Atomic Masses of the Elements

Element	Symbol	Atomic number	Atomic mass
Actinium	Ac	89	227.028
Aluminum	Al	13	26.9815
Americium	Am	95	(243)
Antimony	Sb	51	121.757
Argon	Ar	18	39.948
Arsenic	As	33	74.9216
Astatine	At	85	(210)
Barium	Ba	56	137.327
Berkelium	Bk	97	(247)
Beryllium	Be	4	9.0122
Bismuth	Bi	83	208.9804
Boron	B	5	10.811
Bromine	Br	35	79.904
Cadmium	Cd	48	112.411
Calcium	Ca	20	40.078
Californium	Cf	98	(251)
Carbon	C	6	12.011
Cerium	Ce	58	140.115
Cesium	Cs	55	132.9054
Chlorine	Cl	17	35.4527
Chromium	Cr	24	51.9961
Cobalt	Co	27	58.9332
Copper	Cu	29	63.546
Curium	Cm	96	(247)
Dysprosium	Dy	66	162.50
Einsteinium	Es	99	(252)
Erbium	Er	68	167.26
Europium	Eu	63	151.965
Fermium	Fm	100	(257)
Fluorine	F	9	18.9984
Francium	Fr	87	(223)
Gadolinium	Gd	64	157.25
Gallium	Ga	31	69.723
Germanium	Ge	32	72.61
Gold	Au	79	196.9665
Hafnium	Hf	72	178.49
Helium	He	2	4.0026
Holmium	Ho	67	164.9303
Hydrogen	H	1	1.0079
Indium	In	49	114.818
Iodine	I	53	126.9045
Iridium	Ir	77	192.22
Iron	Fe	26	55.847
Krypton	Kr	36	83.80
Lanthanum	La	57	138.9055
Lawrencium	Lr	103	(260)
Lead	Pb	82	207.2
Lithium	Li	3	6.941
Lutetium	Lu	71	174.967
Magnesium	Mg	12	24.3050
Manganese	Mn	25	54.9380
Mendelevium	Md	101	(258)
Mercury	Hg	80	200.59
Molybdenum	Mo	42	95.94
Neodymium	Nd	60	144.24
Neon	Ne	10	20.1797
Neptunium	Np	93	237.0482
Nickel	Ni	28	58.6934
Niobium	Nb	41	92.9064
Nitrogen	N	7	14.0067
Nobelium	No	102	(259)
Osmium	Os	76	190.23
Oxygen	O	8	15.9994
Palladium	Pd	46	106.42
Phosphorus	P	15	30.9738
Platinum	Pt	78	195.08
Plutonium	Pu	94	(244)
Polonium	Po	84	(209)
Potassium	K	19	39.0983
Praseodymium	Pr	59	140.9076
Promethium	Pm	61	(145)
Protactinium	Pa	91	231.0359
Radium	Ra	88	226.025
Radon	Rn	86	(222)
Rhenium	Re	75	186.207
Rhodium	Rh	45	102.9055
Rubidium	Rb	37	85.4678
Ruthenium	Ru	44	101.07
Samarium	Sm	62	150.36
Scandium	Sc	21	44.9559
Selenium	Se	34	78.96
Silicon	Si	14	28.0855
Silver	Ag	47	107.8682
Sodium	Na	11	22.9898
Strontium	Sr	38	87.62
Sulfur	S	16	32.066
Tantalum	Ta	73	180.9479
Technetium	Tc	43	(98)
Tellurium	Te	52	127.60
Terbium	Tb	65	158.9253
Thallium	Tl	81	204.3833
Thorium	Th	90	232.0381
Thulium	Tm	69	168.9342
Tin	Sn	50	118.710
Titanium	Ti	22	47.88
Tungsten	W	74	183.84
Unnilennium	(Une)	109	(266)
Unnilhexium	(Unh)	106	(263)
Unniloctium	(Uno)	108	(265)
Unnilpentium	(Unp)	105	(262)
Unnilquadium	(Unq)	104	(261)
Unnilseptium	(Uns)	107	(262)
Uranium	U	92	238.0289
Vanadium	V	23	50.9415
Xenon	Xe	54	131.29
Ytterbium	Yb	70	173.04
Yttrium	Y	39	88.9058
Zinc	Zn	30	65.39
Zirconium	Zr	40	91.224

INTRODUCTORY CHEMISTRY

2ND

ALTERNATE

Victor S. Krimsley

HARTNELL COLLEGE

Brooks/Cole Publishing Company

I(T)P™ An International Thomson Publishing Company

Pacific Grove • Albany • Bonn • Boston • Cincinnati • Detroit • London • Madrid • Melbourne
Mexico City • New York • Paris • San Francisco • Singapore • Tokyo • Toronto • Washington

SPONSORING EDITOR: *Lisa J. Moller*
EDITORIAL ASSOCIATE: *Beth Wilbur*
PRODUCTION EDITOR: *Ellen Brownstein*
PRODUCTION SERVICE: *Phyllis Niklas*
MANUSCRIPT EDITOR: *Phyllis Niklas*
PERMISSIONS EDITOR: *May Clark*
INTERIOR AND COVER DESIGN: *Vernon T. Boes*
COVER PHOTO: *Clark Dunbar, Uniphoto*
PHOTO EDITOR: *Larry Molmud*
TYPESETTING: *TSI Graphics, St. Louis, MO*
COVER PRINTING: *Color Dot*
PRINTING AND BINDING: *R. R. Donnelley, Crawfordsville, IN*

For more information, contact:

BROOKS/COLE PUBLISHING COMPANY
511 Forest Lodge Road
Pacific Grove, CA 93950
USA

International Thomson Publishing
Berkshire House 168–173
High Holborn
London WC1V 7AA
England

Thomas Nelson Australia
102 Dodds Street
South Melbourne, 3205
Victoria, Australia

Nelson Canada
1120 Birchmount Road
Scarborough, Ontario
Canada M1K 5G4

International Thomson Editores
Campos Eliseos 385, Piso 7
Col. Polanco
11560 México D. F. México

International Thomson Publishing Gmbh
Königwinterer Strasse 418
53227 Bonn
Germany

International Thomson Publishing Asia
221 Henderson Road #05–10
Henderson Building
Singapore 0315

International Thomson Publishing—Japan
Hirakawacho-cho Kyowa Building, 3F
2-2-1 Hirakawacho-cho
Chiyoda-ku, Tokyo 102
Japan

Printed in the United States of America.

10 9 8 7 6 5 4 3 2 1

Library of Congress Cataloging-in-Publication Data

Krimsley, Victor S.
Introductory chemistry / Victor S. Krimsley.—2nd ed.
p. cm.
Includes index.
ISBN 0-534-25315-6
1. Chemistry. I. Title
QD33.K85 1994
540—dc20 94-32107
CIP

This text is dedicated to my friends and colleagues at Hartnell College, for the many years of friendship, support, and inspiration that they have provided for me,
and to my wife,
Sylvia,
and my children,
David and Deborah.

BRIEF CONTENTS

CONTENTS

3 MEASUREMENT 29

QUANTITATIVE DESCRIPTION OF MATTER 79

THE PERIODIC TABLE 163

CHEMICAL BONDING 189

15 *Working with Acids and Bases* 448

16 *Oxidation–Reduction* 479

List of Summary Boxes

PREFACE

As with the first edition of *Introductory Chemistry,* this text is designed to prepare students with no background in chemistry for a general college chemistry course. It is also a suitable text for the first half of some two-semester sequences, such as those required for certain allied health majors. This text develops the concepts of chemistry in an orderly fashion. Ideas are explained step-by-step, carefully avoiding any omissions in the logical sequence required for comprehension.

LOGICAL ORGANIZATION AND COVERAGE

Introductory Chemistry is designed to build knowledge one step at a time. The first four chapters present a basic description of matter, concluding with a quantitative description in Chapter 4, where the mole concept is introduced. Chapters 5–7 examine the structure of the atom and the relationship of this structure to that of the periodic table. Chapter 8 (Chemical Bonding) and Chapter 9 (Chemical Nomenclature) complete the introductory portion of the text. Following this preliminary material, Chapters 10 (Chemical Changes and Stoichiometry), 11 (The Gas Laws), 13 (Solutions), 14 (The Reactions of Aqueous Solutions), and 15 (Working with Acids and Bases) complete the topics usually considered essential for an introductory course. Some instructors will also wish to include Chapter 12 (Properties of the Liquid and Solid States) in this core material.

The temptation to cover certain topics early for convenience sake has been deliberately resisted. For example, while nomenclature skills would be desirable early in the course, our discussion of nomenclature is delayed until a complete description of ions has been given. Similarly, while equation balancing is a useful skill, it too is delayed until the reader is ready to study stoichiometry.

Broad Topic Selection

There is a considerable diversity of thought about the appropriate contents for an introductory chemistry course. A wide variety of topics has been included to enable instructors to shape courses to meet their individual needs. Chapters on oxidation–reduction and kinetics and equilibrium provide a range of subjects with which to round out the course.

Flexibility

Chapters have been written with an internal integrity that permits some rearrangement of the order of topics. In particular, the chapters on stoichiometry (Chapter 10), gas laws (Chapter 11), and solutions (Chapter 13) may be covered any time following Chapter 4. In addition, nomenclature (Chapter 9) may be moved forward to any point following Chapter 5. There is also little difficulty in inverting the order of Chapter 2 (The Classification of Matter) and Chapter 3 (Measurement). The details of tailoring such rearrangements are spelled out in the *Instructor's Guide.*

Aids to Learning

This text includes many features to aid students in their development of the crucial skills needed to understand chemistry: problem solving and vocabulary building.

Examples and Practice Problems Explanations of concepts are accompanied by worked-out examples of typical problems students are likely to encounter. These are followed immediately by problems for students to work on, using the worked-out examples as models for problem solving. Solutions to these problems are presented at the end of the text, enabling the reader to establish a pattern of: (a) reading a section, (b) working the related problems, and (c) checking the solutions before proceeding to the next section.

End-of-Chapter Problem Sets Problem sets at the end of chapter provide problems similar to the practice problems within the chapter. In addition, a section of general problems provides more challenging exercises designed to test the student's ability to link together or extend concepts introduced in the chapter. Solutions to these problem sets are provided in the *Solutions Manual.*

Writing Exercises Chemistry students are usually asked to demonstrate their proficiency through problem-solving exercises, rather than through written explanations of concepts and ideas. However, students benefit from writing about new material they have learned. Consequently, a series of questions requiring students to practice their writing skills in the context of chemical concepts accompanies each chapter. These exercises take a number of forms, such as writing a letter on a subject addressed in the chapter, or preparing an informal explanation or summary of a concept or idea.

Keyed-in Objectives Learning objectives are listed at the beginning of each section. This helps focus the student's attention on the important concepts presented in each section.

Vocabulary Development Aids Much of what students learn in their first chemistry course is a new language—that of the chemist. This text provides four aids to help students assimilate the many new terms presented:

Key terms are set in boldface type where they are first introduced.

Definitions of all key terms are repeated in an abbreviated form in the margin where the term is introduced.

A list of key terms appears at the end of every chapter.

A comprehensive Glossary at the end of the book provides a valuable reference tool.

Summary and Reinforcement Features Students benefit from seeing ideas presented in a number of different ways. The following features are used to reinforce important concepts:

In-text Summaries outline the important steps in problem-solving techniques.

Unique Summary Diagrams are used to show the links between a number of concepts of problem-solving methods. Some of the Summary Diagrams present flowcharts as an additional aid to solving problems.

See the List of Summary Boxes following the Contents.

COMPLETE SUPPLEMENT PACKAGE

To aid both the student and the instructor, a complete supplement package has been developed to accompany this text. The package includes a *Solutions Manual, Laboratory Manual, Study Guide,* and *Instructor's Guide.* In addition, transparencies and a computerized test bank are available to assist instructors.

Solutions Manual

Complete solutions to all problems presented in the text are included in this manual, including both the in-chapter problems and the Additional Problems at the end of each chapter. While responses to the Writing Exercises are not included in the *Solutions Manual,* general suggestions for dealing with these questions are provided in the *Instructor's Guide.*

Laboratory Manual

The *Laboratory Manual* provides a wide range of well-tested experiments that have been used successfully with beginning students. There is a sufficient range of experiments to provide the instructor with considerable choice. The experiments are designed to enhance the concepts developed within the text. The *Laboratory Manual* allows instructors to use either a journal-type notebook or tear-out report sheets, and a set of prelab questions accompanies each experiment. Experiments have been carefully written to meet concerns over environmental issues. (Suggestions for the disposal of laboratory wastes are included in the *Instructor's Guide.*)

Study Guide

The *Study Guide* presents a unique diagnostic approach in which students analyze their own areas of weakness. They do so with the aid of multiple-choice questions that are designed to check for common errors and provide the reader with further insights through annotations to their incorrect choices.

Instructor's Guide

The *Instructor's Guide* facilitates the administration and teaching of the course. It fully supports the use of the text by providing a variety of materials to aid instruction. The contents and philosophy of each chapter is summarized, enabling instructors to understand how each chapter fits into the educational plan, and workable alternate arrangements for covering the order of the chapters are presented. A test bank of questions is included, providing problems for worksheets, quizzes, and exams. The *Instructor's Guide* also covers the accompanying *Laboratory Manual,* providing information for coordinating the laboratory program with the lecture material, including estimates of laboratory time and a list of materials needed for each experiment. Suggestions for the management of laboratory wastes are also included.

ACKNOWLEDGMENTS

Many individuals have contributed to this text, including reviewers, colleagues at Hartnell College, and the many individuals at Brooks/Cole who worked on this project.

I wish to thank the following reviewers whose helpful criticisms and comments have aided in the development of this text:

Caroline L. Ayers, East Carolina University
James Coke, University of North Carolina/Chapel Hill
Gordon Ewing, New Mexico State University
Mitchel Fedak, Community College of Allegheny County
Robert Fremland, San Diego Mesa University
Donald Glover, Bradley University
Christine Kerr, Montgomery College
Karen Long, Diablo Valley College
Michael O'Connor
Susan Thornton, Montgomery College
Ronald Tjeerdema, University of California, Santa Cruz
Bruce Toder, State University of New York/Brockport
Patricia Wilde
David Williamson, California Polytechnic State University
Donald Young, Ashville-Buncombe Community College

I also wish to thank my colleagues in the Science Department at Hartnell College for their support and friendship throughout my career: Richard Ajeska, Jesse Cude, Larry Elder, Ed Mercurio, Bette Nybakken, Ray Puck (emeritus), Keith Simmons, and Darold Skerritt. Their unwavering encouragement has provided me with a constant source of motivation to put forth the time and effort necessary to complete a project such as this. The encouragement of the administration of Hartnell College is also acknowledged, including John Totten (Emeritus Vice-President of Instruction), Vearl Gish (Emeritus Dean of Science and Mathematics), and Gustavo Valadez-Ortiz (Dean of Science and Mathematics).

A very special thanks must go to two of my local colleagues who have also been involved in writing the supplement package: Beverly Harrison of Robert Louis Stevenson School in Pebble Beach, California, and my very patient office partner, Darold Skerritt of Hartnell College. Both of these individuals always find the time to act as sounding boards for new ideas and for assisting me with the development of new materials and the evaluation of reviewers' comments.

Of course, none of this would be possible without the very able and helpful assistance of the many individuals at Brooks/Cole Publishing Company who participated in this project. In particular, I wish to acknowledge the efforts of Lisa Moller, Beth Wilbur, Ellen Brownstein, Vernon T. Boes, Larry Molmud, Faith Stoddard, Phyllis Niklas, and Harvey Pantzis. Each of these individuals provided me with the encouragement and support necessary to complete this project, and for that, I thank them.

Finally, no acknowledgment would be complete without thanking my loving wife, Sylvia, for the sacrifices she endures during each of my publishing ventures; my son, David, who provided considerable technical assistance; and my daughter, Deborah, who takes it all in stride.

Victor S. Krimsley

INTRODUCTION

1.1 WELCOME TO THE WORLD OF CHEMISTRY

OBJECTIVE 1.1 Define the term *chemistry.* State three contemporary problems that fall within the scope of chemistry.

Inviting you to enter the world of chemistry is much like welcoming the living to life. As you are about to discover, the world that surrounds us is composed of chemicals, and most of what we commonly refer to as the nature of the universe belongs (in part at least) to the world of chemistry.

Chemistry: The study of the nature of matter and the changes it undergoes.

Chemistry is formally defined as the study of the nature of matter—its structure, its composition, and the transformations it undergoes. As we look out on the world about us, everything we see is composed of matter: the mountains, the streams and rivers, the clouds above, even the sea of air we walk through. Indeed, all that we can touch and feel (as well as much that we cannot perceive through our senses) is composed of matter.

Chemistry concerns itself with some of the most pressing problems facing our society today: worldwide food shortage, dwindling energy resources, pollution of our environment, and all sorts of medical issues. The chemical industries furnish the raw materials we need for buildings and machinery both on the job and at home. Pharmaceutical companies provide us with medicines to help combat disease and to ease the suffering caused by illness. Many of the household products we rely on for cleaning, cooking, and gardening are also products of our very diverse chemical industry. Our daily reliance on chemistry is immense!

On the other hand, the prominence of chemistry in our modern world often seems to be a mixed blessing. While the chemical industries provide us with useful products to enhance our comfort, the unsafe disposal of toxic by-products frequently threatens our environment. The morning newspaper teems with endless references to such toxic substances as PCB's, dioxin, and nuclear wastes. The problem of "acid rain" has also been attributed to the discharge of gaseous waste products by various chemical industries.

It would be easy to condemn these industries, but it is likely that the good we derive from the application of chemistry far outweighs the bad. Furthermore, solutions to the problems of chemical pollution will ultimately come from our understanding of chemistry itself. Currently, there is great emphasis on the control and reduction of pollution generated both in the workplace and at home. The substitution of so-called "green" (or nontoxic) materials for more toxic ones represents one approach to the problem. As a classic example of this approach, our society benefits from the availability of synthetic detergents. In the 1960s, however, such detergents were not readily biodegradable and were fouling our nation's waters. Through a knowledge of chemistry, newer, biodegradable products that can be used safely were developed. This quest for safe, nontoxic products continues today and requires a knowledge of basic chemical principles.

Finally, there is an additional unexpected "payoff" that most students find quite exciting as they pursue their study of chemistry. A great number of phenomena that we encounter every day, but may never have expected to understand, become comprehensible! Here are just a few of the questions you will be able to answer after completing this course in chemistry: What is combustion? Why does natural gas burn? What is color? What is table salt made of? What is sugar? How does a pressure cooker work? How does a barometer work? Why does ice melt? Why does a flag wave in the breeze? How does a battery work? Why doesn't water flow uphill? What is boiling? And why do most people float?

PROBLEM 1.1 List three newsworthy problems whose solutions are related to chemistry. How can a knowledge of chemistry be applied to each?

1.2 WHAT DO CHEMISTS DO?

OBJECTIVE 1.2 Describe what academic chemists do. Describe what industrial chemists do. List and describe each of the five branches of chemistry.

We have just begun to explore in the most general way what chemistry is. As a practical matter, it might also be said that chemistry *is* what chemists *do*. Unfortunately, there is no precise description of this, because chemists pursue an enormous variety of activities. Nevertheless, let us examine some of the categories that chemists use to describe their work.

Most chemists consider themselves either academic or industrial chemists. Other chemists are hired by government agencies to work in such areas as public health or defense. Teachers of chemistry are considered academic chemists. In addition to their teaching duties, college and university professors often engage in research. Industrial chemists work for private companies or corporations.

Roughly one-fifth of all chemists enter the teaching profession; about three-fifths are employed by the chemical industries. The work of an industrial chemist is usually related to the sale of products. Examples include flavorings and fragrances, drugs (over 20,000 tons of aspirin are produced annually!), synthetic fabrics, and petroleum products such as gasoline and Vaseline. Industrial chemists may engage in either the research and development (R&D) aspect of this work or in the production of goods for the marketplace. For example, a researcher at Du Pont was responsible for the development of nylon in the late 1930s. Now this textile is manufactured in many parts of the world, and its production requires the participation of chemical engineers and quality-control chemists. So, research goes on in both academic and industrial settings.

Chemistry itself, as it is studied in school, is generally divided into the following branches:

1. **Analytical chemistry** concerns itself with either finding out what is present in a given sample (qualitative analysis) or finding out how much of something is present (quantitative analysis). The other branches of chemistry rely heavily on the techniques of analytical chemistry.

2. **Organic chemistry** is the study of carbon-containing substances. Animal and vegetable matter are rich sources of these substances. Early workers coined the term *organic* because the materials they studied could be derived from living (or formerly living) organisms. Now the term *organic* refers to any substance whose primary component is carbon. Organic substances deserve a branch of their own because there are so many of them. In addition to the countless numbers of biological substances (proteins, carbohydrates, fats), petroleum and coal are sources of organic substances, as are synthetic organic materials prepared in the laboratory.
3. **Inorganic chemistry** involves the study of substances other than those classified as organic. Unlike organic chemicals, these substances are generally derived from mineral sources. Water, which is probably our most important chemical, is also classified as inorganic.
4. **Physical chemistry** concerns the physical properties of chemicals and the physics of chemical processes. This branch of chemistry is the most mathematical in nature.
5. **Biochemistry** is the study of the chemistry of biological processes in all living cells, including those of humans, other animals, plants, and bacteria. The study of viruses also falls within this branch.

The branches overlap considerably with each other, and all chemists practice a mixture of the various branches, depending on the exact nature of their work.

PROBLEM 1.2

Match each of the following tasks with the branch of chemistry with which it is associated.

	Task Carried Out	Branch of Chemistry Employed
______	**(a)** A polyester is prepared from petroleum products.	**1.** Analytical chemistry
______	**(b)** The structure of hemoglobin, the oxygen-carrying component of red blood, is determined.	**2.** Organic chemistry
______	**(c)** An ore is determined to be 72.4% iron and 27.6% oxygen in composition.	**3.** Inorganic chemistry
______	**(d)** Table salt is produced by purification from seawater.	**4.** Physical chemistry
______	**(e)** The pressure of the gas in a helium tank is found to increase if the tank is heated.	**5.** Biochemistry

1.3 SCIENTIFIC INQUIRY AND THE SCIENTIFIC METHOD

OBJECTIVE 1.3 List and describe five major components of the scientific method. Explain the importance of scientific literature to scientific progress. Explain the role of models in the scientific method.

It is through scientific research that scientific discovery and progress are made. Let us examine how the researcher goes about his or her business. In so doing, we will describe the scientific method. Although there is no universal agreement on exactly what the steps of the **scientific method** are, its most important components are observation, hypothesis, controlled experiment, law, and theory.

Scientific method: Set of steps used in scientific inquiry.

Observation: An event that has been witnessed.

Scientific inquiry is generally sparked by some **observation** that requires explanation. Often the observation is a chance event. Such was the case when Alexander Fleming discovered a mold growing on a bacterial culture he had prepared. Fleming noticed that the bacteria were not growing in the vicinity of the mold on the culture plate. So he began to wonder what was inhibiting the growth of the bacteria. Two elements described here are involved in the opening phases of *all* scientific inquiry: alert observation and an attitude of curiosity.

Before seeking an explanation for a particular observation, the scientist often tries to determine whether the phenomenon being observed exhibits some regularity. In Fleming's case, the absence of bacterial growth in the vicinity of the mold was such a regularity. Often such a set of observed regularities may be stated in a concise verbal or mathematical statement known as a **law.** Fleming's observation was never formally stated in a law, but similar types of regularity have led to laws. For example, Sir Isaac Newton's observation that an apple, when released from its stem, always falls from the tree to the earth led to the formulation of the laws of gravity.

Law: A set of observed regularities expressed in a concise verbal or mathematical statement.

The next step is to ponder possible explanations for the observed phenomenon. Fleming made a well-educated guess: Some chemical agent produced by the mold might be responsible for killing the bacterial culture in its vicinity. We call such a guess a **hypothesis.** It is our first attempt to explain the newly observed behavior.

Hypothesis: An educated guess used to explain an observation.

Controlled experiment: An experiment in which only one variable (or factor) is changed at a time.

Next we must test the hypothesis. This requires devising **controlled experiments** that will tend either to support our guess or discredit it. A controlled experiment is generally designed to alter only one variable (or factor) at a time, such as the temperature or the type of bacteria being tested. It is important to understand that our experiments can never *prove* a hypothesis. We can only gather data that tend to support it. When a considerable body of knowledge is gathered in support of a particular hypothesis, it is frequently referred to as a **theory.** If it turns out that we gather information contrary to the hypothesis, it is absolutely disproven; but absolute *positive* proof is never possible.

Theory: An explanation for an observation or series of observations that is substantiated by a considerable body of evidence.

As Fleming began to experiment with the mold, he soon found that it would inhibit the growth of several types of bacteria, which were usually associated with infectious diseases. These results were all consistent with his hypothesis that some chemical agent was responsible for the bactericidal action of the mold. However, it was not until several years later that an independent group of researchers, reading of Fleming's findings, was able to purify the substance involved and test it in human subjects. That substance, the widely used antibiotic penicillin, has been instrumental in our control of bacterial infection ever since.

In order to put Fleming's work in proper perspective, we should keep in mind that his chance discovery took place while he was studying bacterial growth. In other words, he was already at work experimenting, actively engaged in research, and *prepared* to learn. For this reason, chance findings are quite common in scientific research. The activity of scientific inquiry leads the researcher to make careful observations and to be on the lookout for the unusual. Furthermore, our story does not end with Fleming's discovery of penicillin, nor with its subsequent isolation. This discovery prompted further experimentation in the search for other antibiotic substances. Thus, experimentation begets further experimentation.

Another important facet of this classic story is the role of reporting experimental findings. As we have said, Fleming was never able to isolate penicillin himself, and for years his discovery went unnoticed. However, he did report his findings concerning the penicillin mold in a medical journal. A decade later, a biochemist named Ernst Chain, who was working in the laboratories of Howard Florey at Oxford University, came across Fleming's article as he searched the literature for promising leads to an antibacterial agent. It was Chain's skill in purifying penicillin, coupled with Florey's dedication to the task of pioneering a useful drug for combating infectious diseases, that ultimately led to the development of penicillin as the miracle drug of its era. Were it not for the work of Florey and Chain, it is likely that Fleming's discovery would have remained an obscure finding that never amounted to much. Consequently, the 1945 Nobel prize for medicine, awarded for the discovery of penicillin, was shared by Fleming, Florey, and Chain.

To summarize our discussion of the scientific method, observations may lead investigators to discern regularities (laws, perhaps) that foster hypotheses (explanations). Hypotheses suggest further experimentation, which may lead to new observations. Periodically, a hypothesis or related hypotheses lead to a theory.

Model: A mental image used to describe observations in concrete terms.

As you pursue your study of chemistry you will often find that hypotheses, theories, and laws are more easily thought of in terms of a simple mental picture. We refer to such a visual image as a **model.** When we discuss gases, for example, we find it convenient to picture their behavior as that of tiny billiard ball-like particles flying about and colliding with each other and with objects in their paths. Of course, gases are not made of tiny billiard balls, but this model serves us very well in describing gaseous behavior. Good models render the explanations of our hypotheses and laws more concrete.

PROBLEM 1.3

Explain the importance of reporting experimental findings in the scientific literature. How was this facet of the scientific method illustrated in the development of penicillin as an effective antibiotic?

CHAPTER SUMMARY

Chemistry *is the study of the nature of matter and the changes it undergoes.*

Chemists are generally classified as either academic or industrial chemists. Academic chemists teach, and they may also carry out scientific research. Industrial chemists are generally involved in the production of products for the marketplace.

The major branches of chemistry are **analytical chemistry, organic chemistry, inorganic chemistry, physical chemistry,** and **biochemistry.**

Chemical investigations, like those of the other sciences, are carried out in accordance with the **scientific method.** The primary components of the scientific method are **observation, hypothesis, controlled experiment, law,** and **theory.** Scientific inquiry begins with observation. If there is regularity in the observed behavior, it can sometimes be stated in a law. A hypothesis is a possible explanation for the observation. Controlled experiments, in which the variables are altered systematically, are used to test a hypothesis. When a considerable body of knowledge is gathered in support of a hypothesis, the hypothesis may be referred to as a theory. A **model** may be used to help explain such a theory.

KEY TERMS

Review each of the following terms, which are discussed in this chapter and defined in the Glossary at the end of the text.

chemistry
analytical chemistry
organic chemistry
inorganic chemistry
physical chemistry
biochemistry
scientific method
observation
hypothesis
controlled experiment
law
theory
model

ADDITIONAL PROBLEMS

SECTION 1.1

1.4 Define the term *chemistry.*

SECTION 1.2

1.5 Describe what academic chemists do. Describe what industrial chemists do.

1.6 What are the five branches of chemistry? Describe each.

SECTION 1.3

1.7 What are the components of the scientific method? Describe each.

1.8 What is a model?

1.9 Explain how a hypothesis becomes a theory.

GENERAL PROBLEMS

1.10 The presence of chemistry in our lives is unavoidable. We constantly come into contact with hundreds of chemical products. Look around you and find ten plastic products that you use regularly.

1.11 When an anthropologist discovers paintings in a cave, the paintings can be analyzed for style and content to ascertain their age. Paintings also can be dated by analyzing the chemical composition of the paint. Using similar techniques of observation, it is possible to date the clothing in your wardrobe. Explain how it would be possible to date your wardrobe by the intensity of the colored dyes in your clothing. Describe some other observations that could be used to establish the age of your clothing.

WRITING EXERCISES

1.12 Suppose you are the public relations officer for Medical Devices, a plastics company that manufactures prosthetic hip replacements. Your company wishes to establish a plant in the town of Sunshine, a retirement community. However, local citizens are concerned about having a chemical manufacturing plant located in their community. Write a letter to the city council, explaining why they should approve a facility use permit for your plant.

1.13 Suppose you are the chapter secretary of Living Earth, a local environmental group that is concerned with the establishment of a plastics manufacturing plant in your community. Write a letter to the Chief Executive Officer of Medical Devices, explaining your concern for the welfare of the local environment in your area. (See Problem 1.12.)

1.14 Find an article in the daily newspaper that involves chemistry. Write a paragraph summarizing the contents of the article.

2 THE CLASSIFICATION OF MATTER

Matter: Anything that has mass and occupies space.

In the first chapter, chemistry was defined as the study of the nature of matter and the changes it undergoes. But what do we mean by matter? **Matter** may be formally defined as anything that has mass and occupies space. In the next chapter we discuss the meaning of mass in some detail, but for the moment let us say that this book is composed of matter and that the amount of matter in it is the book's mass.

If our eyes could examine matter in its finest detail, we would see that matter is actually composed of tiny particles. Thus, although the page you are reading appears to be smooth and continuous, it is actually composed of tiny discrete (separate) particles. This chapter examines the structure of matter and introduces you to the symbols and language used to describe its composition.

2.1 PHASES OF MATTER

OBJECTIVE 2.1 Name the three common physical states of matter. Describe the characteristics of each state, and give two examples of each.

Physical state or phase: The solid, liquid, or gaseous state of matter.

We can classify matter according to its three **physical states:** solid, liquid, and gas. These states are frequently referred to as the **phases** of matter. You are undoubtedly familiar with each of these phases. For example, wood is a solid, drinking water is a liquid, and air is a gas. (Scientists also recognize a fourth state of matter known as *plasma,* which is found in stars. In this book, however, we will concentrate on solids, liquids, and gases.)

The particles that make up the solid and liquid states are packed tightly together (Fig. 2-1a and b). By contrast, the particles making up a gas are relatively far away from one another. Because their component particles are so close to one another, solids and liquids are not readily compressed, and we refer to these states as the "condensed" phases of matter. On the other hand, the space between gaseous particles allows gases to be compressed with relative ease (Fig. 2-1c).

Solid: One of the physical states of matter, characterized by resistance to changes in shape and volume.

A **solid** has a definite shape and volume and tends to resist any change in that shape. If we move a solid from one container to another, its shape and volume remain

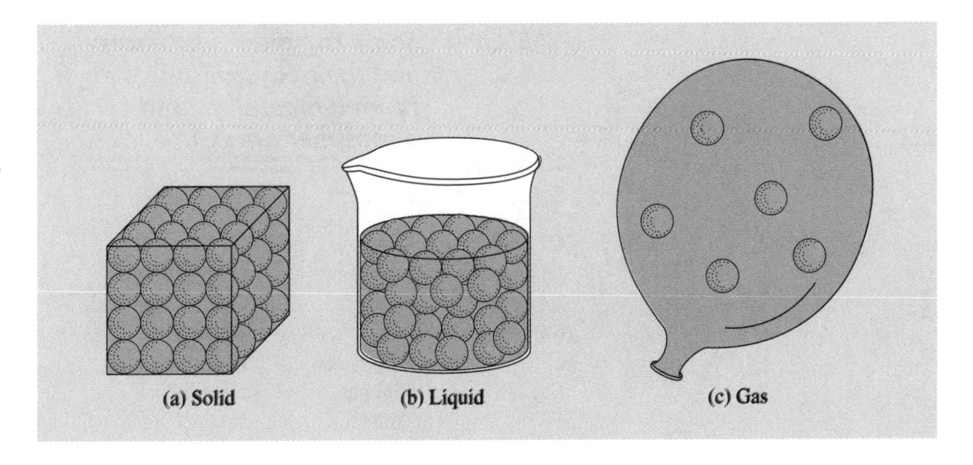

FIGURE 2-1 (a) An orderly arrangement of particles in the solid state leads to resistance to any change in shape or volume. (b) In the liquid state, particles move randomly over one another, allowing liquids to flow and change shape. (c) In the gaseous state, particles spread apart or move closer, allowing gases to fit their container exactly.

the same. For example, a piece of ice (a solid) does not alter its shape to fit through the neck of a vessel if the opening is too small. Some additional examples of solids are iron, marble, and diamond. (Although granulated solids, such as sugar, salt, and sand, might *appear* to change shape when they are transferred from one container to another, individual granules maintain a constant shape.)

Liquid: One of the physical states of matter, characterized by an ability to flow freely but exhibiting a fixed volume.

A **liquid** tends to flow freely, having a definite volume but no definite shape. Because a liquid flows freely, it assumes the shape of the container in which it is placed. However, its volume remains unchanged when we change its container. Unlike a cube of solid ice, liquid water alters its shape to fit through the neck of a vessel. Some examples of liquids are water, gasoline, kerosene, vinegar, oil, and turpentine.

Gas: One of the physical states of matter, characterized by the ability to expand or contract to fit its container.

A **gas** is a substance in which the particles of matter are so far away from each other that it has no definite shape whatsoever and no definite volume. Instead, a gas expands (or contracts) to fill the container in which it is placed. With the exception of gases that are colored (such as chlorine gas, which is green), gases appear to be invisible. However, their presence can be observed indirectly. For example, air is composed of gases. Although we cannot actually see the air in front of us, when we watch a tree blowing in the wind, we are actually seeing the effect of these gases striking the branches and leaves of the tree. By contrast, when American astronauts landed on the moon, they found it necessary to place a rigid flag in front of the television camera. In the absence of air, there was no wind on the moon's surface to hold the flag up. Some examples of gases are natural gas, carbon dioxide, carbon monoxide, neon, helium, and steam.

You may have noticed that water, in one of its various forms, appeared as an example for each of the three physical states. At room temperature, water exists as a liquid. When water is cold enough, it freezes to form solid ice. When hot enough, it vaporizes (forms a gas) to become steam. This is a general characteristic of most substances: Depending on the temperature, a given substance may exist as a solid, a liquid, or a gas. For the present, we will discuss substances in the phase they exhibit at room temperature (about 20°C, or 68°F). Pressure can also affect the physical state that matter exhibits. We normally discuss substances at atmospheric pressure, the normal pressure found at sea level on most days. Table 2-1 lists some common substances according to their physical states at room temperature and atmospheric pressure.

TABLE 2-1 Some common substances and their physical states at room temperature and atmospheric pressure

Solids	Liquids	Gases
Copper	Gasoline	Air
Diamond	Kerosene	Carbon dioxide
Iron	Mercury	Carbon monoxide
Salt	Oil	Helium
Sand	Turpentine	Hydrogen
Sugar	Vinegar	Natural gas
Wood	Water	Neon

PROBLEM 2.1

Describe each of the three states of matter. How do these states differ in terms of changes in shape? In which of these states can matter be compressed? Which are referred to as the "condensed" phases?

2.2 HETEROGENEOUS AND HOMOGENEOUS MATTER

OBJECTIVE 2.2 Distinguish between heterogeneous matter and homogeneous matter.

Matter may be classified as heterogeneous or homogeneous. Suppose we pour some sand into a glass of water. We can see that this sand–water mixture is not uniform throughout. If we take a portion from near the bottom of the glass, we are likely to get more sand than we would if we took our sample from the top of the glass. Such a visibly nonuniform mixture is said to be **heterogeneous.**

Heterogeneous mixture: A visibly nonuniform mixture.

Now suppose we pour a small amount of sugar into another glass of water. If we stir the sugar with a spoon, eventually all of the sugar dissolves and we are no longer able to see the sugar. Nevertheless, we still have a mixture of sugar and water, just as we had a mixture of sand and water before. However, the characteristics of these mixtures are quite different. The sugar–water mixture is visibly uniform throughout. If we remove a portion of the mixture, it will be the same in composition as any other portion we might choose to remove. Such a uniform mixture is said to be **homogeneous.**

Homogeneous mixture: A visibly uniform mixture.

In distinguishing heterogeneous from homogeneous matter, we often say that distinct *phase boundaries* exist in heterogeneous matter. For example, when sand is suspended in water, we are actually able to see the boundary where a solid particle of sand ends and the liquid water begins. Similarly, if we shake up oil and vinegar salad dressing, it is possible to see the liquid oil droplets suspended in the vinegar. Again, we can see the boundary where the oil droplet ends and the vinegar begins. A glass of water with an ice cube in it is also heterogeneous, even though both the solid and liquid phases are of the same substance. In this case, we can see the boundary where the solid ice ends and the liquid form of water begins.

In contrast, homogeneous matter does not have any distinct phase boundaries that are visible to the eye. A number of homogeneous mixtures of metals, known as alloys, are commonly used. Steel, bronze, and brass are examples of these. Bronze is a homogeneous mixture of copper and tin. Brass is a mixture of copper and zinc. In the case of bronze, we cannot *see* any phase boundary that separates the copper from the tin. Likewise, we cannot see any phase boundary separating copper from zinc in a piece of brass.

PROBLEM 2.2

Identify each of the following as heterogeneous or homogeneous.

(a) A piece of brass
(b) A glass of ice water with ice cubes in it
(c) Clean air
(d) Fog (Fog contains water droplets suspended in air.)

(e) Smog (Smog contains solid particles suspended in air.)
(f) A glass of bubbling soda water
(g) Oil and vinegar salad dressing
(h) A cup of clear tea
(i) Mud (Mud contains solid particles suspended in water.)

PROBLEM 2.3 What characteristics distinguish homogeneous matter from heterogeneous matter?

2.3 MIXTURES AND PURE SUBSTANCES

OBJECTIVE 2.3 Describe a pure substance; distinguish between a pure substance and a homogeneous mixture.

Matter may be subdivided into mixtures and pure substances. Any time we mix two or more different substances together, we have a mixture. In the last section, we examined the nature of a heterogeneous mixture (sand–water) and a homogeneous mixture (sugar–water). As a general rule, heterogeneous matter is a mixture, rather than a pure substance. (There are exceptions, however, as is the case when an ice cube is placed in a glass of liquid water. Both the ice and the liquid are different forms of the same pure substance—water. For simplicity, we will not concern ourselves with this type of exception.) Homogeneous matter may be either a mixture (as in a sugar–water mixture) or a pure substance (as we are about to describe).

Pure substance: A substance having a fixed and definite composition.

A **pure substance** is one that is uniform throughout and has a fixed and definite chemical composition and characteristic properties. It may seem that our definition of a pure substance is the same as that of a homogeneous mixture. However, unlike our homogeneous sugar–water mixture (in which the proportions of sugar and water can be varied), a pure substance has a fixed and definite composition. Water, for example, is a pure substance. It always has the same chemical composition. A sugar–water mixture may vary in its composition, depending on how it was prepared. These various classifications of matter are summarized in Fig. 2-2.

PROBLEM 2.4 What characteristics distinguish a pure substance from a homogeneous mixture?

2.4 COMPOUNDS AND ELEMENTS

OBJECTIVE 2.4 Define an element and give two examples. Define a compound and give two examples.

Compound: A pure substance composed of more than one element; can be decomposed into simpler substances.

Pure substances are divided into two categories: compounds and elements (see Fig. 2-2). **Compounds** are pure substances that can be decomposed into simpler substances.

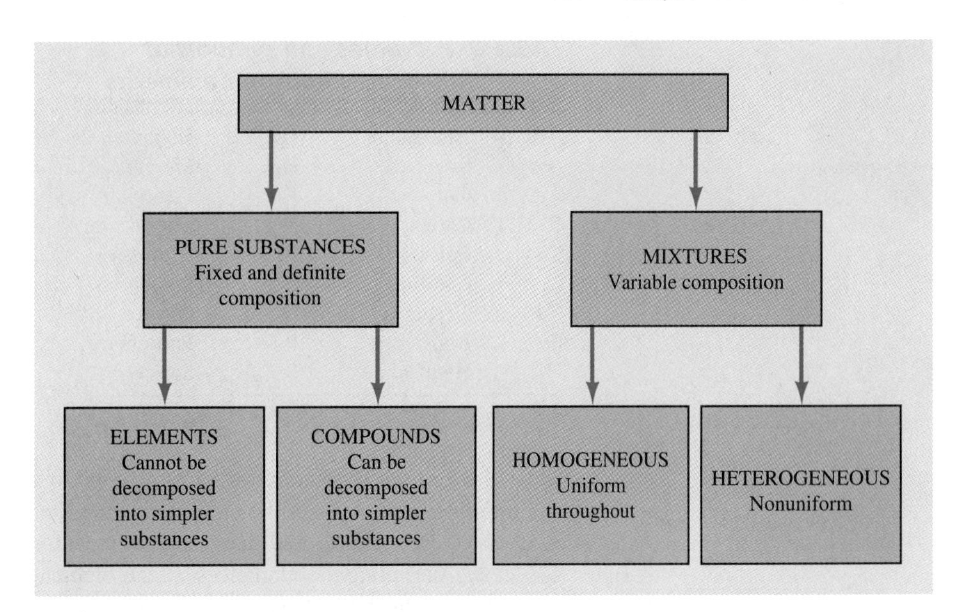

FIGURE 2-2 The classification of matter.

For example, water can be broken down into hydrogen and oxygen. Thus, water must be a compound. A pure substance that cannot be broken down into simpler substances by ordinary chemical means is called an **element.** Neither hydrogen nor oxygen (the constituents of water) can be further decomposed, so each is classified as an element.

Element: A pure substance that cannot be decomposed into simpler substances by ordinary means.

Do not be misled by these examples, however. On heating, calcium carbonate (limestone) can be decomposed into calcium oxide and carbon dioxide. All that this tells us is that calcium carbonate must be a compound, because it decomposes into simpler compounds. We may *not* conclude that calcium oxide and carbon dioxide are elements. Indeed, they are not. Both calcium oxide and carbon dioxide can be further decomposed, indicating that they too are compounds.

Also, note that a compound is not a mixture, though both can be broken down into simpler substances. Remember that a pure substance is distinguished from a mixture by its fixed composition. A compound is made up of two or more kinds of elements, but they are always combined in the same fixed proportions. (The distinctions we have made so far between mixtures and pure substances were summarized in Fig. 2-2.) Since compounds and elements are pure substances, they are homogeneous.

(Later in this chapter, you will learn a much easier way to distinguish compounds from elements by examining their formulas. So, do not be concerned if these definitions seem a little mysterious at the moment. Their meanings will become much clearer as we proceed.)

By 1989, the first 109 elements had been discovered; these are tabulated on the inside of the front cover of this book. You will also find these elements displayed in a chart found on the inside front cover of the book. This chart is called the **periodic table.** Its format is designed for ease of reference, and we will discuss the reasons for its organization in later chapters. Some of the most common elements are listed in Table 2-2 along with the symbols used to represent them. Notice that each symbol given in Table 2-2 is composed of one or two letters. In each case the first letter is always capitalized, whereas the second letter (if there is one) is lowercase. You may also notice that

Periodic table: The table of the elements.

TABLE 2-2 Names and symbols of some important elements

Al	Aluminum	He	Helium
Ag	Silver	Hg	Mercury
Au	Gold	K	Potassium
C	Carbon	Mg	Magnesium
Ca	Calcium	N	Nitrogen
Cl	Chlorine	Na	Sodium
Cu	Copper	O	Oxygen
Fe	Iron	P	Phosphorus
H	Hydrogen	S	Sulfur

some of the symbols do not seem to correspond to the names of their elements. The reason for this is that many of the elements derive their names from either Latin or Greek. A few of these symbols and their original meanings are included in Table 2-3. Figure 2-3 shows the relative abundances of the elements as they occur in our earthly surroundings and in our bodies.

Following World War II, the practice of naming new elements after prominent scientists or places of scientific importance became popular. Unfortunately, with the discovery of elements 104 and 105, disagreement arose over who should be credited with the discovery of each, and the names to be given to those elements have been in dispute ever since. This impasse led to a proposal that new elements be named in a systematic fashion until official agreement is reached. By this system, each new element is given a three-letter symbol that corresponds to its three-digit number. For example, element 104 is named Unnilquadium (Unq): un = 1, nil = 0, quad = 4. Element 105 is Unnilpentium (Unp), 106 is Unnilhexium (Unh), 107 is Unnilseptium (Uns), 108 is Unniloctium (Uno), and 109 is Unnilennium (Une).

PROBLEM 2.5 Distinguish between an element and a compound.

TABLE 2-3 Selected symbols and their origins

Symbol	English name	Language	Original name	Meaning
H	Hydrogen	Greek	Hydros and genes	Water-forming
Au	Gold	Latin	Aurum	Shining dawn
Ag	Silver	Latin	Argentum	Silver
Cu	Copper	Latin	Cuprum	From island of Cyprus
Ca	Calcium	Latin	Calx	Lime
Sn	Tin	Latin	Stannum	Tin
Fe	Iron	Latin	Ferrum	Iron
O	Oxygen	Greek	Oxys and genes	Acid-forming
S	Sulfur	Latin	Sulfur	Sulfur
Cl	Chlorine	Greek	Chloros	Greenish-yellow

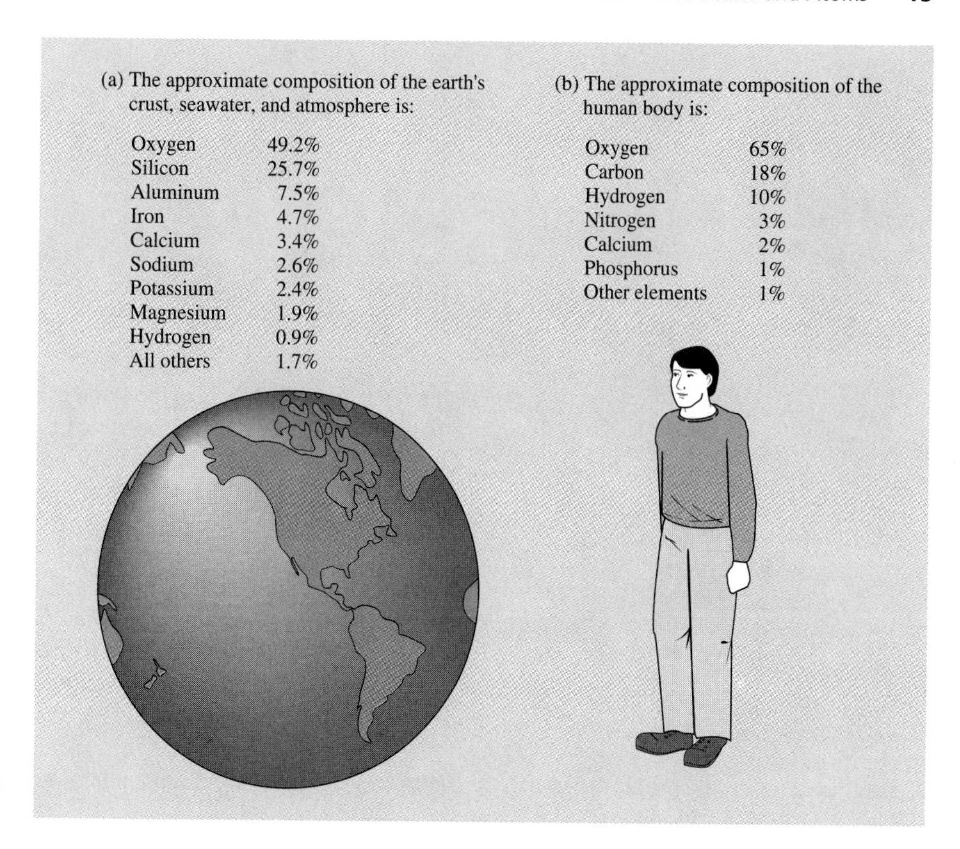

FIGURE 2-3 The distribution of elements in (a) the earth's crust, seawater, and atmosphere and (b) the human body.

PROBLEM 2.6

On heating, substance A is decomposed to substances B and C. Substance C may also be made by combining substances D and E. Substance D cannot be decomposed into simpler substances by ordinary chemical means. Classify each of the substances, A through E, as follows: (1) an element, (2) a compound, or (3) there is insufficient information given in the problem to tell.

2.5 MOLECULES AND ATOMS

OBJECTIVE 2.5 Draw a picture of a molecule of water and identify the atoms present. Distinguish between a molecule and an atom of an element.

In the first half of this chapter, we examined matter by looking at some of its gross, or *macroscopic*, characteristics. The definitions and distinctions we made concentrated on those characteristics that can be seen with our eyes. However, we are going to find it much easier to understand the nature of matter if we look more closely at the *microscopic* characteristics of matter, those characteristics that cannot be seen directly.

Earlier, we mentioned that matter is composed of tiny discrete particles. Let us now examine the composition of those particles. Suppose we take a glass of water and

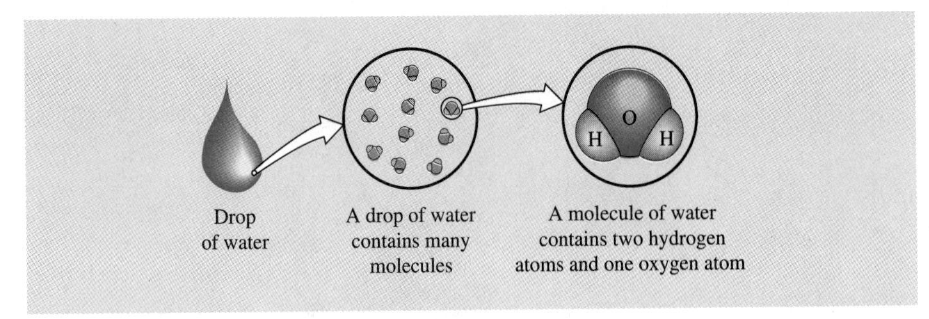

FIGURE 2-4 A sample of water contains many individual molecules. The molecule is the smallest unit of water. Each molecule is made up of atoms.

Molecule: The smallest unit of a pure substance that can exist independently and exhibit all of the properties of the substance.

Atom: A building block from which molecules are made.

divide its contents in half. Each half of the original sample still possesses all the characteristics, or *properties,* we normally associate with water. Suppose we could divide the contents of the sample again and again. At some point we would reach the smallest unit of matter that could still be identified as water. We refer to this unit, which is far too small to be seen directly, as a **molecule** of water.

However, a molecule is still not the smallest particle we use in our description of matter. If we could examine our single molecule of water even more closely, we would find that it is composed of three particles known as **atoms** (Fig. 2-4). In the case of water, each molecule contains two atoms of hydrogen and one atom of oxygen. Note that the names of the atoms in water are the names of elements. The same is true of the names of *all* atoms. In fact, there are as many kinds of atoms as there are elements.

Figure 2-5 illustrates several molecules, including water and oxygen. We have divided the figure into elements and compounds. Note that each of the compounds shown

FIGURE 2-5 Examples of elements and compounds. An element contains only one kind of atom; a compound contains two or more kinds of atoms.

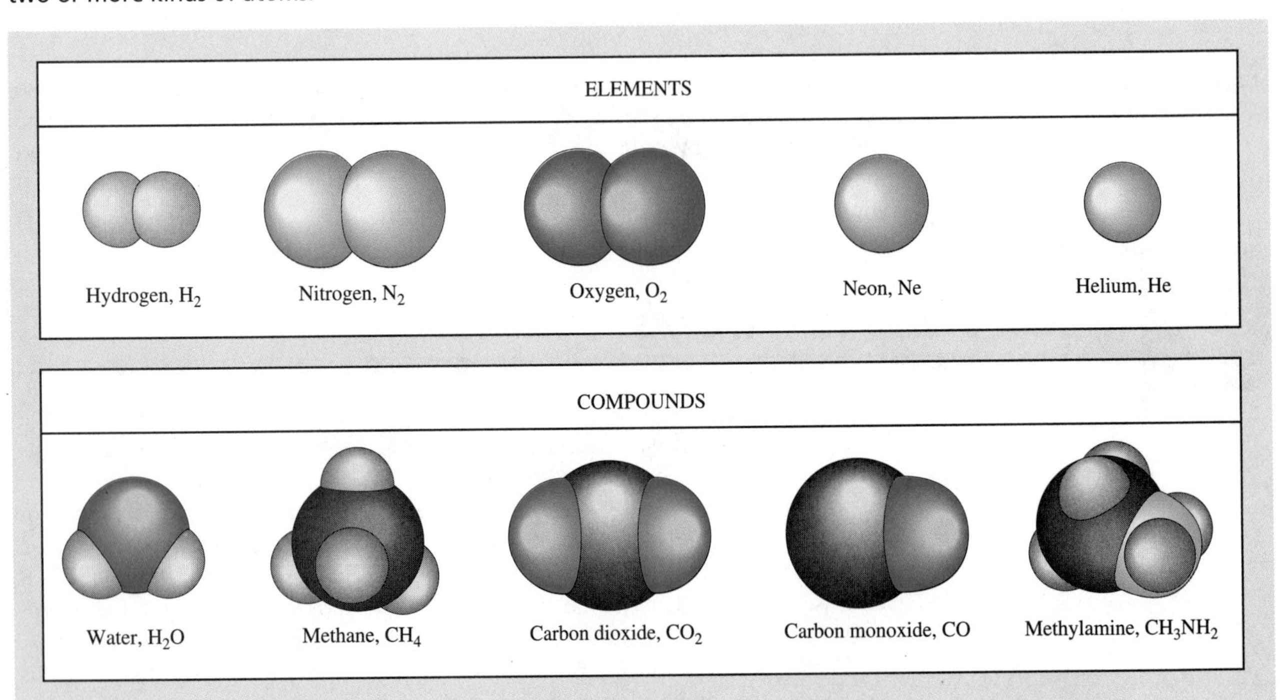

has more than one kind of atom per molecule. For example, a molecule of methane (a compound) contains one carbon atom and four atoms of hydrogen. On the other hand, molecules of elements contain only one type of atom. For instance, a molecule of oxygen (an element) is made exclusively of oxygen atoms. In the case of oxygen, there are two atoms in each molecule (O_2). The oxygen in the air we breathe is made up of these molecules. Similarly, hydrogen (H_2) and nitrogen (N_2) contain two atoms per molecule. (The subscript "2" in each of these formulas tells us the number of atoms present. We will have more to say about writing chemical formulas in the next section.) We say that these substances are **diatomic,** meaning they contain two atoms per molecule. Note that the compound carbon monoxide (CO) is also diatomic. Thus, there are diatomic compounds as well as diatomic elements. Molecules with three atoms, such as water and carbon dioxide, are said to be **triatomic,** whereas the term **polyatomic** is used to describe molecules with many atoms, such as methylamine (see Fig. 2-5). Note that the smallest unit of helium contains only one atom. Although a single helium atom fits our basic definition of a molecule, many chemists reserve the term *molecule* for species that contain two or more atoms. Thus, we generally refer to helium *atoms* rather than helium *molecules*. Because it exists in nature as independent atoms, helium is a **monatomic** element, and on occasion we will use the term *molecule* when referring to monatomic substances such as helium because their behavior is that of molecules. Note that each molecule shown in Fig. 2-5 possesses a fixed composition of atoms, regardless of whether the substance is an element or a compound.

Diatomic: Having two atoms.

Triatomic: Having three atoms.

Polyatomic: Having many (two or more) atoms.

Monatomic: Having one atom.

PROBLEM 2.7

(a) Distinguish between a molecule and an atom.
(b) Distinguish between an element and a compound in terms of the atoms present.
(c) What is the meaning of the term *diatomic?* What does the term *monatomic* mean? Give an example of a diatomic element and an example of a monatomic element. Give an example of a diatomic compound. Is it possible to have a monatomic compound?

2.6 CHEMICAL FORMULAS

OBJECTIVE 2.6 Write the formula of a substance, given its composition (names and numbers of atoms in the correct order).

In Section 2.5, we examined molecules of several compounds and elements. In order to express the composition of each substance, we found that the most important information for us to convey is what kind of atoms are present and how many of each are present. This information is expressed in the **chemical formula** of the substance. For example, the chemical formula for water must convey the fact that each molecule of water contains two atoms of hydrogen and one atom of oxygen. To do this, we write the chemical symbol for each atom in the molecule, and then we indicate how many of each type of atom are present. Thus, the chemical formula for water is:

Chemical formula: A symbolic representation of the composition of a pure substance.

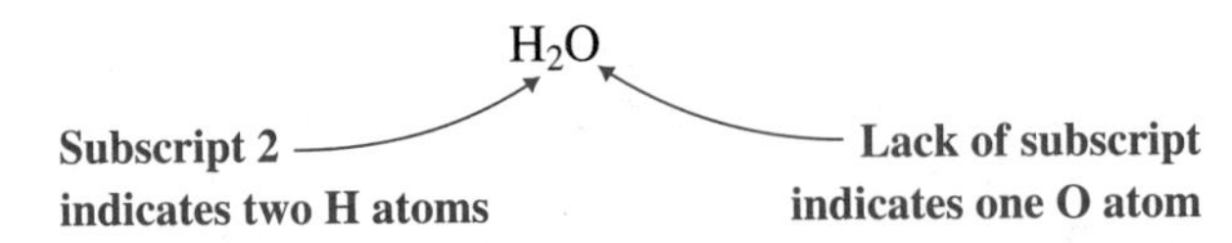

In the formula we have indicated the number of hydrogen atoms by placing the subscript 2 at the lower right of the symbol for hydrogen, H. When no subscript appears next to an element, it means that only one atom of that particular element is present. Thus, we did not write H_2O_1 but simply omitted a subscript after the oxygen. Let us do a few more examples to illustrate this principle and to convey a few more rules.

EXAMPLE 2.1

Write the formula for the compound called potassium nitrate. It contains one atom of potassium (K), one atom of nitrogen (N), and three atoms of oxygen (O).

SOLUTION

$$KNO_3$$

No subscript was written after the symbols for potassium and nitrogen, because only one of each of these kinds of atoms is present. Occasionally we refer to a substance by its formula rather than by its chemical name. This substance would be called "K-N-O-three."

EXAMPLE 2.2

Octane, a major component of gasoline, contains eight atoms of carbon (C) and eighteen atoms of hydrogen (H). Write its molecular formula.

SOLUTION

$$C_8H_{18}$$

This formula is read "C-eight-H-eighteen."

EXAMPLE 2.3

Potassium chromate contains two atoms of potassium (K), one atom of chromium (Cr), and four atoms of oxygen (O). Write its formula.

SOLUTION

$$K_2CrO_4$$

This formula is read "K-two-C-r-O-four."

Note that in Example 2.3, the symbol for chromium has a capital C and a lowercase r. Recall from our earlier discussion of the elements that the symbols of the first 103 elements consist of either one or two letters, where the first letter is capitalized and the second letter (if there is one) is lowercase. The importance of this rule is apparent when we compare the chemical formulas of CO and Co. CO is carbon monoxide. It consists of one atom of the element C and one atom of the element O. Carbon monoxide is a colorless and odorless gas at room temperature. Co is cobalt, a metallic solid under the same conditions. CO and Co are two quite different substances. Note that CO is a compound, whereas Co is an element. In order to communicate correctly, it is essential to write the symbols of elements properly, using capital and lowercase letters appropriately.

Certain groups of atoms are found together in many different substances. When one of these groups appears more than once in a chemical formula, parentheses are used to maintain the identity of the group. For example, the formula for aluminum hydroxide is $Al(OH)_3$. The group inside the parentheses is called the *hydroxide* group. The subscript 3 outside the parentheses indicates that there are three hydroxide groups in the formula. Because each hydroxide group has one atom of oxygen and one atom of hydrogen, the total formula consists of one aluminum atom, three oxygen atoms, and three hydrogen atoms. Similarly, the formula of magnesium nitrate, $Mg(NO_3)_2$, denotes one magnesium atom, two nitrogen atoms, and six oxygen atoms.

PROBLEM 2.8

Write the chemical formula for each of the following compounds. In each case, the correct order of the atoms is the order in which they are given. (The name of each is provided for your information.)

(a) One atom of hydrogen and one atom of chlorine (hydrogen chloride)
(b) Two atoms of nitrogen and five atoms of oxygen (dinitrogen pentoxide)
(c) Six atoms of carbon and six atoms of hydrogen (benzene)
(d) Two atoms of hydrogen, one atom of sulfur, and four atoms of oxygen (sulfuric acid)
(e) One atom of sodium and one atom of chlorine (sodium chloride)
(f) One atom of neon (neon)
(g) One atom of magnesium and two atoms of bromine (magnesium bromide)
(h) Two atoms of aluminum and three atoms of sulfur (aluminum sulfide)

PROBLEM 2.9

How many oxygen atoms are present in each of the following chemical formulas?
(a) K_2SO_4 **(b)** $Al_2(SO_4)_3$ **(c)** $Ba(OH)_2$ **(d)** $Fe(NO_3)_3$

Now that we know how to interpret chemical formulas, we review what we have learned about compounds and elements, molecules and atoms, and mixtures and pure substances in the Summary that follows.

SUMMARY: Compounds and Elements, Molecules and Atoms, Mixtures and Pure Substances

1. A **compound** is a pure substance that can be broken down into simpler substances. We recognize a compound by the fact that there is more than one kind of atom present in its chemical formula. We are able to distinguish a compound from a mixture by its fixed composition, as reflected in its chemical formula. The formula states the ratio of atoms present, such as the 2:1 ratio of hydrogen atoms to oxygen atoms in water.
2. An **element** is a pure substance that cannot be broken

down into simpler substances by ordinary chemical means. We recognize an element by the fact that there is only one type of atom present in its formula.

3. A **molecule** is the smallest unit of any pure substance that possesses all of the properties of that substance. We can tell what atoms are present in a molecule of a substance by examining its chemical formula. The formula tells us what kinds of atoms are present (what elements) as well as how many of each.
4. An **atom** is a building block from which molecules are made. There are as many different kinds of atoms as there are elements. Although some elements contain more than one atom per molecule, an atom is the simplest unit that can still be identified with a particular element.
5. A **pure substance** has a fixed and definite composition and is uniform throughout. A single chemical formula can be used to describe a pure substance. The formula states the fixed composition of atoms. Elements and compounds are both pure substances.
6. A **mixture** can be separated into components. Unlike a pure substance, which has a fixed composition, a mixture may be prepared in any proportions. More than one chemical formula is needed to describe a mixture fully. For example, to describe a mixture of table sugar ($C_{12}H_{22}O_{11}$) and water (H_2O), we need the formulas of both substances.

2.7 PHYSICAL CHANGES

OBJECTIVE 2.7 Explain the nature of a physical change and give an example.

In Section 2.1 we described the three phases of matter in terms of some of their observable properties. We saw that the particles of matter in the solid state (Fig. 2-1) were very orderly, occupying fixed locations. Although you might think that the particles of matter in a solid are stationary, they actually vibrate about their fixed positions. By comparison, the particles of matter in the liquid state have more freedom of movement as they tumble over one another. Finally, the particles of matter in the gaseous state have even greater freedom of movement, being separated from one another and flying about in various directions.

Phase change: A change in the physical state of a substance, such as occurs during melting or boiling.

Physical change: A change in a substance that does not alter its fixed composition.

As a substance changes from one state to another, it is said to undergo a **phase change.** For example, when ice melts, it undergoes a phase change from solid to liquid. Or when water boils, a phase change from liquid to gas occurs. These changes belong to a more general class of change, referred to as *physical change.* A **physical change** is one that involves no change in the fixed composition of the substance in question. Thus, when water freezes or boils, each molecule is still composed of two atoms of hydrogen and one atom of oxygen. However, the molecules are arranged differently with respect to one another. Figure 2-6 shows how a monatomic substance might look to us if we could observe its phase changes at the atomic level.

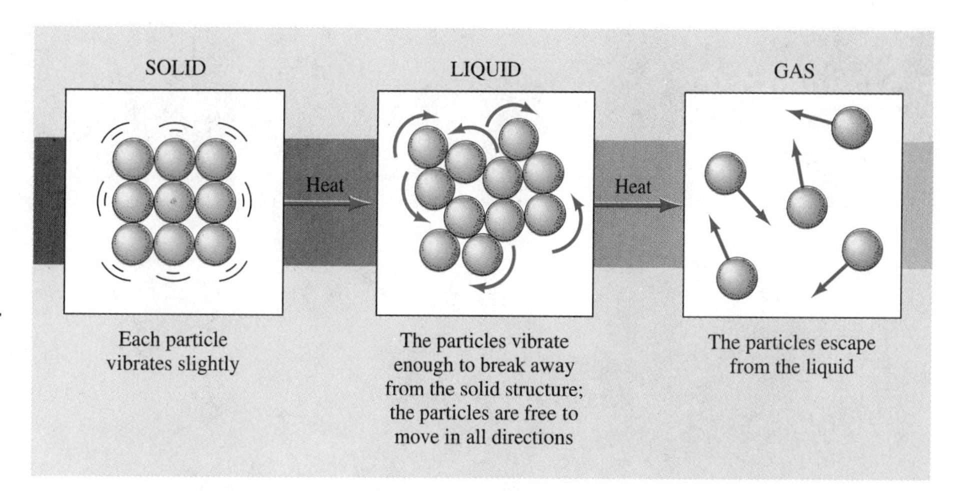

FIGURE 2-6 **Changes of state. When a phase change occurs, the molecules are not changed. However, when a substance passes from the solid to the liquid state or from the liquid to the gaseous state, the molecules acquire greater freedom of movement.**

Melting point: The temperature at which a substance melts.

Boiling point: The temperature at which a substance boils.

Sublimation: A phase change in which a substance passes directly from the solid to the gaseous state.

The **melting point** of a substance is the temperature at which a transition occurs between the solid and the liquid state. For example, the temperature at which ice melts is its melting point. When a liquid solidifies (or freezes), the reverse process is occurring. This reverse process occurs at the same temperature (the melting point). The **boiling point** of a liquid is the temperature at which a transition occurs between the liquid and the gaseous state. Some substances pass directly from the solid to the gaseous state, without ever forming a liquid. This process is known as **sublimation,** and we call the transition temperature the *sublimation point*. Probably the most familiar example of this phenomenon occurs with dry ice, which is solid carbon dioxide. Dry ice sublimes when left out in a room. (If you have ever watched a block of dry ice sublime and seen a puddle of liquid form around the block, you may have been fooled. The liquid you saw was water condensation from the moisture in the air!) The Summary Diagram on page 22 illustrates the various changes of state.

A phase change is just one type of physical change. As another example, consider how the tungsten filament in a light bulb glows when an electrical current is passed through it. This physical change in appearance involves no alteration of the chemical composition of the tungsten wire. The physical change is readily reversed when the current is turned off.

Another familiar example of a physical change occurs when a substance dissolves in a liquid, as in the case of making a homogeneous sugar–water mixture. The molecules of sugar, which are packed together in their crystalline form, become separated from one another and are dispersed among the water molecules. Since the composition of each sugar molecule remains unchanged in the process of being dissolved, this is classified as a physical change. Furthermore, sugar crystals can be regenerated by evaporation of the water.

PROBLEM 2.10 Explain why a phase change is a physical change.

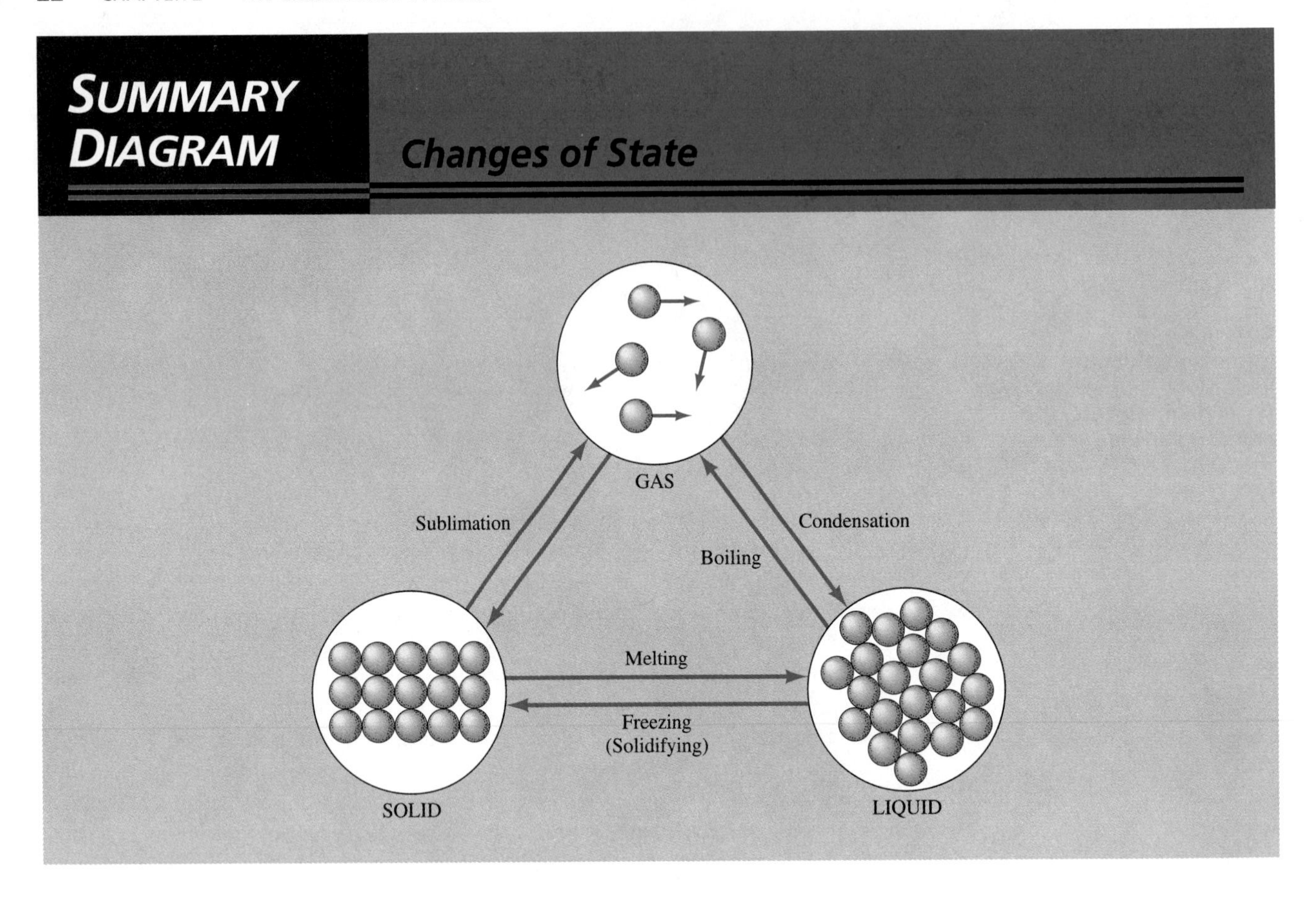

2.8 CHEMICAL CHANGES

OBJECTIVE 2.8A Explain the nature of a chemical change and give an example.

Imagine a log burning in a fireplace. While you watch its cheerful light, heat from the fire warms you. As it burns, you may also hear the sound of gases escaping from the wood. After a period of time the fire dies down and is extinguished. A pile of ashes is all that remains. You have just witnessed one of the oldest chemical changes known to us.

The combustion of a log involves the chemical transformation of substances composed primarily of carbon and hydrogen into water vapor and carbon dioxide. This process requires the presence of oxygen, which also undergoes a change. The original substances undergoing the chemical change (in this example, the components of the log plus oxygen) are referred to as *reactants;* the newly formed substances (the water and carbon dioxide) are called *products*. The term *react* is often employed when discussing the transformation of *react*ants, so the phrase *chemical reaction* has the same meaning as chemical change. A **chemical change** is one that involves the transformation (or reaction) of one or more substances, referred to as **reactants,** into one or more different substances, known as **products.**

Chemical change: A change in the fixed composition of a substance, such as occurs during a chemical reaction.

Reactant: A substance that undergoes a chemical change.

Product: A substance that results from a chemical change.

Chemical equation: A symbolic representation of a chemical change.

We may represent the transformation that occurs during a chemical change by writing a **chemical equation,** a symbolic description of the change. For example, the following chemical equation describes the reaction that occurs when sodium hydrogen carbonate ($NaHCO_3$, also known as sodium bicarbonate or, more commonly, as baking soda) reacts with hydrochloric acid (HCl). The products formed are sodium chloride (NaCl), water (H_2O), and carbon dioxide (CO_2).

$$\underbrace{NaHCO_3 + HCl}_{\textbf{Reactants}} \rightarrow \underbrace{NaCl + H_2O + CO_2}_{\textbf{Products}}$$

The arrow may be taken to mean "form" or "react to form," and it separates the reactants (which are usually written on the left) from the products (which are written on the right). This equation says, "Sodium hydrogen carbonate and hydrochloric acid react to form sodium chloride, water, and carbon dioxide."

PROBLEM 2.11 What is the difference between a physical change and a chemical change?

Conservation of Mass

OBJECTIVE 2.8B Explain the Law of Conservation of Mass.

Although the products of a chemical reaction are chemically different from the reactants, the total amount of matter produced during a reaction equals the amount present before the reaction took place. For example, consider what happens when we heat a sample of the compound silver oxide, Ag_2O, which is made up of silver and oxygen. As we heat the compound, we observe that the brownish-black silver oxide has changed to a silvery, metallic-looking species (substance), indicating that a new substance has been formed. If we weigh our sample prior to heating, and then reweigh it afterward, we find that the original sample has lost mass. What has happened? If we carry out the experiment under carefully controlled conditions, we find that a gas is given off. Analysis of the gas reveals that its mass is equal to the mass lost from the solid starting material. Thus, the total mass of the products formed equals the total mass of the reactants.

The Law of Conservation of Mass: Mass can be neither created nor destroyed.

This fundamental principle is a guide for much of what comes in the following chapters. Known as the **Law of Conservation of Mass,** the principle states that matter can be neither created nor destroyed during a chemical change.

Twentieth-century investigations have demonstrated a relationship between matter and energy, such that the two may be interconverted under certain conditions. This modification of nineteenth-century theories is incorporated into a more complete relationship known as the *Law of Conservation of Mass and Energy*. This law takes into account the fact that extremely small amounts of matter may be lost in the production of chemical energy. However, the error introduced by this deviation from the Law of Conservation of Mass is so slight that it is undetectable under normal laboratory conditions.

Chemical Energy

OBJECTIVE 2.8C Give an example of a chemical change in which energy is given off. Give an example of a chemical change in which energy is absorbed.

Chemical energy: Energy liberated or absorbed during a chemical change.

Earlier in this section, we saw that heat and light are given off by a burning log. Heat and light are both forms of energy that we refer to as **chemical energy,** because they accompany a chemical change. Chemical transformations are almost always accompanied by some form of energy change. Sometimes energy is liberated, as it is when a log burns. At other times it is absorbed from the surroundings. Such is the case when green plants use energy, in the form of sunlight, to produce sugar from atmospheric carbon dioxide and water. In the study of chemistry, energy plays a central role in many of the concepts we encounter.

2.9 PHYSICAL AND CHEMICAL PROPERTIES

OBJECTIVE 2.9 Explain the nature of a physical property and give three examples. Explain the nature of a chemical property and give two examples.

Just as a person may be described by his or her physical and personality attributes, such as height, eye color, and temperament, so can all substances be described by characteristics known as *physical* and *chemical properties.* Every pure substance has its own unique set of chemical and physical properties, and these can be used to distinguish it from other substances.

Physical property: A property of a substance that can be observed without changing its fixed composition.

Physical properties are those properties that can be observed or measured without changing the fixed composition of a substance. The many physical properties include color, taste, odor, hardness, melting point, and boiling point. Some of these properties describe a characteristic that involves a physical change, such as the melting point or the boiling point. Other physical properties, such as color or hardness, do not involve any apparent physical change. All, however, may be observed without altering the fixed composition of the substance.

Physical properties enable us to distinguish substances from one another. Iron is a hard, silvery metal that is practically tasteless and odorless and cannot be dissolved in water. It is a reasonably good conductor of heat and electricity and is attracted to a magnet. Table salt, by contrast, can be readily dissolved in water, is odorless, and tastes "salty." It is not a good conductor of electricity in its normal solid state, nor is it attracted to a magnet. Thus, it would not be difficult to distinguish iron from salt by its physical properties.

Similarly, physical properties also enable us to distinguish iron from copper. Like iron, copper is a hard, metallic substance that is tasteless, odorless, and a good conductor of electricity. However, unlike iron, copper has a reddish-gold color and is not attracted to a magnet. Furthermore, iron characteristically melts at 1535°C, whereas copper melts at 1083°C. The physical properties of iron and copper enable us to distinguish between these two substances with relative ease.

Chemical property: A property associated with a change in the fixed composition of a substance.

The various elements and compounds can also be distinguished from one another by their chemical properties. The **chemical properties** of a substance describe its tendencies to undergo a chemical change. For example, when iron rusts, a chemical change takes place in which the iron combines with oxygen to form a new substance containing both iron and oxygen. Thus, the statement "iron rusts" describes a chemical property of iron. Likewise, the description of gasoline as a "flammable" substance refers to the chemical change in which gasoline combines with oxygen to form two new substances, carbon dioxide and water.

In contrast to the measurement of a physical property, which requires no change in the fixed composition of a substance, the measurement of a chemical property requires that the substance undergo a change in its composition. Thus, to measure the amount of heat produced when a gram of ethyl alcohol burns (a chemical property), it is necessary to burn the ethyl alcohol (a chemical change). By contrast, the composition of each ethyl alcohol molecule remains unchanged when measuring its boiling point (a physical property), even though the liquid form is converted to the gaseous state (a physical change).

PROBLEM 2.12

Identify each of the following as either a physical property or a chemical property.

(a) Hydrogen sulfide smells like rotten eggs.
(b) Ice melts at 0°C. (0°C is the same temperature as 32°F.)
(c) Chlorine is a pale green gas.
(d) Natural gas burns.
(e) Diamond is hard.
(f) Wood will rot.
(g) Sugar dissolves in water.

CHAPTER SUMMARY

Matter may exist in one of three **physical states** (or **phases**): **solid, liquid,** or **gas.** If we could examine matter in its finest detail, we would find that it is composed of tiny particles. In the solid and liquid states (the "condensed phases"), these particles are closely packed. Particles of gaseous matter are spread out, permitting expansion or compression of gases. Solids resist any change in shape, whereas liquids flow freely, but both have fixed volumes. Gases assume the volume of the container in which they are placed.

Matter may be described as **homogeneous** or **heterogeneous.** Homogeneous matter is uniform throughout, whereas distinct boundaries exist within portions of heterogeneous matter. Matter may also be classified as a mixture or a **pure substance.** A pure substance has a fixed composition; the composition of a mixture is variable.

Pure substances may be classified as **elements** or **compounds.** An element cannot be decomposed into simpler substances, whereas compounds can be broken down. A **molecule** is the smallest identifiable unit of any pure substance. A molecule is composed of **atoms.** Elements are substances composed of only one type of atom, whereas a compound contains at least two different types of atoms. There are the same number of types of atoms as there are elements. The elements are organized in the **periodic table.**

The composition of a pure substance is generally expressed in a **chemical formula,** which gives the number of atoms of each element present in a molecule (or other unit) of the substance. Many substances are also characterized as **monatomic, diatomic, triatomic,** or **polyatomic,** according to the number of atoms per molecule.

Pure substances may be described in terms of their physical and chemical properties. A **physical property,** such as color or melting point, is one that can be determined without any change in the fixed composition of the substance. A **phase change** is an example of a **physical change.** When a substance *melts, boils,* or *sublimes* (all phase changes), its composition remains the same. We refer to the respective temperatures at which these phase changes occur as the **melting point, boiling point,** or **sublimation point** of the substance.

A **chemical property,** such as the ability to burn, involves a change of the substance into one or more new substances. We refer to such a change as a **chemical change.** When a chemical change takes place, one or more substances, known as **reactants,** are transformed into one or more new substances, known as **products.** This process is generally described by a **chemical equation,** which represents the **chemical reaction** that occurs. All chemical changes obey the **Law of Conservation of Mass,** which states that matter can be neither created nor destroyed in a chemical reaction. Chemical changes are usually accompanied by the liberation or absorption of **chemical energy.**

KEY TERMS

Review each of the following terms, which are discussed in this chapter and defined in the Glossary.

matter
physical state
phase
solid
liquid
gas
heterogeneous matter
homogeneous matter
pure substance
compound
element
periodic table
molecule
atom
monatomic
diatomic
triatomic
polyatomic
chemical formula
physical change
phase change
melting point
boiling point
sublimation
sublimation point
chemical change
chemical reaction
reactant
product
chemical equation
Law of Conservation of Mass
chemical energy
physical property
chemical property

ADDITIONAL PROBLEMS

SECTION 2.1

2.13 Name each of the three physical states, describe the characteristics of each, and give two examples of each.

2.14 Classify each of the following as solid, liquid, or gas.

(a) Marble **(b)** Oil **(c)** Safety pin
(d) Air **(e)** Kerosene **(f)** Helium

SECTION 2.2

2.15 Identify each of the following as heterogeneous or homogeneous.

(a) Iced tea with nothing in it
(b) Iced tea with undissolved sugar in it
(c) Iced tea with sugar that is completely dissolved
(d) Iced tea with ice cubes

2.16 Identify each of the following as heterogeneous or homogeneous.
(a) A salt–pepper mixture
(b) An alloy of tin and lead (known as solder)
(c) Fog
(d) A gasoline–water mixture
(e) An alcohol–water mixture
(f) Milk
(g) A carbonated soft drink

SECTION 2.3

2.17 Identify each of the following as either a pure substance or a homogeneous mixture.
(a) Water (b) Salt (c) Salt water
(d) Iron (e) Copper (f) Bronze
(g) Tea (h) Oxygen (i) Carbon dioxide
(j) Clean air

SECTION 2.4

2.18 Define the terms *element* and *compound.*

2.19 What symbol is used to represent each of the following elements?
(a) Helium (b) Silicon (c) Silver
(d) Argon (e) Uranium (f) Sulfur
(g) Copper (h) Potassium (i) Gold
(j) Iron

2.20 What are the names of the elements that are symbolized as follows?
(a) N (b) C (c) Na (d) Cl (e) S
(f) H (g) O (h) Mg (i) Ca (j) Pb

2.21 Which of the following are elements? Which are compounds?
(a) C_2H_6 (b) N_2 (c) Xe (d) CO (e) S_8
(f) $KMnO_4$ (g) Br_2 (h) P_4

SECTION 2.5

2.22 Draw a picture of a molecule of water and identify the atoms present.

2.23 Referring to Fig. 2-5, draw a picture of a molecule of each of the following. In every case, list the type and number of each kind of atom present per molecule.
(a) Helium (b) Methane (c) Carbon dioxide
(d) Oxygen

SECTION 2.6

2.24 How many atoms of each element are in each of the following molecules?
(a) H_3PO_4 (b) $C_{12}H_{22}O_{11}$ (c) HNO_3
(d) P_4O_{10} (e) $C_3H_2F_6$

2.25 Write the chemical formula that corresponds to each of the following compositions.
(a) Two atoms of lithium, one atom of carbon, and three atoms of oxygen (lithium carbonate)
(b) Two atoms of potassium and one atom of sulfur (potassium sulfide)
(c) Two atoms of aluminum and three atoms of oxygen (aluminum oxide)
(d) Twelve atoms of carbon, twenty-two atoms of hydrogen, and eleven atoms of oxygen (sucrose, table sugar)
(e) One atom of carbon, three atoms of hydrogen, one atom of nitrogen, and two atoms of oxygen (nitromethane)

2.26 How many oxygen atoms are present in each of the following formulas?
(a) KNO_3 (potassium nitrate)
(b) $Mg(NO_3)_2$ (magnesium nitrate)
(c) $Al(NO_3)_3$ (aluminum nitrate)
(d) What atoms must be present in the nitrate group?

SECTION 2.7

2.27 Explain the nature of a physical change and give an example.

SECTION 2.8

2.28 Explain the nature of a chemical change and give an example.

2.29 State the Law of Conservation of Mass and explain why it is true for chemical changes.

2.30 Give an example of a chemical change in which energy is given off. Give an example of a chemical change in which energy is absorbed.

SECTION 2.9

2.31 Explain the nature of a physical property and give three examples.

2.32 Explain the nature of a chemical property and give two examples.

2.33 The following properties describe magnesium. Classify each as either a chemical or a physical property.
(a) Solid at room temperature
(b) Conducts electricity
(c) Gray
(d) Metallic
(e) Forms a white powdery substance when exposed to hydrogen chloride gas
(f) Burns in air with an intense white flame
(g) Melts at 651°C

GENERAL PROBLEMS

Note: Problems indicated by an asterisk (*) in this and future chapters are more challenging than other problems.

2.34 Marble is made up of calcium carbonate, $CaCO_3$. When treated with acid, marble gives off a well-known gas composed of elements contained in the marble. What do you suppose the gas might be?

***2.35** When mothballs are placed in a closet or drawer, they eventually disappear.
(a) Is this phenomenon a violation of the Law of Conservation of Mass? Explain your answer.
(b) Give a possible explanation (a hypothesis) of what becomes of the mothballs. [*Hint:* When water evaporates, water molecules in the liquid state escape into the gaseous state at a temperature below the boiling point of the liquid.]
(c) What evidence do you have to support your hypothesis? [*Hint:* Think about your various senses, and remember what you experience when you open a closet or drawer that contains mothballs.]
(d) Based upon your hypothesis, is the disappearance of the mothballs the result of a chemical change or a physical change?

***2.36** The names and formulas of five compounds follow. Use the pattern of combining capacities to predict the missing formula of the sixth compound.
(a) Potassium nitrate, KNO_3
(b) Potassium sulfate, K_2SO_4
(c) Magnesium nitrate, $Mg(NO_3)_2$
(d) Magnesium sulfate, $MgSO_4$
(e) Aluminum nitrate, $Al(NO_3)_3$
(f) Aluminum sulfate, ?

2.37 If the reactants in a chemical reaction are mixed together in a sealed container, will the mass at the end of the reaction be greater than, equal to, or less than it was at the beginning of the reaction? Explain your answer.

***2.38** When making pottery, there is a period of time known as the curing period. The clay is heated and then cooled. When the clay is cooled back to room temperature, it assumes different physical properties. Although only a change in temperature was applied to the clay, is the process a physical or a chemical change? Explain your answer.

2.39 If trapped in a dark elevator, why is it a bad idea to light a candle?

WRITING EXERCISES

2.40 A friend who has never studied chemistry has asked you to explain how chemical formulas are written. Write an explanation for your friend. Be sure to explain the correct manner for writing the symbols for the elements, as well as the use of subscripts and parentheses.

2.41 Water in the ocean is warmed by the sun, evaporates, and is carried over the mountains where it falls as snow on the mountain peaks. Eventually, the snow melts and flows down the mountains, back to the ocean. Write a paragraph describing the various phase changes that occur during this hydrologic cycle.

2.42 Write a paragraph describing what occurs when a campfire burns.

MEASUREMENT

In this chapter we begin to quantify our descriptions of matter. With this goal in mind, we first discuss the uncertainty of measured numbers and the use of units in carrying out calculations. We then consider several properties of matter that a chemist is likely to measure. However, before proceeding further with this chapter, you should review the following mathematical skills, which are discussed in Appendixes A and B at the end of the text: (a) rounding off numbers, (b) expressing fractions in decimal notation, (c) powers of 10, (d) expressing numbers in scientific notation, (e) working with numbers in scientific notation, and (f) using a calculator to multiply, divide, add, and subtract.

3.1 A Word About Problem-Solving

As the introductory remarks indicate, we begin to tackle numerical problems in this chapter. Every attempt has been made in this book to facilitate your development as a problem-solver. In most sections of the text, there are one or more worked-out examples that apply the concepts developed in that section. These examples are always followed by similar problems for you to solve. Many of the problems are drill-type problems that require you to apply a new concept or definition. Others are word problems that require you to combine a group of concepts or definitions in order to arrive at a solution.

As you approach a problem, the following general sequence of steps will help you work through it:

1. Read the problem carefully.
2. Determine what it is that you are trying to find out. In other words, what is the *unknown* you are looking for?
3. List the *known* information you are given. For example, if you are asked how many inches are in 5 miles, the information that you are given is *5 miles.*
4. Map out a strategy for attacking the problem. In the problem just stated, you might wish to find the number of feet in 5 miles and then find the number of inches from the number of feet you have calculated. Lay out a road map of the strategy, for example,

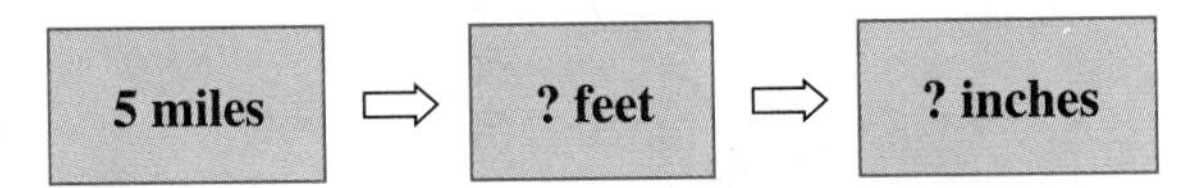

5. Write down the information that you need to complete the solution. In this case, you need to know that there are 5280 feet in a mile and 12 inches in a foot:

$$1 \text{ mile} = 5280 \text{ ft} \quad \text{and} \quad 1 \text{ ft} = 12 \text{ in.}$$

Sometimes a problem may contain some extra (or extraneous) information that you do not need. While you should certainly consider carefully which information is needed and which is not, do not be sidetracked by information that does not fit into your solution. For example, suppose you are told that a man who is 6 ft tall and weighs 170 lb with blond hair and blue eyes robbed a bank at 1:45 P.M. on June 15, 1994. If you are asked to give the sex of the suspect, there is no problem ignoring the extraneous information in order to answer that the suspect was male. In a similar fashion, you may need to sort through information provided in a problem to find the facts that are relevant to the question you are asked.

6. Calculate the answer to the problem. In this chapter we introduce you to a technique for solving problems with units. In addition to requiring you to use units in your calculations, you must show a *setup* and an *answer.* Be sure to use this technique from the very start. With practice, it will become second nature to you, and it will be a great aid to you in your problem-solving.
7. Check your setup. Reexamine your solution to be sure that you have used all of the numbers the way you intended. For example, suppose you divided where you meant to multiply. See that all of your units cancel properly.
8. Examine your answer for *reasonableness.* In other words, does your answer make sense? Suppose that in the problem discussed in Steps 3–5, you calculated an answer of 18 in. Since there are over 5000 ft in a mile, and a foot is larger than an inch, an answer of 18 in. is unreasonably small for a distance of 5 miles. You would know that you must have made an error. You must go back and check your calculations until you find that error.

In the worked-out examples, we model the procedure you are to follow by laying out road maps and discussing strategies for attacking problems. Do not try to memorize a routine for solving problems. Instead, try to follow the *logic* that is presented in the examples. As you come to understand the logic behind the solutions to the examples, you will learn what you need to know in order to attack new problems that are stated somewhat differently.

To facilitate your progress in approaching problems, many of the examples are broken down into one or more of the following categories:

1. *Symbolic statement.* Often the solution of a problem becomes more obvious if we can reduce it to a simple mathematical statement indicating what we are given and what we are trying to find.
2. *Planning the solution.* Once we have determined what we wish to find out, we lay out a stepwise strategy that leads us to our goal.
3. *Solution.* This is the actual calculation or set of calculations that we make, using our strategy as a guide.
4. *Checking the answer.* Whenever you finish a calculation, you should check over your numbers to see that you have used them the way you intended, and that your answer looks reasonable for the numbers you multiplied, divided, added, or subtracted. Limitations of space make it impossible for us to model that procedure in our examples. However, we will occasionally check the answer in an example to see if it makes sense. For example, it would seem unreasonable if the average height of basketball players in the National Basketball Association were calculated at $5\frac{1}{2}$ ft. Reason would tell us that most professional basketball players are at least 6 ft tall, and many are close to 7 ft in height. Although it is certainly *possible* for $5\frac{1}{2}$ ft to be a correct average, the nature of the specific problem itself (the group is restricted to professional basketball players) would lead us to expect an answer of around $6\frac{1}{2}$ ft. We will occasionally check an answer for this type of reasonableness.

In each chapter, there is a variety of problems with varying degrees of difficulty. Within the various sections of each chapter, there are very routine problems following the examples. These represent the basics that you need to know. A second, parallel set of problems appears at the end of each chapter in the Additional Problems, listed by section. However, you may find some of these problems to be more challenging than those found within the body of the chapter. The General Problems section at the end of the Additional Problems presents some problems that have not been covered directly, but they, too, are within your reach with a little thought and application of ideas already covered. Among these problems are several that are more challenging than the others. These are designated by an asterisk (*).

The key to success as you pursue this and the coming chapters is practice, practice, practice! Do not attempt to cram. Instead, if you work steadily, covering a few sections every day without fail, you will be amazed at how smoothly your problem-solving skills develop, and you will be rewarded with success in your study of chemistry.

3.2 UNCERTAINTY IN NUMBERS

OBJECTIVE 3.2 Distinguish between an exact number and a measured number. Explain what is meant by uncertainty. Explain the meaning of precision and accuracy.

Exact number: A number with no uncertainty.
Measured number: A number that has been determined by measurement and therefore has uncertainty associated with it.
Uncertainty: The degree to which a measured number is in doubt.

In every scientific inquiry it is necessary to gather facts and to organize them into meaningful relationships. In the process, chemists generally work in a laboratory where they make careful measurements related to the investigation at hand. The use of measurements enables the chemist to gather information that is accurate enough to use and transmit to others.

Before we examine the basic units of measure used in scientific work, it is helpful to learn a little more about numbers themselves. First, we can distinguish between an **exact number** and a **measured number.** When we buy a dozen eggs at the supermarket, the number of eggs in the carton is exactly 12. No uncertainty is involved. No matter how often we count the eggs, we always get 12. On the other hand, we often obtain numbers by using some type of measuring device, such as a tape measure or a bathroom scale. *Numbers obtained by measurement are not exact.* If we take repeated measurements, it is likely that we will get a range of values rather than the same value each time. There is **uncertainty** associated with measured numbers.

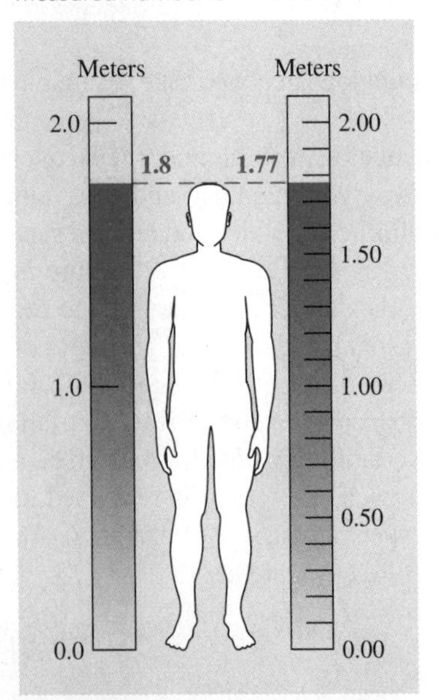

FIGURE 3-1 A comparison of two measuring devices. Because the one on the left shows less detail, a greater degree of uncertainty is associated with readings taken on it.

Suppose we wish to measure the height of the person pictured in Fig. 3-1. If we use the measuring stick shown on the left, we can determine with certainty that this person's height is somewhere between 1 and 2 meters. (In Section 3.6 we will see that a meter is slightly longer than a yard.) Looking more carefully, we can further estimate the height to be about 1.8 m. We can see that it is probably somewhere between 1.7 and 1.9 m, but we do not know exactly where. Because we are reasonably confident that the person's height is within 0.1 m of 1.8 m, we describe the uncertainty in our measurement as ±0.1 m.

Now suppose that we repeat the measurement using the measuring stick on the right. This time we can determine with certainty that the person's height is somewhere between 1.7 and 1.8 m, and we can further estimate it to be somewhere between 1.76 and 1.78 m. We assign it a value of 1.77 m. Using the second device enables us to determine a greater number of digits than the first. However, it is still necessary to estimate the last digit. In this case the uncertainty is ±0.01 m. No matter what measuring device we use, there will always be uncertainty in the last digit we read.

Precision: The agreement between two or more measured quantities.

Accuracy: The degree to which a measured number agrees with the correct value.

Related to uncertainty are two additional terms: *precision* and *accuracy.* **Precision** refers to how closely repeated measurements agree with one another. **Accuracy** is how close a measurement is to the correct value (Fig. 3-2). In carrying out scientific work, we repeat measurements to help catch careless errors. If our precision is high (the results agree with one another), we are more confident that we have measured correctly. Even so, it is possible that faulty equipment or a systematic error in making a series of measurements may lead us to a result that is not accurate. When chemists work in the laboratory, they strive for both accuracy and precision.

PROBLEM 3.1

Which of the following quantities are exact numbers? Which are measured numbers?

(a) A gross of pencils contains 144 pencils.

(b) The height of the Statue of Liberty is 45.3 meters.

(c) A dollar is worth 100 pennies.

(d) The distance from San Francisco to New York is 3000 miles.

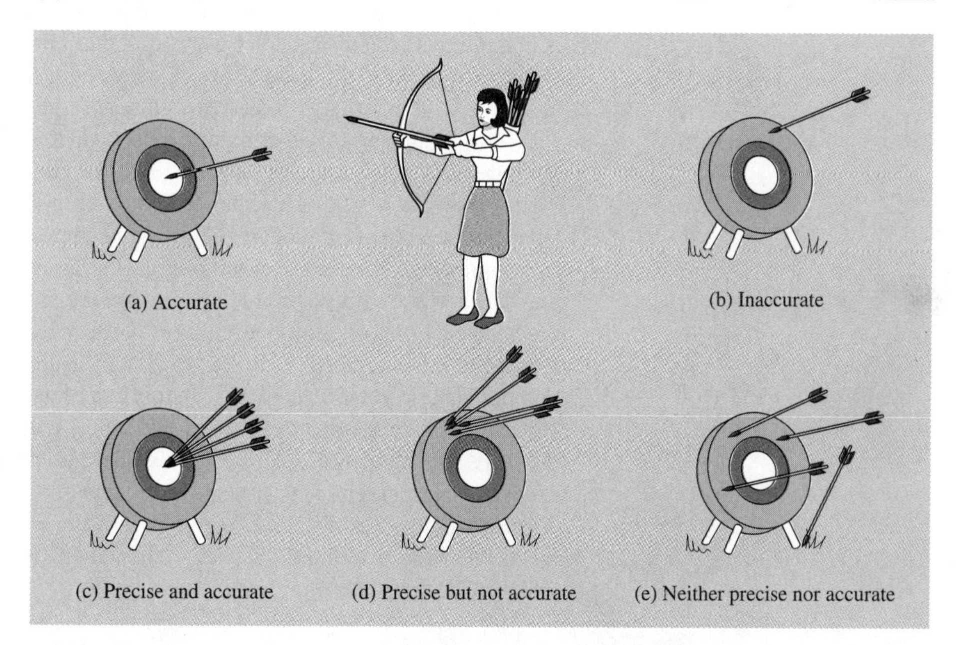

FIGURE 3-2 Like arrows aimed at the center of a target, a laboratory result is accurate when it is close to the correct value and inaccurate when it is far from that value. Precision exists when repeated results are close to one another.

3.3 SIGNIFICANT FIGURES

OBJECTIVE 3.3A Determine the number of significant figures in any number.

In scientific work it is important that we report only digits that we have actually measured. Suppose that, in our work with Fig. 3-1, we report our first measurement as 1.80 m rather than 1.8 m. Because we included a zero in the hundredths place, an outsider reading the measurement assumes that we know the person's height to be somewhere between 1.79 m and 1.81 m (an uncertainty of ±0.01 m). In fact, the best we were able to determine was that the height was somewhere between 1.7 m and 1.9 m (an uncertainty of ±0.1 m), so adding the extra zero amounts to misrepresenting our data. (For the purpose of simplicity in this text, we assume that *estimated* digits vary by ±1. Thus, in our example, we assume that the number 1.8 varies from 1.7 to 1.9. In practice, the actual variation depends on the measuring device and may differ from ±1.)

Any time we report a measured number, it is assumed that all of the digits in the number are accurately known except the last digit, which represents an estimate. For example, if we measure an object on a balance and report a mass of 12.34 g, it means that we are certain of the first three digits and have estimated the fourth. We refer to each of these digits as a **significant digit** or **significant figure.** We are frequently interested in the *number* of significant figures. To determine the number of significant figures, we simply count the digits present. For example, 12.34 contains four significant figures, 1.8 contains two significant figures, and 1.77 contains three significant figures. It sounds simple and it is. You need only one more rule to assign correctly the number of significant figures that a number contains. *Any zero whose sole function is to hold a decimal point is not a significant digit.*

Significant figures (or significant digits): In a measured number, any of the digits that are known exactly, plus the last digit, which has been estimated.

For example, in the number 0.04, the zeros are needed in order to show that the 4 is in the hundredths place. They serve no other purpose than to locate the decimal and, therefore, are not significant. Thus, there is only one significant figure in 0.04. Similarly, the number 0.00853 has three significant figures.

A certain amount of ambiguity exists when we write a number like 500, as in the measurement of the mass of an object that weighs 500 g. Remember that a significant digit is one that has been measured. If we were to measure the mass of an object on a balance that reads to the nearest 100 g, our measurement would tell us that the mass was somewhere between 400 and 600 g. In other words, only the 5 is a significant figure. The zeros are needed to locate the decimal and they tell us that the 5 is in the hundreds place. Suppose, however, that we had a more sensitive balance that is capable of reading to the nearest 10 g. In this case, a measurement of 500 g would mean that the number is between 490 and 510 g. In this case, the zero in the tens place *has* been measured and *is* significant, but the zero in the units place is just holding the decimal. Here, there are two significant figures. Finally, if we used a balance that can read to the nearest gram, a measurement of 500 g would mean that the mass is between 499 and 501 g. The zero in the units place has been measured, and all three digits are significant. Thus, when we write a number such as 500, without more information, we do not know whether the zeros are significant.

In the absence of additional information (such as was given in the three cases we just considered), we generally assume that the zeros holding the decimal are *not* significant. (For example, there are two significant figures in the number 2500.) If we wish to show that a particular zero is significant in a number such as 500, the only universally accepted method is to use scientific notation, as follows:

Number of significant figures	Scientific notation	Uncertainty	Number is between
1	5×10^2	±100	400 and 600
2	5.0×10^2	±10	490 and 510
3	5.00×10^2	±1	499 and 501

When using scientific notation, the number of significant figures is determined *only* by the factor that precedes the power of 10. (If you are not familiar with scientific notation or need to review it, refer to Appendix A at the end of the text.)

In practice, it is unlikely that a chemist would express a volume such as 500 mL in scientific notation; yet most common measuring equipment is accurate to the nearest ±1 mL. Similarly, in the measurement of temperature, laboratory thermometers are generally accurate to the nearest ±1°C, and gas pressure can easily be measured to the nearest ±1 torr. (The torr is a unit of pressure.) For the sake of simplicity in this text, we will assume that all volumes expressed in units of milliliters are accurate to at least the nearest ±1mL, all temperatures are accurate to the nearest ±1°C, and all pressures expressed in torr are accurate to ±1 torr. Thus, we interpret a pressure of 1700 torr to correspond to 1.700×10^3 torr, having four significant figures. However, when interpreting other nonscientific numbers, we will assume that zeros whose only *apparent* function is to hold the decimal point are not significant.

The following examples should help to illustrate the rules for assignment of significant figures.

EXAMPLE 3.1

How many significant figures are there in each of the following numbers?

113	Three
1.7254	Five
0.012	Two (Both zeros shown are merely holding a decimal point.)
12.0	Three (The zero is not needed to hold the decimal point, indicating that it must be significant.)
37,800	Three (Without being given information about how this number was measured, we must assume that the zeros were not measured and are therefore not significant. In other words, the number is somewhere between 37,700 and 37,900.) This number can be expressed as 3.78×10^4.
37,800.0	Six (Since the last zero is not needed to hold the decimal point, it is significant, and therefore all of the zeros preceding it must be significant.) In scientific notation, this is 3.78000×10^4.
0.00002	One (All of the zeros are holding the decimal point.) This is expressed in scientific notation as 2×10^{-5}.

1.00002	Six (Since the first digit is not a zero, it must have been measured. Since the last digit was also measured, all of the zeros in between must have been measured and are therefore significant.)
7.6×10^3	Two (This number corresponds to 7600 and has been measured to the nearest hundred.)
7.600×10^3	Four (This number corresponds to 7600 and has been measured to the nearest unit.)

Students often wonder why zeros are not significant when their sole function is to hold the decimal point. Let us consider the number we obtained when we made our first measurement of the person in Fig. 3-1. We decided that the height was 1.8 m, somewhere between 1.7 m and 1.9 m. Now suppose that for some reason we wish to express this measurement in centimeters. There are 100 centimeters (cm) in each meter, so the height is expressed as 180 cm. Someone who presumes the zero in this number to be significant, however, will conclude that we knew the height to be between 179 and 181 cm, when in fact our actual estimate was between 170 and 190 cm. Similarly, if we wish to express the height in kilometers (1000 m = 1 km), it is 0.0018 km. Our actual measurement contained only two significant figures, and it is unreasonable to believe that converting from meters to kilometers changes the uncertainty of our original measurement. Using zeros to move a decimal point does not change the uncertainty of the original measurement, and these zeros are simply holding the decimal point, so they are not significant.

PROBLEM 3.2

How many significant figures are there in each of the following numbers?

(a) 412 **(b)** 2.6 **(c)** 130 **(d)** 10.00 **(e)** 4.63
(f) 0.463 **(g)** 0.0039 **(h)** 0.0390 **(i)** 14.00 **(j)** 12,300
(k) 1,560,000.0 **(l)** 4.000×10^3 **(m)** 4.0×10^3 **(n)** 4.00×10^{-5}

PROBLEM 3.3

Round off each of the following numbers to the number of significant figures indicated. (If you are not familiar with the rules for rounding off, see Appendix A, Section A.1.)

(a) 17.6432 to 3 significant figures **(b)** 0.006267 to 3 significant figures
(c) 35.973 to 3 significant figures **(d)** 35.973 to 2 significant figures
(e) 35.973 to 1 significant figure **(f)** 6.9951×10^{-6} to 2 significant figures

PROBLEM 3.4

If the number 61,000 is measured to the nearest tens, how many significant figures does it have? Express the number in scientific notation.

Working With Significant Figures

OBJECTIVE 3.3B Express the answer to any addition, subtraction, multiplication, or division problem to the correct number of significant figures.

In carrying out mathematical calculations, it is important that we obey two very simple rules about significant figures.

Addition and Subtraction *Any time two or more numbers are added or subtracted, the final answer is rounded off to the same decimal place as the number having its last significant figure (the estimated digit) the furthest to the left.* This corresponds to the number having its final measured digit in the largest decimal place. (If you are not familiar with the rules for rounding off, refer to Appendix A, Section A.1.)

For example, suppose we add 1.12 and 2.3:

```
  1.12
+ 2.3     ←Estimated digit is furthest to left
  3.42   rounded off to 3.4
```

Our answer must be 3.4, not 3.42, because 2.3 has its last significant figure (its estimated digit) the furthest to the left. Since our knowledge of the number 3.4 ends with the tenths place (which is larger than the hundredths place), it is impossible for us to know anything about the hundredths place in our sum.

Let us examine this problem more closely. Suppose we are combining the lengths (in centimeters) of two objects. The object that was measured to be 2.3 cm must be somewhere between 2.2 and 2.4 cm in length. The other object must be somewhere between 1.11 and 1.13 cm in length. It is possible that their sum is as little as 3.31 cm or as much as 3.53 cm.

```
The sum is   as small as    or   as large as
               1.11 cm             1.13 cm
             + 2.2  cm      or   + 2.4  cm
               3.31 cm             3.53 cm
```

If we were to report the sum as 3.42 cm, we would be claiming to know their sum with an uncertainty of ±0.01 cm. This would misrepresent our true uncertainty, because the sum could be anywhere from 3.31 to 3.53 cm. In rounding off our answer to 3.4 cm, we are claiming that the sum lies between 3.3 and 3.5 cm. This is a much more accurate picture of our uncertainty.

EXAMPLE 3.2

Add the following numbers.

```
   343.25
    26.417
43,321.12
     4.5
```

Planning the solution

First we will add up all of the numbers. Then we must round off the answer to the same place as 4.5, because that number has its estimated digit the furthest to the left.

SOLUTION

```
   343.2|5
    26.4|17
43,321.1|2
     4.5|
43,695.2|87   rounded off to 43,695.3
```

EXAMPLE 3.3

Add the following numbers.

```
 26.4
  0.453
275.
```

Planning the solution

We must round this off to the units place, the same column as the last significant digit in 275. The other numbers have digits to the right of that column.

SOLUTION

```
 26.|4
  0.|453
275.|
301.|853   rounded off to 302
```

EXAMPLE 3.4

Add the following numbers.

```
12,400
   230
```

Planning the solution

The last significant figure (the estimated digit) in 12,400 is the 4 in the hundreds place. The last significant figure in 230 is the 3 in the tens place. We round off our answer to the hundreds place since that column is the furthest to the left.

SOLUTION

```
12,4|00   ←Hundreds column has estimated
   2|30     digit furthest to the left
12,6|30   rounded off to 12,600
```

EXAMPLE 3.5

Subtract the following.

```
 476.534
−  5.00
```

Planning the solution

The rule for subtraction is the same as that for addition. The answer is rounded off to the same place as 5.00, the number with the fewest digits (two) to the right of the decimal point.

SOLUTION

```
 476.53|4
−  5.00|
 471.53|4   rounded off to 471.53
```

PROBLEM 3.5

Express each of the following sums and differences to the correct number of significant figures.

(a) 40.2 + 3.4 **(b)** 32 + 11.7 + 1.58 **(c)** 14.00 − 1.3
(d) 0.943 − 0.1 **(e)** 226.113 + 30 + 1582

Multiplication and Division *Any time we multiply or divide, our answer must be rounded off to the same number of significant figures as the number entered into the calculation that contains the least number of significant figures.*

For example, suppose we divide 2.7 by 3.61.

$$\frac{2.7}{3.61} = 0.7479 \quad \text{rounded off to } 0.75$$

2 significant figures (2.7)
3 significant figures (3.61)
2 significant figures (0.75)

The answer is rounded off to two significant figures, the same number of significant figures as there are in 2.7.

Once again, the reason for this rule is quite simple. The quotient of this division could be as small as

$$\frac{2.6}{3.62} = 0.7182$$

or as large as

$$\frac{2.8}{3.60} = 0.7778$$

Certainly we have no right to report our answer to more than the second significant figure. As you can see by the wide variation in possible answers (from 0.7182 to 0.7778), even our system of significant figures is somewhat less than perfect. However, it will keep our calculations reasonably close to the precision we are entitled to report.

EXAMPLE 3.6

Carry out the following operations.

(a) $(3.74)(2.653) = ?$ **(b)** $\frac{276.54}{47.3} = ?$ **(c)** $\frac{(274.6)(4.3)}{74.3125} = ?$

Planning the solution

We must round off each product or quotient to the same number of significant figures as the factor with the fewest significant figures.

SOLUTION

(a) The number 3.74 has three significant figures, whereas 2.653 has four significant figures. We must round off the answer to three significant figures.

$$(3.74)(2.653) = 9.9222 \quad \text{rounded off to } 9.92$$

(b) The number 47.3 has the fewest significant figures—three. We must round off the answer to three significant figures.

$$\frac{276.54}{47.3} = 5.8465 \quad \text{rounded off to } 5.85$$

(c) The number 4.3 has two significant figures—the fewest of all the factors in the calculation. Thus, the answer must be rounded off to two significant figures.

$$\frac{(274.6)(4.3)}{74.3125} = 15.889 \quad \text{rounded off to } 16$$

PROBLEM 3.6

For each of the following, select the correct form for the final answer.

(a) (5.24)(3.65) = ? **(i)** 19.126 **(ii)** 19.13 **(iii)** 19.1 **(iv)** 19

(b) $\frac{52.78}{13.6} = ?$ **(i)** 3.8809 **(ii)** 3.881 **(iii)** 3.88 **(iv)** 3.9

(c) $\frac{6.744}{3.000} = ?$ **(i)** 2.248 **(ii)** 2.25 **(iii)** 2.2 **(iv)** 2

(d) (47.6)(24) = ? **(i)** 1142.4 **(ii)** 1142 **(iii)** 1140 **(iv)** 1100

(e) $\frac{16}{8.000} = ?$ **(i)** 2.000 **(ii)** 2.00 **(iii)** 2.0 **(iv)** 2

(f) (4)(1.235) = ? **(i)** 4.940 **(ii)** 4.94 **(iii)** 4.9 **(iv)** 5

Occasionally, it is necessary to combine addition or subtraction with multiplication or division, as in the following example.

EXAMPLE 3.7

Evaluate the following expression.

$$(3.18 + 1.2) \times (4.85 - 1.827) = ?$$

Planning the solution

Examination of the expression reveals that we must first carry out the addition and subtraction, followed by the multiplication.

SOLUTION

First we add and subtract the numbers within the parentheses.

$$(3.18 + 1.2) \times (4.85 - 1.827) = ?$$
$$(4.4) \times (3.02) = ?$$

Then we multiply the factors together.

$$(4.4) \times (3.02) = 13$$

In practice, the expression should first be evaluated as shown to determine the correct number of significant figures for the final answer. Then the calculation should be repeated without rounding off until the final answer:

$$(3.18 + 1.2) \times (4.85 - 1.827) = ?$$
$$(4.38) \times (3.023) = 13.24074 \quad \text{rounded off to } 13$$

PROBLEM 3.7

Round off the following answers in accord with the rules for significant figures.

(a) $(12.623 - 1.00) \times (0.338 + 0.21) = 6.369404$

(b) $\frac{123.45 - 65.65}{210.0 - 111.0} = 0.5838384$

3.4 DIMENSIONAL ANALYSIS: THE USE OF UNITS IN CALCULATIONS

OBJECTIVE 3.4 Use the dimensional analysis technique to convert a quantity expressed in terms of one unit to an equivalent quantity expressed in terms of another unit, given one or more appropriate conversion factors.

Dimensional analysis: Problem-solving technique that relies on cancellation of units.

In our scientific work, it is frequently necessary to convert our measured quantities from one unit to another. For example, suppose we wish to know how many inches there are in 5 ft. This problem is so simple that you can probably do it in your head. However, we can use it to illustrate a method of problem-solving known as **dimensional analysis.** First we identify the *unknown* we are looking for, as well as the *known,* which is what we are given:

$$? \text{ in.} = 5 \text{ ft}$$

This equation states that we wish to find the number of inches in 5 ft. To do so, we use the equivalency relationship between feet and inches:

$$1 \text{ ft} = 12 \text{ in.}$$

Because both sides of the relationship represent exactly the same distance, we can make the following mathematical statement:

$$\frac{12 \text{ in.}}{1 \text{ ft}} = 1 \qquad (3\text{-}1)$$

Conversion factor: A relationship between quantities that permits conversion from one unit to another.

This statement merely says that the numerator and denominator are equal. Just as $\frac{6}{6} = 1$ and $\frac{223}{223} = 1$, any time the numerator and denominator are equal, the fraction has a value of 1. We refer to the fraction in equation (3-1) as a **conversion factor.** Like all conversion factors, it allows us to convert a quantity from one unit (such as feet) to another (such as inches).

To find the number of inches in 5 ft, we multiply the known quantity by this conversion factor:

$$? \text{ in.} = 5\,\cancel{\text{ft}}\left(\frac{12 \text{ in.}}{1\,\cancel{\text{ft}}}\right)$$

Because 12 in./1 ft is equal to 1, multiplying by it does not change the length represented by 5 ft. However, when these two quantities are multiplied together, the unit of feet in the numerator cancels the unit of feet in the denominator.

$$? \text{ in.} = \frac{(5\,\cancel{\text{ft}})(12 \text{ in.})}{1\,\cancel{\text{ft}}} = (5)(12 \text{ in.}) = 60 \text{ in.}$$

The distance originally expressed as 5 ft is now expressed in inches. We have used the equivalency of 12 in. = 1 ft to calculate the number of inches in 5 ft.

Suppose we had chosen to write our conversion factor with a numerator of 1 ft and a denominator of 12 in., another correct way to state the relationship between inches and feet:

$$\frac{1\text{ ft}}{12\text{ in.}} = 1$$

When we multiply 5 ft by this conversion factor, the units no longer cancel.

$$?\text{ in.} = 5\text{ ft}\left(\frac{1\text{ ft}}{12\text{ in.}}\right) = \frac{(5\text{ ft})(1\text{ ft})}{12\text{ in.}} = \frac{5\text{ ft}^2}{12\text{ in.}}$$

There is no agreement between the units we wish to find (in.) and the units we have calculated ($\text{ft}^2/\text{in.}$). We have not found the number of inches in 5 ft.

For any equivalency relationship, there are always two and only two conversion factors possible. For example, the relationship between feet and miles,

$$1\text{ mi} = 5280\text{ ft}$$

leads to the following conversion factors:

$$\frac{1\text{ mi}}{5280\text{ ft}} = 1 \quad \text{and} \quad \frac{5280\text{ ft}}{1\text{ mi}} = 1 \qquad (3\text{-}2)$$

To determine which conversion factor to use in a given problem, ask yourself, "Which unit do I want to cancel? Which unit do I want to end up with? Where do these units have to be in the conversion factor to cancel properly?" Only one of the two conversion factors correctly converts the units; the other always leads to an answer in units that make no sense. Let us solve a few more sample problems to illustrate this method further.

EXAMPLE 3.8

Calculate the number of feet in 2.50 miles.

Symbolic statement

First we identify the *unknown* we are looking for, as well as the *known,* which we are given. We are given 2.50 miles. We wish to determine how many feet this equals. We can state this information symbolically as follows:

$$?\text{ ft} = 2.50\text{ mi}$$

Planning the solution

This problem requires the equivalency relationship between feet and miles:

$$1\text{ mi} = 5280\text{ ft}$$

Using this relationship, we can convert directly from miles to feet:

2.50 mi ⇨ **? ft**

SOLUTION

The two possible conversion factors for the relationship between feet and miles are

$$\frac{1\text{ mi}}{5280\text{ ft}} = 1 \quad \text{and} \quad 1 = \frac{5280\text{ ft}}{1\text{ mi}}$$

We wish to convert miles to feet, so we choose the conversion factor on the right. It cancels the units of miles, leaving an answer with units of feet:

$$? \text{ ft} = 2.50 \cancel{\text{mi}} \left(\frac{5280 \text{ ft}}{1 \cancel{\text{mi}}} \right) = 13{,}200 \text{ ft}$$

If we had accidentally chosen the wrong conversion factor, the resulting calculation would have given units of mi^2/ft rather than ft:

$$? \text{ ft} = 2.50 \text{ mi} \left(\frac{1 \text{ mi}}{5280 \text{ ft}} \right) = \frac{2.50 \text{ mi}^2}{5280 \text{ ft}}$$

This would have told us that we had chosen the wrong conversion factor, and we would have gone back and reworked the problem using the correct one.

Checking the answer

Before concluding this example, let us check our answer for reasonableness. We were asked the number of feet in 2.50 miles. A mile is slightly more than 5000 ft. So 2 miles would be slightly more than 10,000 ft; and 2.50 miles would be somewhat larger yet. An answer of 13,200 ft is in good agreement with what we would expect.

EXAMPLE 3.9

Calculate the number of inches in 1.40 miles.

Symbolic statement

In this example, the known quantity is 1.40 miles. The unknown we wish to find is the number of inches it equals.

$$? \text{ in.} = 1.40 \text{ mi}$$

Planning the solution

Unlike our last example, we do not have a direct conversion factor between inches and miles. However, we do have the ability to convert miles to feet and to convert feet to inches. We can map out our strategy for solving this problem schematically as follows:

1.40 mi ⇨ **? ft** ⇨ **? in.**

SOLUTION

The relationships we need in order to carry out these conversions are

$$1 \text{ mi} = 5280 \text{ ft} \quad \text{and} \quad 1 \text{ ft} = 12 \text{ in.}$$

The first step is to convert miles to feet, as we did in Example 3.8, by using one of the following conversion factors:

$$\frac{1 \text{ mi}}{5280 \text{ ft}} = 1 \quad \text{or} \quad 1 = \frac{5280 \text{ ft}}{1 \text{ mi}}$$

We use the conversion factor on the right, thus canceling out the units of miles:

$$? \text{ in.} = 1.40 \cancel{\text{mi}} \left(\frac{5280 \text{ ft}}{1 \cancel{\text{mi}}} \right) \tag{3-3}$$

Although we could evaluate the number of feet at this point, that is not necessary. Instead, we actually multiply this expression by a second conversion factor to convert feet to inches:

$$\frac{12 \text{ in.}}{1 \text{ ft}} = 1 \quad \text{or} \quad 1 = \frac{1 \text{ ft}}{12 \text{ in.}}$$

The conversion factor on the left cancels the units of feet.

$$? \text{ in.} = 1.40 \cancel{\text{mi}}\left(\frac{5280 \cancel{\text{ft}}}{1 \cancel{\text{mi}}}\right)\left(\frac{12 \text{ in.}}{1 \cancel{\text{ft}}}\right) = 88{,}700 \text{ in.} \quad (\text{or } 8.87 \times 10^4 \text{ in.}) \qquad (3\text{-}4)$$

Note that each step of the solution requires a conversion factor. Note also that the final answer has been rounded off to three significant figures, the same number of significant figures as there are in 1.40 miles. (Your calculator would have given an answer of 88,704.) Although some of the numbers in our conversion factors *appear* to have fewer than three significant figures, the relationships 1 ft = 12 in. and 1 mi = 5280 ft are *exact* relationships and may be considered to have an unlimited number of significant figures. We say more about exact versus measured relationships in Section 3.5.

As you work problems, you will undoubtedly be using a calculator. When carrying out multistep calculations, such as in the preceding example, wait until your final answer before you round off the digits displayed on your calculator to the correct number of significant figures. We apply dimensional analysis to a wide variety of problems throughout the text, and it is important that you become skilled in using it.

PROBLEM 3.8

Use dimensional analysis to convert each of the following quantities to the indicated units. (The relationships 1 ft = 12 in. and 1 yd = 3 ft are exact relationships.)
(a) 3.0 ft to inches **(b)** 54 in. to feet **(c)** 2.0 yd to inches

3.5 SI AND ENGLISH UNITS

OBJECTIVE 3.5A Identify the basic SI unit of measure for each of the following: length, mass, volume, temperature, pressure, and energy. State the numerical meaning of the following prefixes: mega, kilo, deci, centi, milli, micro, and nano.

English system: System of measurement used in the United States.

International System (SI): System of measurement used in most countries as well as for all scientific work.

In this chapter we explore two systems of measurement: the **English system** and the **International System (SI).** (SI stands for *le Système International d'Unités.*) The English system is a system of measurement that is still used in the United States, but it has been replaced in most parts of the world by the International System, which is the modern-day version of the metric system. The six types of measurement you will encounter in this course are measurements of length, mass, volume, temperature, pressure, and energy. Table 3-1 summarizes the basic SI and English system units for these quantities.

TABLE 3-1 SI and English units of measure

Quantity	SI unit	Related units used in scientific work	English unit
Length	meter (m)		foot (ft)
Mass	kilogram (kg)		pound (lb)
Volume	cubic meter (m^3)	liter (L)	quart (qt)
Temperature	Kelvin (K)	degree Celsius (°C)	degree Fahrenheit (°F)
Pressure	Pascal (Pa)	torr, atmosphere (atm)	pound per square inch (psi)
Energy	joule (J)	calorie (cal)	British thermal unit (Btu)

In addition to the basic SI units shown in Table 3-1, we use many other units that are either multiples or fractions of these basic units. For example, the *kilo*meter is a length equal to 1000 meters (10^3 m). A *centi*meter is equal to one one-hundredth ($\frac{1}{100}$ or 0.01) of a meter (10^{-2} m). As you can see, each of these units is related to the meter by some power of 10. The name of each consists of a prefix that tells us which power of 10, followed by the word *meter.* These simple relationships based on powers of 10 make it relatively easy for us to convert from one SI unit to another. Table 3-2 lists the various prefixes used with SI units. The most important of these have been italicized.

TABLE 3-2 Prefixes used with SI units

Prefix	Symbol	Multiple or fraction	Meaning
giga	G	1 000 000 000 = 10^9	one billion times
mega	M	1 000 000 = 10^6	one million times
kilo	k	1 000 = 10^3	one thousand times
hecto	h	100 = 10^2	one hundred times
deka	da	10 = 10^1	ten times
deci	d	0.1 = 10^{-1}	one-tenth of
centi	c	0.01 = 10^{-2}	one-hundredth of
milli	m	0.001 = 10^{-3}	one thousandth of
micro	μ	0.000 001 = 10^{-6}	one-millionth of
nano	n	0.000 000 001 = 10^{-9}	one-billionth of
pico	p	0.000 000 000 001 = 10^{-12}	one-trillionth of

The advantages of working with SI units rather than English units will become more obvious as you work with them. The SI units are systematically defined, and they are *all* related to one another by powers of 10. By contrast, the English units were defined much earlier, and no consistent relationship links all of them. The foot, for instance, was roughly equivalent to the length of the human foot. The mile represented 1000 double steps by a Roman legionary during Caesar's time. The yard was the distance from King Henry I's royal nose to the tips of his fingers. The cumbersome relationships between these units make them difficult to use. For example, there are 12 inches in a foot, 3 feet in a yard, and 5280 feet in a mile. Thus, converting from one English unit to another involves more complicated arithmetic than converting between SI units.

PROBLEM 3.9

If 1 cm equals one one-hundredth of a meter, how many centimeters are there in a meter? If 1 mL equals one one-thousandth of a liter, how many milliliters are there in a liter?

PROBLEM 3.10

State the relationship between each of the following.

(a) kilograms (kg) and grams (g)
(b) millimeters (mm) and meters (m)
(c) centiliters (cL) and liters (L)
(d) kilojoules (kJ) and joules (J)
(e) microseconds (μsec) and seconds (sec)
(f) milligrams (mg) and grams (g)
(g) megawatts (MW) and watts (W)
(h) microliters (μL) and liters (L)

Exact Versus Measured Relationships

OBJECTIVE 3.5B Predict whether a given equivalency is an exact relationship or a measured relationship.

As you proceed with your work, you may occasionally wonder how to tell whether a relationship is an exact one or whether it is measured. As a general rule, if the relationship is a *defined* relationship, it is exact. For example, the definition of kilo is 1000. Using this prefix we find that 1 km = 1000 m. Similarly, centi means $\frac{1}{100}$. Thus, 1 cm = $\frac{1}{100}$ m or 1 m = 100 cm. All these relationships are exact. In a similar fashion, the following relationships are also defined, and are therefore exact: 1 mile = 5280 ft, 1 ft = 12 in., 1 yd = 3 ft. As long as we stay within the same system of measurement (the SI system or the English system), the relationships are exact. When an exact relationship is used in a calculation, it may be considered to have an infinite, or unlimited, number of significant figures, because there is no uncertainty in the relationship.

However, when we wish to compare the units of one system with those in another, the comparison requires measurement and is generally not defined. Thus, when you go to the gas station, you may see a sign that tells you that 1 gal equals 3.785 L. This relationship is not exact and can be determined to a greater number of significant figures, such as 3.7854 L. (In fact, both of these values have been rounded off from values with an even greater number of significant figures.) Depending upon the value given, when a relationship *between* systems is used in a calculation, the number of significant figures in the relationship must be considered in determining the number of significant figures in the answer. Thus, the relationship 1 gal = 3.785 L has four significant figures (the number of significant figures in 3.785), whereas the relationship 1 gal = 3.7854 L has five significant figures (the number of significant figures in 3.7854). The number "1" that is associated with the units of gallons should be considered to have as many significant figures as the relationship itself. Thus, when the relationship 1 gal = 3.785 L is used in a calculation, the answer is limited to no more than four significant figures.

For the most part, the relationships *between* systems are measured and the value given has some uncertainty, as in the case of 3.785 L. However, occasionally, scientists will define the relationship between systems, giving an exact relationship to some conversion factor between systems. For example, an international agreement has *defined* the relationship between yards and meters as 1 yd = 0.9144 m. (Since there are exactly

36 in. in a yard, 1 in. is exactly $\frac{1}{36}$ of 0.9144 m, which is 0.0254 m or 2.54 cm.) As we proceed through the rest of this chapter, we will indicate which relationships are exact (and may therefore be ignored in determining the number of significant figures in a calculation) and which are measured or have been rounded off (thereby limiting the number of significant figures to which an answer may be reported).

PROBLEM 3.11

Predict whether each of the following relationships is *exact* or *measured.*
(a) 1 qt = 4 cups **(b)** 1 gal = 3.785 L **(c)** 1 gal = 4 qt
(d) 1 lb = 16 oz **(e)** 1 lb = 454 g

3.6 LENGTH

OBJECTIVE 3.6 Convert quantities of length expressed in either SI or English units to equivalent quantities in other SI or English units.

The basic SI unit of length is the *meter:* it was initially based on a measurement of the distance from the North Pole to the Equator. Division of this distance by 10 million gives the meter. Later, the meter became the distance between two engraved lines on a bar made of a platinum–iridium alloy. This bar is still kept at a specified temperature in a vault near Paris, but today the meter is defined even more precisely. It is related to a characteristic wavelength of light emitted by a substance known as krypton-86.

Conversion Between SI Units

Table 3-3 shows the most commonly encountered SI units of length. Each of these units is related to the meter by a power of 10. As a result, converting from one unit to another simply involves moving a decimal point. For example, suppose we want to convert 72.3 m to kilometers:

$$? \text{ km} = 72.3 \text{ m}$$

We use the equivalency 1 km = 1000 m as follows:

$$\frac{1 \text{ km}}{1000 \text{ m}} = 1 \quad \text{or} \quad 1 = \frac{1000 \text{ m}}{1 \text{ km}}$$

Choosing the conversion factor on the left leads to correct cancelation of the units:

$$? \text{ km} = 72.3 \cancel{\text{m}}\left(\frac{1 \text{ km}}{1000 \cancel{\text{m}}}\right) = 0.0723 \text{ km}$$

Note that we could have obtained the same result simply by moving the decimal point in 72.3 m three places to the left:

$$72.3 \text{ m} = 0.0723 \text{ km}$$

TABLE 3-3 SI units of length

Unit	Relation to meter		
1 kilometer (km)	= 1000 meter (m)	= 10^3 m	All are exact relationships
1 meter (m)			
1 decimeter (dm)	= 0.1 m	= 10^{-1} m	
1 centimeter (cm)	= 0.01 m	= 10^{-2} m	
1 millimeter (mm)	= 0.001 m	= 10^{-3} m	
1 micrometer (μm)		= 10^{-6} m	
1 nanometer (nm)		= 10^{-9} m	
1 picometer (pm)		= 10^{-12} m	
The following non–SI unit is also frequently encountered:			
1 angstrom (Å)	= 10^{-10} m	= 10^{-8} cm	
Some frequently used relationships:			
	1 km =	1000 m	
	100 cm =	1 m	
	1000 mm =	1 m	
	1,000,000 μm =	1 m	
	1 nm =	10 Å	

The fact that multiplication or division of a number by any power of 10 involves just moving the decimal point simplifies the conversion between SI units.

Two of the units expressed in Table 3-3 are worthy of mention. The *angstrom* is a unit that previously enjoyed widespread use, because most atoms are approximately 1 Å in diameter. However, this unit is not derived from the meter by using one of the SI prefixes, so it is not considered an SI unit. In recent years the nanometer (1 nm = 10 Å) has been more widely used to express length in this size range. The micrometer (μm) used to be called the *micron,* and you may come across that term in other books.

Conversion from English to SI Units

Having just demonstrated the simplicity of conversion between SI units, we can hope for the day when the English system will be an oddity. Everyone will know how long a meter is, and few people will care how many inches or feet it equals. However, for the present we live in a world with two systems, and it is useful for us to be able to convert measurements in the English system to their SI equivalents. Figure 3-3 compares inches and centimeters. There are slightly more than $2\frac{1}{2}$ cm in an inch. The following equivalencies are used to convert between English and SI units:

1 in. = 2.54 cm } **This is an exact relationship**

39.4 in. = 1 m
0.621 mi = 1 km } **These relationships have been rounded off and are accurate to three significant figures**

You need to learn only the first of these, because it can be used to calculate the others.

FIGURE 3-3 Size comparison between centimeters and inches: 2.54 cm = 1 in.

EXAMPLE 3.10

Calculate the number of centimeters in 45.8 in.

Symbolic statement

$$? \text{ cm} = 45.8 \text{ in.}$$

Planning the solution

Since we have an equivalency between inches and centimeters, we can make this conversion in one step:

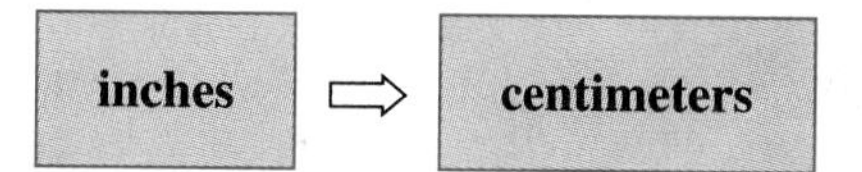

SOLUTION

Because 1 in. = 2.54 cm,

$$\frac{1 \text{ in.}}{2.54 \text{ cm}} = 1 \quad \text{and} \quad 1 = \frac{2.54 \text{ cm}}{1 \text{ in.}}$$

We use the conversion factor on the right.

$$? \text{ cm} = 45.8 \cancel{\text{in.}} \left(\frac{2.54 \text{ cm}}{1 \cancel{\text{in.}}} \right) = 116 \text{ cm}$$

The answer of 116 cm has been rounded off to three significant figures, the same number of significant figures contained in 45.8 in. Recalling our earlier discussion of exact versus measured relationships, 1 in. = 2.54 cm is an exact relationship and may be considered to have an unlimited number of significant figures.

EXAMPLE 3.11

Calculate the number of feet in 1280 cm (1.28×10^3 cm).

Symbolic statement

$$? \text{ ft} = 1280 \text{ cm}$$

Planning the solution

We do not have a direct equivalency between centimeters and feet. However, we do have relationships for converting centimeters to inches and inches to feet. We can combine these in the following strategy:

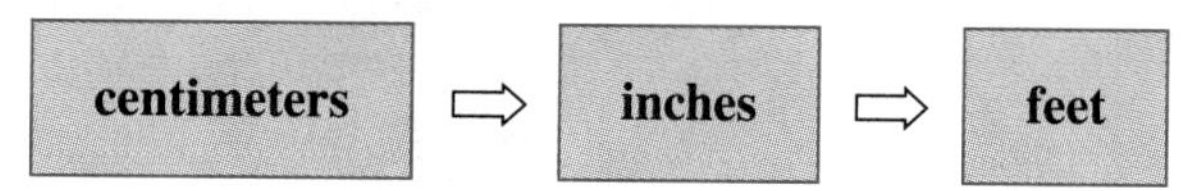

SOLUTION

We need the following relationships:

$$2.54 \text{ cm} = 1 \text{ in.} \qquad 12 \text{ in.} = 1 \text{ ft}$$

$$? \text{ ft} = 1280 \cancel{\text{cm}}\left(\frac{1 \cancel{\text{in.}}}{2.54 \cancel{\text{cm}}}\right)\left(\frac{1 \text{ ft}}{12 \cancel{\text{in.}}}\right) = \frac{1280 \text{ ft}}{(2.54)(12)} = 42.0 \text{ ft}$$

Our final answer has been rounded off to three significant figures—the same number of significant figures present in 1280 cm. Since the relationships 1 ft = 12 in. and 2.54 cm = 1 in. are exact, they have an unlimited number of significant figures.

PROBLEM 3.12

Use the dimensional analysis method to carry out the following one-step conversions.

(a) 42.6 cm = ______ m **(b)** 1.5 km = ______ m
(c) 502 m = ______ km **(d)** 0.372 m = ______ cm
(e) 943 mm = ______ m **(f)** 10.0 cm = ______ in.
(g) 3.86 in. = ______ cm **(h)** 15.0 m = ______ cm

PROBLEM 3.13

Use the dimensional analysis method to carry out the following multistep conversions. (Plot out a strategy before you begin each part.)

(a) 3.23 km = ______ mm **(b)** 1.5 km = ______ cm
(c) 1.40 ft = ______ cm **(d)** 125 cm = ______ ft
(e) 3.20 yd = ______ cm **(f)** 10.0 ft = ______ m

PROBLEM 3.14

Suppose that the only relationship given for converting between English and SI units of length was 1 in. = 2.54 cm. Use dimensional analysis to verify the following relationships. [*Hint:* In part (a), begin with 1.00 m and convert to inches; in part (b), convert 1.00 km to miles.]

(a) 1 m = 39.4 in. **(b)** 1 km = 0.621 mi

3.7 MASS

OBJECTIVE 3.7 Distinguish between mass and weight. Convert quantities of mass from one unit to another.

Mass: A quantity of matter.

In Chapter 2 matter was defined as anything that has mass and occupies space. At that time we deferred a detailed description of **mass,** simply describing it as the amount of matter in a substance.

To illustrate the concept of mass, suppose you are helping a friend move and are given a choice between carrying a box of books upstairs and carrying up an identical-sized box of sweaters. Depending on how close your friend is, you will probably choose to carry the box of sweaters; they have the smaller mass.

Weight: The force with which gravity acts on an object.

At this point it may seem that mass and weight are the same thing. However, there is a significant distinction between the two. Whereas mass is a measure of the quantity of matter present, the **weight** of an object refers to the force of gravity acting upon it. Since the force of gravity acting upon an object varies with location, the weight of an object can change, depending upon its location. For example, the force of gravity on the moon is approximately one-sixth the gravitational attraction on earth. Thus, the weight of an object on the moon is roughly one-sixth its weight on earth. However, the mass of the object does not change, because it is still composed of the same quantity of matter.

To further illustrate this distinction, let us compare the manner in which mass and weight are measured. Mass is determined using a *balance*. Chemists originally used balances similar to the one shown in Fig. 3-4. When the pans come to rest as shown, the masses of the objects in both pans must be equal. To determine the mass of an object, it is compared to an object of known mass. Thus, the unknown object in Fig. 3-4 has a 1 kg mass, the same as that of the known object. Although modern electronic balances are not quite as simple as this, they too are calibrated using known masses. As long as we calibrate a balance in the location where it is used, it will give the correct mass, regardless of location.

On the other hand, weight is measured using a *scale,* such as the spring scale shown in Fig. 3-5. When an object is hung from the scale, the force of gravity pulls the object toward the earth's surface, stretching the spring and enabling us to measure the weight. If we transport our equipment to the moon, we find that the spring does not stretch as far, because the force of gravity pulling the object toward the moon's surface is only one-sixth what it is on earth. The mass of the object has not changed (there is still the same quantity of matter), but the weight is only one-sixth what it was on earth. In a similar fashion, the gravitational field is not uniform everywhere on earth, and the weight of an object may vary from one location to another. However, its mass will always be constant, regardless of location. Because chemists are concerned with the *amount of matter* they are dealing with, they are always interested in measuring mass rather than weight.

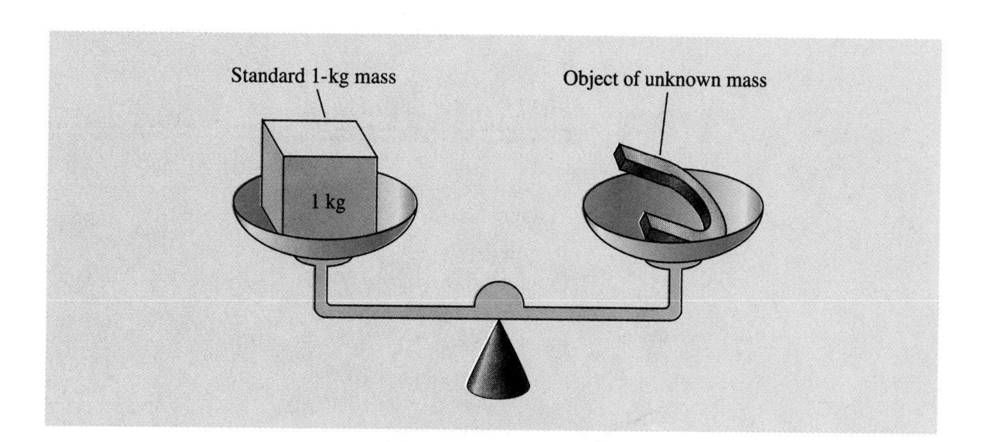

FIGURE 3-4 A balance compares the masses of objects.

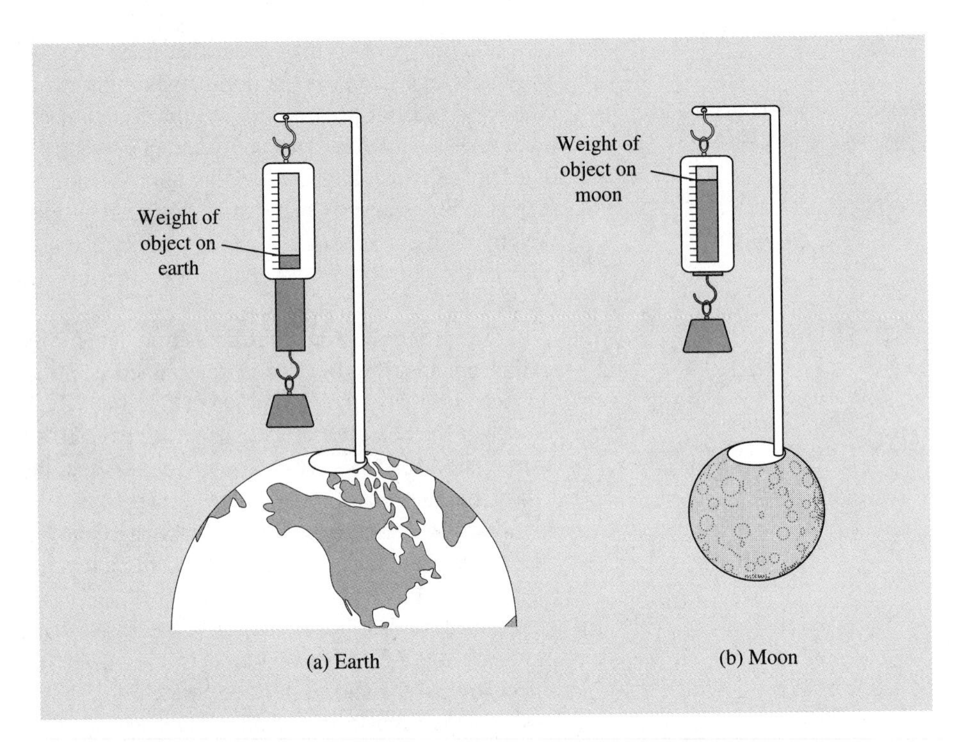

FIGURE 3-5 Comparison of an object's weight (a) on the earth and (b) on the moon.

In actual practice, chemists often use the terms *mass* and *weight* interchangeably. Nevertheless, it is important to be aware of the distinction between them. The SI unit of mass is the *kilogram,* a multiple of the gram. Other masses related to these are shown in Table 3-4. The most commonly used units of mass are the gram (g), kilogram (kg), and milligram (mg).

The following relationships may be used to convert English units of mass to SI units:

$$1 \text{ lb} = 454 \text{ g}$$
$$1 \text{ kg} = 2.20 \text{ lb}$$

These measured relationships have three significant figures

It is sufficient to memorize the first of these; the second can be calculated using the first.

TABLE 3-4 SI units of mass

Unit	Relation to gram	
1 kilogram (kg)	= 1000 gram (g) = 10^3 g	**All are exact relationships**
1 gram (g)		
1 milligram (mg)	= 0.001 g = 10^{-3} g	
1 microgram (μg)	= 10^{-6} g	
Useful relationships:		
1 kg	= 1000 g	
1 g	= 1000 mg	

PROBLEM 3.15

Use the equivalency 1 lb = 454 g to calculate the number of pounds in 1.00 kg.

PROBLEM 3.16

Use the dimensional analysis method to carry out the following conversions.

(a) 1235 g = ______ kg **(b)** 3.45 g = ______ mg
(c) 598 mg = ______ g **(d)** 3.00 lb = ______ g
(e) 7.00 kg = ______ lb **(f)** 3.50 lb = ______ kg
(g) 0.620 lb = ______ mg

3.8 VOLUME

OBJECTIVE 3.8 Convert quantities of volume from one unit to another.

It is often more convenient to measure the volume of a substance than to measure its mass. This is particularly true when the substance to be measured is in the liquid state. For example, when we purchase gasoline for our cars, the amount of gasoline is measured as a volume.

Although the cubic meter is the basic SI unit of volume, this volume is so large that most chemistry work is done in terms of the liter, a volume that is one one-thousandth as large. One *liter* (L) is the volume contained within a cubic box that measures 10 cm on each side (Fig. 3-6). The volume of this box is equal to 1000 cubic centimeters (1000 cm^3). The liter is most commonly subdivided into 1000 *milliliters* (mL). Thus, a milliliter is the same as a cubic centimeter, the volume contained within a cube measuring 1 cm on each side. When a doctor calls for "5 cc's" of a medication, a volume of 5 mL is needed. Table 3-5 summarizes the most important relationships between metric volumes.

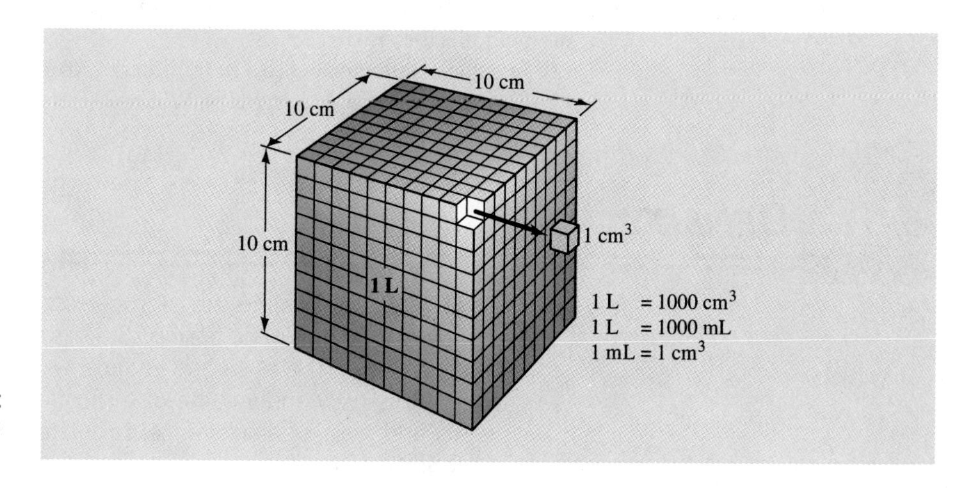

FIGURE 3-6 The liter is a metric unit of volume. It is equal to 1000 mL or 1000 cm^3. (The liter is related to the cubic meter, the SI unit of volume, as follows: 1 m^3 = 1000 L.)

TABLE 3-5 SI and related metric units of volume

1 cubic meter	(m^3)	= 1000 L	All are exact relationships
1 cubic decimeter	(dm^3)	= 1 liter (L)	
1 milliliter	(mL)	= 0.001 L	
1 centiliter	(cL)	= 0.01 L	
1 deciliter	(dL)	= 0.1 L	
1 microliter	(μL)	= 10^{-6} L	
Useful relationships:			
	1 L	= 1000 mL	
	1 L	= 1000 cm^3	
	1 mL	= 1 cm^3	

The conversion factors that are most useful in relating the English and SI units of volume are

$$1 \text{ qt} = 946 \text{ mL}$$
$$1 \text{ gal} = 3.785 \text{ L}$$

These are measured relationships

In recent years, the beverage industries and oil companies have begun to market their products in metric units of measure. Since a quart is less than a liter (1 qt = 0.946 L), it follows that a liter is slightly larger than a quart. (Section 3.13 examines the units of volume in greater detail.)

PROBLEM 3.17

Convert the following.

(a) 525 mL = ______ L **(b)** 4.95 L = ______ mL
(c) 1.32 qt = ______ mL **(d)** 2.25 qt = ______ L
(e) 10.8 gal = ______ L **(f)** 45.3 L = ______ gal

PROBLEM 3.18

A rectangular tin box has the dimensions 12.0 cm by 8.00 cm by 9.50 cm. What volume of liquid fills this box:
(a) In cubic centimeters? **(b)** In milliliters? **(c)** In liters?
[*Remember:* Volume = Length × Width × Height]

3.9 DENSITY

OBJECTIVE 3.9 Define the term *density* and write a mathematical equation expressing this relationship. State the mathematical meaning of the word *per.* Given the mass and volume of a substance, calculate its density. Given the density and volume of a substance, calculate its mass. Given the density and mass of a substance, calculate its volume. Explain the meaning of specific gravity.

An old riddle asks, "Which weighs more, a pound of iron or a pound of cork?" The unwary subject is tempted to answer that the iron weighs more, despite being told that both weigh a pound. The person being tricked is confusing density with mass! Both have equal masses, but a pound of iron occupies a much smaller volume than a pound of cork. If we compare equal volumes, however, we find that the iron has a greater mass.

In our scientific work, we frequently encounter relationships between two measured quantities such as this. This relationship between mass and volume is known as **density (*d*)** and is defined as the mass (*m*) per unit volume (*V*) of a substance. The word *per* tells us to divide. Replacing the word *per* with the phrase *divided by* tells us exactly how to calculate density.

Density: Mass per unit volume.

$$\text{Density} = \frac{\text{Mass}}{\text{Volume}} \qquad d = \frac{m}{V} \qquad \textbf{“Per” means “divided by”}$$

Thus, if a sample of a liquid has a mass of 135 g and occupies a volume of 60.0 mL, its density is found as follows:

$$\text{Density} = \frac{\text{Mass}}{\text{Volume}} = \frac{135 \text{ g}}{60.0 \text{ mL}} = 2.25 \text{ g/mL}$$

In other words, every milliliter (unit of volume) has a mass of 2.25 g.

Perhaps a more familiar example of a "per" relationship is that of velocity, the distance we travel in a given period of time. If we drive 100 miles in 2 hours, our velocity is 50 miles per hour (miles/hr). To obtain this result, we divide the distance (100 miles) by the time (2 hours). Replacing the word *per* with the phrase *divided by* in the units of "miles per hour" tells us to divide the miles by the hours.

$$\text{Velocity} = \frac{\text{Miles}}{\text{Hours}} = \frac{100 \text{ mi}}{2 \text{ hr}} = 50 \text{ mi/hr}$$

Strictly speaking, when we make a statement such as "50 miles per hour," we are really saying "50 miles per 1 hour." Thus, we are stating the distance for each *unit* of time (1 hour). Similarly, the density we just calculated told us that there are 2.25 g in each milliliter (unit of volume). Looking back to our discussion of conversion factors, we realize that they too express "per" relationships, such as 2.54 cm/in., 946 mL/qt, and 454 g/lb.

EXAMPLE 3.12

A 15.0-mL sample of an unknown liquid is found to have a mass of 12.4 g. Calculate the density of the liquid.

Symbolic statement

$$\text{Density} = \ ?\ \text{g/mL}$$

Planning the solution

Since density is the mass per unit volume, we must divide the mass by the volume.

SOLUTION

$$\text{Density} = \frac{\text{Mass}}{\text{Volume}}$$

$$d = \frac{12.4 \text{ g}}{15.0 \text{ mL}} = 0.827 \text{ g/mL}$$

Checking the answer

Since there are more milliliters (15.0) than there are grams (12.4), the mass of a milliliter must be less than 1 g, in agreement with our calculated density of 0.827 g/mL.

PROBLEM 3.19

(a) A 4.0-g sample of a substance has a volume of 2.5 mL. Calculate the density of the substance.

(b) A 2.50-mL sample of mercury has a mass of 34.0 g. Calculate the density of mercury.

(c) A 59.2-g sample of ethyl alcohol occupies a volume of 75.0 mL. What is the density of ethyl alcohol?

(d) Bromine is a red liquid that is one of the two elements that exist as liquids at room temperature (20°C). A 5.00-mL sample of bromine has a mass of 15.6 g. Calculate the density of this element.

(e) Acetone is a solvent used for a large number of chemical applications. If 11.0 mL of acetone has a mass of 8.69 g, what is its density?

To get a more qualitative sense of density, consider why some objects sink to the bottom of a glass of water whereas others float to the top. Objects that sink (such as coins, paper clips, and stones) have a greater density than water. Objects that float (such as corks and pencils) are less dense than water. If you are familiar with oil and vinegar salad dressing, you know that no matter how many times you shake it up, when the oil separates from the vinegar, it is always on top—never on the bottom. That is because the oil is less dense than the vinegar.

Most substances tend to expand when they are heated. Hence, the density of most liquids decreases as their temperature rises. As a result, most densities of substances are reported at a specific temperature. The densities of several common chemicals are shown in Table 3-6. Like a melting point or boiling point, the density of a substance is a physical property of a substance and may be used to describe, or *characterize,* a substance.

Recalling that molecules in the gaseous state are much farther apart than molecules in the liquid and solid states, we should not be surprised to find that the densities of gases are much lower than those of liquids or solids. As a result, when gases are bubbled through liquids, they rise to the top. You have undoubtedly observed this

TABLE 3-6 Densities of some common substances

Solids and liquids		Gases	
Alcohol (ethyl)	0.789 g/mL (20°C)	Carbon dioxide	0.00183 g/mL (20°C)
Benzene	0.879 g/mL (20°C)	Oxygen	0.00133 g/mL (20°C)
Carbon tetrachloride	1.59 g/mL (20°C)	Hydrogen	0.0000832 g/mL (20°C)
Chloroform	1.48 g/mL (20°C)		
Ether	0.714 g/mL (20°C)		
Iron	7.86 g/mL (20°C)		
Lead	11.3 g/mL (20°C)		
Mercury	13.6 g/mL (20°C)		
Water	1.000 g/mL (4°C)		

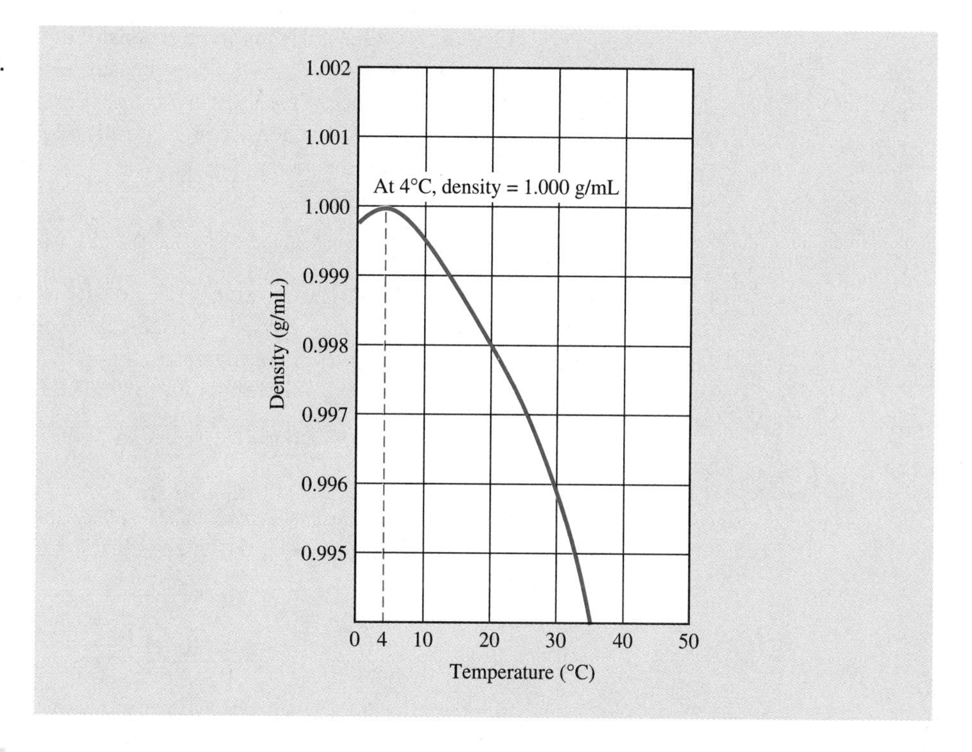

FIGURE 3-7 The density of water varies with temperature. Water reaches its maximum density at 4°C.

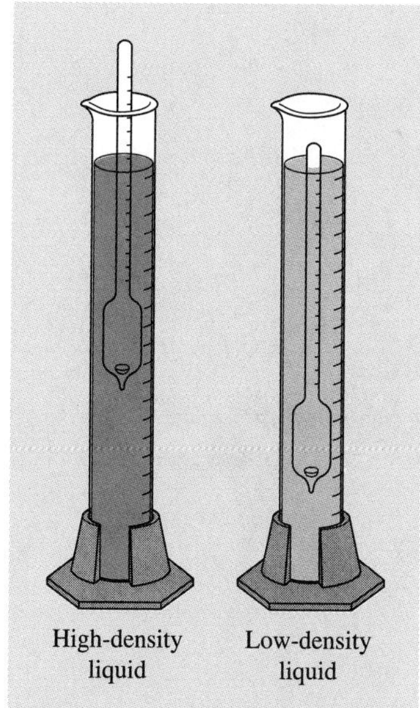

FIGURE 3-8 A hydrometer is used to measure the density of liquids. The greater the density of the liquid, the higher the hydrometer floats.

phenomenon if you have ever watched the carbon dioxide bubbles in a soft drink rising to the top of the glass.

The density of water varies with temperature, but its density is unusual in two ways. Water is unlike most substances in that the density of liquid water is greater than that of the solid (ice). As a result, ice cubes always float in a soft drink. Similarly, as a lake freezes, the layer of ice that forms remains on the surface rather than sinking. In addition, the density of liquid water increases as its temperature rises from 0°C (32°F, the freezing point of water) to 4°C (39°F, see Fig. 3-7). Above 4°C, the density of water decreases in a fashion similar to that of most substances. Thus, at 4°C water has its maximum density: d = 1.000 g/mL. In fact, the gram was originally defined as the mass of 1 mL of water at 4°C.

A *hydrometer* is an instrument that allows the direct measurement of the density of a solution (Fig. 3-8). The mass and volume of a hydrometer do not change, so the hydrometer has a fixed density. In a solution of high density the hydrometer floats higher than in less dense solutions. There is a calibrated scale on the hydrometer that enables the user to measure the density of the solution by noting where the scale crosses the surface of the liquid.

A brewer's hydrometer is used in the production of beer or wine. In either process, sugar is fermented to make ethyl alcohol. The density of the alcohol produced is less than that of the initial sugar–water mixture, so the density of the fermenting mixture decreases as fermentation proceeds. By utilizing tables that correlate density and alcoholic content, brewmasters and winemakers use the hydrometer to determine the appropriate time to halt the fermentation of sugar in their brew.

One of the values of knowing the density of a substance is that it makes it possible to obtain the mass of a liquid by measuring its volume, rather than by actually weighing the substance. Since this is often easier to do, a chemist who wishes to use 12 g of a substance with a density of 2.0 g/mL might choose to measure out 6.0 mL of the substance, rather than weighing out 12 g:

$$12\text{ g} = 6.0\,\cancel{\text{mL}}\left(\frac{2.0\text{ g}}{1\,\cancel{\text{mL}}}\right)$$

As this example shows, we can use density as a conversion factor. That is reasonable; the density relates the mass of a substance to its volume. Saying that a substance has a density of 2.0 g/mL is the same as saying that 1 mL = 2.0 g for the substance in question. Consequently, either of the following could be used as a conversion factor:

$$\frac{2.0\text{ g}}{1\text{ mL}} = 1 \qquad \text{or} \qquad 1 = \frac{1\text{ mL}}{2.0\text{ g}}$$

If we want to know the mass of 16 mL of this substance, we multiply 16 mL by the conversion factor on the left (which is just the density itself). This process converts volume to mass.

$$?\text{ g} = 16\,\cancel{\text{mL}}\left(\frac{2.0\text{ g}}{1\,\cancel{\text{mL}}}\right) = 32\text{ g}$$

On the other hand, suppose we want to calculate the number of milliliters occupied by 8.0 g of this substance. In this case, multiplication by the conversion factor on the right (the reciprocal of the density) converts grams into milliliters.

$$?\text{ mL} = 8.0\,\cancel{\text{g}}\left(\frac{1\text{ mL}}{2.0\,\cancel{\text{g}}}\right) = 4.0\text{ mL}$$

EXAMPLE 3.13

What is the mass of 25.0 mL of ether? (d = 0.714 g/mL)

Symbolic statement

$$?\text{ g} = 25.0\text{ mL}$$

Planning the solution

The unknown we wish to find is the mass of ether. The information we are given is the volume of ether and its density. Density can be used to relate the mass and volume. In this case:

$$1\text{ mL} = 0.714\text{ g}$$

We can use this to convert from milliliters to grams:

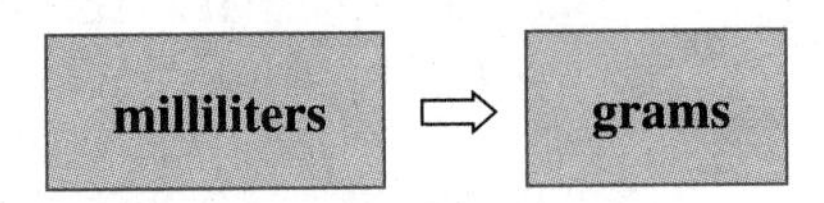

SOLUTION

Our density gives us two possible conversion factors:

$$\frac{0.714\text{ g}}{1\text{ mL}} = 1 \qquad \text{or} \qquad 1 = \frac{1\text{ mL}}{0.714\text{ g}}$$

We will use the conversion factor on the left, thereby canceling the units of milliliters:

$$? \text{ g} = 25.0 \not{\text{mL}}\left(\frac{0.714 \text{ g}}{1 \not{\text{mL}}}\right) = 17.8 \text{ g}$$

Checking the answer

Let us see if the answer is reasonable. Since the density of ether is less than 1 g/mL, the number of grams must be less than the number of milliliters (25.0). Our answer of 17.8 g seems quite reasonable.

EXAMPLE 3.14

A chemist needs 225 g of methylene chloride for use in a chemical extraction. She decides to measure it out by volume, rather than by weighing it out on a balance. The density of methylene chloride is 1.33 g/mL. What volume does she need?

Symbolic statement

$$? \text{ mL} = 225 \text{ g}$$

Planning the solution

In this case, the unknown we wish to find is the volume of methylene chloride. We are given the mass and the density of methylene chloride. The density will enable us to convert grams to milliliters:

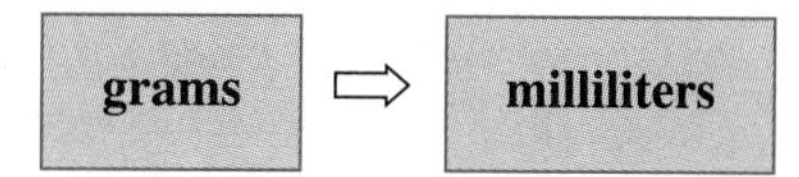

SOLUTION

A density of 1.33 g/mL leads to the following pair of conversion factors:

$$\frac{1.33 \text{ g}}{1 \text{ mL}} = 1 \quad \text{or} \quad 1 = \frac{1 \text{ mL}}{1.33 \text{ g}}$$

We will choose the conversion factor on the right, canceling the units of grams.

$$? \text{ mL} = 225 \not{\text{g}}\left(\frac{1 \text{ mL}}{1.33 \not{\text{g}}}\right) = 169 \text{ mL}$$

Checking the answer

The density of methylene chloride is greater than 1 g/mL, so the number of milliliters must be less than the number of grams (225 g). Our answer of 169 mL is reasonable.

Examples 3.13 and 3.14 can also be solved by substitution into the formula for density:

$$d = \frac{m}{V}$$

In Example 3.13, since we wish to know the mass, we can solve the formula for m:

$$d \cdot V = m = \left(\frac{0.714 \text{ g}}{\not{\text{mL}}}\right)(25.0 \not{\text{mL}}) = 17.8 \text{ g}$$

Note that the final setup looks essentially the same as the setup we obtained using dimensional analysis in Example 3.13.

Similarly, Example 3.14 may be solved using the substitution method. Since we wish to know the volume, we will solve the formula for V:

$$d = \frac{m}{V}$$

$$V = \frac{m}{d} = \frac{225\ \text{g}}{\frac{1.33\ \text{g}}{\text{mL}}}$$

To divide, we must invert the denominator and then multiply it by the numerator:

$$V = (225\ \cancel{\text{g}})\left(\frac{\text{mL}}{1.33\ \cancel{\text{g}}}\right) = 169\ \text{mL}$$

Again, note the similarity between the final setup here and the one obtained using dimensional analysis. In general, we use dimensional analysis throughout the text. However, substitution into a formula is also an acceptable method of problem-solving, as long as the substitution includes units and shows how cancellation of those units takes place.

PROBLEM 3.20

(a) What is the mass of 11.5 mL of a substance with density 0.85 g/mL?
(b) Carbon tetrachloride has a density of 1.59 g/mL. If a chemist wishes to use 12 g of carbon tetrachloride, what volume should be measured out?
(c) The density of benzene is 0.879 g/mL. Calculate the mass of 65.0 mL of benzene.
(d) What volume of chloroform (d = 1.48 g/mL) has a mass of 27.0 g?
(e) What is the volume (in milliliters) of a pound (1.00 lb) of mercury? (d = 13.6 g/mL)
(f) What is the mass of 245 mL of ethyl alcohol? (d = 0.789 g/mL)

Density and Specific Gravity

Specific gravity: The ratio of the density of a substance to the density of water at 4°C.

You may occasionally hear people refer to the specific gravity of a substance. The **specific gravity** of a substance is the ratio of the density of that substance to the density of water at 4°C (the maximum density of water, 1.000 g/mL):

$$\text{Specific gravity (of substance X)} = \frac{\text{Density of substance X}}{\text{Density of water at 4°C}}$$

For example, the density of mercury is 13.6 g/mL. It follows that the specific gravity (sp gr) of mercury is 13.6:

$$\text{sp gr} = \frac{13.6\ \cancel{\text{g/mL}}}{1.00\ \cancel{\text{g/mL}}} = 13.6$$

There are two things to note about this specific gravity. First, the units of density in the numerator of the setup cancel those in the denominator. Thus, the specific gravity is a *unitless* quantity; that is, it is just a ratio. Second, the numerical value of the specific gravity is equal to the density of the substance, *if the density is expressed in grams per milliliter.* (The reason that this is true is because the density of water is 1.000 g/mL.)

Thus, if we know that the specific gravity of a substance is 3.65, it follows that its density is 3.65 g/mL. Conversely, if we determine that the density of a substance is 0.895 g/mL, then its specific gravity is 0.895.

It might seem that the specific gravity of a substance is an unnecessary quantity. In fact, if we are working with units of grams per milliliter, it is an extraneous term. However, if one is working with English system units, where the density of water is 62.4 lb/ft^3, or with any units other than grams per milliliter, the numerical value of the specific gravity will not equal that of the density. For example, since the specific gravity of mercury is 13.6, it follows that the density of mercury is 13.6 times as large as that of water. Thus, the density of mercury must be 13.6 times 62.4 lb/ft^3, or 849 lb/ft^3.

PROBLEM 3.21

Answer the following questions about specific gravity. (Remember, the density of water is 62.4 lb/ft^3.)

(a) What is the specific gravity of bromine (d = 3.12 g/mL)?

(b) What is the density of bromine in pounds per cubic foot (lb/ft^3)?

(c) The density of carbon tetrachloride is 99.5 lb/ft^3. Calculate its specific gravity.

(d) What is the density of carbon tetrachloride in grams per milliliter?

3.10 TEMPERATURE

OBJECTIVE 3.10 Convert between temperatures expressed in the Kelvin and Celsius scales; convert between temperatures expressed in the Fahrenheit and Celsius scales; convert between temperatures expressed in the Fahrenheit and Kelvin scales.

We all have some kind of qualitative sense about hot and cold, but in scientific work we need to be more specific about the meaning of the term *temperature.* In Chapter 11 we consider the physical meaning of temperature. For now, let us look at the temperature scales we are likely to encounter in our daily lives and in our scientific work. The English system employs the **Fahrenheit** temperature scale, which is still the most commonly used scale in the United States. Increasingly, this scale is being replaced by the **Celsius** temperature scale, which employs a metric unit of temperature. The International System (SI) utilizes the **Kelvin** temperature scale.

Fahrenheit temperature: English system temperature scale.

Celsius temperature: Metric temperature scale used widely in scientific work.

Kelvin temperature: SI temperature scale; often called the *absolute* temperature scale.

Figure 3-9 compares the Fahrenheit, Celsius, and Kelvin temperature scales. On the Fahrenheit scale the freezing point of water is 32°F, and its boiling point is 212°F. The distance between these temperatures represents 180 equal steps, each of which is called a Fahrenheit degree. This thermometric scale is not employed in scientific work; instead, we use the Celsius and Kelvin scales.

The Celsius scale (often referred to as the "centigrade" scale) was established by defining 0°C as the freezing point of water and 100°C as the boiling point of water. (As stated in Chapter 2, boiling points depend on atmospheric pressure. This is the boiling point at 1 atm pressure. We will discuss this further in Chapter 11.) There are 100 equal divisions between these points, each of which represents one Celsius degree.

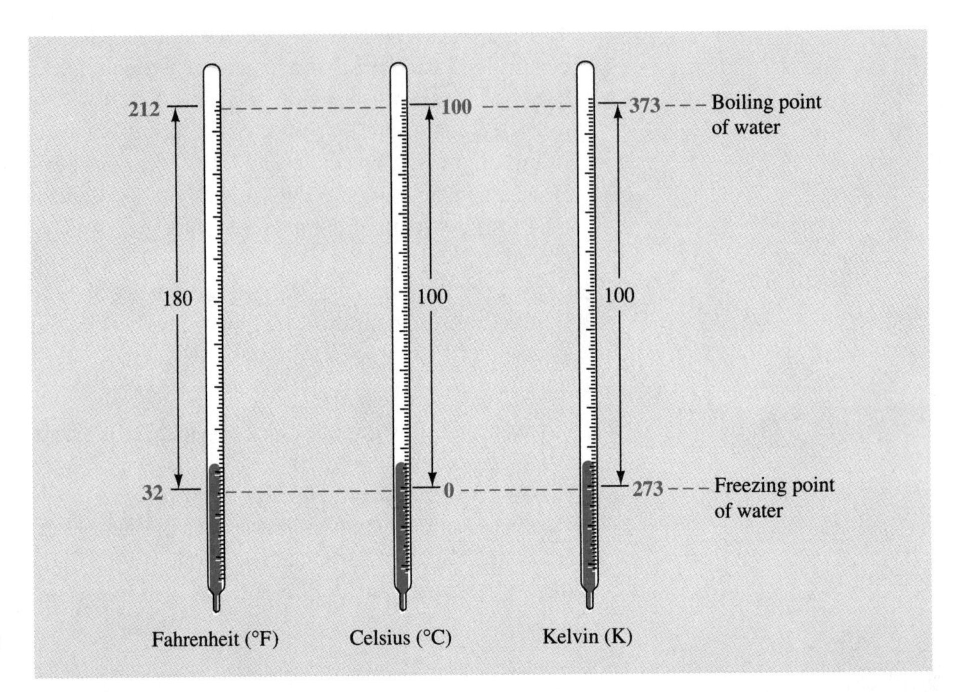

FIGURE 3-9 **Comparison of Fahrenheit, Celsius, and Kelvin temperature scales.**

The Kelvin scale is often referred to as the "absolute" temperature scale. The zero point on this scale is referred to as "absolute zero," because no substance can be cooled below that point. The size of each degree on the Kelvin scale is the same as that on the Celsius scale, and for both scales there are 100 degrees between the freezing point and the boiling point of water. The unit of Kelvin temperature is called the *kelvin,* K. (No "degree" sign is used.) The zero point for the Kelvin scale is 273.15 degrees lower than that for the Celsius scale. Therefore, the relationship between the Kelvin and Celsius scales is

$$\text{Kelvins} = \text{Degrees Celsius} + 273.15$$

or

$$T_K = T_C + 273.15$$

However, since most of our routine work in the laboratory is done with thermometers that measure temperature to the nearest degree, we will simplify this equation to

$$T_K = T_C + 273$$

and round off all calculations in this section to the nearest degree. Table 3-7 shows several temperature readings in Celsius degrees and in kelvins.

EXAMPLE 3.15

What Kelvin temperature corresponds to room temperature, 20°C?

Planning the solution

We simply substitute our Celsius temperature into the following relationship:

$$T_K = T_C + 273$$

SOLUTION

$$T_K = 20 + 273 = 293\text{ K}$$

TABLE 3-7 Relationship between Kelvin and Celsius temperatures

Degrees Celsius + 273	= Kelvins	°C	K
0	273	0°C	273 K
20	293	20°C	293 K
100	373	100°C	373 K
–40	233	–40°C	233 K
–273	0	–273°C	0 K

EXAMPLE 3.16

What Kelvin temperature corresponds to the sublimation point of dry ice, –78°C?

SOLUTION

$$T_K = -78 + 273 = 195 \text{ K}$$

PROBLEM 3.22

Convert each of the following temperatures to the indicated unit.
(a) 20°C to K **(b)** 145°C to K **(c)** –78°C to K
(d) –223°C to K **(e)** 298 K to °C **(f)** 577 K to °C
(g) 100 K to °C **(h)** 23 K to °C

The relationship between the Fahrenheit and Celsius scales is expressed mathematically as follows:

$$T_F = 1.8T_C + 32 \tag{3-5}$$

We use the factor 1.8 because each Celsius degree is 1.8 times as large as each Fahrenheit degree. (Note that between the melting point and boiling point of water, there are exactly 100 Celsius degrees and 180 Fahrenheit degrees, leading to the factor of 1.8.) The number 32 comes from the difference in zero points between the two scales. In applying this equation, the number 1.8 is *exact,* and therefore does not limit the number of significant figures when used to multiply or divide. Similarly, the number 32 is also exact. For simplicity, you may round off all temperatures in this section to the nearest degree.

To convert from Celsius to Fahrenheit temperature, we substitute appropriately into equation (3-5), as shown in the following examples.

EXAMPLE 3.17

What Celsius temperature corresponds to 212°F?

Planning the solution

We simply substitute the Fahrenheit temperature into the following relationship:

SOLUTION

$$T_F = 1.8T_C + 32$$
$$212 = 1.8T_C + 32$$
$$180 = 1.8T_C$$
$$T_C = \frac{180}{1.8} = 100°\text{C}$$

We could have predicted this easily, because 212°F is the boiling point of water (see Fig. 3-9).

EXAMPLE 3.18

What Celsius temperature corresponds to 113°F?

SOLUTION

$$T_F = 1.8T_C + 32$$
$$113 = 1.8T_C + 32$$
$$81 = 1.8T_C$$
$$T_C = \frac{81}{1.8} = 45°C$$

This temperature might be experienced during a summer day on the Mojave Desert.

EXAMPLE 3.19

Convert 25°C to its Fahrenheit equivalent.

SOLUTION

$$T_F = 1.8T_C + 32$$
$$= 1.8(25) + 32$$
$$= 45 + 32 = 77°F$$

EXAMPLE 3.20

What Kelvin temperature corresponds to 5°F?

Planning the solution

We need to do this calculation in two steps. First we convert from Fahrenheit to Celsius, and then from Celsius to Kelvin:

T_F ⇨ T_C ⇨ T_K

SOLUTION

Step 1. $T_F = 5 = 1.8T_C + 32$

$$-27 = 1.8T_C$$
$$\frac{-27}{1.8} = T_C$$
$$T_C = -15°C$$

Step 2. $T_K = T_C + 273$

$$= -15 + 273$$
$$T_K = 258 \text{ K}$$

PROBLEM 3.23

Convert the following Fahrenheit temperatures into Celsius temperatures.
(a) 87°F **(b)** 68°F **(c)** −40°F **(d)** 0°F
(e) −48°F **(f)** −104°F **(g)** 932°F **(h)** 400°F

PROBLEM 3.24

Convert each of the Celsius temperatures you calculated in Problem 3.23 to its equivalent in Kelvin temperature.

PROBLEM 3.25

Convert the following Celsius temperatures into Fahrenheit temperatures.
(a) 95°C **(b)** 60°C **(c)** 20°C **(d)** –40°C
(e) –70°C **(f)** 37°C

PROBLEM 3.26

Convert the following Fahrenheit temperatures into Kelvin temperatures.
(a) 59°F **(b)** –13°F **(c)** 257°F

3.11 ENERGY

OBJECTIVE 3.11A Distinguish between kinetic energy and potential energy. State the Law of Conservation of Energy. List three forms of energy.

Energy: The ability to do work.

Kinetic energy: The energy of motion.

Potential energy: The energy due to the position of an object in a force field, such as a gravitational field; stored energy.

Law of Conservation of Energy: Energy can be neither created nor destroyed.

Energy is defined as the ability to do work. Work is done any time we apply a force through some distance. For example, when we lift an object, such as a book, a force is applied through the distance the book is lifted.

Energy exists as either kinetic energy or potential energy. **Kinetic energy** is energy associated with the *motion* of an object. A baseball traveling at a velocity of 80 miles per hour has a certain kinetic energy associated with it. On the other hand, objects may *store* energy in the form of **potential energy.** A book on a shelf has potential energy. We can release this potential energy by removing the shelf, thereby allowing the book to go into motion (exhibiting kinetic energy) as it falls toward the floor.

One of the fundamental principles concerning energy is the **Law of Conservation of Energy:** Energy can be neither created nor destroyed. However, it *is* possible to convert energy from one form to another. For example, the behavior of a pendulum is an excellent illustration of how kinetic and potential energy can be interconverted. When a pendulum is at rest, the bob (weight) has no kinetic or potential energy (Fig. 3-10a, page 66). To start the pendulum in motion, the bob must be pulled to one side, thereby raising it (as shown in Fig. 3-10b). Raising the bob requires that we invest energy, which is stored in the bob in the form of potential energy. On being released, the bob begins its swing, picking up velocity and thereby gaining kinetic energy (Fig. 3-10c). As it passes through the lowest point in its swing, the bob has given up all of the potential energy it possessed at the moment of release (Fig. 3-10d). That potential energy has all been converted to kinetic energy, the energy of motion. As the bob continues upward in its swing, it slows down, losing kinetic energy and regaining potential energy. Before stopping to reverse direction, the pendulum rises to the same height from which it was initially dropped (Fig. 3-10e). Because it stops momentarily at this point, it has no velocity in any direction and hence no kinetic energy. The kinetic energy it possessed at the bottom of its swing has all been reconverted to potential energy. The pendulum would continue to swing back and forth forever, continually interconverting kinetic to potential energy and vice versa, if frictional forces did not act to eventually bring the pendulum to a halt. (Friction causes energy to be dissipated to the surroundings in the form of heat. Thus, there is no violation of conservation of energy.)

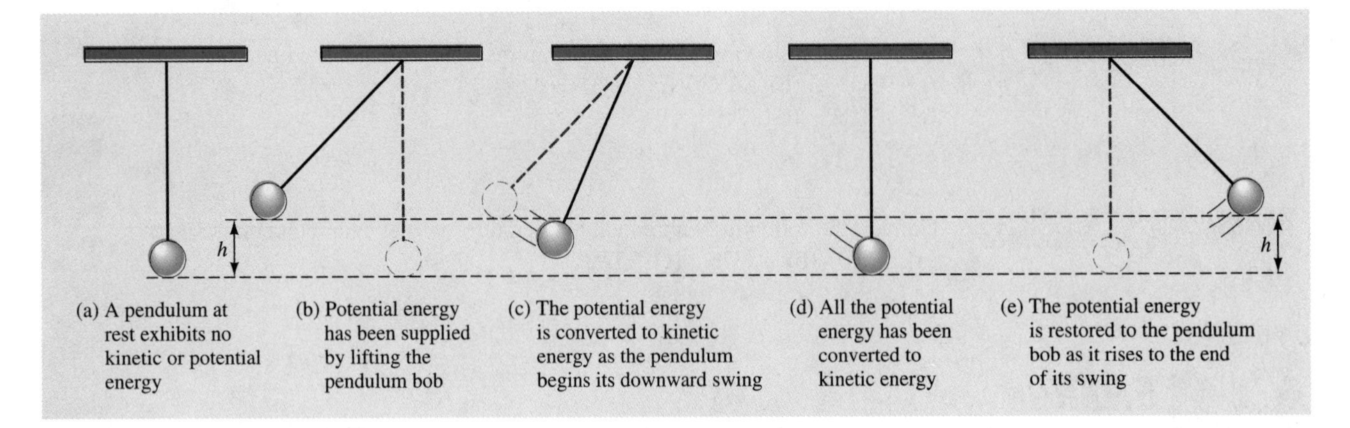

FIGURE 3-10 Interconversion of potential energy and kinetic energy.

The concept of energy is a thread that weaves its way through our entire presentation of chemistry. Indeed, energy is the primary factor responsible for the behavior of the physical universe. Energy may be encountered in a variety of forms—including heat, light, electrical energy, mechanical energy, chemical energy, and nuclear energy—and these forms can generally be interconverted. For example, most conventional power plants (Fig. 3-11) produce electrical energy by burning fossil fuels, thereby releasing their chemical energy in the form of heat. The heat is then used to boil water, producing steam. The steam drives large turbines (mechanical energy in action), which convert this mechanical energy into electrical energy.

Units of Energy

OBJECTIVE 3.11B Define the calorie in terms of the heat required to raise the temperature of water; convert quantities of energy from calories to joules, and from joules to calories.

Calorie: The amount of heat required to raise the temperature of 1 g of water by 1°C.

The SI unit of energy is the *joule* (J). Traditionally, however, chemists have worked with energy in terms of the **calorie,** the amount of energy required to raise the temperature of 1 g of water by 1°C. It is not necessary here to define the joule in its precise physical terms; instead, let us describe it in comparison to the calorie:

$$4.184 \text{ J} = 1 \text{ cal} \quad \text{This is an exact relationship}$$

The kilojoule (kJ) and kilocalorie (kcal) are multiples of these two energy units that we encounter frequently:

$$\left.\begin{aligned} 1 \text{ kJ} &= 1000 \text{ J} \\ 1 \text{ kcal} &= 1000 \text{ cal} \end{aligned}\right\} \text{These are exact relationships}$$

The nutritional "Calorie" discussed with respect to diet is actually a kilocalorie. Nutritionists sometimes refer to this as a "big" Calorie (with a capital "C") to distinguish it from our definition of the "little" calorie. The caloric content of a food represents

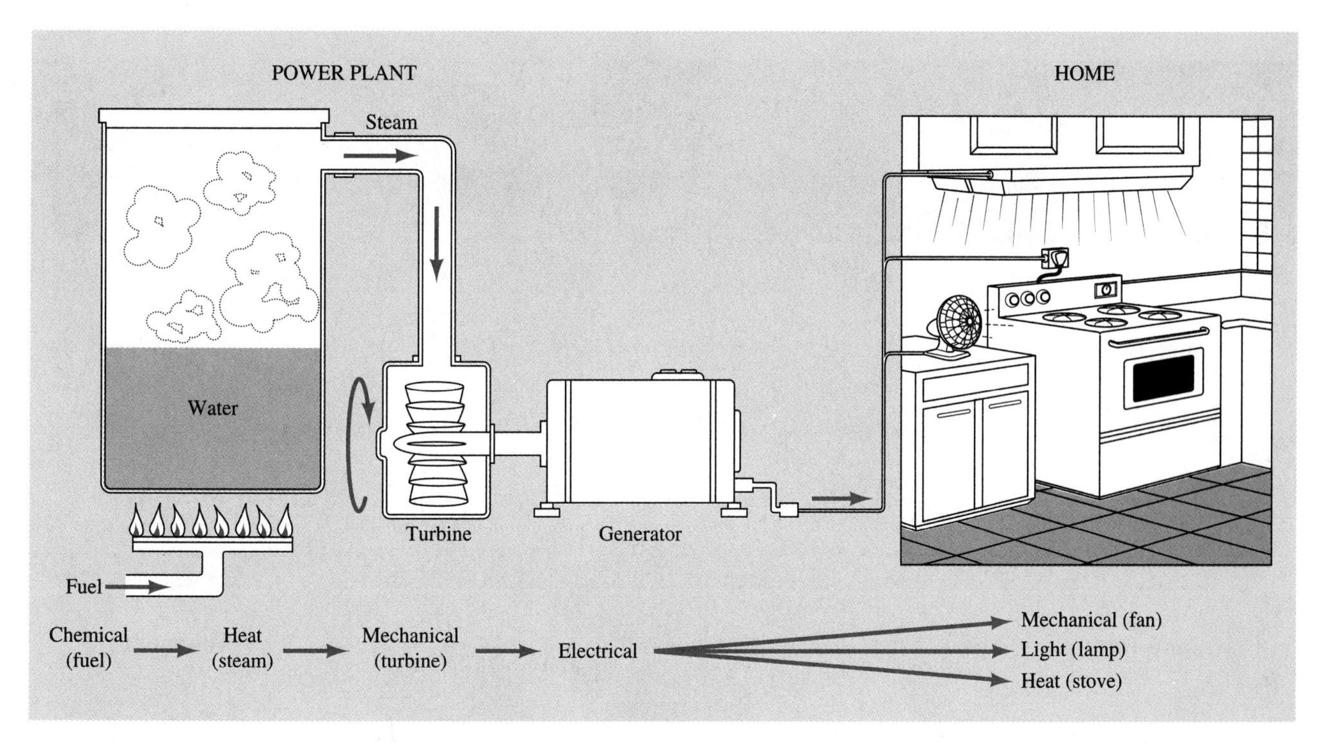

FIGURE 3-11 Interconversion of energy forms.

the number of kilocalories of energy it is capable of liberating upon digestion and metabolism.

The Summary Diagram on page 68 gives the conversions we have discussed in this chapter.

PROBLEM 3.27

Convert each of the following to the indicated unit.

(a) 25.0 cal = ______ J **(b)** 0.575 cal = ______ J **(c)** ______ cal = 1.43 J
(d) 1.75 kcal = ______ kJ **(e)** 325 cal = ______ kJ

3.12 HEAT AND TEMPERATURE

OBJECTIVE 3.12 Distinguish between heat and temperature.

Heat: A form of energy related to molecular motion.

Heat is a form of energy created by molecules in motion, so it is a form of kinetic energy. (There is more discussion about the motion of molecules later in the text.) Heat may be transferred from one object to another, but it always flows from a hotter object to a colder one. For example, when a cup of hot coffee cools, heat from the coffee is transferred to the surroundings. A cold spoon placed in the cup of coffee will get hot, absorbing heat from the coffee.

It is important that we distinguish between heat and temperature. You probably think of temperature as a measure of the *hotness* or *coldness* of an object or its surroundings. However, it is important not to confuse "hot" with "heat." When we

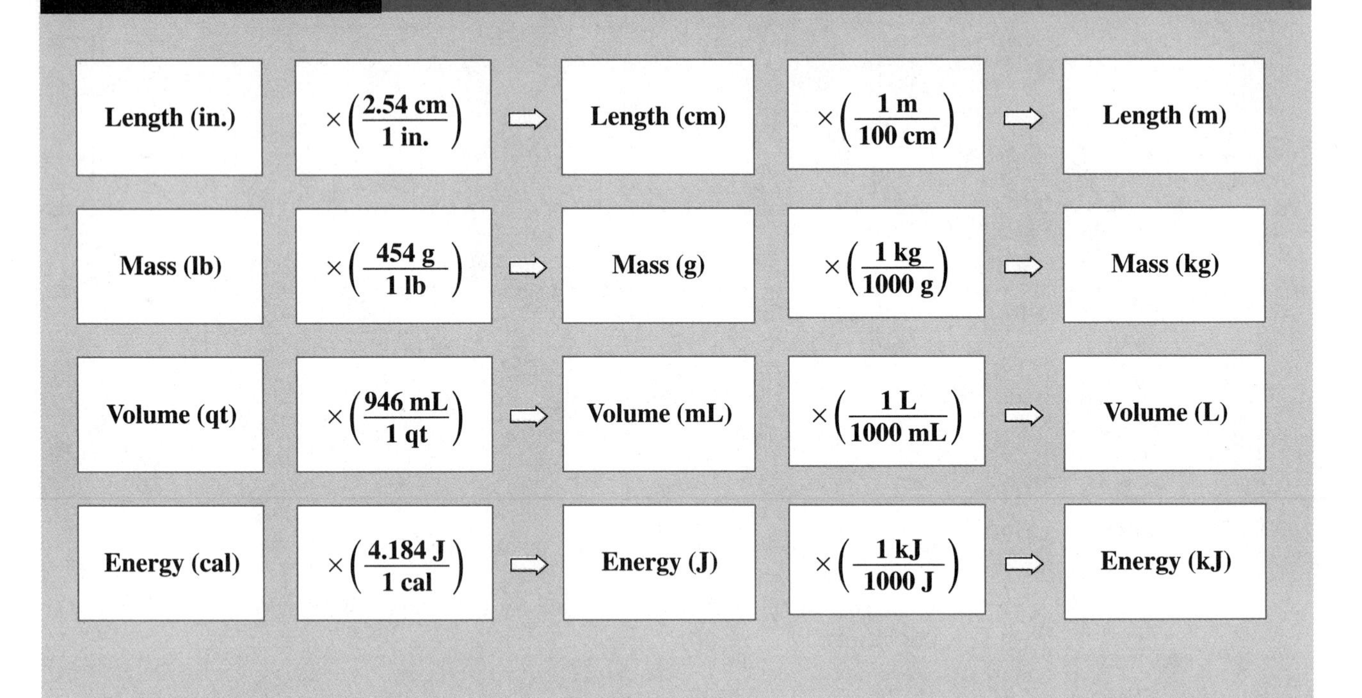

Temperature: The average kinetic energy of a sample.

measure heat, we are measuring a *quantity* of kinetic energy. When we measure **temperature,** we are measuring the *average* kinetic energy present in a sample. To illustrate the difference, suppose we place two identical beakers over identical burners on the kitchen stove (Fig. 3-12). In one beaker we put 1 L of water, and in the other, 2 L of water. Next we measure the temperature of both beakers of water and find that they are at the same temperature, 20°C. Now let us turn on both burners so that the flames are the same height and let the water heat for 10 minutes. When we return, we find that the beaker containing 2 L of water is at 30°C, whereas the beaker containing 1 L of water is at 40°C. Both beakers absorbed the same quantity of heat. However, the one containing less water (fewer water molecules) ended up with a greater *average* energy (more heat per molecule) and hence a higher temperature.

PROBLEM 3.28

Complete the following paragraph by inserting the word *heat* or the word *temperature,* as appropriate, in each blank.

A 5-g gold coin and a 1-kg gold block are both placed in an oven at 150°C for 1 hr. At the end of this time, both objects have the same ______, although the gold block has absorbed more ______. If both objects were to absorb the same quantity of ______, the gold coin would have the higher ______.

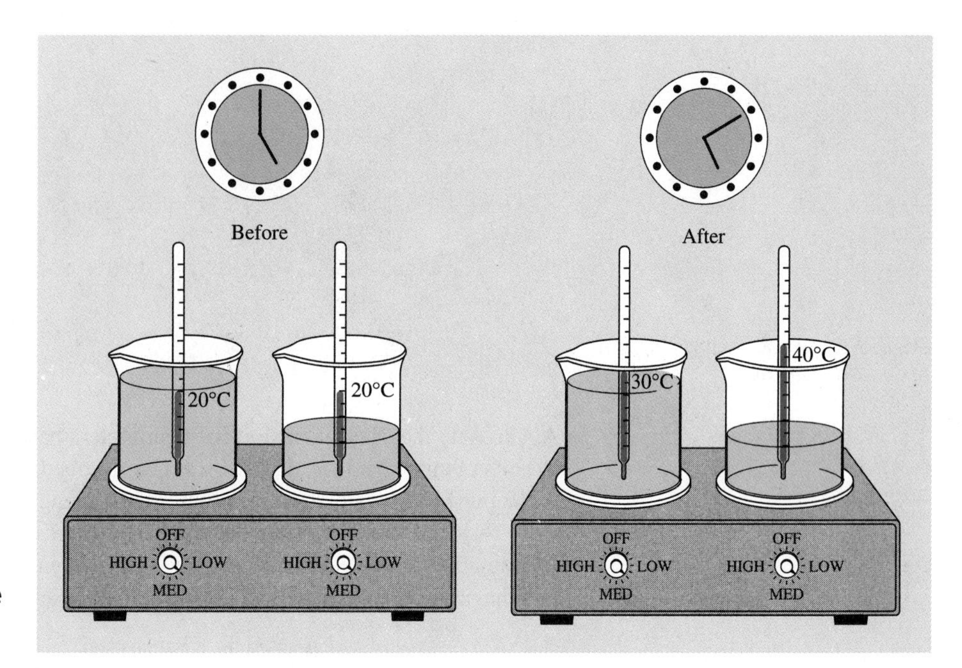

FIGURE 3-12 The distinction between heat and temperature. Both beakers of water absorb the same amount of heat. Because the beaker containing the smaller sample has less water to distribute the heat to, it has a higher average energy and thus a higher temperature.

3.13 MORE ABOUT DIMENSIONS

Areas and Volumes

OBJECTIVE 3.13A Convert between units of area expressed as square lengths and between units of volume expressed in cubic lengths.

The area of a rectangle is measured by multiplying the length times the width:

$$A = l \times w$$

Thus, an area that measures 16.0 cm by 12.0 cm is calculated as follows:

$$\begin{aligned} A &= l \times w \\ &= (16.0\text{ cm})(12.0\text{ cm}) = 192\text{ cm}^2 \end{aligned}$$

Note that the dimensions of centimeters are multiplied together, leading to units of area in square centimeters, or centimeters squared. The units of area have dimensions of length to the second power, because we are multiplying a length times a length.

In a similar fashion, the volume of a regular solid object (one with corners that are all right angles) is calculated by measuring the length times the width times the height: $V = l \times w \times h$. To calculate the volume of a regular object that measures 8.00 cm by 6.00 cm by 12.0 cm, we multiply the dimensions as follows:

$$\begin{aligned} V &= l \times w \times h \\ &= (8.00\text{ cm})(6.00\text{ cm})(12.0\text{ cm}) = 576\text{ cm}^3 \end{aligned}$$

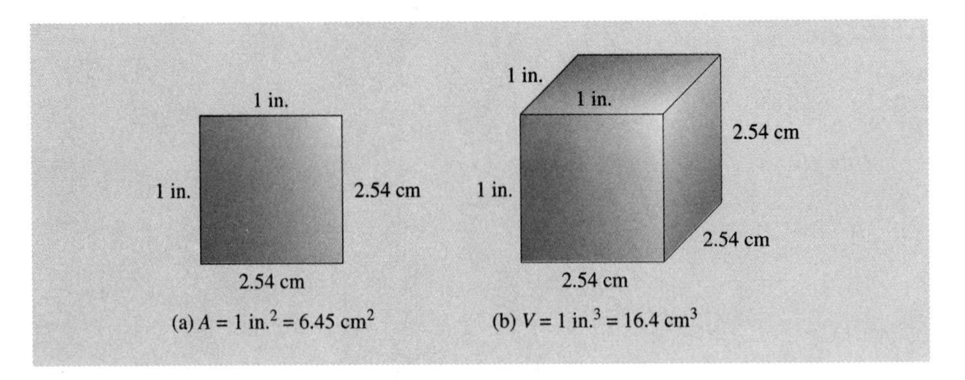

(a) $A = 1\text{ in.}^2 = 6.45\text{ cm}^2$ (b) $V = 1\text{ in.}^3 = 16.4\text{ cm}^3$

FIGURE 3-13 Relating units of area and volume.

Again, note that the dimensions of centimeters have been multiplied together, leading to units of cubic centimeters, or centimeters cubed. Volumes have dimensions of length to the third power, since we have multiplied a length times a length times a length.

Suppose we wish to convert between English units of area and SI units of area. Examination of a square inch (Figure 3-13a) reveals that it is equal to 6.45 cm^2. To obtain that area, we had to square 2.54 cm, the SI equivalent of an inch:

$$A = (1\text{ in.})(1\text{ in.}) = (2.54\text{ cm})(2.54\text{ cm})$$
$$A = (1\text{ in.})^2 = (2.54\text{ cm})^2$$
$$A = 1^2\text{ in.}^2 = 2.54^2\text{ cm}^2$$
$$A = 1\text{ in.}^2 = 6.45\text{ cm}^2$$

By the same logic (Fig. 3-13b), it follows that a volume of $1\text{ in.}^3 = 16.4\text{ cm}^3$:

$$V = (1\text{ in.})^3 = (2.54\text{ cm})^3$$
$$V = 1^3\text{ in.}^3 = 2.54^3\text{ cm}^3$$
$$V = 1\text{ in.}^3 = 16.4\text{ cm}^3$$

Other relationships between units of area and volume may be similarly determined. The important thing to remember is that it is necessary to raise *both* the number in the equivalency and its unit to the appropriate power. Thus, to calculate the number of square kilometers in a square mile, we must square the entire relationship:

$$1\text{ mile} = 1.61\text{ km}$$
$$(1\text{ mile})^2 = (1.61\text{ km})^2$$
$$1^2\text{ mile}^2 = 1.61^2\text{ km}^2$$
$$1\text{ mile}^2 = 2.59\text{ km}^2$$

We can use relationships derived in this fashion as conversion factors. However, it is possible to derive the conversion factors within the setup itself by selecting the correct conversion factor of length and then squaring it or cubing it, depending on the units to be canceled. The next example shows how this is done.

EXAMPLE 3.21

Convert 18.0 ft^3 to cubic meters.

Symbolic statement

$$?\text{ m}^3 = 18.0\text{ ft}^3$$

Planning the solution

To convert from cubic feet to cubic meters, we cube each of the linear relationships between feet and inches, inches and centimeters, and centimeters and meters, as follows:

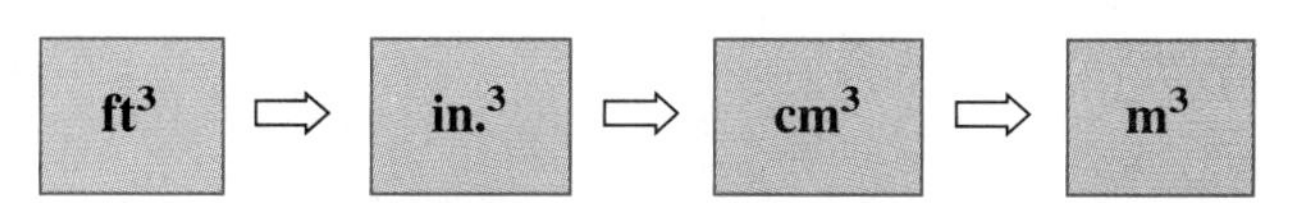

SOLUTION

$$? \text{ m}^3 = 18.0 \text{ ft}^3 \left(\frac{12 \text{ in.}}{1 \text{ ft}}\right)^3 \left(\frac{2.54 \text{ cm}}{1 \text{ in.}}\right)^3 \left(\frac{1 \text{ m}}{100 \text{ cm}}\right)^3$$

$$= 18.0 \text{ ft}^3 \left(\frac{12^3 \text{ in.}^3}{1^3 \text{ ft}^3}\right)\left(\frac{2.54^3 \text{ cm}^3}{1^3 \text{ in.}^3}\right)\left(\frac{1^3 \text{ m}^3}{100^3 \text{ cm}^3}\right) = 0.510 \text{ m}^3$$

Earlier, we learned that 1 mL = 1 cm^3. Thus, we can convert from units of volume expressed as a cubic length to those of liters, as shown in the next example.

EXAMPLE 3.22

Calculate the number of liters in a box that measures 11.0 in. by 8.50 in. by 2.00 in.

Symbolic statement

$$? \text{ L} = (11.0 \text{ in.})(8.50 \text{ in.})(2.00 \text{ in.})$$

Planning the solution

First we determine the volume of the box in cubic inches (in.^{3}) and then apply the following strategy:

in.^{3} ⇨ **cm^3** ⇨ **mL** ⇨ **L**

The key to the strategy is the conversion of cubic centimeters to milliliters using the relationship 1 cm^3 = 1 mL.

SOLUTION

We can calculate the result directly, without actually computing the number of cubic inches.

$$? \text{ L} = (11.0 \text{ in.})(8.50 \text{ in.})(2.00 \text{ in.})\left(\frac{2.54 \text{ cm}}{1 \text{ in.}}\right)^3 \left(\frac{1 \text{ mL}}{1 \text{ cm}^3}\right)\left(\frac{1 \text{ L}}{1000 \text{ mL}}\right)$$

$$? \text{ L} = (11.0 \text{ in.})(8.50 \text{ in.})(2.00 \text{ in.})\left(\frac{2.54^3 \text{ cm}^3}{1^3 \text{ in.}^3}\right)\left(\frac{1 \text{ mL}}{1 \text{ cm}^3}\right)\left(\frac{1 \text{ L}}{1000 \text{ mL}}\right)$$

$$? \text{ L} = 187 \text{ in.}^3 \left(\frac{2.54^3 \text{ cm}^3}{1^3 \text{ in.}^3}\right)\left(\frac{1 \text{ mL}}{1 \text{ cm}^3}\right)\left(\frac{1 \text{ L}}{1000 \text{ mL}}\right) = 3.06 \text{ L}$$

This is roughly the volume of a standard ream of paper.

PROBLEM 3.29

Carry out each of the following conversions.

(a) 4.50 yd^2 to ft^2

(b) 3.50 m^3 to dm^3

(c) 3.50 m^3 to L [Looking back to part (b), what is the relationship between liters and cubic decimeters?]

(d) 7.25 ft^3 to L

PROBLEM 3.30

Use the following strategy to calculate the number of gallons in a cubic foot: ? gal = 1.00 ft^3. [*Remember:* 1 gal = 4 qt]

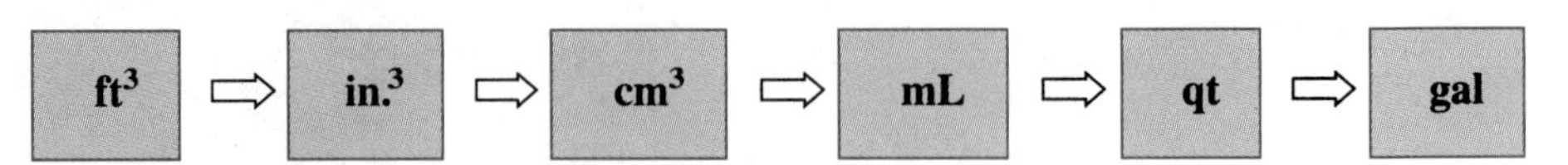

Working With Units in the Denominator

OBJECTIVE 3.13B Convert between various units found in the denominator of a quantity.

Frequently we wish to convert units that appear in a denominator. For example, suppose that we are traveling at a velocity of 45 miles per hour (miles/hr), and we wish to know our velocity in miles per minute (miles/min):

$$\frac{?\ \text{mi}}{\text{min}} = \frac{45\ \text{mi}}{\text{hr}}$$

We can convert the denominator from hours to minutes by using the relationship 1 hr = 60 min:

$$\frac{60\ \text{min}}{1\ \text{hr}} = 1 \qquad \text{or} \qquad 1 = \frac{1\ \text{hr}}{60\ \text{min}}$$

Since the unit of hours appears in the denominator of 45 miles/hr, we select the conversion factor with hours in the numerator (the one on the right), thereby canceling hours and giving us units of miles per minute:

$$\frac{?\ \text{mi}}{\text{min}} = \frac{45\ \text{mi}}{\cancel{\text{hr}}}\left(\frac{1\ \cancel{\text{hr}}}{60\ \text{min}}\right) = \frac{0.75\ \text{mi}}{\text{min}}$$

Similarly, we can convert units in both the numerator and the denominator of the same setup, as follows.

EXAMPLE 3.23

Convert the velocity of 60.0 miles/hr to units of feet per second (ft/sec).

Symbolic statement

$$\frac{?\ \text{ft}}{\text{sec}} = \frac{60.0\ \text{mi}}{\text{hr}}$$

Planning the solution

In this case we need to convert the numerator from miles to feet:

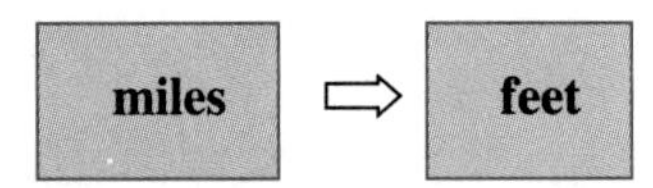

and the denominator from hours to seconds:

SOLUTION

$$\frac{?\ \text{ft}}{\text{sec}} = \frac{60.0\ \cancel{\text{mi}}}{\cancel{\text{hr}}}\left(\frac{5280\ \text{ft}}{1\ \cancel{\text{mi}}}\right)\left(\frac{1\ \cancel{\text{hr}}}{60\ \cancel{\text{min}}}\right)\left(\frac{1\ \cancel{\text{min}}}{60\ \text{sec}}\right) = \frac{88.0\ \text{ft}}{\text{sec}}$$

The first conversion factor converted the numerator from miles to feet, and the next two conversion factors converted the denominator from hours to seconds.

PROBLEM 3.31 The density of oxygen is 0.00133 g/cm^3. What is its density in kilograms per cubic meter (kg/m^3)?

PROBLEM 3.32 Convert 60.0 miles/hr to meters per second (m/sec).

PROBLEM 3.33 The density of titanium is 0.163 pound per cubic inch. Convert this density to grams per cubic centimeter (g/cm^3).

CHAPTER SUMMARY

Measurements may be made in either the **English system** or the **International System (SI),** though scientists universally use the International System. The SI units of measure employed by chemists include the *meter* (length), *kilogram* (mass), *liter* (volume), *kelvin* (temperature), and *joule* (energy). Multiples or fractions of these basic units of measurement are related to one another by powers of 10. In general, the names of these derived units consist of a prefix that tells the power of 10 by which to multiply or divide, followed by the name of the basic unit, as in the *milli*liter.

Numbers themselves may either be **exact** or **measured.** Measured numbers have some **uncertainty** associated with them, and the degree of uncertainty depends on the measuring instrument used. We may describe the degree of uncertainty in a number by referring to the number of **significant figures** (or **significant digits**) in the number. In any measured number, all digits are significant except for those zeros whose sole function is to hold the decimal point. When adding or subtracting numbers, we must round off the final answer to the same decimal place as the number (among those we started with) that has the greatest degree of uncertainty. When multiplying or dividing numbers, we must round off the final answer to the same number of significant figures as the factor that has the least number of significant figures. Measurements are often repeated in scientific work. **Precision** refers to how closely repeated measurements agree with one another. **Accuracy** refers to how closely a measurement agrees with the accepted (or correct) value.

Units can aid calculations when we use a method known as **dimensional analysis.** With this technique, equivalency relationships expressed as **conversion factors** are used to convert a given quantity to a different desired unit of measure. Scientific work often involves the relationship between two measured quantities. **Density,** an example of this, is the mass per unit volume of a substance. The density of a substance can be used as a conversion factor to convert between the mass of a substance and its volume.

Among the various quantities that chemists measure frequently are mass, volume, temperature, and energy. It is important not to confuse mass with weight. **Mass** is a measure of the quantity of matter. **Weight** refers to the gravitational forces acting on an object. Chemists are interested in the quantity of matter, so we measure mass. It is also important to distinguish heat from temperature. **Heat** is a quantity of energy that can be transferred from one object to another. For example, a **calorie** is the amount of heat required to raise the temperature of 1 g of water by 1°C. **Temperature** is a measure of the average energy of the matter in an object or substance.

Two common temperature scales are employed in scientific work, the **Celsius** scale and the **Kelvin** scale. The Kelvin, or absolute, temperature is the SI unit of measure, although the Celsius scale is more widely used. The relationship between the two is $T_K = T_C + 273$. The English **Fahrenheit** scale is related to the Celsius scale as follows: $T_F = 1.8T_C + 32$.

Energy is required any time a force is applied through a distance. Energy may exist as either **kinetic energy,** the energy of motion, or **potential energy,** which is stored energy. The **Law of Conservation of Energy** tells us that energy can be neither created nor destroyed. However, energy can be converted from one form to another.

KEY TERMS

Review each of the following terms, which are discussed in this chapter and defined in the Glossary.

exact number
measured number
uncertainty
precision
accuracy
significant figures (or digits)
dimensional analysis
conversion factor
International System (SI)
English system
mass
weight
density
Kelvin temperature
Celsius temperature
Fahrenheit temperature
energy
kinetic energy
potential energy
Law of Conservation of Energy
calorie
heat
temperature

ADDITIONAL PROBLEMS

SECTION 3.2

3.34 Answer the following questions about numbers themselves.

(a) Distinguish between an exact number and a measured number.
(b) Explain what is meant by uncertainty.
(c) Explain thc mcanings of prccision and accuracy.

3.35 Three chemists carry out an experiment designed to determine the percentage of iron in an alloy. Each chemist repeats the experiment three times and averages the results obtained. The actual percentage of iron in the alloy is 57.2%. The chemists' results are shown in the following table:

	Trial #1	Trial #2	Trial #3	Average
Chemist #1	57.1%	57.3%	57.2%	57.2%
Chemist #2	53.5%	53.3%	53.4%	53.4%
Chemist #3	55.0%	59.1%	57.5%	57.2%

(a) Which chemist had good precision but poor accuracy?
(b) Which chemist had good accuracy but poor precision?
(c) Which chemist had both good accuracy and good precision?

SECTION 3.3

3.36 How many significant figures are in each of the following numbers?

(a) 42.0 **(b)** 1864 **(c)** 0.920
(d) 1010 **(e)** 456,000 **(f)** 512,000.0
(g) 14.06 **(h)** 0.0035 **(i)** 4.5×10^{-3}
(j) 7.000×10^{6}

3.37 Round off each of the following numbers to three significant figures.

(a) 47.654 **(b)** 0.0082147 **(c)** 19.982
(d) 17,465 **(e)** 6.5621×10^{-4}

3.38 Round off each of the following answers to the correct number of significant figures.

(a) $1.032 + 4.7 + 62.43 = 68.162$
(b) $73.45 - 5 + 17.8 = 86.25$
(c) $521.6 + 130 + 216 = 867.6$
(d) $47.3 \times 2.40 = 113.52$

(e) $\dfrac{0.19}{4.68} = 0.0405983$

(f) $\dfrac{122 + 3.7}{243.7 - 8.12} = 0.5335767$

(g) $(1.076 + 3.25) \times (16.77 + 8) = 107.15502$

3.39 The precise definition of the meter is 1,650,763.73 wavelengths of orange-red light emitted by the krypton-86 atom. How many significant figures are in the number 1,650,763.73?

SECTION 3.4

3.40 Use dimensional analysis and the relationships between units of time to carry out the following conversions. (All are exact relationships except for the first one, which is accurate to three significant figures.)

1 yr = 365 days; 1 day = 24 hr; 1 hr = 60 min;
1 min = 60 sec

(a) How many hours are there in 1.5 days?
(b) How many seconds are there in 0.110 hr?
(c) How many seconds are there in 21.0 days? Express your final answer in scientific notation.
(d) How many years are there in 15.0 min? Express your final answer in scientific notation.

3.41 The following exact relationships describe an antiquated set of British liquid units.

1 hogshead = 7 firkin
18 pottle = 1 firkin
140 pottle = 1 puncheon
504 pottle = 1 tun

Use dimensional analysis to determine each of the following.
(a) The number of puncheons in 42 pottle
(b) The number of firkins in 8.00 tun
(c) The number of tuns in 144 hogshead

3.42 If apricots are selling for 69¢ per pound and the average pound contains six apricots, how many apricots can be purchased for $2.07? (Use dimensional analysis to set this up.)

SECTION 3.5

3.43 State the numerical meaning of each of the following prefixes: mega, kilo, deci, centi, milli, micro, and nano.

3.44 Identify the basic SI unit of measure for each of the following: length, volume, mass, temperature, energy, and pressure.

SECTION 3.6

3.45 Use dimensional analysis to carry out each of the following conversions.
(a) 14.9 in. to cm
(b) 0.427 in. to mm
(c) 31.0 ft to m
(d) 1.25 yd to m
(e) 246 miles to km (Use 1 in. = 2.54 cm to convert from English to SI units.)
(f) 4.52 in. to nm

3.46 Calculate your height in meters.

SECTION 3.7

3.47 Distinguish between mass and weight.

3.48 The gravitational force on the planet Jupiter is 2.8 times that on earth.
(a) What would be the mass of a 50.0-kg woman on Jupiter?
(b) What would be the weight of a 150-lb man on Jupiter?

3.49 Complete the blank spaces in the following sentence with either the word *balance* or the word *scale*.

A ______ is used to measure mass, whereas a ______ measures weight.

3.50 Use dimensional analysis to carry out the following conversions.
(a) 3.47 lb to g
(b) 0.260 lb to mg
(c) 4.38 oz to g (1 lb = 16 oz)
(d) 5.20 ton to kg (1 ton = 2000 lb)
(e) 3.50 ton to Mg
(f) 0.653 mg to μg

3.51 Calculate your weight in kilograms.

SECTION 3.8

3.52 Use dimensional analysis to carry out the following conversions.
(a) 2.40 qt to L
(b) 43.0 L to gal
(c) 12.0 oz to L (32 oz = 1 qt)

3.53 Most modern kitchenware is graduated in both English and SI units. Keeping in mind that there are exactly 4 cups in a quart, calculate the number of milliliters in a cup. (Check a measuring cup to see if your answer is correct.)

SECTION 3.9

3.54 Answer the following questions concerning density.

(a) Define the term *density* and write a mathematical equation expressing the relationship.

(b) How is the density of *most* substances affected by raising the temperature?

(c) Why are the densities of gases much less than those of liquids and solids?

(d) Why do gas bubbles rise to the surface of a liquid?

(e) Chloroform has a density of 1.48 g/mL. Like gasoline, it does not dissolve in water. Will chloroform also act like gasoline by floating on top of water, or will it sink to the bottom like a coin?

3.55 A 25.0-mL sample of a liquid is found to have a mass of 18.3 g. Calculate its density.

3.56 The density of mercury is 13.6 g/mL. What is the mass of 35.0 mL of mercury?

3.57 The density of benzene is 0.879 g/mL. What is the mass of 1.32 L of benzene?

3.58 The density of chloroform is 1.48 g/mL. What volume of chloroform has a mass of 654 g?

3.59 The density of ether is 0.714 g/mL. What volume of ether has a mass of 2.44 kg?

3.60 The volume of an irregular solid object may be determined by measuring its displacement in water. This is often done in a graduated cylinder, a common piece of laboratory equipment for measuring volumes. Suppose a graduated cylinder is filled to the 35.0-mL mark with water and a nugget of copper weighing 98.1 g is immersed in the cylinder, raising the level of water to the 46.0-mL mark. What is the volume of the copper nugget? What is the density of copper?

3.61 The density of a regular solid object can be determined by weighing the object and determining its volume by measurement of its dimensions. Calculate the density of a block of aluminum with sides measuring 5.42 cm by 8.12 cm by 6.57 cm and with a mass of 781 g. Would you expect aluminum to sink or float in water?

3.62 Answer the following questions about specific gravity.

(a) What is specific gravity?

(b) How is the specific gravity of a substance related to its density?

(c) If the specific gravity of a substance is greater than 1.00, will the substance sink or float in water? Why?

3.63 The density of lead is 708 lb/ft^3.

(a) What is its specific gravity? [*Remember:* The density of water is 62.4 lb/ft^3.]

(b) What is the density of lead in grams per milliliter?

SECTION 3.10

3.64 Freon is a coolant used in refrigerators. Its boiling point is –20.2°F. Express this temperature in Celsius degrees. What Kelvin temperature does this equal?

3.65 Ethyl alcohol freezes at –179°F. Express this temperature in Celsius degrees. What Kelvin temperature does this equal?

3.66 Normal body temperature for human beings is 98.6°F. To what Celsius temperature does this correspond? To what Kelvin temperature?

3.67 What Fahrenheit temperature is the same as –40°C?

3.68 What Fahrenheit temperature corresponds to absolute zero, –273°C?

3.69 Superconductors are materials that conduct electricity with virtually no electrical resistance.

(a) Until recently, known superconductors could do so only at temperatures below 34 K. To what Fahrenheit temperature does this correspond?

(b) In 1987, Ching-Wu Paul Chu of the University of Houston made an enormous breakthrough in the field of superconductors by developing a ceramic material that superconducts below temperatures of 93 K. To what Fahrenheit temperature is this equivalent?

(c) The boiling point of liquid nitrogen is –196°C. Will liquid nitrogen cool Chu's ceramic adequately for use as a superconductor?

(d) The sublimation point of dry ice is –78°C. Is this cold enough to allow Chu's ceramic to superconduct?

(e) By 1989 Chu had improved his ceramic to superconduct at temperatures below –143°C. How many Celsius degrees higher is this than his original ceramic, reported in 1987?

SECTION 3.11

3.70 Answer each of the following questions concerning energy.

(a) Distinguish between kinetic and potential energy.

(b) State the Law of Conservation of Energy.

(c) List three forms of energy.

(d) Define the term *calorie.*

3.71 Use dimensional analysis to carry out the following conversions.

(a) 11.2 cal to J (b) 7.31 J to cal
(c) 893 cal to kJ (d) 6.53 kcal to kJ

SECTION 3.12

3.72 Distinguish between heat and temperature.

SECTION 3.13

3.73 An acre is an area of 43,560 ft^2. Calculate the area of an acre in each of the following.
(a) Square yards (b) Square miles
(c) Square meters (d) Square kilometers
(e) How many acres are there in a rectangular piece of property that measures 106 ft by 82 ft?

3.74 Hartnell College has an Olympic-size swimming pool that is 50.0 *meters* long by 25.0 *yards* wide, with an average depth of 5.00 *feet.* Calculate the volume of the pool in:
(a) Cubic feet (b) Cubic meters
(c) Liters (d) Gallons

3.75 A wood block has dimensions of 10.7 cm by 11.2 cm by 10.1 cm. If its mass is 983 g, calculate its density in grams per milliliter. [*Hint:* What is the relationship between mL and cm^3?]

3.76 The density of gold is 18.88 g/cm^3. What is the volume of a 5.00-lb block of gold?

3.77 The velocity of light is 3.00×10^8 m/sec. What is this velocity equal to in:
(a) Kilometers per hour?
(b) Miles per hour?

3.78 What is the density of water in pounds per cubic foot? (d = 1.00 g/mL at 4°C)

GENERAL PROBLEMS

***3.79** Water is most dense at 4°C. In a frozen lake, what conclusion can be drawn about the temperature of the (liquid) water at the top of the lake (just below the frozen surface) compared to the temperature at the bottom of the lake?

***3.80** When liquid water is added to a glass with ice in it, the ice will float to the top of the liquid water. If enough liquid is added, a portion of the solid (ice) can float higher than the rim of the glass. Suppose enough water is added to bring the liquid level to the rim of the glass. Will liquid water overflow the rim of the glass as the ice melts? Explain your answer.

***3.81** Body temperature was originally found to be 37°C to two significant figures. When the value was converted from the Celsius scale to the Fahrenheit scale, it was given an extra significant figure (98.6°F). Given that 37°C could mean any value between 36.5°C and 37.5°C, what is the range in Fahrenheit for body temperatures? (There are two ways to solve this problem—one is much easier than the other. See if you can identify both ways to do it.)

***3.82** At what value does the temperature on the Kelvin scale equal the value for the temperature on the Fahrenheit scale?

3.83 If picking up an aluminum rod at 50°C causes less pain than picking up a glass rod of the same mass at 50°C, what can be said about the relative amounts of heat in the two rods?

Problems 3.84–3.86 involve your understanding of the meaning of *per.*

3.84 Answer each of the following questions.
(a) State the mathematical meaning of the word *per.*
(b) Seven apples cost 91 cents. What is the cost per apple?
(c) If the seven apples weigh 1.82 lb, what is the cost per pound?
(d) What is the weight per apple?

3.85 One pound equals 16 oz. A pound is also equal to 454 g. What is the number of grams per ounce?

3.86 A car travels 225 miles on 9.62 gal of gas.
(a) Calculate the fuel efficiency in miles per gallon (miles/gal).
(b) Calculate the fuel efficiency in gallons per mile.
(c) Use the equivalencies of 1 km = 0.621 mi and 1 gal = 3.785 L to convert the answer from part (a) to kilometers per liter. You will have to convert the units in both the numerator and the denominator.

3.87 Atmospheric pressure is often expressed in English units as 14.7 lb/in.^{2}. Convert this measurement to:
(a) g/cm^2 (b) kg/m^2

3.88 The density of mercury is 13.6 g/mL. Convert this density to English units of pounds per gallon (use 1 gal = 3.785 L). Would you need help lifting a gallon of mercury?

***3.89** In parts of the country, drought conditions have required water rationing. Under severe conditions, residents may be restricted to 50 gal per person per day, but their water meters read in cubic feet. Using dimensional analysis, calculate the number of cubic feet that corresponds to 50.0 gal.

Problems 3.90–3.97 are designed to sharpen your skills in dimensional analysis, as well as your critical thinking skills. The last three problems have some extraneous information, so be careful in selecting the information you actually need to solve each of those problems.

3.90 How many dollars will it cost to fill an 18.0-gal gas tank if gasoline is selling for 32.9 cents per liter?

3.91 An economy-minded student traveling on limited funds has budgeted \$125 to spend on gasoline during a sight-seeing trip she plans to take. If her automobile gets 37.0 miles per gallon and the average cost of gasoline is \$1.33 per gallon, what is the maximum total round-trip distance she can expect to travel?

3.92 The owner's manual of a popular foreign automobile says that the capacity of the gas tank is 35.0 L. The car averages 42.3 miles per gallon on the open road. What is the maximum distance the car can travel on a tank of gas?

***3.93** A box of saltine crackers sells for \$1.29. There are exactly 80 crackers in a box. The average weight of each cracker is 0.100 oz. What is the price of crackers per pound? (1 lb = 16 oz)

***3.94** Each car on the sky tram at a popular amusement park makes a round trip in 10 min and 15 sec. A car leaves each departure station every 8.2 sec. The distance between cars is 24 ft. What is the one-way distance (half the round-trip) between sky tram stations?

3.95 A 10-lb bag of potatoes contains 41 potatoes. The price of the bag is \$1.89. How many pounds of potatoes can a shopper purchase for \$5.67?

3.96 Cherries cost \$1.19/lb. A consumer counts 58 cherries to a pound. How many grams does a cherry weigh?

3.97 Red bell peppers cost \$1.89 per pound. Green bell peppers cost 89¢ per pound. A shopper purchases three red bell peppers for \$2.35 and two green bell peppers for 74¢. How many grams does a green bell pepper weigh?

Writing Exercises

3.98 Describe the difference between the English and International systems. Your answer should include a discussion of the features that make the International System easy to use and the features that make the English system difficult to use.

3.99 Summarize the meaning of significant figures. Your summary should include an explanation of what are significant figures and how to determine which digits in a number are significant. In addition, explain how to add or subtract numbers and how to multiply or divide numbers in accord with the rules for significant figures.

3.100 Explain how a pendulum works. Your explanation should include a description of kinetic and potential energy, as well as how a pendulum interconverts these energy forms.

3.101 A friend has asked you to explain the concept of density. Write a note to your friend explaining this concept, and describe how you would measure the density of an irregular solid object.

QUANTITATIVE DESCRIPTION OF MATTER

In Chapter 2 you were introduced to the nature of matter in terms of its many classifications, its properties, and the changes it undergoes. In Chapter 3 we studied the various units of measure used to describe the properties of matter. Now we are prepared to undertake a quantitative description of matter. Since many of the calculations require the use of chemical formulas, we begin this chapter by examining some of the types of formulas you will encounter.

4.1 MORE ABOUT CHEMICAL FORMULAS

OBJECTIVE 4.1 Distinguish between a molecular formula and an empirical formula. Given a molecular formula, determine the empirical formula.

Molecular formula: Shows the total number of atoms of each element present per molecule of a substance.

Empirical formula: Shows the simplest whole-number ratio of atoms present in a substance.

In Chapter 2 we began to write chemical formulas. Many of the formulas we wrote were molecular formulas. A **molecular formula** indicates the total number of each type of atom present in a molecule of the substance. For example, the molecular formula of octane is C_8H_{18}. This means that every molecule of octane contains eight atoms of carbon and eighteen atoms of hydrogen. In certain situations, however, we have only enough information to write another type of formula, known as an empirical formula. The **empirical formula** is the simplest whole-number ratio of atoms in a compound. For octane, the simplest ratio of atoms of carbon to hydrogen is 4 : 9, so its empirical formula is C_4H_9. Water has a molecular formula of H_2O. The molecular formula also happens to represent the simplest whole-number ratio of atoms. Hence, the empirical formula of water is also H_2O.

Not all substances form distinct molecules. For many such *nonmolecular* substances, the term *formula unit* is used to describe the formula. Sodium chloride, NaCl, potassium carbonate, K_2CO_3, and magnesium nitrate, $Mg(NO_3)_2$, are examples you will encounter in the next section. You will learn more about the meaning of a formula unit when we study chemical bonding in Chapter 8.

With the exception of certain elements that form diatomic molecules (for example, oxygen, O_2, or nitrogen, N_2), most elements are nonmolecular. We express the formulas of nonmolecular elements as the symbol for one atom, as in the case of silicon, Si, or copper, Cu. Thus, most of the formulas you are likely to encounter are either atomic formulas, molecular formulas, or formula units.

EXAMPLE 4.1

What is the empirical formula of ethene, C_2H_4?

SOLUTION

The ratio of carbon to hydrogen atoms is 2 : 4. The simplest whole-number ratio corresponding to 2 : 4 is 1 : 2. Hence, the empirical formula is CH_2.

PROBLEM 4.1

Write the empirical formula corresponding to each of the following molecular formulas.

(a) C_4H_{10} (butane) **(b)** CCl_4 (carbon tetrachloride) **(c)** B_2H_6 (diborane)
(d) $C_6H_{12}O_6$ (glucose) **(e)** N_2H_4 (hydrazine)

4.2 ATOMIC MASS

OBJECTIVE 4.2 Define the term *atomic mass,* and explain the meaning of relative mass. Find the atomic mass of any element from the periodic table, and round off the value to the nearest tenth. State the appropriate unit to use when describing the atomic mass of one atom.

Atomic mass: The relative mass of an atom, as compared to some standard.

Relative mass: The mass of an object, as compared to some standard.

We mentioned in Section 2.5 that the atoms of each element differ from those of every other element. One of the ways in which the atoms of two elements differ from each other is in their **atomic mass.** The periodic table (which appears inside the front cover) displays the atomic mass of each element directly beneath its symbol. When we speak of atomic mass, we are actually referring to the *relative* mass of an atom, as compared to some standard. (The atomic mass used to be called the *atomic weight,* and you may encounter this term in your readings elsewhere. But note that the word *weight* is so entrenched in our language that we continue to use the verb *weigh* when referring to the process of measuring mass. Thus, we *weigh* a substance on a balance to measure its *mass.*)

To explain what we mean by a **relative mass,** consider Mr. Goldsmith, a world-renowned jeweler, who wishes to compare the masses of several pieces of jewelry: a pair of gold earrings, a gold tie clasp, and the famed Goldsmith golden nugget, a treasured family heirloom. Although Mr. Goldsmith does not have a balance capable of providing numerical values of each object's mass, he does have a two-pan balance capable of *comparing* the masses of his jewelry. To begin the comparisons, Mr. Goldsmith places one earring on each of the pans and observes that the balance comes to rest as expected (Fig. 4-1a). Although he does not know the actual mass of either earring, he concludes that the two earrings are of equal mass.

Next he places the gold tie clasp on one of the pans and both of the earrings on the other, and again the pans come to rest in the balanced position (Fig. 4-1b). Mr.

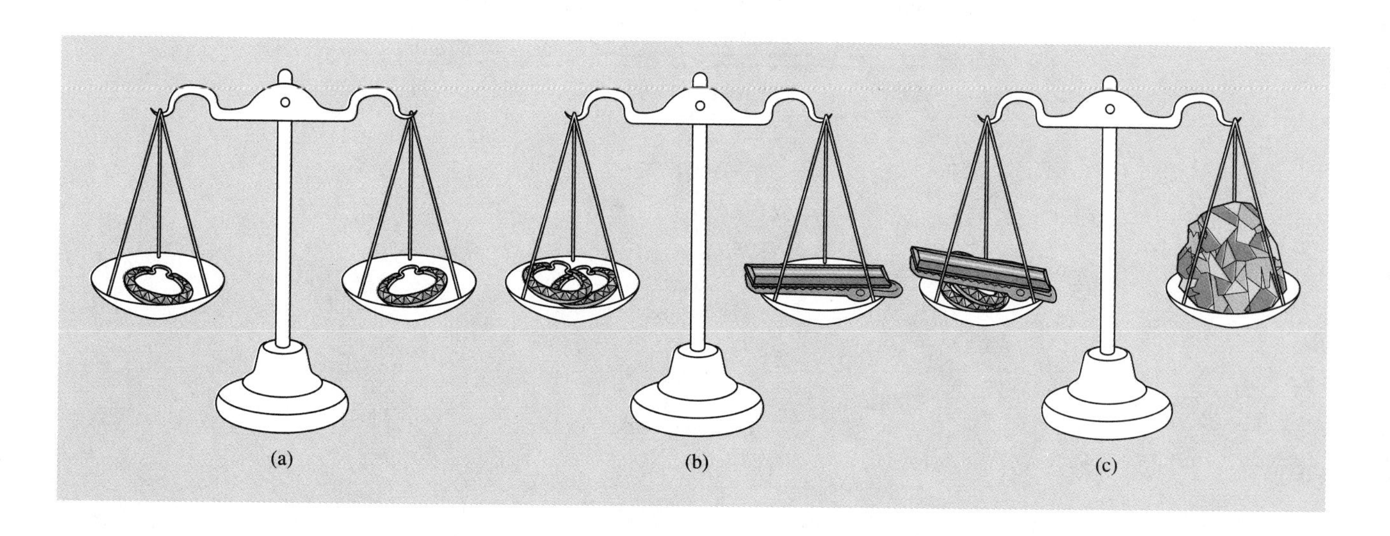

FIGURE 4-1 Relative masses of several objects. The tie clasp weighs twice as much as each earring. The gold nugget weighs four times as much as each earring and twice as much as the tie clasp.

TABLE 4-1 Relative masses of Goldsmith's jewelry

Item	Sets of relative masses		
One gold earring	1	$\frac{1}{2}$	$\frac{1}{4}$
Gold tie clasp	2	1	$\frac{1}{2}$
Goldsmith nugget	4	2	1

Goldsmith concludes that the tie clasp must have twice the mass of each earring, although he does not know its actual mass.

Finally, he decides to match these objects against his prized gold nugget. He places both earrings, along with the tie clasp, on one pan and the Goldsmith nugget on the other. When the pans come to rest in the balanced position, he concludes that the nugget's mass is twice that of the tie clasp and four times that of each earring (Fig. 4-1c). At this point, Mr. Goldsmith does not know the actual mass of any of his jewelry; however, he does know the *relative* mass of each piece. These are summarized in Table 4-1.

If we assign each earring a mass of 1, the tie clasp must have a mass of 2 and the nugget a mass of 4. These values are shown in the first column of numbers in Table 4-1. However, assignment of the earring as 1 is purely arbitrary. The other columns in Table 4-1 also express the same information about the relative masses of the jewelry. Throughout our discussion, we have used the earrings as the *standard* of comparison. However, it would have been just as valid to make the tie clasp or the nugget the standard. Finally, when we assign each earring a relative mass of 1, we do not specify any units, because we do not know its actual mass. We know only its mass *relative* to one of the other objects.

In much the same way, early chemists were able to gain information about the relative masses of atoms, even though they could not actually determine the masses of individual atoms. The atomic masses presented in the periodic table are just a set of these

TABLE 4-2 Atomic masses of some common elements

Element	Precise atomic mass	Rounded-off mass
H	1.0079	1.0
He	4.0026	4.0
C	12.011	12.0
N	14.0067	14.0
O	15.9994	16.0
F	18.9984	19.0
Na	22.9898	23.0
Mg	24.3050	24.3
S	32.066	32.1
Cl	35.453	35.5
K	39.0983	39.1
Ca	40.08	40.1

relative masses. Table 4-2 lists the atomic masses of several frequently encountered elements. For simplicity in our calculations and discussions in this text, we adopt the practice of rounding off atomic masses to the nearest tenth. Table 4-2 shows the precise atomic masses, as well as the rounded-off values we will use for these elements.

When we assign carbon an atomic mass of 12.0 and hydrogen an atomic mass of 1.0, we are simply stating that the mass of a carbon atom is roughly twelve times greater than that of a hydrogen atom. It does not mean that individual carbon atoms have masses of 12.0 g or that hydrogen atoms have masses of 1.0 g. It does mean, however, that if we have 12.0 g of carbon and 1.0 g of hydrogen, we must have an equal number of atoms of each. Today we can determine the masses of individual atoms, and we know that most hydrogen atoms have a mass of 1.67×10^{-24} g. However, chemists continue to use atomic masses in most of their work because they rarely work with one atom at a time, and it is simpler to work with numbers like 1.0 and 12.0 than with numbers the size of 1.67×10^{-24}.

The standard to which each atom is compared has changed periodically as our technology has advanced, but, as our jewelry example demonstrates, the relative masses are not dependent on the actual standard of comparison. The current standard is a substance known as carbon-12.

Atomic Mass Units

The relative masses (atomic masses) expressed in the periodic table are actually unitless, because each represents a ratio of masses. However, scientists have adopted a unit known as the *atomic mass unit* to facilitate working with one atom at a time. The SI symbol used to abbreviate atomic mass units is "u," and we use that symbol in this text. (You are also likely to encounter the symbol "amu" in your readings elsewhere. Both symbols have the same meaning.) Thus, you will frequently see the atomic mass of hydrogen reported as 1.0 u and that of carbon as 12.0 u. When we use this symbol, we are talking about the mass of one atom of hydrogen, carbon, or whatever. Note that the ratio of the relative mass of one atom of carbon to that of one atom of hydrogen is still 12.0 : 1.0. Chemists rarely work with only one atom at a time, so we rarely encounter the atomic mass unit in our practical work in chemistry.

PROBLEM 4.2

Find the atomic mass of each of the following elements. Round off each atomic mass to the nearest tenth.

(a) Kr **(b)** Rb **(c)** U **(d)** Rh **(e)** Pt
(f) Br **(g)** Ca **(h)** Al **(i)** I **(j)** P

4.3 MOLECULAR MASS

OBJECTIVE 4.3 Distinguish between molecular mass and formula mass. Calculate the molecular mass or formula mass of any substance. State the appropriate unit to use when describing the molecular mass of one molecule.

Just as every atom has a mass, so does every molecule. To find the mass of a molecule, we merely add up the masses of all of the atoms in the molecule. Since the mass of each atom is its atomic mass, this corresponds to adding up the atomic masses. For example, a molecule of water contains two atoms of hydrogen and one atom of oxygen. Therefore, the mass of a molecule of water can be calculated as follows:

$$\begin{array}{rcl} 2\text{ H atoms} = 2 \times 1.0 = & 2.0 & \leftarrow \textbf{Mass of H in } \mathbf{H_2O} \\ 1\text{ O atom} = 1 \times 16.0 = & 16.0 & \leftarrow \textbf{Mass of O in } \mathbf{H_2O} \\ \hline H_2O = & 18.0 & \end{array}$$

Molecular mass: The sum of the atomic masses of all the atoms in a molecule.

We refer to the mass of a molecule, calculated in this fashion, as its molecular mass. Thus, the **molecular mass** of any substance is the sum of the atomic masses of all the atoms in the molecule. When the molecular mass is expressed in atomic mass units, it represents the mass of one molecule of the substance. Thus, one molecule of water has a mass of 18.0 u. As a practical matter, we generally work with collections of molecules, rather than with one molecule at a time. Consequently, we do not include units in the molecular mass calculations in this text. Instead, we compute the molecular mass as shown above and add appropriate units as necessary.

Formula mass: The sum of the atomic masses of all the atoms in a chemical formula.

For nonmolecular substances, the term **formula mass** is often used in place of molecular mass. (Examples 4.3 and 4.4 represent such compounds.) Although it is too soon for you to predict which substances belong in each category (molecular or nonmolecular), this poses no difficulty since the calculation of each is carried out in identical fashion: simply sum the atomic masses of all atoms in the chemical formula. We will use the term formula mass when it is specifically called for; however, be aware that many chemists use the term molecular mass to refer to the formula mass of all compounds, and we likewise do so when we are speaking in general terms.

EXAMPLE 4.2

Calculate the molecular mass of glucose, $C_6H_{12}O_6$.

Planning the solution

For each element, we must look up its atomic mass on the periodic table and then multiply it by the number of atoms to obtain the contribution of that element to the total molecular mass. After doing that for all of the elements, we simply add them up.

SOLUTION

$$\begin{array}{rcl} 6 \times C = 6 \times 12.0 = & 72.0 \\ 12 \times H = 12 \times 1.0 = & 12.0 \\ 6 \times O = 6 \times 16.0 = & 96.0 \\ \hline C_6H_{12}O_6 = & 180.0 \end{array}$$

EXAMPLE 4.3

Calculate the formula mass of potassium carbonate, K_2CO_3.

Planning the solution

Even though this is a nonmolecular compound, our approach is exactly the same as it was in Example 4.2.

SOLUTION

$$\begin{array}{rcl} 2 \times K = 2 \times 39.1 = & 78.2 \\ 1 \times C = 1 \times 12.0 = & 12.0 \\ 3 \times O = 3 \times 16.0 = & 48.0 \\ \hline K_2CO_3 = & 138.2 \end{array}$$

EXAMPLE 4.4

Calculate the formula mass of magnesium nitrate, $Mg(NO_3)_2$.

Planning the solution

To calculate a formula mass when parentheses appear, we rewrite the formula in terms of the number of atoms of each element: MgN_2O_6. Then we proceed as usual.

SOLUTION

$$
\begin{aligned}
1 \times Mg &= 1 \times 24.3 = 24.3 \\
2 \times N &= 2 \times 14.0 = 28.0 \\
6 \times O &= 6 \times 16.0 = 96.0 \\
\hline
Mg(NO_3)_2 &= 148.3
\end{aligned}
$$

PROBLEM 4.3

Calculate the molecular (or formula) mass for each of the following compounds.

(a) Na_2CO_3, sodium carbonate

(b) $BiCl_3$, bismuth(III) chloride

(c) $KMnO_4$, potassium permanganate

(d) $NaHCO_3$, sodium hydrogen carbonate

(e) C_2H_5Cl, ethyl chloride

(f) $Ca(OH)_2$, calcium hydroxide

(g) $Al_2(SO_4)_3$, aluminum sulfate

4.4 Percentage Composition: The Law of Constant Composition

OBJECTIVE 4.4A Calculate the percentage composition of any substance, given its chemical formula.

The Law of Constant Composition: A pure substance has a fixed and definite composition.

Percentage composition: The composition of a substance expressed as a percentage of each element by mass.

Water is a compound made up of two atoms of hydrogen and one atom of oxygen per molecule. Furthermore, every molecule of water has this exact composition. Early chemists expressed this fact in a law. The **Law of Constant Composition** states that every pure substance has a fixed and definite composition. We can express the composition of a pure substance by the mass percentage of each of the elements making up the substance. This is known as the **percentage composition.** (If you need to review percentages, see Section A.4 of Appendix A.) For example, we found the molecular mass of water as follows:

$$
\begin{aligned}
2 \times H &= 2 \times 1.0 = 2.0 \quad \leftarrow \text{Mass of H in } H_2O \\
1 \times O &= 1 \times 16.0 = 16.0 \quad \leftarrow \text{Mass of O in } H_2O \\
\hline
H_2O &= 18.0
\end{aligned}
$$

The percentage of hydrogen in a molecule of water must be

$$
\begin{aligned}
\%H &= \frac{\text{Mass H}}{\text{Mass } H_2O} \times 100\% \\
&= \frac{2.0}{18.0} \times 100\% = 11\%
\end{aligned}
$$

And the percentage of oxygen in water must be

$$
\begin{aligned}
\%O &= \frac{\text{Mass O}}{\text{Mass } H_2O} \times 100\% \\
&= \frac{16.0}{18.0} \times 100\% = 89\%
\end{aligned}
$$

The percentages add up to 100%, as we would expect. The percentage composition of any substance can be calculated in exactly the same way.

EXAMPLE 4.5

Calculate the percentage composition of each element in sodium chloride (table salt), NaCl.

Planning the solution

The percentage composition of an element is the portion (as a percent) of the total formula mass contributed by that element. To calculate that portion, we need to know the total mass of all atoms of that element in the formula and also the total formula mass. Both of these quantities are determined in a formula mass calculation (see Example 4.3). Thus, we begin by calculating the formula mass. Following this, we obtain the percentage of each element by dividing its contribution to the formula mass by the formula mass itself and then multiplying the fraction obtained by 100%.

SOLUTION

First we calculate the formula mass:

$$\begin{aligned} 1 \times \text{Na} &= 1 \times 23.0 = 23.0 \\ 1 \times \text{Cl} &= 1 \times 35.5 = 35.5 \\ \hline \text{NaCl} &= 58.5 \end{aligned}$$

Then we find the percentage of each element.

$$\%\text{Na} = \frac{\text{Mass Na}}{\text{Mass NaCl}} \times 100\% = \frac{23.0}{58.5} \times 100\% = 39.3\%$$

$$\%\text{Cl} = \frac{\text{Mass Cl}}{\text{Mass NaCl}} \times 100\% = \frac{35.5}{58.5} \times 100\% = 60.7\%$$

EXAMPLE 4.6

Calculate the percentage composition of each element in DMSO (dimethyl sulfoxide), C_2H_6SO.

Planning the solution

Although there are more elements and more atoms of each, we proceed exactly as in the last example.

SOLUTION

$$\begin{aligned} 2 \times \text{C} &= 2 \times 12.0 = 24.0 \\ 6 \times \text{H} &= 6 \times 1.0 = 6.0 \\ 1 \times \text{S} &= 1 \times 32.1 = 32.1 \\ 1 \times \text{O} &= 1 \times 16.0 = 16.0 \\ \hline C_2H_6SO &= 78.1 \end{aligned}$$

$$\%\text{C} = \frac{\text{Mass C}}{\text{Mass } C_2H_6SO} \times 100\% = \frac{24.0}{78.1} \times 100\% = 30.7\%$$

$$\%\text{H} = \frac{\text{Mass H}}{\text{Mass } C_2H_6SO} \times 100\% = \frac{6.0}{78.1} \times 100\% = 7.7\%$$

$$\%\text{S} = \frac{\text{Mass S}}{\text{Mass } C_2H_6SO} \times 100\% = \frac{32.1}{78.1} \times 100\% = 41.1\%$$

$$\%\text{O} = \frac{\text{Mass O}}{\text{Mass } C_2H_6SO} \times 100\% = \frac{16.0}{78.1} \times 100\% = 20.5\%$$

EXAMPLE 4.7

Calculate the percentage composition of each element in calcium nitrate, $Ca(NO_3)_2$.

Planning the solution

First we rewrite the chemical formula in terms of the atoms present: CaN_2O_6. Then we proceed as usual.

SOLUTION

$$\begin{aligned} 1 \times Ca &= 1 \times 40.1 = 40.1 \\ 2 \times N &= 2 \times 14.0 = 28.0 \\ 6 \times O &= 6 \times 16.0 = 96.0 \\ \hline Ca(NO_3)_2 &= 164.1 \end{aligned}$$

$$\%Ca = \frac{\text{Mass Ca}}{\text{Mass } Ca(NO_3)_2} \times 100\% = \frac{40.1}{164.1} \times 100\% = 24.4\%$$

$$\%N = \frac{\text{Mass N}}{\text{Mass } Ca(NO_3)_2} \times 100\% = \frac{28.0}{164.1} \times 100\% = 17.1\%$$

$$\%O = \frac{\text{Mass O}}{\text{Mass } Ca(NO_3)_2} \times 100\% = \frac{96.0}{164.1} \times 100\% = 58.5\%$$

PROBLEM 4.4

Calculate the percentage composition of each element in each of the following compounds.

(a) MnO_2 **(b)** C_3F_8 **(c)** $CuBr_2$ **(d)** H_3PO_4
(e) CH_2BrF **(f)** $Zn(NO_3)_2$ **(g)** $Mg(CN)_2$

Calculating the Mass of an Element in a Given Quantity of a Compound

OBJECTIVE 4.4B Determine the mass of an element in a given quantity of a pure substance.

As a practical matter, we often wish to know the quantity of a particular element in a sample of a substance. For example, suppose that as a dietary supplement you are taking calcium. The calcium-containing substance in the dietary supplement is not pure calcium (the element), but is instead a compound that contains calcium. A very commonly used compound is calcium carbonate, $CaCO_3$. In addition to the calcium present, there is also carbon and oxygen. Suppose you wish to know how much calcium is in a 0.500-g (500-mg) tablet. The next example shows you how this can be determined.

EXAMPLE 4.8

Calculate the mass of calcium in a 0.500-g sample of calcium carbonate, $CaCO_3$.

Symbolic statement

$$? \text{ g Ca} = 0.500 \text{ g } CaCO_3$$

Planning the solution

One way that you could determine the mass of calcium present is to calculate the percentage of calcium in calcium carbonate (just as we have been doing), and then take that percentage of 0.500 g. However, we save ourselves a step by using the formula mass calculation to obtain a conversion factor for converting grams of calcium carbonate to grams of calcium.

g $CaCO_3$ ⇨ g Ca

SOLUTION

Let us begin by determining the formula mass in the usual fashion:

$$\begin{array}{lcl} 1 \times \text{Ca} = 1 \times 40.1 = & 40.1 \\ 1 \times \text{C} = 1 \times 12.0 = & 12.0 \\ 3 \times \text{O} = 3 \times 16.0 = & 48.0 \\ \hline \text{CaCO}_3 = & 100.1 \end{array}$$

To calculate the percentage of calcium, we would create a fraction, obtained by dividing the contribution from calcium by the total formula mass (40.1/100.1), and then multiply by 100%:

$$\%\text{Ca} = \frac{40.1}{100.1} \times 100\%$$

Fraction of Ca in $CaCO_3$

The fraction tells us that there are 40.1 g of calcium in a 100.1-g sample of $CaCO_3$. By placing units on our fraction, we can use it as a conversion factor to convert grams of calcium carbonate to grams of calcium:

$$\frac{40.1 \text{ g Ca}}{100.1 \text{ g CaCO}_3} = 1$$

g $CaCO_3$ ⇨ g Ca

$$? \text{ g Ca} = 0.500 \text{ g } \cancel{\text{CaCO}_3} \left(\frac{40.1 \text{ g Ca}}{100.1 \text{ g } \cancel{\text{CaCO}_3}} \right) = 0.200 \text{ g Ca}$$

Note that the fraction we use can be interpreted in any other units of mass that happen to be appropriate to the problem, such as milligrams.

$$? \text{ mg Ca} = 5.00 \times 10^2 \text{ mg } \cancel{\text{CaCO}_3} \left(\frac{40.1 \text{ mg Ca}}{100.1 \text{ mg } \cancel{\text{CaCO}_3}} \right) = 2.00 \times 10^2 \text{ mg Ca}$$

PROBLEM 4.5

Calculate the mass of the element indicated in each of the following samples.

(a) Na in 3.50 g NaCl
(b) Ag in 5.65 g $AgNO_3$
(c) N in 454 g NH_4NO_3 (careful!)
(d) Fe in 125 mg $FeSO_4$

4.5 THE MOLE: AVOGADRO'S NUMBER

OBJECTIVE 4.5 Define the mole in terms of its number of particles. State Avogadro's number.

Avogadro's number: 6.02×10^{23}.

Mole: Avogadro's number of particles.

Our culture has found it useful to represent certain quantities, as we do when we use the word *dozen,* which means 12. Thus, we invariably ask for a dozen eggs or doughnuts, rather than for 12. Likewise, businesses generally order pencils and pens by the *gross,* which equals 144. In a similar fashion, chemists find it useful to work with atoms and molecules in groups of 6.02×10^{23}. This rather large number, known as **Avogadro's number,** represents the number of particles in a **mole** (abbreviated "mol"). Just as a dozen equals 12, a mole equals 6.02×10^{23}; and just as we can have a dozen eggs, doughnuts, or golf balls, we can have a mole of atoms, a mole of molecules, or a mole of any other particles. Thus, a mole of carbon atoms is simply 6.02×10^{23} atoms of C, while a mole of water is 6.02×10^{23} molecules of H_2O. As we pointed out in the last section, it is not possible for us to work with one atom or molecule at a time. Instead, our practical work with substances will be done in multiples of Avogadro's number of atoms and molecules.

4.6 The Mole: Substances With Monatomic Formulas

OBJECTIVE 4.6A Define the mole in terms of formula mass. Determine the mass of 1 mole of any monatomic substance.

We have just defined the mole in terms of Avogadro's number of particles. There is also an older description of the mole that is useful for our practical work. *A mole of any substance is equal to its formula mass (the mass of all the atoms in its formula) taken in grams.* If the substance is a molecular substance, a mole equals its molecular mass taken in grams. If the substance is a nonmolecular compound, a mole equals its formula mass taken in grams. On the other hand, for substances with just one atom in the formula, a mole is equal to its atomic mass taken in grams. (In this section we restrict ourselves to such atomic substances.) By this older definition, 1 mole of helium (atomic mass = 4.0) has a mass of 4.0 g. Similarly, 1 mole of carbon (atomic mass = 12.0) must be 12.0 g. Because each of these quantities represents a mole, each contains 6.02×10^{23} atoms.

Quantity	Abbreviation	
4.0 grams of helium	4.0 g He	All express the same quantity
1 mole of helium atoms	1 mol He	
6.02×10^{23} atoms of helium	6.02×10^{23} atoms He	

$$1 \text{ mol He} = 4.0 \text{ g He} = 6.02 \times 10^{23} \text{ atoms He}$$

Quantity	Abbreviation	
12.0 grams of carbon	12.0 g C	All express the same quantity
1 mole of carbon atoms	1 mol C	
6.02×10^{23} atoms of carbon	6.02×10^{23} atoms C	

$$1 \text{ mol C} = 12.0 \text{ g C} = 6.02 \times 10^{23} \text{ atoms C}$$

Let us see whether our two definitions of the mole are consistent with each other. Recalling our earlier discussion of atomic mass, the ratio of the masses of one carbon atom to one helium atom is 12.0 : 4.0, the ratio of their atomic masses. If we take two atoms of each, the ratio of the masses is still 12.0:4.0. If we take 100 atoms of each, the ratio remains 12.0 : 4.0. In fact, as long as we have equal numbers of the two atoms, their ratio of masses is always 12.0:4.0, as listed below:

Number of atoms of each	Mass of C : Mass of He	Ratio of masses also equals
1 atom of each	12.0 u : 4.0 u	12.0 : 4.0
2 atoms of each	24.0 u : 8.0 u	12.0 : 4.0
3 atoms of each	36.0 u : 12.0 u	12.0 : 4.0
10 atoms of each	120 u : 40 u	12.0 : 4.0
100 atoms of each	1200 u : 400 u	12.0 : 4.0
6.02×10^{23} atoms of each	12.0 g : 4.0 g	12.0 : 4.0

It follows that, if we take 12.0 g of carbon and 4.0 g of helium, the samples must contain equal numbers of atoms. The number of atoms they contain is Avogadro's number, 6.02×10^{23}. In other words, we have a mole of each.

Students often wonder why it is necessary to work with moles. Atoms and molecules are so small and have so little mass that we cannot weigh out one molecule of water or one atom of copper. In the laboratory we must deal with quantities that we can see and measure conveniently. Defining the mole as a quantity equal to the atomic or molecular mass in grams provides masses that are easy to work with under normal laboratory conditions.

Molar mass: The mass of 1 mole of a substance.

Because it is necessary to carry out most of our practical work in mole-size quantities, we have a special name for the mass of 1 mole of a substance: the **molar mass.** In this case, the word *molar* means "per mole"; that is, the molar mass is the mass per mole. Numerically, the molar mass is equal to the formula mass of the substance with units of grams per mole. Thus, the molar mass of carbon is 12.0 g/mol, while that of helium is 4.0 g/mol.

Converting Between Grams and Moles of Monatomic Substances

OBJECTIVE 4.6B Convert between the mass and the number of moles of any monatomic substance.

We have just demonstrated that when we have an equal number of moles of two monatomic substances, we must also have an equal number of atoms. As chemists, we are more interested in the relationships between *numbers of atoms* than we are in their masses. For example, stating that a molecule of water is made up of two atoms of hydrogen and one atom of oxygen is much more meaningful than stating that the mass of each molecule is two-eighteenths (11%) hydrogen and sixteen-eighteenths (89%) oxygen. Because the number of atoms in a sample is related to the number of moles, it is very useful for us to be able to calculate the number of moles in a given mass of a substance. The following examples illustrate how this is done. [*Remember:* The symbol mol is used to abbreviate the mole.]

EXAMPLE 4.9

Calculate the number of moles in 24.0 g of carbon.

Symbolic statement

$$? \text{ mol C} = 24.0 \text{ g C}$$

Planning the solution

The atomic mass of carbon is 12.0, so

$$1 \text{ mol C} = 12.0 \text{ g C}$$

Hence, 24.0 g of carbon must be equal to 2.00 mol. Although this problem was simple enough to do in our heads, let us set it up by using dimensional analysis to demonstrate a method for solving more complex problems. The relationship 1 mol C = 12.0 g C may be used to create a conversion factor (just like those we discussed in Chapter 3) for converting from grams to moles.

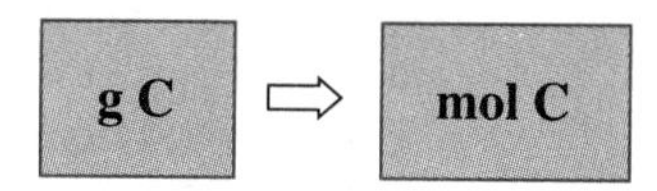

SOLUTION

The conversion factor we use is

$$\frac{1 \text{ mol C}}{12.0 \text{ g C}} = 1$$

Then we use the same method we have used before.

$$? \text{ mol C} = 24.0 \text{ g}\cancel{\text{C}}\left(\frac{1 \text{ mol C}}{12.0 \text{ g}\cancel{\text{C}}}\right) = 2.00 \text{ mol C}$$

EXAMPLE 4.10

Calculate the number of moles in 10.0 g of copper, Cu.

Symbolic statement

$$? \text{ mol Cu} = 10.0 \text{ g Cu}$$

Planning the solution

g Cu ⇨ mol Cu

SOLUTION

$$1 \text{ mol Cu} = 63.5 \text{ g Cu}$$

$$? \text{ mol Cu} = 10.0 \text{ g}\cancel{\text{Cu}}\left(\frac{1 \text{ mol Cu}}{63.5 \text{ g}\cancel{\text{Cu}}}\right) = 0.157 \text{ mol Cu}$$

We can also use the relationship between grams and moles to determine the mass of a given number of moles, as shown in Example 4.11.

EXAMPLE 4.11

Calculate the mass of 0.567 mol of aluminum, Al.

Symbolic statement

$$? \text{ g Al} = 0.567 \text{ mol Al}$$

Planning the solution

We again need to make use of the relationship between moles and atomic mass, but in this case, we are converting from moles to grams.

mol Al ⇨ **g Al**

SOLUTION

$$1 \text{ mol Al} = 27.0 \text{ g Al}$$

$$? \text{ g Al} = 0.567 \cancel{\text{mol Al}} \left(\frac{27.0 \text{ g Al}}{1 \cancel{\text{mol Al}}} \right) = 15.3 \text{ g Al}$$

PROBLEM 4.6

Calculate the number of moles in each of the following.
(a) 6.00 g C (carbon) **(b)** 4.19 g Kr (krypton)
(c) 47.6 g U (uranium) **(d)** 8.37 g Fe (iron)
(e) 8.82 g Ca (calcium) **(f)** 316 g Mg (magnesium)

PROBLEM 4.7

Calculate the mass of each of the following.
(a) 2.50 mol Ne (neon) **(b)** 0.452 mol Na (sodium)
(c) 0.125 mol K (potassium) **(d)** 0.742 mol Hg (mercury)
(e) 0.0531 mol Au (gold)

Converting from Moles of a Monatomic Substance to Atoms

OBJECTIVE 4.6C Convert between the number of moles and the number of atoms of any monatomic substance.

If we know the number of moles of a given monatomic substance, we can easily calculate the number of atoms involved, because 1 mol of any monatomic substance must contain Avogadro's number of atoms.

EXAMPLE 4.12

Calculate the number of atoms in 0.25 mol C (carbon).

Symbolic statement

$$? \text{ atoms C} = 0.25 \text{ mol C}$$

Planning the solution

To find the number of atoms in 0.25 mol of carbon, we need a relationship between moles and atoms. One mole of a monatomic substance (such as carbon, C) contains Avogadro's number of atoms:

$$1 \text{ mol C} = 6.02 \times 10^{23} \text{ atoms C}$$

We may use this relationship to convert from moles of carbon to atoms of carbon.

mol C ⇨ **atoms C**

SOLUTION

$$? \text{ atoms C} = 0.25 \text{ mol C}\left(\frac{6.02 \times 10^{23} \text{ atoms C}}{1 \text{ mol C}}\right) = 1.5 \times 10^{23} \text{ atoms C}$$

Thus, to convert from moles of a monatomic substance to atoms, we simply multiply the number of moles by Avogadro's number.

EXAMPLE 4.13

Calculate the number of atoms in 8.05 g Na (sodium).

Symbolic statement

$$? \text{ atoms Na} = 8.05 \text{ g Na}$$

Planning the solution

We do not have a direct relationship between grams and atoms. However, we can convert grams to moles and then moles to atoms:

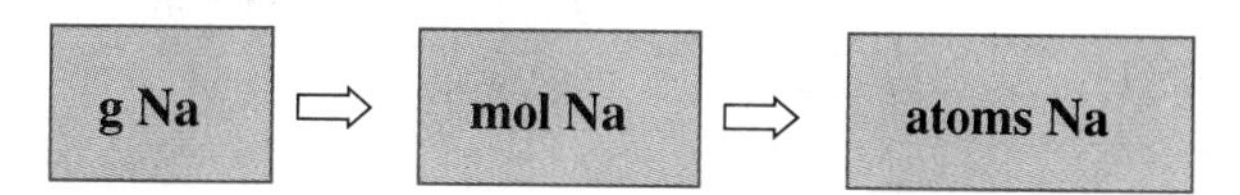

The relationships we need are:

$$1 \text{ mol Na} = 23.0 \text{ g Na}$$
$$1 \text{ mol Na} = 6.02 \times 10^{23} \text{ atoms Na}$$

SOLUTION

$$? \text{ atoms Na} = 8.05 \text{ g Na}\left(\frac{1 \text{ mol Na}}{23.0 \text{ g Na}}\right)\left(\frac{6.02 \times 10^{23} \text{ atoms Na}}{1 \text{ mol Na}}\right)$$
$$= 2.11 \times 10^{23} \text{ atoms Na}$$

PROBLEM 4.8

How many atoms are there in each of the following?
(a) 0.25 mol He **(b)** 3.50 mol Fe **(c)** 0.0155 mol Au
(d) 6.35 g Cu **(e)** 253 g Na **(f)** 10.0 g S

4.7 THE MOLE: SUBSTANCES CONTAINING MORE THAN ONE ATOM

OBJECTIVE 4.7A Convert between the mass and the number of moles of any polyatomic substance.

In Section 4.6 we defined a mole of any substance to be the formula mass taken in grams. Since we confined our discussion to monatomic substances (such as carbon or helium), the mass of a mole was equal to the atomic mass. Now let us expand that discussion to include substances that contain more than one atom, such as water or carbon dioxide. For these *polyatomic* substances, we can determine the mass of a mole from the molecular (or formula) mass. For example, water has a *molecular mass* of 18.0. Thus, a mole of water has a mass of 18.0 g. Similarly, since the molecular mass of carbon dioxide (CO_2) is 44.0, a mole of carbon dioxide has a mass of 44.0 g. Furthermore, there are 6.02×10^{23} molecules of water in 1 mol (18.0 g) of water and 6.02×10^{23} molecules of carbon dioxide in 1 mol (44.0 g) of CO_2.

$1 \text{ mol } H_2O$
$18.0 \text{ g } H_2O$
6.02×10^{23} molecules of H_2O
} **All express the same quantity**

$1 \text{ mol } CO_2$
$44.0 \text{ g } CO_2$
6.02×10^{23} molecules of CO_2
} **All express the same quantity**

Just as we used *atomic* masses to obtain the molar masses of monatomic substances, we use *molecular* (or *formula*) masses to obtain the molar masses of polyatomic substances. Furthermore, just as the molar mass of an atomic substance may be used to convert between grams and moles, the molar mass of a polyatomic substance may be used to make similar conversions. The following examples illustrate how this is done.

EXAMPLE 4.14

How many moles are there in 11.0 g of carbon dioxide, CO_2?

Symbolic statement

$$? \text{ mol } CO_2 = 11.0 \text{ g } CO_2$$

Planning the solution

We work these problems exactly as we worked Examples 4.9–4.11 in Section 4.6. However, here we have polyatomic substances, so the first step is to find the molecular (or formula) mass rather than the atomic mass. By taking the molecular mass in grams, we obtain the molar mass, which can be used as a conversion factor between grams and moles.

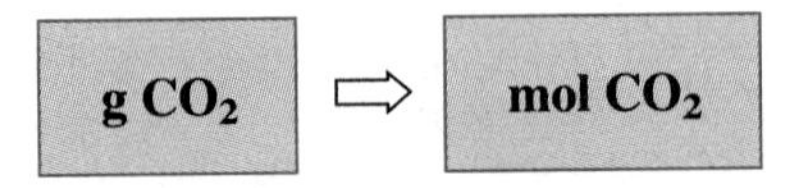

SOLUTION

To compute the molecular mass of carbon dioxide, we simply add up the atomic masses of all of the atoms in the molecule:

$$\begin{aligned} 1 \times C &= 1 \times 12.0 = 12.0 \\ 2 \times O &= 2 \times 16.0 = 32.0 \\ \hline CO_2 &= 44.0 \end{aligned}$$

Thus, $1 \text{ mol } CO_2 = 44.0 \text{ g } CO_2$.

In order to find the number of moles in 11.0 g of CO_2, we use the relationship we have just derived as a conversion factor:

$$? \text{ mol } CO_2 = 11.0 \cancel{\text{g } CO_2} \left(\frac{1 \text{ mol } CO_2}{44.0 \cancel{\text{g } CO_2}} \right) = 0.250 \text{ mol } CO_2$$

EXAMPLE 4.15

How many moles are there in 7.45 g of NH_3?

Symbolic statement

$$? \text{ mol } NH_3 = 7.45 \text{ g } NH_3$$

Planning the solution

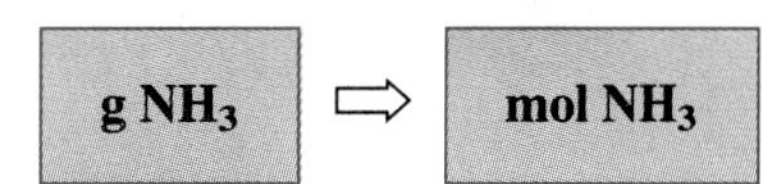

SOLUTION

The molecular mass of ammonia is calculated as follows:

$$\begin{aligned} 1 \times N &= 1 \times 14.0 = 14.0 \\ 3 \times H &= 3 \times \ \ 1.0 = \ \ 3.0 \\ \hline NH_3 &= 17.0 \end{aligned}$$

Thus, 1 mol NH_3 = 17.0 g NH_3.

$$? \text{ mol } NH_3 = 7.45 \cancel{\text{ g } NH_3}\left(\frac{1 \text{ mol } NH_3}{17.0 \cancel{\text{ g } NH_3}}\right) = 0.438 \text{ mol } NH_3$$

EXAMPLE 4.16

What is the mass of 0.164 mol of aluminum hydroxide, $Al(OH)_3$?

Symbolic statement

$$? \text{ g } Al(OH)_3 = 0.164 \text{ mol } Al(OH)_3$$

Planning the solution

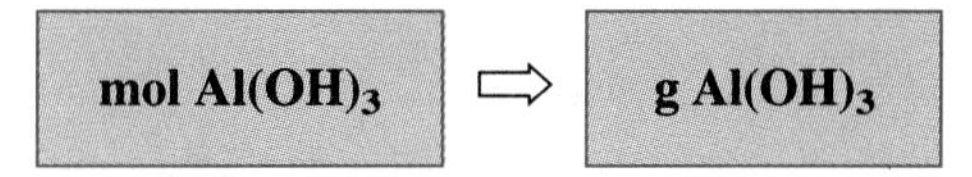

SOLUTION

To calculate the formula mass of aluminum hydroxide, we first rewrite its formula as AlO_3H_3:

$$\begin{aligned} 1 \times Al &= 1 \times 27.0 = 27.0 \\ 3 \times O &= 3 \times 16.0 = 48.0 \\ 3 \times H &= 3 \times \ \ 1.0 = \ \ 3.0 \\ \hline Al(OH)_3 &= 78.0 \end{aligned}$$

Thus, 1 mol $Al(OH)_3$ = 78.0 g $Al(OH)_3$.

$$? \text{ g } Al(OH)_3 = 0.164 \cancel{\text{ mol } Al(OH)_3}\left(\frac{78.0 \text{ g } Al(OH)_3}{1 \cancel{\text{ mol } Al(OH)_3}}\right) = 12.8 \text{ g } Al(OH)_3$$

PROBLEM 4.9

Calculate the number of moles in each of the following.

(a) 1.56 g carbon monoxide, CO **(b)** 15.0 g carbon dioxide, CO_2
(c) 2.50 g nitrogen dioxide, NO_2 **(d)** 12.5 g potassium bromide, KBr
(e) 20.0 g potassium carbonate, K_2CO_3 **(f)** 12.1 g sulfur trioxide, SO_3

PROBLEM 4.10

Calculate the mass of each of the following.

(a) 2.50 mol CH_4 **(b)** 0.550 mol KCl
(c) 0.315 mol NaOH **(d)** 0.114 mol H_2SO_4

Working With the Diatomic Elements

OBJECTIVE 4.7B Distinguish between the molecular and atomic forms of the elements. List the seven common elements that form diatomic molecules.

You may recall that, in our discussion of molecules and atoms (Section 2.5), we cited several diatomic elements. Among these were hydrogen (H_2) and oxygen (O_2). Since molecular hydrogen is diatomic, its molecular mass is 2.0. Thus, a mole of molecular hydrogen has a mass of 2.0 g and contains 6.02×10^{23} *molecules:*

$$1 \text{ mol } H_2 = 2.0 \text{ g } H_2 = 6.02 \times 10^{23} \text{ molecules } H_2$$

This is in contrast to atomic hydrogen, H. One mole of hydrogen atoms (atomic mass = 1.0) has a mass of 1.0 g and contains 6.02×10^{23} *atoms* of hydrogen. Similarly, the mass of a mole of oxygen atoms is 16.0 g, whereas that of a mole of molecular oxygen (O_2) is 32.0 g.

Atomic substances

$$1 \text{ mol H (atoms)} = 6.02 \times 10^{23} \text{ atoms H} = 1.0 \text{ g H (atoms)}$$

$$1 \text{ mol O (atoms)} = 6.02 \times 10^{23} \text{ atoms O} = 16.0 \text{ g O (atoms)}$$

Molecular substances

$$1 \text{ mol } H_2 = 6.02 \times 10^{23} \text{ molecules } H_2 = 2.0 \text{ g } H_2$$

$$1 \text{ mol } O_2 = 6.02 \times 10^{23} \text{ molecules } O_2 = 32.0 \text{ g } O_2$$

The other diatomic elements include nitrogen (N_2), fluorine (F_2), chlorine (Cl_2), bromine (Br_2), and iodine (I_2), giving us seven diatomic elements altogether. You will find it helpful to memorize this list, since the formulas of almost all other elements are monatomic. As an aid in remembering which of the elements are diatomic, note their locations in the periodic table. Hydrogen is the very first element (atomic number 1), and the other six form a right angle in the upper right-hand corner of the periodic table, beginning with nitrogen, oxygen, and fluorine (across the second row) and continuing down from fluorine to chlorine, bromine, and iodine.

When we work with the diatomic elements, we generally wish to know the number of moles of molecules. Thus, if asked for the number of moles of oxygen in a 5.00-g sample, we calculate the number of moles of molecular oxygen, O_2. However, we occasionally wish to know the number of moles of atoms, as is the case in the next section. The following examples illustrate how to handle each situation.

EXAMPLE 4.17

How many moles are there in 5.00 g of molecular oxygen?

Symbolic statement

$$? \text{ mol } O_2 = 5.00 \text{ g } O_2$$

Planning the solution

Since we are dealing with molecular oxygen, we must use the molecular mass of O_2. If the problem does not specifically state *molecular* oxygen, we nevertheless assume that it is molecular oxygen, O_2, since that is the way oxygen occurs in nature. We use the atomic mass of oxygen only if we are asked to calculate the moles of oxygen *atoms*.

g O_2 ⇨ **mol O_2**

SOLUTION

The molecular mass of oxygen is calculated as follows.

$$2 \times \text{O} = 2 \times 16.0 = 32.0$$

$$1 \text{ mol } O_2 = 32.0 \text{ g } O_2$$

$$? \text{ mol } O_2 = 5.00 \cancel{\text{g } O_2}\left(\frac{1 \text{ mol } O_2}{32.0 \cancel{\text{g } O_2}}\right) = 0.156 \text{ mol } O_2$$

EXAMPLE 4.18

How many moles of nitrogen atoms are in 6.23 g of nitrogen?

Symbolic statement

$$? \text{ mol N} = 6.23 \text{ g N}$$

Planning the solution

In this case, we have been asked specifically for the moles of nitrogen atoms, so we use the atomic mass of nitrogen and interpret its formula as N (rather than as N_2).

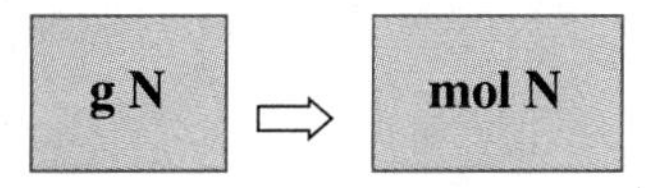

SOLUTION

$$1 \text{ mol N} = 14.0 \text{ g N}$$

$$? \text{ mol N} = 6.23 \cancel{\text{g N}}\left(\frac{1 \text{ mol N}}{14.0 \cancel{\text{g N}}}\right) = 0.445 \text{ mol N}$$

The Summary Diagram on page 98 shows how to convert between masses and moles of monatomic and polyatomic substances.

PROBLEM 4.11

Calculate the number of moles indicated in each of the following.

(a) 5.0 g hydrogen, H_2 **(b)** 20.0 g chlorine gas, Cl_2
(c) 11.3 g iodine atoms, I **(d)** 11.3 g molecular iodine, I_2
(e) 1.40 g argon, Ar **(f)** 17.0 g fluorine

Converting Between Moles and Molecules

OBJECTIVE 4.7C Convert between the number of moles and the number of molecules of any molecular substance.

Earlier, we said that 1 mole of any monatomic substance contains Avogadro's number of atoms: 6.02×10^{23}. Similarly, a mole of a molecular substance must contain Avogadro's number of molecules. Let us see whether this is reasonable.

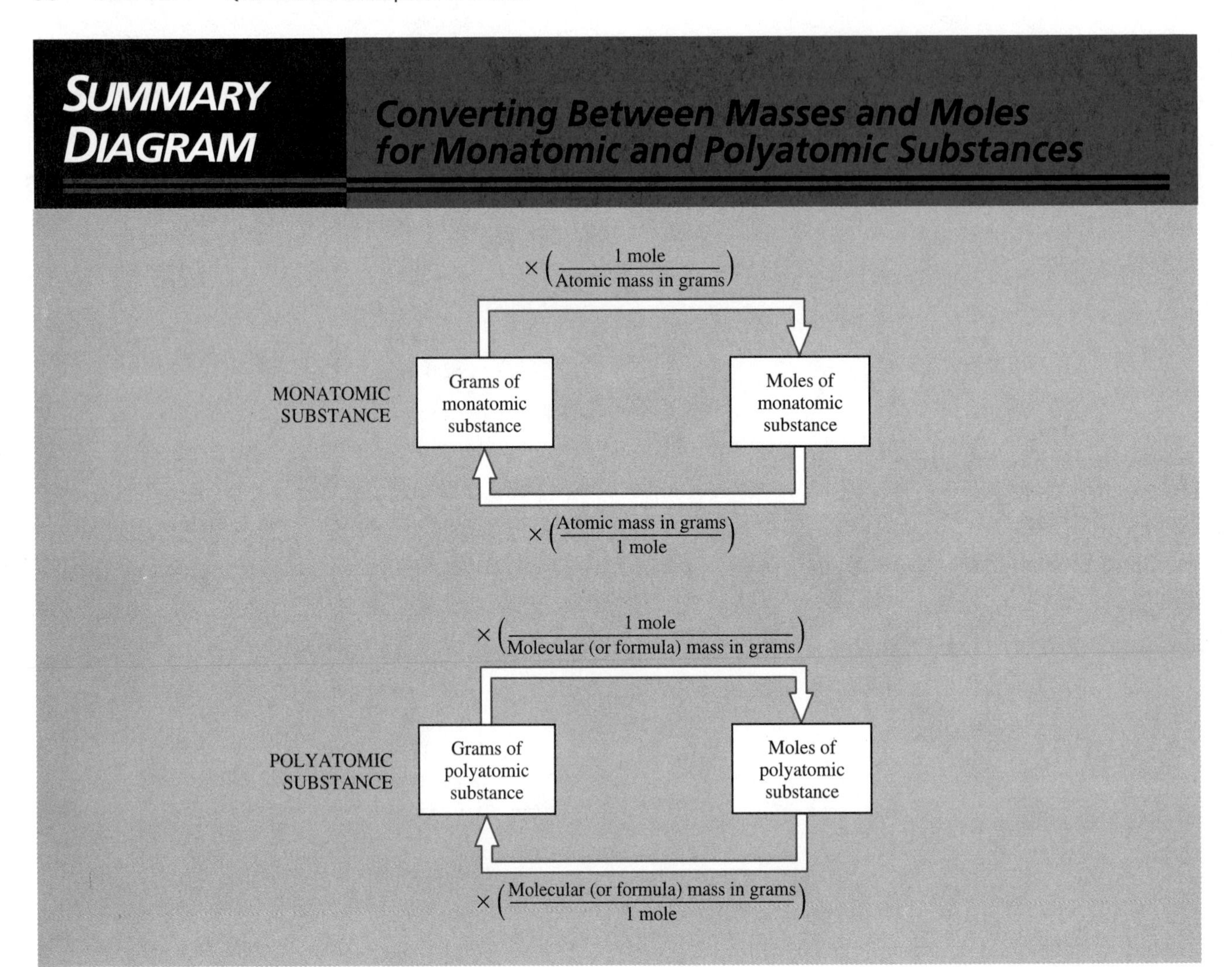

If a mole of water, H_2O, contains Avogadro's number of molecules, it follows that it must contain twice Avogadro's number of hydrogen atoms plus Avogadro's number of oxygen atoms.

$$\underbrace{6.02 \times 10^{23} \text{ molecules of } H_2O}_{1 \text{ mole of } H_2O}$$

contains

$$\underbrace{2 \times 6.02 \times 10^{23} \text{ atoms of H}}_{2 \text{ moles of H atoms}} \quad \text{and} \quad \underbrace{1 \times 6.02 \times 10^{23} \text{ atoms of O}}_{1 \text{ mole of O atoms}}$$

Twice Avogadro's number of hydrogen atoms is 2 mol of hydrogen atoms. Avogadro's number of oxygen atoms is 1 mol of oxygen atoms.

$$2\ \cancel{\text{mol H}}\left(\frac{1.0\ \text{g H}}{1\ \cancel{\text{mol H}}}\right) = 2.0\ \text{g H}$$

$$1\ \cancel{\text{mol O}}\left(\frac{16.0\ \text{g O}}{1\ \cancel{\text{mol O}}}\right) = 16.0\ \text{g O}$$

$$1\ \text{mol H}_2\text{O} = 18.0\ \text{g H}_2\text{O}$$

We already know that 1 mol of water has a mass of 18.0 g, so our assumption that a mole of water contains Avogadro's number of molecules must be correct. Let us use Avogadro's number to calculate the number of molecules in a given sample.

EXAMPLE 4.19

How many molecules are there in 0.400 mol of carbon dioxide, CO_2?

Symbolic statement

$$?\ \text{molecules CO}_2 = 0.400\ \text{mol CO}_2$$

Planning the solution

This problem is handled exactly like Example 4.12 except that we are dealing with molecules rather than atoms. Thus, using 1 mol CO_2 = 6.02×10^{23} molecules CO_2, we convert from moles to molecules.

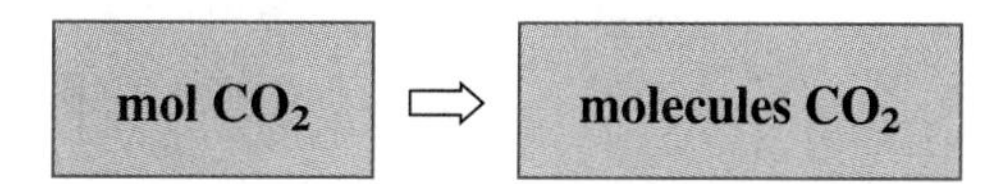

SOLUTION

$$?\ \text{molecules CO}_2 = 0.400\ \cancel{\text{mol CO}_2}\left(\frac{6.02 \times 10^{23}\ \text{molecules CO}_2}{1\ \cancel{\text{mol CO}_2}}\right)$$

$$= 2.41 \times 10^{23}\ \text{molecules CO}_2$$

EXAMPLE 4.20

How many molecules are there in 12.0 g of methane, CH_4?

Symbolic statement

$$?\ \text{molecules CH}_4 = 12.0\ \text{g CH}_4$$

Planning the solution

Using a strategy similar to that employed in Example 4.13, we can convert grams to moles and moles to molecules.

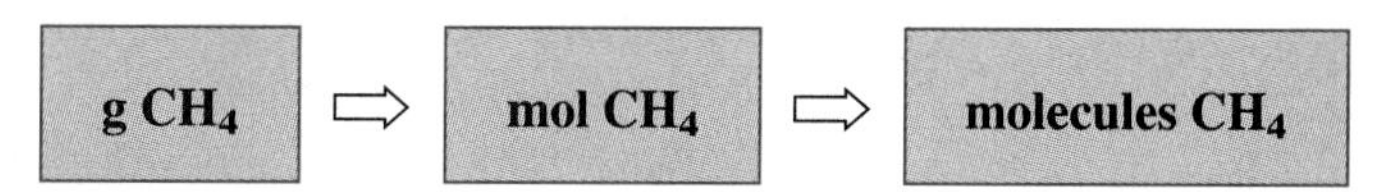

SOLUTION

We use the following relationships.

$$1\ \text{mol CH}_4 = 16.0\ \text{g CH}_4$$ ← **Obtained from the molecular mass**

$$1\ \text{mol CH}_4 = 6.02 \times 10^{23}\ \text{molecules CH}_4$$

$$?\ \text{molecules CH}_4 = 12.0\ \cancel{\text{g CH}_4}\left(\frac{1\ \cancel{\text{mol CH}_4}}{16.0\ \cancel{\text{g CH}_4}}\right)\left(\frac{6.02 \times 10^{23}\ \text{molecules CH}_4}{1\ \cancel{\text{mol CH}_4}}\right)$$

$$= 4.52 \times 10^{23}\ \text{molecules CH}_4$$

PROBLEM 4.12 Calculate the number of molecules in each of the following.
(a) 4.75 mol NH_3 **(b)** 0.450 mol CCl_4 **(c)** 0.0135 mol F_2
(d) 14.7 g C_3H_6O **(e)** 8.55 g N_2

4.8 DETERMINING EMPIRICAL FORMULAS

OBJECTIVE 4.8 Given the percentage composition of a substance or the masses of the elements present in a sample of the substance, calculate its empirical formula.

In Section 4.1 you learned to reduce a molecular formula to its corresponding empirical formula. You might wonder why a chemist is interested in the empirical formula of a molecular substance. After all, a molecular formula is much more informative; it tells us exactly what is in the molecule. However, sometimes the experimental information gathered in the laboratory enables us to determine the empirical formula but not the molecular formula. For example, let us consider two very common substances, benzene and acetylene. Benzene has the molecular formula C_6H_6. If we calculate the percentage composition of benzene, we get

$$\%C = \frac{72.0}{78.0} \times 100\% = 92.3\% \qquad \%H = \frac{6.0}{78.0} \times 100\% = 7.7\%$$

Acetylene has the molecular formula C_2H_2. If we calculate the percentage composition of acetylene, we find it is exactly the same as that of benzene! The reason for this is that both have a 1 : 1 ratio of carbon atoms to hydrogen atoms. Hence, if we determine the percentage composition of an unknown compound to be 92.3% C and 7.7% H, we have no way of knowing the molecular formula of the compound without some additional information. However, the percentage composition enables us to calculate the simplest ratio of atoms and to determine the empirical formula to be CH. Let us see how.

Suppose we have a sample of benzene. The ratio of carbon to hydrogen is the same whether we have a sample weighing 50, 60, or 100 g; let us assume that we have a 100.0-g sample. The percentage of carbon is 92.3%, so the sample must contain 92.3 g of carbon (92.3% of 100.0 g). Similarly, it must contain 7.7 g of hydrogen (7.7% of 100.0 g). Knowing the mass of each element is of very little use to us, but we can use this information to find the simplest ratio of atoms in the molecule. In Section 4.6 we saw that equal numbers of moles of monatomic substances contain equal numbers of atoms. Hence, if we know the ratio of *moles* of carbon atoms to *moles* of hydrogen atoms, we also know the ratio of the atoms themselves. Let us calculate the number of moles of carbon atoms and hydrogen atoms in our 100.0-g sample:

$$? \text{ mol C} = 92.3 \cancel{\text{g C}}\left(\frac{1 \text{ mol C}}{12.0 \cancel{\text{g C}}}\right) = 7.7 \text{ mol C}$$

$$? \text{ mol H} = 7.7 \cancel{\text{g H}}\left(\frac{1 \text{ mol H}}{1.0 \cancel{\text{g H}}}\right) = 7.7 \text{ mol H}$$

1:1 ratio of C:H

The simplest ratio of moles of carbon atoms to moles of hydrogen atoms is 1 : 1, hence the empirical formula must be CH. We obtain an identical result for acetylene, because it has the same percentage composition as benzene.

EXAMPLE 4.21

Calculate the empirical formula for the following compound: 50.1% S, 49.9% O.

Symbolic statement

We wish to determine the simplest whole-number ratio in the formula $S_?O_?$.

Planning the solution

We assume we have a 100.0-g sample and calculate the number of moles of sulfur atoms and oxygen atoms present. Once the number of moles of each has been established, we can determine the simplest whole-number ratio between them.

SOLUTION

The sample is 50.1% S, so 100.0 g must contain 50.1 g S. We convert grams of sulfur to moles of sulfur atoms using the following relationship:

$$32.1 \text{ g S} = 1 \text{ mol S}$$

$$? \text{ mol S} = 50.1 \text{ g S}\left(\frac{1 \text{ mol S}}{32.1 \text{ g S}}\right) = 1.56 \text{ mol S}$$

Likewise, there must be 49.9 g O in this sample. To convert grams of oxygen to moles of oxygen atoms, we use the relationship

$$16.0 \text{ g O} = 1 \text{ mol O}$$

$$? \text{ mol O} = 49.9 \text{ g O}\left(\frac{1 \text{ mol O}}{16.0 \text{ g O}}\right) = 3.12 \text{ mol O}$$

As a general rule, the easiest way to find the simplest whole-number ratio of atoms is to divide the smallest number of moles into every other number of moles. For example,

$$\frac{1.56 \text{ mol S}}{1.56} = 1.00 \text{ mol S}$$

$$\frac{3.12 \text{ mol O}}{1.56} = 2.00 \text{ mol O}$$

Finally, we must write an empirical formula that expresses this ratio.

Empirical formula: SO_2

We outline a general procedure for calculating empirical formulas from the percentage composition in the Summary on page 102.

EXAMPLE 4.22

Find the empirical formula for the substance with composition 52.9% Al, 47.1% O.

Symbolic statement

$Al_?O_?$

Planning the solution

We follow the general procedure just outlined.

SOLUTION

Step 1. 52.9 g Al, 47.1 g O in a 100.0-g sample

SUMMARY Calculating Empirical Formulas from the Percentage Composition

Step 1. Assume a 100.0-g sample of the substance.

Step 2. Determine the number of moles of each element in the 100.0-g sample.

Step 3. Find the simplest ratio of moles. (It is equal to the simplest ratio of atoms.) To find the simplest ratio of moles, divide the smallest number of moles from Step 2 into every other number. If this does not result in a whole-number ratio, multiply by an appropriate integer to obtain a whole-number ratio. (For example, multiplication of the ratio 1.5 : 1 by 2 gives the whole-number ratio 3 : 2.)

Step 4. Write the empirical formula for the ratio you find.

Step 2.

$$? \text{ mol Al} = 52.9 \text{ g Al}\left(\frac{1 \text{ mol Al}}{27.0 \text{ g Al}}\right) = 1.96 \text{ mol Al}$$

$$? \text{ mol O} = 47.1 \text{ g O}\left(\frac{1 \text{ mol O}}{16.0 \text{ g O}}\right) = 2.94 \text{ mol O}$$

Step 3.

$$\frac{1.96 \text{ mol Al}}{1.96} = 1.00 \text{ mol Al}$$

$$\frac{2.94 \text{ mol O}}{1.96} = 1.50 \text{ mol O}$$

In this case, division by the smallest number does not lead directly to a whole-number ratio, but multiplication of each element by 2 gives us a 2 : 3 ratio.

Step 4. The empirical formula is Al_2O_3.

As Example 4.22 illustrates, our procedure does not always lead to a whole-number ratio; however, it usually leads to a ratio with a recognizable fraction. The following fractional ratios can be converted to whole-number ratios as shown:

If the ratio from the calculation is	Its fractional equivalent is	Multiply the ratio by	To give a whole-number ratio
1.50 : 1.00	$1\frac{1}{2}:1$ or $\frac{3}{2}:1$	2	3 : 2
1.33 : 1.00	$1\frac{1}{3}:1$ or $\frac{4}{3}:1$	3	4 : 3
1.25 : 1.00	$1\frac{1}{4}:1$ or $\frac{5}{4}:1$	4	5 : 4
1.67 : 1.00	$1\frac{2}{3}:1$ or $\frac{5}{3}:1$	3	5 : 3

We will not consider fractions more complicated than these.

The empirical formula of a substance may also be determined when the actual masses of the various elements present in a sample are known. The following example shows how this is done.

EXAMPLE 4.23

A 5.36-g sample of a mineral contains 2.02 g Na, 1.23 g Si, and 2.11 g O. Calculate the empirical formula of this substance.

Symbolic statement

$$Na_?Si_?O_?$$

Planning the solution

Since we are given the number of *grams* of each element (rather than its *percentage*), we can calculate the actual number of moles of each element in the sample. Once the number of moles of each element is known, we can calculate the simplest whole-number ratio as in Examples 4.21 and 4.22.

SOLUTION

First, find the number of moles of each element:

$$2.02\ \cancel{\text{g Na}}\left(\frac{1\ \text{mol Na}}{23.0\ \cancel{\text{g Na}}}\right) = 0.0878\ \text{mol Na}$$

$$1.23\ \cancel{\text{g Si}}\left(\frac{1\ \text{mol Si}}{28.1\ \cancel{\text{g Si}}}\right) = 0.0438\ \text{mol Si}$$

$$2.11\ \cancel{\text{g O}}\left(\frac{1\ \text{mol O}}{16.0\ \cancel{\text{g O}}}\right) = 0.132\ \text{mol O}$$

Next, divide each by the smallest (in this case, by 0.0438):

$$\frac{0.0878\ \text{mol Na}}{0.0438} = 2.00\ \text{mol Na}$$

$$\frac{0.0438\ \text{mol Si}}{0.0438} = 1.00\ \text{mol Si}$$

$$\frac{0.132\ \text{mol O}}{0.0438} = 3.01\ \text{mol O}\ (\approx 3\ \text{mol O})$$

The empirical formula is Na_2SiO_3.

Of course, we also could have determined the percentage composition from the masses given, and then proceeded as in Examples 4.21 and 4.22.

PROBLEM 4.13

Find the empirical formula for each of the following substances.

(a) K = 52.4%, Cl = 47.6%
(b) Na = 16.2%, Mn = 38.6%, O = 45.2%
(c) H = 3.1%, P = 31.5%, O = 65.4%
(d) K = 56.6%, C = 8.7%, O = 34.7%
(e) C = 10.04%, H = 0.84%, Cl = 89.12%

PROBLEM 4.14

A 4.63-g sample of a solid is found to contain 1.67 g Ca and 2.96 g Cl. Calculate the empirical formula of this compound.

PROBLEM 4.15

The composition of a 0.864-g sample of a fluorinated hydrocarbon contains 0.432 g C, 0.090 g H, and 0.342 g F. Determine the empirical formula.

4.9 DETERMINING MOLAR MASSES AND MOLECULAR FORMULAS

OBJECTIVE 4.9 Given the mass and number of moles of a substance, calculate its molar mass. Given the empirical formula and molecular mass of a substance, determine its molecular formula.

Determining Molar Masses

In Examples 4.14–4.16, we used the molecular (or formula) masses of substances to obtain their molar masses for use as conversion factors. For example, 1 mol CO_2 = 44.0 g CO_2. We could restate the information contained in this equivalency by stating that the molar mass of carbon dioxide is 44.0 g/mol. In other words, the molar mass is equal to the molecular (or formula) mass, but with units of grams per mole. It follows that if we know the molar mass of a substance, we also know its molecular (or formula) mass.

We can obtain the molar mass of a substance experimentally by determining the number of moles present in a given mass of the substance. Recalling that the meaning of "per" is "divided by," we can obtain the molar mass by dividing the number of grams by the number of moles:

$$\text{Molar mass} = \frac{\text{Number of grams}}{\text{Number of moles}}$$

Remember: **"per" means "divided by"**

The following examples illustrate how this is done.

EXAMPLE 4.24

Acetone is a colorless liquid that is used widely as a solvent in the chemical industry. A 2.24-g sample of acetone is also 0.0386 mol. Calculate the molar mass of acetone.

Symbolic statement

$$\text{Molar mass} = \text{? g/mol}$$

Planning the solution

Since the molar mass is the number of grams per mole, we obtain the molar mass by dividing the number of grams given in the problem by the number of moles it equals.

SOLUTION

$$\text{Molar mass} = \frac{2.24\ \text{g}}{0.0386\ \text{mol}} = 58.0\ \text{g/mol}$$

Checking the answer

Since the lightest element (hydrogen) has an atomic mass of 1.0, no molecular mass (or its corresponding molar mass) can be less than 1.0. Our answer of 58.0 g/mol is reasonable.

EXAMPLE 4.25

Tetrachloroethene (also known as perchloroethylene, or "PERC") is a solvent used in the dry cleaning industry. A 0.0545-mol sample of tetrachloroethene has a mass of 9.05 g. Determine the molar mass of this substance.

Symbolic statement

$$\text{Molar mass} = \text{? g/mol}$$

SOLUTION

$$\text{Molar mass} = \frac{9.05\text{ g}}{0.0545\text{ mol}} = 166\text{ g/mol}$$

PROBLEM 4.16

Calculate the molar mass of each of the following substances.
(a) A 1.25-mol sample has a mass of 165 g. **(b)** A 14.5-g sample represents 0.320 mol.
(c) A 2.35-g sample is 0.0155 mol.

Determining Molecular Formulas

Once the molar mass of a substance has been determined, this information can be combined with the empirical formula to provide the molecular formula. For example, suppose the empirical formula of a substance has been determined to be CH_2O. This empirical formula could correspond to quite a few commonly known substances—including formaldehyde, CH_2O (molecular mass = 30.0); lactic acid, $C_3H_6O_3$ (molecular mass = 90.0); and glucose, $C_6H_{12}O_6$ (molecular mass = 180.0).

If, in addition to the empirical formula, we know that the molar mass is 90.0 g/mol, then the molecular mass is 90.0. It follows that the molecular formula has to be $C_3H_6O_3$. We can obtain this result in a systematic fashion.

Step 1. Add the atomic masses in the empirical formula:

$$\begin{aligned} 1 \times C &= 1 \times 12.0 = 12.0 \\ 2 \times H &= 2 \times 1.0 = 2.0 \\ 1 \times O &= 1 \times 16.0 = 16.0 \\ \hline CH_2O &= 30.0 \end{aligned}$$

This provides us with an *empirical formula mass.* The molecular mass must be a whole-number multiple of this value.

Step 2. Divide the molecular mass by the empirical formula mass:

$$\frac{90.0}{30.0} = 3 \qquad \leftarrow \textbf{Number of multiples of } \mathbf{CH_2O} \textbf{ in a molecule}$$

The molecular formula must contain three multiples of the empirical formula.

Step 3. Multiply each atom in the empirical formula by the multiple you determined in Step 2:

$$3 \times CH_2O \qquad \text{leads to} \qquad C_3H_6O_3$$

The procedure used to determine molecular formulas is shown in the Summary Diagram on page 106.

EXAMPLE 4.26

The empirical formula of a substance is determined to be C_2H_3F. If 0.634 mol of the substance has a mass of 87.5 g, what is the molecular formula of the substance?

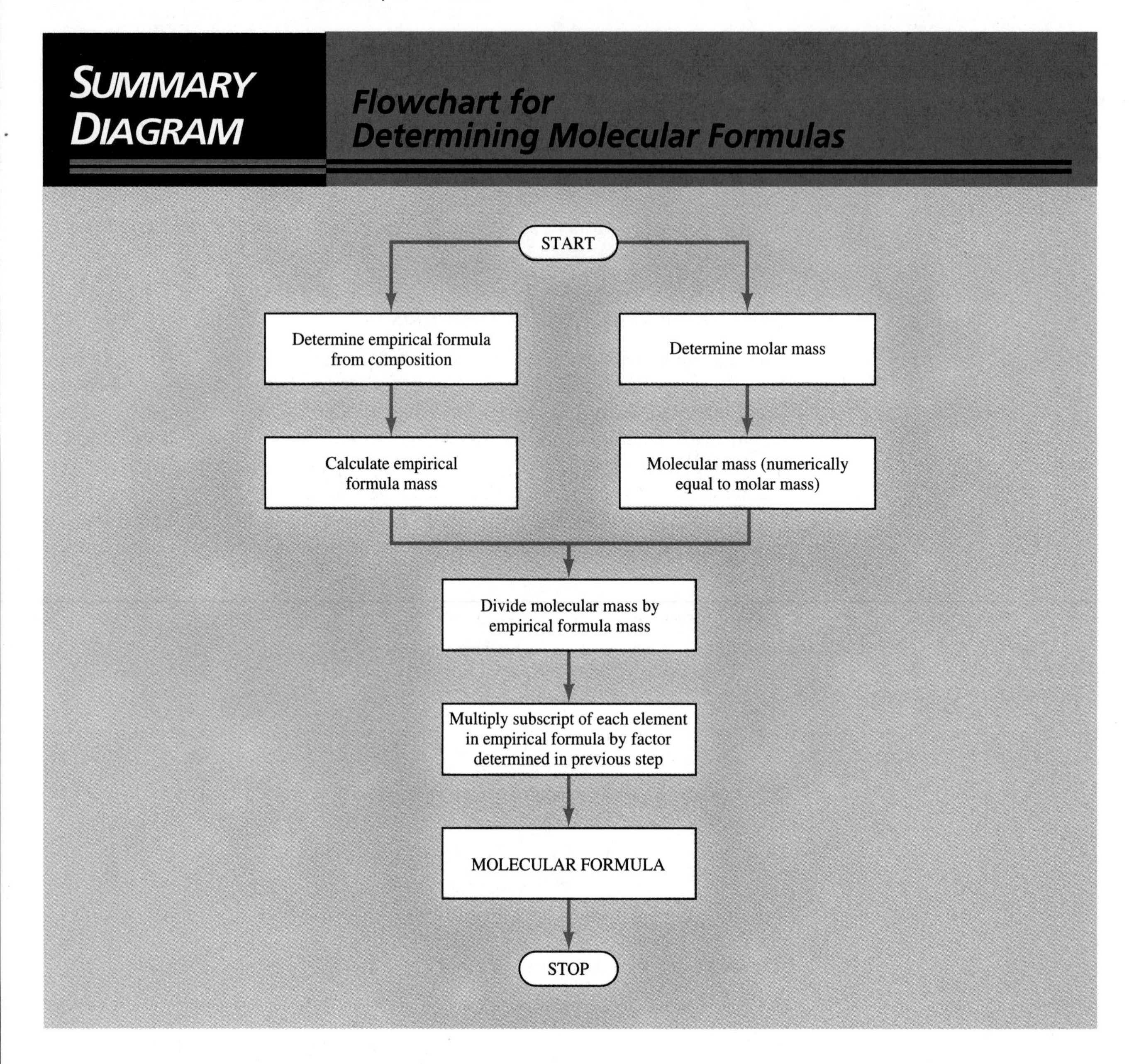

Planning the solution

To determine the molecular formula of the substance, we need both its molecular mass and its empirical formula mass. Division of the molecular mass by the empirical formula mass will then tell us the number of multiples of the empirical formula present in a molecule. Multiplication of the empirical formula by this multiple gives the molecular formula.

SOLUTION

First we calculate the molecular mass by determining the molar mass as follows:

$$\text{Molar mass} = \frac{87.5 \text{ g}}{0.634 \text{ mol}} = 138 \text{ g/mol}$$

Since the molar mass is numerically equal to the molecular mass, it follows that the molecular mass must be 138.

To determine the empirical formula mass, we simply add up the atomic masses in the empirical formula:

$$\begin{aligned} 2 \times C &= 2 \times 12.0 = 24.0 \\ 3 \times H &= 3 \times 1.0 = 3.0 \\ 1 \times F &= 1 \times 19.0 = 19.0 \\ \hline C_2H_3F &= 46.0 \end{aligned}$$

Dividing the molecular mass by the empirical formula mass tells us that the molecular formula contains three multiples of the empirical formula:

$$\frac{138}{46.0} = 3$$

Multiplication of the empirical formula (C_2H_3F) by 3 now leads to the molecular formula: $C_6H_9F_3$.

EXAMPLE 4.27

A compound is found to contain 30.4% nitrogen and 69.6% oxygen. If it has a molecular mass of 92.0, what is its molecular formula?

Planning the solution

This problem involves two steps. First we must determine the empirical formula, as we did in Section 4.8. Then we can proceed as in Example 4.26.

SOLUTION

Assume a 100.0-g sample.

$$? \text{ mol N} = 30.4 \text{ g N} \left(\frac{1 \text{ mol N}}{14.0 \text{ g N}} \right) = 2.17 \text{ mol N}$$

$$? \text{ mol O} = 69.6 \text{ g O} \left(\frac{1 \text{ mol O}}{16.0 \text{ g O}} \right) = 4.35 \text{ mol O}$$

Next divide the moles of each by 2.17 (the smaller number):

$$\frac{2.17 \text{ mol N}}{2.17} = 1.00 \text{ mol N}$$

$$\frac{4.35 \text{ mol O}}{2.17} = 2.00 \text{ mol O}$$

The empirical formula is NO_2.

The empirical formula mass is

$$\begin{aligned} 1 \times N &= 1 \times 14.0 = 14.0 \\ 2 \times O &= 2 \times 16.0 = 32.0 \\ \hline NO_2 &= 46.0 \end{aligned}$$

Dividing the molecular mass by the empirical formula mass tells us that there are two multiples of the empirical formula in the molecular formula:

$$\frac{92.0}{46.0} = 2$$

The molecular formula must be N_2O_4.

PROBLEM 4.17

Determine the molecular formula for each of the following.

Empirical formula	Molecular mass	Molecular formula
(a) CH	52.0	______
(b) CH	78.0	______
(c) C_2H_3F	92.0	______
(d) NaO	78.0	______
(e) C_5H_5N	79.0	______
(f) C_2HNO_2	213.0	______

CHAPTER SUMMARY

Matter may be described quantitatively. The atoms of each element have a characteristic **atomic mass,** which is the **relative mass** of an atom compared to some standard. The standard for atomic mass is a substance known as carbon-12. The atomic mass unit (u) is the unit of mass that we use with atomic mass when we are referring to the mass of one atom.

The **molecular mass** or **formula mass** of a substance can be obtained by summing the atomic masses of all the atoms in the chemical formula of the substance. The mass of one molecule of a molecular substance equals its molecular mass in atomic mass units. We indicate the **percentage composition** of each element in a compound by expressing its contribution to the molecular mass as a percentage of the total molecular mass. The **Law of Constant Composition** tells us that a pure substance always has the same percentage composition. The percentage composition may be used to determine the mass of an element in a given quantity of pure substance.

The **mole** (mol) provides a large enough collection of atoms or molecules to work with under laboratory conditions. A mole refers to a specific number of particles. This number of particles, known as **Avogadro's number**, is equal to 6.02×10^{23}. One mole of a substance is also equal to its atomic, molecular, or formula mass in grams; this is known as the **molar mass.** The molar mass enables us to write a conversion factor between moles and the corresponding mass of any monatomic or polyatomic substance. We can convert among mass, moles, and atoms or molecules by using relationships derived from these definitions of the mole.

Under certain circumstances, we have only enough information to write an **empirical formula,** the simplest whole-number ratio of atoms in a compound. The procedure for calculating empirical formulas from percentage composition is as follows: (1) assume a 100.0-g sample; (2) calculate the moles of atoms for each element; and (3) determine the simplest ratio of atoms.

The relationship between grams and moles also can be used to determine the molar mass of a substance. Because 1 mole equals the atomic, molecular, or formula mass in grams, it follows that the molar mass is numerically equal to the atomic, molecular, or formula mass. Consequently, a substance's molar mass can be determined by dividing the mass of the substance by the number of moles it equals. The **molecular formula** of a substance can be determined by dividing the molecular mass by the empirical formula mass. This tells us how many multiples of the empirical formula there are in each molecule. Multiplying the empirical formula by this multiple yields the molecular formula.

KEY TERMS

Review each of the following terms, which are discussed in this chapter and defined in the Glossary.

molecular formula	relative mass	Law of Constant Composition	Avogadro's number
empirical formula	molecular mass	percentage composition	mole
atomic mass	formula mass		molar mass

ADDITIONAL PROBLEMS

SECTION 4.1

4.18 Distinguish between a molecular formula and an empirical formula.

4.19 Determine the empirical formula corresponding to each of the following molecular formulas.

(a) H_2O_2 (hydrogen peroxide)
(b) C_2H_6O (ethyl alcohol)
(c) $C_4H_6O_6$ (tartaric acid)
(d) $C_6H_6O_2$ (hydroquinone, photographic developer)
(e) $C_3H_6O_3$ (lactic acid)

SECTION 4.2

4.20 What is an *atomic mass?* Explain the meaning of relative mass.

4.21 Use the periodic table to find the atomic mass of each of the following elements. Round off the precise value to the nearest tenth.

(a) Li **(b)** Ne **(c)** B **(d)** S **(e)** Cl
(f) Ar **(g)** Se **(h)** Fe **(i)** Cu **(j)** Zn

SECTION 4.3

4.22 Calculate the molecular (or formula) mass for each of the following substances.

(a) NaOCl (sodium hypochlorite)
(b) C_4H_{10} (butane)
(c) HNO_3 (nitric acid)
(d) $Al(OH)_3$ (aluminum hydroxide)
(e) $Zn_3(PO_4)_2$ (zinc phosphate)

SECTION 4.4

4.23 Calculate the molecular (or formula) mass and the percentage composition of each of the following.

(a) N_2O (nitrous oxide, or laughing gas)
(b) C_2H_6O (ethyl alcohol)
(c) $C_{14}H_9Cl_5$ (DDT)
(d) $C_7H_7O_2N$ (*para*-aminobenzoic acid, a sunscreen)
(e) $C_9H_{13}O_3N$ (adrenalin)
(f) $KC_4H_5O_6$ (potassium hydrogen tartrate, cream of tartar)
(g) $NaC_5H_8O_4N$ (monosodium glutamate, MSG)
(h) $Mg(C_2H_3O_2)_2$ (magnesium acetate)
(i) $(NH_4)_2SO_4$ (ammonium sulfate)

4.24 Calculate the mass of the element indicated in each of the following samples.

(a) I in 1.25 g KI **(b)** Ca in 0.750 g CaC_2O_4
(c) P in 3.85 g Na_3PO_4 **(d)** Mg in 275 mg $Mg(OH)_2$

SECTIONS 4.5 AND 4.6

4.25 Define the mole in terms of:

(a) The number of particles **(b)** The mass

4.26 Calculate the number of moles in each of the following.

(a) 4.51 g Kr **(b)** 48.0 g Mg **(c)** 0.273 g Ag

4.27 Calculate the mass of each of the following.

(a) 1.37 mol Fe **(b)** 0.783 mol Si **(c)** 0.0915 mol Ar

4.28 Calculate the number of atoms in each of the following.

(a) 0.853 mol Pb **(b)** 0.637 mol Ni **(c)** 3.11 g Ne
(d) 0.274 g Ag

SECTION 4.7

4.29 Calculate the number of moles in each:

(a) 6.73 g $NaNO_3$ **(b)** 11.4 g C_2H_6O
(c) 0.855 g $CuSO_4$ **(d)** 0.218 g F_2

4.30 Calculate the mass of each of the following.

(a) 0.263 mol Cl_2 **(b)** 3.45 mol $CHCl_3$
(c) 0.513 mol AgCl **(d)** 0.0125 mol $Na_2S_2O_3$

4.31 Calculate the number of molecules in each:

(a) 0.713 mol C_4H_{10} **(b)** 0.473 mol O_2
(c) 2.67 g C_2H_6 **(d)** 1.08 g $C_6H_{12}O_6$

4.32 List the seven common diatomic elements.

4.33 Calculate each of the following quantities for a 3.58-g sample of bromine.

(a) The number of moles of bromine molecules

(b) The number of bromine molecules

(c) The number of moles of bromine atoms

(d) The number of bromine atoms

SECTION 4.8

4.34 Calculate the empirical formula corresponding to the following compound.

%C = 37.2% %H = 7.8% %Cl = 55.0%

4.35 A beautiful mineral crystal, englesite, has the following composition. Determine the empirical formula of englesite.

%Pb = 68.3% %S = 10.6% %O = 21.1%

4.36 A fluorinated hydrocarbon has the following composition. What is the empirical formula of this compound?

%C = 45.0% %H = 7.5% %F = 47.5%

4.37 Three mineral compounds each contain iron and oxygen. Determine the empirical formula of each of these minerals.

(a) Hematite: 69.94% Fe, 30.06% O

(b) Magnetite: 72.36% Fe, 27.64% O

(c) Wuestite: 77.73% Fe, 22.27% O

4.38 A 0.700-g sample of a fluorocarbon (a substance composed of carbon and fluorine) is found to contain 0.122 g C and 0.578 g F. Calculate its empirical formula.

4.39 A 0.846-g sample of an unknown substance is found to contain 0.227 g Ni and 0.619 g Br. What is the empirical formula of this substance?

4.40 A 4.82-g sample of an oxide of chromium is composed of 3.30 g of chromium and 1.52 g of oxygen. Determine the empirical formula of this substance.

SECTION 4.9

4.41 Glycerol, a component of fats, is often used in cosmetic products. A 12.6-g sample of glycerol contains 0.137 mol. What is the molar mass of glycerol?

4.42 Phosgene is a toxic gas that must be handled with great care. If 2.05 mol of this substance has a mass of 203 g, what is its molar mass?

4.43 Sucrose, also known as either cane or beet sugar, is used in our homes as table sugar. If 72.5 g are equal to 0.212 mol, what is the molar mass of this sweetener?

4.44 Vinyl chloride has a molar mass of 62.5 g/mol. What is the molecular mass in terms of atomic mass units per molecule?

4.45 The empirical formula of hydrazine is NH_2. If hydrazine has a molecular mass of 32.0, what is its molecular formula?

4.46 Trifluorobenzene has a molar mass of 132.0 g/mol. Its empirical formula is C_2HF. What is its molecular formula?

4.47 Ethane is a gaseous hydrocarbon.

(a) The percentage composition of this substance is 80.0% C and 20.0% H. Calculate its empirical formula.

(b) If 0.125 mol of ethane has a mass of 3.75 g, what is its molar mass?

(c) Use the answers you obtained in parts (a) and (b) to determine the molecular formula of ethane.

4.48 *para*-Dichlorobenzene is frequently used in mothballs. Its percentage composition is 49.0% C, 2.7% H, and 48.3% Cl.

(a) Calculate the empirical formula of this substance.

(b) A 10.0-g sample of *para*-dichlorobenzene is found to contain 0.0680 mol. What is its molar mass?

(c) What is the molecular formula of this substance?

GENERAL PROBLEMS

4.49 Potassium chromate is a yellow compound that has been used as a pigment in paints. Calculate the number of oxygen atoms in 355 mg of potassium chromate, K_2CrO_4.

4.50 How many mercury atoms are present in 3.45 mL of mercury? The density of mercury is 13.6 g/mL.

***4.51** Xylene is a common organic solvent. A 0.0430-mol sample of xylene has a mass of 4.56 g, which is composed of 4.13 g C and 0.43 g H. What is the molecular formula of this compound?

***4.52** Apply your knowledge of the meaning of *per* to solve the following problems.

(a) A 4.00-g sample of helium (1 mol) contains Avogadro's number of atoms. Calculate the mass of each helium atom in grams per atom.

(b) Because the atomic mass of helium is 4.00, each atom must also have a mass of 4.00 atomic mass units (that is, there are 4.00 u per atom). Use this and the answer you obtained in part (a) to determine the mass of an atomic mass unit in grams per atomic mass unit (g/u).

***4.53** The price of precious metals fluctuates according to the availability of the metal and the industrial need for it. Rhodium, Rh, has a price that fluctuates around $20.00 per gram. Would it be a wise investment to buy rhodium(III) chloride hexahydrate, $RhCl_3 \cdot 6H_2O$, at $8.00 per gram? (The formula mass of the hexahydrate is obtained by adding six times the molecular mass of water to the formula mass of rhodium(III) chloride, $RhCl_3$.)

***4.54** In one process for decaffeinating coffee, methylene chloride, CH_2Cl_2, is added to coffee beans to extract (or dissolve) the caffeine, $C_8H_{10}N_4O_2$. Once the caffeine has been extracted into the methylene chloride, the coffee beans are filtered (thereby removing the caffeine and methylene chloride) and allowed to dry. How many grams of caffeine contain the same mass of carbon as 1.00 g of methylene chloride?

4.55 The recommended daily allowance (RDA) of sodium is given as a range, from 1100 to 3300 mg. Most Americans consume far more than this quantity. A teaspoon (tsp) of table salt (NaCl) has a mass of approximately 7.8 g. A not-too-health-conscious host prepares a 2-lb meatloaf, which he seasons with a tablespoon of salt (1 T = 3 tsp). He and four of his hearty friends consume the entire meatloaf. Assuming that each consumed equal portions of the meatloaf, what percentage of the minimum RDA (1100 mg) did each consume in his portion of meatloaf? What percentage of the maximum RDA (3300 mg) did each consume? (For the sake of your calculation, assume his tablespoon of salt is accurate to 1.00 T.)

4.56 Hydrocarbons are a class of organic compounds that are composed only of carbon and hydrogen. Alkanes are a subclass of hydrocarbons that have the generic formula C_nH_{2n+2}, where n is any positive integer. For what values of n is the molecular formula not equal to the empirical formula? (Your answer should be a generalization, rather than a list of specific numbers.)

4.57 Sucrose (table sugar) is a sugar composed of two smaller sugars, known as glucose ($C_6H_{12}O_6$) and fructose (also $C_6H_{12}O_6$). When glucose combines with fructose to form sucrose, water is given off. What can be said about the mass percentage of carbon in sucrose, compared to the percentage in glucose or fructose? (You may find it interesting that both glucose and fructose have the same chemical formula. Although their formulas are the same, the atoms are connected differently, leading to differences in their physical and chemical properties. Thus, they are different substances.)

4.58 Some combinations of elements are capable of forming several different compounds. An unexpected element that forms several compounds with fluorine is xenon, Xe. Xenon can form XeF_2, XeF_4, and XeF_6. Explain how elemental analysis can be used to distinguish these three compounds.

WRITING EXERCISES

4.59 Your aunt takes potassium supplements for her diet. The bottle that the pharmacist sold her is labeled as "potassium gluconate." When she examined the label carefully, she noted that a 550-mg tablet of potassium gluconate contains only 92 mg of potassium. She is confused by these distinctions and wonders why there is so little potassium. Write a note to your aunt, explaining why there are only 92 mg of potassium in each tablet, and tell her how the amount of potassium in each tablet is calculated.

4.60 In your own words, explain the mole concept and tell why the mole is such an important quantity for chemists.

4.61 A classmate has come across two bottles in the laboratory, each containing a pure substance. One bottle contains a liquid and the other a solid. With each bottle is a laboratory analysis reporting the percentage composition of the contents of the bottle. Your classmate sees that both substances have the same percentage composition and wonders how the composition of two different substances can be the same. Write a note to your classmate explaining this apparent paradox, and describe the limitations on information that can be obtained from a percentage composition. Support your explanation with an example of two different substances that have the same percentage composition.

5

ATOMIC STRUCTURE

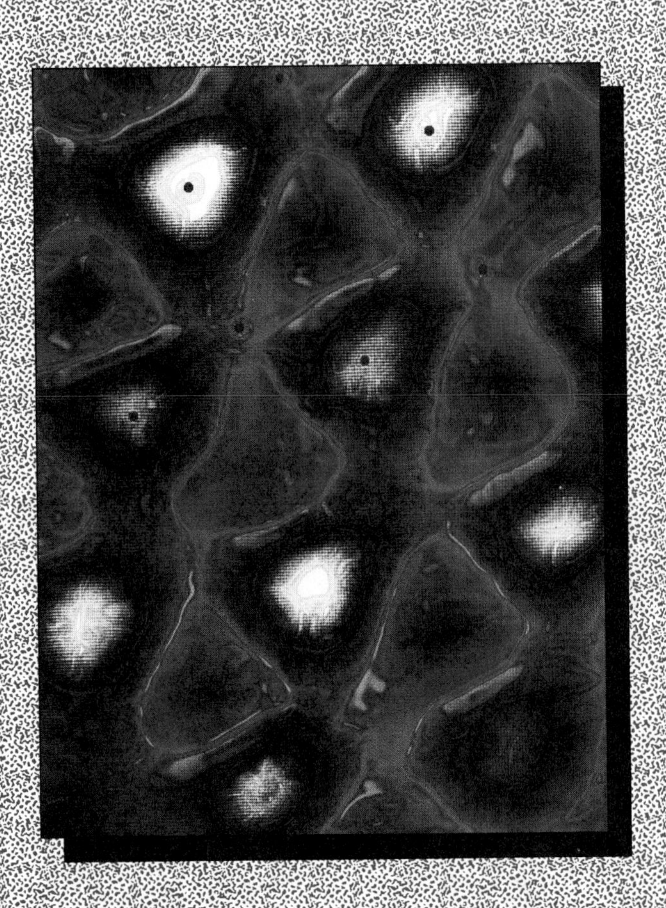

So far we have pictured atoms as billiard balls, noting that 109 different kinds of atoms are displayed in the periodic table. We have referred to the various differences in their chemical and physical properties, including their differences in atomic mass. However, we have not yet been precise about the specific characteristic that distinguishes one kind of atom from every other kind. In this chapter we look at the basic overall structure of atoms and learn about the *defining* characteristic that determines what element a particular atom is.

A great deal is known about atoms, even though no one has ever actually seen one with the naked eye. This is one of science's most remarkable achievements. We have developed the ability to extend the human senses through the use of various instruments. Our equipment and ingenuity have enabled us to gain indirect evidence about atoms without actually seeing them.

What is meant by "extending the senses to gain indirect evidence"? Consider blind people who learn about objects by exploring their shapes and textures with their hands, rather than by actually seeing the objects. These people extend their senses to gain indirect information about the appearance of the objects they touch. In much the same way, scientists have devised instruments capable of probing atoms and molecules in order to provide information about their structural organization.

We must remember that the picture of atomic structure that emerges, however much it helps us to explain the behavior of atoms, is just a model. Again, consider the blind person who develops a mental picture of an object's appearance through the sense of touch. This mental picture (or model) may bear very little resemblance to the actual appearance of the object. And so it is with the models we use in our study of chemistry. The description of atomic and molecular structure that follows in this and succeeding chapters represents the best model (or picture) we can construct on the basis of the indirect evidence we have available today.

5.1 Historical Background

OBJECTIVE 5.1A State two ideas from Dalton's atomic theory that are still accepted today; state two ideas from his theory that had to be modified. Explain Rutherford's major contribution to our description of the atom.

OBJECTIVE 5.1B Name the three fundamental particles composing atoms, stating the charge and approximate mass of each. Identify those fundamental particles that are components of the nucleus. Compare the size and approximate mass of the nucleus to those of the atom as a whole.

In Section 2.5, we described the *atom* as a building block of matter, and we later saw how it acts as a component of molecules. Much of what we have described thus far was first published during the early nineteenth century in a series of papers by an English schoolteacher named John Dalton. The ideas presented in Dalton's atomic theory laid the foundation for our current understanding of atoms. His ideas included the following.

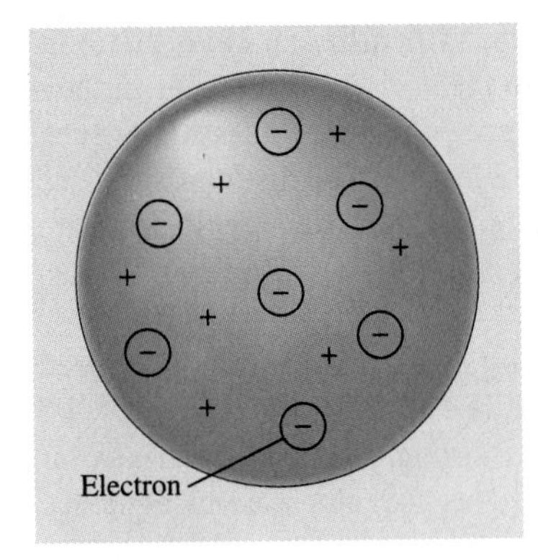

FIGURE 5-1 J. J. Thomson's plum-pudding model of an atom.

1. All matter is composed of tiny particles called atoms.
2. Atoms are indivisible.
3. The atoms of each element all have the same mass and properties.
4. The atoms of each element are different from the atoms of all other elements.
5. Compounds are formed from the combination of atoms of different elements in whole-number ratios.
6. The same elements may combine to form more than one compound, but when they do, the ratio of atoms differs for each compound.
7. When atoms combine to form compounds, they remain unchanged.

Despite several incorrect values, Dalton was also the first to propose a table of atomic masses. Even though later findings revealed that atoms are not indivisible, and that all atoms of the same element do not necessarily have the same mass, Dalton's theory was largely correct. It represents a major contribution to the development of modern chemistry.

In fact, by the late nineteenth century, new discoveries had already led to the belief that atoms were not indivisible but instead were made up of smaller *subatomic* particles that possessed electrical charges. Scientists postulated the existence of very small, negatively charged particles called *electrons* and much more massive particles that bore positive charges. J. J. Thomson proposed a model of the atom that is often referred to as the "plum-pudding model." He suggested that atoms were composed primarily of a very massive, positively charged blob. Embedded in this positively charged blob were negatively charged electrons. The tiny, negatively charged electrons balanced the positive charge, resulting in an electrically neutral atom. Thomson's model is shown in Fig. 5-1.

Thomson's model of the atom was rejected as a result of the work of Ernest Rutherford, who demonstrated that, contrary to the plum-pudding model, the atom is primarily made up of empty space. Rutherford bombarded a very thin gold foil (a thin sheet made up of gold atoms) with a stream of positively charged particles known as *alpha particles.* These particles were of a very high energy, and they were expected to pass right through the foil with little or no deflection, like a bullet passing through a tissue paper. Although most of the particles passed right through the foil as expected, some were deflected at various angles, while others were reflected back in the direction

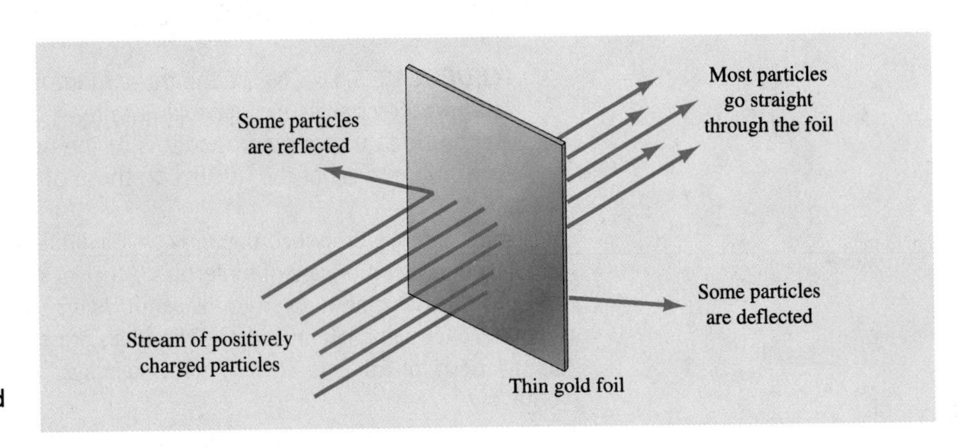

FIGURE 5-2 Rutherford's experiment fired high-energy particles at a thin piece of gold foil.

FIGURE 5-3 (a) Rutherford's model of an atom. (b) How Rutherford's model accounts for the experimental results.

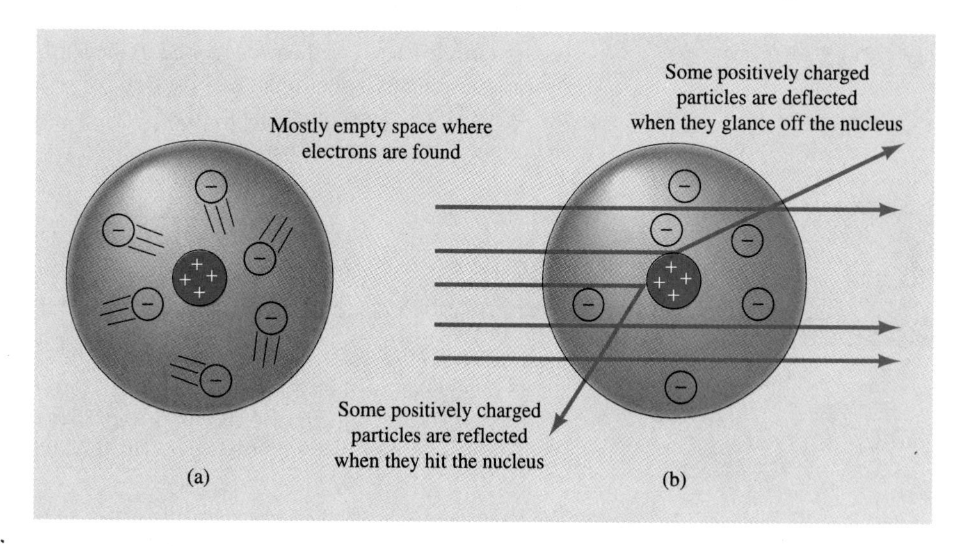

Atom: Building block of matter composed of a very dense positively charged nucleus, which is surrounded by negatively charged electrons.

Nucleus: Central component of atoms, composed of protons and neutrons.

Proton: A subatomic particle found in the nucleus; it has a charge of +1 and a mass of approximately 1 u.

Neutron: A subatomic particle found in the nucleus; it has a zero charge and a mass of approximately 1 u.

Electron: A subatomic particle found outside the nucleus; it has a charge of −1 and a negligible mass compared to that of a proton or neutron.

Fundamental particle: A proton, neutron, or electron.

from which they had come (Fig. 5-2). By determining the various locations at which the particles were deflected (or reflected), Rutherford concluded that an **atom** was made up of a very dense, positively charged **nucleus** surrounded primarily by empty space in which the electrons could be found (Fig. 5-3). Compared with the overall size of the atom, the nucleus is extremely small; it has a radius of about 1/100,000 that of the atom itself. The space that the electrons occupy essentially determines the volume of the atom. Today we know that electrons do not exist in any fixed location but instead move about the nucleus. To illustrate these ideas, a hydrogen atom is shown in Fig. 5-4.

It was later shown by Sir James Chadwick that the nucleus is made up of two kinds of particles, **protons** and **neutrons.** Protons bear a single positive charge (+1), whereas neutrons are neutral (0), which means they have neither a positive nor a negative charge. Thus, an atom is made up of a positively charged nucleus surrounded by negatively charged **electrons,** each of which bears a single negative charge (−1). Because the charge of each proton (+1) cancels that of each electron (−1), a neutral atom (one with no electrical charge) must have an equal number of protons and electrons. The properties of these **fundamental particles** are summarized in Table 5-1.

Note that the masses of both the proton and the neutron are very close to 1 u (recall that "u" is the symbol for atomic mass unit) and that the mass of the electron (which is approximately 1/1800 that of a proton or a neutron) is negligible by comparison. Consequently, we consider the approximate mass of a proton or neutron to equal 1 u, whereas that of an electron is considered to be zero. Thus, to find the approximate mass of any atom, we simply add up the number of protons and neutrons.

FIGURE 5-4 The radius of a hydrogen atom is 100,000 times larger than the radius of the nucleus.

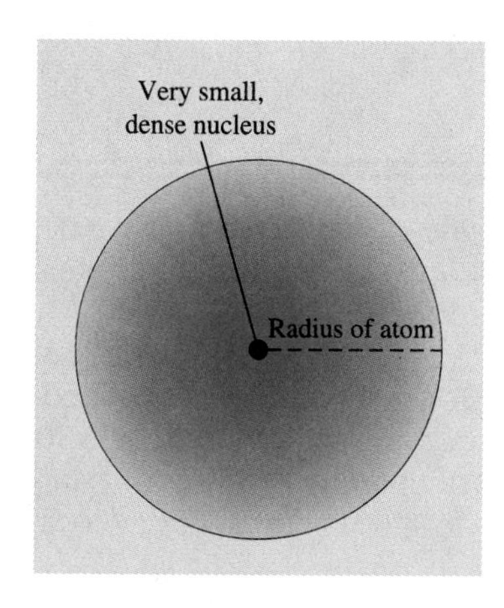

TABLE 5-1 Fundamental particles

Particle	Symbol	Charge	Mass	Approximate mass
Proton	p^+	+1	1.0072765 u	1 u
Neutron	n	0	1.0086649 u	1 u
Electron	e^-	−1	0.0005486 u	0 u

PROBLEM 5.1

Alpha particles, such as those described in the Rutherford experiment, are composed of two protons and two neutrons.
(a) What is the approximate mass of an alpha particle?
(b) What is its electrical charge?

5.2 ATOMIC NUMBER

OBJECTIVE 5.2 Define the term *atomic number.* Find the atomic number of any element in the periodic table. Determine any one of the following for any element, given one of the others: the number of protons in the nucleus, the atomic number of the element, the symbol for the element.

Atomic number: The number of protons in the nucleus of an atom.

We have seen that the various elements differ in atomic mass. However, we have not said what it is that makes one kind of atom different from every other kind. If we examine the periodic table, we find that each element has a whole number (integer) associated with it and that the whole number is different for every element. This number is called the **atomic number,** and it gives the number of protons in the nucleus of an atom. Thus, hydrogen (atomic number = 1) has one proton in its nucleus. Carbon (atomic number = 6) has six protons in the nucleus. Oxygen (atomic number = 8) has eight protons, and uranium (atomic number = 92) has ninety-two. The atomic number determines which element an atom is. Examine the periodic table inside the front cover and be sure that you can find the atomic number for each element. Since an atom is electrically neutral, the atomic number also gives the number of electrons in an atom, because an equal number of electrons is required to balance the charge of the protons.

PROBLEM 5.2

Use the periodic table inside the front cover to find the atomic number for each of the following elements.
(a) Ne **(b)** Cu **(c)** Bi **(d)** N **(e)** Sb **(f)** Hg

5.3 MASS NUMBER: ISOTOPES

OBJECTIVE 5.3A Define the term *mass number.* Write the nuclear symbol for any atom, given: (a) its atomic number, number of protons, or symbol; and (b) its mass number or number of neutrons.

OBJECTIVE 5.3B Define the term *isotope.* Determine the number of neutrons in a specific isotope, given its mass number and either its symbol or its atomic number. Determine the mass number of an isotope, given the number of neutrons plus its symbol, its atomic number, or its number of protons.

Mass number: The sum of the protons and neutrons in the nucleus of an atom.

Another number also gives valuable information about atoms. The **mass number** of an atom is the sum of its protons and neutrons. For example, an atom having 13 protons and 14 neutrons in its nucleus has a mass number of 27. The nuclei of aluminum atoms found in nature have this composition. In addition to its protons and neutrons, an aluminum atom must have 13 electrons to balance the charge of the nucleus. The following relationships are used to describe the composition of a particular atom.

$$\left.\begin{aligned} \text{Mass number} &= \text{Number of protons} + \text{Neutrons} \\ \text{Atomic number} &= \text{Number of protons} \\ \text{Mass number} - \text{Atomic number} &= \text{Number of neutrons} \end{aligned}\right\} \textbf{Nuclear composition}$$

$$\left.\text{Atomic number} = \text{Number of electrons}\right\} \textbf{Surrounding the nucleus}$$

When we recall that the approximate mass of a proton or neutron is 1 u, whereas that of an electron is zero, it becomes clear that the approximate mass of a particular atom is very nearly equal to its mass number.

Scientists use either of two symbolic representations to specify the complete nuclear composition of an atom. The simplest of these is the name of the element followed by its mass number. For example, the aluminum atom we just considered is denoted *aluminum-27.* Another representation is the nuclear symbol, which is obtained by writing the symbol of the element with the atomic number as a subscript in the lower left-hand corner and the mass number as a superscript, also on the left. For aluminum-27, the nuclear symbol is

$$\begin{matrix} \textbf{Mass number} \rightarrow \\ \textbf{Atomic number} \rightarrow \end{matrix} {}^{27}_{13}\text{Al}$$

The symbols A (mass number), Z (atomic number), X (symbol for the element), p (proton), and n (neutron) are often used to designate the various nuclear quantities.

$$\begin{matrix} (\mathbf{p} + \mathbf{n}) \rightarrow \\ (\mathbf{p}) \rightarrow \end{matrix} {}^{A}_{Z}X \qquad \begin{aligned} n &= A - Z \\ n &= (p + n) - p \end{aligned}$$

Though atoms of the same element always have the same atomic number (number of protons), they may differ in the number of neutrons in their nuclei. For example, the nuclei of most hydrogen atoms contain one proton and no neutrons (1p, 0n). This form of hydrogen is sometimes called *protium.* However, there are two other forms of hydrogen that have one and two neutrons, respectively, per nucleus. These additional forms of hydrogen, known as *deuterium* (1p, 1n) and *tritium* (1p, 2n), must also have one proton in the nucleus or they would not be hydrogen atoms (Fig. 5-5, p. 118). (Remember, the number of protons determines which element we have, and all hydrogen nuclei *must* have one proton.) In addition, an atom of each has one electron to balance the nuclear charge.

Isotopes: Atoms of the same element that differ in mass number.

We refer to these various forms of hydrogen as isotopes. **Isotopes,** then, are atoms of the same element that differ in mass number. Applying the definition of mass number to the three isotopes of hydrogen, we find that the most common form of hydrogen has a mass number of 1, deuterium has a mass number of 2, and tritium has a mass number of 3.

FIGURE 5-5 Isotopes of hydrogen: (a) protium; (b) deuterium; (c) tritium.

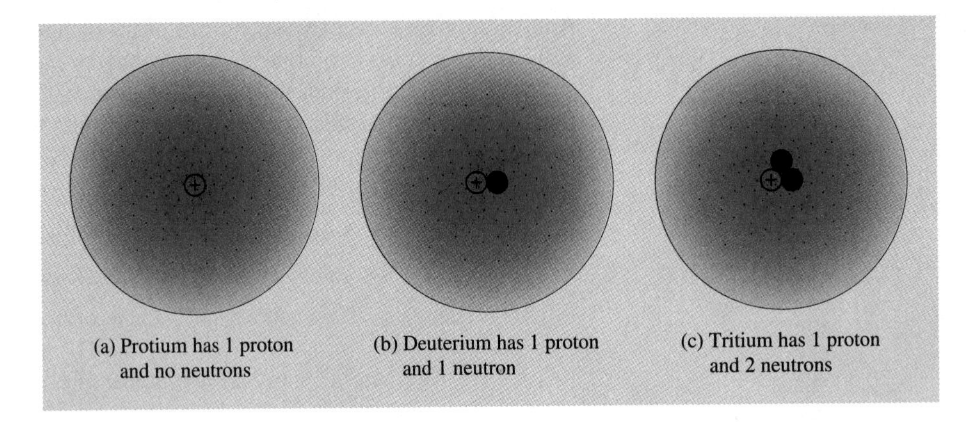

	Nuclear composition	Electrons
Hydrogen-1:	1 proton + 0 neutrons (protium)	1 electron
Hydrogen-2:	1 proton + 1 neutron (deuterium)	1 electron
Hydrogen-3:	1 proton + 2 neutrons (tritium)	1 electron

Naturally occurring uranium consists of several isotopes, including uranium-235 and uranium-238. Uranium-235 is used as a fuel in nuclear power plants. Similarly, the carbon found in nature consists of carbon-12, carbon-13, and carbon-14. Carbon-14 is used as a tool for dating some types of fossils and ancient relics. The nuclear symbols for these various isotopes are

$$\begin{array}{r} \text{Mass number:} \\ \text{Atomic number:} \end{array} \quad {}^{235}_{92}\text{U} \quad {}^{238}_{92}\text{U} \quad {}^{12}_{6}\text{C} \quad {}^{13}_{6}\text{C} \quad {}^{14}_{6}\text{C}$$

Minor differences between isotopes of the same element do exist, but their behavior is so similar that, for our purposes, we consider isotopes of the same element to have the same physical and chemical properties.

TABLE 5-2 Structure of some isotopes on the first three elements

Name	Symbol	Atomic number	Protons	Electrons	Neutrons	Approximate mass	Precise mass
Hydrogen-1 (protium)	${}^{1}_{1}\text{H}$	1	1	1	0	1 u	1.007825 u
Hydrogen-2 (deuterium)	${}^{2}_{1}\text{H}$	1	1	1	1	2 u	2.01410 u
Hydrogen-3 (tritium)	${}^{3}_{1}\text{H}$	1	1	1	2	3 u	3.01605 u
Helium-3	${}^{3}_{2}\text{He}$	2	2	2	1	3 u	3.01603 u
Helium-4	${}^{4}_{2}\text{He}$	2	2	2	2	4 u	4.00260 u
Lithium-6	${}^{6}_{3}\text{Li}$	3	3	3	3	6 u	6.01514 u
Lithium-7	${}^{7}_{3}\text{Li}$	3	3	3	4	7 u	7.01600 u

Special names for isotopes, such as deuterium and tritium, are the exception. Most isotopes are known by the name of the element and its mass number. Table 5-2 presents information on the isotopes of the first three elements. The distinction between elements and isotopes is outlined in the Summary Diagram.

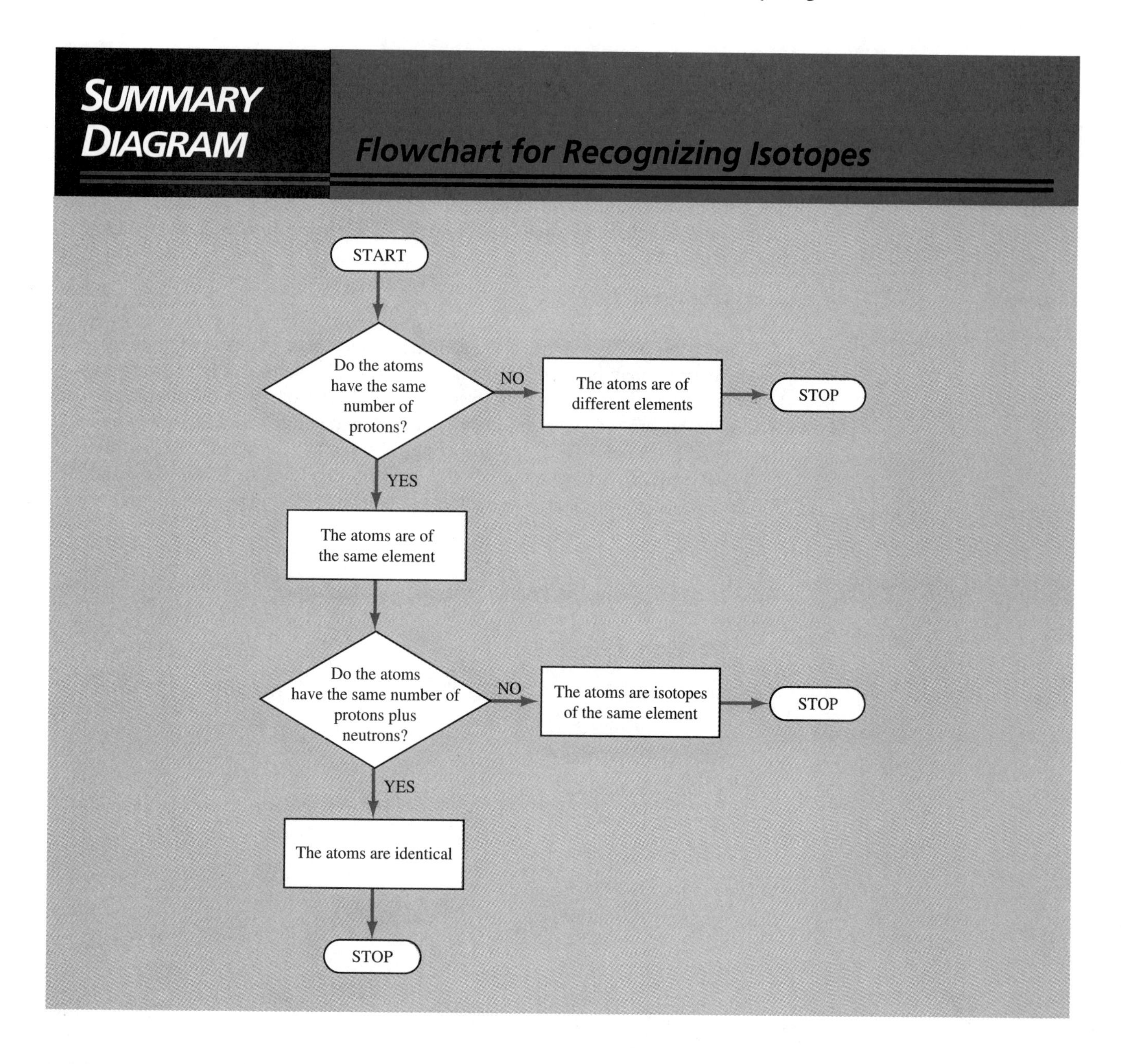

EXAMPLE 5.1

How many protons, neutrons, and electrons are in an atom of rubidium-85: $^{85}_{37}Rb$?

SOLUTION

The atomic number (37) equals the number of protons, so there must be 37 protons. The mass number (85) equals the sum of the protons and neutrons. Therefore, the number of neutrons must equal the mass number minus the atomic number: 85 − 37 = 48 neutrons. The number of electrons in an atom equals the number of protons, so there are 37 electrons.

Answer

37 protons, 48 neutrons, 37 electrons

EXAMPLE 5.2

Write a nuclear symbol for the isotope that has 79 protons and 118 neutrons in its nucleus.

SOLUTION

The periodic table tells us that gold, Au, is the element with atomic number 79. The atomic number equals the number of protons: 79. The mass number equals the protons plus neutrons: 79 + 118 = 197.

Answer

The isotope is gold-197: ${}^{197}_{79}Au$.

Since the atomic number is the defining characteristic of an element, the symbol for any element may be found in the periodic table by looking up its atomic number. Likewise, the atomic number may be found for any symbol. Since the atomic number and its symbol are redundant information, the atomic number is occasionally omitted from nuclear symbols. Thus, you might find the symbol for rubidium-85 written as ${}^{85}Rb$ (see Example 5.1), while gold-197 could be simplified to ${}^{197}Au$ (see Example 5.2). You should use the complete nuclear symbol when solving problems in this text.

PROBLEM 5.3

Write the name and nuclear symbol for each of the following nuclei. Part (a) is done for you.

Nuclear composition	Name	Nuclear symbol
(a) 4 protons, 5 neutrons	Beryllium-9	${}^{9}_{4}Be$
(b) 6 protons, 6 neutrons	______	______
(c) 8 protons, 10 neutrons	______	______
(d) 17 protons, 18 neutrons	______	______
(e) 92 protons, 146 neutrons	______	______
(f) 103 protons, 154 neutrons	______	______
(g) 6 protons, 7 neutrons	______	______
(h) 17 protons, 20 neutrons	______	______

PROBLEM 5.4

Which of the nuclei in Problem 5.3 are isotopes of one another?

PROBLEM 5.5

Fill in the following table. Assume that each line refers to an uncharged atom. Use the tables provided inside the front and back covers of this book for reference.

Name	Symbol	Atomic number	Number of protons	Number of electrons	Number of neutrons	Mass number
Lithium	Li	3	3	3	4	7
Carbon	____	____	____	____	7	____
____	N	____	____	____	____	15
____	Au	____	____	____	118	____
Lead	____	____	____	____	____	206
____	____	____	92	____	143	____
____	____	____	____	92	____	238
Neptunium	____	____	____	____	144	____
____	Pu	____	____	____	____	242

5.4 *Atomic Mass Revisited: The Average Mass of an Element*

OBJECTIVE 5.4A Given the percentage abundances of the naturally occurring isotopes of an element, estimate the atomic mass of the element.

Atomic mass unit (u): One-twelfth the mass of an atom of carbon-12. 1 u = 1.66×10^{-24} g.

Atomic mass: The average mass of the naturally occurring isotopes of an element.

In Chapter 4 we described the atomic masses of the elements as a set of relative masses. We also mentioned that the reference standard is a substance known as carbon-12, though we were not prepared to give a detailed description of that standard. In fact, carbon-12 is an isotope of carbon composed of six protons and six neutrons. In establishing the values shown in the periodic table, we arbitrarily assign a mass of *exactly* 12 u to each atom of carbon-12, and every atomic mass represents a relative mass compared to that standard. Consequently, 1 **atomic mass unit** represents exactly one-twelfth the mass of an atom of carbon-12 (which corresponds to a mass of 1.66×10^{-24} g).

This being so, you may wonder why the atomic mass of carbon shown in the periodic table is a fractional number, 12.011. The **atomic mass** of each element given in the periodic table represents the average mass per atom of the various isotopes as they occur in nature. Recalling that the mass of an atom is approximately equal to its mass number, we can estimate the atomic mass of an element if we know its isotopic composition. For example, the carbon found in nature is actually a mixture of several isotopes: carbon-12, carbon-13, and carbon-14. The abundances of these isotopes are

Carbon-12: 98.90%

Carbon-13: 1.10%

Carbon-14: Negligibly small by comparison

Because carbon is 98.90% carbon-12 and only 1.10% carbon-13, it seems quite reasonable that the average mass of carbon is close to 12 u; most of the atoms have masses of 12 u. However, the contribution of a small number of atoms with masses of approximately 13 u (the mass of a carbon-13 atom is 13.00335 u) increases the average mass to 12.011 u, a value slightly larger than 12 u. (We will show how this type of average mass is calculated at the end of this section.)

Table 5-3 shows the natural abundances of isotopes of the first 20 elements. Referring to the periodic table, we can see how the atomic mass of each element reflects the percentage composition of the isotopes that comprise it. We have just seen how the isotopic composition of carbon results in an atomic mass of 12.011. Lithium is composed of 92.5% lithium-7 and 7.5% lithium-6. We would predict the atomic mass of lithium to be very close to 7 u (the approximate mass of most of the lithium atoms), but somewhat less than 7 u (due to the presence of lithium atoms with masses of approximately 6 u). In agreement with our expectations, the atomic mass of lithium is 6.941 u.

TABLE 5-3 Distribution of protons, electrons, neutrons, mass numbers, and abundances of naturally occurring isotopes of the first 20 elements

Name	Symbol	Atomic number	Protons	Electrons	Neutrons	Mass number	% Natural abundance[a]
Hydrogen-1	$^{1}_{1}H$	1	1	1	0	1	99.985
Hydrogen-2	$^{2}_{1}H$	1	1	1	1	2	0.015
Hydrogen-3	$^{3}_{1}H$	1	1	1	2	3	[b]
Helium-3	$^{3}_{2}He$	2	2	2	1	3	0.000137
Helium-4	$^{4}_{2}He$	2	2	2	2	4	99.999863
Lithium-6	$^{6}_{3}Li$	3	3	3	3	6	7.5
Lithium-7	$^{7}_{3}Li$	3	3	3	4	7	92.5
Beryllium-9	$^{9}_{4}Be$	4	4	4	5	9	100
Boron-10	$^{10}_{5}B$	5	5	5	5	10	19.9
Boron-11	$^{11}_{5}B$	5	5	5	6	11	80.1
Carbon-12	$^{12}_{6}C$	6	6	6	6	12	98.90
Carbon-13	$^{13}_{6}C$	6	6	6	7	13	1.10
Carbon-14	$^{14}_{6}C$	6	6	6	8	14	[b]
Nitrogen-14	$^{14}_{7}N$	7	7	7	7	14	99.634
Nitrogen-15	$^{15}_{7}N$	7	7	7	8	15	0.366
Oxygen-16	$^{16}_{8}O$	8	8	8	8	16	99.762
Oxygen-17	$^{17}_{8}O$	8	8	8	9	17	0.038
Oxygen-18	$^{18}_{8}O$	8	8	8	10	18	0.200
Fluorine-19	$^{19}_{9}F$	9	9	9	10	19	100
Neon-20	$^{20}_{10}Ne$	10	10	10	10	20	90.48
Neon-21	$^{21}_{10}Ne$	10	10	10	11	21	0.27
Neon-22	$^{22}_{10}Ne$	10	10	10	12	22	9.25
Sodium-23	$^{23}_{11}Na$	11	11	11	12	23	100
Magnesium-24	$^{24}_{12}Mg$	12	12	12	12	24	78.99
Magnesium-25	$^{25}_{12}Mg$	12	12	12	13	25	10.00
Magnesium-26	$^{26}_{12}Mg$	12	12	12	14	26	11.01
Aluminum-27	$^{27}_{13}Al$	13	13	13	14	27	100
Silicon-28	$^{28}_{14}Si$	14	14	14	14	28	92.23
Silicon-29	$^{29}_{14}Si$	14	14	14	15	29	4.67

TABLE 5-3 (continued)

Name	Symbol	Atomic number	Protons	Electrons	Neutrons	Mass number	% Natural abundance[a]
Silicon-30	$^{30}_{14}Si$	14	14	14	16	30	3.10
Phosphorus-31	$^{31}_{15}P$	15	15	15	16	31	100
Sulfur-32	$^{32}_{16}S$	16	16	16	16	32	95.02
Sulfur-33	$^{33}_{16}S$	16	16	16	17	33	0.75
Sulfur-34	$^{34}_{16}S$	16	16	16	18	34	4.21
Sulfur-36	$^{36}_{16}S$	16	16	16	20	36	0.02
Chlorine-35	$^{35}_{17}Cl$	17	17	17	18	35	75.77
Chlorine-37	$^{37}_{17}Cl$	17	17	17	20	37	24.23
Argon-36	$^{36}_{18}Ar$	18	18	18	18	36	0.337
Argon-38	$^{38}_{18}Ar$	18	18	18	20	38	0.063
Argon-40	$^{40}_{18}Ar$	18	18	18	22	40	99.600
Potassium-39	$^{39}_{19}K$	19	19	19	20	39	93.2581
Potassium-40	$^{40}_{19}K$	19	19	19	21	40	0.0117
Potassium-41	$^{41}_{19}K$	19	19	19	22	41	6.7302
Calcium-40	$^{40}_{20}Ca$	20	20	20	20	40	96.941
Calcium-42	$^{42}_{20}Ca$	20	20	20	22	42	0.647
Calcium-43	$^{43}_{20}Ca$	20	20	20	23	43	0.135
Calcium-44	$^{44}_{20}Ca$	20	20	20	24	44	2.086
Calcium-46	$^{46}_{20}Ca$	20	20	20	26	46	0.004
Calcium-48	$^{48}_{20}Ca$	20	20	20	28	48	0.187

[a]The relative natural isotope abundance values are expressed as percentages of total number of atoms.
[b]Radioactive isotope with negligible abundance compared to other naturally occurring isotopes.

(Examples 5.3 and 5.4 can be solved without using Table 5-3.)

EXAMPLE 5.3

Boron is 19.9% boron-10 and 80.1% boron-11. Which of the following would you predict the atomic mass of boron to be?
(a) 10.0 u **(b)** Greater than 10.0 u but less than 10.5 u
(c) 10.5 u **(d)** Greater than 10.5 u but less than 11.0 u **(e)** 11.0 u

SOLUTION

The answer is (d). If boron were 100% boron-11, its atomic mass would be approximately 11.0 u. If boron were a 50 : 50 mixture of the two isotopes, its atomic mass would be approximately 10.5 u. However, there is about four times as much boron-11 as boron-10, so the atomic mass of boron is greater than 10.5 u but less than 11.0 u. In fact, the atomic mass of boron is 10.811.

EXAMPLE 5.4

Bromine is 50.69% bromine-79 and 49.31% bromine-81. Which of the following would you predict the atomic mass of bromine to be?
(a) 79 u **(b)** Slightly less than 80 u
(c) 80 u **(d)** Slightly more than 80 u **(e)** 81 u

SOLUTION

The answer is (b). This is very close to a 50 : 50 mixture, so we would expect the average mass to be near 80 u (the average of 79 u and 81 u). Because there are slightly more atoms of bromine-79, however, we would expect the atomic mass of bromine to be slightly less than 80 u (closer to 79 u than to 81 u). In accordance with our predictions, the atomic mass of bromine is 79.904 u.

Calculating the Atomic Mass of an Element

OBJECTIVE 5.4B Given the percentage abundances of the naturally occurring isotopes of an element and the mass of each isotope, calculate the atomic mass of the element.

In Examples 5.3 and 5.4, we estimated the atomic masses of boron and bromine. Let us revisit these elements to see how their atomic masses are actually calculated. In practice, the atomic mass of an element is obtained by multiplying the mass of each of the isotopes found in nature by the fraction of that isotope and then adding the results. In Example 5.4, we saw that bromine is composed of 50.69% bromine-79 and 49.31% bromine-81. To convert these percentages to decimal fractions, we simply divide each by 100. Thus, 50.69% equals 0.5069 and 49.31% equals 0.4931. The precise mass of each isotope must be determined experimentally. For bromine, the naturally occurring isotopes have the following masses and fractional compositions:

	Mass	Fraction
Bromine-79:	78.9183 u	0.5069
Bromine-81:	80.9163 u	0.4931

For each isotope, we multiply the precise mass by the fraction of that isotope. The atomic mass is obtained by adding up the contribution from each isotope as follows:

$$0.5069 \times 78.9183\ \text{u} = 40.00\ \text{u} \quad \leftarrow \textbf{Contribution from bromine-79}$$
$$0.4931 \times 80.9163\ \text{u} = 39.90\ \text{u} \quad \leftarrow \textbf{Contribution from bromine-81}$$
$$\text{Atomic mass} = 79.90\ \text{u}$$

This value agrees with the estimate made in Example 5.4 that the atomic mass of bromine is slightly less than 80 u. It also corresponds reasonably well to the value of 79.904 given in the periodic table.

EXAMPLE 5.5

Boron is composed of 19.9% boron-10 and 80.1% boron-11. The precise masses of these isotopes are 10.0129 u and 11.0093 u, respectively. Calculate the atomic mass of boron.

Planning the solution

To calculate the atomic mass, we multiply the mass of each isotope by its fraction. Once this has been done for all of the isotopes, we add them up. To convert a percentage to a decimal fraction, we simply divide by 100. Thus, 19.9% = 0.199 and 80.1% = 0.801. We may summarize the data as follows:

	Mass	Fraction
Boron-10:	10.0129 u	0.199
Boron-11:	11.0093 u	0.801

SOLUTION For each isotope, we multiply the precise mass by the fraction and then add the results:

$$0.199 \times 10.0129 \text{ u} = 1.99 \text{ u}$$
$$0.801 \times 11.0093 \text{ u} = 8.82 \text{ u}$$
$$\text{Atomic mass} = 10.81 \text{ u}$$

This value agrees well with the value of 10.811 given in the periodic table.

Before concluding this discussion, it is useful to review certain distinctions between atomic number, mass number, and atomic mass. First, it is important not to confuse the mass number with the atomic mass. A mass number represents the sum of the protons and neutrons of a specific isotope. As such, it is approximately equal to the mass of *that particular* isotope. But be careful not to confuse it with atomic mass, which represents the average mass per atom of a *collection* of isotopes.

Because the atomic number and atomic mass are both found in the periodic table, it is also important not to confuse these two terms. We can correctly identify these two numbers if we keep their definitions in mind. The atomic number represents the number of protons in the nucleus, so it makes no sense to have a fractional number for an atomic number. On the other hand, it is unlikely that the average mass of a collection of isotopes would be a whole number. Thus, in the case of carbon, the value 12.011 could not represent the number of protons, nor could the number of protons in a chlorine nucleus be 35.4527. These fractional quantities must be the atomic masses, and the whole numbers 6 and 17 must be the atomic numbers of these elements. These various terms are reviewed in the Summary.

SUMMARY — Distinctions Between Atomic Number, Mass Number, and Atomic Mass

Term	Definition	Can it refer to a single atom?	Can it refer to a collection of atoms?
Atomic number	Number of protons	Yes	Yes
Mass number	Sum of protons plus neutrons	Yes	No
Atomic mass	Average mass of the naturally occurring isotopes of an element	No	Yes

PROBLEM 5.6

Use Table 5-3 to answer each of the following questions.

(a) What is the approximate ratio of chlorine-35 atoms to chlorine-37 atoms?

(b) In naturally occurring chlorine, how many chlorine-35 atoms are there in every 100,000 atoms of chlorine?

(c) How many calcium-46 atoms are present in 1,000,000 atoms of calcium?

(d) Approximately how many nitrogen-15 atoms are present in every 300 atoms of nitrogen? What is the most abundant isotope of nitrogen? Does this explain its atomic mass of 14.0067?

(e) Identify five elements that are composed of only one naturally occurring isotope. For each of these, how does the atomic mass compare with the mass number of the isotope?

(f) What makes potassium-40 different from calcium-40?

(g) It is possible for isotopes of two different elements to have the same number of neutrons. Give five examples.

PROBLEM 5.7

Calculate the atomic mass of copper from the following data.

	Mass	Percentage
Copper-63:	62.9296 u	69.17%
Copper-65:	64.9278 u	30.83%

PROBLEM 5.8

Calculate the atomic mass of silver from the following data.

	Mass	Percentage
Silver-107:	106.9051 u	51.84%
Silver-109:	108.9048 u	48.16%

5.5 IONS

OBJECTIVE 5.5 Define the term *ion*. Describe the difference between a monatomic ion and a polyatomic ion. Write the symbol for any monatomic ion, given the number of protons and electrons; or given the symbol for any monatomic ion, determine the number of protons and electrons.

Ion: An atom or group of atoms bearing an electrical charge other than zero.

In our discussion so far, we have assumed the number of protons and the number of electrons of an atom to be equal. However, much of our study of chemistry involves chemical species known as ions. An **ion** is an atom (or group of atoms) that bears an electrical charge other than zero. For example, suppose a magnesium nucleus (which must have 12 protons) is surrounded by only 10 electrons. The total positive charge (+12) exceeds the total negative charge (−10), creating a net charge of +2.

$$\begin{matrix} 12(+1) & + & 10(-1) & = & +2 \\ \begin{pmatrix}\text{Nuclear} \\ \text{charge}\end{pmatrix} & + & \begin{pmatrix}\text{Electronic} \\ \text{charge}\end{pmatrix} & = & \begin{pmatrix}\text{Net electrical} \\ \text{charge}\end{pmatrix} \end{matrix}$$

Recalling that the defining characteristic of an element is the number of protons (the atomic number), we name this ion accordingly. We have been describing a magnesium ion. We write its formula as Mg^{2+}. Note that the plus-two charge is indicated in the upper right-hand corner.

It is possible for an ion to have a negative charge if the number of electrons exceeds the number of protons. Such is the case in a sulfide ion, which has 16 protons in the nucleus and 18 electrons surrounding the nucleus:

$$\begin{matrix} 16(+1) & + & 18(-1) & = & -2 \\ \begin{pmatrix}\text{Nuclear} \\ \text{charge}\end{pmatrix} & + & \begin{pmatrix}\text{Electronic} \\ \text{charge}\end{pmatrix} & = & \begin{pmatrix}\text{Net electrical} \\ \text{charge}\end{pmatrix} \end{matrix}$$

We represent the sulfide ion as S^{2-}. Again, the charge is shown in the upper right-hand corner.

Cation: A positively charged ion.

Anion: A negatively charged ion.

The formulas of several ions are shown in Table 5-4. Note that, in every case, we show both the sign and the magnitude of the charge. If the charge in question is either +1 or −1, this is designated by a simple + or – sign. A positively charged ion (such as Na^+ or Mg^{2+}) is known as a **cation** (pronounced CAT′-ION); a negatively charged ion (such as F^- or S^{2-}) is called an **anion** (pronounced AN′-ION). The Summary Diagram on the next page indicates how cations and anions are formed. The names of the cations listed in Table 5-4 are the same as the names of the elements from which they are derived. The anions have names that are based on the names of their parent elements, but they take an *-ide* ending.

TABLE 5-4 Some simple ions

Name	Formula	Protons	Electrons
Hydrogen	H^+	1	0
Lithium	Li^+	3	2
Sodium	Na^+	11	10
Potassium	K^+	19	18
Magnesium	Mg^{2+}	12	10
Calcium	Ca^{2+}	20	18
Strontium	Sr^{2+}	38	36
Barium	Ba^{2+}	56	54
Aluminum	Al^{3+}	13	10
Hydride	H^-	1	2
Fluoride	F^-	9	10
Chloride	Cl^-	17	18
Bromide	Br^-	35	36
Iodide	I^-	53	54
Oxide	O^{2-}	8	10
Sulfide	S^{2-}	16	18

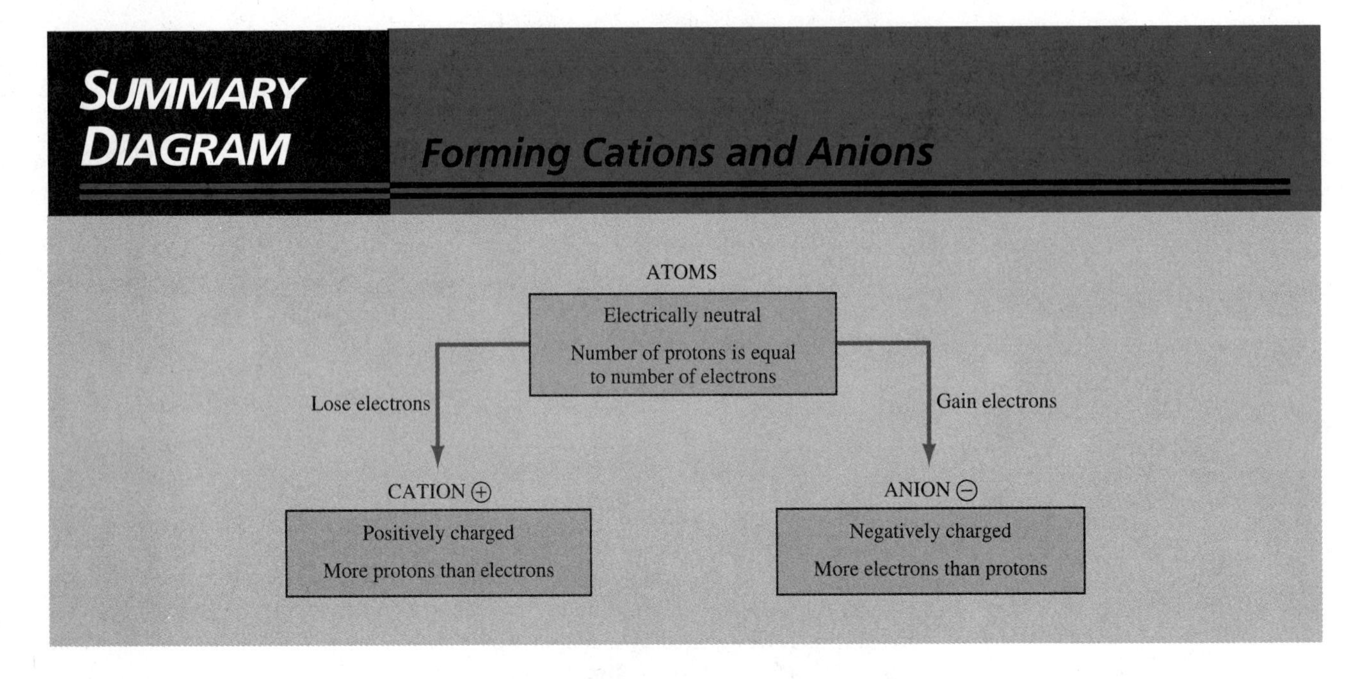

In Chapters 6–8, we discuss the reasons a neutral atom gains or loses electrons to become an ion. For the moment, we simply observe that atoms that gain electrons obtain those additional electrons from atoms that lose them, in a process called *electron transfer.*

$$\underset{\text{Neutral atom}}{\underset{\begin{pmatrix}12\ p^+\\12\ e^-\end{pmatrix}}{Mg}} + \underset{\text{Neutral atom}}{\underset{\begin{pmatrix}16\ p^+\\16\ e^-\end{pmatrix}}{S}} \rightarrow \underset{\text{Cation}}{\underset{\begin{pmatrix}12\ p^+\\10\ e^-\end{pmatrix}}{Mg^{2+}}} + \underset{\text{Anion}}{\underset{\begin{pmatrix}16\ p^+\\18\ e^-\end{pmatrix}}{S^{2-}}}$$

Monatomic ion: An ion composed of one atom.

Polyatomic ion: An ion composed of more than one atom.

The ions we have just discussed result from only one atom, so they are called **monatomic ions.** There is another category of ions that are composed of more than one atom. These ions are known as **polyatomic ions,** because they are composed of two or more atoms. Some of the more important polyatomic ions are listed in Table 5-5.

TABLE 5-5 Some polyatomic ions

Name	Formula
Hydroxide	OH^-
Cyanide	CN^-
Nitrate	NO_3^-
Sulfate	SO_4^{2-}
Carbonate	CO_3^{2-}
Phosphate	PO_4^{3-}
Ammonium	NH_4^+

The formula for cyanide, CN^-, a polyatomic anion, tells us that a carbon atom and a nitrogen atom are held together and have gained an extra electron, giving this ion a −1 charge.

$$C + N + e^- = CN^-$$

Similarly, the formula for sulfate, SO_4^{2-}, tells us that a sulfur atom and four oxygen atoms are bonded together, having also gained two extra electrons.

$$S + 4\,O + 2\,e^- = SO_4^{2-}$$

Polyatomic cations also exist. The ammonium ion, NH_4^+, is made up of a nitrogen atom and four hydrogen atoms, but these atoms have managed to lose one electron, leaving the polyatomic ion with a net charge of +1.

$$N + 4\,H - e^- = NH_4^+$$

In Chapter 8 we discuss chemical bonding, the "glue" that holds atoms together in chemical compounds. The same type of bonding that holds the atoms of a water molecule together is responsible for holding the atoms together in a polyatomic ion. However, a polyatomic ion is unlike a water molecule, which has equal numbers of protons and electrons and so is uncharged. *Polyatomic ions have unequal numbers of protons and electrons, leaving them with a net charge.*

EXAMPLE 5.6

How many protons and electrons are in a Ca^{2+} ion?

SOLUTION

Calcium has atomic number 20, so there are 20 protons. Since the charge is +2, there are two *less* electrons than protons, or there are 18 electrons:

$$20(+1) + 18(-1) = +2$$

Answer

20 protons, 18 electrons

EXAMPLE 5.7

Write a symbol for the ion having 53 protons and 54 electrons.

SOLUTION

Iodine (atomic number = 53) has 53 protons. Since there is one *more* electron than the number of protons, the charge is −1.

$$53(+1) + 54(-1) = -1$$

Answer

The formula of this ion is I^-.

PROBLEM 5.9

Determine the number of protons and the number of electrons in each of the following monatomic ions.

(a) Ag^+ **(b)** Fe^{3+} **(c)** Zn^{2+} **(d)** Sn^{4+} **(e)** Se^{2-} **(f)** P^{3-}

PROBLEM 5.10

Write a symbol for each of the following monatomic ions. Part (a) is done for you.

(a) 80 protons, 78 electrons Hg^{2+}

(b) 29 protons, 28 electrons ______

(c) 37 protons, 36 electrons ______

(d) 28 protons, 26 electrons ______

(e) 79 protons, 76 electrons ______

(f) 52 protons, 54 electrons ______

CHAPTER SUMMARY

Much of what we accept today about the nature of **atoms** was put forth in John Dalton's atomic theory, which proposed that matter is composed of indivisible atoms that combine in whole-number ratios to form compounds. Dalton also suggested that all atoms of a particular kind have the same mass and that, when they combine, each remains unchanged.

Later researchers demonstrated that atoms are divisible, being made up of positively and negatively charged particles. Thomson's plum-pudding model proposed that matter is composed of a positively charged "pudding" in which negatively charged electrons are embedded. Rutherford demonstrated that an atom is actually composed of a small but very dense, positively charged **nucleus** surrounded by a volume in which negatively charged **electrons** can be found. Compared to its effective volume, most of an atom is empty space. However, the mass of an atom is almost entirely located in the nucleus. Chadwick later discovered that the nucleus itself is made up of both positively charged **protons** and uncharged **neutrons.** Thus, atoms are composed of protons, neutrons, and electrons, which are known as **fundamental particles** and have the following charges and approximate masses:

Proton	+1	1 u	Neutron	0	1 u
Electron	−1	0 u			

The defining characteristic of an atom is its **atomic number,** which gives the number of protons in the nucleus. Not all atoms of the same element have the same number of neutrons in the nucleus. Thus, they may differ in **mass number,** which is the sum of the protons and neutrons. **Isotopes** are atoms of the same element that differ in mass number.

It is important to distinguish mass number from atomic mass. An **atomic mass** represents the average mass of the isotopes of an element as they occur naturally, compared to the mass of one carbon-12 atom, which is assigned a mass of exactly 12 u. (Thus, an **atomic mass unit** is exactly one-twelfth the mass of a carbon-12 atom.)

When the number of electrons in an atom equals the number of protons in the nucleus, the atom has no charge (it is electrically neutral). However, if these two are not equal, there is a net charge and the resulting species is known as an **ion.** When the ion bears a positive charge, it is known as a **cation.** When the ion is negatively charged, it is known as an **anion.** Ions containing only one atom are known as **monatomic ions. Polyatomic ions** are those in which a group of atoms bonded together bears a net charge.

KEY TERMS

Review each of the following terms, which are discussed in this chapter and defined in the Glossary.

atom	electron	isotope	cation
nucleus	fundamental particle	atomic mass unit	anion
proton	atomic number	atomic mass	monatomic ion
neutron	mass number	ion	polyatomic ion

ADDITIONAL PROBLEMS

SECTION 5.1

5.11 State two ideas from Dalton's atomic theory that are still accepted today; state two ideas from his theory that had to be modified.

5.12 Explain Rutherford's major contribution to our understanding of atomic structure. Compare the relative size of the nucleus to that of the atom as a whole. Compare the mass of the nucleus to that of the atom as a whole.

5.13 List the three fundamental particles that compose atoms, and describe each in terms of charge and approximate mass.

5.14

(a) What is the approximate mass of an atom composed of 6 protons, 6 neutrons, and 6 electrons?

(b) What is the approximate mass of an atom composed of 6 protons, 8 neutrons, and 6 electrons?

(c) What is the approximate mass of an atom composed of 7 protons, 7 neutrons, and 7 electrons?

(d) Which two of these atoms have the same nuclear charge?

SECTIONS 5.2 AND 5.3

5.15 Define each of the following:
(a) Atomic number **(b)** Mass number **(c)** Isotope

5.16 Write the name and nuclear symbol for each of the following nuclei. Part (a) is done for you. Which of these nuclei are isotopes of one another?

Nuclear composition	Name	Nuclear symbol
(a) 3 protons, 4 neutrons	Lithium-7	${}^{7}_{3}Li$
(b) 6 protons, 8 neutrons	______	______
(c) 9 protons, 10 neutrons	______	______
(d) 16 protons, 17 neutrons	______	______
(e) 26 protons, 30 neutrons	______	______
(f) 94 protons, 144 neutrons	______	______
(g) 26 protons, 32 neutrons	______	______
(h) 16 protons, 15 neutrons	______	______

5.17 Fill in the chart shown at the bottom of this page. Assume that each line refers to an uncharged atom. Use the tables provided inside the front and back covers of this book for reference.

SECTION 5.4

Do not use a periodic table or any other table of atomic masses for Problems 5.18–5.20.

5.18 Naturally occurring nitrogen is 99.634% nitrogen-14 and 0.366% nitrogen-15. Which of the following would you predict the atomic mass of nitrogen to be?
(a) Slightly less than 14.0 u
(b) Slightly more than 14.0 u
(c) 14.5 u
(d) Slightly less than 15.0 u

5.19 Naturally occurring chlorine is 75.77% chlorine-35 and 24.23% chlorine-37. Which of the following would you predict the atomic mass of chlorine to be?
(a) Between 35.0 u and 36.0 u
(b) Between 36.0 u and 37.0 u
(c) 36.0 u
(d) None of these

5.20 The relative abundances of naturally occurring magnesium are magnesium-24, 78.99%; magnesium-25,

Name	Symbol	Atomic number	Number of protons	Number of electrons	Number of neutrons	Mass number
______	B	______	______	______	6	______
______	______	8	______	______	______	17
______	______	______	17	______	18	______
______	P	______	______	______	16	______
Calcium	______	______	______	______	______	48
______	______	______	50	______	69	______
Polonium	______	______	______	______	______	210
______	______	86	______	______	136	______
______	______	______	102	______	______	254

10.00%; and magnesium-26, 11.01%. Which of the following would you predict the atomic mass of magnesium to be?
(a) Between 24.0 u and 25.0 u
(b) Between 25.0 u and 26.0 u
(c) 25.0 u
(d) None of these

5.21 Gallium is an element with properties similar to those of aluminum. Its melting point of 30°C and boiling point of 2403°C give it the widest liquid temperature range of any element. Use the data below to calculate the atomic mass of gallium.

	Mass	Percentage
Gallium-69:	68.9256 u	60.108%
Gallium-71:	70.9247 u	39.892%

5.22 Rubidium is a soft silvery white metallic element that reacts readily with oxygen and violently with water. Calculate the atomic mass of rubidium from the following data.

	Mass	Percentage
Rubidium-85:	84.9118 u	72.165%
Rubidium-87:	86.9092 u	27.835%

SECTION 5.5

5.23 Determine the number of protons and electrons in each of the following monatomic ions.
(a) Rb^+ **(b)** Co^{3+} **(c)** Hg^{2+}
(d) Pb^{4+} **(e)** Te^{2-} **(f)** N^{3-}

5.24 Write a symbol for each of the following monatomic ions. Part (a) is done for you.

(a) 30 protons, 28 electrons	Zn^{2+}
(b) 24 protons, 21 electrons	______
(c) 31 protons, 28 electrons	______
(d) 55 protons, 54 electrons	______
(e) 78 protons, 74 electrons	______
(f) 51 protons, 54 electrons	______

5.25 Define the term *ion*. Explain the difference between a monatomic ion and a polyatomic ion.

GENERAL PROBLEMS

***5.26** The diameter of a uranium-238 atom is approximately 10^{-10} m, and its nucleus is roughly 10^{-14} m in diameter. If a marble is used to model the nucleus of the atom, what would be a good model for the atom as a whole? The diameter of a marble is roughly 2 cm. [*Hint:* Calculate the number of yards required for the diameter of the model to be to scale with the nucleus, and think of a structure that would be roughly that size.]

5.27 In biological research, compounds containing phosphorus (such as DNA and ATP) can be enriched with phosphorus-32, a radioactive isotope of phosphorus. How many neutrons are present in this isotope?

***5.28** Dibromomethane, CH_2Br_2, is used as a pesticide in the fumigation of buildings. In Section 5.4, we saw that bromine is very nearly a 50 : 50 mixture of bromine-79 and bromine-81. Assuming that all of the hydrogen is hydrogen-1 and all of the carbon is carbon-12, determine the *approximate* masses of the various molecules found in a sample of dibromomethane and their approximate percentages of each.

***5.29** *Heavy water* is water in which the hydrogen atoms are all deuterium (hydrogen-2). Chemists often use D as the symbol for deuterium and write heavy water as D_2O. How much would a 130-lb woman weigh if all of the H_2O in her body were replaced by D_2O? Assume her body is 70% (0.70) water by mass.

5.30 Determine the number of protons and electrons for each of the following ions.
(a) K^+ **(b)** Cl^- **(c)** S^{2-} **(d)** Ca^{2+}
(e) What do all of these have in common?

5.31 Iron is found in hemoglobin, a complex molecule used to carry oxygen in the bloodstream. The iron in hemoglobin is found to exist with a charge of either +2 or +3, depending upon whether it has oxygen attached. If the iron in a specific hemoglobin molecule is derived from an iron-56 atom and exhibits a +2 charge, what is the number of protons, neutrons, and electrons in this iron atom?

5.32 Determine the number of protons, neutrons, and electrons for each of the following chemical species.
(a) ${}^{41}_{19}K^+$ **(b)** ${}^{32}_{16}S^{2-}$ **(c)** ${}^{27}_{13}Al^{3+}$ **(d)** ${}^{127}_{53}I^-$

***5.33** Write symbols for each of the following species. Place the atomic numbers, mass numbers, and charges (if any), as you did in Sections 5.3 and 5.5.
(a) 26 p^+, 30 n, 23 e^- **(b)** 27 p^+, 32 n, 27 e^-
(c) 26 p^+, 30 n, 24 e^- **(d)** 27 p^+, 31 n, 24 e^-
(e) Which two are isotopes of each other?
(f) Which two are ions of the same atom?
(g) Which two have the same charge?
(h) Which is not an ion?

5.34 Ionic compounds form from the simplest whole-number ratio of cations and anions that results in a zero charge. What ratio of ions is required for each of the following combinations?
(a) Na^+ and Br^- **(b)** Ca^{2+} and O^{2-}
(c) Fe^{3+} and Cl^- **(d)** Mg^{2+} and N^{3-}

5.35 Classify each of the following molecules or ions as either (i) an element, (ii) a compound, (iii) a monatomic ion, or (iv) a polyatomic ion. (By itself, a monatomic ion is not considered an element, nor is a polyatomic ion considered a compound.)
(a) Br^- **(b)** Br_2 **(c)** SO_3 **(d)** SO_3^{2-}
(e) O^{2-} **(f)** O_2^{2-} **(g)** CO **(h)** Co

Writing Exercises

5.36 A blind friend has asked you to describe the structure of an atom. Since you cannot use drawings to illustrate your description, you have decided to write out your explanation and then read it aloud. Write a description of the atom, including such things as the relative sizes of the nucleus and the atom as a whole, the locations and masses of the various fundamental particles, and the charge characteristics of the various components of the atom.

5.37 A friend has seen a story in the newspaper about the dangers of radon in homes. The article mentioned that the offending species is an isotope known as radon-222. Your friend wishes to gain a better understanding of the problem and has asked you to explain what an isotope is and what radon-222 is. Write a note to your friend, explaining the meaning of isotopes, and illustrate this definition by describing the composition of an atom of radon-222. Your explanation should include definitions of atomic number and mass number.

5.38 Astronauts have landed on a distant planet where the composition of the elements differs from that on earth. As a member of the monitoring team on earth, your assignment includes the task of developing a periodic table for this newly discovered planet. You plan to delegate the task of computing the various atomic masses to one of your subordinates. One of the astronauts has just sent you data on the masses and abundances of the various isotopes of the first element investigated. Write a memorandum to your subordinate, explaining what an atomic mass is and how to calculate the atomic mass of the element from the data you have just received.

6 Electronic Configuration

In Chapter 5 we examined the structure of the atom, concentrating primarily on the composition of the nucleus. In that discussion, we carefully avoided any detailed description of the location of the electrons, saying only that they are located outside of the nucleus. In this chapter, we look at the organization of electrons within atoms, using the results of quantum mechanics, a very sophisticated branch of physics. Although the mathematics of quantum mechanics is quite advanced, it is not difficult to use its results to describe the locations of electrons within the various atoms we encounter.

Important discoveries that took place during the early portion of this century revealed that the electrons in atoms exist in certain *energy levels.* Associated with the various energy levels are regions of space, known as *orbitals,* in which the electrons may be found. Much of what we examine in this chapter is a description of the various orbitals and of their energies, as well as a bookkeeping method, known as *electronic configuration,* for cataloging the locations of the electrons in the various orbitals.

The chapter is divided into two portions. In the first portion of the chapter we present a relatively simple description of the various orbitals and their energies. The sections appearing at the end of the chapter present some of the experimental evidence that led to the development of quantum mechanics.

6.1 ELECTRONIC ENERGY LEVELS

OBJECTIVE 6.1 Describe what is meant by the principal energy levels of electrons in atoms. Explain the meaning of *ground state* and *excited state.* Tell what is meant by a sublevel. Determine the number of sublevels within a given principal energy level.

Imagine yourself at the top of a hill, about to roll a ball down the hill. The ball has more potential energy than it will have when it gets to the bottom of the hill. If you allow the ball to roll downhill, it will roll smoothly down the hill. As it descends, it loses potential energy in a continuous fashion until it reaches the bottom of the hill (Fig. 6-1a, p. 136). If we wish, we can stop the ball at any point along the hill. It is capable of existing at any potential energy level between the top of the hill and the bottom of the hill. If we allow the ball to roll to the bottom of the hill, it will be at its lowest possible energy state. We refer to this as its *ground state.*

On the other hand, imagine yourself rolling the same ball down a set of steps that runs next to the hill (Fig. 6-1b). In this case the ball descends step by step, losing potential energy in increments. As it descends, the ball cannot be stopped in between steps but may only rest at certain *discrete* energy levels, which correspond to the various steps. Thus, the increments of potential energy that are lost correspond to the potential energy differences between the various steps. If allowed to continue its downward path, the ball will eventually descend to the bottom of the hill, as in the previous situation.

The electrons in atoms exist in discrete energy levels, much like the ball on the set of steps. However, unlike a set of steps, where the spacing between levels is roughly equal, the spacing between electronic energy levels is varied. Figure 6-2 pictures several of the electronic energy levels of a hydrogen atom. The various levels are referred to as **principal energy levels.** (An older term that you may come across is *shell.* A shell

Principal energy level: One of the discrete electronic energy levels, *n*.

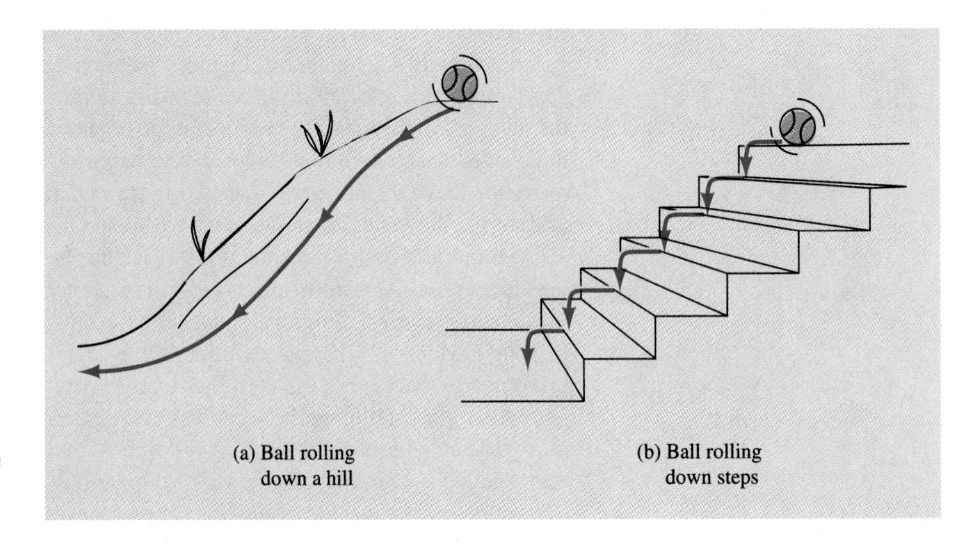

FIGURE 6-1 **(a) A ball rolling down a hill makes continuous changes in energy. (b) A ball rolling down a set of steps can rest only at certain discrete energy levels.**

is the same as a principal energy level.) The principal energy levels are numbered from the lowest level upward. We use the letter n to refer to a particular principal energy level, starting with $n = 1$ for the lowest level. Note that as n increases, the spacing between levels decreases. As a practical matter, we will be concerned with the first seven levels only. A hydrogen atom has only one electron, and when the electron is in the first principal energy level ($n = 1$), hydrogen is in its **ground state.** If the electron is not in the lowest possible energy state, it is said to be in an **excited state.**

Ground state: The lowest possible energy state of the electrons in an atom.

Excited state: An electronic energy state other than the ground state.

If we examine the various levels more carefully, we find that within each level, there may exist one or more **sublevels** (also known by the older term *subshell*). It turns out that for the first principal energy level there is only one sublevel, but for the second principal energy level there are two sublevels, for the third principal energy level there are three sublevels, and so forth. The first four of the various sublevels are designated *s*,

Sublevel: One of the subdivisions of a principal energy level; usually *s, p, d,* or *f.*

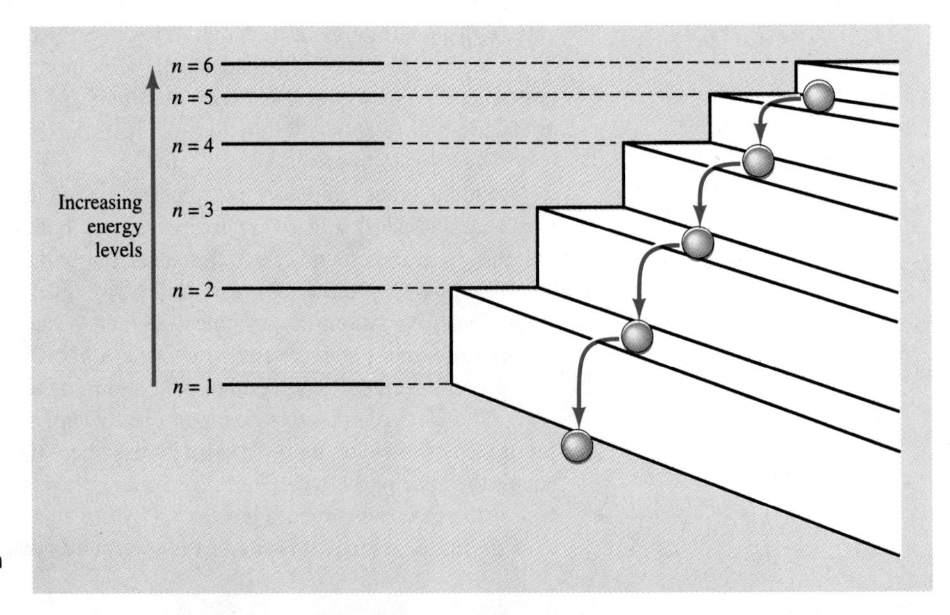

FIGURE 6-2 **Electronic energy levels within a hydrogen atom. Like a ball on a set of steps, an electron can exist only at certain energy levels and not in between.**

TABLE 6-1 Principal energy levels and their sublevels

Principal level	Sublevel
$n = 1$	1*s*
$n = 2$	2*s*, 2*p*
$n = 3$	3*s*, 3*p*, 3*d*
$n = 4$	4*s*, 4*p*, 4*d*, 4*f*
$n = 5$	5*s*, 5*p*, 5*d*, 5*f*, 5*g*
$n = 6$	6*s*, 6*p*, 6*d*, 6*f*, 6*g*, 6*h*
$n = 7$	7*s*, 7*p*, 7*d*, 7*f*, 7*g*, 7*h*, 7*i*

p, d, and *f.* Since each sublevel is in a particular principal energy level, it is designated by both a number and a letter (Table 6-1). Thus, for $n = 1$, the only possible sublevel is 1*s*. For $n = 2$, the two possible sublevels are 2*s* and 2*p;* for $n = 3$, there are the 3*s,* 3*p,* and 3*d* levels; and so forth. (Beyond the first four sublevels, lettering is alphabetical and continuous: *g, h, i,* and so forth.)

The description of the electronic energy levels in hydrogen represents a particularly simple situation, because there is only one electron in the atom. Furthermore, all of the sublevels in a given principal energy level have the same energy. However, the situation becomes slightly more complicated for elements that have more than one electron (which is everything *except* hydrogen). In these cases, the energies of the sublevels within each principal level are in the order $s < p < d < f$. Thus, for $n = 3$, the 3*s* sublevel is lower in energy than the 3*p* sublevel, and the 3*p* sublevel is below the 3*d* sublevel. Imagine a housing complex on a hill (Fig. 6-3) in which there is a one-story

FIGURE 6-3 Like the stories of a house, each principal energy level may be divided into sublevels.

building at the bottom of the hill, a two-story building a little way up the hill, a three-story building farther up, and buildings with increasing numbers of stories as we continue on up the hill. The buildings themselves represent the principal energy levels, and the stories within each building represent the sublevels.

PROBLEM 6.1

Indicate whether each of the following statements is true (T) or false (F).

(a) A ground state represents the lowest possible energy state.
(b) The principal energy level having $n = 1$ represents the highest possible energy state.
(c) The $5p$ sublevel is lower in energy than the $5f$ sublevel.
(d) There is no $2d$ sublevel.
(e) The lowest possible principal energy level is $n = 0$.
(f) The following sublevels are correctly arranged in order of increasing energy:

$$4s < 4p < 4d < 4f$$

6.2 ATOMIC ORBITALS

OBJECTIVE 6.2 Describe an orbital. Explain how the behavior of an electron in an orbital differs from the behavior of macroscopic particles. State the number of orbitals in the various types of sublevels (*s, p, d,* and so forth).

The size of an electron is so small that its behavior does not correspond to the usual behavior we associate with objects we are used to encountering. For example, when the moon orbits the earth, it travels in a path that we feel confident we can describe. Because of its motion and its position relative to the earth, the moon has a certain energy associated with it. Like the moon, an electron has energy (one of the energy levels we discussed in the preceding section). However, the motion of an electron is not readily described.

Our best theories about electronic behavior allow us to describe only the *probability* of finding an electron in a given region of space. The electrons within atoms do not exist in any fixed location or well-defined path, as in the case of the moon orbiting the earth. Instead, electrons occupy indefinite locations within regions surrounding the nucleus, called **orbitals.** An orbital defines a volume in which we are most likely to find an electron. Quantum mechanics tells us that the *exact* location of electrons within the orbitals cannot be known, nor is it important.

Orbital: Energy state describing a region of space within which one or two electrons may be found.

We often use models or pictures such as those shown in Fig. 6-4 to describe the shapes of the orbitals, likening the location of an electron in an orbital to a cloud. Note that the various designations assigned to the orbitals shown in Fig. 6-4 correspond to the various energy levels we examined in Section 6.1. *Each orbital represents an energy state and can hold two electrons.* The two electrons that fill an orbital differ in a property known as *electron spin.* We represent electrons of opposite spin with the symbols ↿ and ⇂ (Fig. 6-5, p. 140).

Examination of the orbitals shown in Fig. 6-4 reveals several important principles. Note that the $1s$ orbital is smaller than the $2s$ orbital, which, in turn, is smaller than the $3s$ orbital. As the principal energy level increases, so does the size of the orbital. In

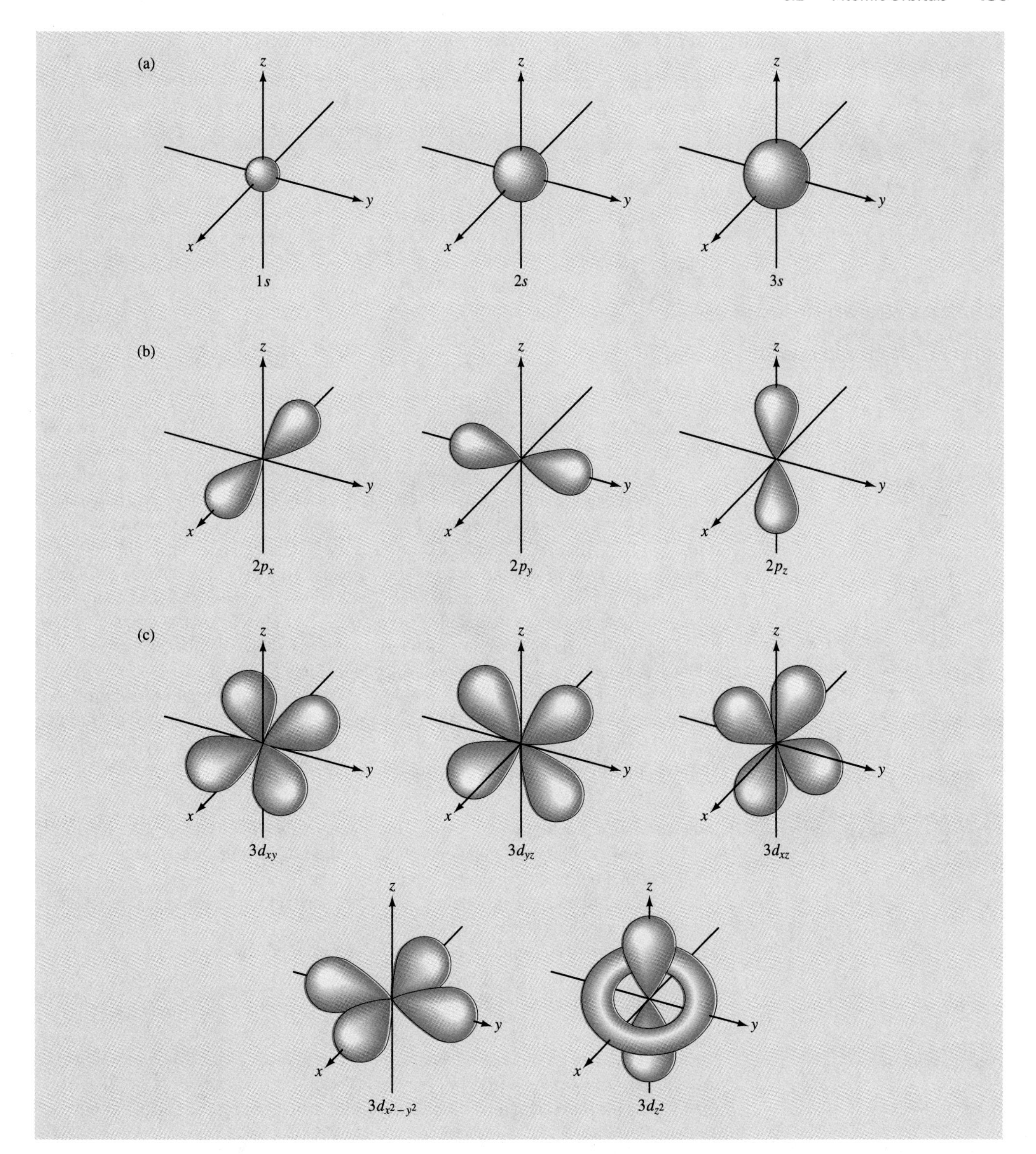

FIGURE 6-4 The shapes of representative orbitals. (a) The 1*s*, 2*s*, and 3*s* orbitals. The size of orbitals increases with increasing principal energy level. (There is only one orbital in any *s* sublevel.) (b) The three 2*p* orbitals. (c) The five 3*d* orbitals.

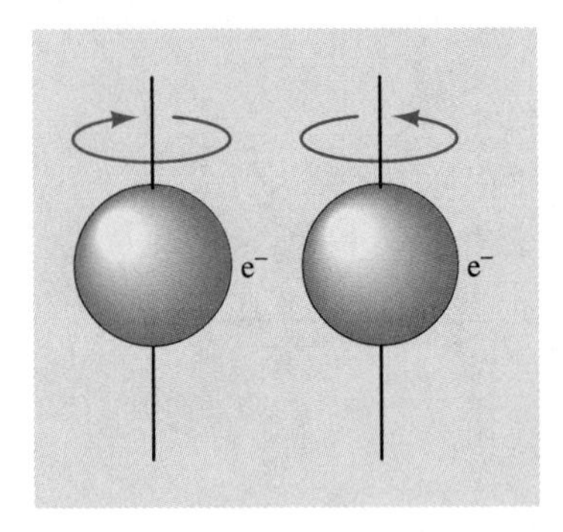

FIGURE 6-5 Like a child's top, an electron can spin in either of two directions, represented as ↑ and ↓.

TABLE 6-2 Principal energy levels, sublevels, and orbitals

Principal energy level (n)	Sublevels and their orbitals			
1	○ 1s			
2	○ 2s	○○○ 2p		
3	○ 3s	○○○ 3p	○○○○○ 3d	
4	○ 4s	○○○ 4p	○○○○○ 4d	○○○○○○○ 4f

other words, the probability of finding the electron farther from the nucleus increases with increasing principal energy level.

Figure 6-4 also reveals a general pattern about the number of orbitals in a given sublevel. We can see that only one 1s orbital is shown. (Similarly, there is only one 2s orbital and one 3s orbital.) However, there are three 2p orbitals and five 3d orbitals. The three 2p orbitals are all of equal energy, but each is oriented along a different axis. Likewise, the five 3d orbitals are of equal energy, although their shapes and spatial orientations also differ from one another. The various s sublevels always contain one orbital, p sublevels always contain three orbitals, d sublevels always contain five orbitals, and, although they are not pictured here, the f sublevels always contain seven orbitals. A comprehensive pattern is displayed in Table 6-2.

Recalling that each orbital can hold 2 electrons, the maximum number of electrons that can occupy an s sublevel is 2, for a p sublevel it is 6 electrons, for a d sublevel it is 10 electrons, and for an f sublevel it is 14 electrons. The Summary lists what we know about principal energy levels, sublevels, orbitals, and electron capacities.

PROBLEM 6.2

Indicate whether each of the following statements is true (T) or false (F).

(a) Electrons circulate about the nucleus in fixed orbits having specific radii.
(b) An orbital represents an energy state.
(c) An orbital represents an indefinite location, surrounding the nucleus, in which an electron may be found.
(d) The quantum-mechanical model tells us precisely where an electron may be found at any instant.
(e) An orbital can hold an indefinite number of electrons.

PROBLEM 6.3

What is the maximum number of electrons that can fill each of the following sublevels?
(a) 5d **(b)** 7s **(c)** 6p **(d)** 4f **(e)** 6h **(f)** 5g
(g) What is the maximum total number of electrons that can fill the fourth principal energy level?

SUMMARY Principal Energy Levels, Sublevels, Orbitals, and Electron Capacities

1. *Principal energy levels:*
   These are represented by integers and increase in energy as n increases: $n = 1, 2, 3, \ldots$
2. *Sublevels:*
   (a) The number of sublevels is the same as the integer associated with the principal energy level.
   (b) For a given principal energy level, the energies of the sublevels are (from lowest to highest) $s < p < d < f$. For example, $4s < 4p < 4d < 4f$.
3. *Orbitals:*
   (a) The number of orbitals in each *type* of sublevel increases in the following pattern: one *s* orbital, three *p* orbitals, five *d* orbitals, seven *f* orbitals.
   (b) An orbital can hold two electrons, which are represented as ↿ and ⇂.

Principal energy level	Number of orbitals in each sublevel							Number of orbitals in principal energy level
	s	*p*	*d*	*f*	*g*	*h*	*i*	
$n = 1$	1							1
$n = 2$	1	3						4
$n = 3$	1	3	5					9
$n = 4$	1	3	5	7				16
$n = 5$	1	3	5	7	9[a]			25
$n = 6$	1	3	5	7[a]	9[a]	11[a]		36
$n = 7$	1	3[a]	5[a]	7[a]	9[a]	11[a]	13[a]	49

[a]No known element has ground-state electrons in this sublevel.

Principal energy level	Maximum number of electrons in each sublevel							Maximum number of electrons in principal energy level
	s	*p*	*d*	*f*	*g*	*h*	*i*	
n = 1	2							2
n = 2	2	6						8
n = 3	2	6	10					18
n = 4	2	6	10	14				32
n = 5	2	6	10	14	18[b]			50[c]
n = 6	2	6	10	14[b]	18[b]	22[b]		72[c]
n = 7	2	6[b]	10[b]	14[b]	18[b]	22[b]	26[b]	98[c]

[b]No known element has electrons in this sublevel.
[c]No known element has filled this principal energy level.

6.3 AN ELECTRONIC ENERGY-LEVEL DIAGRAM

OBJECTIVE 6.3 State the Aufbau principle. List the order in which the electronic sublevels fill, starting with 1*s* and ending with 6*d*.

Aufbau principle: Electrons fill the lowest available energy levels first.

In the next section we will see how the electrons of atoms are arranged in the various orbitals we have been discussing. First, let us examine the order in which electrons fill these orbitals. It should make sense that electrons fill the orbitals of lowest energy first, and as the lower-level orbitals fill, the electrons then fill orbitals of progressively higher energy. This principle of electrons filling the orbitals in order of increasing energy is known as the **Aufbau principle.** Applying this principle, it is no surprise to find the lone electron of a hydrogen atom in a 1*s* orbital, since it is the lowest-energy orbital available. Before we apply the Aufbau principle to other elements, let us take a careful look at the order in which the orbitals are filled.

Figure 6-6a combines the information discussed in Sections 6.1 and 6.2, showing the relative energies of the various sublevels of each principal energy level and the orbitals in each sublevel. (Some of the higher sublevels have been omitted for principal levels 5, 6, and 7, since we are not concerned with them.) Figure 6-6a is organized with the sublevels of each principal energy level grouped vertically.

Figure 6-6b reorganizes the information in Fig. 6-6a so that all of the *s* sublevels are in the same vertical column, and similarly for the *p, d,* and *f* sublevels. Figure 6-6b shows the relative energies of the various sublevels as they become filled with electrons. The arrows show the order in which the various sublevels fill. Note that this is the same as the order of increasing energy. (The rectangles placed around the various groups of sublevels are used to relate electronic configuration to the periodic table in Chapter 7.)

You should spend a few minutes examining Fig. 6-6 *carefully.* Let us review what you should find there, starting with part (a). Note that for each principal energy level, the sublevels fall in the order $s < p < d < f$. Next, note that for each type of sublevel (for example, the *s* sublevels), the energy increases with increasing principal energy level ($1s < 2s < 3s$, and so forth). Finally, note that *s* sublevels contain one orbital, *p* sublevels contain three orbitals, *d* sublevels five orbitals, and *f* sublevels seven orbitals.

Now let us turn our attention to part (b). Part (b) represents the relative energies as electrons fill the various sublevels. Note that the 4*s* sublevel lies below the 3*d* sublevel and therefore fills first. You might wonder how this is possible. Recalling our housing-complex analogy, imagine that you live up the hill (have a higher principal energy level) from a neighboring building (Fig. 6-7, p. 144). Your apartment is on the first floor (an *s* sublevel) of your building, and as you look out of your window, you see that you are below the third story (*d* sublevel) of the neighboring (downhill) building. Although your building (the principal level) has a higher energy level, your particular story (the sublevel) is below one of the stories of the neighboring building. As you examine Fig. 6-6b, you can find other examples of this phenomenon. For example, the 6*s* sublevel lies below the 4*f* and the 5*d* sublevels.

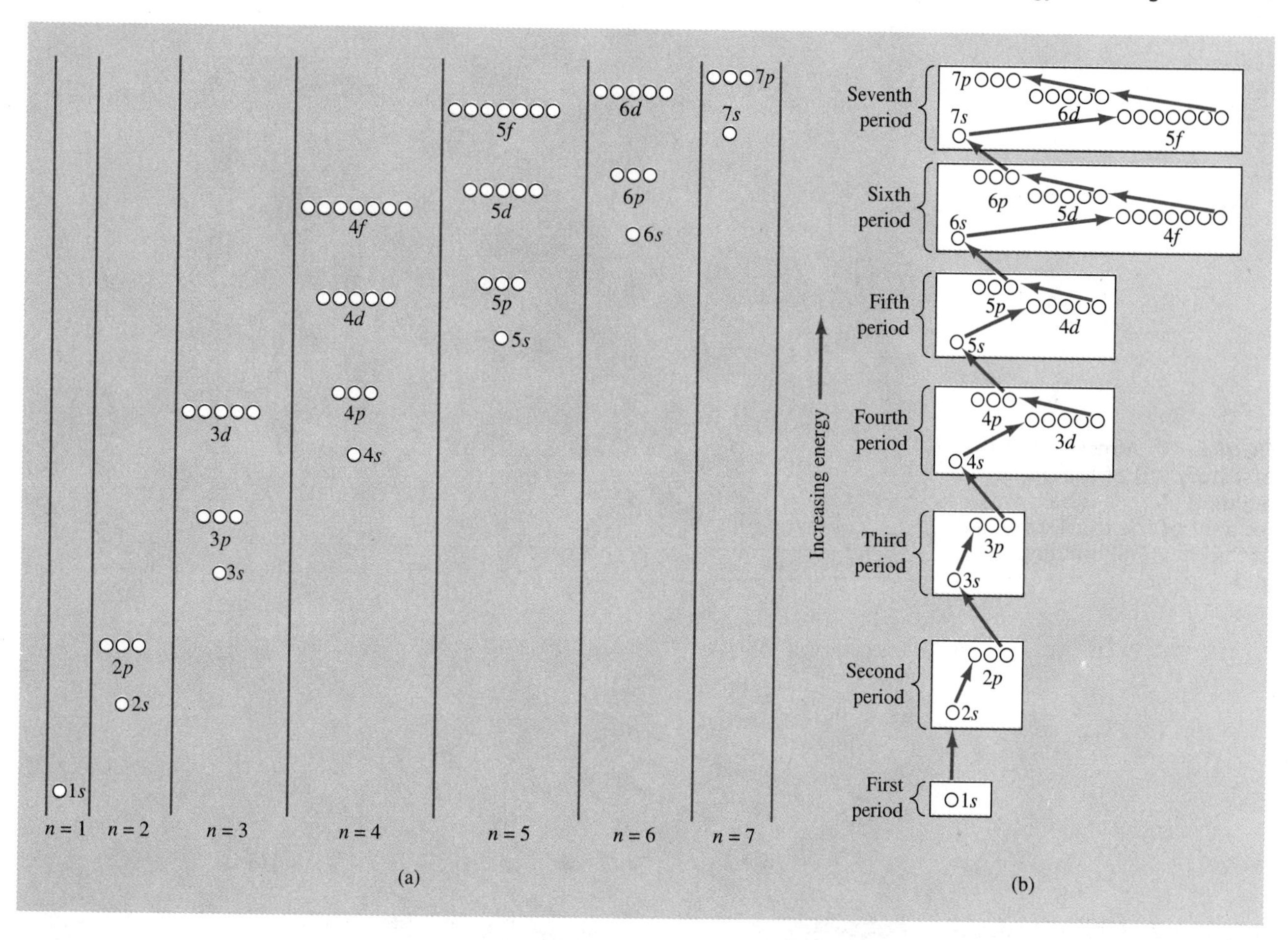

FIGURE 6-6 Relative energies of the orbitals as they fill. (Each circle represents an orbital. Each orbital can accommodate up to two electrons.) (a) The orbitals are grouped according to principal energy level. Reorganization leads to (b), which shows the order in which the orbitals fill. Each rectangular block of sublevels corresponds to a row (or period) of the periodic table.

A simplified version of Fig. 6-6b is presented in Fig. 6-8, which is a useful memory device for determining the order in which the sublevels fill. Either diagram can be used to assign electronic configurations in the next section. Though these diagrams are a useful tool in understanding electronic configuration, many elements have electronic configurations that differ slightly from what we would predict using these diagrams. The actual configurations of the elements are listed in Appendix D.

PROBLEM 6.4

Referring to Fig. 6-6b, which sublevel is filled after 3*d*? After 5*s*? After 6*s*? After 4*d*? After 4*f*?

FIGURE 6-7 A resident of the first story (4*s*) of the uphill building ($n = 4$) is below a resident of the third story (3*d*) of the downhill building ($n = 3$).

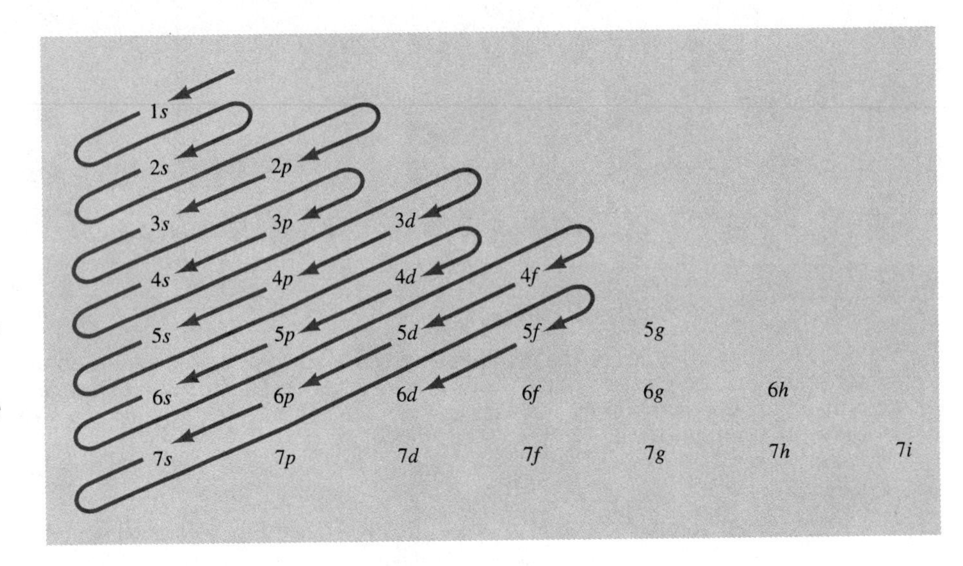

FIGURE 6-8 A memory device for determining the order in which sublevels fill. To reconstruct the diagram, write out the possible sublevels in each principal energy level, then draw diagonal lines as shown. Known elements do not exceed the 6*d* sublevel.

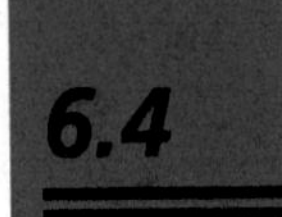

6.4 ELECTRONIC CONFIGURATION

OBJECTIVE 6.4 State Hund's rule. Predict the electronic configuration of any element with the aid of Fig. 6-6b or a similar diagram. Use the core symbols for the noble gases to abbreviate electronic configurations.

Electronic configuration: A description of the orbital locations of the electrons in an atom.

The **electronic configuration** of an element is a shorthand method for telling where the electrons in an atom are located. For example, applying the Aufbau principle to a hy-

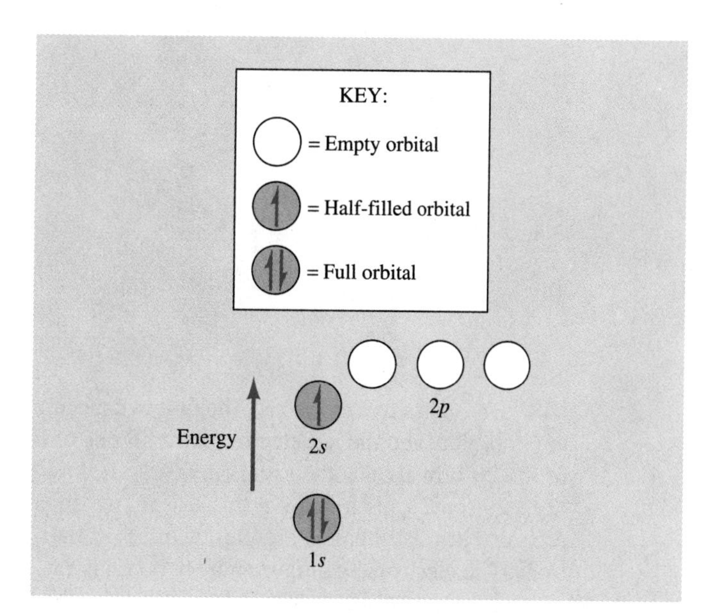

FIGURE 6-9 The electronic configuration of lithium: $1s^2 2s^1$.

drogen atom (atomic number 1), we predict that its lone electron must be found in the 1*s* orbital, since this orbital is in the sublevel having the lowest energy. We represent this as $1s^1$. The superscript 1 tells us that there is one electron in the 1*s* orbital. This configuration represents the ground state of hydrogen. As we continue to apply the Aufbau principle, we predict only ground-state electronic configurations.

Recalling that an orbital has a capacity for two electrons, helium (atomic number 2) has both of its electrons in the 1*s* orbital, hence its configuration is $1s^2$. The superscript 2 tells us that both electrons are in the 1*s* orbital. Note that in helium, the 1*s* orbital is filled, and the first block in Fig. 6-6b is likewise completed.

Lithium (atomic number 3) has three electrons. Referring to Fig. 6-9, we see that lithium has its first two electrons in the 1*s* orbital. That fills the 1*s* orbital, so the third electron must occupy the 2*s* orbital, the next lowest available orbital. The electronic configuration of lithium is $1s^2 2s^1$. This representation shows that there are two electrons in the 1*s* orbital and one electron in the 2*s* orbital. (Remember that the two electrons that fill an orbital differ in their electron spin, and we represent electrons of opposite spin with ↿ and ⇂.)

Beryllium has four electrons. The first two electrons fill the 1*s* orbital; the next two fill the 2*s* orbital. The electronic configuration of beryllium is $1s^2 2s^2$ (Fig. 6-10).

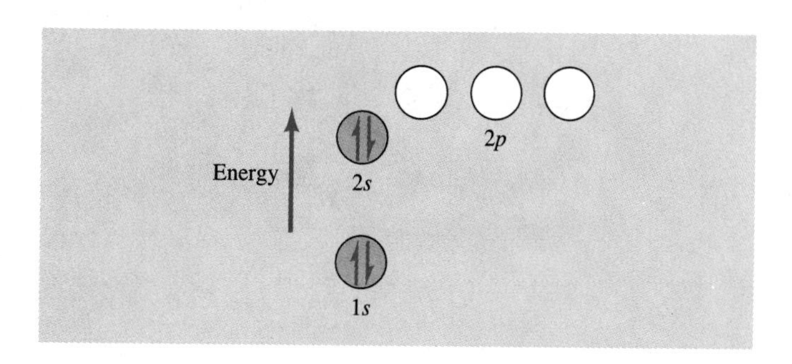

FIGURE 6-10 The electronic configuration of beryllium: $1s^2 2s^2$.

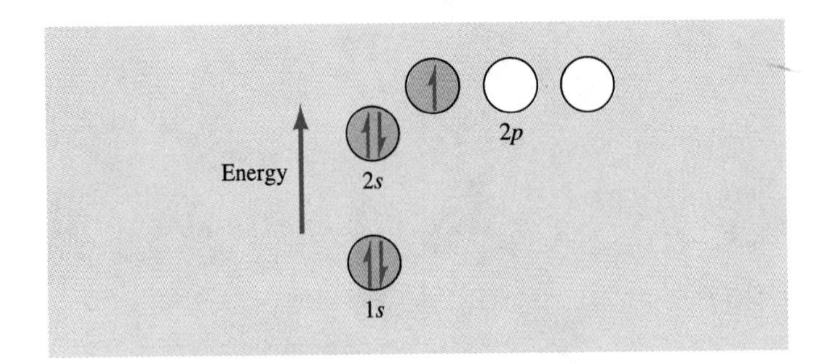

FIGURE 6-11 The electronic configuration of boron: $1s^22s^22p^1$.

Boron has five electrons. The first two electrons fill the $1s$ orbital, the next two fill the $2s$ orbital, and the last electron must fill one of the three equivalent $2p$ orbitals. Each of the $2p$ orbitals has the same energy, so it does not matter which of the $2p$ orbitals ($2p_x$, $2p_y$, or $2p_z$) is filled. For this reason, we do not distinguish among the various $2p$ orbitals but treat them as a group of three orbitals having a total capacity of six electrons. The electronic configuration for boron is $1s^22s^2p^1$ (Fig. 6-11).

Hund's rule: Electrons half-fill each of the orbitals in a sublevel before pairing.

Carbon has six electrons. It fills the $1s$ orbital with two electrons and the $2s$ orbital with two electrons. The remaining two electrons go in the various $2p$ orbitals. At this point you might wonder whether to put the two remaining electrons in the same $2p$ orbital or in two different $2p$ orbitals, since the energy of each of the $2p$ orbitals is the same. **Hund's rule** tells us that when electrons have more than one equivalent orbital available, they will half-fill each of the equivalent orbitals before filling the second half of each. Thus, the last two electrons in a carbon atom go into separate $2p$ orbitals, as shown in Fig. 6-12. By half-filling the orbitals of a sublevel in this fashion, repulsive forces that exist between the electrons are reduced. The electronic configuration of carbon is $1s^22s^22p^2$.

Nitrogen has seven electrons. Its electronic configuration is shown in Fig. 6-13. Once again, the three electrons that fill the $2p$ orbitals spread out evenly, half-filling each of the $2p$ orbitals. The electronic configuration of nitrogen is $1s^22s^22p^3$. Notice that the three electrons in the p sublevel are not distinguished as $2p_x{}^12p_y{}^12p_z{}^1$; instead, we simply write $2p^3$. We do the same whenever we fill the equivalent orbitals in a sublevel.

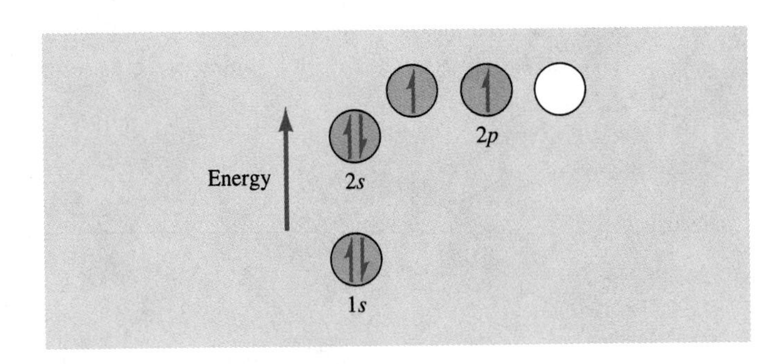

FIGURE 6-12 The electronic configuration of carbon: $1s^22s^22p^2$.

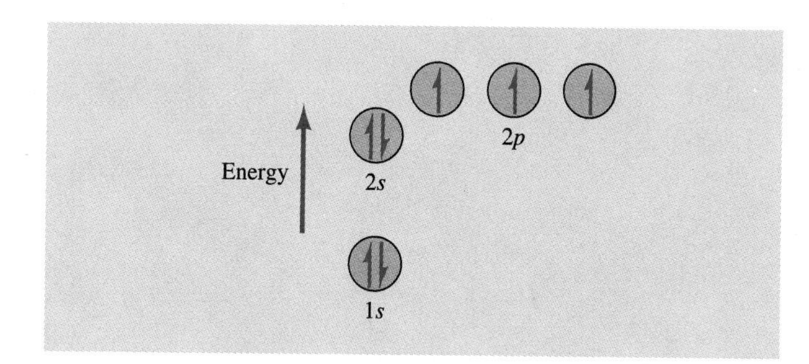

FIGURE 6-13 The electronic configuration of nitrogen: $1s^2 2s^2 2p^3$.

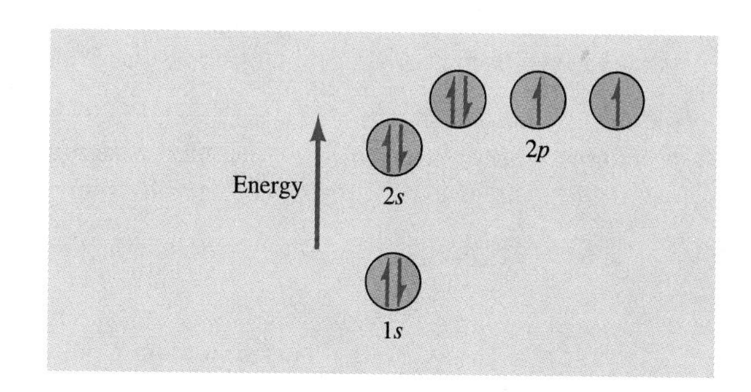

FIGURE 6-14 The electronic configuration of oxygen: $1s^2 2s^2 2p^4$.

Oxygen has eight electrons, and its electronic configuration is $1s^2 2s^2 2p^4$, as shown in Fig. 6-14. In accordance with Hund's rule, the four $2p$ electrons half-fill each orbital before pairing up.

Fluorine, with nine electrons, has the electronic configuration $1s^2 2s^2 2p^5$ (Fig. 6-15). Neon, with ten electrons, has the configuration $1s^2 2s^2 2p^6$ (Fig. 6-16, p. 148). Note that the electronic configuration of neon represents the complete filling of the second block of sublevels in Fig. 6-6b. Earlier we saw that helium represented the complete filling of the first block. Both helium and neon are found in the last column of the periodic table (group VIIIA). These and the other elements in that column represent a very special group of elements known as the *noble gases*.

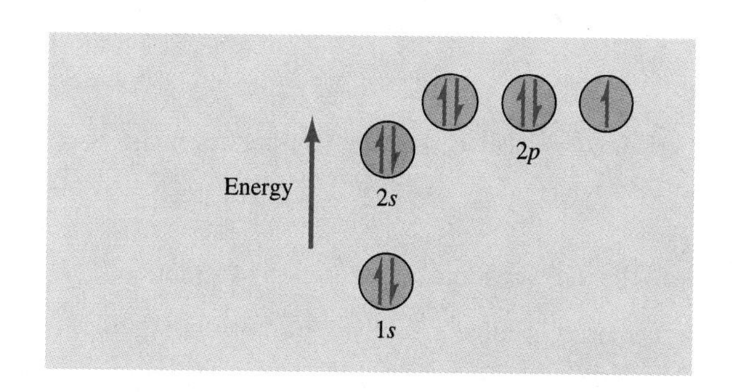

FIGURE 6-15 The electronic configuration of fluorine: $1s^2 2s^2 2p^5$.

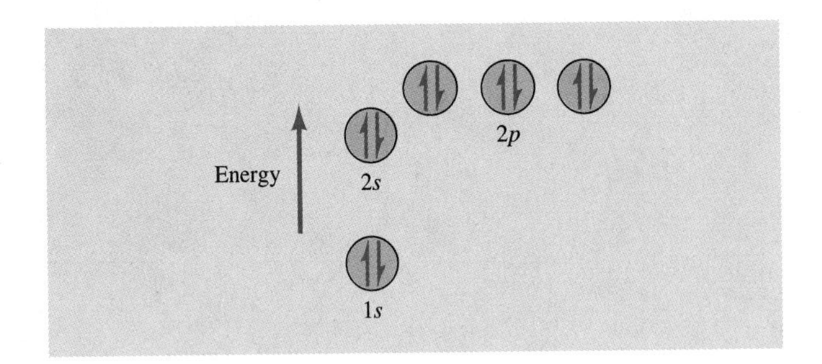

FIGURE 6-16 The electronic configuration of neon: $1s^2 2s^2 2p^6$.

EXAMPLE 6.1

Write the electronic configuration of manganese (atomic number 25).

Planning the solution

Manganese has atomic number 25, so there must be 25 electrons. We use the Aufbau principle to place the electrons in the appropriate orbitals, starting with the 1*s* sublevel and working up, using Fig. 6-6b (or Fig. 6-8) to give us the order of filling.

SOLUTION

The first two electrons go into the 1*s* sublevel:

$$1s^2 \ldots$$

Electron count: $2 = 2$

Next comes the 2*s* sublevel (two electrons), followed by the 2*p* sublevel (six electrons). This brings the total to 10 electrons:

$$1s^2 2s^2 2p^6 \ldots$$

Electron count: $2+2+6 = 10$

We continue with the 3*s* sublevel (two electrons) and the 3*p* sublevel (six electrons), bringing the total to 18 electrons:

$$1s^2 2s^2 2p^6 3s^2 3p^6 \ldots$$

Electron count: $2+2+6+2+6 = 18$

We have seven electrons remaining. The next two electrons fill the 4*s* sublevel, leaving five electrons, which can all be accommodated by the 3*d* sublevel. Thus, the electronic configuration of manganese is as follows:

$$1s^2 2s^2 2p^6 3s^2 3p^6 4s^2 3d^5$$

Electron count: $2+2+6+2+6+2+5 = 25$

Figure 6-17 shows the orbitals filled for manganese.

EXAMPLE 6.2

Write the electronic configuration of xenon (atomic number 54).

SOLUTION

The orbitals filled in xenon are shown in Fig. 6-18. The electronic configuration is

$$1s^2 2s^2 2p^6 3s^2 3p^6 4s^2 3d^{10} 4p^6 5s^2 4d^{10} 5p^6$$

Electron count: $2+2+6+2+6+2+10+6+2+10+6 = 54$

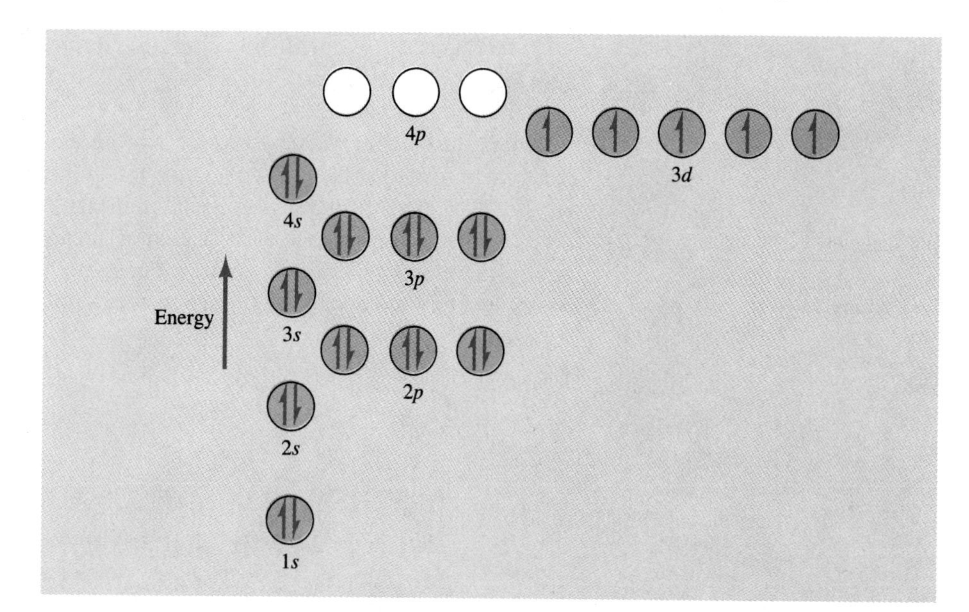

FIGURE 6-17 The electronic configuration of manganese: $1s^22s^22p^63s^23p^64s^23d^5$.

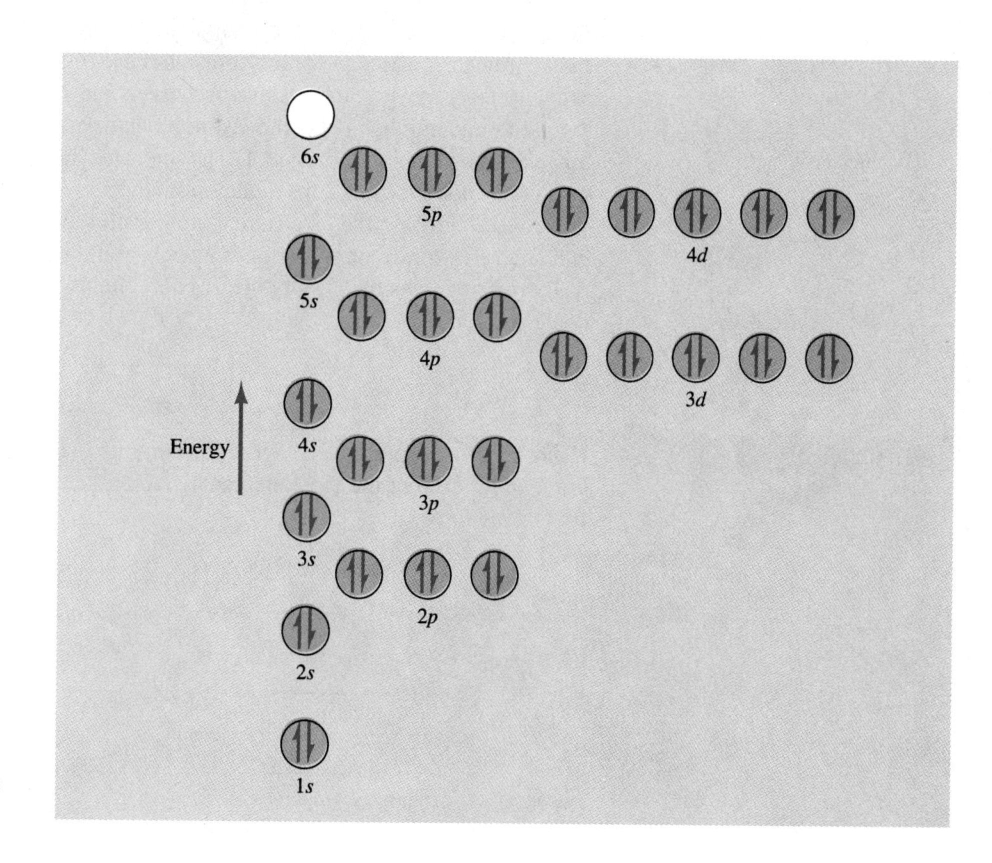

FIGURE 6-18 The electronic configuration of xenon: $1s^22s^22p^63s^23p^64s^23d^{10}4p^65s^2 4d^{10}5p^6$.

In Chapter 7, we will show you how to predict electronic configurations using only the periodic table (rather than Fig. 6-6b or Fig. 6-8) as your guide.

Before completing our discussion of electronic configuration, we introduce a commonly used abbreviated form for writing electronic configurations. We have already mentioned that the elements in the last column of the periodic table (He, Ne, Ar, Kr, Xe, and Rn) constitute a family of elements known as the noble gases. The electronic configuration for each of these elements represents an exceptionally stable condition for an atomic species. In writing electronic configurations, we frequently find it convenient to write one of the following **core symbols** in place of all of the electrons it represents.

Core symbol: An abbreviation used to represent the electronic configuration of one of the noble gases.

$$[\mathrm{He}] = 1s^2$$
$$[\mathrm{Ne}] = 1s^2 2s^2 2p^6$$
$$[\mathrm{Ar}] = 1s^2 2s^2 2p^6 3s^2 3p^6$$
$$[\mathrm{Kr}] = 1s^2 2s^2 2p^6 3s^2 3p^6 4s^2 3d^{10} 4p^6$$
$$[\mathrm{Xe}] = 1s^2 2s^2 2p^6 3s^2 3p^6 4s^2 3d^{10} 4p^6 5s^2 4d^{10} 5p^6$$
$$[\mathrm{Rn}] = 1s^2 2s^2 2p^6 3s^2 3p^6 4s^2 3d^{10} 4p^6 5s^2 4d^{10} 5p^6 6s^2 4f^{14} 5d^{10} 6p^6$$

In Example 6.1, we wrote the electronic configuration of manganese as $1s^2 2s^2 2p^6 3s^2 3p^6 4s^2 3d^5$. Rather than write out the entire electronic configuration, we can abbreviate it using one of the core symbols. To find the appropriate core symbol, we examine the periodic table to find the last noble gas before manganese. In this case, argon (atomic number 18) is the last noble gas before manganese (atomic number 25). Since the core symbol for argon, [Ar], represents the 18 electrons from $1s^2$ through $3p^6$, we simply continue with the remaining seven electrons beyond that noble gas core. Thus, we abbreviate the electronic configuration of manganese as $[\mathrm{Ar}]4s^2 3d^5$. An abbreviated electronic configuration is always written as the previous noble gas core (generally referred to as the "last noble gas core") plus the remaining electrons beyond that noble gas core.

PROBLEM 6.5

Using Fig. 6-6b (or Fig. 6-8), write the complete *and* the abbreviated electronic configurations for each of the eight elements given in the third row of the periodic table. Sulfur is done for you.

	Complete configuration	Abbreviated configuration
(a) Na	______	______
(b) Mg	______	______
(c) Al	______	______
(d) Si	______	______
(e) P	______	______
(f) S	$1s^2 2s^2 2p^6 3s^2 3p^4$	$[\mathrm{Ne}]3s^2 3p^4$
(g) Cl	______	______
(h) Ar	______	______

PROBLEM 6.6

Use Fig. 6-6b (or Fig. 6-8) to predict the electronic configuration of each of the following elements. Write out the entire configuration first, then abbreviate it with the appropriate core symbol.
(a) P **(b)** Ni **(c)** Sr **(d)** Hg **(e)** I **(f)** Eu

PROBLEM 6.7

Use Hund's rule to show how the indicated electrons are distributed in each of the following sublevels. For example, a correct arrangement for $4f^{10}$ is:

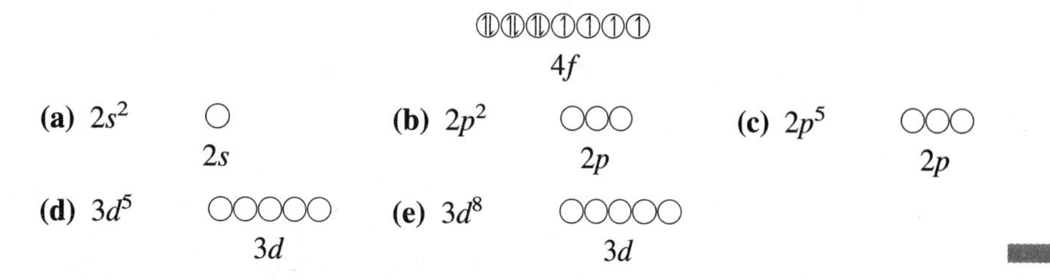

6.5 LIGHT AND THE NATURE OF WAVES

OBJECTIVE 6.5 Describe the relationships among wavelength, frequency, and velocity of a light wave. Describe the relationship between the frequency of a light wave and the energy of each photon.

In the first portion of the chapter, you learned that the electrons in atoms exist in discrete energy levels. You also learned that the behavior of electrons differs from that of larger objects. Let us now look at some of the evidence that led scientists to describe electron behavior in this fashion. To do so, we must first consider some of the properties of light, because so much of what we know about the behavior and properties of electrons was derived from observing how atoms interact with light.

When we go to the beach, we can observe the action of waves coming in against the shore. The waves have crests, and they come in to shore in a regular rhythm. In general, a wave is a periodic (that is, regular) disturbance that carries energy, usually through some medium. In the familiar example of water waves at the beach, we can easily recognize such a periodic disturbance. And ocean waves obviously carry energy, as evidenced by the powerful effect they can have on objects in their path.

If we could position ourselves some distance from the shore and watch the waves come past us, a wave might look something like the one in Fig. 6-19 (p. 152). The highest point of each wave is called a *crest;* the low point is referred to as a *trough.* The distance between the crests of any two consecutive waves is called the **wavelength.** We use the Greek letter lambda (λ) as the symbol for wavelength.

Wavelength: The distance between the crests of two neighboring waves.

If we stand in one place, we can count the number of waves that pass a given point in a given period of time. This number is called the **frequency.** For example, if we count 10 waves passing by in each minute, the frequency can be expressed as 10 waves per minute. Rather than talk about "waves," however, scientists refer to frequency in

Frequency: The number of cycles per second.

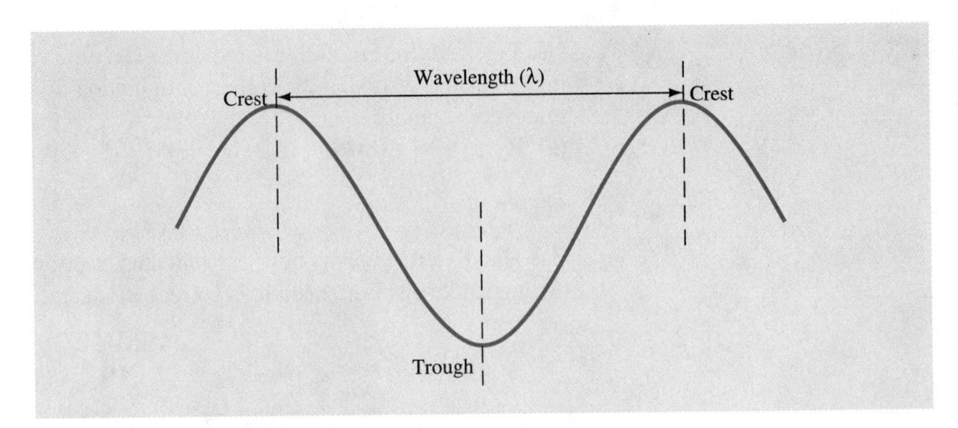

FIGURE 6-19 A cross-sectional view of a typical wave.

terms of "cycles" per unit time—usually cycles per seconds (cps), which is given the unit hertz (Hz). The Greek letter nu (ν) is used to symbolize frequency.

Velocity of a wave: The velocity with which a point on a wave travels.

We can also describe the velocity with which waves travel. Suppose we pick out one crest and follow its motion in toward the shore. If the crest moves 20 m in each minute, it has a velocity of 20 m/min. The **velocity of a wave,** represented by the symbol v, is the rate at which a given point on the wave travels from one place to another.

The velocity (v), frequency (ν), and wavelength (λ) are related to one another by the following equation.

$$v = \lambda\nu \tag{6-1}$$

Velocity of light: The velocity at which light travels: 3.0×10^8 m/sec

Like the waves of the ocean, light travels as a wave and carries energy. Equation (6-1) works as well with light waves as with other types of waves. However, light has a definite velocity. The **velocity of light** is 3.0×10^8 m/sec, which is given the symbol c. Consequently, a special form of equation (6-1) is used whenever the wave nature of light is being discussed:

$$c = \lambda\nu \tag{6-1a}$$

Electromagnetic radiation: A light wave; consists of an electrical component and a magnetic component.

Light is actually made up of two wavelike components: an electrical component and a magnetic component. These two waves oscillate in planes that are perpendicular to one another (Fig. 6-20). Because it is made up of both an electrical and a magnetic component, light is referred to as an **electromagnetic radiation.**

Normal sunlight, or "white light," is made up of light of many wavelengths. The various wavelengths correspond to different colors of visible light. Passing sunlight through a prism separates it into its component colors, resulting in a small "rainbow" (Fig. 6-21). We call this separated band of colors a *spectrum.* The red light at one end of the spectrum has a wavelength of about 660 nm (= 6.6×10^{-7} m), whereas the violet light at the other end of the spectrum has a wavelength of about 410 nm (= 4.1×10^{-7} m). The wavelengths of the colors increase as we scan the spectrum from violet to red. Since the velocity of light is a constant, each wavelength corresponds to a specific frequency, which can be calculated by substitution into equation (6-1a). Thus, either frequency or wavelength may be used to describe a particular light wave. Examining equation (6-1a) reveals that, as the wavelength increases, the corresponding frequency decreases, and vice versa (Fig. 6-22, p. 154).

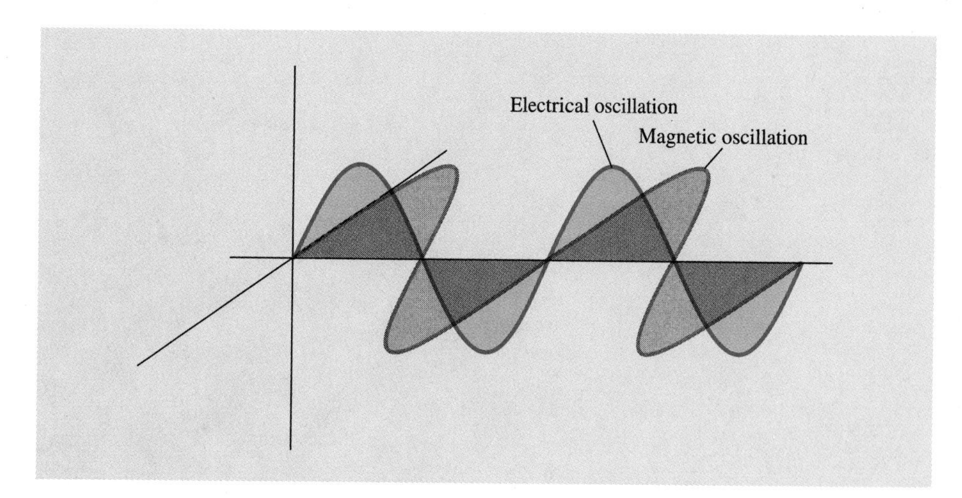

FIGURE 6-20 Light is an electromagnetic radiation. Notice that the two components lie in planes that are perpendicular to each other.

Photon: A single bundle of light energy.

Like other waves, light waves also carry energy. During the early part of this century, experimental work carried out independently by Albert Einstein and Max Planck demonstrated that light behaves as though it is made up of little bundles of energy, known as **photons.** The energy (E) associated with a particular photon of light is related to the frequency (ν) of that light wave by the relationship

$$E = h\nu \tag{6-2}$$

where h is a physical constant known as *Planck's constant.* This relationship tells us that *the energy associated with a given light wave increases as the frequency of the light wave increases.*

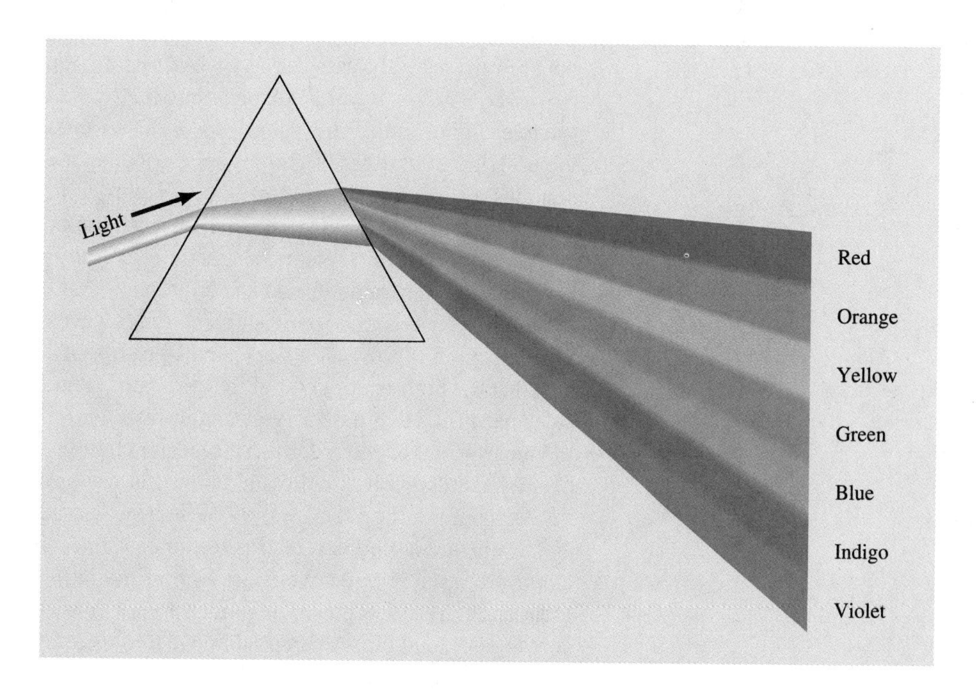

FIGURE 6-21 Visible light can be separated into the many colors that comprise it.

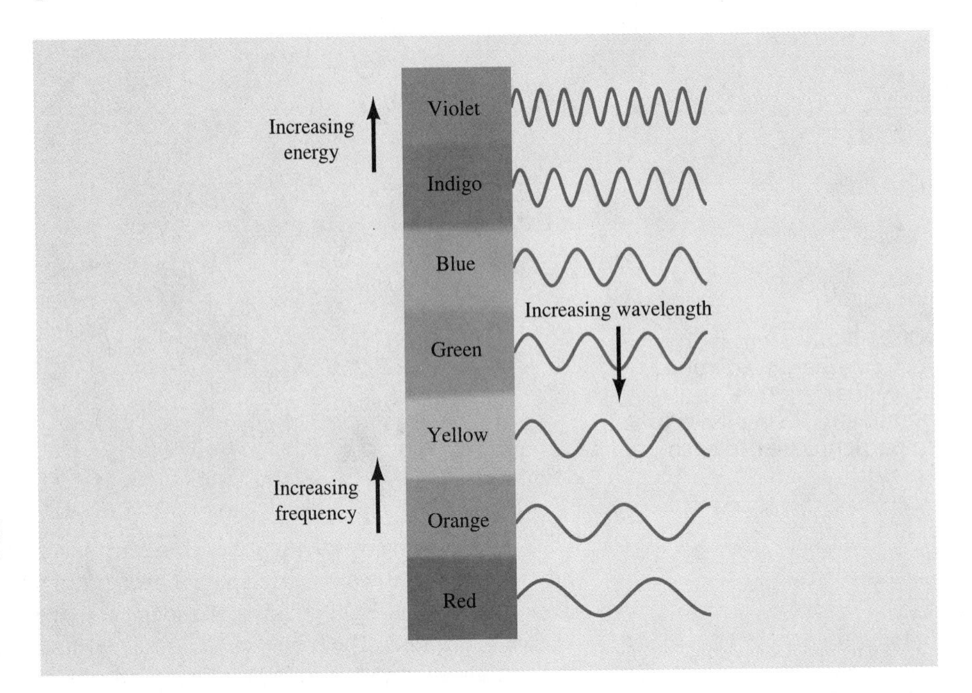

FIGURE 6-22 Moving from red toward violet light, the energy and frequency of the light wave increases, whereas the corresponding wavelength decreases.

Light is not restricted to the colors that we can see with our eyes, which is generally referred to as *visible light.* Instead, any electromagnetic radiation can be considered a form of light. You have probably heard the terms *infrared* and *ultraviolet.* If we were to extend our spectrum from Fig. 6-21 beyond the visible region, we would find electromagnetic radiations of lower frequencies than red. The radiations just beyond the visible region are referred to as infrared (meaning "below red") light. Although we cannot see infrared light with our eyes, like visible light, it is an electromagnetic radiation. However, it is of a lower frequency than our eyes can detect. In much the same way, ultraviolet light exists at the other end of the visible spectrum, just beyond (or "above") violet. Ultraviolet light is of a higher frequency and energy than visible light. Sunlight contains a considerable amount of ultraviolet light. It is this high-energy radiation that causes sunburn and damage to the eyes if we look directly at the sun.

The spectrum of electromagnetic radiation (often referred to as the *electromagnetic spectrum*) is rather broad and consists of many different types of radiation, including microwaves, X rays, television and radio waves, and radar (Fig. 6-23). Note that the visible region represents a very small portion of the electromagnetic spectrum. Note also that the high-energy end of the spectrum includes radiations that we generally consider harmful, such as X rays, cosmic rays, and gamma rays. On the other hand, the low-energy end of the spectrum includes radiations that we generally do not think of as health hazards, such as radio and television waves or radar.

You might find the location of microwave radiation on the low-energy end of the electromagnetic spectrum confusing, since we need to be careful about microwave radiations from microwave ovens. In fact, all radiations can be harmful if the intensity is great enough. Although we generally do not consider sound a harmful radiation, exposure to very loud noises can do damage to the eardrums. When the intensity of sound is extreme, the energy carried by the waves can pose a health hazard. In a similar fashion,

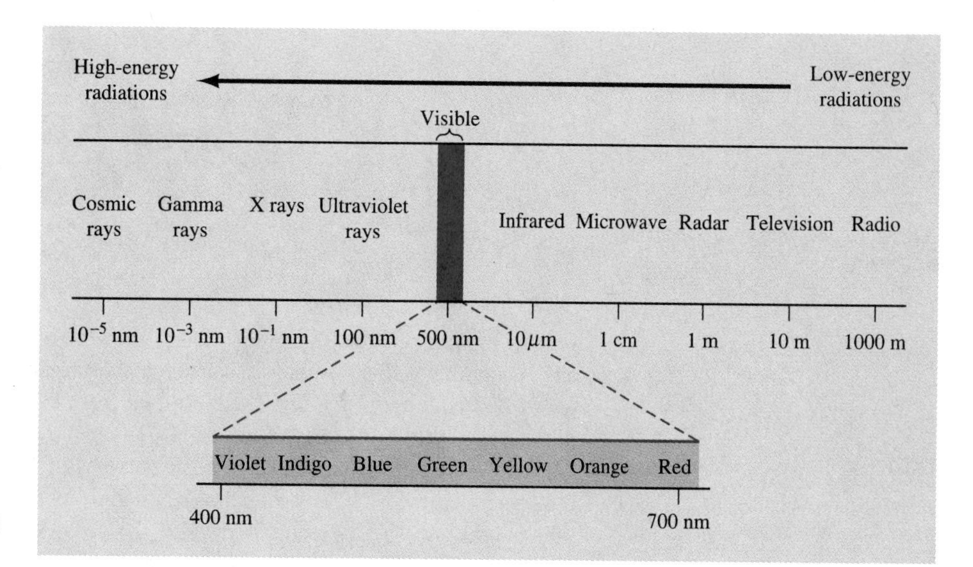

FIGURE 6-23 The electromagnetic spectrum. Visible light is only a narrow portion of the electromagnetic spectrum.

the fact that microwave radiation is at the low-energy end of the electromagnetic spectrum means that the energy *per photon* is low. However, if the intensity is great enough (meaning that there are a great many photons), the total energy dose can do severe damage to living tissue.

Although we cannot see nonvisible forms of electromagnetic radiation with our eyes, scientists have developed instruments to detect all forms of electromagnetic radiations. These devices make possible the beneficial use of electromagnetic radiation, such as radios (which detect radio waves), X-ray film (which is used for many types of medical applications), and radar scanners (used to locate airplanes and ships at sea). Most forms of electromagnetic radiation have useful properties if handled properly. However, all can have devastating health effects when misused.

PROBLEM 6.8

Fill in the blanks in each statement with one of the following: (1) increases, (2) decreases, or (3) remains the same.

(a) As the frequency of a light wave increases, its wavelength ______.
(b) As the wavelength of a light wave increases, its frequency ______.
(c) As the frequency of a light wave increases, the energy of each photon ______.
(d) As the wavelength of a light wave increases, the energy of each photon ______.
(e) As the frequency of a light wave increases, its velocity ______.
(f) As the wavelength of a light wave increases, its velocity ______.

6.6 THE HYDROGEN SPECTRUM

OBJECTIVE 6.6 Explain the difference between a continuous and a discontinuous spectrum. Account for the discontinuous spectrum of hydrogen in terms of specific electronic energy levels.

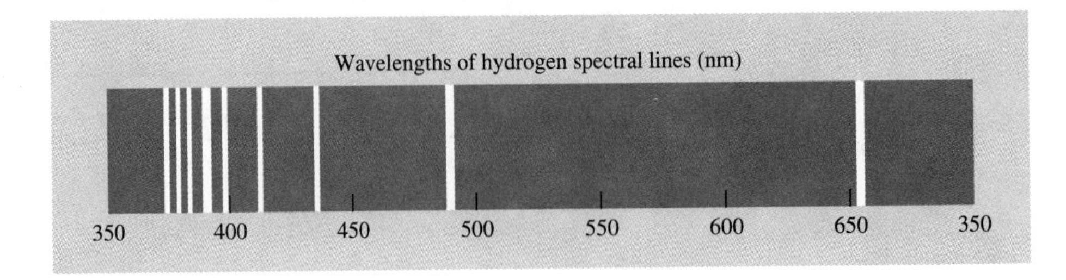

FIGURE 6-24 The hydrogen spectrum.

Continuous spectrum: A continuous band of light waves.

Discontinuous spectrum: A band of light waves that contains only certain frequencies.

During the late nineteenth century, scientists were observing very exciting but inexplicable phenomena. They found that, when an electric current was passed through a glass tube filled with hydrogen, light was emitted from the tube. (We observe a similar effect when we see the glow of a neon light.) When the light from the hydrogen tube was passed through a prism, it gave a spectrum very different from the "rainbow" spectrum resulting from white light. Instead of a **continuous spectrum,** such as that produced by white light, only a few separated bands of color were given off in a **discontinuous spectrum** (Fig. 6-24). Furthermore, when the same experiment was tried with other gases, such as helium and sodium vapor, they too produced spectra that consisted of separated lines of color. (Under the experimental conditions used, any diatomic molecules are separated into individual atoms. Thus, the various spectra represent the behavior of individual atoms, rather than that of molecules.)

Two questions needed to be answered in order to understand these spectra. First, why was light given off by these substances? And second, why was the emitted light of only certain frequencies? To answer these questions, scientists eventually postulated that the electrons in atoms exist in definite energy levels—those we examined in Section 6.1. Just as a ball on a staircase can come to rest only on a step and not in between,

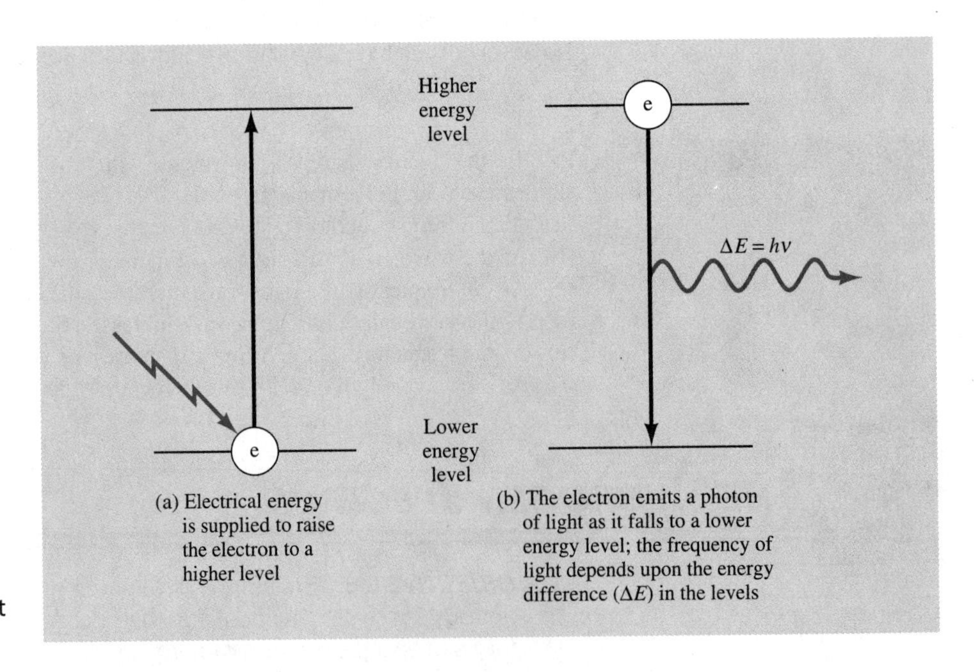

FIGURE 6-25 Emission of light by an electron.

an electron in a hydrogen atom can exist only at one of the various energy levels but not in between.

Why was light given off by the hydrogen tube? Light is a form of energy that an electron may absorb when going to higher energy levels or may emit when falling to a lower energy level. In our hydrogen tube experiment, the electron in each hydrogen atom starts out in the lowest possible energy level ($n = 1$) and absorbs electrical energy from the high-voltage source, which "promotes" it to a higher level (Fig. 6-25a). Having been raised above its lowest possible energy state, the electron spontaneously falls back down to a lower level, emitting its energy in the form of light (Fig. 6-25b). Thus, the light given off by the hydrogen tube represents the energy lost by electrons that are falling from higher to lower energy levels.

To understand why this light contains only certain frequencies, remember that only certain energy levels are available to the electron. Whenever an electron changes from one level to another, the amount of energy absorbed or given off must match exactly the energy difference, ΔE, between the levels. When a photon of light is given off by an electron falling to a lower level, its frequency corresponds to the energy difference, ΔE, between the two levels:

$$\Delta E = h\nu \qquad (6\text{-}2a)$$

Thus, the only possible frequencies of light that the electron can emit are those that correspond to the exact differences between the various energy levels. Figure 6-26 shows a simplified energy diagram for a system with only three energy levels. Note that transitions may occur among any of the three energy states. In this case, only three different frequencies of light are possible. Note, too, that it is not necessary for a promoted electron to return directly to the lowest level.

PROBLEM 6.9

Draw each of the following four-level energy diagrams (similar to Fig. 6-26, but with four levels rather than three) and answer the accompanying questions.

(a) Construct the levels so that the distance between the first and second levels is twice

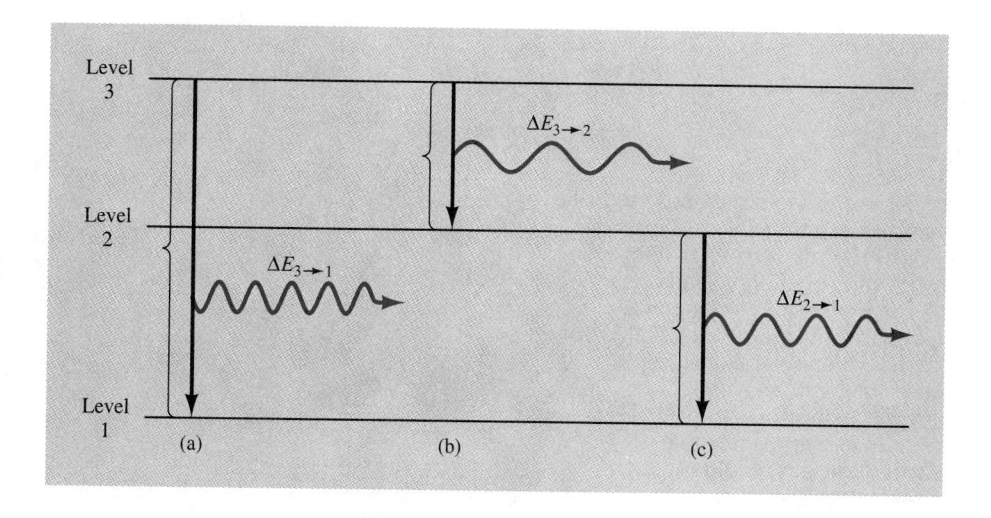

FIGURE 6-26 Three different frequencies of light may be given off by electrons in this three-level system. (a) An electron falls from level 3 to level 1. (b) An electron falls from level 3 to level 2. (c) An electron falls from level 2 to level 1.

that between the second and third levels, which in turn is twice that between the third and fourth levels. How many different frequencies (or energies) of emitted light are possible from this system?

(b) Construct the levels so that the distances between the first and second levels, the second and third levels, and the third and fourth levels are all equal. How many different frequencies (or energies) of emitted light are possible from this system?

6.7 THE BOHR MODEL VERSUS THE QUANTUM-MECHANICAL DESCRIPTION OF ELECTRONS

OBJECTIVE 6.7 Describe the Bohr model of electronic behavior. Explain what important feature of this model is correct. Explain what feature of this model is incorrect.

During the early part of this century, a Danish physicist named Niels Bohr presented a model designed to explain the existence of the various energy levels in atoms. The Bohr model pictured the electron orbiting the nucleus in much the same way that a planet orbits the sun (Fig. 6-27a). According to the Bohr model, when an electron changed energy levels, it also changed the radius at which it orbited the nucleus. Thus, electrons in higher energy levels orbited at greater distances from the nucleus. Bohr's model explained a great number of the facts known at that time, including the discontinuous hydrogen spectrum. However, it is inconsistent with known facts about the nature of charged particles in motion. Of particular importance is the fact that a negatively charged electron orbiting a positively charged nucleus gives off radiant energy. This loss of energy would cause its orbit to decay, and the electron would spiral very rapidly into the nucleus (Fig. 6-27b).

Eventually, the Bohr model gave way to the current description of electrons in orbitals, in which the location of the electron is described in terms of a probability of find-

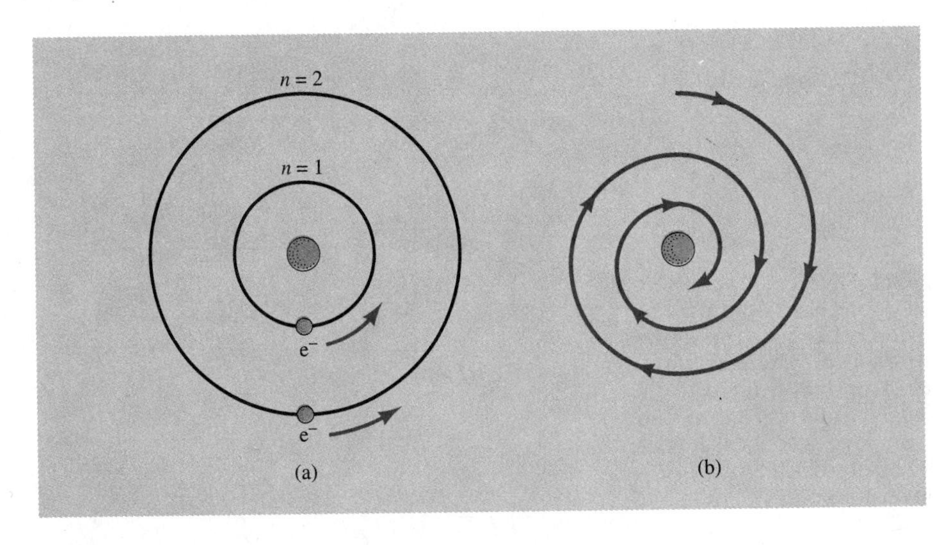

FIGURE 6-27 (a) The Bohr model of the atom pictured electrons orbiting the nucleus with fixed radii corresponding to the various energy levels. (b) The Bohr model is known to be inconsistent with several physical phenomena, including the behavior of charged particles in motion. An electron orbiting a positively charged nucleus would rapidly spiral into the nucleus.

ing it in some region of space. Although his description of electrons in orbits was not correct, the Bohr model correctly explained the role of electrons in giving rise to the discontinuous hydrogen spectrum. At the time of its presentation, Bohr's model represented an enormous step forward in our understanding of electronic behavior.

PROBLEM 6.10

In both the Bohr model and the quantum-mechanical model, electrons in higher energy levels are located at greater average distances from the nucleus than electrons in lower energy levels. However, the models differ greatly in how these greater average distances are achieved. Compare the two models with respect to average electronic distances from the nucleus, and explain how both models give rise to the same general conclusion.

CHAPTER SUMMARY

The electrons in atoms can exist in various discrete energy levels, known as **principal energy levels,** designated by integers: $n = 1, 2, 3$, and so forth. Within each principal energy level are one or more **sublevels,** designated by the letters *s, p, d,* and *f* for the first four sublevels. Within a given principal energy level, the energies of sublevels follow the general pattern $s < p < d < f$. There are as many sublevels within a given principal energy level as the integer associated with that principal level.

Within each sublevel, the location of an electron may be described by its **orbital,** which is an energy state that describes a given region of space within which the electron may be found. The precise location of an electron within an orbital cannot be known but, rather, it is described in terms of the probability of locating the electron at any given point. Each orbital has a capacity for a maximum of two electrons. The number of orbitals in a given sublevel follows a set pattern of one *s* orbital, three *p* orbitals, five *d* orbitals, and seven *f* orbitals. The energies of the orbitals within a given sublevel are all equal.

Electrons fill the various orbitals in an atom according to the **Aufbau principle,** which states that electrons always fill the available orbitals of lowest energy first. The order in which the orbitals fill does not strictly correspond to their principal energy level; for example, the 4*s* sublevel fills prior to the 3*d* sublevel. When orbitals of equivalent energy are available for filling, as in a given sublevel, electrons half-fill these orbitals before the pairing of electrons takes place. This is known as **Hund's rule.** When all of the electrons in an atom occupy the orbitals of the lowest possible energy, the atom is in its **ground state.** However, if any electron is in an orbital other than those of the lowest possible energy, the atom is in an **excited state.**

We can describe the location of electrons in an atom by writing an **electronic configuration,** which is a shorthand way to tell which orbitals are occupied. We can abbreviate an electronic configuration by substituting noble gas **core symbols** for all of the electrons up to the last core.

Our knowledge of the behavior of electrons is based, in part, upon the interaction of atoms with light. Light travels as a wave, having a constant velocity, *c*, of 3.0×10^8 m/sec. For any given light wave, the **velocity of light** (c) equals the **wavelength** (λ) times the **frequency** (ν): $c = \lambda\nu$. When passed through a prism, visible light is separated into a rainbow of colors in a **continuous spectrum.** Each color corresponds to a different frequency of light. Light also behaves as though it is made up of little bundles of energy, called **photons.** The energy of a given photon is proportional to the frequency of the light wave and may be calculated by multiplying the frequency by *Planck's constant* (h): $E = h\nu$.

When gaseous atoms are subjected to a high-voltage source of energy, light is emitted. When this emitted light is passed through a prism, a **discontinuous spectrum** results, indicating that only specific frequencies of light have been emitted. The electrons in atoms may exist in various specific energy states. Ordinarily, the electrons assume the lowest energy possible, known as the ground state. However, when electrical energy is supplied to an electron, it may rise to a higher energy level, in which case the atom is said to be in an excited state. When the electron falls back down to one of the lower energy levels, light is emitted, and the frequency of light corresponds to the energy difference between the higher and the lower energy levels.

Several theories have been proposed to explain the various electronic energy states. Among these was the Bohr model, which eventually gave way to the quantum-mechanical model. The Bohr model correctly postulated the existence of electronic energy levels, but it incorrectly described the location of electrons in well-defined circular orbits rather than in terms of the probability of finding each electron at given locations in space.

KEY TERMS

Review each of the following terms, which are discussed in this chapter and defined in the Glossary.

principal energy level
sublevel
orbital
ground state
excited state
Aufbau principle
electronic configuration
Hund's rule
core symbol
wavelength
frequency
velocity of a wave
velocity of light (3.0×10^8 m/sec)
photon
continuous spectrum
discontinuous spectrum

ADDITIONAL PROBLEMS

SECTION 6.1

6.11 Indicate whether each of the following statements is true (T) or false (F).

(a) The electrons in atoms may exist only at certain discrete energy levels.
(b) With increasing principal energy level n, the spacing between energy levels increases.
(c) The lowest principal energy level is $n = 1$.
(d) The number of sublevels in a given principal energy level is equal to the integer that designates that level. (For example, there are three sublevels in principal energy level $n = 3$.)
(e) The letters *s, p, d,* and *f* may be used to designate principal energy levels.
(f) When the lone electron in a hydrogen atom is in the second principal energy level ($n = 2$), it is in an excited state.

SECTION 6.2

6.12 Answer each of the following questions.

(a) How many sublevels are there in principal energy level $n = 5$?
(b) What are the sublevels in the fourth principal energy level?
(c) How many orbitals are in the 5*f* sublevel?
(d) What is the maximum number of electrons that can fill a 5*d* orbital?
(e) What is the maximum number of electrons that can fill the 5*d* sublevel?
(f) What is the maximum number of electrons that can fill the third principal energy level?

6.13 Based on the pattern established for the number of orbitals in the *s, p, d,* and *f* sublevels (see Table 6-2), how many orbitals do you predict are in a *g* sublevel? What is the maximum number of electrons that can fill this sublevel?

SECTION 6.3

6.14 State the Aufbau principle.

6.15 Reconstruct from memory the Aufbau diagram shown in Fig. 6-8.

6.16 Explain why the 5*s* sublevel fills before the 4*d* sublevel.

SECTION 6.4

6.17 Use Fig. 6-6b to predict the electronic configurations of the following elements. In each case write out the complete electronic configuration, and then abbreviate it using the appropriate noble gas core symbol.

(a) Na **(b)** Ga **(c)** Ca **(d)** V **(e)** Pm **(f)** Zr

6.18 What is Hund's rule? The electronic configuration of Eu is $[Xe]6s^24f^7$. How must the electrons in the 4*f* sublevel be distributed?

6.19 Use Hund's rule to show how five electrons could

be distributed in each of the following sublevels.

(a) $4p$ sublevel: ○○○ $4p$

(b) $4d$ sublevel: ○○○○○ $4d$

(c) $4f$ sublevel: ○○○○○○○ $4f$

SECTION 6.5

6.20 What is the relationship among the wavelength, frequency, and velocity of a light wave? What relationship exists between the frequency of a light wave and the energy of each photon?

6.21 During World War II, it was common practice to have enlisted men clean radar sending units while they were in operation. This practice was suspended when it was found that soldiers were becoming ill. Explain why the soldiers became ill despite the relatively low energy of individual radar photons.

SECTION 6.6

6.22 Explain the difference between a continuous spectrum and a discontinuous spectrum. How can the discontinuous spectrum of hydrogen be explained in terms of specific electronic energy levels?

SECTION 6.7

6.23 Describe the Bohr model of electronic behavior. Explain what major feature of his model forms the basis of our current understanding of electronic behavior. Explain what feature of his model is incorrect.

6.24 Describe an orbital. Explain how the behavior of an electron in an orbital differs from that predicted by the Bohr model.

GENERAL PROBLEMS

6.25

(a) Write the electronic configurations for Li, Na, and K.

(b) What similarity do these electronic configurations exhibit?

(c) What do you notice about the location of these three elements in the periodic table?

(d) Repeat parts (a), (b), and (c) for O, S, and Se.

6.26

(a) Write the electronic configuration for Na.

(b) Sodium can lose one electron to form a sodium ion, Na^+. What is the electronic configuration of this ion?

(c) What element has the same electronic configuration as the sodium ion?

(d) Chlorine atoms often gain one electron to form chloride ions, Cl^-. What is the electronic configuration of the chloride ion?

(e) What element has the same electronic configuration as the chloride ion?

(f) What do you notice about the elements in parts (c) and (e)?

(g) Using your answer to part (f), predict the number of electrons that a magnesium atom is likely to lose. What would be the resulting charge on a magnesium ion?

(h) Sulfur atoms form ions by gaining electrons. How many electrons is a sulfur atom likely to gain, and what would be the resulting charge?

(i) Repeat part (g) for aluminum.

6.27 What do you notice about the principal energy level of each sublevel that follows a filled p sublevel? What relationship does this have to the structure of the periodic table?

6.28 Fireworks function by causing metals to absorb excess heat released from the reaction of pyrotechnics (fireworks chemicals). The thermal energy released in the reaction excites electrons from their ground-state configurations to excited states. When the excited electrons return to the ground state, the excess energy is given off as photons of light having colors that correspond to the energy loss. Strontium and barium are commonly used in fireworks to produce the colors red and green, respectively. Compare the energies of the transitions for these two elements.

***6.29** What change occurs in the radius of a selenium atom when a photon excites the atom from a ground-state electronic configuration of $[Ar]4s^23d^{10}4p^4$ to an excited state of $[Ar]4s^23d^{10}4p^35s^1$?

***6.30** The ability of substances to be attracted by a

magnetic field is attributed to unpaired electrons in their electronic energy levels. The more unpaired electrons in an element, the greater the ability to be attracted by a magnetic field. For example, iron has the electronic configuration [Ar]$4s^2 3d^6$, which corresponds to four unpaired electrons in the $3d$ sublevel. Iron is readily attracted by magnetic fields. List the transition metals that should be incapable of attraction by a magnetic field. (Refer to Fig. 7-4 on p. 170 to identify the transition metals.)

6.31 The sun is 93 million miles from earth. How many minutes does it take for light to reach the earth?

***6.32** A light-year is the distance that light can travel in 1 year. Assuming that a year is 365 days and the velocity of light is 3.00×10^8 m/sec, calculate the quantities requested in the following questions.

(a) How many kilometers are in a light-year?

(b) How many kilometers is it to Alpha Centauri, the nearest constellation of stars beyond the sun? Alpha Centauri is 4.3 light-years away.

(c) How many light-years is the sun from earth? The sun is 93 million miles from earth.

***6.33** Radio and television signals broadcast here on earth travel at the speed of light across the galaxy. These signals, if received by intelligent life forms on distant planets, would be capable of describing our society to those distant creatures. How far away is a planet that has just received a live broadcast announcing the assassination of President John F. Kennedy on November 22, 1963?

WRITING EXERCISES

6.34 Your mother has always been concerned about the warnings that accompany the use of microwave ovens. Recently, she was looking through your chemistry book and noticed that microwaves are shown on the low-energy side of the electromagnetic spectrum. Now she is confused and wants to know more about the electromagnetic spectrum and the dangers associated with microwave ovens. Write her a note describing the electromagnetic spectrum, and clear up her confusion about the dangers of microwave radiations.

6.35 A friend has just visited San Francisco, where mercury lamps are widely used to illuminate the local bridges because of their ability to penetrate the night fog. Your friend wants to know why the light given off by these lamps is an orange-pink color. Write your friend a note explaining the nature of electronic energy levels and why different elements are observed to give off different colors of light.

6.36 In your own words, describe briefly the currently accepted theory of electronic behavior. Your description should include the behavior of electrons in orbitals, as well as the arrangement of electrons in energy levels and sublevels.

THE PERIODIC TABLE

Earlier, you learned that there are more than 100 known elements and that advances in research will undoubtedly lead to the discovery of new elements. The study of chemistry is focused on learning about the elements, so it is logical for us to attempt to organize them with respect to their various properties. In this chapter you will learn how the periodic table is organized so that you can obtain information from it at a glance.

7.1 HISTORICAL DEVELOPMENT OF THE PERIODIC TABLE

OBJECTIVE 7.1 Describe how the elements in a period of the periodic table are arranged. Describe how the elements in a group or chemical family are arranged. Determine whether atomic mass or atomic number is currently used as one of the principal organizing features of the periodic table.

During the early part of the nineteenth century, chemists began to notice regularities in the properties of the various elements that had been discovered. J. W. Dobereiner (1790–1849) noticed that certain groups of elements having similar properties seemed to come in groups of three, such that the atomic mass of the middle one was approximately equal to the average mass of the heaviest and the lightest. The same was true of other properties, such as density. These various groups became known as *Dobereiner's triads.*

Later, John A. Newlands (1837–1898) noticed that when the elements were arranged in order of increasing atomic mass, the properties of every eighth element were similar. Newlands' *Law of Octaves* was so named because of its resemblance to a musical octave, which occurs every eighth note.

Shortly after this, both Dmitri Mendeleev and Lothar Meyer (working independently) proposed that, when the elements were arranged in order of increasing atomic mass, recurring patterns appeared in their properties. These regular, or periodic, trends led Mendeleev to organize the elements into a table in order of their atomic masses, with the elements that exhibit similar properties appearing next to one another. Eventually, it was shown that *atomic number,* rather than atomic mass, should be used to arrange the elements. In our modern version of the periodic table, each row represents one cycle through the recurring (or periodic) pattern of properties. Each (horizontal) row of the periodic table is called a **period.** The (vertical) columns, which consist of elements having similar chemical properties, are referred to as either **groups** or (in some cases) **chemical families.** The periodic recurrence of the properties of elements, when they are arranged according to atomic number, is referred to as the *Periodic Law.*

Period: A horizontal row of the periodic table.

Chemical family or group: A vertical column of the periodic table.

PROBLEM 7.1

There are several places in the modern periodic table where the atomic mass of an element is less than that of the preceding element. One of the elements involved is iodine.

(a) How is it possible for iodine (atomic number 53, atomic mass 126.9) to have a smaller atomic mass than tellurium (atomic number 52, atomic mass 127.6)?

(b) Find two other elements in the first five rows of the periodic table whose atomic masses are less than those of the preceding element.

Mendeleev's periodic table is an excellent illustration of how the scientific method works in practice. First of all, to construct his table Mendeleev utilized information already available. However, in order to arrange the elements so that those of similar properties were within the same group, it was necessary for him to leave several blank spaces. He predicted that these spaces would eventually be filled by elements yet to be discovered. His various hypotheses turned out to be correct: the missing elements were eventually discovered, and they exhibited properties corresponding to those he predicted.

As events of the twentieth century unfolded, many more elements were added to the table, and the electronic structure of each element became known. Today we know that the periodic relationships expressed in the periodic table are the result of the various electronic configurations of the elements. Consequently, as we pursue our examination of the periodic table in this chapter, we will draw particular attention to the relationships among electronic configuration, properties of the elements, and the overall organization of the periodic table.

The periodic table continues to be a living document, with standards and formats that are occasionally revised to meet the needs of the times. For example, the carbon-12 standard currently used for atomic masses was adopted in 1961. This required a change from an earlier standard based on oxygen. The system of three-letter symbols, used to designate newly discovered elements, is a far more recent invention.

An even more recent proposal has changed the numerical designations of the columns of the periodic table. Examination of the periodic table inside the front cover reveals two sets of column headings, designated *IUPAC* and *traditional.* Under the older, traditional system used in the United States, the various families (or groups) are designated by a Roman numeral and the letter A or B, starting with IA at the left and continuing as shown. The newer system, adopted by the International Union of Pure and Applied Chemistry (IUPAC), numbers the columns consecutively from 1 to 18, as also shown in the table. Because you are likely to encounter both systems on wall charts and texts, we have included both designations in our periodic table, and in our discussions we use the traditional (American) designations with the IUPAC numbers in parentheses. (A third system, used in Europe, is similar to the traditional system shown here, except that most of the A and B designations are reversed. We will not use this system.)

7.2 IONIZATION ENERGY AND ELECTRON AFFINITY

OBJECTIVE 7.2 Explain what electrostatic forces are, including the effects of like and unlike charges on one another. Define the term *ionization energy.* Explain why energy must be supplied to bring about ionization of an atom. Write an equation for the ionization of any element. Define the term *electron affinity.* Explain why energy is released when an electron is captured by an atom.

Among the various trends within the periodic table, two are intimately related to the underlying basis for the organization of the table: trends in ionization energy and electron

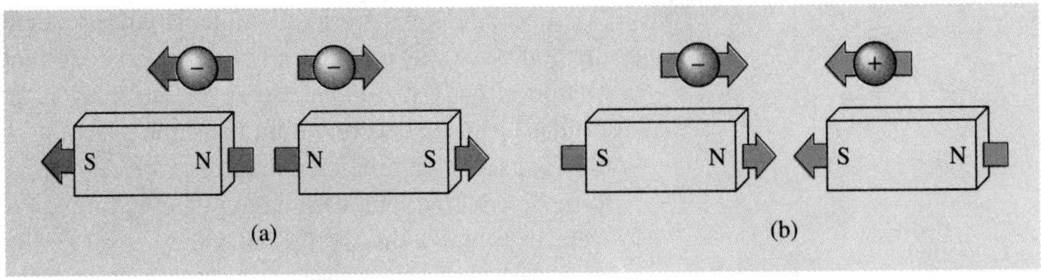

FIGURE 7-1 Comparison of electrical charges and magnetic poles. (a) Like charges and like poles repel. (b) Unlike charges and unlike poles attract.

Ionization energy: Energy required to remove an electron from an atom.

Electron affinity: Energy liberated when an electron is captured by an atom.

affinity. An element's **ionization energy** is the amount of energy required to remove an electron from an uncharged atom in the gaseous state. The **electron affinity** of an element is the amount of energy released when an atom in its gaseous state gains an electron. The values for both of these properties differ from element to element.

In order to understand these terms, recall that an uncharged atom is made up of a positively charged nucleus and enough negatively charged electrons outside the nucleus to balance the electrical charges. From the laws of physics, we know that particles of like charge repel each other, whereas particles of opposite charge attract one another. This behavior is similar to that of two magnets whose like poles repel each other, whereas their unlike poles are attracted (Fig. 7-1). We refer to the electrical forces of attraction or repulsion between charged objects as **electrostatic forces.**

Electrostatic forces: Attractive or repulsive forces between electrically charged particles.

The magnitude of an electrostatic force depends upon the charges on the two objects as well as the distance between them (Fig. 7-2). As we might expect, the electrostatic force of attraction (or repulsion) increases when the magnitude of either charge (whether positive or negative) increases. However, the electrostatic force decreases as the distance between the objects increases. Conversely, the electrostatic force between charged bodies increases with decreasing distance. Again, note the similarity to the behavior of magnets, whose attraction (or repulsion) increases as they are brought closer together.

You may be wondering why the nuclei of atoms that have more than one proton do not fly apart. After all, if particles with like charges repel each other, we might pre-

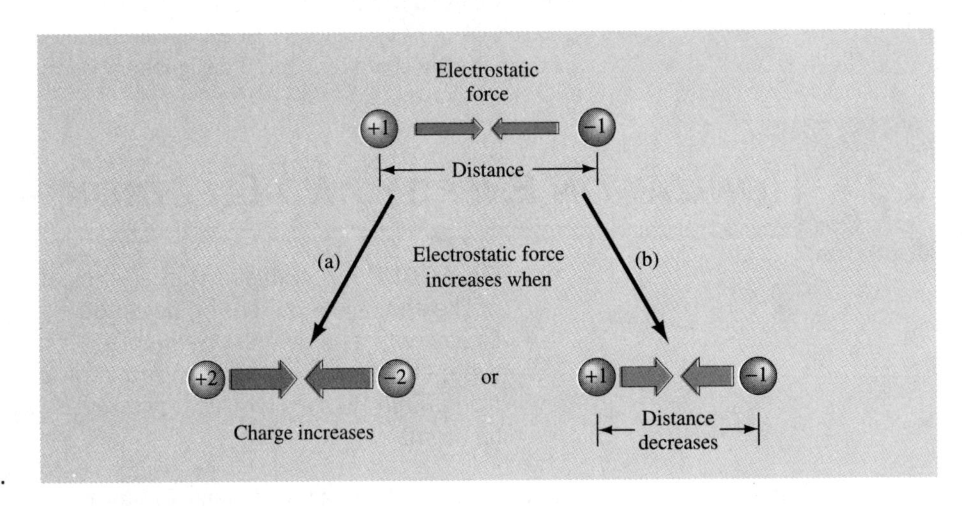

FIGURE 7-2 The electrostatic force between two charged objects increases when (a) the charge on either object increases or (b) the distance between the objects decreases.

dict that the repulsive forces within the nucleus would cause the nucleus to be broken apart. Apparently there are forces within the nuclei of these atoms holding the various nuclear particles together. These forces result from a property known as the *binding energy.*

Just as it takes energy to pull two magnets apart, it requires energy to remove an electron from an atom. Here we must work against the electrostatic attraction between the negatively charged electron and the positively charged nucleus. It is possible to remove an electron from an atom by supplying enough energy to overcome the electrostatic force of attraction between the electron and the nucleus. The amount of energy supplied represents the ionization energy. You may recall that we defined an ion as an atom (or group of atoms) with an electrical charge other than zero. Removal of an electron from an atom creates a positive ion, because the number of electrons that remain is one fewer than the number of protons. It is for this reason that the energy required for removal is referred to as the ionization energy.

If the atom in question is a hydrogen atom, its ionization may be represented by the following chemical equation:

$$\underset{(1p^+,\,1e^-)}{H} \rightarrow \underset{(1p^+,\,0e^-)}{H^+} + \underset{(1e^-)}{e^-}$$

We interpret this chemical equation to mean that a hydrogen atom forms a positively charged hydrogen ion plus an electron.

Just as it is possible for an atom to lose an electron through the process of ionization, it is also possible for an atom to gain an electron. The energy released when a gaseous atom gains an electron is its *electron affinity.* For example, a chlorine atom can accept an electron to form a chloride ion:

$$\underset{(17p^+,\,17e^-)}{Cl} + \underset{(1e^-)}{e^-} \rightarrow \underset{(17p^+,\,18e^-)}{Cl^-}$$

This equation says that a chlorine atom plus an electron forms a chloride ion.

To understand why energy is released, we must reverse the logic that explained why energy is required to remove an electron. When we remove an electron, energy must be supplied to separate the negatively charged electron from the positively charged nucleus. Reversing this procedure, when an electron is attracted by a nucleus, energy is given off.

PROBLEM 7.2

(a) Define the term *ionization energy;* define the term *ion.*

(b) Write a chemical equation (similar to the one given for hydrogen) for the ionization of potassium. Show the locations of the protons and electrons on both sides of the equation. Write a similar equation for rubidium.

(c) Define the term *electron affinity.*

(d) Write a chemical equation (similar to the one given for chlorine) to show the capture of one electron by a bromine atom to form a bromide ion. Show the locations of the protons and electrons on both sides of the equation.

7.3 PERIODIC TRENDS IN IONIZATION ENERGY AND ELECTRON AFFINITY

OBJECTIVE 7.3A State the feature of electronic configuration that is common to all noble gases except helium. Tell what generalization can be made about the ionization energies of the noble gases. Explain why the alkali metals have relatively low ionization energies. Explain why halogens have relatively high electron affinities.

OBJECTIVE 7.3B Classify any element that is a member of the alkali metals, alkaline earth metals, chalcogens, halogens, or noble gases according to its chemical family. If the element forms ions, predict the charge that an ion of that element is most likely to bear.

OBJECTIVE 7.3C Classify any element that is a transition metal or a rare earth metal. State the characteristic of electronic configuration that is common to all transition metals. State the characteristic of electronic configuration that is common to all rare earth metals.

The ionization energy is not the same for every element. Rather, there is considerable variation among the elements. Figure 7-3 shows how the ionization energy varies with atomic number. A high ionization energy indicates that atoms of that element are resistant to the loss of electrons. This means that the electronic configuration of the atom

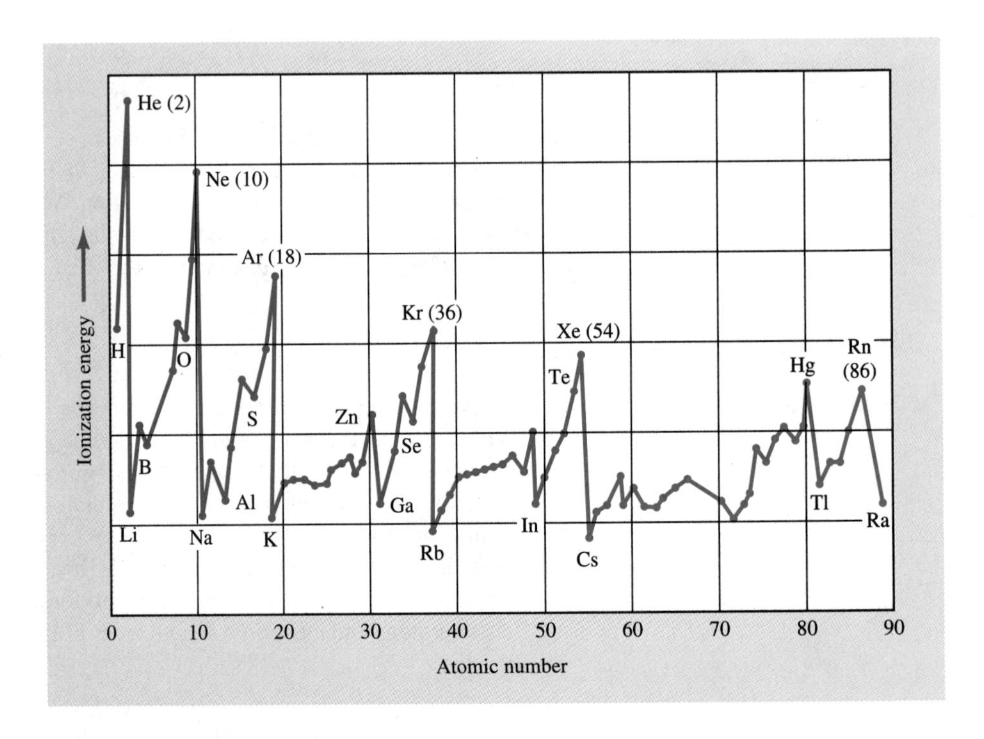

FIGURE 7-3 Ionization energies.

must be quite stable. On the other hand, a low ionization energy indicates that an electron is easily removed from the atom.

The Noble Gases

Noble gases: The elements in group VIIIA (18).

Certain features of Fig. 7-3 stand out. The elements with atomic numbers 2, 10, 18, 36, 54, and 86 seem to have unusually high ionization energies compared to the elements on either side of them. This group of elements is known as the **noble gases.** Their electronic configurations are as follows:

He	$1s^2$
Ne	$1s^22s^22p^6$
Ar	$1s^22s^22p^63s^23p^6$
Kr	$1s^22s^22p^63s^23p^64s^23d^{10}4p^6$
Xe	$1s^22s^22p^63s^23p^64s^23d^{10}4p^65s^24d^{10}5p^6$
Rn	$1s^22s^22p^63s^23p^64s^23d^{10}4p^65s^24d^{10}5p^66s^24f^{14}5d^{10}6p^6$

The unusually high ionization energies of these elements (among other considerations) lead us to state that *there is a special stability associated with the electronic configurations of the noble gases.*

Let us see whether we can account for this special stability. Note that the electronic configuration of each of the noble gases (except helium) ends in a p^6 configuration. Table 7-1 displays the relationships of the various sublevels to the periodic table. For example, the elements in the first period (H and He) are those that fill the $1s$ sublevel (which consists of just the $1s$ orbital). The elements in the second period (Li to Ne) have a [He] core and fill the $2s$ and $2p$ sublevels going across the period. Similarly, the elements in the third period (Na to Ar) fill the $3s$ and $3p$ sublevels; those in the fourth period (K to Kr) fill the $4s$, $3d$, and $4p$ sublevels; and so on.

Each time a p sublevel is completed, the very next electron must go into the next principal energy level. Consequently, completion of a p sublevel represents the end of a period, and a special stability accompanies the electronic configuration of the noble gas involved. In the case of helium, a noble gas configuration is achieved on completing the

TABLE 7-1 Sublevels filled across each period of the periodic table

Period	Elements	Last noble gas core	Order in which sublevels are filled
1	H to He	None	$1s$
2	Li to Ne	[He]	$2s \rightarrow 2p$
3	Na to Ar	[Ne]	$3s \rightarrow 3p$
4	K to Kr	[Ar]	$4s \rightarrow 3d \rightarrow 4p$
5	Rb to Xe	[Kr]	$5s \rightarrow 4d \rightarrow 5p$
6	Cs to Rn	[Xe]	$6s \rightarrow 4f \rightarrow 5d \rightarrow 6p$
7	Fr to [a]	[Rn]	$7s \rightarrow 5f \rightarrow 6d \rightarrow 7p$

[a]Element number 118 (not yet known)

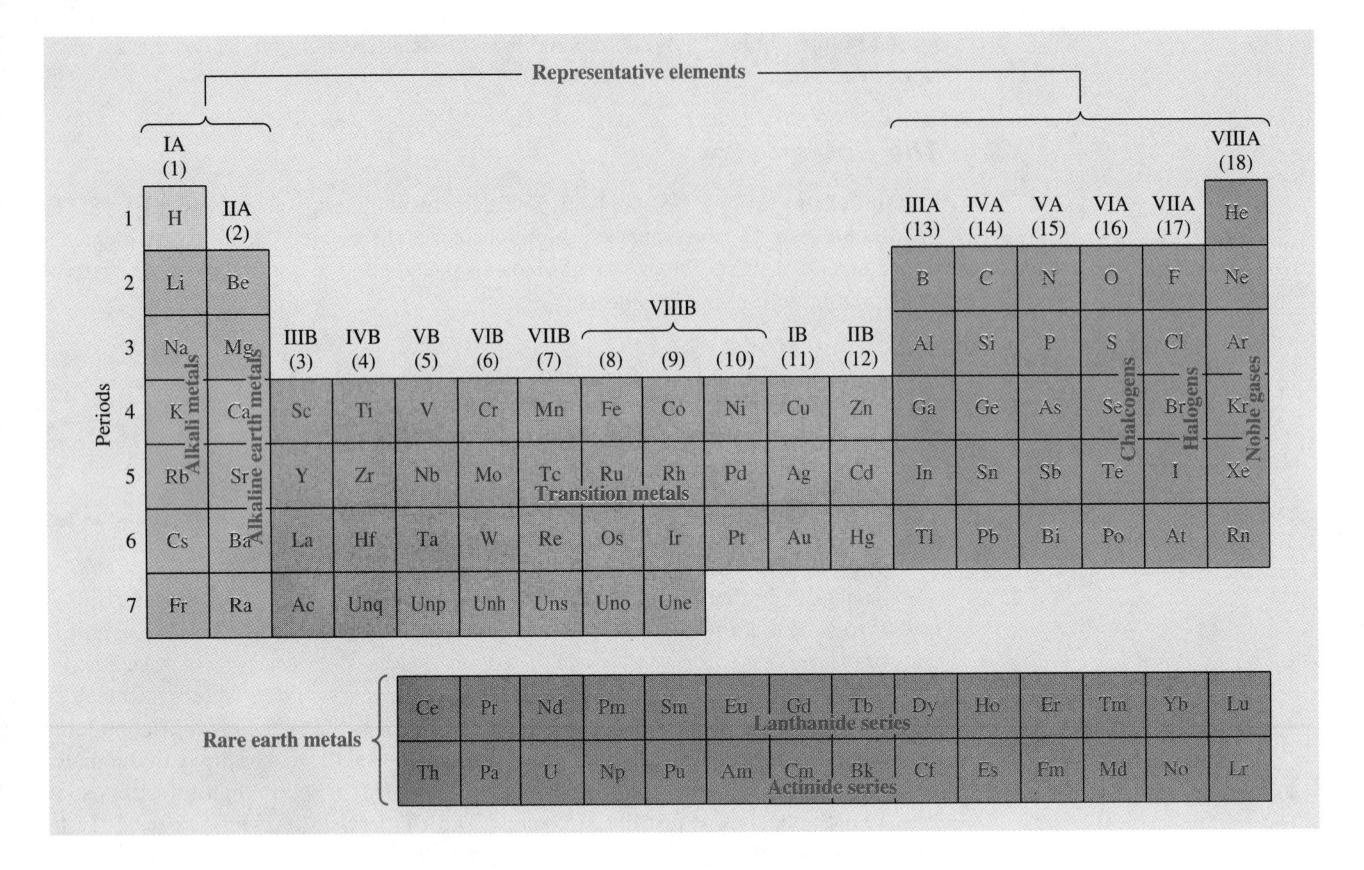

FIGURE 7-4 Organization of the periodic table. Note the locations of the transition metals, the lanthanide series, and the actinide series.

1*s* sublevel, because the next electron must likewise go into the next principal energy level (the 2*s* sublevel). The element following each noble gas begins to fill the next principal energy level, so it corresponds to the beginning of the next period.

Chemists have long been aware of a general lack of chemical reactivity among the noble gases. It is only under special conditions that these elements can be made to undergo chemical change. This lack of reactivity is a direct result of the exceptional stability of their electronic configurations. Because the noble gases exhibit similar chemical behavior, they constitute a chemical family that makes up the last column (group VIIIA or 18) of the periodic table (Fig. 7-4).

Prior to the 1960s, the noble gases were not thought to react with other elements and were known as the *inert* gases. However, Neil Bartlett, now at the University of California, Berkeley, successfully prepared several compounds of xenon and krypton, causing the chemical community to rename this chemical family the noble gases.

Alkali Metals

Further examination of Fig. 7-3 shows that each of the elements immediately following a noble gas (atomic numbers 3, 11, 19, 37, 55, and 87) has an unusually low ionization energy. Thus, lithium (atomic number 3) loses an electron readily. In the process, a lithium ion with only two electrons is formed:

$$\underset{(3p^+, 3e^-)}{\text{Li}} \rightarrow \underset{(3p^+, 2e^-)}{\text{Li}^+} + \underset{(1e^-)}{e^-}$$

This is the same number of electrons as helium, one of the noble gases.

Similarly, when sodium (atomic number 11) loses an electron, the sodium ion formed has 10 electrons. This is the same as neon, another of the noble gases.

$$\underset{(11p^+, 11e^-)}{\text{Na}} \rightarrow \underset{(11p^+, 10e^-)}{\text{Na}^+} + \underset{(1e^-)}{e^-}$$

The elements with atomic numbers 19, 37, 55, and 87 also lose one electron with ease, each forming an ion with a noble gas configuration. For example, potassium ion has the same number of electrons as the noble gas argon.

$$\underset{(19p^+, 19e^-)}{\text{K}} \rightarrow \underset{(19p^+, 18e^-)}{\text{K}^+} + \underset{(1e^-)}{e^-}$$

Alkali metals: The elements in group IA (1)

Lithium, sodium, and potassium are members of the chemical family known as the **alkali metals,** found in the first column (group IA or 1) of the periodic table (Fig. 7-4). Examination of the electronic configurations of the alkali metals reveals that each of these elements has one electron beyond the last noble gas core.

Li	$[\text{He}]2s^1$	or	$1s^22s^1$
Na	$[\text{Ne}]3s^1$	or	$1s^22s^22p^63s^1$
K	$[\text{Ar}]4s^1$	or	$1s^22s^22p^63s^23p^64s^1$
Rb	$[\text{Kr}]5s^1$	or	$1s^22s^22p^63s^23p^64s^23d^{10}4p^65s^1$
Cs	$[\text{Xe}]6s^1$	or	$1s^22s^22p^63s^23p^64s^23d^{10}4p^65s^24d^{10}5p^66s^1$
Fr	$[\text{Rn}]7s^1$	or	$1s^22s^22p^63s^23p^64s^23d^{10}4p^65s^24d^{10}5p^66s^24f^{14}5d^{10}6p^67s^1$

Loss of the outermost *s* electron leads to a noble gas configuration in each case. Thus, *the alkali metals are very susceptible to loss of one electron.* This property leads to similarities in chemical behavior among the elements of this family. These ionization studies, as well as other data gathered by removing more than one electron from elements of other families, all provide additional support for our earlier conclusion that *there is some special stability associated with an atom or ion when it contains the same number of electrons as one of the noble gases.*

In our discussion of the alkali metals, we have ignored hydrogen, another element that has one electron in its highest occupied *s* sublevel: $1s^1$. Like the alkali metals, hydrogen can form a +1 ion by losing its only electron. However, hydrogen has such unique chemical and physical properties that it is often considered a family by itself. To begin with, hydrogen is not even a metal; it is a gas. In addition, there are many known compounds in which hydrogen behaves as though it has *gained* an electron, becoming a negative ion with a helium-like electronic configuration:

$$\underset{(1p^+, 1e^-)}{\text{H}} + \underset{(1e^-)}{e^-} \rightarrow \underset{(1p^+, 2e^-)}{\text{H}^-}$$

We refer to the H^- ion as a hydride ion, thereby distinguishing it from a hydrogen ion, H^+.

Alkaline Earth Metals

Next, we consider the elements in the second column of the periodic table (group IIA or 2). Experimental results have demonstrated that these elements (atomic numbers 4, 12, 20, 38, 56, and 88) tend to lose two electrons to form +2 ions with noble gas configurations. (The energy associated with the loss of the first electron is known more specifically as the *first ionization energy.* The energy associated with the loss of the second electron is known as the *second ionization energy,* and similarly for successive electrons.)

$$\underset{(4p^+,\,4e^-)}{Be} \rightarrow \underset{(4p^+,\,2e^-)}{Be^{2+}} + \underset{(2e^-)}{2e^-}$$

$$\underset{(12p^+,\,12e^-)}{Mg} \rightarrow \underset{(12p^+,\,10e^-)}{Mg^{2+}} + \underset{(2e^-)}{2e^-}$$

$$\underset{(20p^+,\,20e^-)}{Ca} \rightarrow \underset{(20p^+,\,18e^-)}{Ca^{2+}} + \underset{(2e^-)}{2e^-}$$

Alkaline earth metals: The elements in group IIA (2).

This group of elements, known as the **alkaline earth metals,** consists of beryllium, magnesium, calcium, strontium, barium, and radium. Analysis of the electronic configurations of these elements reveals that they all have two electrons (s^2) in their highest occupied sublevel:

Be [He]$2s^2$
Mg [Ne]$3s^2$
Ca [Ar]$4s^2$

Hence, each of these metals achieves a noble gas configuration by losing *two* electrons.

Halogens

Halogens: The elements in group VIIA (17).

Next, let us examine the group VIIA (17) elements, which are known as the **halogens.** The halogen family consists of the elements fluorine, chlorine, bromine, iodine, and astatine. Each of these elements has a p^5 configuration for the highest occupied sublevel, one *less* than the number required to achieve a noble gas configuration.

F [He]$2s^22p^5$
Cl [Ne]$3s^23p^5$
Br [Ar]$4s^23d^{10}4p^5$
I [Kr]$5s^24d^{10}5p^5$

Because the addition of an electron results in a noble gas configuration, the halogens have very high electron affinities, and they *gain* electrons rather than losing them. In the case of fluorine (atomic number 9), for example, addition of an electron leads to the same electronic configuration as neon (atomic number 10):

$$\underset{(9p^+,\,9e^-)}{F} + \underset{(1e^-)}{e^-} \rightarrow \underset{(9p^+,\,10e^-)}{F^-}$$

Note that the addition of electrons leads to a negatively charged ion, because there are more electrons (–1 charges) than protons (+1 charges). Chlorine, bromine, and iodine also form –1 ions.

$$\underset{(17p^+,\ 17e^-)}{Cl} + \underset{(1e^-)}{e^-} \rightarrow \underset{(17p^+,\ 18e^-)}{Cl^-}$$

$$\underset{(35p^+,\ 35e^-)}{Br} + \underset{(1e^-)}{e^-} \rightarrow \underset{(35p^+,\ 36e^-)}{Br^-}$$

$$\underset{(53p^+,\ 53e^-)}{I} + \underset{(1e^-)}{e^-} \rightarrow \underset{(53p^+,\ 54e^-)}{I^-}$$

Chalcogens

Chalcogens: The elements in group VIA (16).

The elements in group VIA (16) are known as the **chalcogens.** Among the elements in this group are oxygen, sulfur, and selenium. They achieve stable noble gas configurations by gaining two electrons, thereby forming –2 ions:

$$\underset{(8p^+,\ 8e^-)}{O} + \underset{(2e^-)}{2e^-} \rightarrow \underset{(8p^+,\ 10e^-)}{O^{2-}}$$

$$\underset{(16p^+,\ 16e^-)}{S} + \underset{(2e^-)}{2e^-} \rightarrow \underset{(16p^+,\ 18e^-)}{S^{2-}}$$

Table 7-2 summarizes the families we have discussed thus far and the ionic charges they are most likely to form.

TABLE 7-2 Chemical families and the charges of their ions

Family name	Charge of most likely ion	Location in periodic table	Most common examples
Alkali metals	+1	IA (1)	Li^+, Na^+, K^+
Alkaline earth metals	+2	IIA (2)	Mg^{2+}, Ca^{2+}, Sr^{2+}, Ba^{2+}
Chalcogens	–2	VIA (16)	O^{2-}, S^{2-}
Halogens	–1	VIIA (17)	F^-, Cl^-, Br^-, I^-
Noble gases	(Do not form ions)	VIIIA (18)	He, Ne, Ar, Kr, Xe, Rn

Transition Metals and Rare Earth Metals

For each of the families we have considered, the elements are located in the same vertical column and possess similar endings to their electronic configurations. This general principle is true throughout the periodic table. For example, consider the elements found in group IVB (4) of the periodic table: titanium, zirconium, and hafnium. These elements end in a d^2 configuration.

Ti $[Ar]4s^23d^2$

Zr $[Kr]5s^24d^2$

Hf $[Xe]6s^24f^{14}5d^2$

Transition metals: The elements that fill *d* sublevels.

Rare earth metals: The elements that fill *f* sublevels.

Lanthanide series: The sixth row series of rare earth metals beginning with lanthanum.

Actinide series: The seventh row series of rare earth metals beginning with actinium.

Representative elements: The elements that fill *s* and *p* sublevels.

Figure 7-5 shows an *idealized* picture of the organization that the periodic table would have if the electronic configurations of all the elements were exactly those predicted using the energy diagram presented in Figure 6-6b. We have selected our examples carefully, in order to illustrate the important fundamental principles surrounding electronic configuration and its relationship to the organization of the periodic table. However, nature is not quite so simple as our drawing, and the outermost configurations of some of the elements filling *d* and *f* sublevels may differ slightly from this. (See Appendix D for the actual configurations of the elements.) Nevertheless, Fig. 7-5 is essentially correct in its presentation of the organization of the periodic table. We refer to the elements that fill *d* sublevels as the **transition metals;** and to the elements that fill *f* sublevels as the **rare earth metals,** which are composed of the **lanthanide series** and the **actinide series.** The elements that fill the *s* and *p* sublevels (the "A" groups) are known as the **representative elements.** The locations of these various groups of elements are included in Fig. 7-4.

Summary of Trends in Ionization Energies and Electron Affinities

As we have seen (Fig. 7-3), the ionization energies of the elements *generally* increase from left to right across each row of the periodic table. This makes sense, since elements at the left of each row (the alkali metals) can achieve noble gas configurations most readily by losing an electron. The noble gases (at the extreme right) are resistant

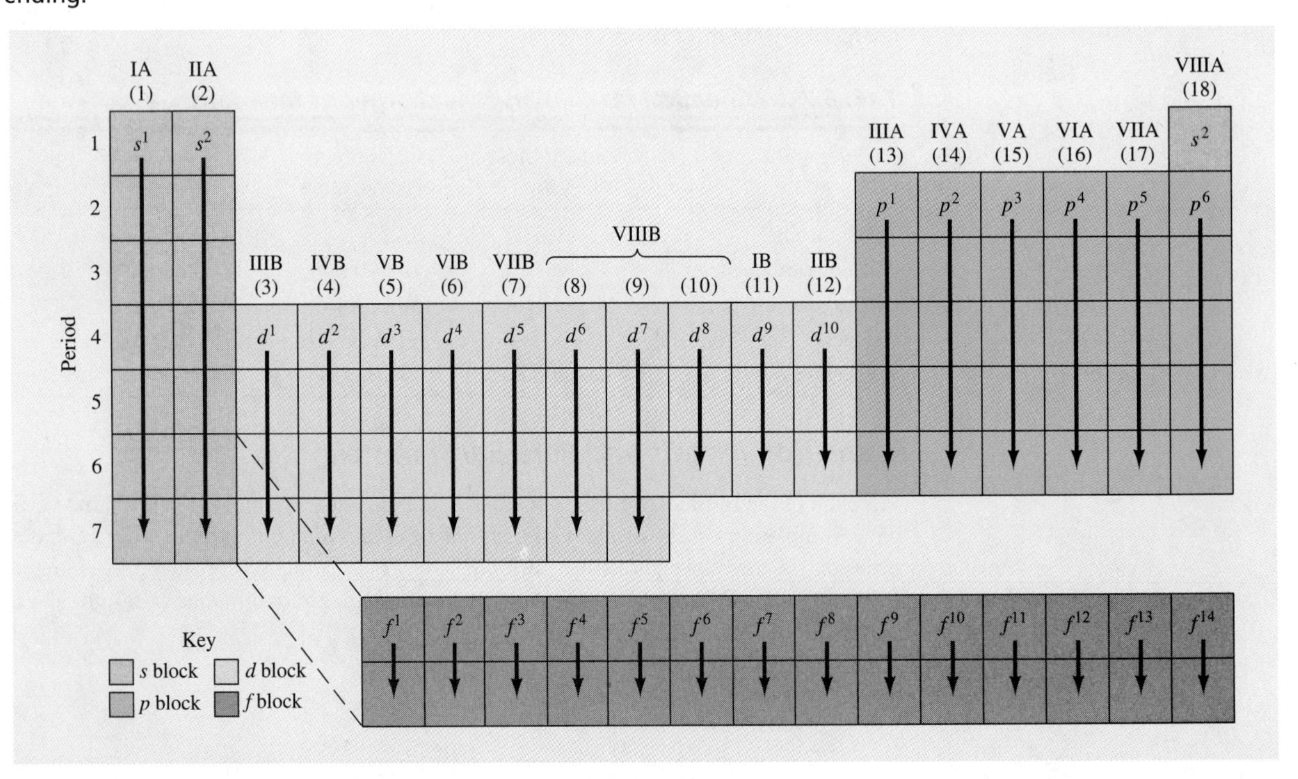

FIGURE 7-5 An idealized organization of the periodic table (by families or groups) that shows how members of the same family are related by electronic configuration ending.

to the loss of electrons, and thus exhibit the highest ionization energies. Figure 7-3 also reveals that ionization energies decrease from top to bottom of a given family.

Trends for electron affinity are not as simple to analyze. Within the various periods, electron affinities are generally largest for the halogens (which need to gain one electron to achieve noble gas configurations). Alkali metals have little tendency to gain an electron (since their tendency is toward loss of an electron). Thus, electron affinities increase from left to right across each period, except that the noble gases have little tendency to gain electrons, and therefore exhibit electron affinities close to zero.

PROBLEM 7.3

Predict the most stable ion formed from each of the following. (If the element forms no ion, write "no ion.")

(a) K **(b)** F **(c)** Ca **(d)** Se **(e)** Ar **(f)** Cs
(g) O **(h)** Xe **(i)** Te

PROBLEM 7.4

State the name of the chemical family for each of the following elements.

(a) Ca **(b)** Ne **(c)** Br **(d)** Li **(e)** Te **(f)** Ar
(g) I **(h)** K **(i)** Sr **(j)** Na **(k)** S

PROBLEM 7.5

Use statements 1–6 below to describe atoms (a)–(h). More than one statement may be correct for each atom.

1. Forms a stable −1 ion
2. A noble gas
3. Has the second principal energy level as its highest occupied energy level
4. Has 36 electrons
5. An alkaline earth metal
6. None of the above

(a) Ne **(b)** Cl **(c)** Xe **(d)** Ba **(e)** F **(f)** Br
(g) Na **(h)** Al

7.4 DETERMINING ELECTRONIC CONFIGURATION FROM THE PERIODIC TABLE

OBJECTIVE 7.4 Predict the complete or abbreviated electronic configuration of any element using only the periodic table as a guide.

In the last section we showed an idealized version of the periodic table (Fig. 7-5). Examination of Fig. 7-5 reveals that there are distinct blocks of sublevels: an *s* block, a *p* block, a *d* block, and an *f* block. Using this organization, we can check the electronic configuration of an element. For example, tin (atomic number 50) is in group IVA (14). This is the second column from the left within the *p* block. Thus, we expect tin to have a configuration that ends in p^2. Using Fig. 6-6b (or Fig. 6-8), we can generate the complete electronic configuration for tin,

$$\text{Sn} \quad 1s^22s^22p^63s^23p^64s^23d^{10}4p^65s^24d^{10}5p^2$$

and then abbreviate its electronic configuration as $[\text{Kr}]5s^24d^{10}5p^2$.

However, with a little reliance on memory and an application of the principles we have learned, we can determine the electronic configuration of any element by examining its position in the periodic table. First, let us examine Fig. 7-6, which is yet another way of showing the order in which the various sublevels are filled. Note that each period begins with an *s* sublevel of the same principal energy level as the period. Thus, period 1 begins with 1*s*, period 2 begins with 2*s*, and so forth. Similarly, after the first period, all periods end in a *p* sublevel of the same principal energy level. For example, the second row ends with 2*p*, and the third row with 3*p*, and so forth. Remember that helium, which is grouped with the p^6 noble gases, completes the 1*s* sublevel: $1s^2$.

The principal energy levels for the *d* and *f* sublevels are not the same as the period in which they are located. A *d* sublevel has a principal energy level that is one less than its period and an *f* sublevel has a principal energy level that is two less than its period. Thus, in period 5 we find $5s \rightarrow 4d \rightarrow 5p$. In period 6, the order is $6s \rightarrow 4f \rightarrow 5d \rightarrow 6p$. Remember that the periodic table itself tells us the order of sublevels. For example, in period 3, as we read across the period we first encounter an *s* sublevel, and a *p* sublevel comes next. However, in period 4, a *d* sublevel follows the *s* sublevel, and then comes the *p* sublevel. Similarly, as we read across period 6, an *s* sublevel is followed by an *f* sublevel, then a *d* sublevel, and finally the *p* sublevel.

To use the periodic table to determine the complete electronic configuration of tin, we simply start at the upper left of the periodic table and go from left to right across each period, filling each sublevel with the maximum number of electrons until we get to the sublevel in which tin is located. We determine the number of electrons in the final sublevel by counting from the left of the block, as we did before. In this case, since tin

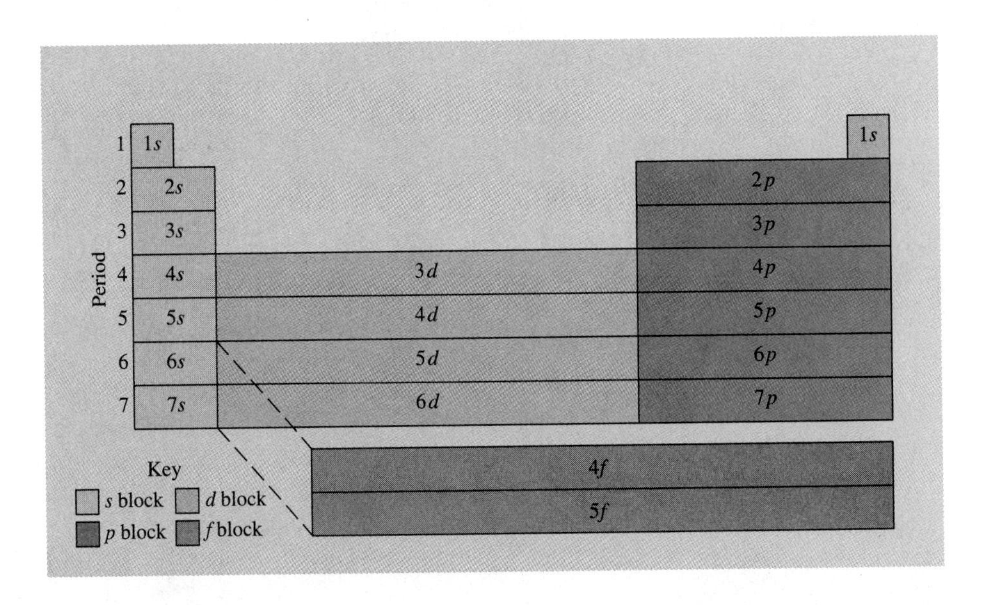

FIGURE 7-6 An idealized organization of the periodic table (by periods) that shows how sublevels fill within each period.

is in the $5p$ sublevel, we fill all of the sublevels through $4d$ (the sublevel preceding $5p$) and then count over two columns, giving us the final $5p^2$.

Determining an abbreviated electronic configuration is even easier. After finding the location of tin in the periodic table, we look for the last noble gas that precedes it; in this case it is krypton, Kr. So our abbreviated electronic configuration begins with the core symbol [Kr]. Since tin is in period 5, we continue with the $5s$ sublevel, then the $4d$ sublevel, and finish up as before with the partially filled $5p$ sublevel: $[Kr]5s^24d^{10}5p^2$.

EXAMPLE 7.1

Using the periodic table as your only guide, determine the complete electronic configuration for phosphorus, P.

Planning the solution

To determine a complete electronic configuration, imagine the periodic table as a page you are about to read. Start at the upper left and, reading across each line, proceed until you get to phosphorus. Phosphorus is in the third line (period 3).

SOLUTION

The first line (period 1) is just the $1s$ sublevel:

$$\text{P} \quad 1s^2 \ldots$$

The second line (period 2) begins with the $2s$ sublevel, followed by the $2p$ sublevel:

$$\text{P} \quad 1s^22s^22p^6 \ldots$$

The third line (period 3) contains the element we are interested in, so we must take care to determine the ending. Phosphorus is in the third column of the p block, so the electronic configuration will end in p^3. First the $3s$ sublevel fills completely, followed by the partially filled $3p$ sublevel.

$$\text{P} \quad 1s^22s^22p^63s^23p^3$$

EXAMPLE 7.2

Using only the periodic table as your guide, determine the abbreviated electronic configuration of iridium, Ir.

Planning the solution

We need to determine three things in order to write the abbreviated electronic configuration: (1) the last noble gas preceding iridium, (2) the period in which iridium is located, and (3) the final sublevel ending.

SOLUTION

Examination of the location of iridium in the periodic table reveals that it is a transition metal (ends in a d sublevel) in period 6. The last noble gas before iridium is xenon. Thus, the electronic configuration begins with the core symbol [Xe]:

$$\text{Ir} \quad [\text{Xe}] \ldots$$

Since iridium is in the sixth period, the electronic configuration continues with the $6s$ sublevel:

$$\text{Ir} \quad [\text{Xe}]6s^2 \ldots$$

In period 6, an f sublevel follows the $6s$ sublevel. (Follow the elements across the period to ascertain which sublevels are filled.) Since the principal energy level of an f sublevel must be two less than the number of the period, it is a $4f$ sublevel:

Ir $[Xe]6s^2 4f^{14} \ldots$

Next comes the *d* sublevel that the electronic configuration ends in. Since the principal energy level of a *d* sublevel is one less than the number of the period, it is a $5d$ sublevel. Iridium is located in the seventh column of the *d* block, so the electronic configuration ends in $5d^7$.

Ir $[Xe]6s^2 4f^{14} 5d^7$

Checking the answer

The method we have applied is its own check. Iridium is in the seventh column of the transition metals. Thus, it must end in a d^7 configuration.

PROBLEM 7.6

Using only the periodic table as your guide, predict the electronic configuration of each of the following elements. Write out the entire configuration first, then abbreviate it with the appropriate core symbol.
(a) P **(b)** Ni **(c)** Sr **(d)** Hg **(e)** I **(f)** Eu

7.5 FEATURES OF THE PERIODIC TABLE

OBJECTIVE 7.5 Locate the metals, semimetals, and nonmetals in the periodic table. Describe three properties of metals.

There are many versions of the periodic table. The most commonly encountered arrangement of the elements is shown in Fig. 7-7. In this version, the rare earth metals (the lanthanide series and actinide series) are separated from the main body of the periodic table. This arrangement gives the table a more compact, less cumbersome width.

The elements may be divided into three broad categories: **metals, semimetals** (or **metalloids**), and **nonmetals.** About 75% of the 109 known elements are metals. Metals are characterized by a variety of features, including luster, malleability (the ability to be pounded and bent into various shapes), and the ability to conduct electricity and heat. Metals also tend to lose electrons, thereby forming cations. Nonmetals may be described as elements that lack these properties. For example, the nonmetals generally exhibit low electrical conductivities and are not shiny. Refer to Fig. 7-7 and note that the metals are located on the left side of the periodic table and the nonmetals on the right. The heavy stair-step line that begins to the left of boron (atomic number 5) and descends to the right represents a rough dividing line between the metals and the nonmetals. A group of elements known as semimetals borders this dividing line. Semimetals are also commonly referred to as *metalloids.* These elements possess some of the properties that are characteristic of metals, but they may lack others. The semimetals generally are solids that are brittle (nonmalleable) and not lustrous. This group of elements consists of boron, silicon, germanium, arsenic, antimony, tellurium, polonium, and astitine. Several of the semimetals, such as silicon and germanium, are used in the production of the semiconductor devices that play such a vital role in the electronics industry.

Metal: An element that is shiny, malleable, and capable of conducting electricity.

Nonmetal: An element that lacks the properties of a metal.

Semimetal (or metalloid): An element that has properties intermediate between those of metals and nonmetals.

Period	IA (1)	IIA (2)	IIIB (3)	IVB (4)	VB (5)	VIB (6)	VIIB (7)	VIIIB (8)	VIIIB (9)	VIIIB (10)	IB (11)	IIB (12)	IIIA (13)	IVA (14)	VA (15)	VIA (16)	VIIA (17)	VIIIA (18)
1	1 H																	2 He
2	3 Li	4 Be											5 B	6 C	7 N	8 O	9 F	10 Ne
3	11 Na	12 Mg											13 Al	14 Si	15 P	16 S	17 Cl	18 Ar
4	19 K	20 Ca	21 Sc	22 Ti	23 V	24 Cr	25 Mn	26 Fe	27 Co	28 Ni	29 Cu	30 Zn	31 Ga	32 Ge	33 As	34 Se	35 Br	36 Kr
5	37 Rb	38 Sr	39 Y	40 Zr	41 Nb	42 Mo	43 Tc	44 Ru	45 Rh	46 Pd	47 Ag	48 Cd	49 In	50 Sn	51 Sb	52 Te	53 I	54 Xe
6	55 Cs	56 Ba	*57 La	72 Hf	73 Ta	74 W	75 Re	76 Os	77 Ir	78 Pt	79 Au	80 Hg	81 Tl	82 Pb	83 Bi	84 Po	85 At	86 Rn
7	87 Fr	88 Ra	**89 Ac	104 Unq	105 Unp	106 Unh	107 Uns	108 Uno	109 Une									

Period														
6	*58 Ce	59 Pr	60 Nd	61 Pm	62 Sm	63 Eu	64 Gd	65 Tb	66 Dy	67 Ho	68 Er	69 Tm	70 Yb	71 Lu
7	**90 Th	91 Pa	92 U	93 Np	94 Pu	95 Am	96 Cm	97 Bk	98 Cf	99 Es	100 Fm	101 Md	102 No	103 Lr

Key: Metal; Semimetal; Nonmetal

FIGURE 7-7 Standard periodic table.

Most of the elements are solids in their normal state at room temperature (20°C). Mercury and bromine are the only liquids at 20°C. Those elements that exist as gases are several of the diatomic elements (H_2, N_2, O_2, F_2, Cl_2) and the noble gases (He, Ne, Ar, Kr, Xe, and Rn).

PROBLEM 7.7

Identify the elements that can be described as follows at 20°C.

(a) A halogen that is a liquid
(b) Two chalcogens that are semimetals
(c) A transition metal that is not a solid
(d) Two halogens that are gases
(e) A group IIIA (13) semimetal
(f) A noble gas that does not end in a p^6 configuration
(g) A diatomic element that is a solid [*Hint:* What diatomic element is *not* mentioned in the preceding paragraph?]

7.6 Atomic Sizes

OBJECTIVE 7.6A Describe the general trend in size, going from top to bottom of any group or family. Explain the factor that is most important in determining this trend.

OBJECTIVE 7.6B Describe the general trend in size, going from left to right across any period. Explain the meaning of effective charge. Explain how effective charge is responsible for the trend in size across each period.

Figure 7-8 illustrates the most important trends in atomic size. *The atomic size increases going from top to bottom in a given family (or group). Atomic size decreases going from left to right across each row.* To explain the first trend, let us recall that orbitals increase in size as the principal energy level increases. Thus, as we go down a given family (or group) of the periodic table, the highest occupied orbital increases in size, and so does the atom. For example,

Lithium: $[He]2s^1$ Sodium: $[Ne]3s^1$

Radius = 152 pm Radius = 186 pm

[1 picometer (pm) = 10^{-12} meter (m)]. The highest occupied orbital in sodium (the $3s$) is larger than that for lithium (the $2s$).

In addition to orbital size, the nuclear charge of an atom may also play a role in determining atomic size. For example, because the nuclear charge is greater in sodium (+11) than in lithium (+3), we might expect the outermost electron in sodium to be drawn closer by the increased electrostatic forces of attraction. However, the $3s$ electron in the sodium atom is shielded from the full impact of the nuclear charge by the

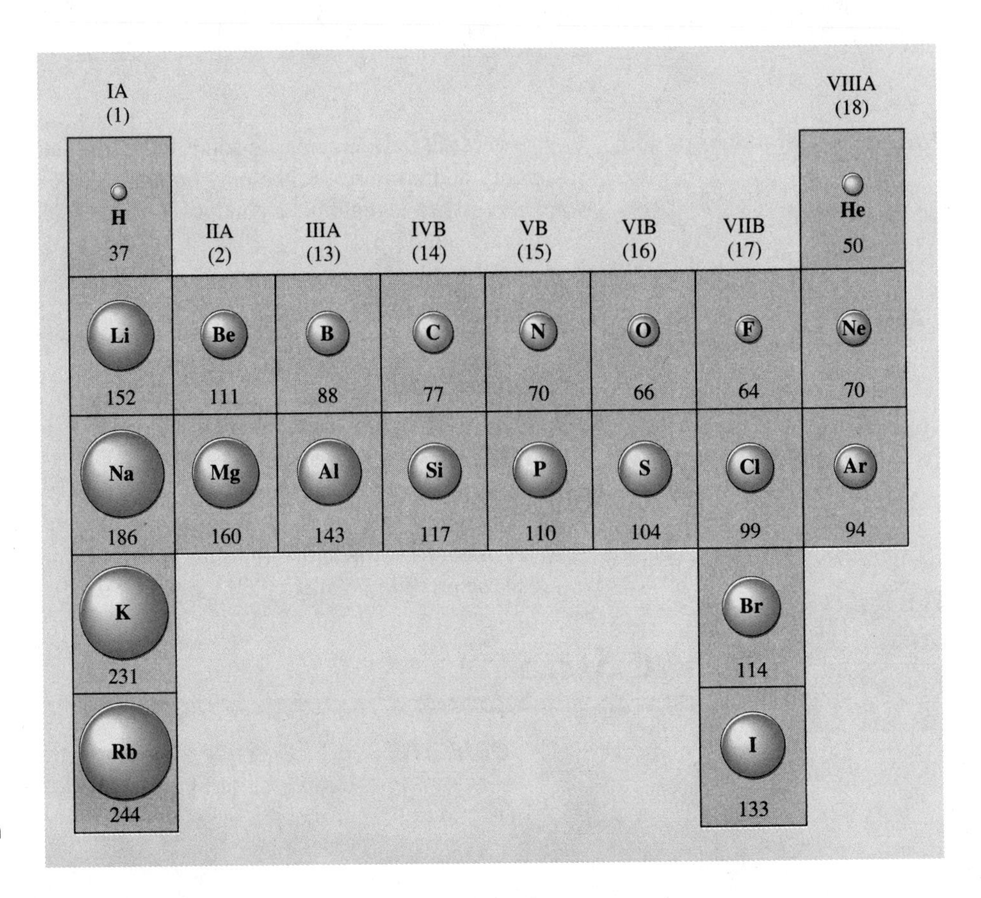

FIGURE 7-8 Selected atomic sizes. Generally, the sizes of atoms increase going down each family (or group) and decrease going across each period from left to right. The radius of each atom is given in picometers (1 pm = 10^{-12} m).

Effective charge: The charge experienced by an outermost electron after subtracting the electrons of the last noble gas core from the nuclear charge.

electrons in smaller orbitals, which lie between the nucleus and the 3*s* electron. These are the electrons of the last complete noble gas core. Thus, in determining electrostatic effects on the size of an atom (or ion), we must consider the effective charge that each electron experiences. The **effective charge** is equal to the nuclear charge minus all of the electrons in the last noble gas core. Thus the outermost electrons in both sodium and lithium experience the same effective charge, so this factor does not affect the trend in atomic size within a given family.

	Nuclear charge	Noble gas core		Effective charge
Lithium:	+3	−2	=	+1
Sodium:	+11	−10	=	+1

However, effective charge *is* an important factor in explaining the decrease in atomic size from left to right across a given period (or row). In general, the sizes of orbitals remain fairly constant across a given period. Thus, we might also expect the corresponding atomic sizes to remain constant. However, the nuclear charge increases going across each row, and this increases the effective charge, leading to increased electrostatic forces of attraction between the electrons and the nucleus. The result is that the electrons are drawn closer to the nucleus, and the size of the atoms decreases across a given period, as in the case of lithium (atomic number 3) and beryllium (atomic number 4):

	Nuclear charge	Noble gas core		Effective charge
Lithium:	+3	−2	=	+1
Beryllium:	+4	−2	=	+2

Lithium: [He]$2s^1$ Beryllium: [He]$2s^2$

Radius = 152 pm Radius = 111 pm

EXAMPLE 7.3

Predict which element in each of the following pairs is smaller:
(a) Ca or Mg **(b)** Na or S

SOLUTION

(a) Mg: Both Ca and Mg are in group IIA (2), but Mg is above Ca. Because the size of atoms increases going down a given family, Mg must be smaller than Ca.
(b) S: Both Na and S are in the third period, but S is to the right of Na. Because the size of atoms decreases from left to right across each row, S must be smaller than Na.

The increasing effective charge going across each period is also responsible for the corresponding increase in ionization energy (within each row) that was observed in Section 7.2. As the effective nuclear charge increases, so does the electrostatic force of attraction, and more energy is required to remove an electron from an atom.

PROBLEM 7.8

Use the principles discussed in this section to predict the larger in each of the following pairs of atoms.
(a) Na or K **(b)** F or Cl **(c)** S or Cl **(d)** K or Ca **(e)** K or Cl

7.7 THE SIZES OF IONS

OBJECTIVE 7.7 Describe how the size of an atom is affected when it loses electrons to become a cation. Describe how the size of an atom is affected when it gains electrons to become an anion. Explain the meaning of the term *isoelectronic*. Describe how the sizes of isoelectronic ions are affected by increasing atomic number. Explain what factor is responsible for this trend.

Ions may be larger or smaller than the parent atoms from which they are derived. For example, when a sodium atom forms an ion, it loses its 3*s* electron. It no longer has electrons in the third principal energy level, so the resulting size of the ion is determined by the smaller second-level orbitals (2*s* and 2*p*). Thus, the sodium ion is considerably smaller than the sodium atom (Fig. 7-9).

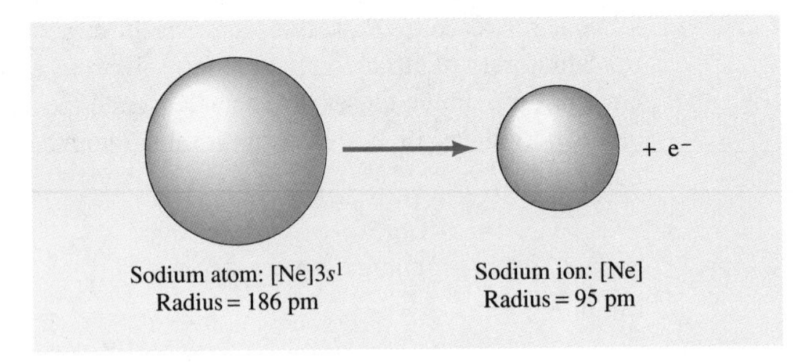

FIGURE 7-9 Relative sizes of the sodium atom and ion.

By contrast, a fluoride ion is larger than a fluorine atom (Fig. 7-10). Here the nuclear charge (+9) on fluorine remains constant, but the number of electrons increases from 9 to 10. The resulting repulsive forces between electrons increases, and the outer-orbital electrons spread out.

Table 7-3 shows how the sizes of several monatomic ions compare with their parent atoms. Cations are smaller than the parent; anions are larger.

We can analyze the sizes of the sodium ion (95 pm) and fluoride ion (136 pm) from an additional viewpoint. Both ions have the same electronic configuration: $1s^22s^22p^6$. (Note that this represents the neon gas core.) Because these two ions have identical electronic configurations, they are said to be **isoelectronic.** However, the sodium ion has a greater nuclear charge (+11) than fluoride ion (+9). We expect this greater nuclear charge to draw the electrons in the sodium ion closer to the nucleus,

Isoelectronic: Having the same electronic configuration.

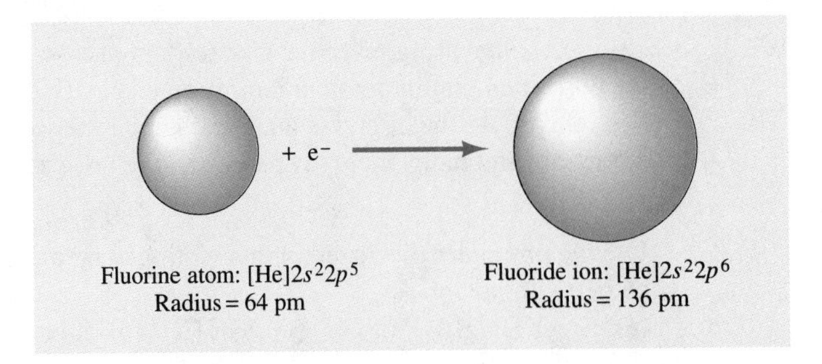

FIGURE 7-10 Relative sizes of the fluorine atom and ion.

TABLE 7-3 Comparison of atomic and ionic size (radii in pm[a])

Atom	Cation	Atom	Anion
Na	Na^+	O	O^{2-}
(186)	(95)	(66)	(140)
Mg	Mg^{2+}	F	F^-
(160)	(65)	(64)	(136)
Al	Al^{3+}	Cl	Cl^-
(143)	(50)	(99)	(181)
K	K^+	S	S^{2-}
(231)	(133)	(104)	(184)

[a] 1 pm = 10^{-12} m

making the sodium ion the smaller of the two. This is in agreement with the actual sizes of the ions. Further analysis of Table 7-3 shows that the following isoelectronic ions decrease in size as the nuclear charge is increased.

Ion:	O^{2-}	F^-	Na^+	Mg^{2+}	Al^{3+}
Nuclear charge:	+8	+9	+11	+12	+13

Decreasing size →

EXAMPLE 7.4

Predict the larger of each of the following pairs.
(a) Li or Li^+ **(b)** S or S^{2-} **(c)** S^{2-} or K^+

SOLUTION

(a) Li: Loss of an electron from an atom leads to a cation, which is smaller than the parent atom. Thus, Li is larger than Li^+.
(b) S^{2-}: Gain of an electron by an atom leads to an anion, which is larger than the parent atom. Thus, S^{2-} is larger than S.
(c) S^{2-}: Both S^{2-} and K^+ have 18 electrons (and are therefore isoelectronic), but the nuclear charge of S (+16) is smaller than that of K (+19). Consequently, the electrons in S^{2-} are not drawn as close to the nucleus, so S^{2-} is larger than K^+.

The Summary lists the trends we have discussed concerning atomic and ionic sizes.

SUMMARY *Atomic and Ionic Sizes*

1. The size of atoms in a family or group *always* increases from the top to the bottom of the column.
2. The size of atoms in a period *generally* decreases from left to right across the row.
3. The size of a cation is smaller than its parent atom.
4. The size of an anion is larger than its parent atom.
5. The size of isoelectronic ions (and atoms) decreases with increasing atomic number.

SUMMARY DIAGRAM

Periodic Trends for (a) Atomic Size, (b) Ionization Energy, (c) Electron Affinity, and (d) Metallic Nature

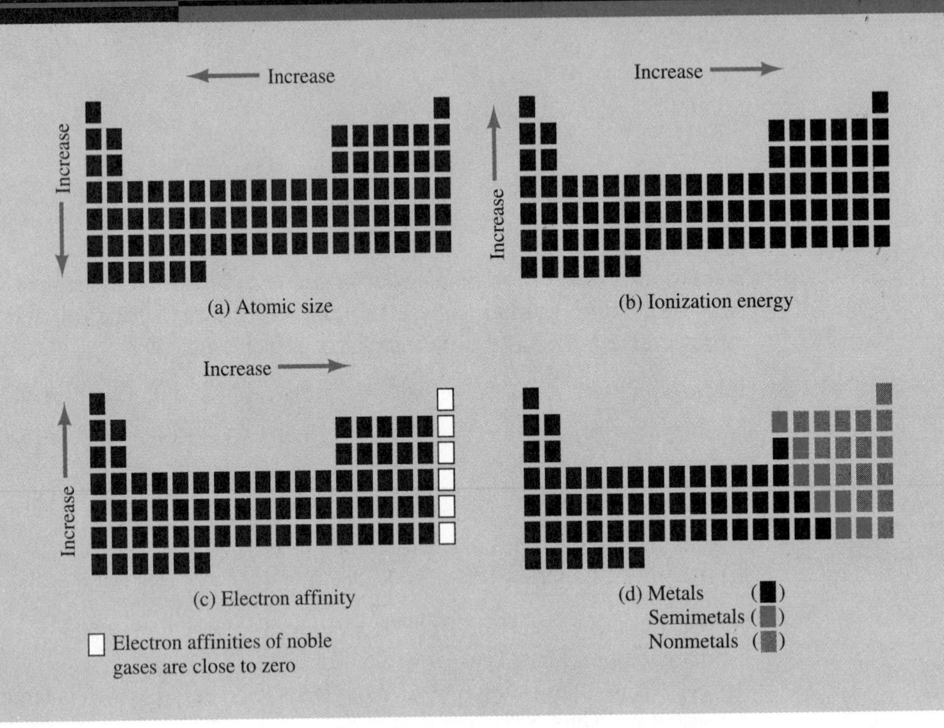

The Summary Diagram illustrates some of these and other trends we have examined in this chapter.

PROBLEM 7.9

Pick out the pair of isoelectronic species in each of the following groups of atoms and ions.
(a) He, Ne, Li^+ **(b)** F^-, S^{2-}, Mg^{2+} **(c)** Ca^{2+}, P^{3-}, Al^{3+}

PROBLEM 7.10

Using the principles discussed in this section, predict the larger of each of the following.
(a) Na^+ or K^+ **(b)** Cl^- or Br^- **(c)** O^{2-} or S^{2-} **(d)** Ne or F^-
(e) Ar or K^+ **(f)** K^+ or Cl^- **(g)** Na^+ or Mg^{2+} **(h)** O^{2-} or F^-

PROBLEM 7.11

Explain the increasing size in the following sequence of ions (see Table 7-3): $K^+ < Cl^- < S^{2-}$.

CHAPTER SUMMARY

The unique organization of the periodic table was originally proposed as a result of recurring physical and chemical properties in the elements. Today we know that these recurring properties reflect recurring patterns in electronic configuration. The rows of the periodic table are referred to as **periods;** the columns represent **groups** or **chemical families.** In general, the sizes of atoms in a given period decrease from left to right. The sizes of atoms in a given family increase going down the column.

Elements are capable of losing or gaining electrons to form ions. Positive ions, known as cations, are formed when neutral atoms lose electrons. Negative ions, called anions, are formed when neutral atoms gain electrons. **Electrostatic forces** exist between charged particles such as electrons and nuclei. Since oppositely charged particles attract, it takes energy to remove an electron from an atom. Conversely, energy is released when an electron is captured by an atom. The **ionization energy** is the energy required to remove an electron from an atom. The **electron affinity** is the energy released when an atom acquires an electron. In general, ionization energies increase from left to right across a given period of the periodic table and decrease from top to bottom of a given family. Trends in electron affinity are not so clear-cut as those for ionization energy, but electron affinities are largest for the halogens. The various ionization energies and electron affinities support the conclusion that there is a special stability associated with noble gas configurations, and neutral atoms tend to form ions having such noble gas configurations.

The elements in a given family have similar chemical properties as a result of their similar electronic configurations. Among the various chemical families, the **noble gases** are resistant to chemical change because of the inherent stability of their electronic configurations. Other elements tend to form ions with noble gas configurations. The other major families and the charges of their ions are: **alkali metals** (+1), **alkaline earth metals** (+2), **chalcogens** (−2), and **halogens** (−1).

The periodic table can be used to predict the electronic configurations of elements. The table is divided up into an *s* block, a *p* block, a *d* block, and an *f* block of elements. The **representative elements** are the elements in the *s* and *p* blocks, the **transition metals** are in the *d* block, and the **rare earth metals** (which can be further subdivided into the **lanthanide series** and **actinide series**) are in the *f* block. Within each block are as many groups (families) as the number of electrons that can fill that sublevel. Each group corresponds to a different configurational ending. By examining an element's position in the periodic table, the electronic configuration can be reconstructed, either in its entirety or in the standard abbreviated form.

The majority of the elements are both solid and metallic in nature. The **metals** are found on the left-hand side of the periodic table, the **nonmetals** on the right. A few **semimetals** separate the two groups.

The sizes of atoms and ions are influenced by both the size of the orbitals that contain the outermost electrons as well as the **effective charge** on those electrons. To summarize the trends, the atoms in a family or group increase in size from the top to the bottom of their column. By contrast, the sizes of atoms in a period generally decrease from left to right across the row. Cations are always smaller than their parent atoms, whereas anions are always larger than their parent atoms. The sizes of **isoelectronic** ions (and atoms) decrease with increasing atomic number.

KEY TERMS

Review each of the following terms, which are discussed in this chapter and defined in the Glossary.

period
group
chemical family
ionization energy
electron affinity
electrostatic force
noble gas
alkali metal
alkaline earth metal
chalcogen
halogen
transition metal
rare earth metal
lanthanide series
actinide series
representative element
metal
semimetal (or metalloid)
nonmetal
effective charge
isoelectronic

ADDITIONAL PROBLEMS

SECTION 7.1

7.12

(a) Are the elements in the periodic table arranged according to atomic mass or atomic number?
(b) What is a period?
(c) What is a chemical family or group?

SECTION 7.2

7.13

(a) What are electrostatic forces?
(b) What are the effects of like and unlike charges on one another?
(c) How are electrostatic forces affected by the magnitude of the charges?
(d) How are electrostatic forces affected by the distance between the charged objects?
(e) What is ionization energy?
(f) Why must energy be supplied to bring about the ionization of an atom?

7.14

(a) Define the term *electron affinity.*
(b) Write an equation for the capture of an electron by a chlorine atom.
(c) Why is energy released when an electron is captured by an atom?

SECTION 7.3

7.15

(a) What feature of electronic configuration is common to all noble gases except helium?
(b) What feature of electronic configuration is common to all noble gases, including helium?
(c) What generalization can be made about the ionization energies of the noble gases?
(d) Why do the alkali metals have relatively low ionization energies?
(e) Why do the halogens have relatively high electron affinities?

7.16 Classify each of the elements in parts (a)–(n) as either an alkali metal, an alkaline earth metal, a chalcogen, a halogen, or a noble gas. If the element forms ions, write the formula for the ion that is most likely to form. If the element is resistant to ion formation, write "no ion."

(a) Li **(b)** Mg **(c)** Cl **(d)** Ba **(e)** S
(f) I **(g)** Rb **(h)** Sr **(i)** Ne **(j)** F
(k) Na **(l)** Br **(m)** Kr **(n)** Po

7.17 Use statements 1–7 to characterize the atoms and ions (a)–(j). More than one statement may be correct for each atom or ion.

1. A noble gas
2. Forms a stable +2 ion
3. Has 18 electrons
4. An alkali metal
5. A halogen
6. The highest occupied sublevel belongs to the third principal energy level
7. None of the above

(a) He **(b)** Li **(c)** Ca **(d)** P **(e)** Ar
(f) Mg **(g)** Fr **(h)** S^{2-} **(i)** Cl^- **(j)** I

7.18 Classify each of the following elements as a transition metal or a rare earth metal.

(a) Ag **(b)** Fe **(c)** U **(d)** Cu **(e)** Eu
(f) Pt **(g)** Pu **(h)** Ho **(i)** Au **(j)** Ni

7.19

(a) What characteristic of electronic configuration do all transition metals have in common?
(b) What characteristic of electronic configuration do all rare earth metals have in common?

SECTION 7.4

7.20 Using only the periodic table as your guide, predict the electronic configurations of the following elements. In each case write out the complete electronic configuration, and then abbreviate it using the appropriate noble gas core symbol.

(a) Na **(b)** Ga **(c)** Ca **(d)** V **(e)** Pm **(f)** Zr

SECTION 7.5

7.21

(a) Describe the three properties of metals.
(b) Classify each of the following as a metal, a semimetal, or a nonmetal: Ge, Al, N_2, As, I_2, Hg.

SECTION 7.6

7.22

(a) Describe the general trend in size going from top to bottom of any group or family.

(b) What factor is most important in determining this trend?

7.23

(a) Describe the general trend in size going from left to right across any period.

(b) What is the meaning of the term *effective charge?*

(c) Explain how effective charge is responsible for the trend in size across each period.

SECTION 7.7

7.24

(a) How is the size of an atom affected when it loses electrons to become a cation?

(b) How is the size of an atom affected when it gains electrons to become an anion?

(c) What is the meaning of the term *isoelectronic?*

(d) How are the sizes of isoelectronic ions affected by increasing atomic number? What is responsible for this trend?

7.25 Using only the periodic table as a guide, arrange each of the following series of atoms and ions in order of size, from smallest (on the left) to largest (on the right).

(a) O, Po, Se **(b)** Fe, Os, Ru **(c)** Al, Cl, Mg
(d) Br, Ca, K **(e)** Cl, Cs, Sn **(f)** As, Br, Sb
(g) Cl^-, K^+, P^{3-} **(h)** Mg, Na, Mg^{2+} **(i)** F, F^-, O^{2-}

GENERAL PROBLEMS

7.26

(a) Judging from its position in the periodic table, predict the charge of an aluminum ion.

(b) Phosphorus gains electrons to become the phosphide ion. Predict the charge of this ion.

7.27 Atoms that form anions gain their extra electrons from atoms that form cations.

(a) How many electrons does a magnesium atom lose to become a cation?

(b) How many electrons does a sulfur atom gain to become an anion?

(c) What ratio of magnesium atoms to sulfur atoms is required to form ions of each with no electrons left over?

(d) Repeat parts (a), (b), and (c) with sodium forming the cation and oxygen the anion.

(e) Repeat parts (a), (b), and (c) with aluminum forming the cation and sulfur the anion.

7.28 The formulas of ionic compounds are obtained by combining the simplest whole-number ratio of ions that results in a neutral charge. For example, K^+ and S^{2-} combine in a 2 : 1 ratio to form K_2S (potassium sulfide). Using this guideline, write formulas for compounds formed between each of the following pairs of ions.

(a) Na^+ and Cl^- (sodium chloride)
(b) Mg^{2+} and Br^- (magnesium bromide)
(c) Al^{3-} and I^- (aluminum iodide)
(d) Al^{3+} and O^{2-} (aluminum oxide)

7.29 Yttrium (atomic number 39) is larger than zirconium (atomic number 40).

(a) To what factor can the size difference between yttrium and zirconium be attributed?

(b) Based upon this information, what can you predict about the relative sizes of scandium (atomic number 21) and titanium (atomic number 22)?

***7.30** A sodium atom is larger than all of the noble gases. What conclusion can be drawn about the importance of the effective nuclear charge compared to the period of an element in determining atomic size? Explain your answer.

***7.31** Mercury is one of the few elemental liquids. Silver will dissolve in mercury liquid. When mixed in the correct ratio, mercury makes the silver more malleable. This mixture, which is known as a silver amalgam, is used by dentists to fill cavities. What is the advantage of adding the mercury to the silver for this purpose?

***7.32** The second ionization energy is defined as the energy required to remove a second electron from an element. The process involves the loss of an electron from E^+ to form E^{2+}, where E is any element. How would you expect the values for the second ionization energies of lithium and beryllium to compare to one another? Explain your prediction.

WRITING EXERCISES

7.33 A friend, who is studying art, has been told that she needs to know about metals and their properties. Write a note to your friend in which you describe the various properties of metals. Compare these properties to those of nonmetals and semimetals. Your note should mention that most of the elements are metals. Be sure to give several important examples of elements that are metals, nonmetals, and semimetals in your note, and suggest some reasons why an artist might need to know the properties of metals.

7.34 In your own words, explain how to write electronic configurations using the periodic table as the only guide. Illustrate your explanation by going through an example (other than any of the textbook examples).

7.35 Explain in general terms how the periodic table is laid out. Your explanation should include the meanings of families and periods, as well as the names and locations of the major families and blocks of elements. The locations of the metals, nonmetals, and semimetals should also be included in your explanation.

CHEMICAL BONDING

8

In Chapter 2, we saw that molecules are usually made up of more than one atom. We wrote formulas for many substances and even drew pictures of some of them. We accepted the fact that atoms can combine to form molecules, but we never really explained how. In this chapter we will explore the nature of the chemical bond and find out what holds these atoms together.

8.1 VALENCE ELECTRONS

OBJECTIVE 8.1 Define the term *valence electrons.* Determine the number of valence electrons for any element in the first three periods of the periodic table. Draw an electron-dot structure for each element.

Valence electrons: The outermost electrons in an atom. For elements in periods 1–3, those electrons beyond the last noble gas core.

As a general rule, the electrons involved in bonding are the outermost electrons in an atom. We refer to these as **valence electrons.** For elements in the first three rows of the periodic table, the valence electrons are those electrons beyond the last noble gas core. Table 8-1 shows the number of valence electrons for each element in the first three rows. We will restrict most of our discussion to these elements in order to avoid complications that can arise for elements having electrons in *d* and *f* sublevels. (Strictly

TABLE 8-1 Valence electrons of elements in the first three periods

Element	Number of valence electrons	Electronic configuration
H	1	$1s^1$
He	2	$1s^2$
Li	1	$[He]2s^1$
Be	2	$[He]2s^2$
B	3	$[He]2s^22p^1$
C	4	$[He]2s^22p^2$
N	5	$[He]2s^22p^3$
O	6	$[He]2s^22p^4$
F	7	$[He]2s^22p^5$
Ne[a]	8	$[He]2s^22p^6$
Na	1	$[Ne]3s^1$
Mg	2	$[Ne]3s^2$
Al	3	$[Ne]3s^23p^1$
Si	4	$[Ne]3s^23p^2$
P	5	$[Ne]3s^23p^3$
S	6	$[Ne]3s^23p^4$
Cl	7	$[Ne]3s^23p^5$
Ar[a]	8	$[Ne]3s^23p^6$

[a]Note that the noble gases neon and argon have eight electrons beyond the previous noble gas. Thus, $[He]2s^22p^6$ and $[Ne]3s^23p^6$ represent noble gas configurations.

speaking, we are limiting our discussion to members of the representative elements. Recall that those are the *s*- and *p*-block elements.)

Lewis structure: A symbolic description of an atom, molecule, or ion using dots to represent the valence electrons.

Dots are often used to represent the valence electrons in atoms and molecules. The structures obtained are referred to as **Lewis structures, electron-dot structures,** or **Lewis electron-dot structures.** These structures are named after G. N. Lewis, a pioneer in our understanding of the relationship between electronic structure and chemical bonding. The following electron-dot representations help us to describe the bonding of the atoms in Table 8-1.

Group Number							
IA (1)	IIA (2)	IIIA (13)	IVA (14)	VA (15)	VIA (16)	VIIA (17)	VIIIA (18)
H·							He:
Li·	·Be·	·B·	·C·	·N·	·O·	:F·	:Ne:
Na·	·Mg·	·Al·	·Si·	·P·	·S·	:Cl·	:Ar:

These electron-dot structures do not all correspond to ground-state configurations. For example, ·C: is the Lewis structure for the ground state of carbon ($[He]2s^2 2p_x^{\ 1} 2p_y^{\ 1}$). The paired electrons represent the 2*s* electrons; the unpaired electrons represent the $2p_x$ and $2p_y$ electrons. For our purposes, however, the structures just given are the most useful.

When writing Lewis structures, we imagine each symbol to be enclosed in a square:

F

The electron-dot structures are obtained by placing the valence electrons, one to a side, until all of the valence electrons are shown. For several of the elements, it is necessary to go around the square more than once. For fluorine, the seven valence electrons are placed as follows:

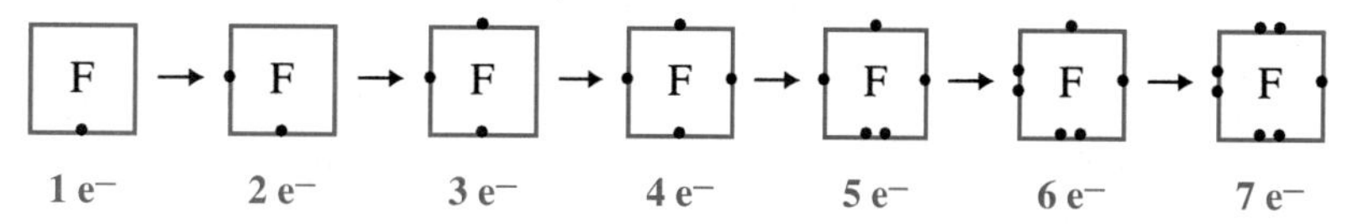

When drawing electron-dot structures, we do not actually draw the square, but instead we use our imagination to picture it. Thus, the electron-dot structure of a fluorine atom is represented as follows:

:F·

Furthermore, the four sides of the square are considered equivalent, so that any of the following structures also represents a fluorine atom.

:F: ·F: :F:

PROBLEM 8.1

Examine the number of valence electrons for each of the atoms in Table 8-1. Next find the location of each atom in the periodic table. What is the relationship between the number of valence electrons and the Roman numeral designation of the group (IA, IIA, or IIIA, for example) in which each element is found? [*Note:* This relationship is restricted to the *representative elements*—the "A" groups—and is true for all representative elements, even those beyond the third row.]

8.2 THE IONIC BOND

OBJECTIVE 8.2 Show the electron transfer that takes place in the formation of an ionic compound between any metal and any nonmetal from the first three periods. Explain what forces hold the ions together in an ionic bond. Give a rule for predicting whether two elements will form an ionic bond with one another.

In Chapter 7 it became clear that a special stability is associated with the electronic configurations of the noble gases. We discussed the tendency of atoms to gain or lose electrons in order to achieve such electronic configurations. For example, in our discussion of ionization energy and electron affinity, we saw that sodium atoms have a tendency to *lose* one electron, thereby achieving a neon core:

$$\underset{(11p^+, 11e^-)}{Na} \rightarrow \underset{(11p^+, 10e^-)}{Na^+} + \underset{(1e^-)}{e^-}$$

By contrast, fluorine atoms have a tendency to *gain* one electron, also achieving a neon core:

$$\underset{(9p^+, 9e^-)}{F} + \underset{(1e^-)}{e^-} \rightarrow \underset{(9p^+, 10e^-)}{F^-}$$

[Note that an ion is formed whenever an (uncharged) atom gains or loses an electron.]

When atoms of sodium and fluorine are brought together, the sodium loses its electron to fluorine, thereby providing *both* with noble gas configurations:

$$Na\cdot + \cdot\ddot{\underset{..}{F}}: \rightarrow Na^+ + :\ddot{\underset{..}{F}}:^-$$

Total electrons:	11 e⁻	9 e⁻	10 e⁻	10 e⁻
Valence electrons (or core):	1 e⁻	7 e⁻	[Ne]	[Ne]

Electron transfer: The transfer of electrons from one chemical species to another.

This process is known as **electron transfer.** As a general rule, oppositely charged ions are formed in this fashion whenever a metal combines with a nonmetal. When the atoms that combine are members of the representative elements, such as sodium and fluorine, the ions that form also achieve noble gas configurations.

Furthermore, as stated in our earlier discussion of electrostatic forces, oppositely charged particles attract one another. In this case, the electrostatic forces of attraction between the oppositely charged ions hold them together in an **ionic bond.** Compounds that contain any ionic bonds are called ionic compounds. *A compound containing both a metal and a nonmetal in its formula is generally an ionic compound.*

Ionic bond: A chemical bond that results from the electrostatic forces of attraction between oppositely charged ions.

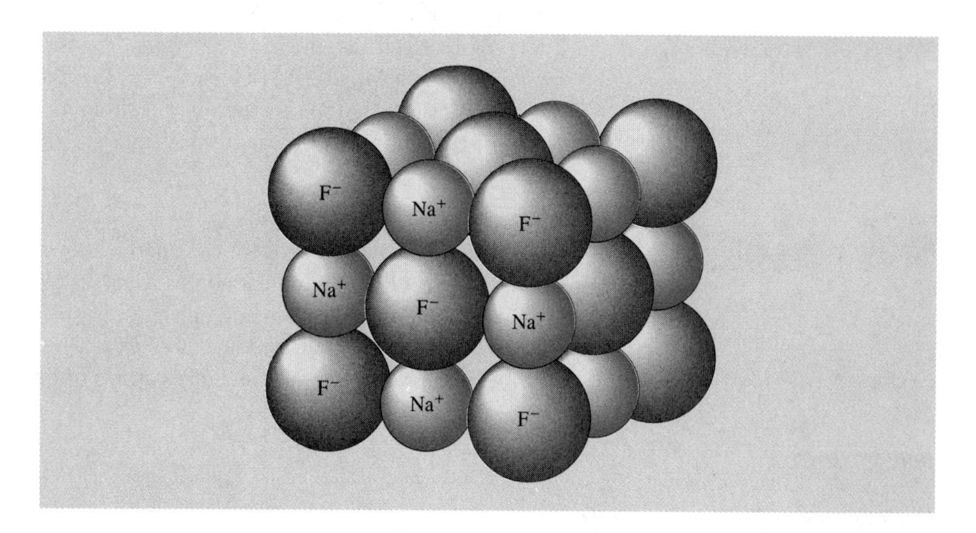

FIGURE 8-1 Sodium fluoride (NaF).

Crystal lattice: The arrangement of the atoms, ions, or molecules in the solid state of a substance.

In the solid state, ionic compounds such as NaF do not actually exist as distinct molecules, as our preceding discussion might suggest. Instead, the ions are arranged in an alternating pattern of positive and negative ions, as shown in Fig. 8-1. This arrangement is called a **crystal lattice.** We can see that this alternating arrangement of sodium and fluoride ions still leads to a neutrally charged substance. Because there are no distinct molecules formed, we write the formulas of ionic compounds as the simplest ratio of ions—in this case 1 : 1. In Chapter 4 we found it necessary to refer to a *formula mass,* rather than a *molecular mass,* when working with compounds such as this. This nonmolecular description of NaF is typical of all ionic compounds.

EXAMPLE 8.1

Atoms may combine in ratios other than 1 : 1. Draw electron-dot structures to show how one atom of magnesium and two atoms of chlorine can combine to form the ionic compound magnesium chloride, $MgCl_2$.

Planning the solution

Magnesium is a member of the alkaline earth metals and therefore tends to form a +2 ion. We expect each atom of magnesium to lose two electrons. Chlorine is a member of the halogens and tends to form a −1 ion. We therefore expect each chlorine atom to gain one electron. If the electrons lost by magnesium are to be gained by chlorine, we will need two chlorine atoms for each magnesium atom.

SOLUTION

$$\underset{(12\ e^-)}{\dot{Mg}\cdot} + \underset{(17\ e^-)}{\cdot\ddot{\underset{..}{Cl}}:} + \underset{(17\ e^-)}{\cdot\ddot{\underset{..}{Cl}}:} \longrightarrow \underset{(10\ e^-)}{Mg^{2+}} \quad \underset{(18\ e^-)}{:\ddot{\underset{..}{Cl}}:^-} \quad \underset{(18\ e^-)}{:\ddot{\underset{..}{Cl}}:^-}$$

Valence electrons: **2** **7** **[Ne]** **[Ar]**

Magnesium, with a total of 12 electrons, loses its 2 valence electrons, leaving it with 10 electrons, the same number as neon. Each chlorine atom has a total of 17 electrons (7 of these are its valence electrons). Each atom adds 1 electron to its valence level, giving it a total of 18 electrons, the same as argon. The resulting ions (one Mg^{2+} plus two Cl^-) combine to form $MgCl_2$. As in the case of NaF, the incentive for forming ions is to obtain noble gas configurations, and the resulting ionic bonds come from the electrostatic forces of attraction between oppositely charged ions.

PROBLEM 8.2

Using electron-dot structures for each atom, show the transfer of electrons between:
(a) One atom of aluminum and three atoms of fluorine
(b) One atom of magnesium and one atom of sulfur
(c) Two atoms of sodium and one atom of oxygen
(d) Two atoms of aluminum and three atoms of oxygen

8.3 THE COVALENT BOND

OBJECTIVE 8.3A Draw a potential-energy curve to describe the energy changes that occur when two atoms are brought together to form a covalent bond. Identify the bond length and the bond energy on the curve.

OBJECTIVE 8.3B Name the ingredients of every covalent bond. Describe the orientation of the electrons and the nuclei with respect to one another in a covalent bond. Describe the arrangement of the orbitals involved in the same covalent bond. Give a rule for predicting whether two elements will form a covalent bond with one another.

FIGURE 8-2 Bonding in molecular hydrogen, H_2. (a) Separated hydrogen atoms. (b) An attraction develops as the atoms move closer together. (c) The bonded atoms form molecular hydrogen, H_2.

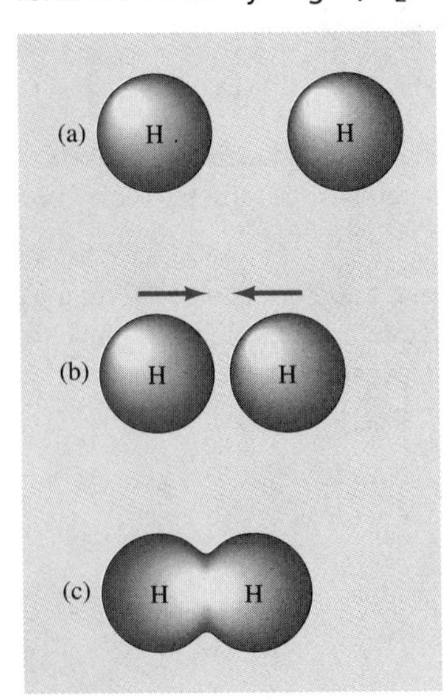

Covalent bond: A chemical bond that arises from the sharing of a pair of electrons between two atoms.

Molecular hydrogen (H_2) is a neutral molecule made up of two hydrogen atoms. Let us consider the manner in which the atoms are held together. Suppose two hydrogen atoms are separated by a large distance, as in Fig. 8-2a. If the two atoms are brought closer together, an attraction will develop between them, and they will become bonded to one another, as shown in Fig. 8-2b and c.

To understand why these two atoms are attracted to one another, recall that electrostatic forces may be either attractive or repulsive. For particles of opposite charge, the forces are attractive, whereas particles of like charge repel one another. As two hydrogen atoms approach one another, an attractive force develops between each electron and the other nucleus. However, repulsive forces also develop between the two nuclei, as well as between the two electrons. The most favorable arrangement of electrons and nuclei exists when the electrons are between the two nuclei (Fig. 8-3). Whether or not two atoms will bond when they approach one another depends on the magnitude of both the attractive and the repulsive forces. If the attractive forces are greater than the repulsive forces, there will be a net attraction and the atoms will bond. Such is the case for the two atoms in molecular hydrogen, H_2. We refer to a bond of this type as a covalent bond. A **covalent bond** results from the sharing of a pair of electrons between two atoms.

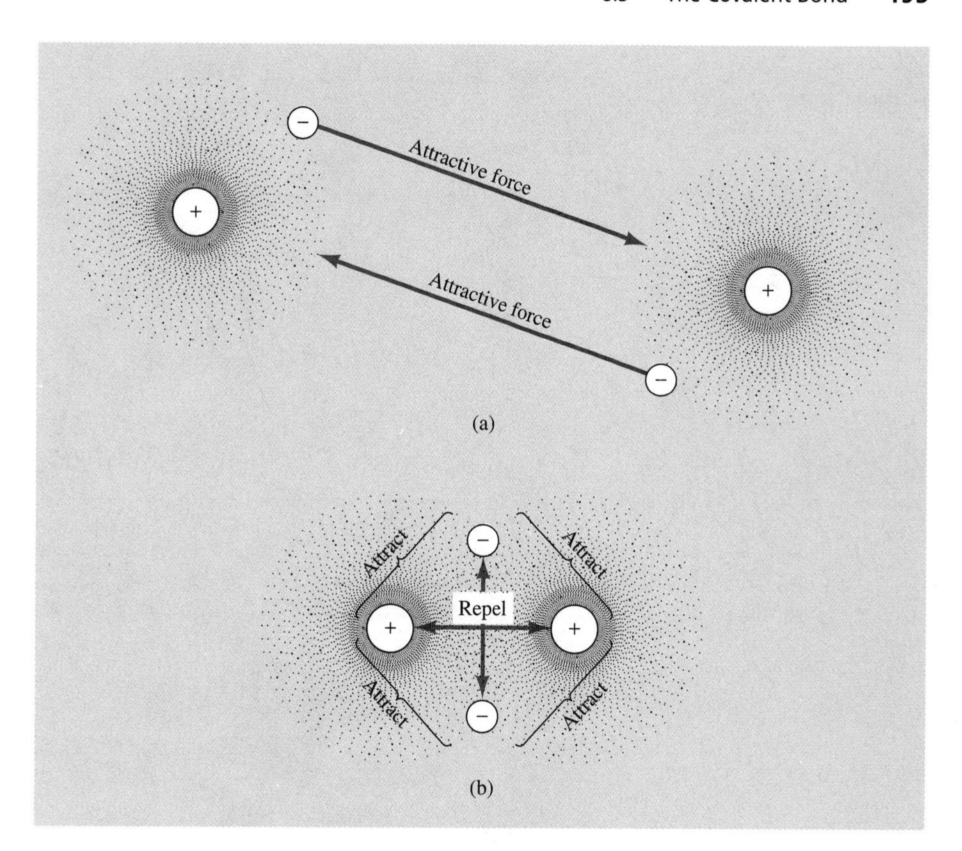

FIGURE 8-3 Electrostatic forces in the covalent bond. (a) As two hydrogen atoms approach one another, an attractive force develops between each electron and the approaching nucleus. Repulsive forces (shown in part b) also develop between the two nuclei and the two electrons. (b) In the bonded state, the electrons orient themselves between the two nuclei in a fashion that maximizes the attractive forces and minimizes the repulsive forces.

Let us consider the energetics of this process. Figure 8-4a (p. 196) shows the potential energy of two hydrogen atoms as their internuclear distance (the distance between the centers of their nuclei) is varied. At very large distances, any attractive or repulsive forces between the atoms are negligible. However, as the atoms are brought closer together, the attractive forces between electrons and nuclei begin to develop, and the potential energy of the system decreases until it reaches a minimum (as shown in Fig. 8-4a). If the atoms are compressed beyond this point, the repulsive forces between the two positively charged nuclei will cause the potential energy to rise sharply.

Just as a ball rolled into a valley (Fig. 8-4b) will come to rest at the bottom of the valley (the potential-energy minimum), the hydrogen atoms will generally exist at the distance that corresponds to the energy minimum in the diagram. This distance is referred to as the bond length. The **bond length** is the distance between the centers of the two nuclei. The energy difference between the separated atoms and the bonded atoms is referred to as the **bond energy.** Thus, the bond energy is the amount of energy required to break the bond. The greater the bond energy between a bonded pair of atoms, the stronger the bond and the more stable the resulting molecule. Table 8-2 gives the bond energies and bond lengths of several diatomic molecules.

Bond length: The distance between the nuclei of two bonded atoms.

Bond energy: The energy required to break the bond between two atoms.

Next let us consider a covalent bond from the viewpoint of the orbitals involved. Each of the hydrogen atoms in H_2 has a 1*s* orbital with a single electron in it (Fig. 8-5a). When bonding occurs, these two orbitals overlap in order to share the two electrons be-

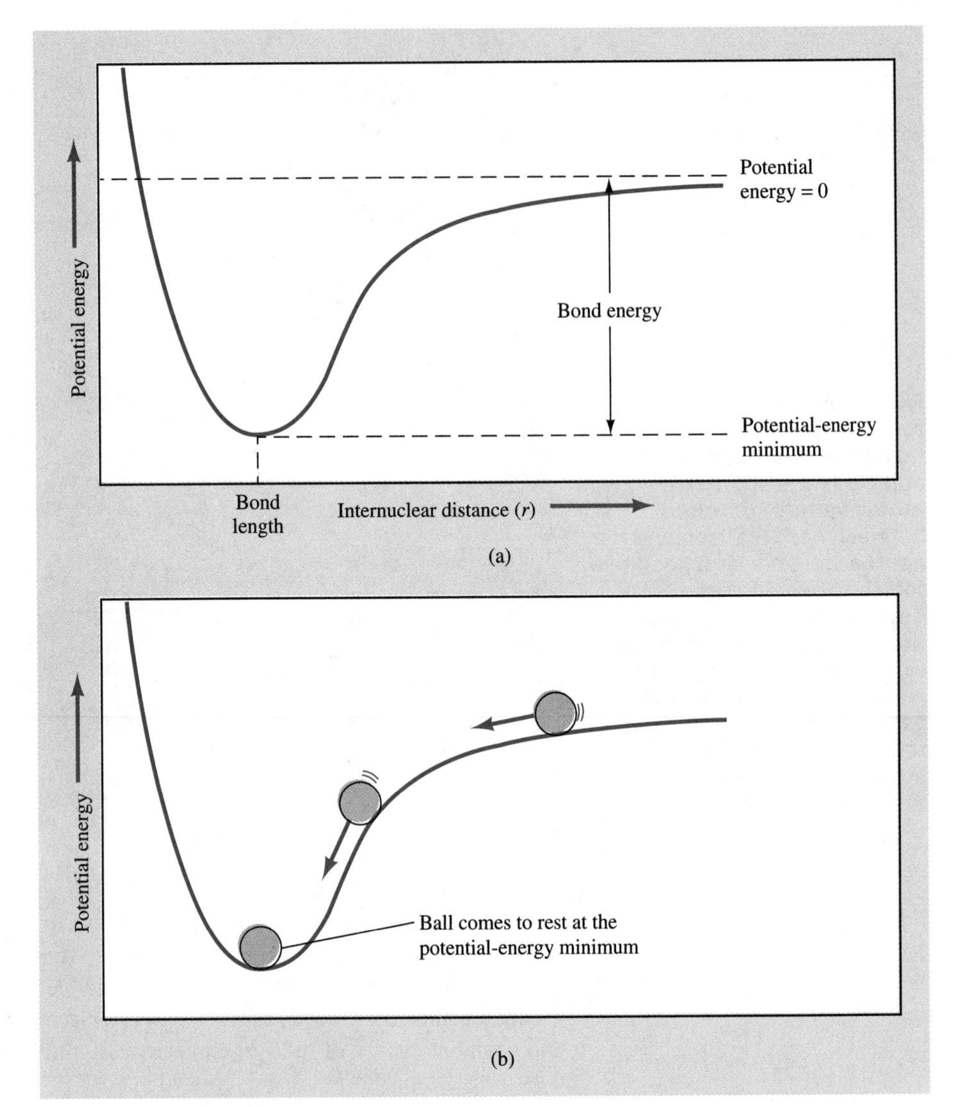

FIGURE 8-4 The energetics of bond formation. (a) Variation of potential energy with internuclear distance between a pair of hydrogen atoms. (b) A ball will come to rest at the bottom of a valley, the potential-energy minimum.

TABLE 8-2 Bond energies[a] and bond lengths of selected diatomic compounds

Bond	Bond energy	Bond length	Bond	Bond energy	Bond length
H–H	436 kJ/mol	74 pm	H–F	570 kJ/mol	92 pm
F–F	159 kJ/mol	141 pm	H–Cl	432 kJ/mol	127 pm
Cl–Cl	243 kJ/mol	199 pm	H–Br	366 kJ/mol	141 pm
Br–Br	193 kJ/mol	228 pm			

[a]These are more precisely known as *bond dissociation energies,* the energy required to break a bond so that each atom comes away with one of the shared electrons (A:B → A· + ·B).

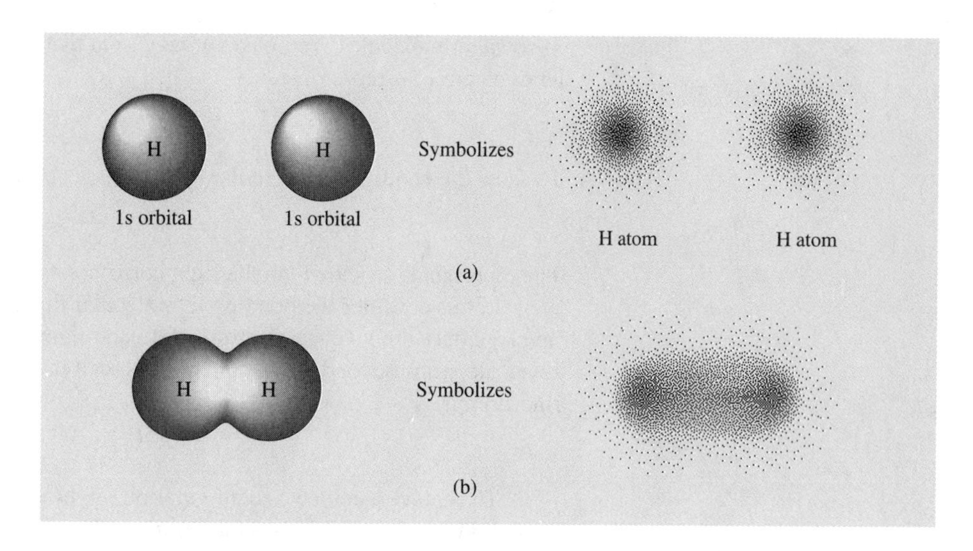

FIGURE 8-5 The overlap of 1*s* orbitals to form the H–H bond. The schematic representation on the left symbolizes the electron densities on the right. (a) Separated hydrogen atoms, each with one electron in its 1*s* orbital. (b) After bond formation, the two 1*s* orbitals overlap, resulting in sharing of the electrons. The region of maximum electron density is between the atoms.

tween them (Fig. 8-5b). The region where the orbitals overlap is the region of greatest electron density, and this corresponds to the general location where the electrons are found most often, as shown in Fig. 8-3. We can see that *a covalent bond involves two atoms sharing two electrons between two overlapping orbitals.*

As with ionic bonding, the electrons involved in covalent bonding are the valence electrons. In the case of hydrogen, this is its lone electron. As a general rule, *covalent bonding takes place between nonmetallic atoms.* Again, the incentive for bonding is to achieve a noble gas configuration, but in the case of covalent bonding, this is accomplished by acquiring electrons through sharing. For this purpose, each atom is considered to have its original valence electrons, plus those it "gains" by sharing with the atoms with which it bonds. In the case of hydrogen, for example, each atom is considered to have two electrons: the one it "donates" plus the one it "gains." This is the same number as helium, one of the noble gases.

PROBLEM 8.3

Which of the following pairs of atoms would you expect to form covalent bonds? Which would be likely to form ionic bonds?
(a) N, H **(b)** Na, S **(c)** S, O **(d)** C, F **(e)** Mg, O **(f)** H, O

8.4 LEWIS STRUCTURES OF MOLECULES

OBJECTIVE 8.4 State the octet rule. Write Lewis structures for selected covalent molecules that satisfy the octet rule and are composed exclusively of atoms from the first three periods.

Earlier, we represented the valence electrons in individual atoms by means of Lewis electron-dot structures. Lewis structures are particularly useful for showing the bonding in covalent molecules. You have already seen that a hydrogen atom, with only one valence electron, is represented as

$$\mathrm{H}\cdot$$

To show the bonding in molecular hydrogen, we bring two hydrogen atoms together:

$$\mathrm{H}\cdot + \cdot\mathrm{H} \rightarrow \mathrm{H}{:}\mathrm{H}$$

The representation shown on the right corresponds to the bonding in Fig. 8.5.

Let us consider the bonding in molecular fluorine, F_2. Because covalent bonding involves only the valence electrons, for each fluorine atom we need consider only the seven electrons beyond the helium core. Recall that the electron-dot structure of a fluorine atom is

$$:\ddot{\underset{\cdot\cdot}{\mathrm{F}}}\cdot$$

To achieve a noble gas configuration (eight valence electrons), each fluorine atom must gain an electron. This situation is similar to the one we just saw for hydrogen. When two fluorine atoms are brought together, a covalent bond forms between the unpaired electrons:

$$:\ddot{\underset{\cdot\cdot}{\mathrm{F}}}\cdot + \cdot\ddot{\underset{\cdot\cdot}{\mathrm{F}}}: \rightarrow :\ddot{\underset{\cdot\cdot}{\mathrm{F}}}{:}\ddot{\underset{\cdot\cdot}{\mathrm{F}}}:$$

By sharing a pair of electrons, each fluorine atom becomes surrounded by eight valence electrons—its original seven valence electrons plus the additional one it has gained through sharing. This represents the same number of valence electrons as neon, the next noble gas.

When drawing electron-dot structures, we want to make it obvious which electrons are involved in bonding and which electrons belong to one atom only. If we imagine that each atomic symbol in the preceding Lewis structure for F_2 is enclosed in a square, it is clear that one pair of electrons is being shared, whereas the remaining six valence electrons on each fluorine atom are not shared.

Let us consider the structure of water, H_2O. Oxygen has six valence electrons, and each hydrogen atom has one. In order to achieve noble gas configurations, oxygen needs two additional electrons and each hydrogen atom needs one additional electron. We can bring the atoms together as follows:

$$\begin{array}{c}\mathrm{H}\cdot\ \cdot\ddot{\mathrm{O}}: \\ \quad\ \ \cdot \\ \quad\ \ \cdot \\ \quad\ \ \mathrm{H}\end{array} \rightarrow \begin{array}{c}\mathrm{H}{:}\ddot{\mathrm{O}}: \\ \quad\ \ \mathrm{\ddot{H}}\end{array}$$

In the final Lewis structure, oxygen is surrounded by eight electrons (the same as neon), and each hydrogen has two electrons (the same as helium). Each atom has a noble gas configuration.

Students often wonder whether they need to be concerned about which sides of the oxygen are bonded to the hydrogens. The answer is: *No.* As a general rule, electron-dot structures are not intended to show geometry. Thus, any of the following structures also would be equally correct for water:

$$\mathrm{H}{:}\ddot{\underset{\cdot\cdot}{\mathrm{O}}}{:}\mathrm{H} \qquad \begin{array}{c}\mathrm{H}\\ :\ddot{\underset{\cdot\cdot}{\mathrm{O}}}:\\ \mathrm{H}\end{array} \qquad \begin{array}{c}\mathrm{H}\\ :\ddot{\underset{\cdot\cdot}{\mathrm{O}}}{:}\mathrm{H}\end{array} \qquad \begin{array}{c}\mathrm{H}\\ \mathrm{H}{:}\ddot{\underset{\cdot\cdot}{\mathrm{O}}}:\end{array} \qquad \begin{array}{c}:\ddot{\underset{\cdot\cdot}{\mathrm{O}}}{:}\mathrm{H}\\ \mathrm{H}\end{array}$$

In Chapter 12, we discuss the shapes of molecules and how to predict their geometries from their Lewis structures.

Because we are restricting ourselves to nonmetallic elements found in the first three periods, the atoms in all our Lewis structures achieve noble gas configurations when eight valence electrons surround each atom. The exception is hydrogen, which needs only two electrons to achieve a noble gas configuration. The **octet rule** directs us to write Lewis structures so that each atom (other than hydrogen) is surrounded by eight electrons; hydrogen must have two electrons (sometimes referred to as a *duet*). Although there are many molecules that do not conform to the octet rule, we restrict our examples and problems to those that do.

Octet rule: Except for hydrogen, the atoms in Lewis structures are generally surrounded by eight electrons.

EXAMPLE 8.2

Draw a Lewis structure for ammonia, NH_3.

Planning the solution

We write the Lewis structures of each of the atoms in ammonia, and then see if they can be attached in a fashion that satisfies the octet rule.

SOLUTION

Nitrogen has five valence electrons:

·N̈·
 ·

Each hydrogen has a lone valence electron:

H· Ḣ ·H

The three unpaired electrons of nitrogen can each form a bond with the unpaired electron from a hydrogen:

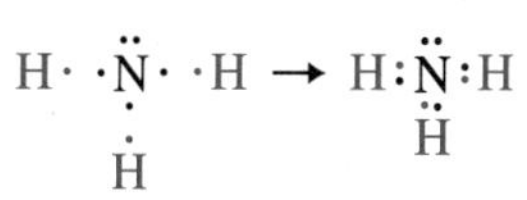

H· ·N̈· ·H → H:N̈:H
 ·H   ··H

The final Lewis structure shows eight electrons around the central nitrogen and two electrons adjacent to each hydrogen, so the octet rule has been satisfied.

EXAMPLE 8.3

Draw a Lewis structure for methane, CH_4.

SOLUTION

Carbon, with four valence electrons, can form four covalent bonds:

·Ċ·
 ·

Each electron on carbon can bond with a lone electron from a hydrogen:

  H·   H
H· ·Ċ· ·H → H:C:H
   ·H   H

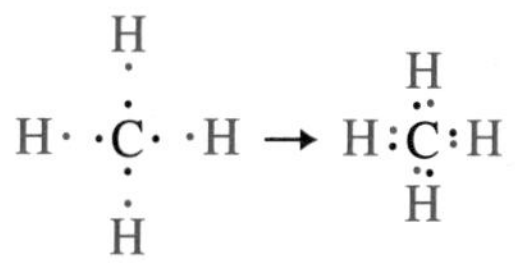

As in the previous example, the octet rule has been satisfied for each atom in the final Lewis structure.

EXAMPLE 8.4

Draw a Lewis structure for phosphorus trichloride, PCl_3.

SOLUTION

Phosphorus has five valence electrons, three of which are unshared:

·P̈·
 ·

Each chlorine atom has seven valence electrons. These elements combine as follows to form phosphorus trichloride:

:C̈l· ·P̈· ·C̈l: → :C̈l:P̈:C̈l:
 :C̈l: :C̈l:

Again, the final Lewis structure satisfies the octet rule; each atom is surrounded by eight electrons.

These examples illustrate several principles that are often helpful in writing electron-dot structures. First, note that phosphorus and nitrogen, which are in the same chemical family, have the same number of valence electrons and so require the same number of electrons to complete the octet. The same is true of chlorine and fluorine. You saw in the last chapter that, with few exceptions, elements within a family have similar outer-level electronic configurations. Thus, they tend to have similar bonding patterns and similar electron-dot structures.

Next, to write a Lewis structure, we must know which atoms are bonded to which. For example, without further information, we might not know whether the arrangement of atoms in sulfur dioxide, SO_2, is OSO or SOO. For many substances the arrangement is not obvious and may have to be determined experimentally. However, the geometries of many simple molecules are found to consist of a central atom of one element surrounded by two or more atoms of another element (or elements). Using this guideline, we would correctly predict the arrangement of atoms in SO_2 to be OSO. As an additional guide, atoms of C, Si, N, P, and S are often found as central atoms. Furthermore, with the exception of H_2O and OCl_2, oxygen is generally *not* the central atom in a molecule (nor are you likely to find two oxygens bonded together, except in oxygen, O_2, hydrogen peroxide, H_2O_2, or ozone, O_3). Finally, remember that hydrogen atoms can form only one bond. Consequently, a hydrogen atom will always be found at the periphery (the outside) of a molecule; never sandwiched between two atoms.

Unfortunately, there are no absolute rules for predicting the arrangement of atoms that will work all of the time. However, as you work your way through the chapter and learn more about typical bonding patterns, your ability to predict the correct arrangement of atoms will improve considerably. In the meantime, we will provide you with the correct arrangement of atoms for problems where the guidelines given here may not be sufficient.

PROBLEM 8.4

Write Lewis structures for each of the following compounds.

(a) NH_3 **(b)** CH_4 **(c)** H_2S **(d)** PH_3 **(e)** NF_3 **(f)** CCl_4
(g) SiH_4 **(h)** Cl_2 **(i)** PF_3 **(j)** $CHCl_3$ **(k)** SiH_2F_2

8.5 DOUBLE AND TRIPLE BONDS

OBJECTIVE 8.5 Write Lewis structures of selected covalent molecules that have multiple bonds, satisfy the octet rule, and are composed exclusively of atoms from the first three periods.

Let us consider the electron-dot structure for carbon dioxide, CO_2. Each oxygen atom has six valence electrons, and the carbon atom has four:

·Ö· ·Ċ· ·Ö·

The atoms have been arranged in a symmetric fashion in agreement with our earlier suggestion. If we attempt to make a bond from carbon to each oxygen atom, the octet rule will not be satisfied:

·Ö:Ċ:Ö·

Each oxygen atom is surrounded by only seven electrons, and the carbon atom is surrounded by only six electrons.

It is possible for each oxygen atom to form an additional bond with carbon by sharing its remaining unpaired electron with one of the unpaired electrons from the carbon:

Ö::C::Ö

Double bond **Double bond**

Double bond: A covalent bond in which two pairs of electrons are shared between two atoms.

Note that each oxygen atom now makes two bonds to carbon. This situation is referred to as double bonding, and each of the carbon–oxygen bonds is called a **double bond.** Both of the bonds in each double bond are covalent bonds; each bond involves the sharing of a pair of electrons between two atoms. When we write an electron-dot structure for a double bond, the two pairs of electrons are shown side by side between the atoms, as in the final Lewis structure above.

With molecular nitrogen, N_2, we encounter a similar situation. Each nitrogen has five valence electrons:

:Ṅ· ·Ṅ:

If we bring the atoms together to form a single bond, the octet rule is not satisfied:

:Ṅ:Ṅ:

Triple bond: A covalent bond in which three pairs of electrons are shared between two atoms.

In this case it is necessary to form three bonds, or a **triple bond,** between the two atoms:

:N:::N: or :N⋮⋮N:

As in double bonding, we show the electron pairs of the three bonds side by side. In addition, each component of the triple bond is a covalent bond formed by sharing a pair of electrons between the atoms.

We can summarize our representation of single, double, and triple bonds as follows:

A:B	D::E	X:::Y or X⁝⁝Y
Single bond	Double bond	Triple bond

Although double and triple bonds are quite common in molecules, bonds of a higher order do not form from the elements we are considering.

PROBLEM 8.5

Draw Lewis structures for each of the following molecules. The arrangement of atoms is given.

(a) CS_2
S C S

(b) HCN
H C N

(c) C_2H_2
H C C H

(d) C_2H_4
H C C H
H H

(e) H_2CO
O
H C H

8.6 LEWIS STRUCTURES: AN EASY WAY TO DRAW THEM

OBJECTIVE 8.6 Write a Lewis structure for any covalent molecule that satisfies the octet rule and is composed exclusively of atoms from the first three periods.

So far we have written Lewis structures by combining the electron-dot structures of individual atoms to form molecules. However, it is often easier to determine Lewis structures by using the following general procedure:

1. Count up all the valence electrons in the molecule.
2. Arrange the atoms on the page.
3. Distribute the valence electrons around the atoms to satisfy the octet rule.

The order of Steps 1 and 2 can be inverted if you prefer. For very simple molecules, this description of the three steps may be sufficient for you to write a Lewis structure. However, a little more explanation will probably be necessary to enable you to work out more complicated structures. We will begin by showing how these three steps can be applied to two very simple examples. Then we will give you a more comprehensive summary of how to apply these steps for more complicated examples.

EXAMPLE 8.5

Draw a Lewis structure for molecular hydrogen, H_2.

Planning the solution

We will follow Steps 1–3.

SOLUTION

Step 1. Molecular hydrogen has two atoms. Each atom has one valence electron, so the total number of valence electrons in the molecule is two:

$$2 \times \text{H} = 2 \times 1\,\text{e}^- = 2\,\text{e}^-$$

Step 2. Since this is a diatomic molecule, the arrangement of the atoms must be side by side:

$$\text{H} \quad \text{H}$$

Step 3. Since the hydrogens must be bonded to one another, we will place a pair of electrons between the two atoms:

$$\text{H:H}$$

This uses up both electrons and places a duet around both hydrogens.

EXAMPLE 8.6

Draw a Lewis structure for methane, CH_4.

SOLUTION

Step 1. Methane contains one carbon atom with four valence electrons and four hydrogen atoms, each having one valence electron. The total number of valence electrons must be eight:

$$\begin{aligned} 1 \times \text{C} &= 1 \times 4\,\text{e}^- = 4\,\text{e}^- \\ 4 \times \text{H} &= 4 \times 1\,\text{e}^- = 4\,\text{e}^- \\ \hline \text{CH}_4 &= 8\,\text{e}^- \end{aligned}$$

Step 2. Next we arrange the atoms with carbon as the central atom and the four hydrogens surrounding it:

$$\begin{array}{ccc} & \text{H} & \\ \text{H} & \text{C} & \text{H} \\ & \text{H} & \end{array}$$

Step 3. Since each hydrogen must be bonded to the carbon, we need to place a pair of electrons between each hydrogen and the carbon:

$$\begin{array}{c} \text{H} \\ \text{H}:\overset{\cdot\cdot}{\underset{\cdot\cdot}{\text{C}}}:\text{H} \\ \text{H} \end{array}$$

This not only uses up all of the eight valence electrons, but it also satisfies the octet rule.

Examples 8.5 and 8.6 were very simple examples, and the examples that follow also can be solved by using this general approach. However, the Summary on the next page is more comprehensive, and Step 3, in particular, is much more detailed in its approach. We illustrate the use of this version of Step 3 in Examples 8.7–8.10.

SUMMARY Procedure for Writing Lewis Structures

1. Count up the total number of valence electrons in the molecule or polyatomic ion. (Polyatomic ions are discussed in Section 8.8. In the case of a polyatomic ion, be sure to add or subtract electrons to account for the charge.)
2. Arrange the atoms on the page with the central atom surrounded by the other atoms. The following information may be helpful in doing this.
   (a) C, Si, N, P, and S are *often* central atoms.
   (b) Oxygen is *rarely* the central atom (except in H_2O and Cl_2O). Nor is oxygen generally bonded to itself (except in O_2, O_3, and H_2O_2).
   (c) Hydrogen is *never* the central atom. Because it can make only one bond, hydrogen is always on the outside of a molecule.
   (d) For a diatomic molecule, just place the atoms side by side.
   (e) Many molecules are symmetrical.
   (f) Use common sense. In a molecule such as N_2H_4, since hydrogen atoms must be on the periphery, the two nitrogen atoms must be bonded to one another (similarly for C_2H_6, C_2H_4, C_2H_2, and H_2O_2).

   (For some chemical formulas, there may be several correct arrangements that are possible. In those cases, we will give you the desired arrangement.)
3. Distribute the valence electrons around the atoms in a fashion that satisfies the octet rule. For simple molecules (such as H_2 or CH_4), Step 3 may be rather obvious. But for more complicated problems, you may find the following procedure helpful. (If the steps seem a little complicated when you first read them, do not be concerned. We will illustrate them in the examples, and once you see them applied, this will serve as a very helpful guide.)
   (a) Place a pair of electrons between the central atom and each atom surrounding it. This is necessary to make sure that all of the atoms are connected. (In more complex structures where several atoms are in a row, it is necessary to place a pair of electrons between every pair of atoms that is bonded together.)
   (b) Count up the number of electrons placed in Step 3a, and subtract that number from the total number of valence electrons determined in Step 1. This represents the number of electrons remaining. You need to place these electrons in a fashion that satisfies the octet rule. Parts (c)–(f) will help you accomplish this task.
   (c) Examine the partial structure and determine the total number of electrons that would be required to complete all of the octets on the atoms that are not yet satisfied. If this number is the same as the number of remaining electrons (determined in Step 3b), simply place those electrons and you are done. In this case, your final structure will have no multiple bonds.
   (d) If the number of electrons calculated in Step 3c is larger than the number of remaining electrons (from Step 3b), there must be multiple bonds (double or triple bonds) in the molecule or ion. For every difference of two electrons, there is one multiple bond. Thus, if the difference (Step 3c minus Step 3b) is 2, there is one double bond. If the difference is 4, there is either one triple bond or two double bonds. And so on.
   (e) If there are multiple bonds in the structure, begin by placing the remaining electrons in pairs around the outer atoms (except H) first. As a general rule, it is helpful to distribute them uniformly (in other words, give one pair to each of the outer atoms before giving a second pair to any of the outer atoms, and so forth) until you run out of electrons.
   (f) At this point, at least one of the atoms in the structure (generally, the central atom) will be short of the electrons it needs to satisfy the octet rule. To correct this, take a pair of *unshared* electrons from a neighboring atom, and use them to make a multiple bond between the two atoms. This will have the overall effect of increasing the number of electrons surrounding the atom that was short of electrons, while leaving the number of electrons on the other atom unchanged. Continue this process until all of the atoms have satisfied the octet rule.

EXAMPLE 8.7

Draw a Lewis structure for PCl_3.

Planning the solution

We use Steps 1–3.

SOLUTION

Step 1. Phosphorus has 5 valence electrons and each chlorine has 7, giving a total of 26 valence electrons:

$$\begin{array}{rcl} 1 \times P = 1 \times 5\,e^- &=& 5\,e^- \\ 3 \times Cl = 3 \times 7\,e^- &=& 21\,e^- \\ \hline PCl_3 &=& 26\,e^- \end{array}$$

Step 2. Since there is only one phosphorus atom and there are three chlorine atoms, we arrange the atoms with phosphorus as the central atom and the chlorine atoms surrounding it:

$$\begin{array}{ccc} Cl & P & Cl \\ & Cl & \end{array}$$

Step 3. **(a)** We begin by placing a pair of electrons between each chlorine and the central phosphorus. (In essence, we are making sure that all of the atoms in the structure are connected.)

$$\begin{array}{ccc} Cl & \!:\!P\!:\! & Cl \\ & \ddot{Cl} & \end{array} \qquad \textbf{Incomplete structure}$$

(b) So far we have placed 6 electrons. Subtracting this from the total number of 26 valence electrons leaves 20 electrons to be placed.

(c) Examination of the partial structure reveals that phosphorus needs 2 more electrons to satisfy its octet, and each of the three chlorine atoms needs 6 more electrons. This is a total of 20 electrons ($2\,e^- + 3 \times 6\,e^- = 20\,e^-$) needed to complete the structure without multiple bonds. Since this is the same as the number of electrons remaining to be placed, we complete the structure by placing the 20 electrons around the atoms to complete each octet:

$$\begin{array}{c} :\ddot{\underset{..}{Cl}}:\ddot{\underset{..}{P}}:\ddot{\underset{..}{Cl}}: \\ :\ddot{\underset{..}{Cl}}: \end{array} \qquad \textbf{Complete structure}$$

EXAMPLE 8.8

Draw a Lewis structure for sulfur trioxide, SO_3.

SOLUTION

Step 1. First we determine the number of valence electrons:

$$\begin{array}{rcl} 1 \times S = 1 \times 6\,e^- &=& 6\,e^- \\ 3 \times O = 3 \times 6\,e^- &=& 18\,e^- \\ \hline SO_3 &=& 24\,e^- \end{array}$$

Step 2. Next we arrange the atoms with sulfur in the center. (Remember, oxygen is rarely in the center, except in H_2O and OCl_2.)

$$\begin{array}{ccc} & O & \\ O & S & O \end{array}$$

Step 3. **(a)** We place a pair of electrons between each oxygen and the sulfur:

```
  O
  ··
O:S:O      Incomplete structure
```

(b) We have used 6 electrons. Since there is a total of 24 valence electrons (Step 1), there are 18 electrons remaining.

(c) To complete the octet around each of the atoms would require an additional 20 electrons (2 for the sulfur atom and 6 each for the three oxygen atoms). Thus, we are 2 electrons short ($20\ e^- - 18\ e^-$).

(d) Since each pair of electrons that we are short represents one multiple bond, SO_3 must have one double bond.

(e) We continue by placing the remaining electrons, one pair at a time around the oxygens, until all of the remaining electrons are used up:

```
   ··
  :O:
 ·· ·· ··
:O:S:O:      Incomplete structure
 ··    ··
```

(f) At this point, the sulfur is short by 2 electrons. We will move one of the unshared pairs of electrons from the top oxygen to a position between the two atoms, thereby creating a double bond between them:

```
  :O:
 ·· ·· ··
:O:S:O:      Complete structure
 ··    ··
```

Now all of the atoms have satisfied the octet rule.

In the example we just completed, you might wonder whether it mattered which oxygen we used to form the double bond. In fact, we could have written either of the following two structures instead:

```
   ··            ··
  :O:           :O:
 ·· ·· ··     ·· ·· ··
 O::S:O:     :O:S::O
 ··    ··     ··    ··
```

All three of these structures are correct Lewis structures for sulfur trioxide. They differ only in the placement of the electrons, yet all satisfy the octet rule. We refer to structures such as these, which differ only in the placement of electrons, as *resonance structures.* You will find that the substances in many of the examples and the problems you solve are capable of multiple resonance structures. We will satisfy ourselves with the ability to write any one of the acceptable resonance forms for a given molecule or polyatomic ion.

EXAMPLE 8.9

Draw the Lewis structure for molecular nitrogen, N_2.

SOLUTION

Step 1. First determine the number of valence electrons:

$$\underline{2 \times N = 2 \times 5\ e^- = 10\ e^-}$$
$$N_2 = 10\ e^-$$

Step 2. For a diatomic molecule, simply arrange the atoms next to one another:

N N

Step 3. **(a)** Place a pair of electrons between the two nitrogens:

N:N **Incomplete structure**

(b) We have used 2 electrons. Since there is a total of 10 electrons, the number of electrons remaining is 8.

(c) Each nitrogen needs an additional 6 electrons to satisfy the octet rule. This is a total of 12 electrons that would be necessary to complete the structure without multiple bonds. We are 4 electrons short.

(d) Since each electron pair that we are short represents one multiple bond, this molecule must have either two double bonds or one triple bond. Obviously, for a diatomic molecule, it must be a triple bond.

(e) We will place the remaining 8 electrons in pairs, one pair per nitrogen, until we run out of electrons:

:N̈:N̈: **Incomplete structure**

(f) Examination of the structure reveals that both nitrogens have only six electrons. We will take an unshared pair from the left-hand nitrogen and place it between the two atoms, creating a double bond:

:N::N̈: **Incomplete structure**

Now the nitrogen on the right has satisfied the octet rule, but the one on the left has not. To solve this problem, we simply repeat the procedure using an unshared pair of electrons from the right-hand nitrogen, thereby creating another multiple bond between the two, which gives us a triple bond:

:N:::N: **Complete structure**

EXAMPLE 8.10

Ozone, O_3, is a form of oxygen that can have both positive and negative environmental effects. When found at the earth's surface, it is considered an air pollutant. However, the ozone layer found in the upper layer of the earth's atmosphere is necessary to shield us from the harmful effects of ultraviolet radiation from the sun. Draw a Lewis structure for ozone.

SOLUTION

Step 1. Since each oxygen atom has 6 valence electrons, there must be a total of 18 valence electrons:

$$3 \times O = 3 \times 6\,e^- = 18\,e^-$$
$$O_3 = 18\,e^-$$

Step 2. Ozone is one of the few examples where oxygen atoms are bonded together or oxygen is the central atom. In this case, the three atoms are bonded in a line:

O O O

Step 3. **(a)** First we place a pair of electrons between each pair of oxygen atoms:

O:O:O **Incomplete structure**

(b) We have used 4 of the electrons. Since there is a total of 18 valence electrons, there are 14 electrons remaining to be placed.

(c) Examination of the partial structure reveals that each of the oxygen atoms on the outside needs 6 more electrons to complete its octet, while the central oxygen needs 4 more electrons. This is a total of 16 electrons. Since there are only 14 electrons left, we are short 2 electrons.

(d) There must be one double bond in this molecule.

(e) We place the electrons in pairs around the outer oxygens first:

:Ö:O:Ö: **Incomplete structure**

This uses only 12 of the 14 remaining electrons. So we assign the remaining 2 electrons to the central oxygen:

:Ö:Ö:Ö: **Incomplete structure**

(f) The central oxygen atom is still short by 2 electrons, so we move an unshared pair of electrons from the left-hand oxygen to create a double bond to the central oxygen:

Ö::Ö:Ö: **Complete structure**

This completes the octet. Note that this is another example where more than one resonance form is possible. Instead of the structure above, we could have written the following structure:

:Ö:Ö::Ö **Resonance structure**

This example illustrates a benefit of using the general approach to writing electron-dot structures presented in this section. If we had tried to assemble the Lewis structure for O_3 from the individual atoms, we might have had difficulty figuring out how to put the pieces together. Attaching the atoms through the unpaired electrons would lead to the following structure:

·Ö· + ·Ö· + ·Ö· → ·Ö:Ö:Ö·

The outer oxygen atoms have only seven electrons. If we had used this method, it would not have been apparent how to satisfy the octet rule. With the technique presented in this section, we simply rearrange the electrons until the octet rule is satisfied. By laying out the atoms first and then adding the electrons, our thinking is not limited by the electron-dot structures of the individual atoms. Instead, we can move electrons around until the octet rule has been satisfied.

PROBLEM 8.6

Draw Lewis structures for each of the following. The arrangements of atoms in parts (f)–(h) are given.

(a) H_2O_2 [*Remember:* Hydrogen must be on the periphery of the molecule.]
(b) N_2H_4 [*Remember:* Hydrogen must be on the periphery of the molecule.]

(c) SO_2 **(d)** CO_2 **(e)** CO
(f) H_2SO_4 **(g)** HNO_3 **(h)** $HClO_4$

```
      O                   O                 O
H  O  S  O  H        H  O  N  O        H  O  Cl  O
      O                                     O
```

8.7 COVALENT MOLECULES: ANOTHER WAY TO DRAW THEM

OBJECTIVE 8.7 Write abbreviated Lewis structures using dashes to represent each bonded pair of electrons.

When we draw structures of covalent molecules, we generally find it too time-consuming and tedious to write in all of the electrons, as we have been doing. Instead, we can use a dash to represent each covalent bond, where a dash represents two electrons.

Electron-dot structure	Abbreviated structure
H:Ö:H (two lone pairs on O)	H—Ö—H (two lone pairs on O)
H:C:H with H above and below C	H—C—H with H bonded above and below C
:Ö::C::Ö:	:Ö=C=Ö:
:N:::N:	:N≡N:
:C̈l:P̈:C̈l: with :C̈l: below P	:C̈l—P̈—C̈l: with :C̈l: bonded below P

The following generalizations are useful for drawing the structures of molecules (but remember that these rules do not always hold).

One bond	Two bonds
H—	—Ö— or :Ö=
:F̈—	—S̈— or :S̈=
:C̈l—	

Three bonds

—N̈— (with a third bond below) or :N≡ or —N̈=

—P̈— (with a third bond below) or :P≡ or —P̈=

Four bonds

—C— (with bonds above and below) or C= (with bonds above and below) or —C≡ or =C=

—Si— (with bonds above and below) or Si= (with bonds above and below) or —Si≡ or =Si=

Using these generalizations, we can easily determine the structures of carbon disulfide, CS_2, and carbon tetrachloride, CCl_4. Since carbon forms four bonds and sulfur forms two bonds, it is likely that each sulfur atom in carbon disulfide forms a double bond to carbon. In the case of carbon tetrachloride, if carbon forms four bonds and chlorine forms one bond, each of the four chlorine atoms must form a single bond to carbon.

$$:\ddot{S}{=}C{=}\ddot{S}: \quad \text{and} \quad \begin{array}{c} :\ddot{Cl}: \\ | \\ :\ddot{Cl}-C-\ddot{Cl}: \\ | \\ :\ddot{Cl}: \end{array}$$

CS_2 $\qquad$ CCl_4

PROBLEM 8.7 Use the generalizations just given and our simplified technique to write structures for each of the molecules in Problems 8.4 and 8.5.

8.8 POLYATOMIC IONS

OBJECTIVE 8.8 Write a Lewis structure for any of the common polyatomic ions composed of elements from the first three periods.

In Section 5.5, we described a set of ions referred to as polyatomic ions. The following are perhaps the most important of them:

NH_4^+	OH^-	NO_3^-	SO_4^{2-}	CO_3^{2-}	PO_4^{3-}
Ammonium	Hydroxide	Nitrate	Sulfate	Carbonate	Phosphate

Let us begin by considering the structure of a nitrate ion, NO_3^-. The formula of this ion tells us that it is made up of one nitrogen atom and three oxygen atoms and that these four atoms have somehow managed to acquire an additional electron, giving the ion a charge of −1.

Let us proceed to draw a Lewis structure, using the same technique introduced in Section 8.6. However, we must account for the additional electron that this group of atoms has acquired. Because it is an additional electron, it must be added to the total number of valence electrons.

$$\begin{array}{lcl} 1 \times N = 1 \times 5\,e^- & = & 5\,e^- \\ 3 \times O = 3 \times 6\,e^- & = & 18\,e^- \\ \underline{1 \times e^- \qquad\qquad\quad} & = & \underline{1\,e^-} \leftarrow \textbf{The additional electron} \\ NO_3^- & = & 24\,e^- \end{array}$$

Like the other ions we have described so far, the nitrate ion gained the extra electron through electron transfer with some other chemical species that must have lost it. In this way, both positive and negative ions are formed.

As before, we place the oxygen atoms symmetrically around the central nitrogen and then insert the 24 valence electrons. This leads to the following Lewis structure, in which the octet rule is satisfied for all atoms:

$$\left[\begin{array}{c} :\mathrm{O}: \\ :\ddot{\underset{..}{\mathrm{O}}}:\ddot{\mathrm{N}}:\ddot{\underset{..}{\mathrm{O}}}: \end{array}\right]^{-}$$

When drawing Lewis structures for polyatomic ions, we place the final structure in brackets and the charge of the ion in the upper right-hand corner, as shown.

Let us draw electron-dot structures for a few more of these polyatomic ions.

EXAMPLE 8.11

Draw a Lewis structure for the hydroxide ion, OH^-.

Planning the solution

Our plan is identical to the one we followed in Section 8.6, except that we must take into account the extra electron responsible for the −1 charge on this ion.

SOLUTION

Oxygen has six valence electrons, hydrogen has one valence electron, and there is one additional electron. Thus, there is a total of eight valence electrons in the hydroxide ion.

$$\begin{array}{rl} 1 \times \mathrm{O} = 1 \times 6\,e^- = & 6\,e^- \\ 1 \times \mathrm{H} = 1 \times 1\,e^- = & 1\,e^- \\ 1 \times e^- = \quad 1\,e^- = & 1\,e^- \\ \hline \mathrm{OH}^- = & 8\,e^- \end{array}$$

Again, we can see that the additional electron this polyatomic ion possesses enables each atom to satisfy the octet rule. The electron-dot structure is as follows:

$$\left[:\ddot{\underset{..}{\mathrm{O}}}:\mathrm{H}\right]^{-}$$

Note that we place the final structure within brackets and show its charge, indicating that it is an ion.

EXAMPLE 8.12

Draw a Lewis structure for the sulfate ion, SO_4^{2-}.

SOLUTION

Sulfur possesses 6 valence electrons, as does each oxygen. In this ion there are 2 extra electrons; hence, the total number of valence electrons is 32.

$$\begin{array}{rl} 1 \times \mathrm{S} = 1 \times 6\,e^- = & 6\,e^- \\ 4 \times \mathrm{O} = 4 \times 6\,e^- = & 24\,e^- \\ 2 \times e^- = \quad 2\,e^- = & 2\,e^- \\ \hline \mathrm{SO_4}^{2-} = & 32\,e^- \end{array}$$

The electron-dot structure for sulfate ion is

$$\left[\begin{array}{c} :\ddot{\mathrm{O}}: \\ :\ddot{\underset{..}{\mathrm{O}}}:\ddot{\underset{..}{\mathrm{S}}}:\ddot{\underset{..}{\mathrm{O}}}: \\ :\underset{..}{\mathrm{O}}: \end{array}\right]^{2-}$$

EXAMPLE 8.13

Draw a Lewis structure for the ammonium ion, NH_4^+.

SOLUTION

The ammonium ion has a +1 charge. This means that it has *lost* one valence electron. Thus, in figuring the total number of valence electrons, we subtract one electron.

$$\begin{aligned} 1 \times N &= 1 \times 5\,e^- = 5\,e^- \\ 4 \times H &= 4 \times 1\,e^- = 4\,e^- \\ -1 \times e^- &= -1\,e^- = -1\,e^- \\ \hline NH_4^+ &= 8\,e^- \end{aligned}$$

$$\left[\begin{array}{c} \mathrm{H} \\ \mathrm{H:\ddot{N}:H} \\ \mathrm{\ddot{H}} \end{array}\right]^+$$

Many compounds have both ionic and covalent bonds present. For example, magnesium sulfate is an ionic compound made up of magnesium ions, Mg^{2+}, and sulfate ions, SO_4^{2-}. Ionic bonds hold the magnesium and sulfate ions together. In Example 8.12 we examined the Lewis structure of the sulfate ion. Thus, there is covalent bonding within the sulfate ion, and ionic bonding between the magnesium and sulfate ions. We show the structure of magnesium sulfate as follows:

$$\mathrm{Mg^{2+}}\left[\begin{array}{c} :\ddot{\mathrm{O}}: \\ :\ddot{\mathrm{O}}:\ddot{\mathrm{S}}:\ddot{\mathrm{O}}: \\ :\ddot{\mathrm{O}}: \end{array}\right]^{2-}$$

This completes our discussion of how to write Lewis structures. Before concluding, it is worth mentioning that not all substances obey the octet rule. For example, in some of its compounds, such as BF_3, boron is surrounded by only six electrons. The result of this electron deficiency is a very high reactivity for these compounds. The reactions they undergo are often those that result in products where boron has achieved a complete octet.

In contrast to this, elements of the third row and beyond are often capable of expanding their octet to ten or twelve electrons. Thus, in addition to PCl_3 (which we used to illustrate the octet rule), phosphorus forms another chloride whose formula is PCl_5. Similarly, sulfur forms a fluoride with the formula SF_6. For our purpose in this chapter, we have been satisfied to limit our skills to those compounds and polyatomic ions that obey the traditional octet rule.

The Summary shown here is an abbreviated version of the Summary presented in Section 8.6. It includes some additional suggestions that incorporate material covered in Section 8.7. Now that you have had some practice in writing Lewis structures, you may find it more efficient to use this Summary rather than the one presented earlier.

PROBLEM 8.8

Draw the electron-dot structure for each of the following polyatomic ions.

(a) Nitrate, NO_3^- **(b)** Hydroxide, OH^- **(c)** Phosphate, PO_4^{3-}
(d) Carbonate, CO_3^{2-} **(e)** Borohydride, BH_4^- **(f)** Azide, N_3^-
(g) Perchlorate, ClO_4^- **(h)** Chlorate, ClO_3^-

SUMMARY Suggestions for Writing Lewis Structures

1. Write down the arrangement of the atoms in the molecule or polyatomic ion you are determining.
   (a) If possible, arrange the atoms symmetrically. Often this means a central atom surrounded by several atoms of another element, as in the previous examples: H_2O, NH_3, CH_4, PCl_3, CO_2, and SO_3.
   (b) Place hydrogen atoms to the outside of the structure. Because hydrogen can form only one bond, it will never be found between two atoms.
   (c) Try to recall familiar molecules that would result from exchanging atoms in the same family. For example, NH_3 and PH_3 have similar Lewis structures because both nitrogen and phosphorus are in the same family. Thus, if you know the Lewis structure or even the arrangement of atoms in NH_3, that information can be used to help write the structure of PH_3.
   (d) Try to arrange the atoms so that it is possible to satisfy the usual bonding patterns discussed in Section 8.7. Although these patterns are not observed in every case, they make a good starting point.
2. Count up the total number of valence electrons in the molecule or polyatomic ion. In the case of a polyatomic ion, be sure to add or subtract electrons to account for the charge. In the case of an anion, be sure to *add* electrons.
3. Arrange the electrons around the atoms in a fashion that satisfies the octet rule, using the following guidelines.
   (a) Begin by placing an electron pair between each pair of atoms that is to be bonded together. Since hydrogen can form only single bonds, the placement of dots around hydrogen atoms is complete at this point.
   (b) Distribute the rest of the electrons in a symmetrical or near-symmetrical fashion around the atoms, attempting to satisfy the octet rule for each atom as the electrons are placed around it.
   (c) For atoms that have too few electrons, rearrange unshared pairs of electrons on neighboring atoms to create multiple bonds between the neighboring atoms and the atoms that are electron-deficient. Pay attention to the bonding patterns discussed in Section 8.7. They can provide important clues to indicate where multiple bonds are needed.

8.9 ELECTRONEGATIVITY

OBJECTIVE 8.9 Define the term *electronegativity.* State the trends in electronegativity from left to right across each period and from top to bottom within a family.

Electronegativity: The tendency of an element to attract electrons in a bonded pair.

So far we have assumed that the electrons are shared equally between two covalently bonded atoms. This is usually not the case, however. Elements differ from one another in a property known as electronegativity. The **electronegativity** of an element is the tendency of that element to attract a bonding pair of electrons. (Do not confuse this with the electron affinity, which measures the tendency of an atom to gain an electron in the gaseous state.) For example, imagine the hypothetical molecule X–Y, in which Y is more electronegative than X. Because of its greater tendency to attract electrons, Y develops a slight negative charge, while X develops a slight positive charge. The symbols δ^+ and δ^- are used to indicate the partial charges on the atoms (δ is the lowercase Greek letter delta).

X has a slightly positive charge → δ^+ δ^- ← **Y has a slightly negative charge**

$$X \;:\; Y$$

We refer to Y as the more *electronegative* of the two elements and to X as the more *electropositive* of the pair.

Metallic elements on the left-hand side of the periodic table have a tendency to lose electrons to form positive ions having noble gas configurations. By contrast, the elements on the right-hand side of the periodic table tend to gain electrons and form negative ions. Thus, as we might expect, the elements on the right-hand side of the periodic table are the most electronegative. Conversely, the metallic elements that are found on the left are the least electronegative.

As a general rule, *the electronegativities of the elements in a given row of the periodic table increase from left to right.* Thus, fluorine has a greater electronegativity than lithium. Similarly, chlorine has a greater electronegativity than sodium. Because the noble gases have neither a tendency to gain nor a tendency to lose electrons, it is not so easy to generalize about their electronegativities, so we shall not discuss them.

Within a family (or group), the electronegativities decrease going down each column of the periodic table. The reason for this trend is that the valence electrons in a large atom are farther from the nucleus, so there is less electrostatic attraction between the nucleus and these electrons. Thus, bromine is less electronegative than fluorine, and francium is less electronegative than lithium. In fact, francium (at the lower left of the periodic table) is the least electronegative element, whereas fluorine (at the upper right) is the most electronegative. Hydrogen has an intermediate electronegativity, comparable to that of phosphorus. Hydrogen's intermediate electronegativity is related to its ability either to gain or lose one electron. The electronegativities of the elements in the first three rows of the periodic table are given in Table 8-3.

PROBLEM 8.9

Use the periodic table to predict the more electronegative element in each of the following pairs.

(a) C, Ge **(b)** N, O **(c)** Se, O **(d)** Mg, Ca **(e)** Al, P **(f)** Sb, Sn

TABLE 8-3 Electronegativities of selected elements

H 2.1						
Li 1.0	Be 1.5	B 2.0	C 2.5	N 3.0	O 3.5	F 4.0
Na 0.9	Mg 1.2	Al 1.5	Si 1.8	P 2.1	S 2.5	Cl 3.0

8.10 ELECTRONEGATIVITY AND BONDING

OBJECTIVE 8.10 State the conditions of electronegativity that result in a nonpolar covalent bond, a polar covalent bond, and an ionic bond.

Let us again imagine a hypothetical diatomic molecule X–Y, in which a pair of electrons is shared by the two elements:

X : Y

We consider three situations:

1. Both atoms are the same and therefore have an identical tendency to attract the electrons (identical electronegativity).
2. One of the atoms has a *slightly* greater tendency to attract the electrons.
3. One of the atoms has a *much* greater tendency to attract the electrons.

Nonpolar Covalent Bonds

Nonpolar covalent bond: A covalent bond with no charge separation.

When two atoms in a covalent bond have identical electronegativities, both atoms have the same tendency to attract electrons. Thus, the electrons are distributed equally between the two atoms (Fig. 8-6a, p. 216), resulting in a net charge of zero on each atom. This type of bond is called a **nonpolar covalent bond** (or simply a **nonpolar bond**). The word *nonpolar* tells us that there is no charge separation in this type of bond. An example is the bond in molecular fluorine, F_2.

F : F **Only the bonding pair of electrons is shown**

Polar Covalent Bonds

Polar covalent bond: A covalent bond in which there is a separation of charge.

In a **polar covalent bond,** one of the atoms has a somewhat greater electronegativity than the other. As a result of this modest difference in their tendencies to attract electrons, the electrons in the bond are pulled closer to the more electronegative atom (Fig. 8-6b). An example of this type of bond is the bond in hydrogen fluoride, HF:

δ^+ δ^-
H : F **Only the bonding pair of electrons is shown**
$\longmapsto$

Dipole: An electrical separation of charges in a bond.

Because the electrons are pulled more closely to the fluorine atoms, there is a partial negative charge on the fluorine and a corresponding partial positive charge on the hydrogen. We say that the bond is polar and possesses a **dipole.** This term refers to the positive and negative electrical poles present in the bond. Despite the charge separation, the hydrogen and fluorine atoms *share* the pair of electrons. Hence, this bond is still classified as covalent. The symbols δ^+ and δ^- indicate that these are only partial charges. The arrow, $\mapsto$, is a symbol that is frequently used instead of (or in addition to) the symbols for partial charges. It points in the direction of the negative pole. (It is often

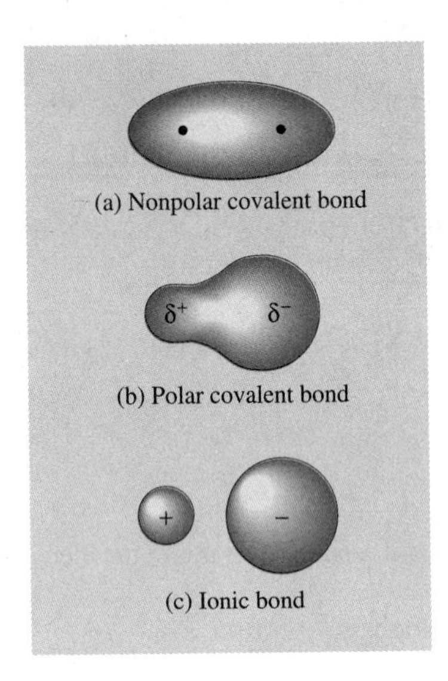

FIGURE 8-6 Three types of bonds.

helpful to think of this symbol as a + sign with an arrow attached; the + sign is found at the positive pole.) Just as we refer to fluorine as the more electro*negative* of the two atoms, we say that hydrogen is the more electro*positive* element in this bond.

Ionic Bonds

The third situation is similar to the second, except that the electronegativity difference is so large that the more electronegative element actually removes the electrons from the less electronegative element and becomes sole possessor of the electron pair (Fig. 8-6c). This corresponds to the ionic bond, which we examined earlier. For example, in sodium fluoride, NaF, an electron is transferred from sodium to fluorine.

(+)		(−)	**Only the bonding pair**
Na	:	F	**of electrons is shown**

We can think of each of these situations as a tug-of-war over a pair of electrons. In the first case, both sides have equal strength and the electrons are shared equally between them. In the second case, one side is stronger, but not strong enough to gain total control of the electrons. The electrons are still shared, though unequally. In the third case, the more electronegative element is so much stronger that it wins the tug-of-war and assumes complete control of the electron pair.

The particular type of bonding that occurs between a given pair of atoms depends on the difference in their electronegativity values (the numbers in Table 8-3). When the electronegativity difference is very small (or zero), the atoms form a nonpolar covalent bond. When the electronegativity difference is very large, an ionic bond results. For intermediate cases, we may observe polar covalent bonds. Chemists like to think of a gradual scale of bond polarities (Fig. 8-7), ranging from nonpolar bonds at one end to ionic bonds at the other. There is no sharp dividing line where polar covalent molecules suddenly become ionic. To distinguish a covalent bond from an ionic bond, we find it most useful to rely on our earlier generalization:

The bond between two nonmetals is covalent.
The bond between a metal and a nonmetal is ionic.

The various types of bonding are shown in the Summary Diagram.

FIGURE 8-7 A scale of electronegativity differences. No sharp dividing line separates bond types.

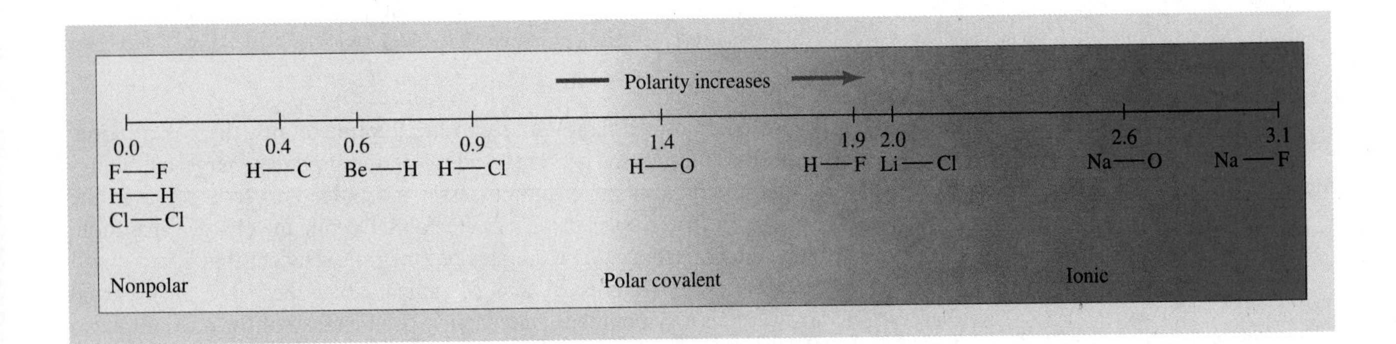

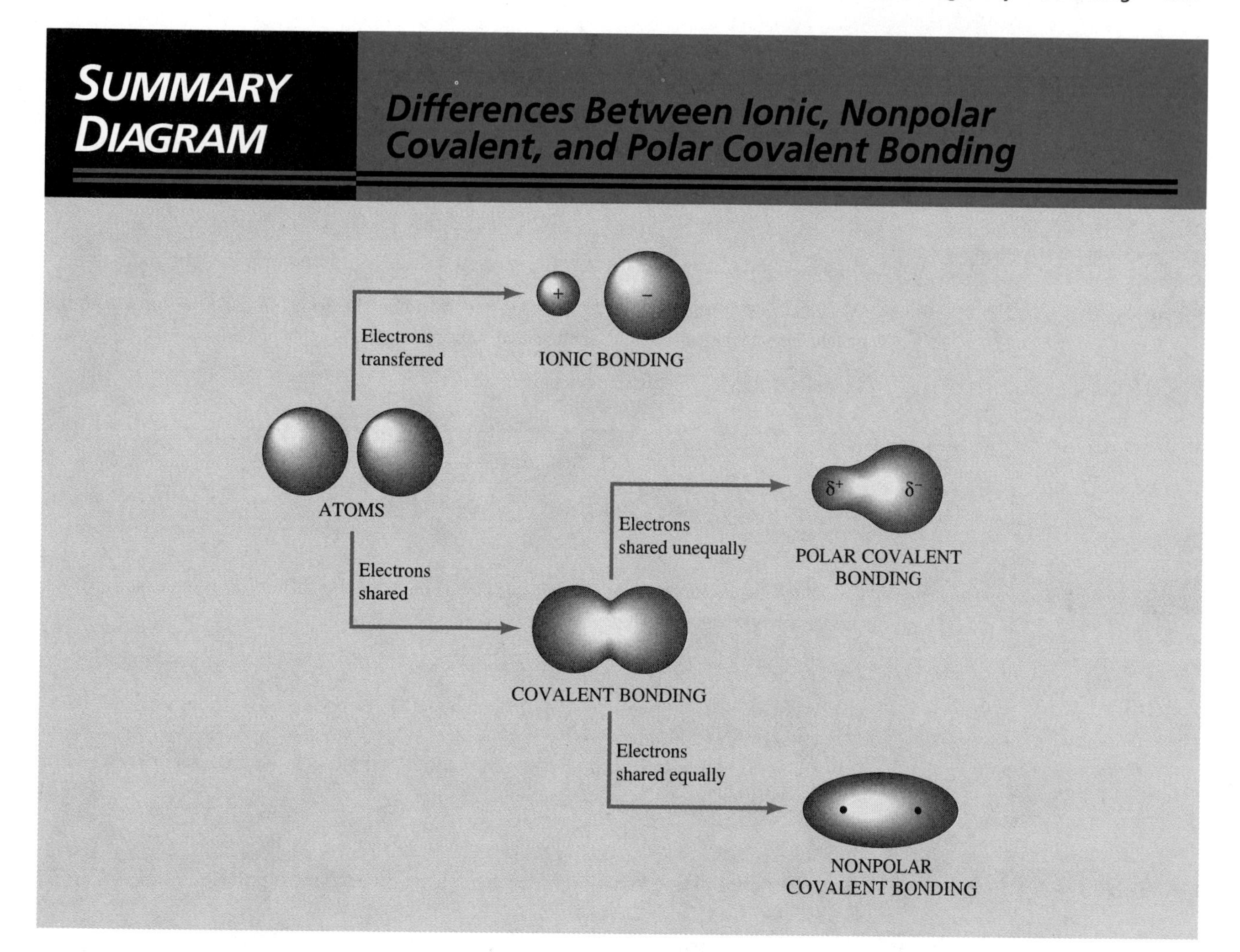

EXAMPLE 8.14

Calculate the electronegativity difference for a C–F bond. Indicate the polarity of the bond by writing δ^+ over the electropositive atom and δ^- over the electronegative atom. Place a dipole arrow below the bond to show the direction of the dipole.

Planning the solution

After looking up the electronegativity values in Table 8-3, we will calculate their difference. The symbol δ^- is written over the element with the greater electronegativity, δ^+ is written over the element with the smaller electronegativity, and the dipole arrow points in the direction of the more electronegative element.

SOLUTION

The electronegativities of the two elements are C = 2.5 and F = 4.0. The electronegativity difference can be determined either by taking the absolute value of the difference in the electronegativities or by subtracting the smaller value from the larger one.

We will use the latter procedure here. The electronegativity difference is 4.0 − 2.5 = 1.5. Fluorine is the more electronegative atom. Thus, we have

$$\overset{\delta^+}{\mathrm{C}} - \overset{\delta^-}{\mathrm{F}}$$
$$\longleftrightarrow$$

EXAMPLE 8.15

Which of the following bonds is the more polar: a P–Cl bond or a Be–H bond?

Planning the solution

We must determine the electronegativity difference for both bonds. The bond with the greater electronegativity difference is the more polar of the two.

SOLUTION

The electronegativity differences are

P–Cl: 3.0 − 2.1 = 0.9
Be–H: 2.1 − 1.5 = 0.6

The P–Cl bond is more polar.

PROBLEM 8.10

Calculate the electronegativity difference for each of the following bonds, and draw a dipole arrow over the bond in the appropriate direction.

(a) H–Cl **(b)** Al–H **(c)** Mg–F **(d)** O–H
(e) Be–S **(f)** P–S **(g)** Li–H
(h) What do you notice about hydrogen when it is bonded to a nonmetal, as in an O–H or H–Cl bond?
(i) What do you notice about hydrogen when it is bonded to a metal, as in an Al–H or Li–H bond?

PROBLEM 8.11

Arrange the following bonds in order, from least polar to most polar: Al–S, C–N, C–H, F–F, H–Cl, Li–Cl, and Si–O.

PROBLEM 8.12

Classify each of the following bonds as ionic or covalent, using the guideline that the bond between a metal and a nonmetal is ionic, whereas one between two nonmetals is covalent.

(a) C–H **(b)** C–F **(c)** Mg–Cl **(d)** Li–O
(e) Cl–Cl **(f)** N–Cl **(g)** Ca–S **(h)** H–Br

8.11 METALLIC BONDING

OBJECTIVE 8.11 Describe metallic bonding.

Metallic bond: Bonding between uncharged metal atoms; characterized by a network of overlapping orbitals that connect an array of inner kernels within a "sea" of electrons.

Before concluding this chapter, let us consider one additional type of bond: the **metallic bond.** This type of bonding is used to describe metals in their uncharged elemental

FIGURE 8-8 Metallic bonding. The nuclei and lower-level electrons make up an orderly array of kernels. The higher-level orbitals overlap, producing a "sea" of very mobile outer electrons. This mobility is responsible for the electrical conductivity of metals.

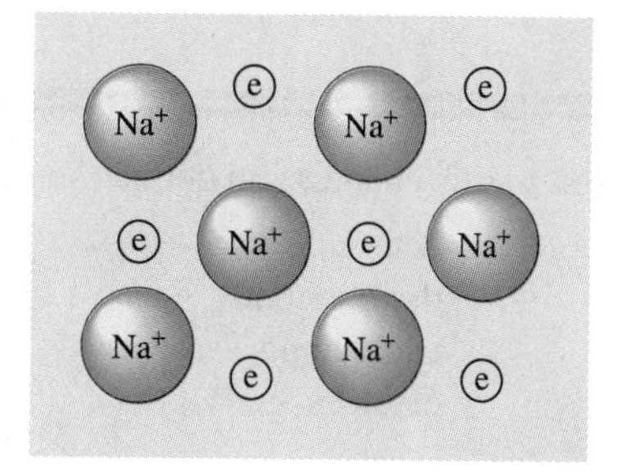

states (as opposed to the ionic bonding we see so frequently when these metals become ions). Each metal atom is considered to be composed of its outer electrons and a *kernel.* The kernel is made up of the nucleus and lower-level electrons. The kernels are arranged in an orderly array, with the outer orbitals overlapping one another. The overlapping network of orbitals contains a "sea" of the outer electrons (Fig. 8-8). This sea of outer electrons is capable of flowing from one atom to the next along the network of kernels. This property accounts for the electrical conductivity that is characteristic of metals.

CHAPTER SUMMARY

Two major kinds of bonds exist: the **ionic bond** and the **covalent bond.** The electrons involved in bond formation are the **valence electrons,** those electrons beyond the last noble gas core. The incentive for bonding is achievement of noble gas configurations by the bonded atoms. In the case of ionic bonding, uncharged atoms form ions through **electron transfer** with other atoms. In the process, both positively and negatively charged ions are formed. The ionic bond results from the electrostatic forces of attraction between such oppositely charged ions. Distinct molecules are not observed for ionic solids. Instead, alternating arrays of oppositely charged ions are held together in a **crystal lattice.**

In the formation of a covalent bond, two orbitals from two atoms overlap and share a pair of electrons. Again, the incentive for bonding is achievement of noble gas configurations, in this case, through the sharing of electrons. Each bond is characterized by a **bond energy** (the energy required to break the bond) and a **bond length** (the distance between the two nuclei). We can represent the bonding in covalently bonded molecules with **Lewis electron-dot structures,** in which each dot represents one of the valence electrons in the molecule.

When atoms bond covalently, they attempt to satisfy the **octet rule** such that each atom is surrounded by eight valence electrons (except for hydrogen, which has only two). It is possible for a pair of atoms to form **double** or **triple bonds** in order to satisfy the octet rule. A useful approach to writing Lewis structures is first to count up all of the valence electrons in the molecule and then to write a structure that satisfies the octet rule.

The atoms in a chemical bond do not always share electrons equally. The **electronegativity** of an element is the tendency of the element to gain electrons in a bonded pair. The electronegativity of the elements increases from left to right across a given period, and it decreases from top to bottom in a given column of the periodic table. Chemical bonds may be classified as **nonpolar covalent, polar covalent,** or **ionic.** The type of bond depends on the difference in electronegativity between the bonded atoms. A **dipole** exists in bonds between atoms of differing electronegativities. In general, a covalent bond forms between two nonmetals, whereas an ionic bond results from the combination of a metal and a nonmetal. A nonpolar covalent bond results from two identical nonmetals, whereas nonequivalent nonmetals often form polar covalent bonds. When nonionic metal atoms combine, they are held together in networks via **metallic bonding.**

KEY TERMS

Review each of the following terms, which are discussed in this chapter and defined in the Glossary.

valence electrons
Lewis electron-dot structure
electron transfer
ionic bond
crystal lattice
covalent bond
bond energy
bond length
octet rule
double bond
triple bond
electronegativity
dipole
nonpolar covalent bond
polar covalent bond
metallic bond

ADDITIONAL PROBLEMS

SECTION 8.1

8.13 Define the term *valence electrons.* What is the relationship between an atom's electronic configuration and its incentive for bonding?

8.14 Determine the number of valence electrons and draw an electron-dot structure for each of the following elements.

(a) Al (b) B (c) Be (d) C (e) Cl (f) F
(g) Li (h) Mg (i) N (j) P (k) S (l) Si

SECTION 8.2

8.15 Show the electron transfer that takes place as the following atoms form ionic compounds.

(a) One atom of lithium and one atom of fluorine
(b) Three atoms of sodium and one atom of nitrogen
(c) Two atoms of lithium and one atom of oxygen
(d) Three atoms of magnesium and two atoms of nitrogen

8.16 State a generalization for predicting whether two elements will form an ionic bond with one another. State a similar generalization for predicting whether two elements will form a covalent bond.

SECTION 8.3

8.17 Draw a potential-energy curve to describe the energy changes that occur as two atoms are brought together to form a covalent bond. Identify the bond length and the bond energy on the curve.

8.18 Show how the two 1*s* orbitals of each hydrogen atom overlap in the formation of a H–H bond. Where is the region of maximum electron density?

SECTIONS 8.4 AND 8.6

8.19 Write Lewis structures for each of the following molecules.

(a) HCl (b) OCl_2 (c) CH_2Cl_2 (d) H_2S_2

SECTIONS 8.5 AND 8.6

8.20
Write Lewis structures for each of the following molecules. The arrangement of atoms is given.

(a) C_2H_3Cl

```
H  C  C  Cl

   H  H
```

(b) C_3H_4

```
   H

H  C  C  C  H

   H
```

(c) HNO_2

H O N O

(d) $HSCN$

H S C N

SECTION 8.7

8.21 Write abbreviated structures for all of the molecules in Problem 8.19, using a dash to represent each bond.

8.22 Write abbreviated structures for all of the molecules in Problem 8.20, using a dash to represent each bond.

SECTION 8.8

8.23 Write Lewis structures for each of the following. You are given the arrangement of the atoms for part (d).

(a) PH_4^+ (b) ClO^- (c) NO_2^-

(d) $C_2O_4^{2-}$

```
[    O  O   ]2-
[ O  C  C  O]
```

SECTION 8.9

8.24 Define the term *electronegativity.* Use the periodic table to predict the more electronegative atom in each of the following pairs.

(a) Na, Al (b) Be, Sr (c) As, Bi
(d) Ge, Br (e) Fr, F (f) Si, O

SECTION 8.10

8.25 Arrange the following bonds in order, from the least polar to the most polar: C–O, Cl–Cl, Mg–O, N–H, Na–H, Na–Cl, P–F.

8.26 Classify each of the following bonds as ionic or covalent, using the guideline that a bond between a metal and a nonmetal is ionic, whereas one between two nonmetals is covalent.

(a) N–H **(b)** Li–Cl **(c)** Mg–O **(d)** P–F
(e) N–S **(f)** C–O **(g)** K–F **(h)** I–Cl

8.27 One method of classifying bond types is according to their electronegativity differences. Some chemists accept the following set of guidelines:

Bond classification	Electronegativity difference
Nonpolar covalent bonds	Less than 0.5
Polar covalent bonds	0.5 to 2.0
Ionic bonds	Greater than 2.0

Use these guidelines to classify each of the following bonds as ionic, polar covalent, or nonpolar covalent.

(a) B–H **(b)** Be–H **(c)** C–F **(d)** H–H
(e) H–F **(f)** Li–O **(g)** Mg–S **(h)** Na–F
(i) P–Cl **(j)** C–H
(k) What do you notice about the polarity of the C—H bond?

GENERAL PROBLEMS

8.28 Write Lewis structures for each of the following molecules or ions. You are given the arrangement of the atoms for part (l).

(a) F_2 **(b)** NCl_3 **(c)** O_3 **(d)** BF_4^-
(e) CH_3CH_3 **(f)** SO_3 **(g)** SO_3^{2-} **(h)** SO_4^{2-}
(i) CN^- **(j)** CH_3NH_2 **(k)** AlH_4^-
(l) $HCONH_2$

```
     O

H  C  N  H

      H
```

8.29 For elements located below the third period of the periodic table, the number of valence electrons is determined by counting the number of electrons beyond the last noble gas core, excluding electrons in filled *d* and *f* sublevels.

(a) How many valence electrons are in each of the following elements: Br, Te, Fr, Sn, Ga, and Bi?

(b) What relationship do you notice between the Roman numeral of the group each element belongs to (traditional designation) and its number of valence electrons?

(c) Use the principle discovered in part (b) to write Lewis structures for Br_2, AsH_3, and SnI_4.

8.30 Element number 117 is yet to be discovered. Answer the following questions about this element.

(a) To what chemical family will element number 117 belong?

(b) What type of bond would two atoms of element 117 form?

(c) What type of bond would it form with a cesium atom (atomic number 55)?

(d) What type of bond would it form with an iodine atom (atomic number 53)?

(e) With which of the following atoms would element 117 form the most polar bond: bromine or chlorine? Explain your answer.

***8.31** The heat absorbed or liberated in a chemical reaction comes from the breaking and making of chemical bonds. The quantity of heat required to break the bonds in H_2, Cl_2, and HCl may be described by the following equations:

$$1 \text{ mol } H_2 + 436 \text{ kJ} \rightarrow 2 \text{ mol H atoms}$$

$$1 \text{ mol } Cl_2 + 243 \text{ kJ} \rightarrow 2 \text{ mol Cl atoms}$$

$$1 \text{ mol HCl} + 432 \text{ kJ} \rightarrow 1 \text{ mol H atoms} + 1 \text{ mol Cl atoms}$$

Use this information to determine the amount of heat liberated when 1 mol of H_2 and 1 mol of Cl_2 react to form 2 mol of HCl.

$$1 \text{ mol } H_2 + 1 \text{ mol } Cl_2 \rightarrow 2 \text{ mol HCl} + ?\text{ kJ}$$

[*Hint:* If 432 kJ are required to break the bonds in a mole of HCl, how much heat is liberated when 2 mol of HCl are formed?]

***8.32** Geometry plays an important role in determining whether a molecule with more than one polar bond has a nonzero net dipole.

(a) For example, a carbon dioxide molecule is linear:

$$\overset{180^\circ}{O{=}C{=}O}$$

Although each C–O bond has a significant dipole, a

carbon dioxide molecule has a net dipole of zero. Suggest a reason for this behavior.

(b) By contrast, water molecules have an angular geometry:

O
H H
105°

The net dipole of water is nonzero. Explain.

(c) A boron trifluoride molecule has a trigonal planar shape:

F
120° 120°
B
F F
120°

Would you expect this molecule to have no net dipole like carbon dioxide, or would it be like water, with a dipole other than zero? Explain your answer.

***8.33** Answer the following questions about the partial charges in a polar diatomic molecule.

(a) How do the electronegativities of the two atoms affect the magnitudes of the partial charges?

(b) Are the magnitudes of the partial charges equal, despite their opposite signs? Explain your answer.

(c) How are the magnitudes of the partial charges related to the distribution of the electrons in the molecule?

***8.34** In Chapter 7, we used the term *isoelectronic* to describe two atoms (or ions) that have the same electronic configuration. For example, a sodium ion and a neon atom are isoelectronic. This term also may be applied to the bonding of chemical species with identical valence electron structures of the bonded atoms in the molecules or ions. For example, CH_4 and NH_4^+ are isoelectronic. To be isoelectronic, the same orbitals must contribute the valence electrons. Thus, CH_4 and SiH_4, although similar in their Lewis structures, are not isoelectronic, since carbon contributes 2*s* and 2*p* electrons to the bonding, whereas silicon donates 3*s* and 3*p* electrons. For each of the following groups of molecules and ions, identify the isoelectronic species. (There is one and only one pair in each group.)

(a) CO, CO_2, and CN^- **(b)** NO_3^-, NO_2^-, and CO_3^{2-}
(c) SO_3, SO_3^{2-}, and PO_3^{3-} **(d)** NCl_3, NF_3, and CCl_3^-
(e) H_3O^+, BH_3, and NH_3

WRITING EXERCISES

8.35 Your friend has always wondered why metals are such good conductors of electricity, and you have agreed to send her a written explanation. She has never taken chemistry, so you need to summarize some of the material covered in the chapter on atomic structure. Begin by explaining how atoms are structured, and then discuss metallic bonding and how it permits the flow of electrons through a network of metal atoms.

8.36 Both sodium and chlorine are considered to be hazardous substances that require great care in handling. Despite the very high reactivity of each of these substances in its pure elemental form, when they are combined as sodium chloride, they are considered relatively harmless. Describe the fate of each element during the formation of sodium chloride, explaining the incentive that exists for each atom to undergo the change that occurs. Then describe the nature of the bond that holds them together. Finally, explain why these substances are dangerous when they exist as individual elements but are harmless once they have combined into sodium chloride.

8.37 In your own words, explain the difference between ionic and covalent bonding. Your explanation should include the incentive that exists for each type of bond to form, how each type of bond is formed, and how to predict which type of bond is most likely to form between a given pair of atoms.

CHEMICAL NOMENCLATURE

9

In this chapter we examine the principles for naming simple inorganic compounds. Chemical nomenclature involves two complementary skills: (1) writing the chemical formula of a substance, given its name, and (2) writing the name of a substance, given its chemical formula.

9.1 COMMON AND SYSTEMATIC NAMES

OBJECTIVE 9.1 Distinguish between a common name and a systematic name.

Common name: Name traditionally associated with a substance; not based on any systematic method of nomenclature.

Systematic name: Derived name; based on a specific set of rules for naming substances.

Many compounds have both a **common name** and a **systematic name.** Some chemicals are encountered so frequently in the laboratory and have been so important traditionally that we recognize them by their common names. Thus, the substance H_2O is commonly referred to as water, although there is no systematic reason for calling it by this name. Similarly, we accept the common name of ammonia for the substance with the formula NH_3. There is no way to learn the common names of substances except by memorization. By this point you probably recognize the formula H_2O as that of water and, as you proceed through your study of chemistry, you will learn to recognize the common names of other important chemicals. Table 9-1 gives the common name, formula, and systematic name for a number of frequently encountered substances.

The large number of possible chemicals makes it impractical to assign a common name to every substance. Consequently, systematic methods for naming compounds have been devised. Once you learn the rules for naming compounds systematically, you will be able to write the correct name for a compound the first time you encounter its formula. Furthermore, that name will communicate the correct formula to anyone else who also knows the rules of nomenclature.

The early sections in this chapter are concerned with writing formulas from the names of compounds. These sections will also further your knowledge of the various

TABLE 9-1 Some common names of chemicals

Common name	Formula	Systematic name
Acetylene	C_2H_2	Ethyne
Baking soda	$NaHCO_3$	Sodium hydrogen carbonate
Cane sugar	$C_{12}H_{22}O_{11}$	Sucrose
Epsom salts	$MgSO_4 \cdot 7H_2O$	Magnesium sulfate 7-hydrate
Grain alcohol	C_2H_5OH	Ethanol
Laughing gas	N_2O	Dinitrogen monoxide
Lye	$NaOH$	Sodium hydroxide
Marble	$CaCO_3$	Calcium carbonate
Milk of magnesia	$Mg(OH)_2$	Magnesium hydroxide
Table salt	$NaCl$	Sodium chloride
Vinegar	$HC_2H_3O_2$	Acetic acid
Washing soda	$Na_2CO_3 \cdot 10H_2O$	Sodium carbonate 10-hydrate
Wood alcohol	CH_3OH	Methanol

ions that are important to us. Later sections deal with writing names from the formulas of substances.

9.2 FORMULAS OF IONIC COMPOUNDS DERIVED FROM MONATOMIC IONS

OBJECTIVE 9.2 Determine the formula of any ionic compound composed of monatomic ions included in Table 9-2, given the name of the compound.

We have seen that many of the elements tend to form ions having noble gas electronic configurations. The ions we have considered thus far are listed in Table 9-2.

In our discussion of the ionic bond, we saw that a sodium ion, Na^+, and a fluoride ion, F^-, combine in a 1 : 1 ratio to form an electrically neutral substance (no net charge) called sodium fluoride, NaF.

In a similar fashion, we might expect a magnesium ion, Mg^{2+}, and a sulfide ion, S^{2-}, to form an ionic compound called magnesium sulfide, MgS. Once again, a 1 : 1 ratio is required for the +2 charge and the −2 charge to balance.

Ionic compounds do not always form in a 1 : 1 ratio. For example, one magnesium ion, Mg^{2+}, combines with two fluoride ions, F^-, to form magnesium fluoride, MgF_2. The total positive charge in the formula is +2 (the charge on one magnesium ion), whereas the total negative charge is −2 (the charge due to two fluoride ions). Although the formula Mg_2F_4 is also electrically neutral, this formula is not correct because it does not represent the simplest ratio possible. *The formula of an ionic compound is written as the simplest whole-number ratio that results in a neutral compound.*

By this rule, one aluminum ion, Al^{3+}, combines with three chloride ions, Cl^-, to give a neutral substance known as aluminum chloride. This 1 : 3 ratio of aluminum ion to chloride ion yields the formula $AlCl_3$.

Finally, let us write the correct formula for aluminum oxide. In this case we need to find the simplest ratio of aluminum ions, Al^{3+}, and oxide ions, O^{2-}, that will balance the charges. A combination of 2 Al^{3+} ions and 3 O^{2-} ions gives a total positive charge of +6 and a total negative charge of −6, resulting in a neutral compound. Thus, the formula of aluminum oxide is Al_2O_3.

In writing the formulas for ionic compounds, we always show the cation (positive ion) first and the anion (negative ion) last. The charges of the ions are not included in the formula of an ionic compound. Thus, we have the following:

Formulas of ions:	Na^+, Ba^{2+}, Al^{3+}	**Charges included**
Formulas of compounds:	NaCl, BaF_2, Al_2S_3	**No charges shown**

TABLE 9-2 Some common ions and their charges

Hydrogen	H^+	Calcium	Ca^{2+}	Chloride	Cl^-
Potassium	K^+	Barium	Ba^{2+}	Bromide	Br^-
Lithium	Li^+	Strontium	Sr^{2+}	Iodide	I^-
Sodium	Na^+	Aluminum	Al^{3+}	Oxide	O^{2-}
Magnesium	Mg^{2+}	Fluoride	F^-	Sulfide	S^{2-}

The Summary outlines a general procedure for writing the formula of any ionic compound from its name.

SUMMARY: Writing the Formula of an Ionic Compound from Its Name

Step 1. Write the formulas of the cation and anion.

Step 2. Determine the simplest whole-number ratio of ions that results in a neutral compound.

Step 3. Use the ratio determined in Step 2 to write the formula of the compound.

EXAMPLE 9.1

Write the formula of potassium oxide.

Planning the Solution

We follow the steps outlined in the Summary.

SOLUTION

We can determine the formulas of the ions from our knowledge of the periodic table (or by looking them up in Table 9-2). Since potassium is an alkali metal (group IA or 1), the potassium ion has a +1 charge: K^+. Since oxygen is a chalcogen (group VIA or 16), its ion has a charge of −2: O^{2-}. A 2 : 1 ratio of these ions results in a neutral compound.

$$2\,K^+ + 1\,O^{2-} \quad \text{becomes} \quad K_2O$$

Thus, the formula of potassium oxide is K_2O.

EXAMPLE 9.2

Write the formula of aluminum sulfide.

SOLUTION

The aluminum ion (group IIIA or 13) has a charge of +3: Al^{3+}. The sulfide ion (group VIA or 16) has a charge of −2: S^{2-}. These two ions must combine in a 2 : 3 ratio.

$$2\,Al^{3+} + 3\,S^{2-} \quad \text{becomes} \quad Al_2S_3$$

The formula of aluminum sulfide is Al_2S_3.

In this section we have restricted ourselves to ions with charges that are readily predicted from the periodic table. In Sections 9.3 and 9.4, we consider ionic compounds composed of ions with formulas that must be memorized (or looked up in a table of ions). The general procedure for writing the formulas of those compounds is exactly the same as that presented in this section.

PROBLEM 9.1

Fill in the following table with the formulas of the compounds obtained by combining each cation with each anion. The formula of sodium oxide is provided as an example.

	Cl^-	O^{2-}	N^{3-}
Na^+		Na_2O	
Ca^{2+}			
Al^{3+}			

PROBLEM 9.2

Write the correct formula for each of the following ionic compounds.

(a) Sodium chloride (table salt) **(b)** Potassium iodide **(c)** Magnesium bromide
(d) Barium oxide **(e)** Lithium oxide **(f)** Potassium sulfide
(g) Aluminum sulfide **(h)** Calcium fluoride **(i)** Strontium oxide

9.3 FORMULAS OF COMPOUNDS CONTAINING POLYATOMIC IONS

OBJECTIVE 9.3 Determine the formula of any ionic compound composed of ions included in Tables 9-2 and 9-3, given the name of the compound.

So far, our discussion of chemical formulas has been restricted to compounds made up of monatomic ions, such as Li^+, Ca^{2+}, F^-, and O^{2-}. In the last chapter, we examined the structures of several polyatomic ions. Their formulas are included in Table 9-3.

Polyatomic ions form compounds in exactly the same way that monatomic ions do: They combine with other ions in the simplest whole-number ratios that give neutral compounds. Thus, the sodium ion, Na^+, and the nitrate ion, NO_3^-, combine in a 1 : 1 ratio to form sodium nitrate:

$$NaNO_3$$

Similarly, the magnesium ion, Mg^{2+}, and the hydroxide ion, OH^-, combine in a 1 : 2 ratio to form a neutral compound. In writing the formula for this compound, we need to show that there are two hydroxide ions. We do this by enclosing the polyatomic ion in parentheses and placing a subscript 2 outside the parentheses:

$$Mg(OH)_2$$

By using parentheses in this fashion, we indicate that the formula includes two hydroxides. If we were to omit the parentheses and write $MgOH_2$, we would be saying that the

TABLE 9-3 Polyatomic ions

Ammonium	NH_4^+	Carbonate	CO_3^{2-}
Hydroxide	OH^-	Sulfate	SO_4^{2-}
Nitrate	NO_3^-	Oxalate	$C_2O_4^{2-}$
Cyanide	CN^-	Chromate	CrO_4^{2-}
Acetate	$C_2H_3O_2^-$	Thiosulfate	$S_2O_3^{2-}$
Hydrogen carbonate	HCO_3^-	Phosphate	PO_4^{3-}
Chlorate	ClO_3^-	Arsenate	AsO_4^{3-}

substance contains one atom of magnesium, one atom of oxygen, and two atoms of hydrogen. The number outside the parentheses indicates that we have two of everything inside the parentheses.

The following examples will further illustrate how to write the formulas of substances containing polyatomic ions.

EXAMPLE 9.3

Write the formula of magnesium carbonate.

Planning the solution

We follow the same procedure we used in Section 9.2.

SOLUTION

Magnesium carbonate is made up of magnesium ion, Mg^{2+}, and carbonate ion, CO_3^{2-}. In order to balance the +2 charge and the −2 charge, we need a 1 : 1 ratio of these ions. Hence the formula for magnesium carbonate is $MgCO_3$.

EXAMPLE 9.4

Write the formula of sodium phosphate.

SOLUTION

Sodium phosphate is made up of sodium ion, Na^+, and phosphate ion, PO_4^{3-}. We need three +1 charges to balance the −3 charge of the phosphate ion. Thus, three Na^+ ions combine with one PO_4^{3-} ion to form a neutral compound, Na_3PO_4.

EXAMPLE 9.5

Write the formula of aluminum sulfate.

SOLUTION

Aluminum sulfate is made up of aluminum ion, Al^{3+}, and sulfate ion, SO_4^{2-}. Since the aluminum ion has a +3 charge and sulfate ion has a −2 charge, it will take a 2 : 3 ratio of aluminum ion to sulfate ion to form a neutral compound. The formula must be written

$$Al_2(SO_4)_3$$

Let us point out a few things about this formula. First, it is not necessary to put monatomic ions such as aluminum in parentheses, because there is no ambiguity about the number of atoms of aluminum in the formula. Second, when we wish to show that there are three sulfate ions, we are not bothered by the fact that each sulfate contains four atoms of oxygen. We merely write the formula for the sulfate ion and place it in parentheses with a subscript 3 to indicate that there are three such ions. Each formula contains two atoms of aluminum, three atoms of sulfur, and twelve ($3 \times 4 = 12$) atoms of oxygen.

EXAMPLE 9.6

Write the formula of aluminum nitrate.

SOLUTION

This compound is composed of aluminum ion, Al^{3+}, and nitrate ion, NO_3^-. In order to balance the charge, the formula must show a 1 : 3 ratio of Al^{3+} to NO_3^-:

$$Al(NO_3)_3$$

This formula contains one atom of aluminum, three atoms of nitrogen, and nine atoms of oxygen.

PROBLEM 9.3

Write the correct formula for each of the following compounds.
(a) Aluminum phosphate **(b)** Lithium sulfate **(c)** Aluminum hydroxide
(d) Barium nitrate **(e)** Ammonium carbonate **(f)** Calcium sulfate
(g) Aluminum chromate **(h)** Strontium phosphate **(i)** Magnesium arsenate
(j) Sodium oxalate

9.4 OXIDATION NUMBERS

OBJECTIVE 9.4 Determine the formula of any ionic compound composed of ions included in a table of ions, given the name of the compound.

Each of the cations we have looked at thus far has a definite charge. For example, the sodium ion always has a +1 charge, and the magnesium ion always has a +2 charge. However, some elements form more than one kind of ion. For example, copper may form both a +1 ion (Cu^+) and a +2 ion (Cu^{2+}). The elements that fall into this category are those found below the third row of the periodic table. The chemistry of these elements is more complicated than that of the elements we have considered so far, and it is often difficult to predict what kinds of ions they are likely to form. In this section we concern ourselves only with how to write the formulas of compounds containing these ions once we know what their charges are.

Oxidation number or oxidation state: The assigned charge of an atom, such as a metal cation.

We call the charge on a monatomic ion its **oxidation number** (or **oxidation state**). For example, the oxidation number of Cu^+ is +1, whereas the oxidation number of Cu^{2+} is +2. We use Roman numerals to write the names of these two ions: Cu^{2+} is named "copper(II)," whereas Cu^+ is named "copper(I)." The table inside the back cover lists the ions we use in this text. If no Roman numeral is included in the name of a particular ion, it means that there is only one common oxidation state for ions of that element. Thus, a zinc ion is always Zn^{2+}. Iron, however, forms two different kinds of ion: iron(II), Fe^{2+}, and iron(III), Fe^{3+}.

When iron(II) and iron(III) combine with anions to form compounds, they combine in different ratios. For example, iron(II), Fe^{2+}, combines with chloride, Cl^-, in a 1 : 2 ratio to give iron(II) chloride, $FeCl_2$. Iron(III), Fe^{3+}, combines with chloride in a 1 : 3 ratio to form iron(III) chloride, $FeCl_3$. Iron(II) chloride and iron(III) chloride are different chemical substances. They have differing chemical compositions, so their physical and chemical properties are different. Because there are two kinds of iron chloride, it is necessary for us to distinguish between them, and we do so when we name them iron(II) chloride and iron(III) chloride.

One ion that is often puzzling at first is the mercury(I) ion. The formula for mercury(I) is $Hg_2{}^{2+}$. Each mercury(I) ion contains two atoms of mercury. In this sense it is similar to the ammonium ion, the nitrate ion, and other polyatomic ions we have considered so far. Because the mercury(I) ion is made up of two identical atoms and bears a total charge of +2, the oxidation state of each individual atom is +1. Thus, we refer to this ion as mercury(I). Let us examine the formula of the following mercury(I) compound:

$$Hg_2Cl_2$$

Mercury(I) chloride

This formula may seem to violate the principle of writing the simplest whole-number ratio of ions; however, the formula really represents a 1:2 ratio of mercury(I), Hg_2^{2+}, to chloride, Cl^-.

PROBLEM 9.4

Write the chemical formula for each of the following compounds.

(a) Copper(I) bromide **(b)** Tin(II) chloride **(c)** Lead(II) nitrate
(d) Cobalt(III) oxide **(e)** Iron(III) hydroxide **(f)** Chromium(III) sulfate
(g) Copper(I) carbonate **(h)** Cobalt(II) chromate **(i)** Mercury(II) fluoride
(j) Mercury(I) fluoride

9.5 THE NAMING OF COMPOUNDS—A WORD OF ADVICE

OBJECTIVE 9.5 State the rule for naming any ionic compound.

The remainder of this chapter is devoted to presenting rules for naming some of the more common types of inorganic compounds. As a practical matter, you will find that the names of most compounds consist of the name of a cation followed by the name of an anion. Consequently, *you should become familiar with the ions listed in the table inside the back cover.* Your instructor may require you to memorize the names and formulas of many of the ions in that table.

You already know how to use the periodic table to predict the charges of monatomic ions derived from some of the representative elements (groups IA, IIA, and so forth). Several of the sections that follow are designed to organize your knowledge of other ions so that the task of learning their names and formulas should be less formidable. As you proceed through the remainder of the chapter, remember that *in order to name any ionic compound, you simply give the name of the cation present followed by the name of the anion.* Thus, NaCl is sodium chloride; $(NH_4)_2SO_4$ is ammonium sulfate; and KOH is potassium hydroxide.

We recommend that you begin by learning the names and formulas of the following polyatomic ions, which you will encounter frequently throughout the text.

Charge	Name	Formula
+1	Ammonium	NH_4^+
−1	Hydroxide	OH^-
	Nitrate	NO_3^-
	Cyanide	CN^-
	Chlorate	ClO_3^-
	Acetate	$C_2H_3O_2^-$
−2	Carbonate	CO_3^{2-}
	Sulfate	SO_4^{2-}
	Chromate	CrO_4^{2-}
	Oxalate	$C_2O_4^{2-}$
−3	Phosphate	PO_4^{3-}

PROBLEM 9.5

State the rule for naming any ionic compound.

9.6 BINARY COMPOUNDS COMPOSED OF A METAL AND A NONMETAL

OBJECTIVE 9.6A Name binary compounds composed of a metal and a nonmetal. Determine the oxidation state of the metal cation from the formula of the compound.

Binary compound: A compound composed of two elements.

A **binary compound** is a compound composed of two different elements. For example, both sodium chloride, NaCl, and aluminum oxide, Al_2O_3, are binary compounds. In each case, the more electronegative element appears second in both the name and the formula. When a binary compound consists of a metal and a nonmetal, the metal is always the less electronegative of the two elements. (Recall that the nonmetals are all on the right-hand side of the periodic table, the same region where we find the elements of highest electronegativity.) Because the bond between a metal and a nonmetal is generally ionic in nature, these compounds must be ionic.

We can subdivide binary compounds of this type into those having a metal with only one oxidation state, such as sodium chloride, NaCl; and those with a metallic element that may exhibit varying oxidation states, such as iron(III) oxide, Fe_2O_3.

Some binary compounds of the first type are the following:

NaCl	Sodium chloride	Ag_2O	Silver oxide
KI	Potassium iodide	$ZnBr_2$	Zinc bromide
MgS	Magnesium sulfide	Al_2O_3	Aluminum oxide
LiH	Lithium hydride	Mg_3N_2	Magnesium nitride

You can see that the name of the metal cation is followed by the name of the nonmetal anion. Also note that the name of the metal cation is the same as the name of the atom from which it was derived. The name of the nonmetal anion is obtained by taking the root of the name and adding an *-ide* ending. Thus, chlor*ine* becomes a chlor*ide* ion. The names of these monatomic anions are included in the table inside the back cover.

When the metallic element in question can exhibit more than one oxidation state, the name must specify the oxidation state, as in the following examples:

CuBr	Copper(I) bromide	HgO	Mercury(II) oxide
FeS	Iron(II) sulfide	$SnCl_4$	Tin(IV) chloride
CuO	Copper(II) oxide	Cu_2S	Copper(I) sulfide
FeP	Iron(III) phosphide	Co_3P_2	Cobalt(II) phosphide

The rules for naming these compounds are the same as those just described. The only difference is that here we include Roman numerals to indicate the oxidation state of the metal cation. This system of naming compounds is known as the *Stock system.*

Earlier in this chapter we wrote the formulas of various ionic compounds, given their names. Now we will name a compound given its formula. Because binary compounds consisting of a metal and a nonmetal are usually ionic in nature, we find it easiest to name such a compound by determining the ions of which it is made.

Suppose we are given the formula ZnI_2. By referring to the table inside the back cover, you can see that the metal cation, zinc, exhibits only one oxidation state, Zn^{2+}.

Thus, the name of the cation is simply zinc (we do not use Roman numerals). The nonmetallic anion is named iodide, I^-. The formula represents zinc iodide.

When the metal cation has more than one common oxidation state, we must determine the oxidation number. For example, for $CuCl_2$ we know that a chloride ion (Cl^-) has a charge of −1, but we must determine the oxidation state of the copper before we can name the compound. To do so, we make use of the fact that the total negative charge in the formula must equal the total positive charge. Let us write down the ions present in the formula:

$$Cu^{?} + 2\,Cl^-$$

There are two chloride ions, so the total negative charge is −2. Hence the total positive charge must be +2. The lone copper ion must bear the entire +2 charge and therefore must be a copper(II) ion, Cu^{2+}. The name of this compound is copper(II) chloride.

EXAMPLE 9.7

Name the compound that has the formula $CoBr_3$.

Planning the solution

Since this compound is composed of a metal and a nonmetal, it is ionic. Thus, its name must consist of the name of the cation followed by the name of the anion. Because cobalt has more than one common oxidation state, we need to determine its oxidation state in order to name the cation.

SOLUTION

The compound must consist of one cobalt cation and three bromide ions:

$$Co^{?} + 3\,Br^-$$

Each bromide ion has a −1 charge, so the total negative charge is −3. Thus, the total positive charge must be +3. The lone cobalt ion must bear the total positive charge and have an oxidation state of +3 (Co^{3+}). The name of the compound is cobalt(III) bromide.

EXAMPLE 9.8

Name the compound that has the formula Fe_2S_3.

SOLUTION

The compound is made up of the following ions:

$$2\,Fe^{?} + 3\,S^{2-}$$

There are three sulfides, so the total negative charge is −6 [= 3 × (−2)]. This means that the total positive charge must be +6. In order for the two iron ions to bear a charge of +6, each one must have a +3 charge, Fe^{3+}. This compound is iron(III) sulfide.

EXAMPLE 9.9

Name the compound that has the formula SnO_2.

SOLUTION

The compound is made up of the following ions:

$$Sn^{?} + 2\,O^{2-}$$

Here, there are two oxide ions, giving a total negative charge of −4. Tin must have an oxidation state of +4. This compound is tin(IV) oxide.

EXAMPLE 9.10

Name the compound that has the formula Mg_3N_2.

SOLUTION

The compound is made up of the following ions:

$$3\ Mg^{2+} + 2\ N^{3-}$$

Since magnesium forms only a +2 ion, we do not include the oxidation number in the name. Consequently, this compound is simply named magnesium nitride. If you are not sure whether a particular metal forms more than one kind of ion, refer to the ion table inside the back cover.

PROBLEM 9.6

Name each of the following binary compounds.

(a) BaS **(b)** $CaCl_2$ **(c)** Ag_3N **(d)** CuF_2 **(e)** HgO **(f)** Hg_2O
(g) $CoCl_2$ **(h)** $CoCl_3$ **(i)** FeS **(j)** Cr_3N_2 **(k)** Zn_3P_2 **(l)** MgH_2

The -ic *and* -ous *Designations—An Older System*

OBJECTIVE 9.6B Translate older *-ic* and *-ous* names into modern systematic nomenclature.

The system of using Roman numerals to designate the oxidation state of a metal is the preferred and currently accepted method of nomenclature. Nevertheless, you may encounter chemicals that have been named according to an older system. In this system, metals of variable oxidation state are distinguished by assigning endings of *-ic* or *-ous* to the Latin, Greek, or English root of the name. The lower oxidation state is indicated by the *-ous* suffix and the higher oxidation state by the *-ic* suffix. Thus, iron(II) is also known as the ferr*ous* ion, and iron(III) is called ferr*ic*. Similarly, copper(I) is known as cupr*ous* and copper(II) is cupr*ic*. Thus, an alternative name for iron(III) chloride is ferric chloride. Similarly, copper(I) iodide is cuprous iodide. It is easy to remember these endings if we keep in mind that *lower* and *-ous* both contain an "o," whereas *higher* and *-ic* both have an "i."

Unfortunately, this system does not tell us the actual oxidation state of the metal. That is why we prefer the use of Roman numerals (which *do* indicate the oxidation state of the cation.) You will find the older names of many of the variable oxidation state metals in the table of ions inside the back cover.

EXAMPLE 9.11

Stannous fluoride is often used as a source of fluoride in toothpaste. Write the preferred name and formula of this substance.

Planning the solution

We must find either the formula or preferred name of the stannous ion. Once we know either, we can proceed as usual. This information can be found in the table of ions.

SOLUTION

The stannous ion corresponds to tin(II): Sn^{2+}. Since fluoride ion (F^-) has a −1 charge, the ratio of tin(II) to fluoride ion is 1 : 2. Therefore, the name of this substance must be tin(II) fluoride, and its formula is SnF_2.

PROBLEM 9.7

Write the preferred name and formula of each of the following compounds.
(a) Mercuric chloride **(b)** Cobaltous iodide **(c)** Stannous bromide
(d) Ferrous phosphide **(e)** Mercurous iodide (mercurous ion is Hg_2^{2+})
(f) Cobaltic nitride **(g)** Plumbous sulfide **(h)** Stannic hydride

Pseudobinary Compounds

Several of the common polyatomic ions form compounds that have names similar to those of binary ionic compounds. These compounds, which are occasionally referred to as *pseudobinary* compounds, include the following ions:

Hydroxide	OH^-
Cyanide	CN^-
Ammonium	NH_4^+

For example, the following compounds contain these ions:

$NaOH$	Sodium hydroxide
KCN	Potassium cyanide
$Ba(OH)_2$	Barium hydroxide
$Hg(CN)_2$	Mercury(II) cyanide
NH_4Cl	Ammonium chloride

As with all ionic compounds, each of their names consists of the name of the cation followed by the name of the anion.

PROBLEM 9.8

Name each of the following compounds.
(a) KOH **(b)** $Cu(CN)_2$ **(c)** $Al(OH)_3$ **(d)** $NaCN$ **(e)** NH_4CN **(f)** $(NH_4)_2S$

9.7 IONIC COMPOUNDS DERIVED FROM OXYANIONS

OBJECTIVE 9.7A Name any ionic compound that includes a polyatomic ion that appears in the table of ions. Determine the oxidation state of any metal cation present.

OBJECTIVE 9.7B Determine the formula of an *-ite* ion, given the formula of its *-ate* parent. Name any *per-* or *hypo-* ions derived from the same parent. Name any ion derived from the combination of one or more hydrogen ions and an oxyanion.

Oxyanion: A polyatomic anion composed of oxygen and another element.

Many compounds possess a polyatomic oxygen-containing anion called an **oxyanion.** These anions are composed of an element other than oxygen plus one or more oxygen atoms. Several oxyanions are listed in Table 9-4.

TABLE 9-4 Some important oxyanions

Nitrate	NO_3^-	Sulfate	SO_4^{2-}	Phosphate	PO_4^{3-}
Nitrite	NO_2^-	Sulfite	SO_3^{2-}	Phosphite	PO_3^{3-}
Perchlorate	ClO_4^-	Carbonate	CO_3^{2-}	Arsenate	AsO_4^{3-}
Chlorate	ClO_3^-	Oxalate	$C_2O_4^{2-}$		
Chlorite	ClO_2^-	Chromate	CrO_4^{2-}		
Hypochlorite	ClO^-	Dichromate	$Cr_2O_7^{2-}$		
Acetate	$C_2H_3O_2^-$	Thiosulfate	$S_2O_3^{2-}$		
Permanganate	MnO_4^-				

You can see that the name of each ion in Table 9-4 ends in *-ate* or *-ite*. Let us compare the formulas of the corresponding oxyanions.

NO_3^-	Nitrate	NO_2^-	Nitrite
SO_4^{2-}	Sulfate	SO_3^{2-}	Sulfite
ClO_3^-	Chlorate	ClO_2^-	Chlorite

In each case, the *-ate* ion contains one more oxygen than its *-ite* counterpart. Thus, there is a systematic assignment of the *-ate* and *-ite* endings according to the number of oxygen atoms. Unfortunately, the system does not tell us how many oxygens go with the *-ate* or *-ite*. Nor does it tell us what the net charge on the ion will be. That is why you are encouraged to learn the list of polyatomic ions presented earlier in Section 9.5.

Like the ionic binary compounds discussed in Section 9.6, the names of ionic compounds containing oxyanions are derived by giving the name of the cation followed by the name of the oxyanion, as in the following examples:

KNO_3	Potassium nitrate	Li_2SO_3	Lithium sulfite
$FeSO_4$	Iron(II) sulfate	$NaClO_2$	Sodium chlorite

For several of the nonmetals, more than two oxyanions exist. We distinguish among these by using the prefixes *per-* and *hypo-*:

ClO_4^-	Perchlorate	ClO_2^-	Chlorite
ClO_3^-	Chlorate	ClO^-	Hypochlorite

per- **Add one oxygen to the *-ate* anion.**

hypo- **Remove one oxygen from the *-ite* anion.**

The prefixes *per-* and *hypo-* alter the number of oxygens as indicated. The prefix *hypo-* means "under," as in a hypodermic syringe (which is injected under the skin). *Per-* may be thought of as a shorter version of the prefix *hyper-*, meaning "over," as in a hyperactive individual.

Oxyanions of the other halogens have formulas and names that are analogous to those of the chlorate series. The parent ions are bromate, BrO_3^-, and iodate, IO_3^-. Thus, the following compounds are named as shown:

$NaBrO_3$	Sodium bromate
KIO_4	Potassium periodate
$NaClO$	Sodium hypochlorite (used to chlorinate swimming pools)

The *per-* prefix is also used to name the peroxide ion, O_2^{2-}, which contains one more oxygen than oxide, O^{2-}. This ion is a component of the common bactericidal agent hydrogen peroxide, H_2O_2. Finally, the *per-* prefix appears in the name of the permanganate ion, MnO_4^-, an ion you will encounter elsewhere in this text. Although the nomenclature of this ion is not as systematic as that of the other ions discussed, it is included here to familiarize you with its name and formula.

EXAMPLE 9.12

What is the name of the ion having the formula PO_2^{3-}?

Planning the solution

This ion is similar to phosphate, PO_4^{3-}, but it has only two oxygens. Its name can be derived from the name *phosphate*.

SOLUTION

The ion with one less oxygen than phosphate would be phosph*ite*. To derive the name of the ion with one less oxygen than an *-ite* ion, we add the prefix *hypo-*. Thus, the name of PO_2^{3-} is *hypophosphite*.

EXAMPLE 9.13

What is the name of the ion having the formula BrO_2^-?

Planning the solution

This ion is similar to chlorate, ClO_3^-. However, the chlorine atom has been replaced by bromine (another halogen), and it has one less oxygen than chlorate. Since bromine and chlorine are both halogens, the name of this ion will be analogous to the name of its chlorine counterpart.

SOLUTION

The name of ClO_3^- is *chlor*ate. Therefore, the name of BrO_3^- must be *brom*ate. Since BrO_2^- has one less oxygen than bromate, its name must be *bromite*.

PROBLEM 9.9

Write the formula of each of the following ions.
(a) Hypobromite **(b)** Periodate **(c)** Iodite **(d)** Hypoiodite

PROBLEM 9.10

Name each of the following.
(a) Cu_2SO_4 **(b)** KNO_2 **(c)** $LiClO_3$ **(d)** $FeSO_3$ **(e)** $Mg(NO_3)_2$ **(f)** $NaClO_2$

PROBLEM 9.11

Name each of the following.
(a) $NaClO_4$ **(b)** $KBrO$ **(c)** $Ca(IO)_2$ **(d)** Na_2O_2 **(e)** $KMnO_4$ **(f)** $AgClO$

A number of oxyanions combine with one or more hydrogen ions, H^+, to give additional anions. For example, the hydrogen carbonate ion may be obtained by combining a hydrogen ion and a carbonate ion:

$$H^+ + CO_3^{2-} \rightarrow HCO_3^-$$

The charge on the resulting ion is equal to the sum of the charges on the ions that were combined.

The following list gives the most common examples of this type of ion.

HCO_3^- Hydrogen carbonate (bicarbonate)
HSO_4^- Hydrogen sulfate (bisulfate)
HSO_3^- Hydrogen sulfite (bisulfite)
HPO_4^{2-} Monohydrogen phosphate
$H_2PO_4^-$ Dihydrogen phosphate

(Several of these ions are also commonly known by the older names given in parentheses.)

Ionic compounds containing these ions are named according to the same general procedure we have used for other ionic compounds. For example, $NaHCO_3$ is composed of sodium ion, Na^+, and hydrogen carbonate ion, HCO_3^-. Thus, the name of this compound is sodium hydrogen carbonate. In this case, the word *hydrogen* is part of the name of the anion. (You may be more familiar with this compound as sodium bicarbonate, an older name that is still frequently used when referring to baking soda.) The names of the following compounds are similarly derived from the name of the cation followed by the name of the anion.

$NaHSO_3$ Sodium hydrogen sulfite
KH_2PO_4 Potassium dihydrogen phosphate
K_2HPO_4 Potassium monohydrogen phosphate

EXAMPLE 9.14

Write the formula of calcium hydrogen sulfate. What is an older name for this compound?

Planning the solution

We will write the formulas of the cation and anion and then take the simplest whole-number ratio that balances the charges.

SOLUTION

The cation is calcium ion, Ca^{2+}, and the anion is the hydrogen sulfate ion, HSO_4^-. A 1:2 ratio of these ions is required:

$$Ca^{2+} + 2\,HSO_4^- \quad \text{becomes} \quad Ca(HSO_4)_2$$

This compound is often referred to as calcium bisulfate.

In this example (as well as others that involve hydrogen-containing anions), care must be taken to recognize that hydrogen is part of the anion and is not an independent cation.

PROBLEM 9.12

Name each of the following ionic compounds.
(a) $KHSO_4$ **(b)** LiH_2PO_4 **(c)** $Mg(HSO_4)_2$ **(d)** $Ca(HCO_3)_2$ **(e)** $MgHPO_4$

9.8 HYDRATED SALTS

OBJECTIVE 9.8 Name the hydrate of any salt composed of ions in the table of ions.

Hydrated salt: A salt that contains a fixed number of moles of water for each mole of salt.

Most of the ionic compounds described thus far are known as *salts*. For now, think of a salt as a substance composed of a metal or ammonium cation and an anion other than oxide, hydroxide, or hydride. Often, when the crystalline form of a salt is obtained, water molecules are found to accompany the salt in a definite ratio. Table 9-1 included two of these salts, which are known as **hydrated salts** because there is water present. The water present is known as the *water of hydration*. The formula of one of these, epsom salts, is $MgSO_4 \cdot 7H_2O$. A dot is used in the formula to separate the salt from the water of hydration. This formula tells us that every mole of magnesium sulfate is accompanied by 7 mol of water. Similarly, washing soda, $Na_2CO_3 \cdot 10H_2O$, has 10 mol of water for each mole of sodium carbonate present. The complete name of a hydrate must include both the name of the salt and the number of waters of hydration present. We may use either a number or a prefix to show how many waters of hydration there are in the formula. The prefixes used are listed in Table 9-5. Thus, our examples may be named in either of two ways.

TABLE 9-5 Prefixes used to indicate numbers

mono-	1	hexa-	6
di-	2	hepta-	7
tri-	3	octa-	8
tetra-	4	nona-	9
penta-	5	deca-	10

$MgSO_4 \cdot 7H_2O$	$Na_2CO_3 \cdot 10H_2O$
Magnesium sulfate 7-hydrate	Sodium carbonate 10-hydrate
or	or
Magnesium sulfate heptahydrate	Sodium carbonate decahydrate

PROBLEM 9.13

Name each of the following hydrated salts, using both of the methods we have described.

(a) $BaCl_2 \cdot 2H_2O$ **(b)** $ZnSO_4 \cdot 7H_2O$ **(c)** $CuSO_4 \cdot 5H_2O$ **(d)** $CoCl_2 \cdot 2H_2O$

9.9 BINARY COMPOUNDS COMPOSED OF TWO NONMETALS

OBJECTIVE 9.9 Name any binary compound composed of two nonmetals, the more electronegative of which is found as a monatomic anion in the table of ions.

In Section 9.6 we stated that most binary compounds composed of a metal and a nonmetal are ionic. By contrast, binary compounds composed of two nonmetals are covalent in nature and require a different system of nomenclature. When naming binary compounds composed of two nonmetals, we use the same prefixes we used for hydrated salts (Table 9-5).

As with binary ionic compounds, the more electronegative element comes last in both the name and formula and also takes an *-ide* ending. In addition, although oxidation numbers can be assigned to both elements, we rarely use oxidation numbers when naming compounds of this type. Instead, the prefixes provide enough information for us to obtain the formula from the name. The following examples should illustrate the use of these prefixes in naming compounds.

CO	Carbon monoxide	SO_3	Sulfur trioxide
CO_2	Carbon dioxide	N_2S_5	Dinitrogen pentasulfide

When an atom appears only once in the formula, it is usually not preceded by a prefix. However, the prefix *mono* may be used when we wish to distinguish between two similar compounds, such as carbon monoxide and carbon dioxide. Those elements that appear more than once in the formula are preceded by the appropriate prefix. Note also that to ease pronunciation, the final letter of the prefix may be dropped, as it was in carbon monoxide (rather than carbon mon*o*oxide).

EXAMPLE 9.15

What is the name of CCl_4?

Planning the solution

For the first element in the formula, we just give the name of the element: carbon. The more electronegative (second) element takes an *-ide* suffix: chloride. Since there are four chlorine atoms, we use the prefix *tetra.*

SOLUTION

Carbon tetrachloride.

EXAMPLE 9.16

What is the name of N_2O_5?

SOLUTION

Dinitrogen pentoxide. Here, the final letter of the prefix *penta-* was dropped, just as in the case of carbon monoxide.

EXAMPLE 9.17

What is the name of CS_2?

SOLUTION

Carbon disulfide.

EXAMPLE 9.18

What is the name of PCl_5?

SOLUTION

Phosphorus pentachloride.

Before going on, let us add the following cautionary note. *Never use prefixes to name ionic compounds.* For the compounds considered in this chapter, prefixes are used only to name binary compounds of two nonmetals. (The only apparent exception occurs when the prefix is part of the name of an ion, as is the case for the monohydrogen phosphate and dihydrogen phosphate ions.)

PROBLEM 9.14

Write the formula for each of the following compounds.
(a) Carbon dioxide (b) Phosphorus trichloride (c) Boron trichloride
(d) Dinitrogen tetroxide (e) Selenium dichloride (f) Sulfur dioxide
(g) Diarsenic pentoxide (h) Nitrogen triiodide (i) Nitrogen dioxide
(j) Xenon tetrafluoride

PROBLEM 9.15

Write the name for each of the following compounds, using the prefixes shown in this section.
(a) SeO_3 (b) OCl_2 (c) P_2O_5 (d) NO (e) TeO_3
(f) $SiBr_4$ (g) NF_3 (h) N_2O (i) XeF_4 (j) SF_6

Binary Compounds Containing Hydrogen

At this point you may be wondering why we have avoided compounds containing hydrogen. Many of the hydrogen-containing compounds of interest to us are acids and are discussed in the next section, where we take up the nomenclature of acids. We may divide the binary compounds of hydrogen into three categories: (1) those composed of a metal and hydrogen; (2) those composed of hydrogen and a nonmetal of group VIA (16) or VIIA (17); and (3) those composed of hydrogen and a nonmetal of group IIIA (13), IVA (14), or VA (15).

1. The binary compounds containing a metal and hydrogen are named according to the principles outlined in Section 9.6. Hydrogen is the more electronegative element, so these compounds are named as metallic hydrides.

LiH	Lithium hydride	NaH	Sodium hydride
BeH_2	Beryllium hydride	MgH_2	Magnesium hydride

2. The common binary compounds containing hydrogen plus another nonmetal are gases in the pure state. *When in their pure gaseous state,* most of those that contain a chalcogen (group VIA or 16) or a halogen (group VIIA or 17) are named as though they were ionic compounds.

HF(g)	Hydrogen fluoride	HBr(g)	Hydrogen bromide
HCl(g)	Hydrogen chloride	H_2S(g)	Hydrogen sulfide

The symbol "(g)" is used here to specify the gaseous state. When dissolved in water, each of these substances forms an acid solution. Although water, too, is composed of a chalcogen and hydrogen, it is rarely called hydrogen oxide.

3. The binary compounds of hydrogen and a nonmetal of group IIIA (13), IVA (14), or VA (15) are named by other rules of nomenclature or are known by their common names. The names and formulas of the most important of these follow. We have included water (group VIA or 16) in the list, since you must memorize its formula too.

BH_3	Borane	NH_3	Ammonia
CH_4	Methane	PH_3	Phosphine
SiH_4	Silane	H_2O	Water

You may have noticed that the formulas for methane and ammonia have the less electronegative element (hydrogen) last. Because these compounds were historically among the earliest substances studied, their formulas have persisted despite having the order of the elements reversed.

9.10 ACIDS AND BASES

OBJECTIVE 9.10A Name any acid derived from a halide ion, the sulfide ion, or the cyanide ion. Give its name in the pure gaseous state.

OBJECTIVE 9.10B Name any acid derived from an oxyanion in the table of ions.

Several of the binary compounds made of hydrogen and an element from group VIA (16) or VIIA (17) form acid solutions when dissolved in water. Hydrogen chloride, HCl, is an example. These compounds are gaseous in their pure state. When dissolved in water they form acid solutions, so they are referred to as *binary acids.* In Chapters 14 and 15 we discuss acids in some detail, and we will see that chemists have several definitions for an *acid.* For the present, let us think of acids as substances that are made up of one or more hydrogen cations, H^+, and some anion (such as Cl^- in the case of hydrogen chloride). When these pure substances are dissolved in water to become acid solutions, their names are changed as follows:

HCl(g)	Hydrogen chloride	becomes	HCl(aq)	*Hydro*chlor*ic acid*
HBr(g)	Hydrogen bromide	becomes	HBr(aq)	*Hydro*brom*ic acid*
H_2S(g)	Hydrogen sulfide	becomes	H_2S(aq)	*Hydro*sulfur*ic acid*
HCN(g)	Hydrogen cyanide	becomes	HCN(aq)	*Hydro*cyan*ic acid*

To name a binary acid, we replace the *-ide* suffix of the anion with *-ic acid* and add the prefix *hydro-*. (Although HCN is not a binary compound, it is named as a binary acid because the anion ends in *-ide.*) The symbols (g) and (aq) refer to the gaseous and aqueous states. A substance in the *aqueous* state is dissolved in water.

Many of the commonly encountered laboratory acids are derived from a combination of hydrogen ion and one of the oxyanions we have discussed. The two categories of oxyanions (*-ate* and *-ite*) result in two categories of acids:

HNO_3	Nitr*ic acid*	contains	Nitr*ate* ion, NO_3^-
H_2SO_4	Sulfur*ic acid*	contains	Sulf*ate* ion, SO_4^{2-}
H_3PO_4	Phosphor*ic acid*	contains	Phosph*ate* ion, PO_4^{3-}
$HClO_4$	Perchlor*ic acid*	contains	Perchlor*ate* ion, ClO_4^-
HNO_2	Nitr*ous acid*	contains	Nitr*ite* ion, NO_2^-
H_2SO_3	Sulfur*ous acid*	contains	Sulf*ite* ion, SO_3^{2-}
H_3PO_3	Phosphor*ous acid*	contains	Phosph*ite* ion, PO_3^{3-}
HClO	Hypochlor*ous acid*	contains	Hypochlor*ite* ion, ClO^-

To name most acids derived from an *-ate* anion: Drop *-ate* from the name of the anion and add *-ic acid.*

To name most acids derived from an *-ite* anion: Drop *-ite* from the name of the anion and add *-ous acid.*

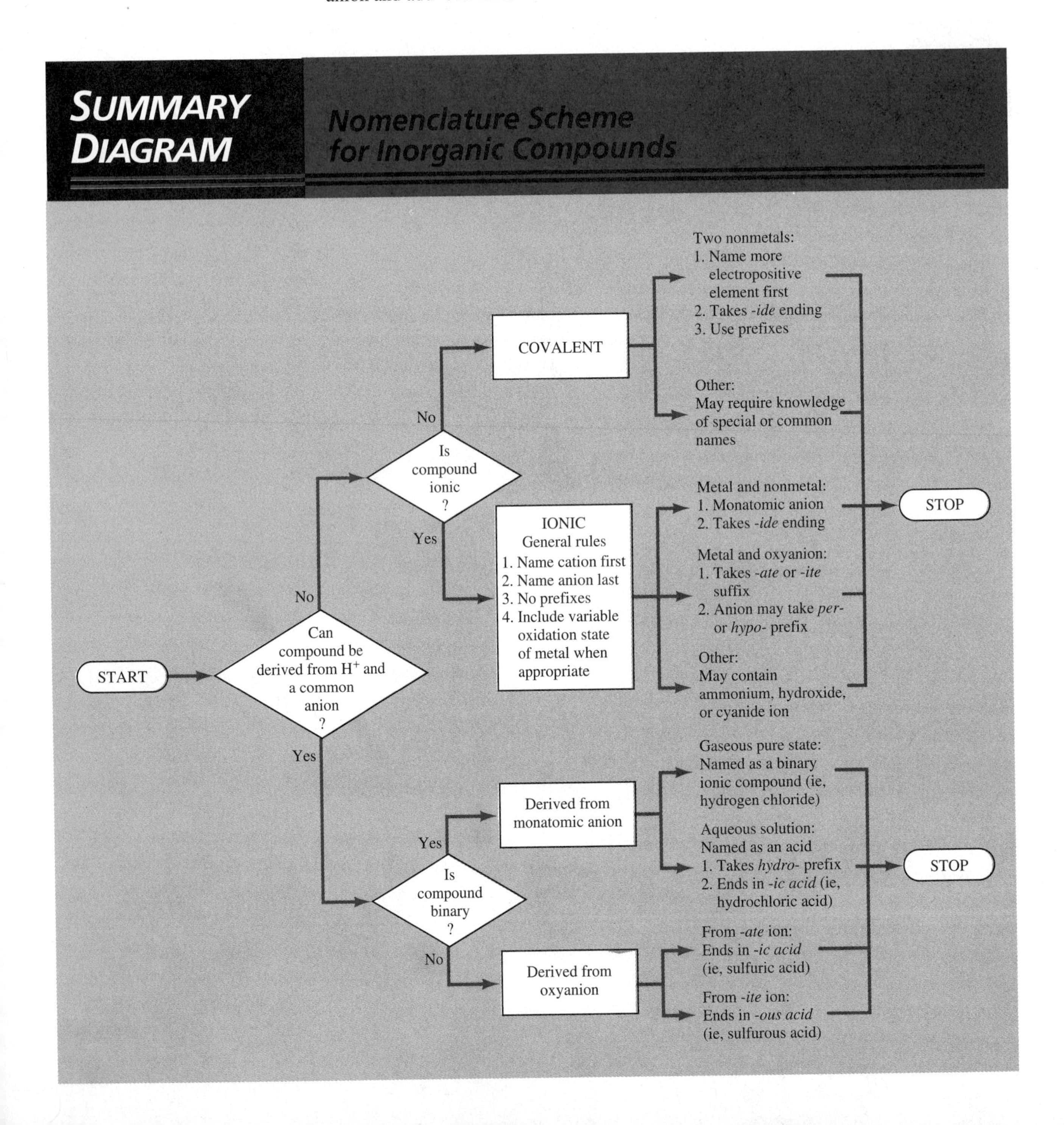

EXAMPLE 9.19

Name the acid whose formula is $H_2C_2O_4$.

Planning the solution

Examine the formula of the anion. If it is an *-ate* ion, the acid takes an *-ic acid* ending. If it is an *-ite* ion, the name will end in *-ous acid.*

SOLUTION

This acid contains the oxal*ate* ion, $C_2O_4^{2-}$. To name the acid, we drop the *-ate* ending and add *-ic acid.* Thus, the name of this acid is oxal*ic acid.*

EXAMPLE 9.20

Name the acid whose formula is $HBrO_2$.

Planning the solution

In Example 9.13 we determined that the BrO_2^- ion is named bromite.

SOLUTION

Since this acid contains the brom*ite* ion, we drop the *-ite* and add *-ous acid.* The name of this acid is brom*ous acid.*

As with the acids, there are several definitions of the term *base.* For the present, we will consider a base to be composed of a cation plus the hydroxide ion, OH^-. The bases encountered most frequently are as follows:

NaOH	Sodium hydroxide	$Ca(OH)_2$	Calcium hydroxide
KOH	Potassium hydroxide	$Ba(OH)_2$	Barium hydroxide

These bases are ionic compounds and are named accordingly by giving the name of the cation followed by the name of the anion. The Summary Diagram reviews the nomenclature discussed in this chapter.

PROBLEM 9.16

Name each of the following acids and bases.

(a) HNO_3 **(b)** H_3PO_3 **(c)** H_3AsO_4 **(d)** H_2CrO_4 **(e)** HNO_2 **(f)** $HC_2H_3O_2$
(g) HI **(h)** HCl **(i)** HCN **(j)** $Al(OH)_3$ **(k)** LiOH **(l)** $Mg(OH)_2$

CHAPTER SUMMARY

Two types of nomenclature exist for naming chemicals: **common names** and **systematic names.** Many frequently encountered substances, such as water, have come to be known by their common names.

Binary compounds are composed of two elements. Most binary compounds composed of a metal and a nonmetal are ionic. The names of all ionic compounds consist of the name of the cation followed by the name of the anion. The cation also precedes the anion in the formula. For binary ionic substances the name of the nonmetallic anion ends in *-ide.* If the cation is a metal of variable **oxidation state,** the name must include the oxidation state, as in iron(III) oxide. However, if the cation is a metal whose ions commonly appear in only one oxidation state, the oxidation number is omitted, as is the case in sodium chloride.

Many ionic substances contain polyatomic ions. **Oxyanions** are polyatomic ions containing an element other than oxygen plus one or more oxygen atoms. Names of oxyanions end in *-ate* or *-ite.* The oxyanion taking the *-ate* ending has one more oxygen than the *-ite* ion. The prefixes *per-* and *hypo-* also may be used in conjunction with oxyanions. Addition of *per-* to the *-ate* anion adds an oxygen; addition of *hypo-* to the *-ite* anion deletes an oxygen. Oxyanions combine with cations to form ionic compounds with names and formulas that are written like those of all ionic compounds: the cation first, followed by the anion.

Some ionic compounds may exist in the crystalline state as **hydrated salts.** A hydrated salt has a definite ratio of salt to water. The ratio of moles of salt to moles of water is included within the chemical formula; an example is $BaCl_2 \cdot 2H_2O$. In naming the salt, we use either numbers or prefixes to specify the number of waters of hydration.

Binary compounds composed of two nonmetals utilize a prefix system and take an *-ide* ending. The less electronegative of the elements comes first in both the formula and the name of the compound. Prefixes, which give the number of atoms of each, precede the names of elements that appear with more than one atom in the formula; an example is phosphorus pentachloride, PCl_5. The prefix *mono-* is used occasionally to distinguish between different binary compounds composed of the same two elements, such as carbon monoxide and carbon dioxide.

Many compounds having hydrogen as the electropositive element are acids and take special names when in aqueous solution. Thus, an *-ide* compound such as hydrogen chloride becomes hydrochloric acid, HCl. Acids derived from oxyanions take new endings, such that *-ate* anions becomes *-ic acids,* as in nitric acid, HNO_3; and *-ite* anions become *-ous acids,* as in nitrous acid, HNO_2.

KEY TERMS

Review each of the following terms, which are discussed in this chapter and defined in the Glossary.

common name
systematic name
binary compound
oxyanion
oxidation number (or state)
hydrated salt

Review the relationships or meanings of the following pairs of prefixes and suffixes.

-ic and *-ous*
-ite and *-ate*
per- and *hypo-*

ADDITIONAL PROBLEMS

SECTION 9.1

9.17 Distinguish between common names and systematic names.

9.18 Match the following chemical formulas with their common names.

______	**(a)** $C_{12}H_{22}O_{11}$	**1.** Ammonia
______	**(b)** $CaCO_3$	**2.** Baking soda
______	**(c)** H_2O	**3.** Lye
______	**(d)** $Mg(OH)_2$	**4.** Marble
______	**(e)** NH_3	**5.** Milk of magnesia
______	**(f)** NaCl	**6.** Table salt
______	**(g)** $NaHCO_3$	**7.** Cane sugar
______	**(h)** NaOH	**8.** Water

SECTION 9.2

9.19 Write the chemical formula for each of the following compounds.

(a) Potassium chloride **(b)** Magnesium sulfide
(c) Barium iodide **(d)** Sodium oxide
(e) Aluminum bromide **(f)** Strontium fluoride
(g) Aluminum oxide

SECTION 9.3

9.20 Fill in the following table with the formulas of the compounds obtained by combining each cation with each anion. The formula of magnesium nitrate is provided as an example.

	Cl^-	NO_3^-	CO_3^{2-}	SO_4^{2-}	PO_4^{3-}
H^+					
Ag^+					
Mg^{2+}		$Mg(NO_3)_2$			
Al^{3+}					
NH_4^+					

9.21 Write the chemical formula for each of the following compounds.

(a) Potassium carbonate **(b)** Strontium sulfate
(c) Magnesium hydroxide **(d)** Aluminum nitrate

(e) Aluminum oxalate (f) Sodium chromate
(g) Ammonium phosphate (h) Aluminum carbonate

SECTION 9.4

9.22 Write the chemical formula for each of the following compounds.
(a) Copper(II) iodide (b) Cobalt(III) sulfide
(c) Iron(II) acetate (d) Cobalt(II) carbonate
(e) Chromium(III) sulfate (f) Tin(IV) chloride
(g) Tin(IV) oxide (h) Copper(II) phosphate

SECTION 9.6

9.23 Name each of the following binary compounds.
(a) CoS (b) LiH (c) CuI_2 (d) Fe_2S_3
(e) Zn_3N_2 (f) PbO_2

9.24 Write the preferred name and a formula for each of the following compounds.
(a) Cuprous chloride (b) Cobaltous hydroxide
(c) Chromic oxide (d) Plumbous nitrate
(e) Ferric sulfate (f) Stannic chromate

9.25 Name each of the following compounds.
(a) NaCN (b) $Mg(OH)_2$ (c) NH_4F (d) CuCN

SECTION 9.7

9.26 Name each of the following.
(a) Cu_2SO_4 (b) $Fe_2(CO_3)_3$ (c) $Mg(ClO_2)_2$
(d) $Cr_3(AsO_4)_2$ (e) $NaHCO_3$ (f) $Cu(NO_2)_2$
(g) $Zn(H_2PO_4)_2$ (h) $Ca(C_2H_3O_2)_2$ (i) $Na_2C_2O_4$

9.27 Name each of the following ionic compounds.
(a) $NaBrO_3$ (b) KIO_2 (c) NH_4ClO (d) $LiIO_4$

9.28 Write a formula for each of the following compounds.
(a) Sodium hydrogen sulfite
(b) Lithium hydrogen carbonate
(c) Magnesium hydrogen carbonate
(d) Calcium monohydrogen phosphate
(e) Calcium dihydrogen phosphate

SECTION 9.8

9.29 Name each of the following hydrated salts.
(a) $CaSO_4 \cdot 2H_2O$ (b) $Na_2SO_4 \cdot 10H_2O$
(c) $CaCO_3 \cdot 6H_2O$ (d) $Ca(ClO)_2 \cdot 3H_2O$
(e) $Hg(NO_3)_2 \cdot 2H_2O$ (f) $Co(C_2H_3O_2)_2 \cdot 4H_2O$
(g) $Al_2(C_2O_4)_3 \cdot 3H_2O$ [*Hint:* What is the name of the $C_2O_4^{2-}$ ion?]
(h) $BaO_2 \cdot 8H_2O$ [*Hint:* What is the name of the O_2^{2-} ion?]

SECTION 9.9

9.30 Write the chemical formula for each of the following compounds.
(a) Dinitrogen pentasulfide (b) Boron trichloride
(c) Oxygen difluoride (d) Silicon tetrafluoride
(e) Dinitrogen monoxide

9.31 Name each of the following compounds.
(a) PCl_5 (b) BF_3 (c) N_2O_4 (d) SO_3 (e) SiO_2

SECTION 9.10

9.32 Name each of the following acids.
(a) HBr (b) H_2SO_4 (c) H_2SO_3 (d) H_3PO_4
(e) $HClO_2$ (f) HIO_4 (g) $HBrO_3$ (h) HF
(i) H_3PO_2 [*Hint:* Derive the name of PO_2^{3-} from the rules discussed in the chapter.]
(j) $H_2C_2O_4 \cdot 2H_2O$ [*Hint:* This hydrated acid is named in the same fashion as a hydrated salt.]

GENERAL PROBLEMS

9.33 Write the formula for each of the following compounds.
(a) Silver acetate (b) Copper(I) phosphate
(c) Mercury(II) nitrate (d) Manganese(II) sulfate
(e) Ammonium dichromate (f) Cadmium chloride
(g) Strontium phosphate
(h) Ammonium hydrogen carbonate
(i) Potassium monohydrogen phosphate
(j) Potassium dihydrogen phosphate
(k) Tin(IV) hydride (l) Chromium(III) oxide
(m) Lead(IV) oxide (n) Silver arsenate
(o) Sodium oxalate (p) Magnesium hydride
(q) Nickel(II) phosphide

9.34 Name each of the following ionic compounds.
(a) LiBr (b) Hg_2O (c) $KMnO_4$
(d) $Na_2S_2O_3$ (e) $NiSO_3$ (f) Sr_3N_2
(g) $(NH_4)_3PO_3$ (h) Na_2O_2 (i) $Sn(C_2H_3O_2)_2$
(j) $Pb(CN)_2$ (k) Ag_2SO_4 (l) $CoCrO_4$
(m) $MgHPO_4$ (n) $Mn_3(AsO_4)_2$

9.35 The rules of nomenclature are not always strictly

applied by chemists and the general public. Below are the names of several substances as they are commonly known. Write the formula of each, and give it another name.
(a) Sodium bicarbonate **(b)** Sodium bisulfite
(c) Manganese dioxide **(d)** Stannous fluoride
(e) Uranium hexafluoride

***9.36** Use the principles you learned in this chapter to name each of the following.
(a) $H_2Se(g)$ **(b)** $H_2Se(aq)$ **(c)** $H_2Te(g)$ **(d)** $H_2Te(aq)$

***9.37** Each of the following pairs of binary compounds is composed of a metal of variable oxidation state and a nonmetal. Use the principles you learned in this chapter to name each pair.
(a) $AuBr$ and $AuBr_3$ **(b)** SbF_3 and SbF_5
(c) CeO_2 and Ce_2O_3
(d) $GaSe$ and Ga_2Se_3 [*Hint:* Judging from its position in the periodic table, what charge would you predict for selenium?]

***9.38** Nitrogen and oxygen combine to form several different compounds. Nitrogen dioxide and dinitrogen tetroxide are two such compounds that are often found in smog. These two substances may be interconverted via a simple chemical reaction. Write a balanced chemical equation to show the formation of dinitrogen tetroxide from nitrogen dioxide.

***9.39** An interesting facet of systematic nomenclature is that names may be derived that correspond to the formulas of substances that have never been observed. Of the possible oxoacids (oxygen-containing acids) that could be derived from fluorine, only HFO has been observed.
(a) What is the name of HFO?
(b) What names correspond to the formulas HFO_2, HFO_3, and HFO_4?

***9.40** Oxidation states may be determined for elements other than metal cations. For oxyanions, the oxidation state of the non-oxygen element may be determined by assigning each oxygen atom a charge of −2 and then calculating the charge on the other element that would cause the sum of the charges to equal the charge on the ion as a whole. For example, the four oxygens in chromate (CrO_4^{2-}) total −8. The sulfur atom must have an oxidation state of +6 for the total charge to add up to −2. Use this method to determine the following.
(a) The oxidation states of N in nitrate and nitrite.
(b) The oxidation states of S in sulfate and sulfite.
(c) The oxidation states of P in phosphate and phosphite.
(d) For each pair of ions, what relationship exists between the oxidation state of the non-oxygen atom and the *-ate* or *-ite* ending of the ion?
(e) When the ions in parts (a), (b), and (c) are found in acids (for example, HNO_3), the oxidation states of the atoms remain the same. For each pair of ions, what relationship exists between the oxidation states of the atoms you just determined and the *-ic acid* or *-ous acid* endings they take when they are found as part of an acid?

WRITING EXERCISES

9.41 One of your classmates is confused about the nomenclature of substances that contain oxyanions. You have agreed to write out an explanation for him. Your explanation should compare the formulas of *-ate* and *-ite* ions, as well as the use of *per-* and *hypo-* prefixes. Finally, it should explain how to name acids that contain oxyanions. Be sure to include examples of acids that contain *-ate* and *-ite* ions, as well as those with *per-* and *hypo-* prefixes.

9.42 Explain the difference between a common and a systematic name. Explain the advantage of systematic nomenclature compared to common names. Give several examples of substances encountered in our daily lives and compare their common and systematic names.

9.43 Explain how to name binary covalent compounds. Be sure to tell how to determine whether a compound should be named by this system.

9.44 Explain how to derive the names of compounds that contain metals of variable oxidation state. Describe the older *-ic/-ous* system, and explain why it is not as informative as the oxidation number method. Compare the names of compounds that contain metals of variable oxidation state to those with metal ions exhibiting only one common oxidation state.

PHOTO QUIZ

1.
The colors displayed by fireworks depend on their chemical composition. The following chemicals produce the colors indicated in parentheses: strontium nitrate (red), copper(II) carbonate (blue), barium nitrate (green), and strontium carbonate (red). What hypothesis can you make as to conditions that will produce a red fireworks display?

2.
The effects of heating a sample of mercury(II) oxide are shown here. The photo on the left shows the mercury(II) oxide prior to heating. The photo on the right shows the cooled contents of the test tube after heating is completed. The balloon (seen in the first two photos) remains inflated after cooling. Do these photos illustrate a physical change or a chemical change? What is your evidence?

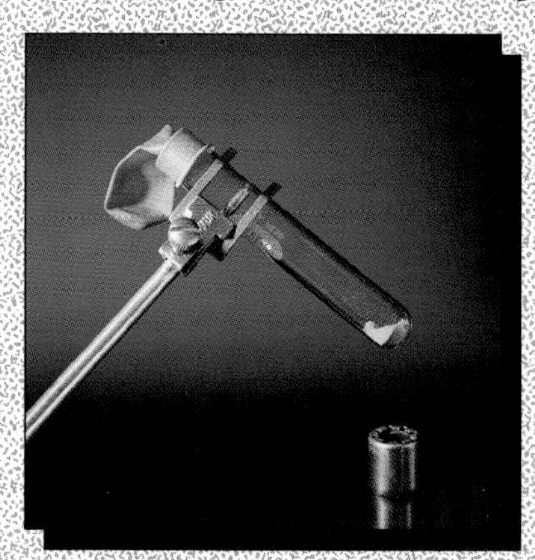

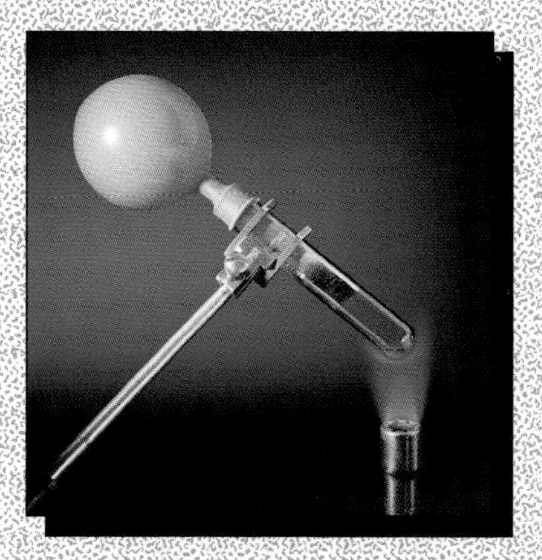

Answers to Photo Quiz Questions can be found following the Solutions to Problems.

PHOTO QUIZ

3.
How many distinct layers can you identify inside the glass cylinder? Why are the substances arranged in the order observed here?

4.
Each of these beakers, left to right, contains 1 lb of the product indicated: $C_{12}H_{22}O_{11}$ (table sugar), NaCl (table salt), and $NaHCO_3$ (baking soda). Which beaker contains the greatest number of moles? Which contains the greatest number of atoms?

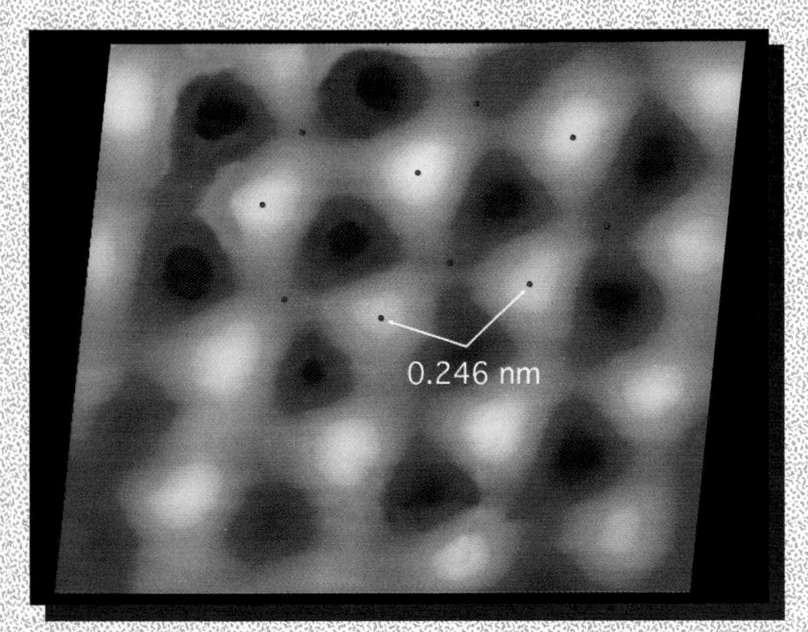

5.
Scanning tunneling microscopy is a technique used to generate a picture of the atoms on the surface of a solid. Shown here is graphite (C), which is generally used as a calibration standard when measuring the dimensions of atoms on other surfaces. The distance between neighboring nuclei in this photo is 0.246 nm. Use this information to calculate the approximate number of carbon atoms that would be on a graphite surface measuring 1 square centimeter ($1 cm^2$).

Answers to Photo Quiz Questions can be found following the Solutions to Problems.

PHOTO QUIZ

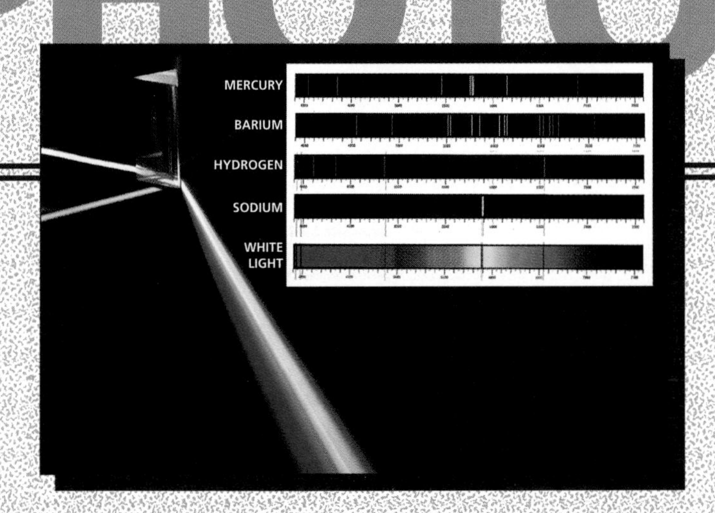

6.
A prism separates the light from a conventional light source (known as "white light") into a continuous rainbow of colors. The strip spectra accompanying the photo show the effects of separating the light given off by a mercury lamp, a barium lamp, a hydrogen lamp, and a sodium lamp. (A strip spectrum of white light is also included for comparison.) Why don't these elements give a continuous spectrum, like that of white light? Why do mercury lamps give off an orange glow?

7.
This young woman is holding onto a Van de Graf generator, which acts as a source of electrons. Why is her hair standing on end?

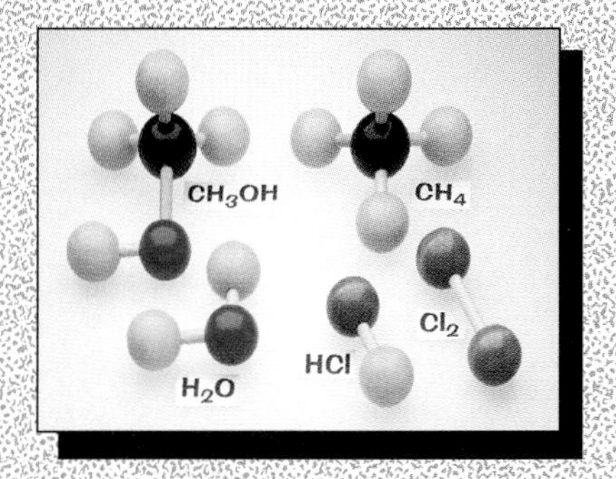

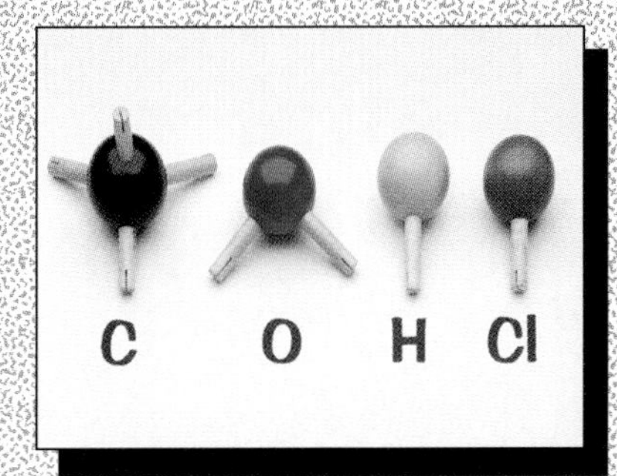

8.
The ball-and-stick models shown here are used to represent molecules. Why are the atoms in a typical ball-and-stick model kit drilled with the following number of holes: C(black) = 4, O(red) = 2, H(yellow) = 1, and Cl(green) = 1? If you were going to make a model of SiH_4, which ball would you use to represent silicon?

Answers to Photo Quiz Questions can be found following the Solutions to Problems.

PHOTO QUIZ

9.
Among other nutrients, plants need nitrogen, phosphorus, and sulfur for growth. This fertilizer contains the ingredients listed in the magnified portion of the label. (Sulfate of potash is an antiquated name for potassium sulfate.) Write the formulas of ammonium nitrate, ammonium phosphate, and potassium sulfate. Which of these compounds would provide nitrogen? Phosphorus? Sulfur? What ambiguity exists with the name *iron sulfate*?

10.
When ammonium dichromate is heated, a mini-volcano of chromium(III) oxide is produced. The reaction that occurs is:

$$(NH_4)_2Cr_2O_7 \rightarrow Cr_2O_3 + N_2 + 4\ H_2O$$

Answers to Photo Quiz Questions can be found following the Solutions to Problems.

PHOTO QUIZ

11.
In order to get a hot air balloon aloft, a fuel such as propane is used to heat the air inside the balloon. How does this enable the balloon to rise? Can you suggest a reason why balloonists prefer to fly in the morning?

12.
A cube of ice floats in liquid water, but solid paraffin sinks when placed in liquid paraffin. Why is there a difference in the behavior of these substances?

WATER AND SODIUM ACETATE

MIXTURE OF SODIUM ACETATE IN WATER AT 90°C

MIXTURE AFTER COOLING TO 30°C

13.
When sodium acetate is mixed with water and the mixture is heated, the solid dissolves completely. Upon cooling, crystals of sodium acetate form. Explain this behavior.

Answers to Photo Quiz Questions can be found following the Solutions to Problems.

PHOTO QUIZ

14.
Marble is composed of calcium carbonate, $CaCO_3$. This marble statue of George Washington shows the effects of years of exposure to rain, wind, heat, and sunlight. How do you account for its severe deterioration?

15.
Shown here are the pH levels of several household products. The pH meter is measuring orange juice. Which of these products is the most acidic?

16.
This photo shows a coil of copper wire in a solution of silver nitrate. What is happening to the portion of the coil that is immersed in the silver nitrate solution? Why does the solution turn blue?

Answers to Photo Quiz Questions can be found following the Solutions to Problems.

PHOTO QUIZ

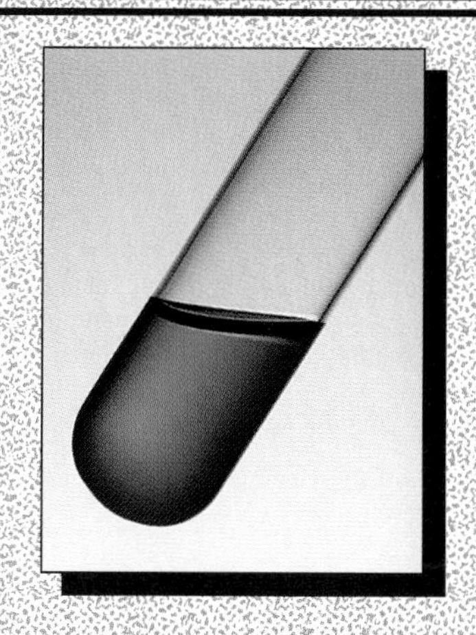
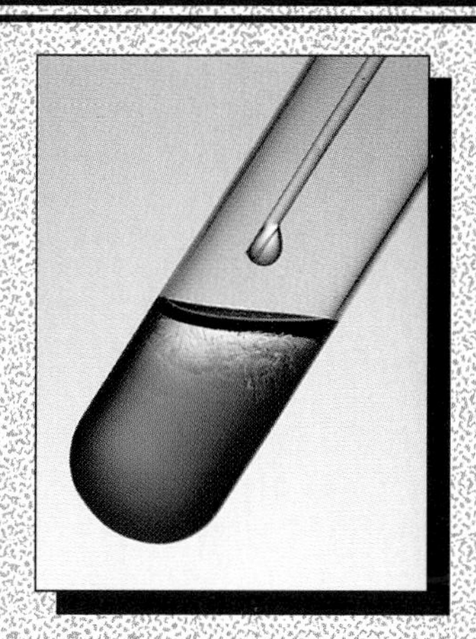
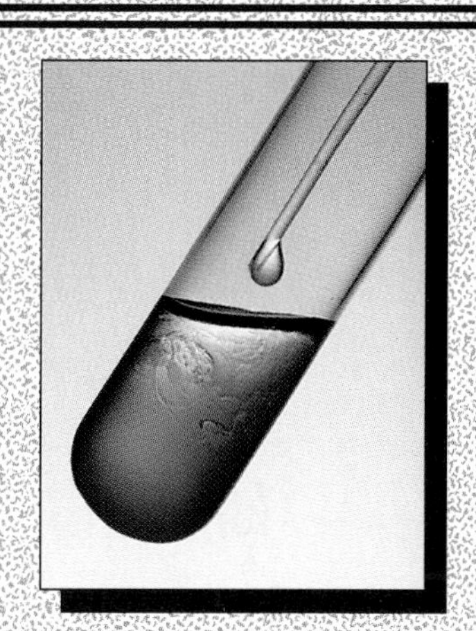
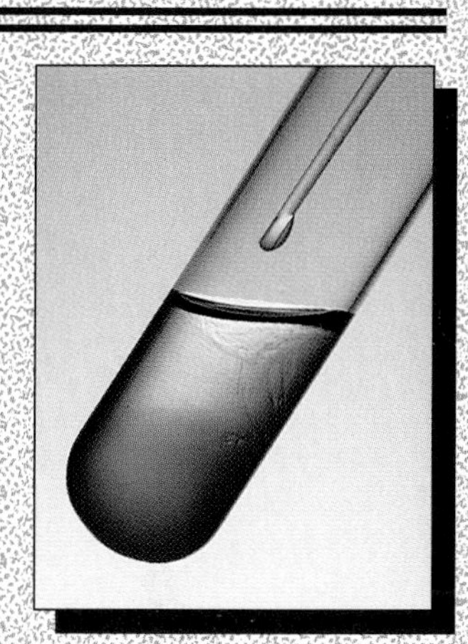

17.

The photo sequence in the top row shows the effect of adding water to a solution containing the blue $CoCl_4^{2-}$ ion. The bottom sequence shows the effect of adding chloride ions to the pink $Co(H_2O)_6^{2+}$ ion. The two reactions are related by the following chemical equation:

$$\underset{\text{blue}}{CoCl_4^{2-}} + 6\,H_2O \rightleftharpoons \underset{\text{pink}}{Co(H_2O)_6^{2+}} + 4\,Cl^-$$

Explain these observations.

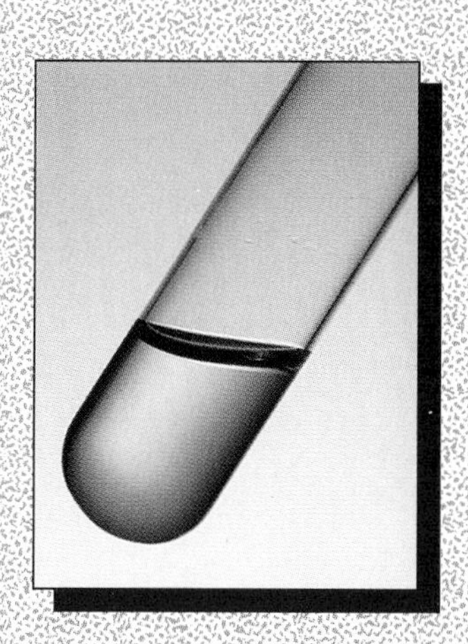
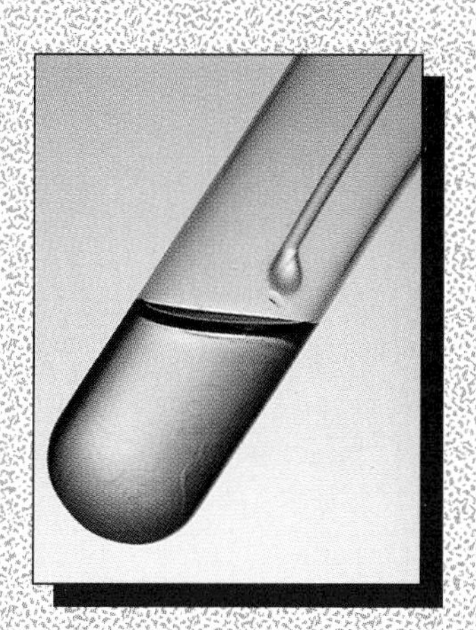
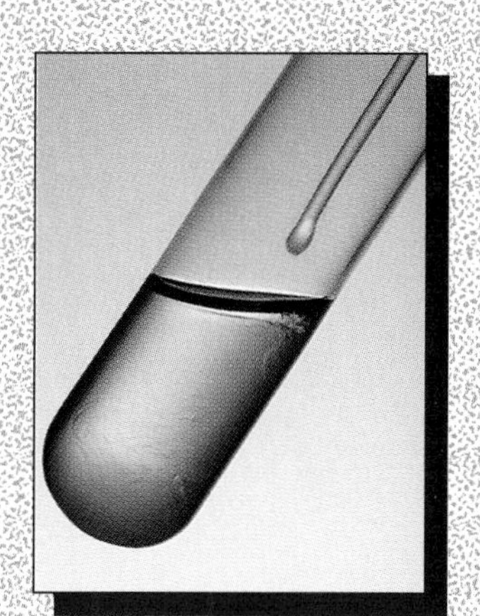
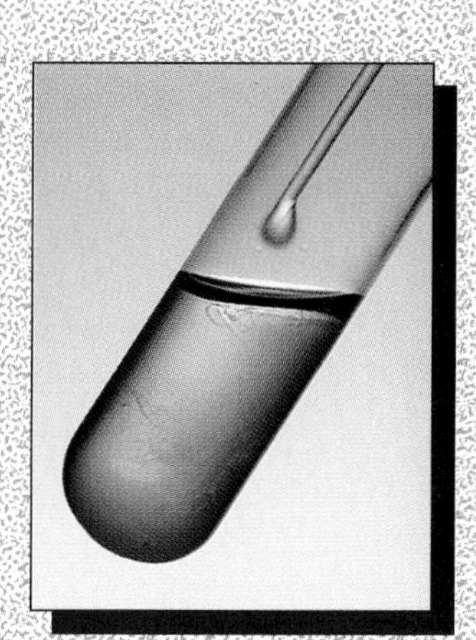

Answers to Photo Quiz Questions can be found following the Solutions to Problems.

PHOTO QUIZ

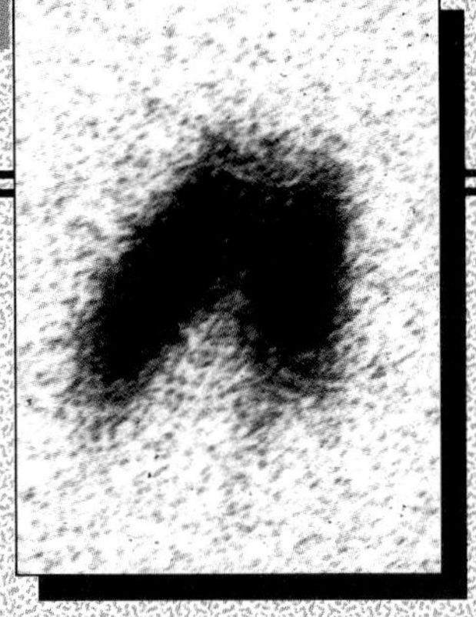

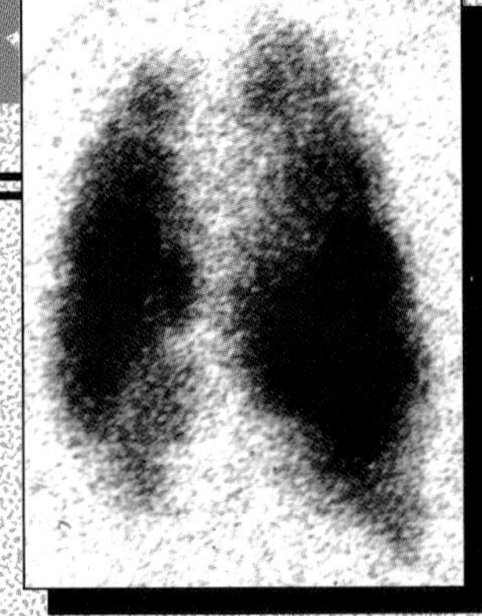

18.
Radioisotopes of iodine are often used in the diagnosis and treatment of disease. Iodine-123, which emits only gamma rays, is used in diagnosis. After being administered to the patient, the radioisotope is absorbed by the body, where its patterns of concentration are determined by exposing a type of photographic film to its radiations. Shown here are a healthy (left) and an enlarged, hyperactive (right) thyroid gland. Iodine-131, which also gives off beta radiations, is used in the treatment of hyperactive thyroid tissue, reducing the number of thyroid cells. Why would a physician prefer the use of iodine-123 for diagnosis and iodine-131 for treatment?

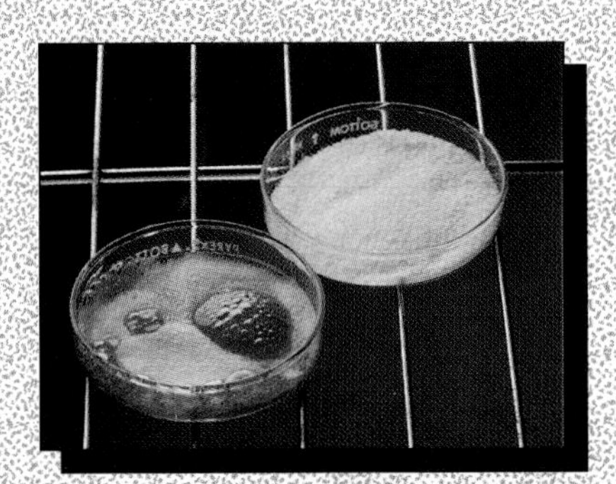

19.
The two substances pictured above are being heated. One is organic and the other is inorganic. Which is which? What evidence do you have to support your hypothesis?

20.
Which two of the following foods are the most similar chemically: sugar, nonfat milk, celery, or butter ?

Answers to Photo Quiz Questions can be found following the Solutions to Problems.

CHEMICAL CHANGES AND STOICHIOMETRY

10

We have defined chemistry as the study of matter and the changes it undergoes. Up to this point, we have been building an understanding of how matter is assembled. At last we are ready to explore the changes it undergoes. In Chapter 2 we mentioned both chemical and physical change. This chapter probes the nature of chemical change in some detail. Perhaps more than any other part of the text, this chapter captures the essence of what chemistry is about.

10.1 THE CHEMICAL EQUATION

OBJECTIVE 10.1 Explain the meaning of the following terms: word equation, reactant, product, unbalanced equation, balanced equation. Explain what happens to the arrangement of atoms during a chemical reaction.

If we were to take a mixture of hydrogen, H_2, and oxygen, O_2, and ignite it with a spark, a rather loud noise would be heard. If large enough quantities of the hydrogen and oxygen were present, the noise would be considered an explosion, and we might also observe a flash of light. Further investigation would reveal that water vapor, H_2O, had been formed during the course of the experiment. We would say that hydrogen and oxygen *reacted* with one another to form water. We might represent this *chemical reaction* with the following expression, which is referred to as a **word equation:**

Word equation: A chemical equation expressed using names, rather than formulas, of the reactants and products.

Hydrogen + Oxygen → Water **Word equation**

This word equation is read, "Hydrogen and oxygen react to form water."

Reactant: Starting material that undergoes change during a chemical reaction.

Product: Substance formed during a chemical reaction.

The substances we mixed initially are known as the **reactants,** and the materials formed are the **products.** Chemical *equations* are generally written from left to right, with the reactants to the left of the arrow and the products to the right. The arrow can mean "form," "react to form," or "yield." We might think of the arrow as a "before and after" sign. The substances appearing on the left of the arrow (the reactants) are "before"; the substances on the right (the products) are "after." The plus sign means either "plus" or "is added to."

Chemical equation: Symbolic representation of a chemical change.

As a general rule, we express a chemical reaction via a **chemical equation,** which employs the chemical formulas of the reactants and products, rather than words. Thus, we can substitute the chemical formulas of hydrogen, oxygen, and water in the word equation above to give the following:

$$H_2 + O_2 \rightarrow H_2O \qquad \textbf{Unbalanced equation}$$

Unbalanced equation: A chemical equation prior to adjustment of the coefficients.

However, this equation is a bit disturbing, because the number of atoms of oxygen is not the same on both sides of the arrow. We call this an **unbalanced equation.** In order to balance the equation, we need to take two molecules of hydrogen for every molecule of oxygen. The result is that two molecules of water will be formed:

$$2\,H_2 + 1\,O_2 \rightarrow 2\,H_2O \qquad \textbf{Balanced equation}$$

4 H atoms + 2 O atoms → **4 H atoms + 2 O atoms**

Balanced equation: A chemical equation in which every atom appearing on the left also appears on the right, and vice versa.

In writing this **balanced equation,** we had to adjust the ratio of hydrogen, oxygen, and water such that every atom of hydrogen that appears on the left is also found

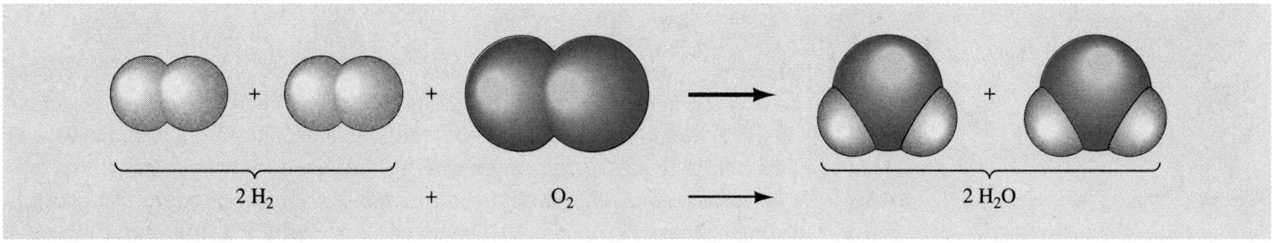

$2 H_2$ + O_2 → $2 H_2O$

FIGURE 10-1 The meaning of a balanced equation. Every atom that appears on the left is also found on the right. The atoms are *rearranged* in a chemical reaction.

on the right. Similarly, every atom of oxygen appearing on the left is also found on the right. (There are four atoms of hydrogen and two atoms of oxygen on both sides of the equation.) We refer to the whole number that appears before each formula as a *coefficient.* The coefficient applies to the entire formula that it precedes. We might read the balanced equation as follows: "Two molecules of hydrogen and one molecule of oxygen react to form two molecules of water." In actual practice, we do not include coefficients of 1. Instead, we write the preceding equation as follows:

$$2 H_2 + O_2 \rightarrow 2 H_2O$$

Figure 10-1 shows the physical meaning of this balanced equation. During the course of this chemical reaction, the atoms have rearranged themselves. On the left side of the equation, each hydrogen atom was bonded to another hydrogen atom. On the right, each hydrogen atom is bonded to an atom of oxygen. *A rearrangement of atoms is the essence of every chemical reaction.*

Let us reconsider the idea that, when a chemical equation is balanced, every atom appearing on the left must also appear on the right (and vice versa). If this were not the case, matter would have to be created or destroyed in the course of a chemical reaction. This would contradict a fundamental principle of nature, the Law of Conservation of Mass.

PROBLEM 10.1

When is a chemical equation balanced? Explain how a balanced chemical equation is consistent with the Law of Conservation of Mass.

10.2 BALANCING CHEMICAL EQUATIONS

OBJECTIVE 10.2 Given an unbalanced equation, balance the equation.

Let us see what steps are required to balance a chemical equation, by reexamining the reaction of hydrogen, H_2, and oxygen, O_2, to form water, H_2O:

$$H_2 + O_2 \rightarrow H_2O \qquad \textbf{Unbalanced equation}$$

Determining the reactants and products of a chemical reaction such as this requires laboratory investigation. We cannot make up the chemical formulas for our own convenience. For example, it might appear that hydrogen, H_2, plus oxygen, O_2, should form H_2O_2, hydrogen peroxide. However, laboratory investigations reveal that hydrogen plus oxygen always react to form water, H_2O—never hydrogen peroxide.

In this chapter, we will tell you what the reactants and products are. Later, you will learn some general principles for predicting products. When presented with an unbalanced chemical equation, the formulas of the reactants and products should not be changed. The task is to *adjust the coefficients of the given reactants and products* in order to balance the atoms of each element. In our early examples, we use question marks to distinguish those coefficients that are still undetermined from those that have already been assigned. Keep in mind that, at first, we do not know the coefficients of *any* of the reactants or products. For example,

$$? H_2 + ? O_2 \rightarrow ? H_2O$$

Let us begin by balancing the atoms of oxygen, because they are clearly unbalanced. There are two atoms of O in each molecule of O_2 and only one atom of O in each H_2O molecule, so we need two molecules of H_2O to balance the oxygen in each molecule of O_2. Consequently, we must place the coefficient 2 in front of the H_2O. Although we do not generally include coefficients of 1 in our balanced equations, we will place a 1 in front of the O_2 in this example to emphasize that this coefficient has been determined and that there are two H_2O molecules for every molecule of O_2.

$$? H_2 + 1 O_2 \rightarrow 2 H_2O$$

Although the oxygen is now balanced, the hydrogen is not. Because we have assigned the coefficient 2 to H_2O, we have four atoms of H on the right. To balance the hydrogen requires two molecules of H_2 on the left. Thus, we must place the coefficient 2 in front of the H_2.

$$2 H_2 + 1 O_2 \rightarrow 2 H_2O$$

Note that we balance an equation by *adjusting the coefficients in front of the formulas.* We cannot change the subscripts in a formula, because that would change the chemical composition of the substance. Remember that each substance has a unique and fixed chemical composition. To balance the preceding equation, we balanced oxygen by placing 2 in front of the H_2O. We did *not* change the subscript on oxygen to H_2O_2, because that would change the water to hydrogen peroxide, an entirely different chemical substance. Since we do not generally write coefficients of 1, the preferred form of the equation we have just balanced is the following:

$$2 H_2 + O_2 \rightarrow 2 H_2O \qquad \textbf{Preferred final form}$$

To illustrate these principles, let us balance a few equations.

EXAMPLE 10.1

Methane, CH_4, is a major constituent of natural gas. When it is burned in the presence of oxygen, O_2, the products formed are carbon dioxide, CO_2, and water, H_2O. Balance the following unbalanced equation for this reaction: $CH_4 + O_2 \rightarrow CO_2 + H_2O$.

Symbolic statement

$$? CH_4 + ? O_2 \rightarrow ? CO_2 + ? H_2O \qquad \textbf{Unbalanced equation}$$

SOLUTION

Because carbon appears in only one formula on each side of the equation, let us begin by assigning a coefficient of 1 to CH_4 and then see which substances can be balanced:

$$1\,CH_4 + ?\,O_2 \rightarrow ?\,CO_2 + ?\,H_2O$$

One molecule of methane contains one atom of carbon and four atoms of hydrogen. If we are to have only one carbon on the right, there must be one CO_2:

$$1\,CH_4 + ?\,O_2 \rightarrow 1\,CO_2 + ?\,H_2O$$ **Partially balanced**

If there are to be four atoms of hydrogen on the right, two molecules of H_2O must be formed:

$$1\,CH_4 + ?\,O_2 \rightarrow 1\,CO_2 + 2\,H_2O$$ **Partially balanced**

Finally, we must balance the oxygen atoms. When we assigned coefficients to the carbon dioxide and water, we established the presence of four oxygen atoms on the right: two from the carbon dioxide molecule and one each from the two water molecules. In order to have four atoms of oxygen on the left, we will need two molecules of O_2:

$$1\,CH_4 + 2\,O_2 \rightarrow 1\,CO_2 + 2\,H_2O$$ **Balanced equation**

or

$$CH_4 + 2\,O_2 \rightarrow CO_2 + 2\,H_2O$$ **Preferred final form**

EXAMPLE 10.2

Propane, C_3H_8, reacts with oxygen, O_2, to form carbon dioxide, CO_2, and water, H_2O. Balance the following unbalanced equation for this reaction:

$$C_3H_8 + O_2 \rightarrow CO_2 + H_2O$$

Symbolic statement

$$?\,C_3H_8 + ?\,O_2 \rightarrow ?\,CO_2 + ?\,H_2O$$ **Unbalanced equation**

SOLUTION

In a problem where there is a somewhat complicated molecule, it is usually easiest to begin by assigning that molecule a coefficient of 1. In this case, we assign a coefficient of 1 to C_3H_8:

$$1\,C_3H_8 + ?\,O_2 \rightarrow ?\,CO_2 + ?\,H_2O$$ **Unbalanced equation**

One molecule of C_3H_8 contains three atoms of carbon and eight atoms of hydrogen. In order to balance both of these elements, we must form three molecules of CO_2 and four molecules of H_2O:

$$1\,C_3H_8 + ?\,O_2 \rightarrow 3\,CO_2 + 4\,H_2O$$ **Partially balanced**

We must now balance the oxygen atoms. There are a total of ten oxygen atoms on the right: six from the three CO_2 molecules and four from the four H_2O molecules. We need five molecules of O_2 on the left to balance the oxygen.

$$1\,C_3H_8 + 5\,O_2 \rightarrow 3\,CO_2 + 4\,H_2O$$ **Balanced equation**

or

$$C_3H_8 + 5\,O_2 \rightarrow 3\,CO_2 + 4\,H_2O$$ **Preferred final form**

EXAMPLE 10.3

In the formation of rust, the elements iron, Fe, and oxygen, O_2, combine to form iron(III) oxide, Fe_2O_3. Balance the following unbalanced equation for this reaction:

$$Fe + O_2 \rightarrow Fe_2O_3$$

Symbolic statement

$$? Fe + ? O_2 \rightarrow ? Fe_2O_3 \qquad \textbf{Unbalanced equation}$$

SOLUTION

We begin balancing this equation by assigning a coefficient of 1 to Fe_2O_3, the most complicated species in the unbalanced equation:

$$? Fe + ? O_2 \rightarrow 1\ Fe_2O_3 \qquad \textbf{Unbalanced equation}$$

With two atoms of iron, it is easy to balance the iron:

$$2\ Fe + ? O_2 \rightarrow 1\ Fe_2O_3 \qquad \textbf{Partially balanced}$$

However, to balance the three oxygen atoms would require a fractional coefficient in front of the O_2:

$$2\ Fe + \frac{3}{2} O_2 \rightarrow 1\ Fe_2O_3 \qquad \textbf{Improperly balanced}$$

Although fractional coefficients are permissible in certain situations, as a general rule, *a properly balanced chemical equation must show the simplest whole-number ratio of molecules with the atoms balanced.*

The easiest way to complete this example is to double the coefficient in front of each substance, thereby eliminating the fractional coefficient:

$$4\ Fe + 3\ O_2 \rightarrow 2\ Fe_2O_3 \qquad \textbf{Balanced equation}$$

EXAMPLE 10.4

Treatment of copper metal, Cu, with a solution of silver nitrate, $AgNO_3$, produces silver metal, Ag, along with a solution of copper(II) nitrate, $Cu(NO_3)_2$. Balance the following unbalanced equation for this reaction: $Cu + AgNO_3 \rightarrow Cu(NO_3)_2 + Ag$.

Symbolic statement

$$? Cu + ? AgNO_3 \rightarrow ? Cu(NO_3)_2 + ? Ag \qquad \textbf{Unbalanced equation}$$

SOLUTION

We assign a coefficient of 1 to $Cu(NO_3)_2$, the most complicated formula in the equation:

$$? Cu + ? AgNO_3 \rightarrow 1\ Cu(NO_3)_2 + ? Ag \qquad \textbf{Partially balanced}$$

Since there is only one copper atom in $Cu(NO_3)_2$, we also assign a 1 to Cu:

$$1\ Cu + ? AgNO_3 \rightarrow 1\ Cu(NO_3)_2 + ? Ag \qquad \textbf{Partially balanced}$$

When working with polyatomic ions such as nitrate, NO_3^-, it is often easier to keep track of the number of ions, rather than their individual atoms. Since there are 2 NO_3^- ions in 1 $Cu(NO_3)_2$, we assign a coefficient of 2 to $AgNO_3$:

$$1\ Cu + 2\ AgNO_3 \rightarrow 1\ Cu(NO_3)_2 + ? Ag \qquad \textbf{Partially balanced}$$

Finally, we balance the silver by assigning a coefficient of 2 to Ag:

$$1\ Cu + 2\ AgNO_3 \rightarrow 1\ Cu(NO_3)_2 + 2\ Ag \qquad \textbf{Balanced equation}$$

$$Cu + 2\ AgNO_3 \rightarrow Cu(NO_3)_2 + 2\ Ag \qquad \textbf{Preferred final form}$$

PROBLEM 10.2

Balance each of the following equations.

(a) $C_5H_{12} + O_2 \rightarrow CO_2 + H_2O$
(b) $H_2 + I_2 \rightarrow HI$
(c) $Li + H_2 \rightarrow LiH$
(d) $S + O_2 \rightarrow SO_2$
(e) $Al + Fe_2O_3 \rightarrow Fe + Al_2O_3$
(f) $Fe_2S_3 + O_2 \rightarrow Fe_2O_3 + SO_2$
(g) $KOH + H_2SO_4 \rightarrow K_2SO_4 + H_2O$
(h) $P_4 + O_2 \rightarrow P_4O_{10}$
(i) $Zn + HCl \rightarrow ZnCl_2 + H_2$
(j) $Al + H_2SO_4 \rightarrow Al_2(SO_4)_3 + H_2$
(k) $C_6H_{12}O_6 + O_2 \rightarrow CO_2 + H_2O$
(l) $Mg_3N_2 + H_2O \rightarrow MgO + NH_3$
(m) $N_2H_4 + O_2 \rightarrow NO_2 + H_2O$

10.3 TRANSLATING WRITTEN STATEMENTS INTO CHEMICAL EQUATIONS

OBJECTIVE 10.3 Given a description of a chemical reaction, write a word equation for that reaction. Translate a word equation into an unbalanced equation and balance the equation.

We are often asked to turn a written sentence into a balanced chemical equation. In Section 10.2, each example included a description of the reaction that corresponded to the equation we were about to balance. In each example, we translated the descriptive statement into an unbalanced equation. In this section, you will learn to do that for yourself.

Let us consider the following statement:

"On heating, calcium carbonate ($CaCO_3$) decomposes to form carbon dioxide (CO_2) and calcium oxide (CaO)."

First, let us convert the sentence into a word equation. The phrase "calcium carbonate *decomposes*" tells us that calcium carbonate is the reactant. The phrase "to *form* calcium oxide and carbon dioxide" tells us that these are the products. Thus,

$$\text{Calcium carbonate} \rightarrow \text{Calcium oxide} + \text{Carbon dioxide} \qquad \textbf{Word equation}$$

Next, we must translate the name of each reactant and product into its corresponding chemical formula. The formula of each substance was given in the statement we are interpreting:

$$?\,CaCO_3 \rightarrow ?\,CaO + ?\,CO_2 \qquad \textbf{Unbalanced equation}$$

Because the number of atoms of each element is the same on both sides, this simple equation is balanced when all of the coefficients are 1.

$$CaCO_3 \rightarrow CaO + CO_2 \qquad \textbf{Balanced equation}$$

The language used to describe chemical reactions is rich and diverse. Phrases such as "treated with," "mixed with," and "allowed to react" tell us that we are dealing with reactants. Phrases such as "evolution of," "are produced," or "are formed" indicate a product.

In each of the examples that follow, we interpret the given statement as a word equation and then translate it into an unbalanced chemical equation. Finally, we balance the chemical equation.

EXAMPLE 10.5

When magnesium (Mg) is treated with hydrochloric acid (HCl), there is an evolution of hydrogen (H_2) accompanied by the formation of magnesium chloride ($MgCl_2$).

SOLUTION

The phrase "magnesium is treated with hydrochloric acid" tells us that these are the reactants. The phrase "evolution of hydrogen" means that hydrogen is formed along with the magnesium chloride, so they are the products:

Magnesium + Hydrochloric acid → Magnesium chloride + Hydrogen **Word equation**

Next, we replace the name of each reactant and product with the formula provided in the problem:

$$Mg + HCl \rightarrow MgCl_2 + H_2$$ **Unbalanced equation**

Using our usual techniques for balancing this equation leads to the following balanced equation:

$$Mg + 2\,HCl \rightarrow MgCl_2 + H_2$$ **Balanced equation**

EXAMPLE 10.6

The formation of sodium sulfate (Na_2SO_4) and water (H_2O) occurs when sulfuric acid (H_2SO_4) is mixed with sodium hydroxide (NaOH).

SOLUTION

Sodium sulfate and water are formed, so they must be the products; since sulfuric acid and sodium hydroxide are mixed together initially, they must be the reactants.

Sulfuric acid + Sodium hydroxide → Sodium sulfate + Water **Word equation**

$$H_2SO_4 + NaOH \rightarrow Na_2SO_4 + H_2O$$ **Unbalanced equation**

$$H_2SO_4 + 2\,NaOH \rightarrow Na_2SO_4 + 2\,H_2O$$ **Balanced equation**

PROBLEM 10.3

Write a word equation, an unbalanced equation, and a balanced equation for each of the following reactions.

(a) Pentane, C_5H_{12}, undergoes combustion with oxygen, O_2, to form carbon dioxide, CO_2, and water, H_2O.

(b) On heating, mercury(II) oxide, HgO, decomposes to mercury, Hg, and oxygen, O_2.

(c) Phosphoric acid, H_3PO_4, and potassium hydroxide, KOH, form potassium phosphate, K_3PO_4, and water, H_2O.

(d) Treatment of magnesium carbonate, $MgCO_3$, with nitric acid, HNO_3, produces magnesium nitrate, $Mg(NO_3)_2$, and water, H_2O. The process is accompanied by the evolution of carbon dioxide, CO_2.

Using Chemical Nomenclature to Translate Word Equations

In Examples 10.5 and 10.6, a chemical formula accompanied the name of each reactant and product. In Examples 10.7 and 10.8 you will have to use your knowledge of nomenclature to translate the word equation.

EXAMPLE 10.7

Potassium reacts with oxygen to form potassium oxide.

SOLUTION

Since potassium reacts with oxygen, these must be the reactants. Potassium oxide is formed, so it must be the product.

Potassium + Oxygen → Potassium oxide **Word equation**

Remember, oxygen is one of the seven common diatomic elements, so its formula is O_2. The formulas of the other elements are generally written in monatomic form; thus, the formula for potassium is K. Potassium oxide is K_2O.

$$K + O_2 \rightarrow K_2O$$ **Unbalanced equation**

Using the techniques of the last section to balance this equation leads to

$$4\,K + O_2 \rightarrow 2\,K_2O$$ **Balanced equation**

EXAMPLE 10.8

Hydrogen and chlorine react to form hydrogen chloride.

SOLUTION

Hydrogen and chlorine are reacting, so these must be the reactants. Since hydrogen chloride is formed, it must be the product.

Hydrogen + Chlorine → Hydrogen chloride **Word equation**

Remember, hydrogen and chlorine are both diatomic elements. So we have the following:

$$H_2 + Cl_2 \rightarrow HCl$$ **Unbalanced equation**

This equation is readily balanced by assigning a coefficient of 2 to HCl:

$$H_2 + Cl_2 \rightarrow 2\,HCl$$ **Balanced equation**

PROBLEM 10.4

Write a word equation, an unbalanced chemical equation, and a balanced chemical equation for each of the following reactions.

(a) Upon heating, magnesium and oxygen form magnesium oxide.
(b) Barium oxide reacts with hydrochloric acid to form barium chloride and water.
(c) Sulfuric acid is produced when sulfur trioxide reacts with water.
(d) Sodium hydroxide and hydrogen are formed when metallic sodium reacts with water.
(e) Sodium carbonate reacts with sulfuric acid to produce sodium sulfate, carbon dioxide, and water.
(f) Reaction of diphosphorus pentoxide with water produces phosphoric acid.
(g) Aluminum and oxygen form aluminum oxide.
(h) Iron(III) oxide and hydrogen form iron and water.

10.4 SYMBOLS USED IN CHEMICAL EQUATIONS

OBJECTIVE 10.4 Explain the meaning of the symbols → and +. Explain the meaning of the state symbols (g), (ℓ), (s), and (aq). Use these symbols in chemical equations.

So far, the only symbols we have used in writing chemical equations are the arrow, →, and the plus sign, +. There are other symbols that we can use when we want to make our equations represent chemical reactions more completely. Table 10-1 lists some of the most commonly used symbols.

The symbols (s), (ℓ), and (g) are handy when we want to show the physical state of a product or reactant. Thus, when hydrogen and oxygen react, if the water produced is in the gaseous state (as water vapor), we write

$$2\,H_2(g) + O_2(g) \rightarrow 2\,H_2O(g)$$

If the water produced is in the liquid state, we write

$$2\,H_2(g) + O_2(g) \rightarrow 2\,H_2O(\ell)$$

Similarly, we can describe the reaction of magnesium (a solid) with bromine (a liquid) to form magnesium bromide (a solid) as follows:

$$Mg(s) + Br_2(\ell) \rightarrow MgBr_2(s)$$

The symbol (aq) is used to show that a particular substance is in aqueous solution. This simply means that it is dissolved in water. For example, hydrogen chloride is a gas in its pure state, HCl(g). When dissolved in water (in aqueous solution), it forms hydrochloric acid, HCl(aq). Treatment of hydrochloric acid with an aqueous solution of sodium hydrogen carbonate results in evolution of carbon dioxide (a gas), water (a liquid), and a solution of aqueous sodium chloride.

$$HCl(aq) + NaHCO_3(aq) \rightarrow CO_2(g) + H_2O(\ell) + NaCl(aq)$$

Precipitate: An insoluble solid product that separates from solution.

Insoluble solid products are called **precipitates,** and reactions that produce them are called *precipitation reactions.* For example, when an aqueous solution of silver

TABLE 10-1 Symbols used in chemical equations

Symbol	Meaning
→	Forms, reacts to form, or yields
+	Plus, or added to
(s)	Solid
(ℓ)	Liquid
(g)	Gas
(aq)	Aqueous, or in aqueous solution
⇌	Equilibrium (discussed in Chapter 17)
ΔH	Heat of reaction, enthalpy of reaction (see Section 10.9)

nitrate is mixed with aqueous sodium chloride, a precipitate (insoluble solid) of silver chloride is formed. Sodium nitrate is also formed, but it is soluble and remains in aqueous solution. We represent this reaction as follows:

$$AgNO_3(aq) + NaCl(aq) \rightarrow AgCl(s) + NaNO_3(aq)$$

In later chapters we discuss the principles of solubility that determine whether a product remains in solution or forms a precipitate. Our purpose here is to illustrate the use of the symbols we are likely to encounter in chemical equations.

State symbol: A symbol used to specify the physical state of a substance.

We frequently refer to the symbols (s), (ℓ), (g), and (aq) as **state symbols,** because they tell us what state the substance is in. A chemical equation that includes the state symbol of each reactant and product can be quite informative and useful. However, for most purposes, a balanced chemical equation showing the reactants and products in their correct proportions is satisfactory. Consequently, we omit state symbols in this text except for those situations that require their use to highlight some important aspect of a reaction or process.

PROBLEM 10.5

Write balanced equations for each of the following reactions, including the appropriate state symbols. [The formulas of reactants and products are provided for you in parts (a)–(c). For parts (d)–(f), you must use your knowledge of nomenclature to figure them out.]

(a) Pentane, C_5H_{12}, is a liquid that burns in the presence of molecular oxygen, O_2, a gas. The products are gaseous carbon dioxide, CO_2, and liquid water, H_2O.
(b) Treating solid magnesium carbonate, $MgCO_3$, with an aqueous solution of nitric acid, HNO_3, results in formation of an aqueous solution of magnesium nitrate, $Mg(NO_3)_2$, gaseous carbon dioxide, CO_2, and liquid water, H_2O.
(c) Solid sodium, Na, reacts with water, H_2O, a liquid, to produce gaseous hydrogen, H_2, and an aqueous solution of sodium hydroxide, NaOH.
(d) On heating, solid mercury(II) oxide decomposes to mercury (a liquid) and molecular oxygen (a gas).
(e) Coal is composed of solid carbon. When coal undergoes incomplete combustion with gaseous oxygen, gaseous carbon monoxide is formed.
(f) When aqueous solutions of lead(II) nitrate and potassium iodide are mixed, a precipitate of solid lead(II) iodide is observed. An aqueous solution of potassium nitrate also forms.

10.5 MOLE RATIOS IN CHEMICAL EQUATIONS

OBJECTIVE 10.5 Interpret a balanced chemical equation in terms of molecules and in terms of moles. Determine the mole ratio between any two substances in a balanced chemical equation. Given a balanced equation and the number of moles of any (known) reactant or product, determine the number of moles of any other (unknown) reactant or product. (Carry out mole–mole problems.)

Suppose we want to prepare 2 moles of water from the reaction of hydrogen and oxygen. Let us see how we can use the balanced chemical equation to determine the quantities of hydrogen and oxygen required.

$$2\,H_2 + O_2 \rightarrow 2\,H_2O$$

The balanced equation tells us that two molecules of hydrogen react with one molecule of oxygen to form two molecules of water. If we took *four* molecules of hydrogen (twice as much), we would need *two* molecules of oxygen (also twice as much); and upon reaction, *four* molecules of water would be produced:

$$4 \text{ molecules } H_2 + 2 \text{ molecules } O_2 \rightarrow 4 \text{ molecules } H_2O$$

Likewise, if we took 200 molecules of hydrogen (100 times as much), 100 molecules of oxygen would be required and 200 molecules of water would be produced:

$$200 \text{ molecules } H_2 + 100 \text{ molecules } O_2 \rightarrow 200 \text{ molecules } H_2O$$

Now suppose we took twice Avogadro's number of hydrogen molecules (remember: Avogadro's number $= 6.02 \times 10^{23}$). Twice Avogadro's number of hydrogen molecules would require Avogadro's number of oxygen molecules. After reacting together, twice Avogadro's number of water molecules would be produced:

$$\underbrace{2 \times 6.02 \times 10^{23} \text{ molecules } H_2}_{\textbf{2 moles } \mathbf{H_2}} + \underbrace{6.02 \times 10^{23} \text{ molecules } O_2}_{\textbf{1 mole } \mathbf{O_2}}$$

$$\downarrow$$

$$\underbrace{2 \times 6.02 \times 10^{23} \text{ molecules } H_2O}_{\textbf{2 moles } \mathbf{H_2O}}$$

However, twice Avogadro's number of hydrogen molecules constitutes 2 moles of hydrogen, and Avogadro's number of oxygen molecules constitutes 1 mole of oxygen. Similarly, twice Avogadro's number of water molecules constitutes 2 moles of water. Thus, we can rewrite the last equation as

$$2 \text{ mol } H_2 + 1 \text{ mol } O_2 \rightarrow 2 \text{ mol } H_2O$$

Looking back to the original balanced equation,

$$2\,H_2 + O_2 \rightarrow 2\,H_2O$$

it is apparent that the coefficients in this equation can refer to either molecules or moles. Thus, the equation can be read, "Two moles of hydrogen react with one mole of oxygen to produce two moles of water." *The same relationships that hold for molecules in a chemical equation also hold for moles.*

It is impossible for us to go into the laboratory to measure out two molecules of hydrogen and one molecule of oxygen to produce two molecules of water. However, there is little difficulty in measuring out 2 moles of hydrogen (mass = 4.0 g) and 1 mole of oxygen (mass = 32.0 g) to produce 2 moles of water. It is for this reason that we define the mole. The mole provides us with quantities that we can work with in the laboratory.

Suppose we wish to produce 8 moles of water. Because we want to produce four times as much water as we did in the preceding example, we need to use four times as much hydrogen and oxygen. We must mix 8 moles of hydrogen and 4 moles of oxygen. An equation to express these ratios is

$$8\ H_2 + 4\ O_2 \rightarrow 8\ H_2O$$

Although this equation is not properly balanced in the sense of representing the simplest whole-number ratio of coefficients, it is balanced from the standpoint of having the same number of atoms of hydrogen and oxygen on each side. (There are 16 atoms of hydrogen and 8 atoms of oxygen on each side.) In fact, there is an infinite number of similarly balanced equations for the reaction of hydrogen and oxygen. For example,

$$6\ H_2 + 3\ O_2 \rightarrow 6\ H_2O$$
$$10\ H_2 + 5\ O_2 \rightarrow 10\ H_2O$$
$$350\ H_2 + 175\ O_2 \rightarrow 350\ H_2O$$

and so forth . . .

Yet each of them expresses the same relationship between the numbers of moles as that expressed in our original equation,

$$2\ H_2 + 1\ O_2 \rightarrow 2\ H_2O$$

In every case, the ratio of moles of hydrogen to moles of oxygen is 2 : 1; the ratio of moles of water to moles of oxygen is 2 : 1; and the ratio of moles of hydrogen to moles of water is 2 : 2 (or 1 : 1).

Let us extend our logic to solve a problem involving fractional numbers. Suppose we wish to produce 1.4 moles of water. Again, the key to the problem lies in examining the ratios involved. The ratio of hydrogen to water is 1 : 1, so we need to use 1.4 moles of hydrogen. And the ratio of oxygen to water is 1 : 2, so we need to use only 0.70 mole of oxygen. In other words,

$$1.4\ \text{mol}\ H_2 + 0.70\ \text{mol}\ O_2 \rightarrow 1.4\ \text{mol}\ H_2O$$

Now, let us apply our dimensional analysis technique to this type of problem. Returning to the original equation,

$$2\ H_2 + O_2 \rightarrow 2\ H_2O$$

Mole ratio: Ratio of the coefficients of two substances in a balanced chemical equation.

we can write each of the following ratios, which are referred to as **mole ratios:**

$$2\ \text{mol}\ H_2 : 1\ \text{mol}\ O_2$$
$$2\ \text{mol}\ H_2 : 2\ \text{mol}\ H_2O$$
$$1\ \text{mol}\ O_2 : 2\ \text{mol}\ H_2O$$

These mole ratios can be interpreted as follows:

"Two moles of hydrogen will react with one mole of oxygen."
"Two moles of hydrogen can produce two moles of water."
"One mole of oxygen can produce two moles of water."

Each mole ratio tells us how a pair of quantities shown in the balanced chemical equation are related to one another. Thus, if we wish to know how many moles of hydrogen are required to produce 1.4 moles of water, we select the mole ratio that relates moles of hydrogen to moles of water:

$$2 \text{ mol } H_2 : 2 \text{ mol } H_2O$$

We can express this ratio as either of the following two fractions, each of which may be used as a conversion factor:

$$\frac{2 \text{ mol } H_2}{2 \text{ mol } H_2O} = 1 \qquad \text{or} \qquad 1 = \frac{2 \text{ mol } H_2O}{2 \text{ mol } H_2}$$

To determine the moles of hydrogen required to produce 1.4 mol H_2O, we will multiply 1.4 mol H_2O by the conversion factor on the left, thereby canceling "mol H_2O."

$$? \text{ mol } H_2 = 1.4 \cancel{\text{mol } H_2O}\left(\frac{2 \text{ mol } H_2}{2 \cancel{\text{mol } H_2O}}\right)$$
$$= 1.4 \text{ mol } H_2$$

Similarly, to find the number of moles of oxygen required, we use the mole ratio that relates moles of water to moles of oxygen:

$$1 \text{ mol } O_2 : 2 \text{ mol } H_2O$$

$$? \text{ mol } O_2 = 1.4 \cancel{\text{mol } H_2O}\left(\frac{1 \text{ mol } H_2}{2 \cancel{\text{mol } H_2O}}\right)$$
$$= 0.70 \text{ mol } O_2$$

In the problem we just solved, we were given the number of moles of water. It was the *known* for the problem. We wished to find the moles of hydrogen and oxygen required. They were the *unknowns*. In each case, we used a mole ratio to convert *moles of known* to *moles of unknown*.

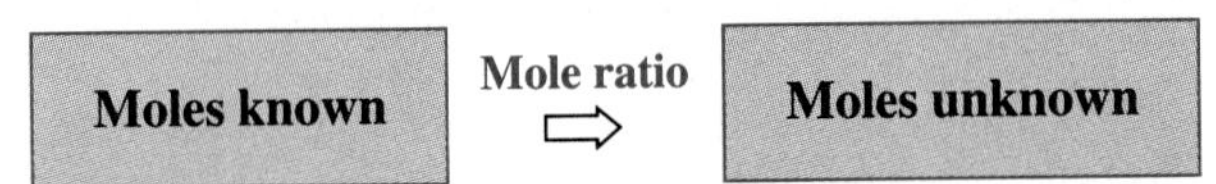

Because we are converting moles of one substance to moles of another, this general problem type is known as a *mole–mole calculation.*

Let us solve a few more examples using this method.

EXAMPLE 10.9

Propane, C_3H_8, burns in the presence of oxygen, O_2, to form carbon dioxide, CO_2, and water, H_2O. How many moles of carbon dioxide are produced by burning 6.40 moles of propane?

Symbolic statement

$$? \text{ mol } CO_2 = 6.40 \text{ mol } C_3H_8$$

Planning the solution

Since we wish to relate the moles of propane (the known) to the moles of carbon dioxide (the unknown), we need a balanced chemical equation to get started. The balanced chemical equation will provide the mole ratio between them, which can then be used to convert moles of propane to moles of carbon dioxide.

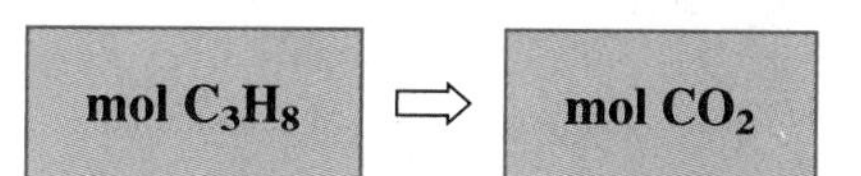

SOLUTION

First we must write and balance an equation for the reaction.

$$C_3H_8 + O_2 \rightarrow CO_2 + H_2O \qquad \textbf{Unbalanced equation}$$
$$C_3H_8 + 5\,O_2 \rightarrow 3\,CO_2 + 4\,H_2O \qquad \textbf{Balanced equation}$$

The problem gives us 6.40 moles of propane as the known. The unknown we are asked to determine is the number of moles of carbon dioxide produced. Thus, we need to find the mole ratio between propane and carbon dioxide from the balanced equation. It is

$$1 \text{ mol } C_3H_8 : 3 \text{ mol } CO_2$$

Multiplying the moles of known by the appropriate conversion factor yields

$$? \text{ mol } CO_2 = 6.40 \text{ mol } C_3H_8 \left(\frac{3 \text{ mol } CO_2}{1 \text{ mol } C_3H_8}\right)$$
$$= 19.2 \text{ mol } CO_2$$

Before we continue working examples, we review the technique we applied in Example 10.9 in the following Summary:

SUMMARY *Using Mole Ratios*

Step 1 Write a balanced equation for the reaction.

Step 2 Determine the known and the unknown.

Step 3 Determine the number of moles of unknown as follows:

(a) Use the balanced equation to determine the mole ratio between the known and the unknown.

(b) Multiply the moles of known by the mole ratio to determine the moles of unknown.

EXAMPLE 10.10

Aluminum, Al, reacts with oxygen, O_2, to produce aluminum oxide, Al_2O_3. How many moles of oxygen are required to react with 2.8 moles of aluminum?

Symbolic statement

$$? \text{ mol } O_2 = 2.8 \text{ mol Al}$$

Planning the solution

The number of moles of known given in the problem is 2.8 mol Al. The unknown we wish to determine is the number of moles of oxygen.

mol Al ⇨ **mol O_2**

SOLUTION

Before we can proceed with the calculation, we must first write a balanced chemical equation for the reaction.

$$? \text{ Al} + ? \text{ O}_2 \rightarrow ? \text{ Al}_2\text{O}_3 \qquad \textbf{Unbalanced equation}$$

$$4 \text{ Al} + 3 \text{ O}_2 \rightarrow 2 \text{ Al}_2\text{O}_3 \qquad \textbf{Balanced equation}$$

To find the moles of oxygen required, we will need the mole ratio between oxygen and aluminum:

$$4 \text{ mol Al} : 3 \text{ mol O}_2$$

Then we use the mole ratio as a conversion factor:

$$? \text{ mol O}_2 = 2.8 \cancel{\text{mol Al}} \left(\frac{3 \text{ mol O}_2}{4 \cancel{\text{mol Al}}}\right)$$

$$= 2.1 \text{ mol O}_2$$

These examples illustrate an important point about mole ratios. We can express the mole ratio between any two substances present in the balanced equation. In our preliminary discussion, we determined the amount of hydrogen and oxygen required to produce a given quantity of water. In that example, the known was a product, while the unknowns were reactants. Then in Example 10.9, we presented a problem where the known was a reactant and we used that to find the amount of a product. In Example 10.10, we used the given amount of one reactant to find the required amount of another reactant. As you might guess, it is also possible to relate the amounts of two products, as shown in the next example.

EXAMPLE 10.11

Copper can react with nitric acid according to the following balanced equation:

$$3 \text{ Cu} + 8 \text{ HNO}_3 \rightarrow 3 \text{ Cu(NO}_3)_2 + 2 \text{ NO} + 4 \text{ H}_2\text{O}$$

How many moles of nitrogen monoxide, NO, are given off in the production of 3.9 mol $Cu(NO_3)_2$?

Symbolic statement

$$? \text{ mol NO} = 3.9 \text{ mol Cu(NO}_3)_2$$

Planning the solution

mol $Cu(NO_3)_2$ ⇨ **mol NO**

SOLUTION

Here the known is $Cu(NO_3)_2$, and the unknown is NO. The balanced equation gives us the mole ratio.

$$3 \text{ mol Cu(NO}_3)_2 : 2 \text{ mol NO}$$

$$? \text{ mol NO} = 3.9 \cancel{\text{mol Cu(NO}_3)_2} \left(\frac{2 \text{ mol NO}}{3 \cancel{\text{mol Cu(NO}_3)_2}}\right)$$

$$= 2.6 \text{ mol NO}$$

PROBLEM 10.6

The following balanced equation was presented in Example 10.11:

$$3\ Cu + 8\ HNO_3 \rightarrow 3\ Cu(NO_3)_2 + 2\ NO + 4\ H_2O$$

Write the mole ratio that relates each of the following pairs of substances found in the equation.

(a) Cu and HNO_3 **(b)** Cu and $Cu(NO_3)_2$ **(c)** HNO_3 and NO
(d) NO and H_2O **(e)** Cu and NO **(f)** $Cu(NO_3)_2$ and H_2O

PROBLEM 10.7

Pentane, C_5H_{12}, burns in the presence of oxygen, O_2, to produce carbon dioxide, CO_2, and water, H_2O.

$$C_5H_{12} + O_2 \rightarrow CO_2 + H_2O \qquad \textbf{Unbalanced equation}$$

(a) Balance the equation.
(b) How many moles of oxygen are required to burn 2.10 moles of pentane?
(c) How many moles of carbon dioxide are produced?
(d) How many moles of water are produced?

PROBLEM 10.8

Aluminum, Al, reacts with hydrochloric acid, HCl, to produce aluminum chloride, $AlCl_3$, and hydrogen, H_2.

$$Al + HCl \rightarrow AlCl_3 + H_2 \qquad \textbf{Unbalanced equation}$$

(a) Balance the equation.
(b) How many moles of hydrochloric acid are required to react with 0.450 mole of aluminum?
(c) How many moles of aluminum chloride are produced?
(d) How many moles of hydrogen form?

PROBLEM 10.9

Hydrazine, N_2H_4, can burn in the presence of oxygen, O_2, to form nitrogen dioxide, NO_2, and water, H_2O.

(a) Write a balanced equation for the reaction.
(b) How many moles of oxygen are required to burn 1.30 moles of hydrazine?
(c) How many moles of nitrogen dioxide are produced?
(d) How many moles of water are produced?

PROBLEM 10.10

Iron(III) sulfide reacts with oxygen to form iron(III) oxide plus sulfur dioxide.

(a) Write a balanced equation for the reaction.
(b) How many moles of iron(III) sulfide are required to produce 0.900 mole of sulfur dioxide?
(c) How many moles of oxygen are required?
(d) How many moles of iron(III) oxide are produced?

10.6 STOICHIOMETRY CALCULATIONS: MASS RELATIONSHIPS

OBJECTIVE 10.6 Given a balanced equation and the number of grams or moles of any (known) reactant or product, calculate the number of grams or moles of any other (unknown) reactant or product. (Solve mass–mass, mole–mass, mass–mole, or mole–mole problems.)

In Section 10.5, we posed questions such as, "How many moles of hydrogen and oxygen are necessary to produce 1.4 moles of water?" Ordinarily, chemists do not ask questions in quite this way. A chemist is more likely to ask, "How many *grams* each of hydrogen and oxygen must be mixed in order to produce 25.2 grams of water?" The reason for this is simple. In the laboratory, we have to measure out the masses of substances using a balance. We do not have devices that measure the number of moles of hydrogen or the number of moles of oxygen. However, if we know the masses equivalent to the numbers of moles we wish to use, we can easily measure out these amounts using a balance. (If you do not recall how to calculate the mass in a given number of moles, or the reverse, review Sections 4.6 and 4.7.) The topic of such mass relationships in chemical reactions is called **stoichiometry.**

Stoichiometry: Quantitative mass relationships between reactants and products in a chemical reaction.

Let us see how we can go about calculating the masses of hydrogen and oxygen required to produce 25.2 g of water. To begin with, we need a balanced equation.

$$\begin{array}{ccccc} 2\,H_2 & + & O_2 & \rightarrow & 2\,H_2O \\ ? & + & ? & \rightarrow & 25.2\text{ g} \end{array} \quad \textbf{Masses of substances}$$

We have been given the mass of water, so this is our known quantity. The unknowns are hydrogen and oxygen. We begin by calculating the mass of hydrogen required. Since hydrogen is an unknown, we will eventually need to use the mole ratio between water (known) and hydrogen (unknown). However, this problem gives us the quantity of water in grams rather than moles. We must first convert the mass of water to moles of water:

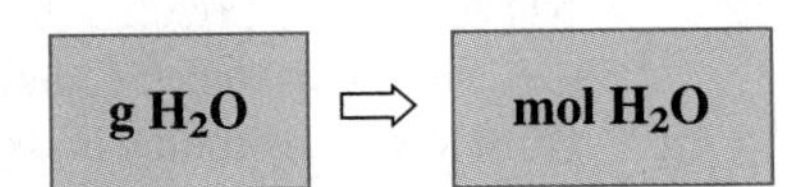

Recalling Section 4.7, we use the molar mass of water to accomplish this:

$$1\text{ mol }H_2O = 18.0\text{ g }H_2O$$

$$?\text{ mol }H_2O = 25.2\text{ g }\cancel{H_2O}\left(\frac{1\text{ mol }H_2O}{18.0\text{ g }\cancel{H_2O}}\right)$$

$$= 1.40\text{ mol }H_2O$$

Now we can use the mole ratio to determine the moles of hydrogen required:

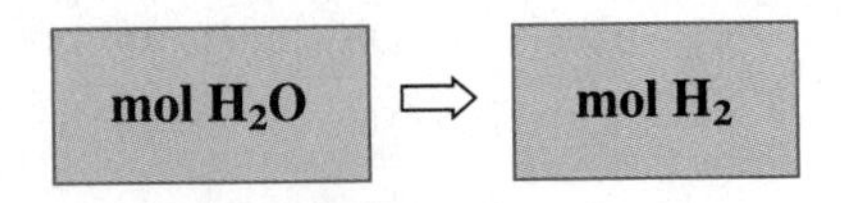

$$2 \text{ mol } H_2 : 2 \text{ mol } H_2O$$

$$? \text{ mol } H_2 = 1.40 \text{ mol } H_2O \left(\frac{2 \text{ mol } H_2}{2 \text{ mol } H_2O} \right)$$

$$= 1.40 \text{ mol } H_2$$

Since we wish to know the *mass* of hydrogen required, we must complete this part of the problem by converting the moles of hydrogen to grams:

mol H_2 ⇨ **g H_2**

We use the molar mass of hydrogen to carry out this conversion:

$$1 \text{ mol } H_2 = 2.0 \text{ g } H_2$$

$$? \text{ g } H_2 = 1.40 \text{ mol } H_2 \left(\frac{2.0 \text{ g } H_2}{1 \text{ mol } H_2} \right)$$

$$= 2.8 \text{ g } H_2$$

Thus, we need 2.8 g of hydrogen.

Looking back over these calculations, there was a very definite strategy that we employed. We can express this strategy symbolically as follows:

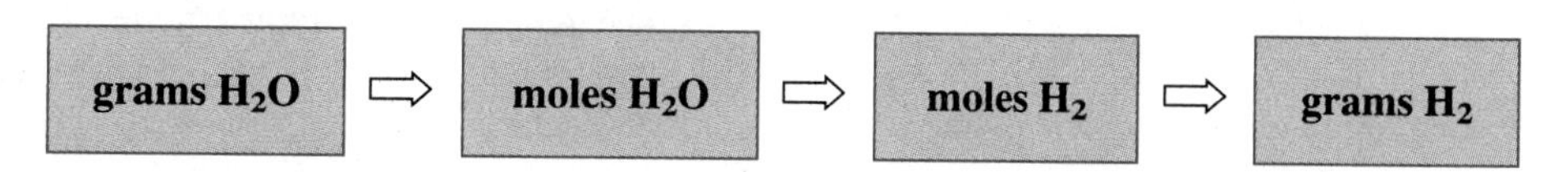

The heart of this strategy involved using the mole ratio to find the number of moles of H_2 from the number of moles of H_2O. This was carried out in exactly the same fashion as in Section 10.5. The other steps were simply conversions between grams and moles.

We can determine the mass of oxygen required by using a similar strategy:

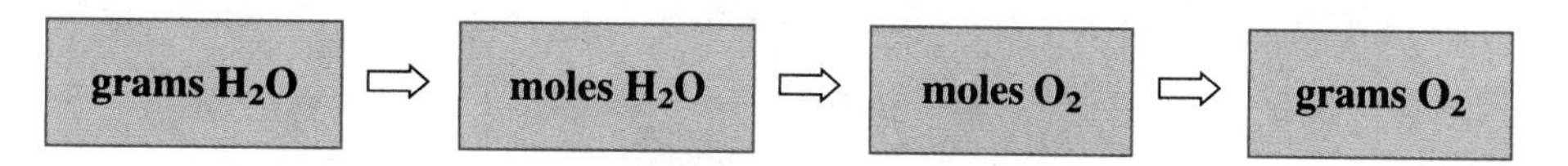

Since we have already determined the moles of H_2O, we skip the first step and start directly with 1.40 mol H_2O. The mole ratio, determined from the chemical equation, will enable us to calculate the moles of oxygen required.

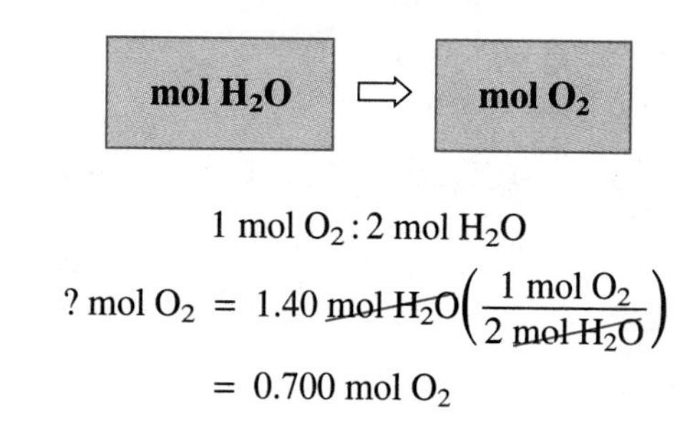

$$1 \text{ mol } O_2 : 2 \text{ mol } H_2O$$

$$? \text{ mol } O_2 = 1.40 \text{ mol } H_2O \left(\frac{1 \text{ mol } O_2}{2 \text{ mol } H_2O} \right)$$

$$= 0.700 \text{ mol } O_2$$

Finally, we use the molar mass of oxygen to calculate the mass of oxygen required in the reaction:

mol O_2 ⇨ g O_2

$$1 \text{ mol } O_2 = 32.0 \text{ g } O_2$$

$$? \text{ g } O_2 = 0.700 \cancel{\text{mol } O_2}\left(\frac{32.0 \text{ g } O_2}{1 \cancel{\text{mol } O_2}}\right)$$

$$= 22.4 \text{ g } O_2$$

Thus, 22.4 g of O_2 are required.

Let us summarize the information we have gathered:

$$\begin{array}{ccccc} 2\,H_2 & + & O_2 & \rightarrow & 2\,H_2O \\ 2.8\text{ g} & + & 22.4\text{ g} & \rightarrow & 25.2\text{ g} \\ 1.40\text{ mol} & + & 0.700\text{ mol} & \rightarrow & 1.40\text{ mol} \end{array}$$

Taking a careful look at these results, we can see that our calculations make a good deal of sense. The moles of substances in the bottom line are in the correct ratio (2 : 1 : 2). Furthermore, the mass of the reactants (2.8 g + 22.4 g) equals the mass of the product (25.2 g). This is consistent with the Law of Conservation of Mass.

Before proceeding to our examples, review the general strategy for working stoichiometry problems outlined in the Summary box on the facing page.

EXAMPLE 10.12

Propane, C_3H_8, undergoes combustion with oxygen, O_2, to produce carbon dioxide, CO_2, and water, H_2O. What mass of carbon dioxide is produced during the combustion of 17.6 g of propane?

Planning the solution

We follow the strategy outlined in the Summary.

Step 1 As in all stoichiometry calculations, we must first write a balanced equation.

$$C_3H_8 + O_2 \rightarrow CO_2 + H_2O \qquad \textbf{Unbalanced equation}$$

$$C_3H_8 + 5\,O_2 \rightarrow 3\,CO_2 + 4\,H_2O \qquad \textbf{Balanced equation}$$

Step 2 Next we must identify the known and unknown specified in the problem. We are given the mass of propane (the known) and asked to determine the quantity of carbon dioxide (the unknown):

$$? \text{ g } CO_2 = 17.6 \text{ g } C_3H_8$$

Step 3 To determine the mass of CO_2 produced, we map out the following strategy:

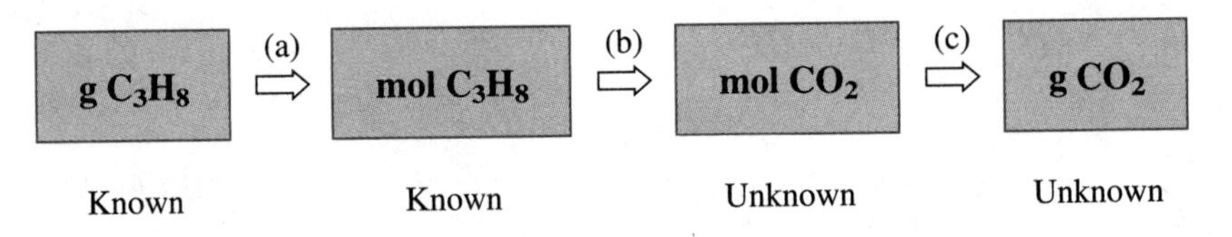

SUMMARY Solving Stoichiometry Problems

Step 1 Write and balance the chemical equation for the reaction involved.

Step 2 Identify the known and unknown (or unknowns) specified in the problem.

Step 3 Employ the following general strategy to calculate the mass of the unknown(s):

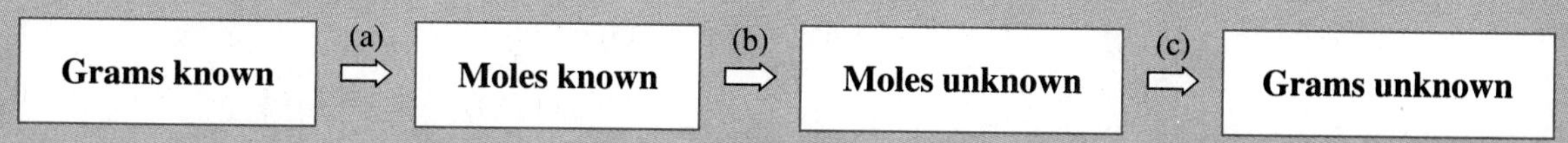

Use molar mass of known Use mole ratio Use molar mass of unknown

(a) Use the molar mass of the known to find the number of moles represented by the given mass. If you are given the number of moles in the problem or have calculated it in a previous part of the problem, this step is unnecessary and you can proceed directly to Step 3b.

(b) Use the appropriate mole ratio from the balanced equation to find the number of moles of unknown.

(c) Use the molar mass of the unknown to convert the moles of unknown (from Step 3b) to grams of unknown.

We use the molar masses of propane and carbon dioxide for Steps 3a and 3c, and the mole ratio between them for Step 3b.

SOLUTION

We will carry out this calculation in stepwise fashion to help reinforce the components of our general strategy.

Step 3a First we determine the number of moles in 17.6 g of propane.

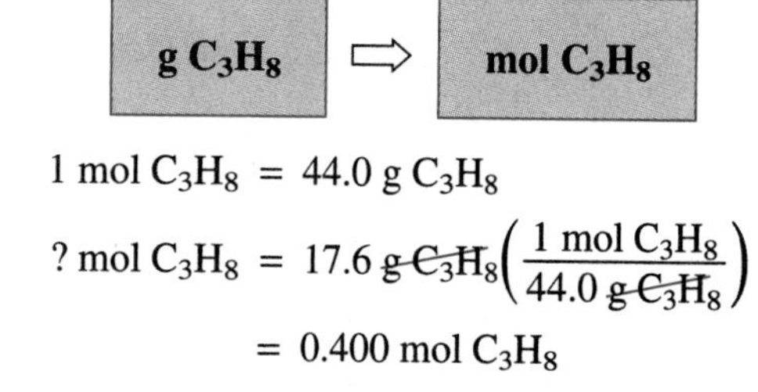

$$1\ \text{mol}\ C_3H_8 = 44.0\ \text{g}\ C_3H_8$$

$$?\ \text{mol}\ C_3H_8 = 17.6\ \cancel{\text{g}\ C_3H_8}\left(\frac{1\ \text{mol}\ C_3H_8}{44.0\ \cancel{\text{g}\ C_3H_8}}\right)$$

$$= 0.400\ \text{mol}\ C_3H_8$$

Step 3b Next we use the mole ratio between propane and carbon dioxide to determine the number of moles of carbon dioxide produced.

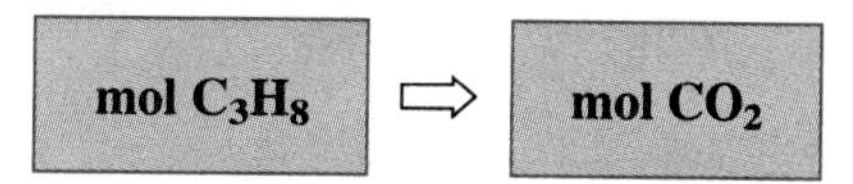

$$1 \text{ mol } C_3H_8 : 3 \text{ mol } CO_2$$

$$? \text{ mol } CO_2 = 0.400 \cancel{\text{mol } C_3H_8}\left(\frac{3 \text{ mol } CO_2}{1 \cancel{\text{mol } C_3H_8}}\right)$$

$$= 1.20 \text{ mol } CO_2$$

Step 3c Finally, we calculate the mass of 1.20 moles of carbon dioxide.

$$\textbf{mol } CO_2 \Rightarrow \textbf{g } CO_2$$

$$1 \text{ mol } CO_2 = 44.0 \text{ g } CO_2$$

$$? \text{ g } CO_2 = 1.20 \cancel{\text{mol } CO_2}\left(\frac{44.0 \text{ g } CO_2}{1 \cancel{\text{mol } CO_2}}\right)$$

$$= 52.8 \text{ g } CO_2$$

Of course, we could have combined all three steps into one setup, as follows:

$$? \text{ g } CO_2 = 17.6 \cancel{\text{g } C_3H_8}\left(\frac{1 \cancel{\text{mol } C_3H_8}}{44.0 \cancel{\text{g } C_3H_8}}\right)\left(\frac{3 \cancel{\text{mol } CO_2}}{1 \cancel{\text{mol } C_3H_8}}\right)\left(\frac{44.0 \text{ g } CO_2}{1 \cancel{\text{mol } CO_2}}\right)$$

$$= 52.8 \text{ g } CO_2$$

However, there are times when we need to use the result of an intermediate calculation, as our next example illustrates.

EXAMPLE 10.13

Find the mass of oxygen required and the mass of water produced in Example 10.12.

SOLUTION

This is just a continuation of the previous example, but now our unknowns are oxygen and water. To find the mass of oxygen we can use the following strategy:

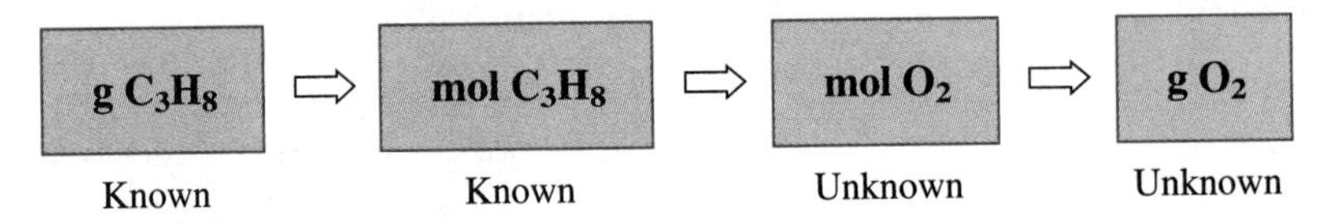

However, we determined the moles of propane (moles of known) in Example 10.12. Thus, we are ready to find the moles of O_2 needed to react with 0.400 mol C_3H_8.

$$? \text{ g } O_2 = 0.400 \text{ mol } C_3H_8$$

$$\textbf{mol } C_3H_8 \Rightarrow \textbf{mol } O_2 \Rightarrow \textbf{g } O_2$$

This time we need the mole ratio between propane and oxygen:

$$1 \text{ mol } C_3H_8 : 5 \text{ mol } O_2$$

$$? \text{ mol } O_2 = 0.400 \cancel{\text{mol } C_3H_8}\left(\frac{5 \text{ mol } O_2}{1 \cancel{\text{mol } C_3H_8}}\right)$$

$$= 2.00 \text{ mol } O_2$$

Converting to grams of oxygen requires the molar mass of oxygen. [*Remember:* Oxygen is diatomic, so its molecular mass is 32.0.]

$$1 \text{ mol } O_2 = 32.0 \text{ g } O_2$$

$$? \text{ g } O_2 = 2.00 \text{ mol } O_2 \left(\frac{32.0 \text{ g } O_2}{1 \text{ mol } O_2}\right)$$

$$= 64.0 \text{ g } O_2$$

Naturally, we can also combine these steps in the usual fashion:

$$? \text{ mol } O_2 = 0.400 \text{ mol } C_3H_8 \left(\frac{5 \text{ mol } O_2}{1 \text{ mol } C_3H_8}\right)\left(\frac{32.0 \text{ g } O_2}{1 \text{ mol } O_2}\right)$$

$$= 64.0 \text{ g } O_2$$

This example illustrates a general principle for solving stoichiometry problems where we need to calculate several unknowns. Once the number of moles of known have been determined, this quantity can become the starting point for determining each of the unknowns. Thus, we also begin with 0.400 mol C_3H_8 to determine the mass of water produced.

$$? \text{ g } H_2O = 0.400 \text{ mol } C_3H_8$$

mol C_3H_8 ⇨ **mol H_2O** ⇨ **g H_2O**

First, we calculate the moles of water:

$$1 \text{ mol } C_3H_8 : 4 \text{ mol } H_2O$$

$$? \text{ mol } H_2O = 0.400 \text{ mol } C_3H_8 \left(\frac{4 \text{ mol } H_2O}{1 \text{ mol } C_3H_8}\right)$$

$$= 1.60 \text{ mol } H_2O$$

Then we determine the mass of 1.60 mol H_2O:

$$1 \text{ mol } H_2O = 18.0 \text{ g } H_2O$$

$$? \text{ g } H_2O = 1.60 \text{ mol } H_2O \left(\frac{18.0 \text{ g } H_2O}{1 \text{ mol } H_2O}\right)$$

$$= 28.8 \text{ g } H_2O$$

Again, we can save time by combining all the conversion factors:

$$? \text{ g } H_2O = 0.400 \text{ mol } C_3H_8 \left(\frac{4 \text{ mol } H_2O}{1 \text{ mol } C_3H_8}\right)\left(\frac{18.0 \text{ g } H_2O}{1 \text{ mol } H_2O}\right)$$

$$= 28.8 \text{ g } H_2O$$

Before proceeding with the next example, let us summarize the information we gathered in working Examples 10.12 and 10.13:

$$\begin{array}{llll} C_3H_8 + & 5\,O_2 \rightarrow & 3\,CO_2 + & 4\,H_2O \\ 17.6 \text{ g} + & 64.0 \text{ g} \rightarrow & 52.8 \text{ g} + & 28.8 \text{ g} \\ 0.400 \text{ mol} + & 2.00 \text{ mol} \rightarrow & 1.20 \text{ mol} + & 1.60 \text{ mol} \end{array}$$

Once again, we see that the total mass of the reactants (17.6 g + 64.0 g = 81.6 g) equals the total mass of the products (52.8 g + 28.8 g = 81.6 g). In addition, the ratios of moles calculated are all in agreement with the balanced equation.

EXAMPLE 10.14

Aluminum, Al, reacts with hydrochloric acid, HCl, to form aluminum chloride, $AlCl_3$, and hydrogen, H_2. How many grams of aluminum should be used to produce 9.0 g of hydrogen?

Symbolic statement

$$? \text{ g Al} = 9.0 \text{ g } H_2$$

SOLUTION

First, we must write a balanced equation:

$$Al + HCl \rightarrow AlCl_3 + H_2 \quad \textbf{Unbalanced equation}$$

$$2\,Al + 6\,HCl \rightarrow 2\,AlCl_3 + 3\,H_2 \quad \textbf{Balanced equation}$$

In this problem, our known is a product (H_2), and we wish to determine the mass of one of the reactants (Al) we need. Our strategy is as follows:

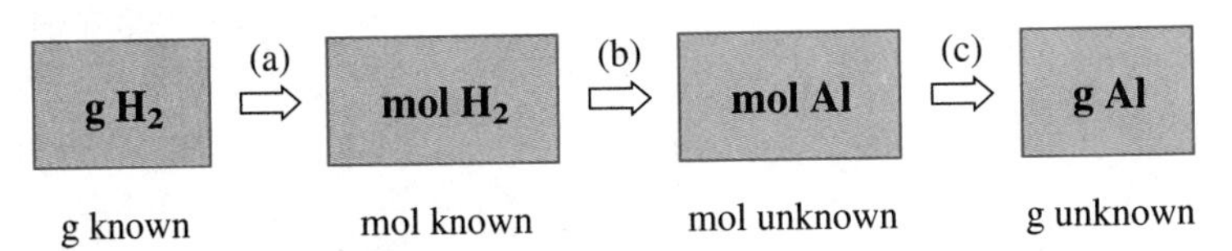

We must use the following relationships:

(a) 1 mol H_2 = 2.0 g H_2
(b) 2 mol Al : 3 mol H_2
(c) 1 mol Al = 27.0 g Al

Since this problem asks for only one unknown, we combine all of our conversions in one setup:

$$? \text{ g Al} = 9.0\,\cancel{\text{g H}_2}\left(\frac{1\,\cancel{\text{mol H}_2}}{2.0\,\cancel{\text{g H}_2}}\right)\left(\frac{2\,\cancel{\text{mol Al}}}{3\,\cancel{\text{mol H}_2}}\right)\left(\frac{27.0\text{ g Al}}{1\,\cancel{\text{mol Al}}}\right) = 81 \text{ g Al}$$

In Examples 10.12 and 10.14, we were given the mass of a known and asked to calculate the mass of one or more unknowns. These problems are known as *mass–mass* problems. However, as Example 10.13 suggests, a problem may provide the number of moles of known as the starting point and ask us to determine the mass of an unknown. Such a problem is called a *mole–mass* problem, and it requires us to use only the following parts of our general strategy:

Similarly, we may be given the mass of known and asked to determine the moles of unknown. This problem type is known as a *mass–mole* problem and uses the following strategy:

Both mole–mass and mass–mole problems are shorter versions of the mass–mass type. The Summary Diagram outlines our strategies for solving these various types of stoichiometry problems.

PROBLEM 10.11

How many moles are present in each of the following?
(a) 14.0 g of carbon monoxide, CO
(b) 25.0 g of calcium carbonate, $CaCO_3$
(c) 30.0 g of sodium sulfate, Na_2SO_4
(d) 62.9 g of ammonium phosphate, $(NH_4)_3PO_4$

PROBLEM 10.12

Calculate the mass of each of the following.
(a) 2.00 mol of carbon dioxide, CO_2
(b) 0.300 mol of sulfur trioxide, SO_3
(c) 0.835 mol of heptane, C_7H_{16}
(d) 0.0175 mol of calcium hydroxide, $Ca(OH)_2$

SUMMARY DIAGRAM

Relationships Among the Four Different Types of Stoichiometry Problems

MASS–MASS: Mass of known $\times \left(\frac{1}{\text{Molar mass of known}}\right)$ → Moles of known $\times \left(\frac{\text{Moles of unknown}}{\text{Moles of known}}\right)$ → Moles of unknown $\times$ Molar mass of unknown → Mass of unknown

MASS–MOLE: Mass of known $\times \left(\frac{1}{\text{Molar mass of known}}\right)$ → Moles of known $\times \left(\frac{\text{Moles of unknown}}{\text{Moles of known}}\right)$ → Moles of unknown

MOLE–MASS: Moles of known $\times \left(\frac{\text{Moles of unknown}}{\text{Moles of known}}\right)$ → Moles of unknown $\times$ Molar mass of unknown → Mass of unknown

MOLE–MOLE: Moles of known $\times \left(\frac{\text{Moles of unknown}}{\text{Moles of known}}\right)$ → Moles of unknown

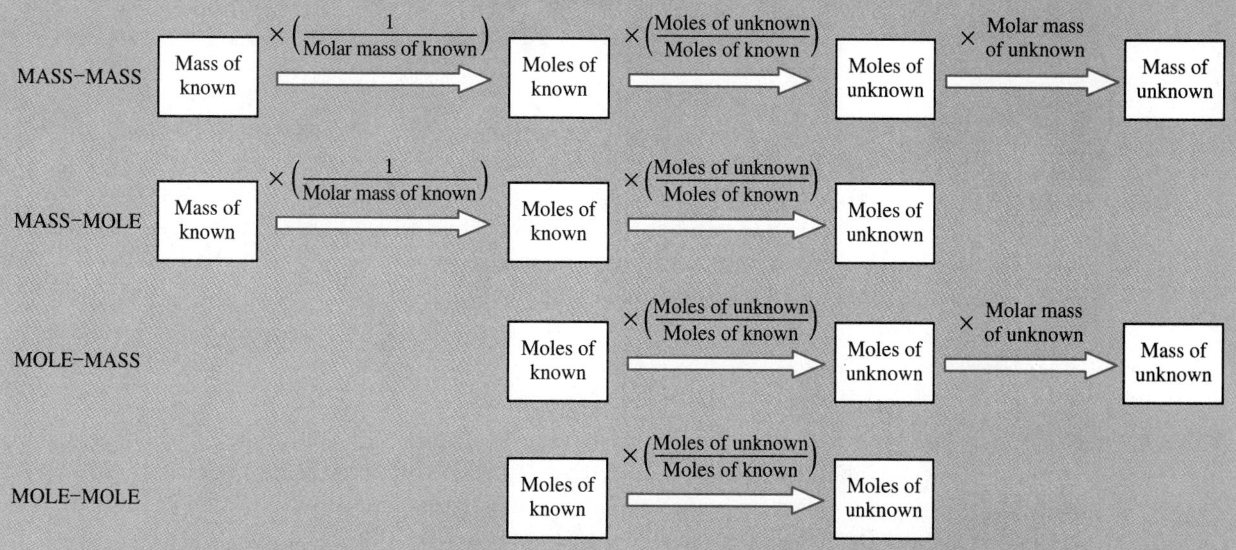

PROBLEM 10.13

In the formation of rust, metallic iron, Fe, reacts with oxygen, O_2, to form iron(III) oxide, Fe_2O_3.

(a) Write a balanced equation for the reaction.

(b) Calculate the mass of iron(III) oxide that results when 5.58 g of iron forms rust.

PROBLEM 10.14

Glucose, $C_6H_{12}O_6$, "burns" in our bodies with oxygen, O_2, to form carbon dioxide, CO_2, and water, H_2O.

(a) Write a balanced equation for the reaction.

(b) Calculate the mass of oxygen required to react completely with 36.0 g of glucose.

(c) What mass of carbon dioxide is produced?

(d) What mass of water results?

(e) Use the results of your calculations to demonstrate the Law of Conservation of Mass.

PROBLEM 10.15

Nitrogen fixation is the process by which molecular nitrogen, N_2, from the atmosphere combines with other elements to form compounds. The Haber process is an example of nitrogen fixation and is used to prepare ammonia, NH_3, from molecular nitrogen, N_2, and hydrogen, H_2.

(a) Write a balanced equation for the reaction.

(b) Calculate the mass of hydrogen required to prepare 85.0 g of ammonia.

(c) What mass of nitrogen is "fixed" (consumed)?

PROBLEM 10.16

Octane, C_8H_{18}, is a major component of gasoline. When octane burns with oxygen, the products are carbon dioxide and water.

(a) Write a balanced equation for the reaction.

(b) A liter of gasoline has a mass of about 702 g. What mass of carbon dioxide is produced when a liter of gasoline burns? (Assume gasoline to be exclusively octane.)

(c) What mass of oxygen is required?

PROBLEM 10.17

Aluminum sulfate, $Al_2(SO_4)_3$, can be produced from the reaction of aluminum, Al, and sulfuric acid, H_2SO_4. Hydrogen, H_2, is also produced in the process.

(a) What mass of aluminum is necessary to produce 455 g of aluminum sulfate?

(b) What mass of sulfuric acid is required?

(c) How many grams of hydrogen are produced?

10.7 PERCENTAGE YIELD

OBJECTIVE 10.7 Given a balanced equation and the mass of reactant used, calculate the theoretical yield of any product. Given the mass of product actually obtained, determine the percentage yield.

Theoretical yield: The quantity of a product predicted on the basis of a stoichiometric calculation.

Percentage yield: The percentage of the theoretical yield actually obtained.

Often when we carry out a reaction, we are not able to isolate all of the product expected. Sometimes this is due to loss in transfer of materials or in purification of the product. At other times, the loss may be due to competing reactions that transform a portion of the reactants into products other than the one we desire. Consequently, we are often interested in describing the efficiency of a particular chemical process, and we usually do this by telling what percentages of the anticipated products are actually obtained. The amount of material expected (based on a stoichiometric calculation) is called the **theoretical yield.** The **percentage yield** is the percentage of the theoretical yield that is actually obtained.

$$\text{Percentage yield} = \frac{\text{Actual yield}}{\text{Theoretical yield}} \times 100\%$$

For example, when aqueous solutions of calcium chloride, $CaCl_2$, and sodium sulfate, Na_2SO_4, are mixed, calcium sulfate, $CaSO_4$, precipitates according to the following equation:

$$CaCl_2(aq) + Na_2SO_4(aq) \rightarrow CaSO_4(s) + 2\ NaCl(aq)$$

The precipitate may be filtered, dried, and weighed. If an excess of sodium sulfate is used, 10.0 g $CaCl_2$ should produce 12.3 g $CaSO_4$:

$$? \text{ g } CaSO_4 = 10.0 \text{ g } \cancel{CaCl_2}\left(\frac{1 \text{ } \cancel{\text{mol } CaCl_2}}{111.1 \text{ } \cancel{\text{g } CaCl_2}}\right)\left(\frac{1 \text{ } \cancel{\text{mol } CaSO_4}}{1 \text{ } \cancel{\text{mol } CaCl_2}}\right)\left(\frac{136.2 \text{ g } CaSO_4}{1 \text{ } \cancel{\text{mol } CaSO_4}}\right)$$
$$= 12.3 \text{ g } CaSO_4$$

The quantity 12.3 g represents the theoretical yield. It is simply the result of a stoichiometric calculation. Suppose, however, that after we carry out this reaction and filter and dry the product, we actually isolate only 10.8 g of $CaSO_4$. The percentage yield (% yield) is then

$$\% \text{ yield} = \frac{\text{Actual yield}}{\text{Theoretical yield}} \times 100\%$$
$$= \frac{10.8 \text{ g}}{12.3 \text{ g}} \times 100\% = 87.8\%$$

EXAMPLE 10.15

Silver chromate, Ag_2CrO_4, can be precipitated by mixing aqueous solutions of silver nitrate, $AgNO_3$, and potassium chromate, K_2CrO_4:

$$2\ AgNO_3(aq) + K_2CrO_4(aq) \rightarrow Ag_2CrO_4(s) + 2\ KNO_3(aq)$$

Suppose an aqueous solution containing 5.32 g $AgNO_3$ is allowed to react with an excess of potassium chromate, and 4.77 g of Ag_2CrO_4 are obtained. Calculate the theoretical yield and the percentage yield.

SOLUTION

The theoretical yield is the amount of silver chromate that should form from 5.32 g $AgNO_3$:

$$? \text{ g } Ag_2CrO_4 = 5.32 \text{ g } AgNO_3$$

We determine this quantity by using a stoichiometric calculation:

g $AgNO_3$ ⇨ mol $AgNO_3$ ⇨ mol Ag_2CrO_4 ⇨ g Ag_2CrO_4

First, we must calculate the theoretical yield:

$$? \text{ g } Ag_2CrO_4 = 5.32 \text{ g } AgNO_3 \left(\frac{1 \text{ mol } AgNO_3}{169.9 \text{ g } AgNO_3}\right)\left(\frac{1 \text{ mol } Ag_2CrO_4}{2 \text{ mol } AgNO_3}\right)\left(\frac{331.8 \text{ g } Ag_2CrO_4}{1 \text{ mol } Ag_2CrO_4}\right)$$

$$= 5.19 \text{ g } Ag_2CrO_4$$

The percentage yield is obtained by taking the actual yield of silver chromate (4.77 g Ag_2CrO_4) and expressing it as a percentage of the theoretical yield (5.19 g Ag_2CrO_4):

$$\% \text{ yield} = \frac{\text{Actual yield}}{\text{Theoretical yield}} \times 100\%$$

$$= \frac{4.77 \text{ g}}{5.19 \text{ g}} \times 100\% = 91.9\%$$

PROBLEM 10.18

The chemical name for aspirin is acetylsalicylic acid, $C_9H_8O_4$. It can be made from salicylic acid, $C_7H_6O_3$, by treatment with an excess of acetic anhydride:

$$C_7H_6O_3 + C_4H_6O_3 \rightarrow C_9H_8O_4 + C_2H_4O_2$$

Salicylic acid	Acetic anhydride	Acetylsalicylic acid	Acetic acid

Suppose the reaction of 3.28 g of salicylic acid produces 3.11 g of aspirin. What is the theoretical yield? The percentage yield?

PROBLEM 10.19

Nitrobenzene, $C_6H_5NO_2$, is produced by the reaction of benzene, C_6H_6, with excess nitric acid:

$$C_6H_6 + HNO_3 \rightarrow C_6H_5NO_2 + H_2O$$

Benzene — Nitrobenzene

A reaction is carried out using 17.5 g of benzene, and 22.6 g of nitrobenzene are isolated. What is the theoretical yield? The percentage yield?

PROBLEM 10.20 Lead(II) chloride can be precipitated by treating lead(II) nitrate with sodium chloride:

$$Pb(NO_3)_2(aq) + 2\ NaCl(aq) \rightarrow PbCl_2(s) + 2\ NaNO_3(aq)$$

When a solution containing 5.36 g of $Pb(NO_3)_2$ is mixed with one containing excess sodium chloride, 3.21 g of $PbCl_2$ are isolated. Calculate the theoretical yield and the percentage yield.

10.8 LIMITING REAGENTS

OBJECTIVE 10.8 Given a balanced chemical equation with two reactants and the masses of each used, determine the limiting reagent and the mass of product formed.

Limiting reagent: A reactant that is completely consumed, thereby limiting the amount of product formed.

In each of the stoichiometry problems we have worked thus far, we have been told the substance on which to base our calculations. For example, in Example 10.12 we calculated the mass of carbon dioxide produced in the combustion of 17.6 g of propane. It was assumed that there was enough oxygen present to react with all of the propane. In other words, there was an *excess* of oxygen. Therefore, the amount of propane, rather than the amount of oxygen, determined the quantities of the products formed. That is, the reaction was not over until all of the propane was used up. Propane limited the reaction; thus, we call it the **limiting reagent.**

Often, we may mix two reactants and not know which one is the limiting reagent. For example, copper and sulfur react with one another, on heating, to form copper(I) sulfide. Suppose we wish to determine the mass of copper(I) sulfide formed when 7.00 g of Cu and 2.50 g of S are combined and allowed to react with each other, as follows:

$$\underset{\mathbf{7.00\ g}}{2\ Cu} + \underset{\mathbf{2.50\ g}}{S} \rightarrow \underset{\mathbf{?}}{Cu_2S}$$

One of the two reactants will be used up completely, and the other will be in excess. Before we can determine the mass of Cu_2S formed, we must determine which of the reactants, copper or sulfur, is the limiting reagent. That is, which of these substances is consumed completely?

To determine the limiting reagent, we will make a wild guess! Let us assume that copper is the limiting reagent. However, now we must check to see whether our guess is correct. We have a 50 : 50 chance of being right. To find out whether our guess is correct, we calculate the amount of sulfur that is actually required to use up 7.00 g of Cu. As in Section 10.6, our strategy is

g Cu ⇨ **mol Cu** ⇨ **mol S** ⇨ **g S**

$$?\ g\ S = 7.00\ \cancel{g\ Cu}\left(\frac{1\ \cancel{mol\ Cu}}{63.5\ \cancel{g\ Cu}}\right)\left(\frac{1\ \cancel{mol\ S}}{2\ \cancel{mol\ Cu}}\right)\left(\frac{32.1\ g\ S}{1\ \cancel{mol\ S}}\right) = 1.77\ g\ S$$

The quantity we have just calculated represents the mass of sulfur that is needed to react with all the copper present. Because 2.50 g of sulfur are actually present, there is more than enough sulfur. In other words, sulfur is in excess and copper *is* the limiting reagent.

Perhaps you are wondering what would have happened if we had guessed that sulfur was the limiting reagent. In that case, we would have calculated the mass of copper required to react with all the sulfur:

g S ⇨ **mol S** ⇨ **mol Cu** ⇨ **g Cu**

$$? \text{ g Cu} = 2.50 \cancel{\text{g S}}\left(\frac{1 \cancel{\text{mol S}}}{32.1 \cancel{\text{g S}}}\right)\left(\frac{2 \cancel{\text{mol Cu}}}{1 \cancel{\text{mol S}}}\right)\left(\frac{63.5 \text{ g Cu}}{1 \cancel{\text{mol Cu}}}\right) = 9.89 \text{ g Cu}$$

The result of this calculation tells us that 9.89 g of Cu are required to consume all the sulfur present. However, only 7.00 g of Cu are actually present. In other words, there is insufficient copper to react with all of the sulfur. So our second assumption must have been incorrect. Copper must be used up first, and therefore it is the limiting reagent. Regardless of which assumption we make, when we check it out, we learn which of the reactants really is the limiting reagent.

We need to perform only one of the preceding calculations. However, once we have found the limiting reagent, we must use it as the *known* for any calculations we perform. Thus, to complete the problem, we determine the mass of Cu_2S in the usual fashion, starting with the mass of copper (the limiting reagent) as our known:

g Cu ⇨ **mol Cu** ⇨ **mol Cu_2S** ⇨ **g Cu_2S**

↑
Limiting reagent

$$? \text{ g Cu}_2\text{S} = 7.00 \cancel{\text{g Cu}}\left(\frac{1 \cancel{\text{mol Cu}}}{63.5 \cancel{\text{g Cu}}}\right)\left(\frac{1 \cancel{\text{mol Cu}_2\text{S}}}{2 \cancel{\text{mol Cu}}}\right)\left(\frac{159.1 \text{ g Cu}_2\text{S}}{1 \cancel{\text{mol Cu}_2\text{S}}}\right)$$
$$= 8.77 \text{ g Cu}_2\text{S}$$

When you examine a limiting reagent problem, do not be misled by the quantities of the reactants. In this problem, the mass of sulfur was considerably smaller than that of copper. However, sulfur was the reagent in excess. Only a stoichiometric calculation will reveal which is the limiting reagent.

EXAMPLE 10.16

Aqueous solutions of silver nitrate, $AgNO_3$, and potassium chloride, KCl, can be mixed to precipitate silver chloride, AgCl, as follows:

$$\underset{\textbf{8.50 g}}{AgNO_3(aq)} + \underset{\textbf{4.75}}{KCl(aq)} \rightarrow \underset{\textbf{?}}{AgCl(s)} + KNO_3(aq)$$

Calculate the mass of silver chloride that is formed in the reaction of 8.50 g $AgNO_3$ with 4.75 g KCl.

Planning the solution

First we must determine whether $AgNO_3$ or KCl is the limiting reagent. Once known, the limiting reagent will become the known for the stoichiometric calculation of our unknown (in this case, AgCl).

SOLUTION

If we assume that $AgNO_3$ is the limiting reagent, we must calculate the mass of KCl required to consume 8.50 g $AgNO_3$:

$$? \text{ g KCl} = 8.50 \cancel{\text{g AgNO}_3}\left(\frac{1 \cancel{\text{mol AgNO}_3}}{169.9 \cancel{\text{g AgNO}_3}}\right)\left(\frac{1 \cancel{\text{mol KCl}}}{1 \cancel{\text{mol AgNO}_3}}\right)\left(\frac{74.6 \text{ g KCl}}{1 \cancel{\text{mol KCl}}}\right)$$
$$= 3.73 \text{ g KCl}$$

Because 4.75 g KCl are present, there is excess KCl, so $AgNO_3$ *is* the limiting reagent.

On the other hand, if we assume that KCl is the limiting reagent, we must calculate the mass of $AgNO_3$ required to consume 4.75 g KCl:

$$? \text{ g AgNO}_3 = 4.75 \cancel{\text{g KCl}}\left(\frac{1 \cancel{\text{mol KCl}}}{74.6 \cancel{\text{g KCl}}}\right)\left(\frac{1 \cancel{\text{mol AgNO}_3}}{1 \cancel{\text{mol KCl}}}\right)\left(\frac{169.9 \text{ g AgNO}_3}{1 \cancel{\text{mol AgNO}_3}}\right)$$
$$= 10.8 \text{ g AgNO}_3$$

Only 8.50 g $AgNO_3$ are actually present, however, so there is *not* enough silver nitrate to react with all of the potassium chloride. Potassium chloride must be in excess. Hence our second assumption is false, and silver nitrate must be the limiting reagent. [*Remember:* We only need to perform one of the two preceding calculations.]

Regardless of which calculation we choose to make, we arrive at the same limiting reagent. Now we may proceed to determine the mass of silver chloride precipitated, and to do so we *must* use $AgNO_3$, the limiting reagent, as our known.

g $AgNO_3$ ⇨ **mol $AgNO_3$** ⇨ **mol AgCl** ⇨ **g AgCl**

$$? \text{ g AgCl} = 8.50 \cancel{\text{g AgNO}_3}\left(\frac{1 \cancel{\text{mol AgNO}_3}}{169.9 \cancel{\text{g AgNO}_3}}\right)\left(\frac{1 \cancel{\text{mol AgCl}}}{1 \cancel{\text{mol AgNO}_3}}\right)\left(\frac{143.4 \text{ g AgCl}}{1 \cancel{\text{mol AgCl}}}\right)$$
$$= 7.17 \text{ g AgCl}$$

PROBLEM 10.21

Magnesium sulfide, MgS, may be produced by allowing the elements to react with one another. Calculate the mass of product obtained when 8.50 g Mg and 9.00 g S are allowed to react according to the following equation:

$$\text{Mg} + \text{S} \rightarrow \text{MgS}$$

PROBLEM 10.22

Barium sulfate, $BaSO_4$, is an insoluble compound that precipitates in the reaction of barium chloride, $BaCl_2$, with sodium sulfate, Na_2SO_4:

$$\text{BaCl}_2(aq) + \text{Na}_2\text{SO}_4(aq) \rightarrow \text{BaSO}_4(s) + 2\ \text{NaCl}(aq)$$

What mass of barium sulfate forms when an aqueous solution containing 4.25 g Na_2SO_4 is mixed with another solution containing 5.75 g $BaCl_2$?

PROBLEM 10.23

When aqueous solutions of lead(II) nitrate, $Pb(NO_3)_2$, and potassium iodide, KI, are mixed, a precipitate of lead(II) iodide, PbI_2, forms:

$$Pb(NO_3)_2(aq) + 2\,KI(aq) \rightarrow PbI_2(s) + 2\,KNO_3(aq)$$

Calculate the mass of precipitate formed when 4.55 g $Pb(NO_3)_2$ are allowed to react with 3.75 g KI.

10.9 HEAT CHANGES THAT ACCOMPANY CHEMICAL REACTIONS

OBJECTIVE 10.9 Explain the meaning of enthalpy of reaction, ΔH. Distinguish between an exothermic and an endothermic reaction. Given the enthalpy of a reaction and the mass of the limiting reagent, calculate the heat liberated or absorbed.

Earlier in the chapter we discussed the formation of water from hydrogen and oxygen,

$$2\,H_2(g) + O_2(g) \rightarrow 2\,H_2O(\ell)$$

Exothermic reaction: A chemical reaction that liberates heat.

This reaction is accompanied by the liberation of energy in the form of heat. A chemical reaction that gives off heat is said to be **exothermic.** Because the heat is given off, we can include it with the products on the right side of the equation:

$$2\,H_2(g) + O_2(g) \rightarrow 2\,H_2O(\ell) + 572\text{ kJ} \qquad (10\text{-}1)$$

[*Remember:* A kilojoule (1 kJ) is 1000 joules.]

Some reactions *absorb* heat, as in the production of glucose, $C_6H_{12}O_6$, from carbon dioxide and water:

$$6\,CO_2(g) + 6\,H_2O(\ell) + 2803\text{ kJ} \rightarrow C_6H_{12}O_6(s) + 6\,O_2(g) \qquad (10\text{-}2)$$

Endothermic reaction: A chemical reaction that absorbs heat.

Enthalpy of reaction: Heat liberated or absorbed during a chemical change.

A reaction that absorbs heat is referred to as **endothermic.** Because heat is absorbed, we have included it with the reactants.

The heat change that accompanies a reaction is known as the **heat of reaction** or **enthalpy of reaction,** and the symbol ΔH is used to designate the enthalpy change. Rather than include the heat term within a reaction, we generally write the value of ΔH to the right of the reaction. If a reaction is exothermic, as in the production of water, the sign of the enthalpy is negative, showing that heat is given off by the reaction:

$$2\,H_2(g) + O_2(g) \rightarrow 2\,H_2O(\ell) \qquad \Delta H = -572\text{ kJ}$$

For an endothermic reaction, the sign of the enthalpy change is positive, showing that heat must be absorbed by the reaction:

$$6\,CO_2(g) + 6\,H_2O(\ell) \rightarrow C_6H_{12}O_6(s) + 6\,O_2(g) \qquad \Delta H = +2803\text{ kJ}$$

We may use the heat of reaction to calculate the amount of heat liberated (or absorbed) during a given reaction. For example, suppose we wish to know how much heat is liberated when 3.00 moles of hydrogen react with oxygen to form water. Equation (10-1) tells us that the reaction of 2 mol H_2 is accompanied by the production of 572 kJ of heat. Just as we are able to derive mole ratios from a balanced equation, we can derive a ratio between any one of the reactants or products and the heat of reaction. In this case, 2 mol H_2 : 572 kJ. (Although the sign of ΔH is negative, we are interested only in the *quantity* of heat given off. Thus, we have taken the absolute value of ΔH and expressed the heat term in our ratio as a positive number.) We can use this ratio to write a conversion factor for converting moles of hydrogen to heat liberated:

mol H_2 ⇨ **kJ**

$$? \text{ kJ} = 3.00 \cancel{\text{mol } H_2}\left(\frac{572 \text{ kJ}}{2.00 \cancel{\text{mol } H_2}}\right) = 858 \text{ kJ}$$

(As in the case of writing the ratio, we are interested in the *quantity* of heat given off and therefore express heat as a positive number in our calculations.)

EXAMPLE 10.17

The heat of combustion of 1 mole of methane, CH_4, is $\Delta H = -890.3$ kJ.

$$CH_4(g) + 2\,O_2(g) \rightarrow CO_2(g) + 2\,H_2O(\ell) \qquad \Delta H = -890.3 \text{ kJ}$$

Calculate the heat liberated in the combustion of 10.0 g of CH_4.

Planning the solution

Since the sign of ΔH is negative, heat is given off. Thus, we may interpret the chemical equation as follows:

$$CH_4(g) + 2\,O_2(g) \rightarrow CO_2(g) + 2\,H_2O(\ell) + 890.3 \text{ kJ}$$

As in the earlier calculations in this chapter, we can begin by converting grams of CH_4 to moles of CH_4. Then we can use the chemical equation to derive a ratio between moles of CH_4 and heat given off:

$$1 \text{ mol } CH_4 : 890.3 \text{ kJ}$$

Our overall strategy is as follows:

g CH_4 ⇨ **mol CH_4** ⇨ **kJ**

SOLUTION

$$? \text{ kJ} = 10.0 \cancel{\text{g } CH_4}\left(\frac{1 \cancel{\text{mol } CH_4}}{16.0 \cancel{\text{g } CH_4}}\right)\left(\frac{890.3 \text{ kJ}}{1 \cancel{\text{mol } CH_4}}\right) = 556 \text{ kJ}$$

PROBLEM 10.24

Identify each of the following reactions as exothermic or endothermic.

(a) $H_2(g) + I_2(s) + 51.8 \text{ kJ} \rightarrow 2\,HI(g)$

(b) $2\,H_2O_2(\ell) \rightarrow 2\,H_2O(\ell) + O_2(g) + 196.4 \text{ kJ}$

(c) $C_2H_5OH(\ell) + 3\,O_2(g) \rightarrow 2\,CO_2(g) + 3\,H_2O(\ell) \qquad \Delta H = -78.1 \text{ kJ}$

(d) $2\,CuO(s) \rightarrow 2\,Cu(s) + O_2(g) \qquad \Delta H = +310.4 \text{ kJ}$

PROBLEM 10.25

Mercury(II) oxide can be decomposed into its elements by heating:

$$2\,HgO(s) + 182 \text{ kJ} \rightarrow 2\,Hg(\ell) + O_2(g)$$

How much heat is required to decompose 50.0 g of mercury(II) oxide?

Chapter Summary

The study of chemistry is the study of matter and the changes it undergoes. One type of change is the chemical reaction. We refer to substances about to undergo change as the **reactants,** whereas the final substances resulting from the change are the **products.** In a chemical reaction, the atoms of the reactants rearrange themselves, resulting in new products.

Chemical reactions may be described by **chemical equations,** which separate the reactants and products by an arrow that indicates the direction of the transformation. The various reactants are separated from one another by plus signs, as are the products. A variety of additional symbols further describe the physical states of the reactants and products. The most important of these are the **state symbols** (s), (ℓ), (g), and (aq), which stand for solid, liquid, gaseous, and aqueous state, respectively. A solid product that separates from an aqueous reaction is known as a **precipitate.**

A **word equation** is often used as an intermediate step between a written description of a chemical reaction and a balanced chemical equation. Substituting the *chemical formulas* of the reactants and products in place of the names of the various substances turns a word equation into an **unbalanced equation,** which may be balanced by adjusting the *coefficients* of the reactants and products so that every atom that appears on the reactant side of the equation is also found on the product side (and vice versa). The coefficients in the **balanced equation** enable us to determine the **mole ratios** between any two substances present in the balanced equation. We may use these mole ratios to carry out **stoichiometric calculations.** These are chemical calculations that relate the masses of reactants and products to one another in accordance with the particular chemical equation involved.

When a reaction is carried out, it is not generally possible to isolate all of the product predicted by our stoichiometric calculations. The amount of product expected from a stoichiometric calculation represents the **theoretical yield.** The amount of product actually isolated represents the actual yield. When we express the actual yield as a percentage of the theoretical yield, we obtain the **percentage yield.**

Often, when two or more substances are allowed to react, one of the substances is totally used up, while the other remains in excess. The substance that reacts completely *limits* the extent of the reaction and is known as the **limiting reagent.**

Chemical reactions are accompanied by energy changes. The **enthalpy of reaction** (or **heat of reaction**) relates the heat liberated or absorbed in a reaction to the moles of reactants and products in the balanced chemical equation. **Exothermic** reactions are those that liberate heat, whereas **endothermic** reactions absorb heat.

Key Terms

Review each of the following terms, which are discussed in this chapter and defined in the Glossary.

chemical equation
word equation
reactant
product
unbalanced equation
balanced equation
precipitate
state symbol
mole ratio
stoichiometry
theoretical yield
percentage yield
limiting reagent
exothermic reaction
endothermic reaction
heat of reaction
enthalpy of reaction

Additional Problems

Section 10.1

10.26 Explain the meaning of the following terms: word equation, reactant, product, unbalanced equation, balanced equation. Explain what happens to the arrangement of atoms during a chemical reaction.

SECTION 10.2

10.27 Balance the following chemical equations.

(a) $H_2 + N_2 \rightarrow NH_3$
(b) $C_4H_{10} + O_2 \rightarrow CO_2 + H_2O$
(c) $Ca + H_2O \rightarrow Ca(OH)_2 + H_2$
(d) $SO_2 + O_2 \rightarrow SO_3$
(e) $C_{12}H_{22}O_{11} + O_2 \rightarrow CO_2 + H_2O$
(f) $K + H_2O \rightarrow KOH + H_2$
(g) $MgO + HNO_3 \rightarrow Mg(NO_3)_2 + H_2O$
(h) $HCl + Al(OH)_3 \rightarrow AlCl_3 + H_2O$
(i) $H_2SO_4 + Al(OH)_3 \rightarrow Al_2(SO_4)_3 + H_2O$
(j) $CCl_4 + O_2 \rightarrow COCl_2 + Cl_2$
(k) $KClO \rightarrow KCl + KClO_3$
(l) $PbO_2 + Pb + H_2SO_4 \rightarrow PbSO_4 + H_2O$

SECTION 10.3

10.28 Interpret each of the following statements as a word equation, translate the word equation into an unbalanced chemical equation, and then balance the equation. In parts (a)–(d) you are provided with the names and formulas of the reactants and products. For parts (e)–(h) you must apply a knowledge of nomenclature to write the formulas of the reactants and products.

(a) When cyclopentane, C_5H_{10}, undergoes combustion with oxygen, O_2, the products formed are carbon dioxide, CO_2, and water, H_2O.
(b) Treatment of zinc carbonate, $ZnCO_3$, with nitric acid, HNO_3, produces zinc nitrate, $Zn(NO_3)_2$, carbon dioxide, CO_2, and water, H_2O.
(c) Treatment of thionyl chloride, $SOCl_2$, with water, H_2O, produces sulfur dioxide, SO_2, and hydrogen chloride, HCl.
(d) In the process of photosynthesis, plants use carbon dioxide, CO_2, and water, H_2O, from the atmosphere to produce glucose, $C_6H_{12}O_6$, and oxygen, O_2.
(e) When phosphoric acid is neutralized with potassium hydroxide, potassium phosphate and water are produced.
(f) Aluminum metal reacts with hydrobromic acid to produce aluminum bromide. The reaction is accompanied by an evolution of hydrogen.
(g) Copper(I) sulfide is formed when copper is heated in the presence of sulfur.
(h) When potassium chlorate is allowed to react in the presence of an appropriate catalyst, potassium chloride and oxygen are produced.

SECTION 10.4

10.29 Explain the meaning of each of the following symbols, which are used in writing chemical equations.

(a) $\rightarrow$ **(b)** + **(c)** (s) **(d)** (ℓ) **(e)** (g) **(f)** (aq)

10.30 Write balanced equations for each of the following reactions, including the appropriate state symbols. [The formulas of reactants and products are provided for you in parts (a)–(c). For parts (d)–(f), you must use your knowledge of nomenclature to figure most of them out.]

(a) Zinc, Zn, is a solid that reacts with liquid bromine, Br_2, to form a new solid, zinc bromide, $ZnBr_2$.
(b) Mixing an aqueous solution of barium hydroxide, $Ba(OH)_2$, with an aqueous solution of sulfuric acid, H_2SO_4, yields a solid precipitate of barium sulfate, $BaSO_4$, and liquid water, H_2O.
(c) Aqueous strontium nitrate, $Sr(NO_3)_2$, reacts with aqueous ammonium oxalate, $(NH_4)_2C_2O_4$, to form a solid precipitate of strontium oxalate, SrC_2O_4, and an aqueous solution of ammonium nitrate, NH_4NO_3.
(d) Complete combustion of gaseous butane, C_4H_{10}, with oxygen (also a gas) produces carbon dioxide gas and liquid water.
(e) Marble is composed of calcium carbonate. Treatment of solid calcium carbonate with aqueous hydrochloric acid produces aqueous calcium chloride, carbon dioxide gas, and liquid water.
(f) Aluminum is a solid that reacts with aqueous copper(II) nitrate to form aqueous aluminum nitrate and solid copper metal.

SECTION 10.5

10.31 Nitroglycerine ($C_3H_5N_3O_9$) is a commonly used explosive. When it explodes, the products formed are carbon dioxide, water, nitrogen, and oxygen. The reaction is also accompanied by a rather large release of energy! The following balanced chemical equation represents the reaction that takes place:

$$4\,C_3H_5N_3O_9 \rightarrow 12\,CO_2 + 10\,H_2O + 6\,N_2 + O_2$$

(a) How many moles of oxygen (O_2) are produced in the explosion of 2.40 moles of nitroglycerine?
(b) How many moles of nitrogen (N_2) are produced in the explosion of 3.60 moles of nitroglycerine?
(c) How many moles of nitroglycerine must explode in order to produce 5.25 moles of water (H_2O)?
(d) How many moles of water accompany the production of 1.80 moles of carbon dioxide (CO_2)?

10.32 Ammonia (NH_3) is a flammable gas capable of undergoing combustion under the proper conditions. When it burns, ammonia reacts with oxygen (O_2) to form nitrogen (N_2) and water (H_2O).

(a) Write a balanced equation for the reaction.
(b) How many moles of O_2 are consumed in the combustion of 0.624 mol NH_3?
(c) How many moles of NH_3 are needed to produce 0.306 mol H_2O?
(d) How many moles of N_2 are produced in the combustion of 0.258 mol NH_3?
(e) How many moles of N_2 accompany the production of 0.738 mol H_2O?

10.33 Diphosphorus pentoxide reacts with water to produce phosphoric acid.

(a) Write a balanced equation for the reaction.
(b) How many moles of diphosphorus pentoxide are needed to produce 3.88 moles of phosphoric acid?
(c) How many moles of water must be used?

SECTION 10.6

10.34 Convert each of the following quantities to the desired unit.

(a) 44.0 g NaOH to moles
(b) 12.6 g $Al(NO_3)_3$ to moles
(c) 4.85 mol H_2SO_4 to grams
(d) 0.250 mol O_3 to grams

10.35 Carbon, C, and oxygen, O_2, react with one another to form carbon dioxide, CO_2.

(a) Write a balanced equation for this reaction.
(b) Calculate the mass of oxygen required to burn 24.0 g of carbon.
(c) Determine the mass of carbon dioxide formed in part (b).
(d) Use the masses you obtained in parts (b) and (c) to demonstrate the Law of Conservation of Mass.

10.36 Acetylene, C_2H_2, can be prepared by allowing calcium carbide, CaC_2, to react with water as follows:

$$CaC_2 + 2\,H_2O \rightarrow C_2H_2 + Ca(OH)_2$$

What mass of calcium carbide is needed to prepare 175 g of acetylene?

10.37 Copper reacts with concentrated nitric acid as follows:

$$3\,Cu(s) + 8\,HNO_3(aq) \rightarrow 3\,Cu(NO_3)_2(aq) + 2\,NO(g) + 4\,H_2O(\ell)$$

(a) How many grams of HNO_3 are required to consume 20.0 g Cu?
(b) What mass of NO is liberated?

10.38 Butane, C_4H_{10}, is a common fuel used by campers who backpack. Butane undergoes combustion with oxygen to produce carbon dioxide and water.

(a) Write a balanced equation for the reaction.
(b) One canister of butane for a backpacking stove contains about 196 g of butane. What mass of oxygen is required to completely burn the butane in a typical canister?
(c) What mass of carbon dioxide is produced?
(d) What mass of water is formed?
(e) Use the masses of reactants and products you calculated in parts (b), (c), and (d) to demonstrate the Law of Conservation of Mass.

10.39 Phosphoric acid can be prepared by treating diphosphorus pentoxide with water.

(a) What mass of diphosphorus pentoxide is required to prepare 75.0 g of phosphoric acid?
(b) What mass of water is needed?

10.40 Ethyl alcohol, C_2H_5OH, can be prepared by fermenting glucose, $C_6H_{12}O_6$ (also known as "grape sugar"), as follows:

$$C_6H_{12}O_6 \xrightarrow{\text{Yeast}} 2\,C_2H_5OH + 2\,CO_2$$

(a) What mass of ethyl alcohol can be produced by fermenting 2.50 moles of glucose?
(b) What mass of carbon dioxide is produced?

10.41 Calcium carbonate, $CaCO_3$, is a relatively insoluble substance frequently found as the "scale" remaining on pots and pans when hard water has been used for cooking. This scale can be removed from pots and pans by treating them with vinegar, which is an aqueous solution of acetic acid, $HC_2H_3O_2$. Calcium carbonate and acetic acid react with each other according to the following equation:

$$2\,HC_2H_3O_2 + CaCO_3 \rightarrow Ca(C_2H_3O_2)_2 + CO_2 + H_2O$$

(a) How many moles of acetic acid are required to "dissolve" 25.0 g of calcium carbonate?
(b) How many moles of calcium carbonate can be "dissolved" by 75.0 g of acetic acid?

SECTION 10.7

10.42 The reaction of salicylic acid, $C_7H_6O_3$, with excess methyl alcohol, CH_3OH, is used to make methyl salicylate, $C_8H_8O_3$, more commonly known as oil of wintergreen:

$$\underset{\text{Salicylic acid}}{C_7H_6O_3} + \underset{\text{Methyl alcohol}}{CH_3OH} \rightarrow \underset{\text{Methyl salicylate}}{C_8H_8O_3} + H_2O$$

Suppose 15.7 g of salicylic acid react in this fashion to produce 10.4 g of methyl salicylate. Calculate the theoretical yield and the percentage yield.

10.43 Magnesium hydroxide, $Mg(OH)_2$, precipitates when solutions of magnesium chloride, $MgCl_2$, and sodium hydroxide, NaOH, are mixed:

$$MgCl_2(aq) + 2\,NaOH(aq) \rightarrow Mg(OH)_2(s) + 2\,NaCl(aq)$$

Suppose a solution containing 9.50 g of $MgCl_2$ is mixed with excess sodium hydroxide, and the product is filtered and dried. The final yield isolated is 5.06 g $Mg(OH)_2$. What is the theoretical yield? The percentage yield?

10.44 Ethyl acetate, $C_4H_8O_2$, is a commonly used solvent that can be made from acetic acid, $HC_2H_3O_2$, and ethyl alcohol, C_2H_5OH:

$$\underset{\text{Acetic acid}}{HC_2H_3O_2} + \underset{\text{Ethyl alcohol}}{C_2H_5OH} \rightarrow \underset{\text{Ethyl acetate}}{C_4H_8O_2} + H_2O$$

Suppose 215 g of ethyl acetate are isolated from the reaction of 176 g of acetic acid with excess ethyl alcohol. What is the theoretical yield? The percentage yield?

SECTION 10.8

10.45 Aluminum sulfide can be prepared from the reaction of its elements. Suppose 5.75 g Al and 8.35 g S are mixed and allowed to react as follows:

$$2\,Al + 3\,S \rightarrow Al_2S_3$$

(a) Which reactant is the limiting reagent?
(b) What mass of aluminum sulfide results?

10.46 Sodium metal reacts vigorously with water to produce sodium hydroxide and hydrogen:

$$2\,Na + 2\,H_2O \rightarrow 2\,NaOH + H_2$$

(a) When 13.5 g Na react with 12.8 g H_2O, which is the limiting reagent?
(b) What mass of hydrogen is produced?

10.47 Sodium hydrogen carbonate (baking soda) is frequently used to neutralize acid spills on highways. In its reaction with sulfuric acid, carbon dioxide is one of the products:

$$2\,NaHCO_3 + H_2SO_4 \rightarrow Na_2SO_4 + 2\,H_2O + 2\,CO_2$$

Would 275 g $NaHCO_3$ provide enough sodium hydrogen carbonate to completely consume 125 g H_2SO_4? If those quantities were mixed, what mass of carbon dioxide would be produced?

SECTION 10.9

10.48 Answer the following questions about heats of reaction.

(a) What is an exothermic reaction? Give an example. What is the sign of ΔH in an exothermic reaction?
(b) What is an endothermic reaction? Give an example. What is the sign of ΔH in an endothermic reaction?

10.49 Gas cylinders containing propane are used by campers to fuel their stoves and lanterns. The heat of combustion for propane is $\Delta H = -2220$ kJ/mol:

$$C_3H_8(g) + 5\,O_2(g) \rightarrow 3\,CO_2(g) + 4\,H_2O(\ell) + 2220\text{ kJ}$$

A typical cylinder of propane contains 0.400 kg (400 g). Calculate the heat delivered during the combustion of this quantity of propane.

10.50 On heating, calcium carbonate ($CaCO_3$) decomposes to calcium oxide (CaO) and carbon dioxide (CO_2). The enthalpy of reaction is $\Delta H = +178.0$ kJ per mole:

$$CaCO_3(s) \rightarrow CaO(s) + CO_2(g) \qquad \Delta H = +178.0\text{ kJ}$$

How much heat is required to decompose 12.5 g of calcium carbonate in this fashion?

GENERAL PROBLEMS

10.51 Two solutions are prepared by dissolving 10.0 g each of sodium chloride and silver nitrate in water. When the solutions are mixed, a reaction occurs, producing silver chloride and sodium nitrate. The silver chloride forms as a precipitate, whereas the sodium nitrate remains in aqueous solution. After the precipitate of silver chloride is filtered, it is dried and weighed, yielding 7.86 g of product.

(a) Write a balanced chemical equation for the reaction, including state symbols.
(b) Determine the limiting reagent.
(c) Calculate the theoretical yield of silver chloride.
(d) Calculate the percentage yield obtained.

10.52 In an experiment similar to that described in Problem 10.51, an aqueous solution containing 8.50 g of lead(II) nitrate is mixed with one containing 6.50 g of potassium iodide. The reaction that occurs results in the precipitation of lead(II) iodide. Potassium nitrate, which is also produced, remains in aqueous solution. After filtering and drying the precipitate, 7.25 g of lead(II) iodide are obtained.

(a) Write a balanced chemical equation for the reaction, including state symbols.
(b) Determine the limiting reagent.
(c) Calculate the theoretical yield of lead(II) iodide.
(d) Calculate the percentage yield obtained.

***10.53** The preparation of isopentyl acetate (banana oil) can be accomplished via the following reaction:

$$\underset{\text{Acetic acid}}{CH_3CO_2H} + \underset{\text{Isopentyl alcohol}}{C_5H_{11}OH} \rightarrow \underset{\text{Isopentyl acetate}}{CH_3CO_2C_5H_{11}} + \underset{\text{Water}}{H_2O}$$

When the reactants are mixed in equal numbers of moles, the reaction typically gives 50–60% yields. Assuming a yield of 55%, what masses of acetic acid and isopentyl alcohol should be mixed to produce 26.0 g of isopentyl acetate?

***10.54** A *bond dissociation energy* represents the amount of energy that must be supplied to break a particular chemical bond. (See Table 8.2, Section 8.3.) It follows that the same amount of energy is given off when a bond forms. The enthalpy of a reaction, ΔH, can be calculated by: (1) summing the energy required to break the bonds that are broken, (2) summing the energy liberated by the bonds that are formed in the reaction, and then (3) subtracting the energy liberated from the energy absorbed. Use the following bond energies to calculate ΔH for the reactions described below: H—H (436 kJ/mol), Cl—Cl (243 kJ/mol), Br—Br (193 kJ/mol), H—Cl (432 kJ/mol), and H—Br (366 kJ/mol).

(a) $H_2(g) + Cl_2(g) \rightarrow 2\,HCl(g)$
(b) $2\,HBr(g) \rightarrow H_2(g) + Br_2(g)$

10.55 Without doing any calculation more complicated than computing a molecular mass, determine which of the following are not possible.

(a) The production of at least 10 g of CO_2 from the combustion of 10 g of CH_4.
(b) The production of at least 10 g of $CaCl_2$ from the reaction of 10 g of Ca with 10 g of Cl_2.
(c) The production of at least 10 g of Hg from heating 10 g of HgO.
(d) The production of at least 10 g of $Al_2(SO_4)_3$ from the reaction of 10 g of Al with 10 g of H_2SO_4.
(e) The production of at least 10 g of C_2H_4 from the dehydration (loss of H_2O) of 10 g of C_2H_5OH.

Explain your reasoning for each case where the desired result is not possible.

10.56 Answer the following question by examining the formulas of the hydrocarbons and applying some logic. (*Do not* do a stoichiometry calculation. You *may* calculate the percentage compositions of the hydrocarbons, although it is not absolutely necessary.)

Which will produce a greater mass of water, the combustion of 25 g of ethane gas (C_2H_6) or the combustion of 25 g of hexane liquid (C_6H_{14})?

Explain your reasoning.

***10.57** Without calculating anything more complicated than the percentage compositions of the hydrocarbons (and even that is not necessary), answer the following question.

Two lamps are filled with 10 g each of propane (C_3H_8) and butane (C_4H_{10}), respectively. Both are placed in identical rooms with excess oxygen. If they both produce carbon dioxide at the same rate, which lamp will burn out first?

Explain your reasoning.

WRITING EXERCISES

10.58 You are working on a rocket that is fueled by liquid hydrogen. The rocket will receive its power by burning the hydrogen with oxygen that will also be carried on board. (As you know, the product of their combustion is water.) One of the engineers has asked you to show him how to calculate the mass of oxygen required to burn each kilogram of hydrogen. Without actually performing the calculation, prepare a written explanation for the engineer, outlining all of the steps required to carry out the desired calculation.

10.59 In your own words, explain how to balance a chemical equation.

10.60 In your own words, explain the meaning of a percentage yield, and tell how to calculate this quantity.

11

The Gas Laws

Earlier we discussed the fact that matter can exist in three states: solid, liquid, and gas. We described the solid and liquid states as "condensed" phases of matter, because in these states the molecules are packed closely together.

By contrast, molecules in the gaseous state are far from one another, resulting in our inability to "see" colorless gases such as nitrogen and oxygen, the main components of air. In Chapter 2 we found that the molecules in a gaseous sample are in constant motion, causing gases to mix rapidly and to expand or contract to fill completely any container they are placed in. In this chapter, we see how this molecular behavior affects the various relationships that exist among the temperature, pressure, and volume of a gaseous sample. In our study, we also reexamine the meaning of temperature in terms of kinetic energy.

11.1 THE KINETIC–MOLECULAR THEORY AND IDEAL GASES

OBJECTIVE 11.1 Describe the kinetic theory of gases in terms of the motion of gaseous molecules, the strength of intermolecular forces, the effect of temperature on molecular velocity, and the conservation of energy when molecules collide.

Kinetic–molecular theory: A model that explains the behavior of gases in terms of particles in motion.

Ideal gas: A gas that conforms to the assumptions of the kinetic–molecular theory.

To help us explain gaseous behavior, we use a model known as the **kinetic–molecular theory.** This model, which is the accepted model of the behavior of gases, pictures each gaseous molecule as a tiny particle that behaves much like a billiard ball. The word *kinetic* refers to the motion of the molecules, which are constantly moving, flying about in all directions (Fig. 11-1). A molecule will collide with any object in its path (such as another molecule or the sides of the container). When it collides, it bounces away in another direction like a billiard ball.

FIGURE 11-1 Molecules of gas inside a container. The molecules are in constant motion, colliding with one another and with the walls of the container.

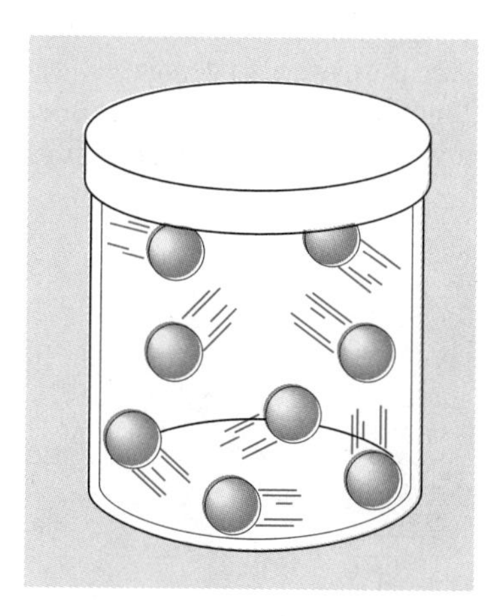

The kinetic theory makes certain assumptions that help us understand why gases behave as they do. A gas that conforms to these assumptions is known as an **ideal gas.** The assumptions we make are as follows:

1. All gases are composed of molecules.
2. The molecules are so far apart that the size of the molecules and the intermolecular forces between them are negligible. Thus, the molecules are so far apart that their size is negligible when compared to the volume the gas occupies as a whole. In addition, the forces of attraction or repulsion between molecules are so small that we can assume no such forces are present.
3. The molecules are in constant linear motion, colliding with one another and with the walls of their container.
4. Molecules move at velocities that depend on their temperature. The higher the temperature, the faster the molecules move; the lower the temperature, the slower they move.
5. All collisions are perfectly elastic. This means that there is no loss of energy when the molecules collide.

In practice, no gas is ideal. Under high pressure or at temperatures near the boiling point of the gaseous substance, our assumptions break down. Under these conditions, intermolecular forces become appreciable and the volumes of the molecules are no longer insignificant. In this chapter, we concern ourselves only with gases under ideal conditions.

PROBLEM 11.1 Why is the word *kinetic* used to describe the behavior of gases?

11.2 PRESSURE

OBJECTIVE 11.2 Define pressure. Explain how gases produce a pressure. Describe how a Torricellian barometer works. Describe how a mercury-filled U-tube can be used to measure pressure. Explain the physical meaning of a vacuum.

Pressure: Force per unit area.

When an object is placed on the floor, it exerts a pressure caused by the force of its weight on the area covered (Fig. 11-2a). The **pressure** (P) is defined as the force (F) per unit area (A):

$$P = \frac{F}{A}$$

Thus, if we double the weight without changing the area it covers, the pressure also doubles (Fig. 11-2b).

In a similar sense, whenever a gaseous molecule collides with the walls of its container, it exerts a force. The greater the number of collisions, the greater the force. Because pressure is the force per unit area, the pressure exerted by a gas is proportional to the number of collisions made with a given area in a given period of time (Fig. 11-3). Thus, when the number of collisions made by a gaseous sample increases, the pressure increases too. When the number of collisions decreases, the pressure also decreases.

Suppose we were to take a glass tube (about 1000 mm long) that is closed at one end and fill it to the top with mercury (Fig. 11-4a). Then very carefully, without letting

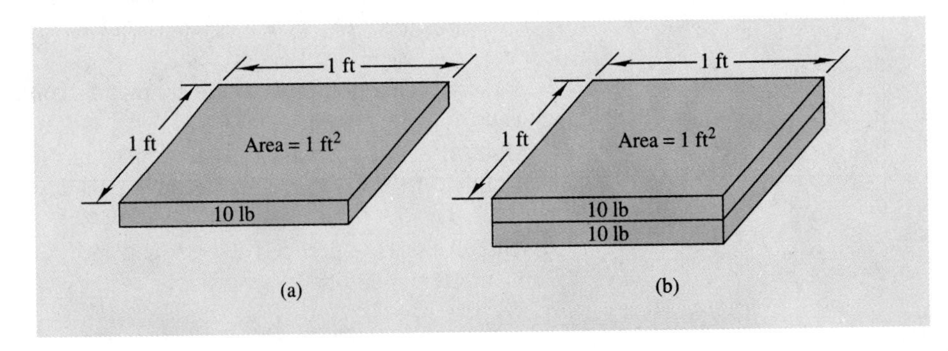

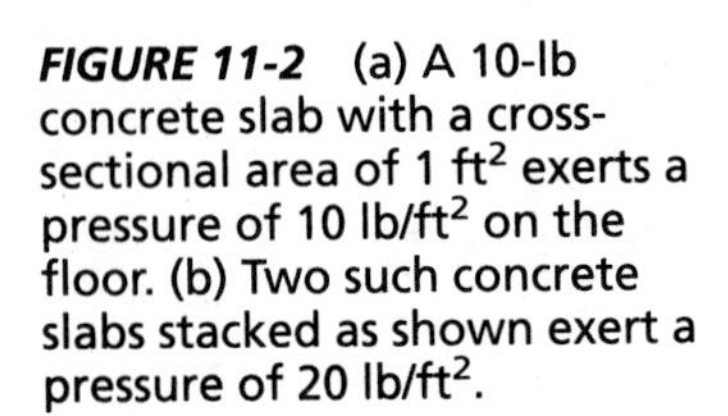
FIGURE 11-2 (a) A 10-lb concrete slab with a cross-sectional area of 1 ft^2 exerts a pressure of 10 lb/ft^2 on the floor. (b) Two such concrete slabs stacked as shown exert a pressure of 20 lb/ft^2.

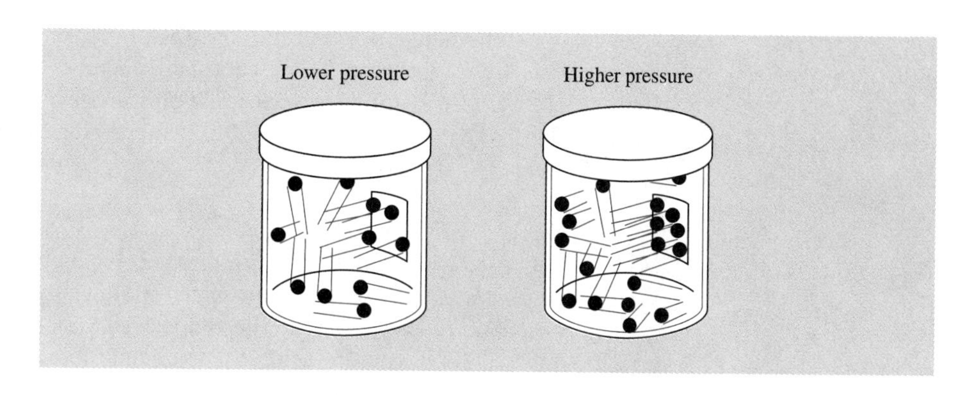

FIGURE 11-3 Gas molecules collide with the wall of a container, exerting a pressure. The greater the number of collisions, the greater the pressure.

any air in, we invert the tube and stand it in a dish of mercury (Fig. 11-4b). We would find that the mercury would not all run out of the tube. Instead, much of it would be supported above the level of mercury in the dish. Furthermore, over a period of days, the height of mercury in the tube would not fluctuate very much.

An Italian physicist, Torricelli, carried out this experiment in the seventeenth century. He concluded that the column of mercury was supported by the pressure of the air outside the tube. The surface of the mercury in the dish has a pressure acting on it from the molecules of air colliding with the mercury (Fig. 11-5, p. 290). This pressure is exactly equal to the pressure exerted by the mass of mercury above the surface inside the glass tube. (Note that there is no air in the tube to exert a pressure on the column of mercury in the tube.) Torricelli found that the height of the mercury in the 1000-mm tube was usually about 760 mm.

FIGURE 11-4 Constructing a Torricellian barometer. (a) Fill a 1000-mm tube with mercury. (b) When the tube is inverted in a pool of mercury, the column in the tube is supported above the level of the pool.

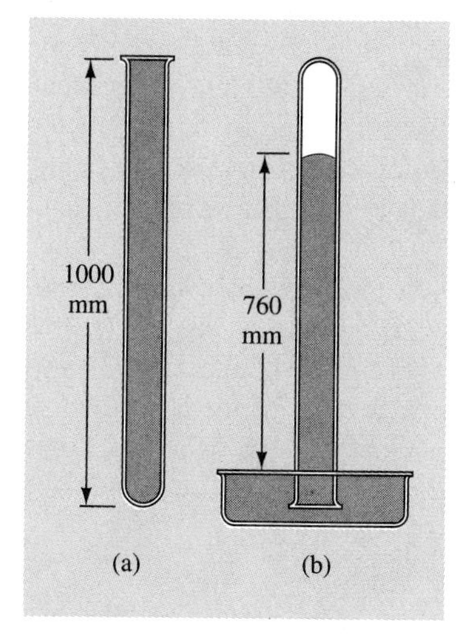

You may wonder whether increasing the diameter of the tube would change the height of the column, inasmuch as the mass of mercury would increase. However, pressure is the force per unit area. If we double the cross-sectional area of the tube, we exactly double the mass of the mercury in the column. Hence the force per unit area remains the same.

If the air pressure outside the tube drops, the height of the mercury column drops just enough so that the pressure inside the tube remains equal to that outside the tube. The higher the column of mercury being supported, the greater the atmospheric pressure must be; and the lower the column of mercury, the lower the pressure. The instrument we have described is called a **Torricellian barometer,** and it is commonly used to measure atmospheric pressure. For many years, the metric unit of pressure was the millimeter of mercury, mm Hg. However, in honor of Torricelli, a unit known as the torr was introduced to represent each millimeter of mercury. At sea level on most days, the height of a column of mercury in a Torricellian barometer is 760 mm, corresponding to a pressure of 760 torr. We refer to this pressure as **standard pressure.**

Torricellian barometer: An instrument used to measure atmospheric pressure.

Standard pressure: A defined reference standard for pressure; equal to 760 torr.

In your daily life, you are likely to encounter pressures expressed in a variety of ways. As we have said, we commonly measure air pressure by determining the height of mercury in a Torricellian barometer. At sea level this is 760 torr (or 760 mm Hg). Most TV weather reports still give the pressure in English system units: 29.9 inches. Pressure is a force per unit area, so it can also be expressed as 14.7 pounds per square inch (psi).

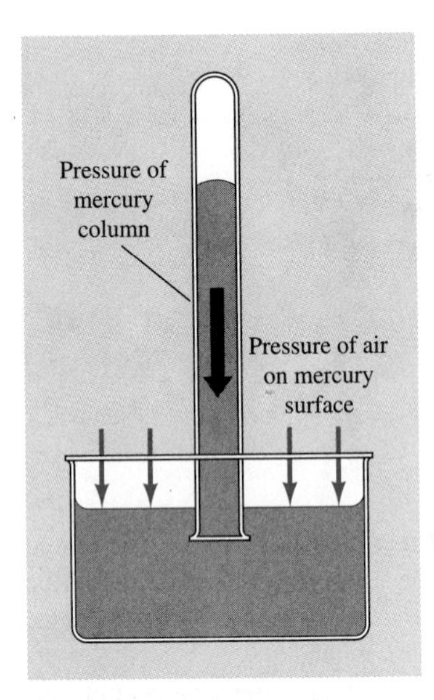

FIGURE 11-5 A Torricellian barometer. The mass of mercury in the tube exerts a pressure on the mercury in the dish. This pressure is equal to the pressure of the air on the mercury surface.

Pascal: The SI unit of pressure.

Atmosphere: A unit of pressure equal to 760 torr (standard pressure).

Vacuum: An absence of gaseous molecules.

The SI unit of pressure is the **pascal (Pa),** a unit that is most useful to the physicist. Rather than use the pascal, many chemists prefer to use either the torr or the **atmosphere (atm)** when measuring pressure. One atmosphere of pressure is equal to standard pressure: 760 torr. (The standard pressures of 1 atm and 760 torr are both exact numbers.) The pascal is related to the atmosphere and the torr as follows:

$$1.013 \times 10^5 \text{ Pa} = 1.000 \text{ atm} = 760.0 \text{ torr}$$

Because so much of our laboratory equipment is calibrated in units of torr or atmospheres, we use these units here. Those who wish may convert to SI units, using the relationships just given. The various values of standard pressure are summarized in Table 11-1. If you are using SI units, note that standard pressure is approximately 100 kPa.

If you have ever traveled in the Sierra Nevada or Rocky Mountain ranges, you may know that, as you go to higher elevations, the atmosphere becomes progressively "thinner." As evidence of this, it becomes more difficult to breathe, particularly above 7000 ft, where there is not as much oxygen available as there is at lower elevations. At these heights the atmosphere becomes less dense, and there are fewer gaseous molecules of all kinds. With fewer molecules there are fewer collisions and a corresponding decrease in atmospheric pressure.

A complete absence of gaseous molecules is known as a perfect **vacuum.** The pressure of gases results from molecular collisions, so a vacuum is always accompanied by a zero pressure. Far away from the earth's atmosphere, in outer space, a negligible number of gaseous molecules are present; hence there is a vacuum. Here on earth we can create a vacuum by evacuating (removing) all gaseous materials, including air, from a vessel (Fig. 11-6). This is generally accomplished by means of a vacuum pump, which "sucks" the gaseous molecules out of their container much as a vacuum cleaner sucks the dirt up from the floor.

In the laboratory setting, the pressure of a gas can be measured by using a combination of a barometer and a U-tube filled with mercury (as pictured in Fig. 11-7). One arm of the U-tube is connected to the gas that is being measured, while the other arm is open to the atmosphere. If the columns of mercury in both arms of the U-tube come to rest at the same height, it means that the pressure of the gas (P_{gas}) is equal to the atmospheric pressure (P_{atm}) in the laboratory (Fig. 11-7a). We can determine the

TABLE 11-1 Selected values of standard pressure

Unit of measure	Standard pressure
Atmosphere	1.00 atm
Torr	760.0 torr
Millimeters of mercury	760.0 mm Hg
Inches of mercury	29.9 in. Hg
Pounds per square inch	14.7 psi
Pascals	1.013×10^5 Pa
Kilopascals	101.3 kPa

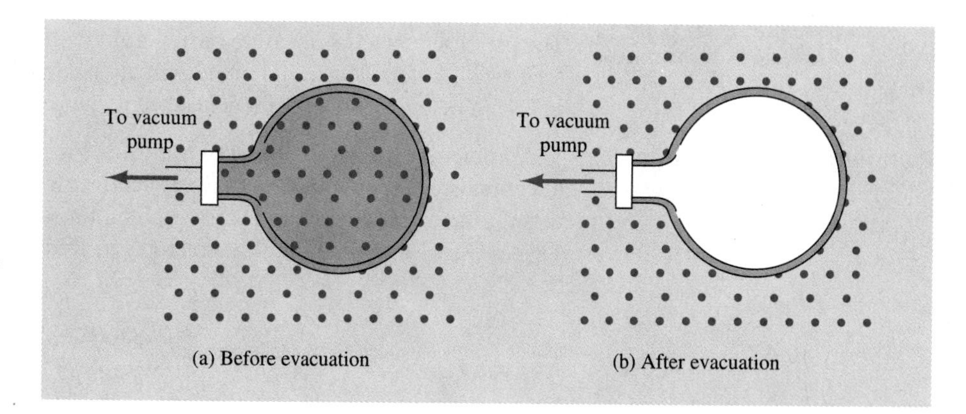

FIGURE 11-6 Creating a vacuum. (a) Before evacuation, the vessel contains gaseous molecules. (b) To create a vacuum, we remove these molecules with a vacuum pump.

atmospheric pressure by reading a barometer placed in the laboratory. If the height of the mercury column in the arm open to the atmosphere is higher than the column in the arm connected to the gas being measured, then the pressure of the gas must be greater than atmospheric pressure (Fig. 11-7b). Since a millimeter of mercury represents 1 torr of gas pressure, the pressure of the gas is determined by adding the height difference in millimeters (P_{diff}) to the atmospheric pressure measured in torr. On the other hand, if the mercury column in the open arm is lower than the column in the other arm, then the pressure of the gas is less than atmospheric pressure and is calculated by subtracting the height difference from the atmospheric pressure measured in the laboratory (Fig. 11-7c).

FIGURE 11-7 Measuring gas pressure with a mercury-filled U-tube. In (a), the gas pressure equals the external pressure (P_{atm}). In (b) and (c), the height difference (in millimeters) represents the pressure difference (P_{diff}) in torr.

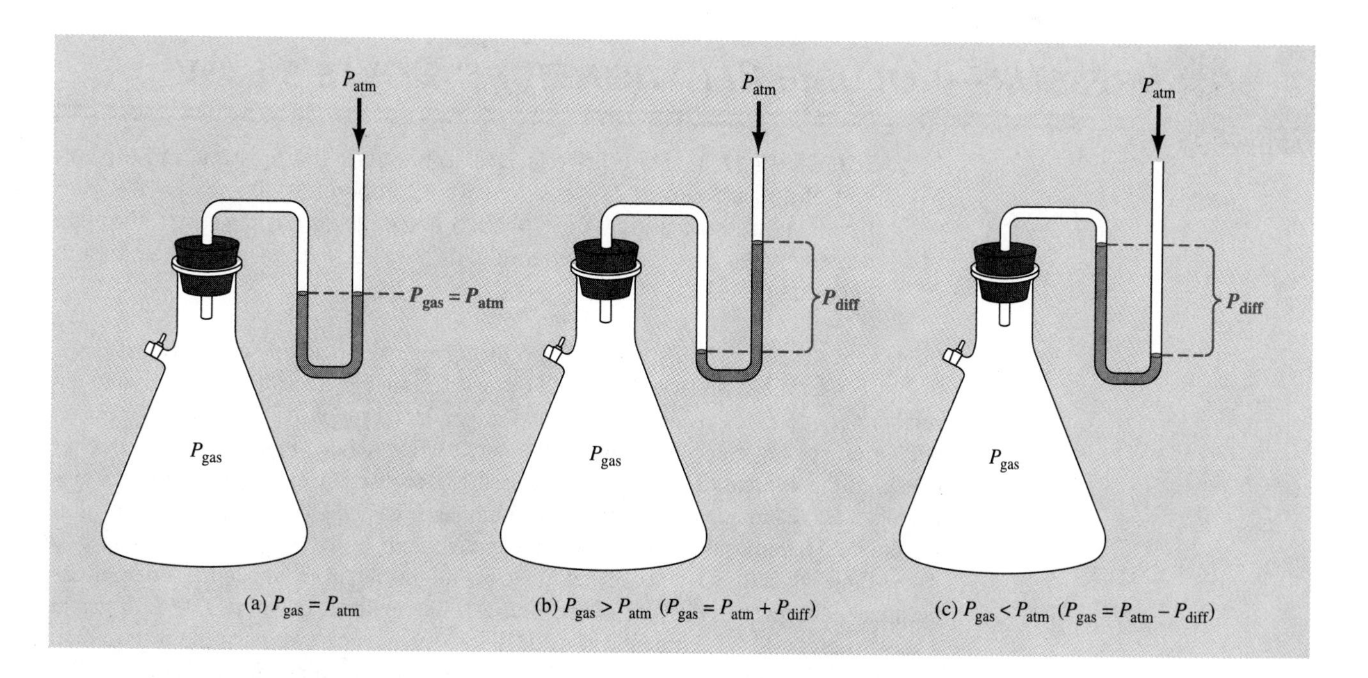

EXAMPLE 11.1

The pressure of a gas is measured using a U-tube. The atmospheric pressure in the laboratory is 743 torr. The height of the mercury column in the arm open to the atmosphere is 11 mm below that in the arm connected to the gas. What is the pressure of the gas?

Planning the solution

The problem describes the situation shown in Fig. 11-7c. Thus, the gas pressure is less than atmospheric pressure. We have to subtract from the atmospheric pressure to correct for the height difference. Since each millimeter of mercury corresponds to 1 torr, 11 mm Hg = 11 torr. We will subtract 11 torr from the atmospheric pressure.

SOLUTION

$$P_{gas} = P_{atm} - P_{diff}$$

$$P_{gas} = 743 \text{ torr} - 11 \text{ torr}$$

$$= 732 \text{ torr}$$

PROBLEM 11.2

The pressure of a gas is measured using a mercury-filled U-tube. The height of the mercury column in the open arm is 7 mm higher than the column in the other arm. The atmospheric pressure is 764 torr. What is the pressure of the gas?

PROBLEM 11.3

When swimmers dive to the bottom of a swimming pool, the pressure on their ears increases as they dive deeper.

(a) What creates the pressure on their ears?

(b) Why does the pressure increase with increasing depth?

11.3 PRESSURE–VOLUME RELATIONSHIPS: BOYLE'S LAW

OBJECTIVE 11.3 State the relationship between the pressure and the volume of a given sample of gas at constant temperature (Boyle's Law). Given the initial pressure and volume of such a sample of gas, calculate the final pressure when the volume is changed, or the final volume when the pressure is changed.

Suppose we have a cylinder with a piston that can move up and down freely, as shown in Fig. 11-8a. The cylinder is fitted with a pressure gauge that enables us to measure the pressure of any gases inside the cylinder. Further, let us suppose that gas molecules can neither escape nor enter the cylinder. We may use this device to examine the effect of changing the volume of the gas on its pressure. Because our experiments will be conducted at a fixed temperature with a fixed number of gas molecules, we will be examining only the relationship between pressure and volume.

Table 11-2 shows a set of data that we might obtain from such a series of experiments carried out with our piston. Initially, the cylinder contains 3.00 L of gas at 1.00 atm pressure (experiment 1, Fig. 11-8a). Now suppose we push the piston in so that the final volume is exactly half of the original volume (experiment 2, Fig. 11-8b).

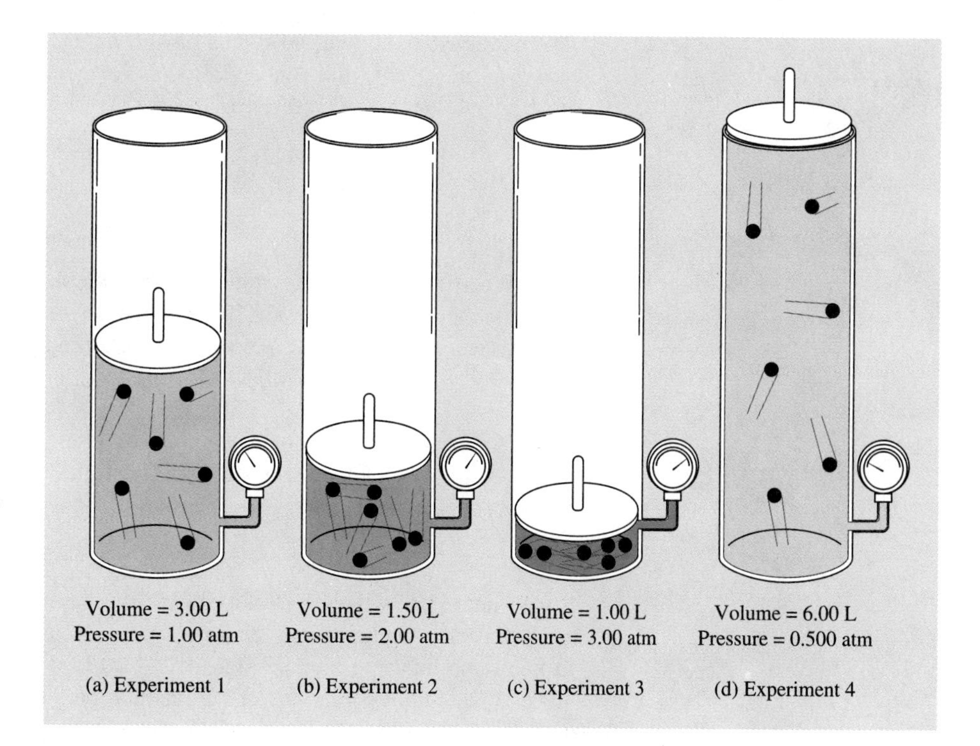

FIGURE 11-8 The relationship between pressure and volume (constant temperature and fixed number of gas molecules). As the volume decreases, the pressure increases. As the volume increases, the pressure decreases.

Being confined to this smaller volume, the molecules collide more often—both with one another and with the walls of the cylinder. In fact, there are twice as many collisions with the walls of the vessel, and the pressure doubles.

Next let us reduce the volume even further, so that it is one-third of its original volume (experiment 3, Fig. 11-8c). The molecules make three times as many collisions as they made originally, and the pressure triples.

You can see from these experiments that the volume of a gas and the pressure it exerts are inversely related. As one increases, the other decreases. As long as the temperature and number of molecules remain fixed, the product of the pressure and volume is a constant (k):

$$PV = k \qquad \textbf{Constant temperature}$$

TABLE 11-2 Variation of pressure with volume (at constant temperature)

Experiment	Volume	Pressure
1	3.00 L	1.00 atm
2	1.50 L	2.00 atm
3	1.00 L	3.00 atm
4	6.00 L	0.500 atm

Boyle's Law: $PV = k$ (constant temperature)

This relationship, known as Boyle's Law, is named after the Irish chemist who discovered it in the seventeenth century. **Boyle's Law** states that, at a fixed temperature, the pressure and volume of a sample of gas are inversely proportional.

$$\text{Experiment 1:} \quad (3.00\text{ L})(1.00\text{ atm}) = 3.00\text{ L}\cdot\text{atm}$$
$$\text{Experiment 2:} \quad (1.50\text{ L})(2.00\text{ atm}) = 3.00\text{ L}\cdot\text{atm}$$
$$\text{Experiment 3:} \quad (1.00\text{ L})(3.00\text{ atm}) = 3.00\text{ L}\cdot\text{atm}$$

When we increase the volume of gas, the molecules make fewer collisions, resulting in a decrease in pressure. Thus, when the volume of the gas is allowed to double (experiment 4, Fig. 11-8d), the pressure is reduced by half.

$$\text{Experiment 4:} \quad (6.00\text{ L})(0.500\text{ atm}) = 3.00\text{ L}\cdot\text{atm}$$

Because the product of the initial pressure (P_i) and volume (V_i) is equal to the product of the final pressure (P_f) and volume (V_f), we can write the following mathematical expression:

$$P_iV_i = P_fV_f$$

If we want to know the effect of a change in volume on the pressure of a sample of gas, we can use the mathematical relationship $P_iV_i = P_fV_f$. For example, suppose we wish to know the final pressure of a gas if a 2.00-L sample at 1.00 atm is compressed to 1.00 L. The initial conditions of volume and pressure are $V_i = 2.00$ L and $P_i = 1.00$ atm. The final volume is $V_f = 1.00$ L. Substitution into the preceding equation allows us to find the final pressure, P_f:

$$P_iV_i = P_fV_f$$
$$(1.00\text{ atm})(2.00\text{ L}) = P_f(1.00\text{ L})$$
$$\frac{(1.00\text{ atm})(2.00\not{\text{L}})}{(1.00\not{\text{L}})} = P_f$$
$$2.00\text{ atm} = P_f$$

However, a little common sense also gets us through problems like this one. Although we use the common-sense approach throughout most of the chapter, you (or your instructor) may prefer the substitution method just illustrated.

As we work the examples in this chapter, we will assume that common laboratory equipment can measure pressures to the nearest torr, volumes to the nearest milliliter, and temperatures to the nearest °C or K. Thus, we will interpret a measurement of 750 torr as 7.50×10^2 torr, 500 mL as 5.00×10^2 mL, 50°C as 5.0×10^1 °C, and 300 K as 3.00×10^3 K.

EXAMPLE 11.2

If 3.00 L of a gas exert a pressure of 0.750 atm, what volume will the gas occupy if the pressure is increased to 1.50 atm?

Planning the solution

We just learned that an inverse relationship exists between the pressure and volume, so when the pressure is increasing from 0.750 atm to 1.50 atm, the volume must be decreasing. More specifically, the pressure is being doubled, so the volume must decrease by one-half:

$$\frac{0.750 \text{ atm}}{1.50 \text{ atm}} = \frac{1}{2} \qquad \leftarrow P_{\text{ratio}} < 1$$

The final volume must equal 1.50 L, half of the initial 3.00 L. In other words, the final volume equals the initial volume times the pressure ratio just determined.

SOLUTION

$$V_f = V_i \times P_{\text{ratio}}$$

$$V_f = 3.00 \text{ L}\left(\frac{0.750 \text{ atm}}{1.50 \text{ atm}}\right) = 1.50 \text{ L}$$

Checking the solution

Since pressure is increasing, the volume must decrease. Our answer of 1.50 L is reasonable. If we had written the pressure ratio as 1.50 atm/0.750 atm = 2.00, we would know that we had made an error, because multiplication by 2 would have caused the volume to increase—and our reasoning told us that the volume should decrease.

Let us solve a few more examples using this common-sense method.

EXAMPLE 11.3

A gas occupies 50.0 mL at 2.40 atm pressure. What volume does the gas occupy at a pressure of 1.60 atm?

Planning the solution

In this case the pressure is decreasing, therefore the volume must increase. Hence, the pressure ratio will have to be a number greater than 1.

$$\text{Pressure ratio} = \frac{2.40 \text{ atm}}{1.60 \text{ atm}} = \frac{3}{2} \qquad \leftarrow P_{\text{ratio}} > 1$$

SOLUTION

$$V_f = V_i \times P_{\text{ratio}}$$

$$V_f = 50.0 \text{ mL}\left(\frac{2.40 \text{ atm}}{1.60 \text{ atm}}\right) = 75.0 \text{ mL}$$

EXAMPLE 11.4

A gas occupies 40.0 mL at a pressure of 654 torr. If the gas is compressed to 30.0 mL, what is the final pressure?

Planning the solution

In this case we wish to find the final pressure, which will equal the initial pressure times a volume ratio. The volume is decreasing, so the pressure must be increasing. We set up the volume ratio to give us a number greater than 1.

$$\text{Volume ratio} = \frac{40.0 \text{ mL}}{30.0 \text{ mL}} = \frac{4}{3} \qquad \leftarrow V_{\text{ratio}} > 1$$

SOLUTION

$$P_f = P_i \times V_{\text{ratio}}$$

$$P_f = 654 \text{ torr}\left(\frac{40.0 \text{ mL}}{30.0 \text{ mL}}\right) = 872 \text{ torr}$$

EXAMPLE 11.5

A gas occupies 50.0 mL at a pressure of 765 torr. What is the final pressure if the gas is expanded to 2.50 L?

Planning the solution

In solving this problem, we need to multiply the initial pressure by a volume ratio. Before we can write our volume ratio, we must write both volumes in terms of the same units. Because 2.50 L = 2500 mL, the volume is increasing, so the pressure must decrease. Thus, the volume ratio must be less than 1.

$$\text{Volume ratio} = \frac{50.0\ \text{mL}}{2500\ \text{mL}} = \frac{0.0500\ \text{L}}{2.50\ \text{L}} \quad \leftarrow V_{\text{ratio}} < 1$$

SOLUTION

$$P_f = P_i \times V_{\text{ratio}}$$

$$P_f = 765\ \text{torr}\left(\frac{50.0\ \text{mL}}{2500\ \text{mL}}\right) = 15.3\ \text{torr}$$

PROBLEM 11.4

Fill in the missing quantities for each of the following gaseous samples. (Assume that all changes occur at a fixed temperature.)

	Initial volume	Initial pressure	Final volume	Final pressure
(a)	30.0 mL	750.0 torr	______	600.0 torr
(b)	1.50 L	500.0 torr	1.00 L	______
(c)	45.0 mL	1.00 atm	______	760.0 torr
(d)	3.60 L	1.20 atm	______	0.800 atm
(e)	84.0 mL	700.0 torr	5.20 L	______
(f)	76.0 mL	0.700 atm	______	480.0 torr
(g)	18.2 L	146 torr	11.4 L	______

11.4 KINETIC THEORY AND ABSOLUTE TEMPERATURE

OBJECTIVE 11.4 Explain the physical meaning of absolute zero in terms of kinetic energy and motion.

Suppose we have an airtight cylinder of fixed volume filled with a gas at room temperature, and suppose that the cylinder is equipped with a device for measuring the pressure inside (Fig. 11-9a). Let us see what effect varying the temperature has on the pressure. As we increase the temperature, we find that the pressure of the gas inside the container increases (Fig. 11-9b). On the other hand, when we decrease the temperature (Fig. 11-9c), the pressure decreases. If we were to vary the temperature over a wide range and measure the pressure corresponding to each temperature, we might obtain a set of data such as that shown in Table 11-3 (p. 298).

If we plot the data obtained, we find that the points we have determined fall on a straight line, as shown in Fig. 11-10. Furthermore, if we extrapolate (or extend) this straight line so that it passes through the point corresponding to zero pressure, the line

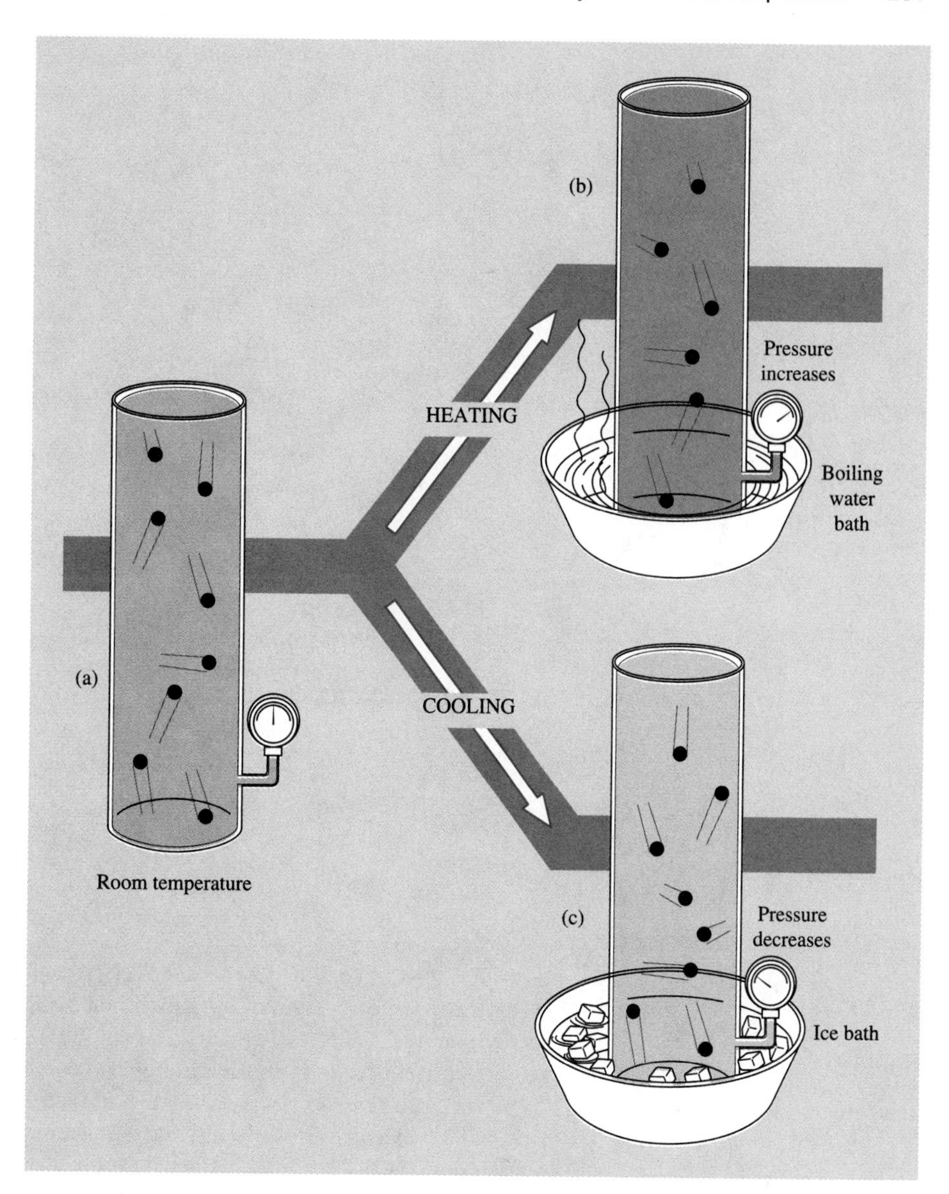

FIGURE 11-9 The effect of temperature on pressure (constant volume and fixed number of gas molecules). As the temperature increases, the pressure increases. As the temperature decreases, the pressure decreases.

intersects the temperature scale at −273°C. In Chapter 3 we referred to this point as absolute zero, and it is the basis for the Kelvin temperature scale, which we described in Section 3.10:

$$T_K = T_C + 273$$

[*Remember:* This equation is more precisely expressed as $T_K = T_C + 273.15$, but for our purposes, the simplified equation shown is adequate.] In this chapter, we will use T to represent Kelvin temperature and t to stand for Celsius temperature. Thus,

$$T = t + 273$$

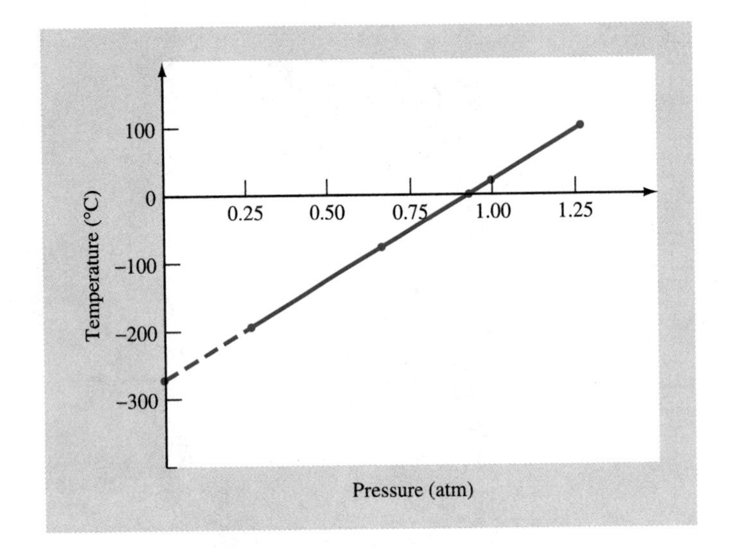

FIGURE 11-10 The relationship between Celsius temperature and pressure. If the graph is extended to zero pressure, it intersects the temperature scale at –273°C.

TABLE 11-3 Variation of pressure with temperature (at constant volume)

Temperature (°C)	Pressure (atm)
100	1.27
20	1.00
0	0.932
–78	0.666
–196	0.263

The size of each degree on the Kelvin scale is the same as that on the Celsius scale. However, each point of the Kelvin scale has a value 273 degrees higher than the corresponding Celsius temperature. The unit of Kelvin temperature is simply called the kelvin, K. (Neither the word nor symbol for degree is included in the unit.)

Absolute temperature: Kelvin temperature.

We often refer to the Kelvin scale as the **absolute temperature** scale, because zero on that scale is the lowest possible temperature. In fact, other absolute temperature scales exist, but the Kelvin scale is the absolute scale for which the size of a degree is the same as it is in the Celsius scale. Because scientists use only Celsius and Kelvin temperatures in their work, the *absolute temperature* is generally taken to mean the Kelvin temperature, and the Kelvin temperature scale is often referred to as the *absolute temperature scale.*

To convert a Celsius temperature to the Kelvin scale, we just add 273.

EXAMPLE 11.6

What is the freezing point of water (0°C) in Kelvin temperature?

SOLUTION

$$T = t + 273 = 0 + 273 = 273 \text{ K}$$

EXAMPLE 11.7

What is the boiling point of water (100°C) in Kelvin temperature?

SOLUTION

$$T = 100 + 273 = 373 \text{ K}$$

EXAMPLE 11.8

Dry ice (solid carbon dioxide) sublimes (passes directly from the solid state to the gaseous state) at –78°C. Express this temperature in kelvins.

SOLUTION

$$T = -78 + 273 = 195 \text{ K}$$

PROBLEM 11.5

Carry out each of the following conversions between the Kelvin and Celsius temperature scales.

(a) 27°C = ______ K **(b)** –40°C = ______ K **(c)** 273°C = ______ K
(d) ______ °C = 157 K **(e)** ______ °C = 415 K **(f)** ______ °C = 0 K

Let us consider the significance of the experimental results we have been describing. Every object in motion has an energy associated with it called its kinetic energy. The kinetic energy of any object can be expressed by the formula

$$\text{KE} = \frac{1}{2} mv^2$$

where m is the mass of the object and v is its velocity.

When objects cease their motion, they must have zero velocity and hence zero kinetic energy. If the objects in question are molecules of a gaseous sample, they will not collide with the walls of their container and will therefore exhibit a pressure of zero. At absolute zero, the molecules of an ideal gas would have no kinetic energy and would therefore exert no pressure. Increasing the temperature of a gas increases the kinetic energy of its molecules. This increase in kinetic energy is reflected in the increased velocities of the molecules, as demonstrated by an increase in pressure. Thus, the Kelvin or absolute temperature of a sample is really a measure of the average kinetic energy of the sample's molecules.

We have repeatedly described temperature as a measure of "average kinetic energy." As this phrase suggests, the molecules in a sample do not all have the same kinetic energy. In fact, some molecules in a sample have high kinetic energy and others have low kinetic energy. Figure 11-11 (p. 300) shows the distribution of energies in a sample at two different temperatures. Note that the high-temperature curve has a much greater proportion of molecules with high kinetic energy. For each curve, however, the temperature is a measure of the average kinetic energy of the collection of molecules represented.

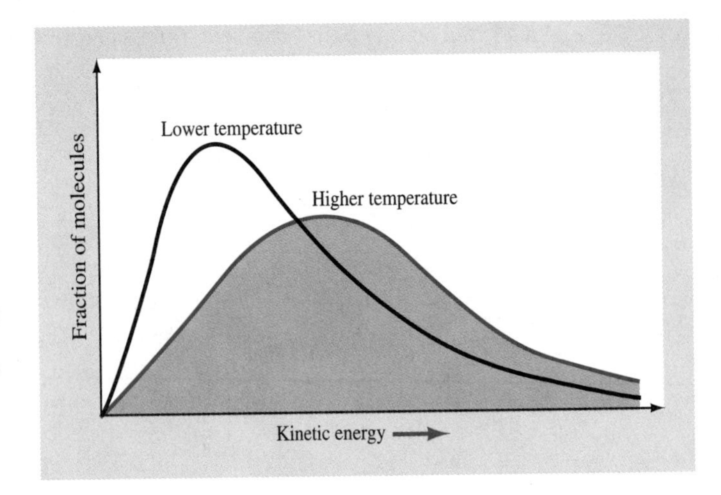

FIGURE 11-11 The distribution of kinetic energies at two different temperatures. At higher temperature there is a greater fraction of high-energy molecules. Thus, the average kinetic energy is higher.

11.5 TEMPERATURE–PRESSURE RELATIONSHIPS: GAY–LUSSAC'S LAW

OBJECTIVE 11.5 State the relationship between the pressure and the absolute temperature of a given sample of gas confined to a constant volume (Gay–Lussac's Law). Given the initial pressure and temperature of such a sample of gas, calculate the final temperature when the pressure is changed, or the final pressure when the temperature is changed.

Gay–Lussac's Law: $P = kT$ (constant volume)

If we redraw Fig. 11-10 using Kelvin instead of Celsius temperatures, we find that as the temperature doubles, so does the pressure (Fig. 11-12). As the temperature is divided in half, so is the pressure. At a fixed volume, the pressure of a gas is directly proportional to the absolute temperature. This statement, known as **Gay–Lussac's Law,** is expressed in the following mathematical relationship:

$$P = kT \qquad \textbf{Constant volume}$$

where T is the Kelvin temperature. *In all calculations involving ideal gases, we must use the Kelvin (or absolute) temperature* rather than the Celsius temperature.

If we want to examine the effect of altering the temperature on the pressure of a gas (at a constant volume), we can modify the preceding mathematical relationship, obtaining

$$\frac{P_i}{T_i} = \frac{P_f}{T_f}$$

where P_i and T_i are the initial temperature and pressure, and P_f and T_f represent the final conditions. However, we can also solve pressure-temperature problems using our common-sense ratio method.

EXAMPLE 11.9

A sample of gas at 200.0 K exerts a pressure of 275.0 torr. If the gas is heated to 400.0 K at a constant volume, what is its final pressure?

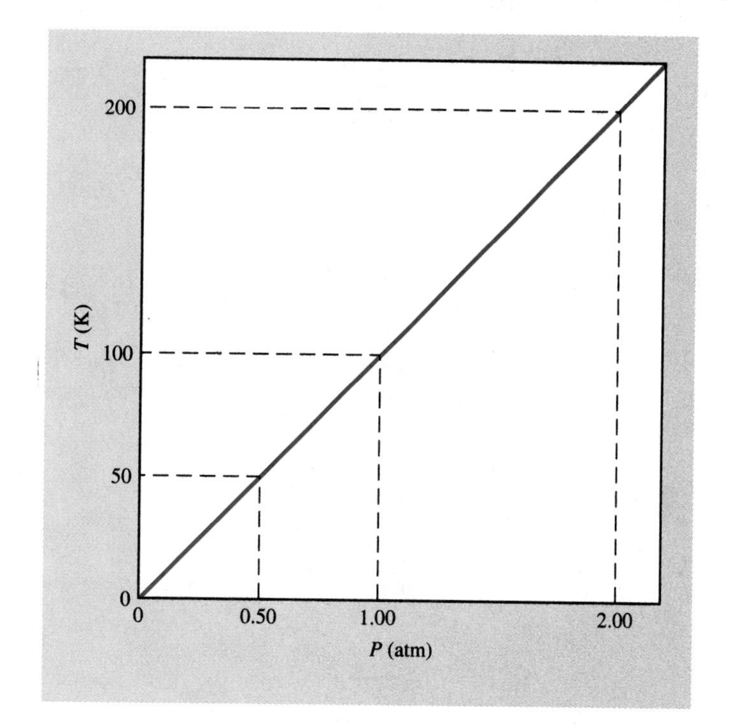

FIGURE 11-12 The relationship between absolute temperature and pressure (constant volume and fixed number of gas molecules). As the temperature doubles, so does the pressure. When the absolute temperature is divided in half, the pressure is reduced by half.

Planning the solution

The temperature of the gas is being doubled. Because the temperature and pressure are directly proportional, the pressure will double too.

$$T_{\text{ratio}} = \frac{400.0\text{ K}}{200.0\text{ K}} = 2.000$$

SOLUTION

$$P_f = P_i \times T_{\text{ratio}}$$

$$P_f = 275.0\text{ torr}\left(\frac{400.0\text{ K}}{200.0\text{ K}}\right) = 550.0\text{ torr}$$

If you cannot recall the effect of increased temperature on pressure, think about what effect increasing the temperature has on the molecules. They move faster, causing more collisions and an increase in pressure.

EXAMPLE 11.10

A sample of gas at 25°C and 755 torr is cooled to –45°C at a constant volume. Calculate the final pressure.

Planning the solution

Before we can begin this problem, we must convert the temperatures to the Kelvin scale:

$$25°\text{C} = 298\text{ K} \qquad -45°\text{C} = 228\text{ K}$$

Since the temperature is decreasing, the molecules will move more slowly, making fewer collisions, and the pressure will decrease. Thus, the final pressure will be equal to the initial pressure times a temperature ratio that is less than 1.

$$T_{\text{ratio}} = \frac{228\text{ K}}{298\text{ K}}$$

SOLUTION

$$P_f = P_i \times T_{\text{ratio}}$$

$$P_f = 755\text{ torr}\left(\frac{228\text{ K}}{298\text{ K}}\right) = 578\text{ torr}$$

If we had written the temperature ratio as 298 K/228 K, we would know that we had made an error because multiplying by this ratio would lead to an increase in pressure.

EXAMPLE 11.11

A sample of gas at 55°C exerts a pressure of 756 torr. To what final temperature must the gas be cooled if it is to exert a pressure of 378 torr at the same volume?

Planning the solution

Once again, we must convert the Celsius temperature to the Kelvin scale:

$$55°\text{C} = 328\text{ K}$$

The final temperature will equal the initial temperature times a pressure ratio. To decrease the pressure, there must be a decrease in the number of collisions. Thus, the temperature must be decreased. The final temperature will equal the initial temperature times a pressure ratio less than 1.

$$P_{\text{ratio}} = \frac{378\text{ torr}}{756\text{ torr}}$$

SOLUTION

$$T_f = T_i \times P_{\text{ratio}}$$

$$T_f = 328\text{ K}\left(\frac{378\text{ torr}}{756\text{ torr}}\right) = 164\text{ K} = -109°\text{C}$$

PROBLEM 11.6

Fill in the missing quantities for each of the following ideal gases. (Assume that all changes occur at a fixed volume.)

	Initial pressure	Initial temperature	Final pressure	Final temperature
(a)	205 torr	212 K	______	636 K
(b)	1.40 atm	324 K	______	216 K
(c)	575 torr	27°C	______	127°C
(d)	2.00 atm	57°C	3.00 atm	______°C
(e)	0.873 atm	25°C	1.00 atm	______°C
(f)	645 torr	100°C	______	0°C
(g)	555 torr	−78°C	1.00 atm	______°C

11.6 VOLUME–TEMPERATURE RELATIONSHIPS: CHARLES'S LAW

OBJECTIVE 11.6 State the relationship between the volume and the absolute temperature of a given sample of gas maintained at a constant pressure (Charles's Law). Given the initial volume and temperature of such a sample of gas, calculate the final volume when the temperature is changed, or the final temperature required to achieve a desired final volume.

Now let us take our gas-filled cylinder and place a fixed weight on top of it, thereby establishing a constant pressure inside the cylinder (Fig. 11-13a, p. 304). Because the piston moves freely, the gas inside the cylinder can expand or contract at a constant pressure. Let us use the cylinder to examine the effect of temperature on the volume of a gas at constant pressure. As we heat the gas, we find that it expands (Fig. 11-13b), whereas cooling causes it to contract (Fig. 11-13c). If we are using Kelvin temperatures, we find that doubling the temperature doubles the volume of the gas and decreasing the temperature by half cuts the volume of the gas in half. We can express these facts in the following statement, which is known as **Charles's Law:** At constant pressure the volume of a gas is directly proportional to its absolute temperature.

Charles's Law: $V = kT$ (constant pressure)

$$V = kT \qquad \textbf{Constant pressure}$$

Again, let us see how this result fits our model of an ideal gas. If we increase the temperature of a gas, the average kinetic energy of the molecules must increase, thereby increasing their velocities. The molecules are moving faster, so they collide more often with the walls of the cylinder—unless we compensate for these increased velocities by allowing the gas to expand within the cylinder. [*Remember:* The piston is movable, permitting expansion and contraction of the gas at a fixed pressure.] Thus, at constant pressure, increasing the temperature causes the gas to expand. Reversing our logic, we can see that cooling a gas causes it to contract.

We can express Charles's Law in the following mathematical relationship:

$$\frac{V_i}{T_i} = \frac{V_f}{T_f}$$

where V_i and T_i are the initial conditions of volume and temperature, and V_f and T_f represent the final conditions.

EXAMPLE 11.12

A 60.0-mL sample of gas is at a temperature of 27°C. What is the volume of the gas at 127°C if there is no change in pressure?

Planning the solution

First we must convert our temperatures to the Kelvin scale:

$$27°\text{C} = 300\text{ K} \qquad 127°\text{C} = 400\text{ K}$$

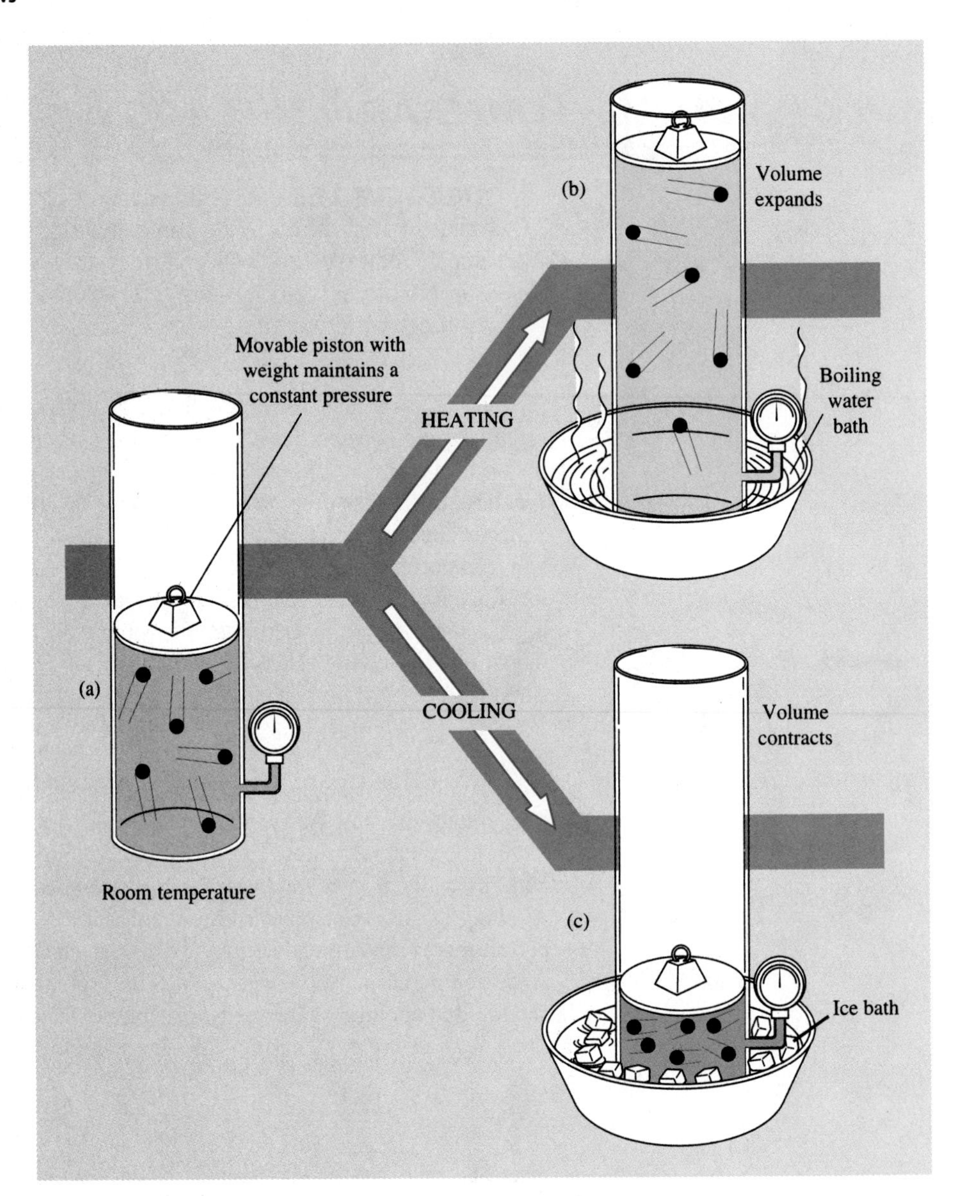

FIGURE 11-13 The effect of temperature on volume (constant pressure and fixed number of gas molecules). As the temperature increases, the volume increases. As the temperature decreases, the volume decreases.

The final volume will equal the initial volume times a temperature ratio. The temperature is increasing, so the volume will increase. Hence, the temperature ratio is greater than 1.

$$T_{\text{ratio}} = \frac{400\text{ K}}{300\text{ K}}$$

SOLUTION

$$V_f = V_i \times T_{\text{ratio}}$$

$$V_f = 60.0\text{ mL}\left(\frac{400\text{ K}}{300\text{ K}}\right) = 80.0\text{ mL}$$

EXAMPLE 11.13

Suppose we wish to reduce the volume of 50.0 mL of a gas at 27°C to 40.0 mL by adjusting the temperature, with no change in pressure. What final temperature is required to achieve the desired volume?

Planning the solution

The initial temperature, T_i, is 27 + 273 = 300 K. In order to decrease the volume of the gas at a constant pressure, it is necessary to cool it. Thus, the temperature must decrease. The final temperature will equal the initial temperature times a volume ratio less than 1.

$$V_{\text{ratio}} = \frac{40.0 \text{ mL}}{50.0 \text{ mL}}$$

SOLUTION

$$T_f = T_i \times V_{\text{ratio}}$$

$$T_f = 300 \text{ K}\left(\frac{40.0 \not{\text{mL}}}{50.0 \not{\text{mL}}}\right) = 240 \text{ K} = -33°\text{C}$$

PROBLEM 11.7

Fill in the missing quantities for each of the following gaseous substances. (In each case, assume that the pressure remains constant.)

	Initial volume	Initial temperature	Final volume	Final temperature
(a)	50.0 mL	27°C	_____	127°C
(b)	3.50 mL	100°C	_____	0°C
(c)	0.400 L	200 K	0.700 L	_____ K
(d)	68.0 mL	25°C	1.02 L	_____ K
(e)	0.250 L	273 K	_____	273°C

11.7 STANDARD TEMPERATURE AND PRESSURE (STP)

OBJECTIVE 11.7 State the specific conditions of standard temperature and pressure (STP).

Standard temperature and pressure (STP): Reference standards of temperature and pressure: 1 atm and 0°C.

In our work with gases, we often find it useful to define a set of conditions known as standard temperature and standard pressure. We have already defined standard pressure as exactly 1 atm pressure. Standard temperature is arbitrarily set at 0°C, or 273 K. When a gas is at **standard temperature and pressure, STP,** it has a temperature of 0°C and a pressure of 1.00 atm (760.0 torr). You will see that if we know the volume of a gas under a given set of conditions of temperature and pressure, we can calculate the volume of that gas at STP.

In working gas law problems, remember that these arbitrary assignments are exact. Thus, a standard pressure of 1 atm may have as many significant figures as we need: 760.000. . . torr or 1.000. . . atm. Similarly, standard temperature is exactly 0.000. . . °C.

11.8 THE COMBINED GAS LAW

OBJECTIVE 11.8 Write a mathematical expression for the Combined Gas Law. Given an initial set of conditions of pressure, volume, and temperature and any two of the final conditions, calculate the third final condition.

Combined Gas Law: $PV/T = k$

The three relationships we have discussed can be combined into the **Combined Gas Law,** which describes the effect of varying the pressure, volume, and temperature of a fixed sample of gas (a constant number of moles). The Combined Gas Law can be expressed as follows:

$$\frac{PV}{T} = k$$

At a fixed temperature, T is a constant, and the Combined Gas Law reduces to Boyle's Law:

$$PV = kT = k' \qquad \text{Constant temperature}$$

At a fixed volume, V is a constant, and we observe Gay–Lussac's Law:

$$\frac{P}{T} = \frac{k}{V} = k'' \qquad \text{Constant volume}$$

At a fixed pressure, P is a constant, and Charles's Law holds:

$$\frac{V}{T} = \frac{k}{P} = k''' \qquad \text{Constant pressure}$$

Thus, the Combined Gas Law is consistent with the three laws we have already learned. In practice, it is often necessary to change more than one condition at a time. For example, we might observe the effect on the volume of a gas when we change the temperature and the pressure. Or we might observe the effect on the pressure of a gas when we change the temperature and the volume. Because the quantity PV/T is equal to a constant, the following relationship holds:

$$\frac{P_iV_i}{T_i} = \frac{P_fV_f}{T_f}$$

where P_i, V_i, T_i, P_f, V_f, and T_f are the initial and final conditions of pressure, volume, and temperature. We can treat these problems in exactly the same manner as our previous problems. When two variables are changed, however, we will have two *independent* variables in our solution. Let us solve a few examples to illustrate this point.

EXAMPLE 11.14

A 30.0-mL sample of gas is measured at 35°C and 745 torr. What is its volume at STP?

Planning the solution

Our problems will be slightly more complicated now, so it will be useful for us to set up a chart like the one that follows. Doing so will help us keep track of the initial and final conditions.

	t (°C)	T (K)	P (torr)	V (mL)
Initial conditions	35	308	745	30.0
Final conditions	0	273	760	V_f

The unknown in this problem is the final volume. The final volume must equal the initial volume times a temperature ratio times a pressure ratio:

$$V_f = V_i \times T_{\text{ratio}} \times P_{\text{ratio}}$$

The temperature is decreasing. This will cause a decrease in the volume, so the temperature ratio must be less than 1.

$$T_{\text{ratio}} = \frac{273\text{ K}}{308\text{ K}}$$

The pressure is increasing. This will cause a further decrease in the volume, so the pressure ratio is also less than 1.

$$P_{\text{ratio}} = \frac{745\text{ torr}}{760\text{ torr}}$$

SOLUTION

$$V_f = 30.0\text{ mL}\left(\frac{273\,\cancel{\text{K}}}{308\,\cancel{\text{K}}}\right)\left(\frac{745\,\cancel{\text{torr}}}{760\,\cancel{\text{torr}}}\right) = 26.1\text{ mL}$$

Checking the answer

Since both ratios are less than 1, it is clear that the final volume must be less than the initial volume. Our calculation agrees with this prediction.

EXAMPLE 11.15

If 0.400 L of a gas at STP is warmed to 100°C and its pressure is increased to 785 torr, what is its final volume?

Planning the solution

	t (°C)	T (K)	P (torr)	V (L)
Initial conditions	0	273	760	0.400
Final conditions	100	373	785	V_f

In this case the temperature is increasing. This will cause the volume to increase, requiring a temperature ratio greater than 1.

$$T_{\text{ratio}} = \frac{373\text{ K}}{273\text{ K}}$$

The pressure is increasing, causing a decrease in the volume. The pressure ratio must be less than 1.

$$P_{\text{ratio}} = \frac{760\text{ torr}}{785\text{ torr}}$$

SOLUTION

$$V_f = V_i \times T_{\text{ratio}} \times P_{\text{ratio}}$$

$$V_f = 0.400\text{ L}\left(\frac{373\,\cancel{\text{K}}}{273\,\cancel{\text{K}}}\right)\left(\frac{760\,\cancel{\text{torr}}}{785\,\cancel{\text{torr}}}\right) = 0.529\text{ L}$$

Checking the answer

Note that in this example, one of the ratios causes an increase in the volume while the other causes a decrease. Without actually carrying out the final calculation, we cannot predict whether the net result of these opposite effects will be an overall increase or decrease. In this case, their combined effects result in a net increase.

EXAMPLE 11.16

What final pressure is necessary to compress 5.00 L of a gas at 100°C and 1.15 atm to 275 mL at –20°C?

Planning the solution

Since the volumes are given in different units, we must first convert one of them to match the units of the other. We will convert 275 mL to 0.275 L.

	t (°C)	T (K)	P (atm)	V (L)
Initial conditions	100	373	1.15	5.00
Final conditions	–20	253	P_f	0.275

In this case we wish to find the final pressure of the gas. The final pressure will equal the initial pressure times a temperature ratio times a volume ratio.

$$P_f = P_i \times T_{\text{ratio}} \times V_{\text{ratio}}$$

The decrease in temperature will cause a decrease in pressure.

$$T_{\text{ratio}} = \frac{253\ \text{K}}{373\ \text{K}}$$

Decreasing the volume will cause the pressure to increase.

$$V_{\text{ratio}} = \frac{5.00\ \text{L}}{0.275\ \text{L}}$$

SOLUTION

$$P_f = 1.15\ \text{atm}\left(\frac{253\ \cancel{\text{K}}}{373\ \cancel{\text{K}}}\right)\left(\frac{5.00\ \cancel{\text{L}}}{0.275\ \cancel{\text{L}}}\right) = 14.2\ \text{atm}$$

EXAMPLE 11.17

A 70.0-mL sample of gas is initially at 27°C and 1.25 atm pressure. The conditions are adjusted so that the final volume and pressure are 35.0 mL and 1.50 atm, respectively. What must be the final absolute temperature?

Planning the solution

	t (°C)	T (K)	P (atm)	V (mL)
Initial conditions	27	300	1.25	70.0
Final conditions	—	T_f	1.50	35.0

The final temperature will equal the initial absolute temperature times a volume ratio times a pressure ratio. In order for the volume to decrease, the temperature must also decrease, so the volume ratio must be less than 1.

$$V_{\text{ratio}} = \frac{35.0\ \text{mL}}{70.0\ \text{mL}}$$

To increase the pressure, the temperature must increase, so the pressure ratio must be greater than 1.

$$P_{\text{ratio}} = \frac{1.50 \text{ atm}}{1.25 \text{ atm}}$$

SOLUTION

$$T_f = T_i \times V_{\text{ratio}} \times P_{\text{ratio}}$$

$$T_f = 300 \text{ K}\left(\frac{35.0 \text{ mL}}{70.0 \text{ mL}}\right)\left(\frac{1.50 \text{ atm}}{1.25 \text{ atm}}\right) = 180 \text{ K}$$

For those who prefer the substitution method for solving problems, the Combined Gas Law can be readily applied to any of the problems in this section. Thus, Example 11.17 can be solved by substituting the initial and final conditions of volume, temperature, and pressure into the equation given at the beginning of this section:

$$\frac{P_iV_i}{T_i} = \frac{P_fV_f}{T_f}$$

$$\frac{(1.25 \text{ atm})(70.0 \text{ mL})}{(300 \text{ K})} = \frac{(1.50 \text{ atm})(35.0 \text{ mL})}{T_f}$$

To solve for T_f, we multiply both sides of the equation by T_f and also by the reciprocal of the left side of the equation:

$$T_f = \frac{(1.50 \text{ atm})(35.0 \text{ mL})(300 \text{ K})}{(1.25 \text{ atm})(70.0 \text{ mL})} = 180 \text{ K}$$

Examination of the final setup in Example 11.17 reveals that both methods lead to the same calculation.

PROBLEM 11.8

Fill in the missing quantities for each of the following ideal gases.

	Initial volume	Initial temperature	Initial pressure	Final volume	Final temperature	Final pressure
(a)	75.0 mL	25°C	732 torr	______	0°C	946 torr
(b)	55.0 mL	37°C	785 torr	60.0 mL	______°C	715 torr
(c)	4.20 L	415 K	1.00 atm	5.00 L	–73°C	______
(d)	17.5 L	111°C	1.40 atm	______	273 K	696 torr
(e)	45.0 mL	–78°C	215 torr	0.900 L	______ K	2.50 atm
(f)	355 mL	273°C	463 torr	0.500 L	–100°C	______

PROBLEM 11.9

A gas measured at 1.23 atm and 57°C has a volume of 0.400 L. What volume will it occupy at STP?

PROBLEM 11.10

A gas measured at 127°C and 695 torr has a volume of 72.0 mL. What volume will it occupy at STP?

PROBLEM 11.11

A gas at 1.00 atm occupies 6.00 L at 77°C. What will be the final pressure if the gas is compressed to 455 mL and cooled to −10°C?

PROBLEM 11.12

The temperature of 40.0 mL of gas is 22°C and its pressure is 666 torr. What final temperature is needed to bring the volume to 50.0 mL at a pressure of 1.00 atm?

11.9 MOLAR VOLUME: AVOGADRO'S LAW

OBJECTIVE 11.9 State Avogadro's Law. State the molar volume of a gas at STP. Given the volume, temperature, and pressure of a sample of gas, calculate the number of moles present by first determining the volume of the gas at STP.

Avogadro's Law: Equal volumes of gases under identical conditions of temperature and pressure contain the same number of molecules.

In the early 1800s, an Italian chemist named Amedeo Avogadro proposed that equal volumes of gases under identical conditions of temperature and pressure also contain equal numbers of molecules. This statement, known as **Avogadro's Law,** holds true for any two samples of gas, regardless of whether the gases are the same or different. And, because equal numbers of molecules represent equal numbers of moles, the equal volumes described must also contain equal numbers of moles.

Since equal volumes of gases contain equal numbers of moles (at the same temperature and pressure), it is apparent that twice the volume of a gas must contain twice the number of moles, or that half the volume must contain half as many moles. In other words, it follows from Avogadro's Law that the volume of a gas is proportional to the number of moles present:

$$V = kn \qquad \textbf{Constant temperature and pressure}$$

where n represents the number of moles.

Molar volume: The volume occupied by 1 mole of any gas. At STP, 1 mole occupies 22.4 L.

It also follows from Avogadro's Law that, if the conditions of temperature and pressure are the same, 1 mole of every gas occupies the same volume as 1 mole of every other gas. This volume is known as the **molar volume;** at standard temperature and pressure (STP), it has a value of 22.4 L. This statement holds true for every gas, regardless of its molecular formula. Therefore, if we know the volume of a gas at STP, we can use the relationship 1 mol = 22.4 L to calculate the number of moles. For example, 5.60 L of a gas at STP contains 0.250 mole:

$$? \text{ mol} = 5.60\cancel{\text{L}}\left(\frac{1 \text{ mol}}{22.4\cancel{\text{L}}}\right) = 0.250 \text{ mol}$$

In Section 11.8 we found that we could calculate the volume of a gas at STP if its volume, temperature, and pressure were known (see Example 11.14). Therefore, if we know the volume of a gas and its temperature and pressure, we can determine the number of moles present by employing the following general strategy:

V (at any T and P) ⇨ **V (at STP)** ⇨ **moles**

Our next example illustrates this strategy.

EXAMPLE 11.18

A 1.50-L sample of gas at 25°C exerts a pressure of 0.874 atm. How many moles are present?

Planning the solution

We will find the volume this gas occupies at STP and then use the molar volume of 22.4 L/mol to determine the number of moles.

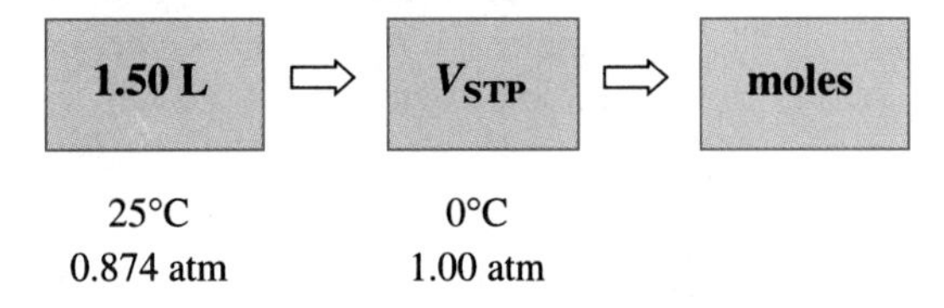

SOLUTION

We must find the volume the gas occupies at STP.

	t (°C)	T (K)	P (atm)	V (L)
Initial conditions	25	298	0.874	1.50
Final conditions	0	273	1.00	V_{STP}

The volume the gas occupies at STP is

$$V_{STP} = 1.50\ \text{L}\left(\frac{273\ \text{K}}{298\ \text{K}}\right)\left(\frac{0.874\ \text{atm}}{1.00\ \text{atm}}\right) = 1.20\ \text{L}$$

Next we make use of the fact that 1 mole of gas occupies 22.4 L at STP,

$$1\ \text{mol} = 22.4\ \text{L}$$

This conversion factor should be used only if the gas is at STP. Under other conditions of temperature and pressure, the volume of 1 mole of a gas usually differs from 22.4 L.

$$?\ \text{mol} = 1.20\ \text{L}\left(\frac{1\ \text{mol}}{22.4\ \text{L}}\right)$$
$$= 0.0536\ \text{mol}$$

PROBLEM 11.13

Calculate the number of moles in each of the following.

(a) 5.60 L of a gas at 0°C and 1.00 atm pressure
(b) 3.20 L of a gas at 25°C and 760.0 torr
(c) 45.0 mL of a gas at –45°C and 745 torr
(d) 5.00 L of a gas at 243°C and 855 torr
(e) 18.0 mL of a gas at –183°C and 17.6 atm

PROBLEM 11.14

Calculate the number of molecules in 1.00 mL of a gas at 1000.0 K and 1.00 torr. [*Remember:* 1 mol contains 6.02×10^{23} molecules.]

11.10 THE IDEAL GAS LAW

OBJECTIVE 11.10 Write a mathematical expression for the Ideal Gas Law. Given any three of the following four variables—moles, volume, temperature, and pressure—calculate the unknown variable.

In Section 11.8 we presented the Combined Gas Law, which describes the relationship among the pressure, volume, and temperature of a fixed sample (constant number of moles) of gas:

$$\frac{PV}{T} = k \qquad \text{or} \qquad PV = kT$$

Then in Section 11.9 we learned that the volume of a gas is proportional to the number of moles present (assuming a constant temperature and pressure):

$$V = kn$$

We can combine these two relationships to give the following expression that relates the number of moles of gas (n) to its volume (V) temperature (T), and pressure (P):

$$PV = nkT$$

Chemists traditionally use the symbol R to stand for the constant in the preceding equation:

$$PV = nRT$$

Ideal Gas Law: $PV = nRT$

Universal gas constant, *R*: Constant used in the Ideal Gas Law: $R = 0.0821\ \text{L}\cdot\text{atm/mol}\cdot\text{K}$

The constant R is called the **universal gas constant,** and the equation $PV = nRT$ the **Ideal Gas Law.** Though R is a constant, its numerical value varies with the particular units selected for pressure and volume. (Temperature is always expressed in kelvins, and n must be the number of moles.) Table 11-4 gives the most common values of R, indicating which units of volume and pressure correspond to each value. Although several values of R exist, we can learn one,

$$\frac{0.0821\ \text{L}\cdot\text{atm}}{\text{mol}\cdot\text{K}}$$

and convert its units of liters or atmospheres to milliliters or torr as needed. For example, to convert 0.0821 L•atm/mol•K to a value of R with units of L•torr/mol•K, we convert atmospheres to torr:

$$\frac{0.0821\ \text{L}\cdot\text{atm}}{\text{mol}\cdot\text{K}}\left(\frac{760\ \text{torr}}{1\ \text{atm}}\right) = \frac{62.4\ \text{L}\cdot\text{torr}}{\text{mol}\cdot\text{K}}$$

Units of liters can be converted to milliliters in a similar fashion. The Summary lists the various gas laws.

Perhaps the most common application of the Ideal Gas Law is to solve problems where three of the four variables (P, V, n, and T) are known and the fourth variable is the desired unknown. Although it is possible to solve such problems using the problem-solving methods illustrated thus far, some problems are more easily solved by substitution into the Ideal Gas Law, as shown in the following examples.

***TABLE 11-4 The universal gas constant,* R**

Volume units	Pressure units	*R*
Liters	Atmospheres	$\frac{0.0821\ L \cdot atm}{mol \cdot K}$
Liters	Torr	$\frac{62.4\ L \cdot torr}{mol \cdot K}$
Milliliters	Atmospheres	$\frac{82.1\ mL \cdot atm}{mol \cdot K}$
Milliliters	Torr	$\frac{6.24 \times 10^4\ mL \cdot torr}{mol \cdot K}$

SUMMARY *The Gas Laws*

Pressure *(P)*	Volume *(V)*	Number of moles *(n)*	Temperature *(T)*
Boyle's Law *($PV = k$):*			
$P = \frac{k}{V}$	$V = \frac{k}{P}$	Constant	Constant
Gay–Lussac's Law *($P = kT$):*			
$P = kT$	Constant	Constant	$T = \frac{P}{k}$
Charles's Law *($V = kT$):*			
Constant	$V = kT$	Constant	$T = \frac{V}{k}$
Combined Gas Law *($PV/T = k$):*			
$P = \frac{kT}{V}$	$V = \frac{kT}{P}$	Constant	$T = \frac{PV}{k}$
Avogadro's Law *($V = kn$):*			
Constant	$V = kn$	$n = \frac{V}{k}$	Constant
Ideal Gas Law *($PV = nRT$):*			
$P = \frac{nRT}{V}$	$V = \frac{nRT}{P}$	$n = \frac{PV}{RT}$	$T = \frac{PV}{nR}$

EXAMPLE 11.19

A 2.00-L gas cylinder contains 1.25 moles of helium, He. What pressure does the gas exert when the contents of the cylinder are at 21°C?

Planning the solution

As in all gas law problems, we need to convert the temperature to kelvins: 21°C = 294 K. Since the volume is given in liters and we are calculating the pressure, we may select either $R = 0.0821$ L•atm/mol•K or $R = 62.4$ L•torr/mol•K. The former value gives pressure units in atmospheres; the latter gives an answer in torr. Let us use the former value.

SOLUTION

$$PV = nRT$$

$$P(2.00\text{ L}) = 1.25\text{ mol}\left(\frac{0.0821\text{ L}\cdot\text{atm}}{\text{mol}\cdot\text{K}}\right)294\text{ K}$$

$$P = \left(\frac{1.25\text{ mol}}{2.00\text{ L}}\right)\left(\frac{0.0821\text{ L}\cdot\text{atm}}{\text{mol}\cdot\text{K}}\right)294\text{ K}$$

$$= 15.1\text{ atm}$$

Note that the units of moles, liters, and kelvins all cancel, leaving an answer with units of atmospheres, an appropriate unit for the unknown pressure.

EXAMPLE 11.20

What volume does 72.0 g of O_2 occupy at 755 torr and 35°C?

Planning the solution

We must first convert 72.0 g of O_2 to moles and 35°C to kelvins (35°C = 308 K). Since the pressure is given in torr and we are calculating the volume, we may use either $R = 62.4$ L•torr/mol•K or 6.24×10^4 mL•torr/mol•K. We will use the former, which gives an answer in liters.

SOLUTION

$$?\text{ mol O}_2 = 72.0\text{ g O}_2\left(\frac{1\text{ mol O}_2}{32.0\text{ g O}_2}\right) = 2.25\text{ mol O}_2$$

$$PV = nRT$$

$$(755\text{ torr})V = 2.25\text{ mol}\left(\frac{62.4\text{ L}\cdot\text{torr}}{\text{mol}\cdot\text{K}}\right)308\text{ K}$$

$$V = \left(\frac{2.25\text{ mol}}{755\text{ torr}}\right)\left(\frac{62.4\text{ L}\cdot\text{torr}}{\text{mol}\cdot\text{K}}\right)308\text{ K}$$

$$= 57.3\text{ L}$$

As with Example 11.19, note that the units of moles, torr, and kelvins cancel, giving a final answer with appropriate units for the quantity sought (in this case, a volume).

The Ideal Gas Law can also be used to solve any of the problems in Section 11.9. Let us repeat Example 11.18.

EXAMPLE 11.21

Use the Ideal Gas Law to determine the number of moles in a 1.50-L sample of gas at 25°C and 0.874 atm.

Planning the solution

Since the volume is given in liters and the pressure in atmospheres, we must use $R = 0.0821\ \text{L} \cdot \text{atm/mol} \cdot \text{K}$. The temperature is 25°C = 298 K.

SOLUTION

$$PV = nRT$$

$$(0.874\ \text{atm})(1.50\ \text{L}) = n\left(\frac{0.0821\ \text{L} \cdot \text{atm}}{\text{mol} \cdot \text{K}}\right)298\ \text{K}$$

Now we must be careful when we solve for n. We multiply both sides by the reciprocal of R and divide both sides by 298 K. We must treat the units of R the same as we would treat numbers.

$$\frac{(0.874\ \cancel{\text{atm}})(1.50\ \cancel{\text{L}})}{(298\ \cancel{\text{K}})}\left(\frac{\text{mol} \cdot \cancel{\text{K}}}{0.0821\ \cancel{\text{L}} \cdot \cancel{\text{atm}}}\right) = n$$

$$n = 0.0536\ \text{mol}$$

PROBLEM 11.15

Use dimensional analysis to convert the value of $0.0821\ \text{L} \cdot \text{atm/mol} \cdot \text{K}$ to units of: **(a)** $\text{mL} \cdot \text{atm/mol} \cdot \text{K}$ **(b)** $\text{L} \cdot \text{torr/mol} \cdot \text{K}$

PROBLEM 11.16

Calculate the volume that 0.275 mole of helium occupies at –78°C and 745 torr.

PROBLEM 11.17

What pressure does a 2.00-mole sample of carbon dioxide exert if it is confined to a 1.00-L cylinder at 22°C?

PROBLEM 11.18

Suppose a 0.640-g sample of oxygen, O_2, occupies a volume of 575 mL and exerts a pressure of 747 torr. What is the Kelvin temperature of the oxygen sample? What is its Celsius temperature?

PROBLEM 11.19

How many moles are present in a 276-mL sample of nitrogen at –24°C and 1.15 atm?

11.11 GAS LAW STOICHIOMETRY

OBJECTIVE 11.11 State Gay–Lussac's Law of Combining Volumes. Carry out stoichiometry problems involving gases.

Avogadro's Law tells us that equal volumes of gases under identical conditions of temperature and pressure contain equal numbers of moles. Thus, it is not too surprising to find out that, when 22.4 L (1 mol) of hydrogen and 22.4 L (1 mol) of chlorine, both at STP, are combined, 44.8 L (2 mol) of hydrogen chloride are produced at STP:

$$H_2(g) + Cl_2(g) \rightarrow 2\,HCl(g)$$

$$\frac{22.4\text{ L}}{1\text{ mol}} + \frac{22.4\text{ L}}{1\text{ mol}} \rightarrow \frac{44.8\text{ L}}{2\text{ mol}}$$

Inspection of this equation shows that the ratios of volumes involved are the same as the mole ratios of the reactants and products. Thus, 1.00 L each of hydrogen and chlorine at STP should lead to production of 2.00 L of hydrogen chloride, because the ratios of volumes of reactants and products are the same as the mole ratios in the balanced equation.

$$\begin{array}{ccccc} 1 & : & 1 & : & 2 \\ H_2(g) & + & Cl_2(g) & \rightarrow & 2\,HCl(g) \\ \frac{1.00\text{ L}}{0.0446\text{ mol}} & + & \frac{1.00\text{ L}}{0.0446\text{ mol}} & \rightarrow & \frac{2.00\text{ L}}{0.0892\text{ mol}} \end{array}$$

The production of ammonia from nitrogen and hydrogen proceeds in accordance with the same underlying principle:

$$\begin{array}{ccccc} 1 & : & 3 & : & 2 \\ N_2(g) & + & 3\,H_2(g) & \rightarrow & 2\,NH_3(g) \\ \frac{1.00\text{ L}}{0.0446\text{ mol}} & + & \frac{3.00\text{ L}}{0.134\text{ mol}} & \rightarrow & \frac{2.00\text{ L}}{0.0892\text{ mol}} \end{array}$$

← **Volumes at STP**

In this case, 2.00 L of ammonia are formed from 1.00 L of N_2 and 3.00 L of H_2. Again, the ratios of volumes (1 : 3 : 2) are the same as the mole ratios in the balanced chemical equation.

Law of Combining Volumes: The volumes of reacting gases and their products are in simple whole-number ratios (at constant temperature and pressure).

Observations such as these led Gay–Lussac to his **Law of Combining Volumes:** When gases react to form gaseous products, the volumes of the reacting gases and their products are in simple whole-number ratios. The ratios, of course, are simply the mole ratios that we examined during our study of stoichiometry. The Law of Combining Volumes also requires that the reactants and products be at the same temperature and pressure, or their molar volumes would differ and the ratios would no longer hold. Let us make use of this law to solve some simple stoichiometry problems.

EXAMPLE 11.22

When propane (C_3H_8) undergoes combustion, it reacts with oxygen (O_2) to form carbon dioxide (CO_2) and water (H_2O). What volume of carbon dioxide is produced by the combustion of 4.50 L of propane if all reactants and products are at STP?

Symbolic statement

$$?\text{ L }CO_2\text{ (at STP)} = 4.50\text{ L }C_3H_8\text{ (at STP)}$$

Planning the solution

As in our previous stoichiometry problems, we need a balanced chemical equation.

$$C_3H_8(g) + 5\,O_2(g) \rightarrow 3\,CO_2(g) + 4\,H_2O(g)$$

Since the reactants and products are at the same temperature and pressure (STP), their volumes are in the same ratio as the mole ratios. Hence, we may use the following volume ratio as a conversion factor:

$$1\ \text{L } C_3H_8 : 3\ \text{L } CO_2$$

L C_3H_8 (at STP) ⇨ **L CO_2 (at STP)**

SOLUTION

$$?\ \text{L } CO_2 = 4.50\ \cancel{\text{L } C_3H_8}\left(\frac{3\ \text{L } CO_2}{1\ \cancel{\text{L } C_3H_8}}\right) = 13.5\ \text{L } CO_2$$

(Because the volume ratio comes from the balanced equation, it is an *exact* relationship and has an infinite number of significant figures.)

EXAMPLE 11.23

Referring to Example 11.22, determine what volume of oxygen, O_2, is required for the combustion.

Planning the solution

Since oxygen is also at STP, we can use the following volume ratio:

$$1\ \text{L } C_3H_8 : 5\ \text{L } O_2$$

L C_3H_8 (at STP) ⇨ **L O_2 (at STP)**

SOLUTION

$$?\ \text{L } O_2 = 4.50\ \cancel{\text{L } C_3H_8}\left(\frac{5\ \text{L } O_2}{1\ \cancel{\text{L } C_3H_8}}\right) = 22.5\ \text{L } O_2$$

In the preceding examples, we simplified the problems by maintaining the volumes of all reactants and products at STP. However, it is possible to extend the topic of gas stoichiometry to cover ideal gases at conditions other than STP, as our next example illustrates.

EXAMPLE 11.24

What volume of carbon dioxide, CO_2, is produced at 315°C and 465 torr by the combustion of 2.80 L of propane, C_3H_8, at 25°C and 755 torr?

$$C_3H_8(g) + 5\ O_2(g) \rightarrow 3\ CO_2(g) + 4\ H_2O(g)$$

Planning the solution

In Examples 11.22 and 11.23, we were able to use a volume ratio derived from the chemical equation, because the reactants and products were at the same conditions of temperature and pressure, namely, at STP. In fact, as long as the conditions of the reactants and products remain the same, it does not matter what conditions of temperature and pressure exist. Thus, if this example had asked us to determine the volume of carbon dioxide produced at 25°C and 755 torr (the initial conditions of the propane), we would have multiplied our volume of 2.80 L of C_3H_8 by the volume ratio derived from the chemical equation, as in the previous examples:

$$3\ \text{L } CO_2 : 1\ \text{L } C_3H_8$$

$$?\ \text{L } CO_2 = 2.80\ \cancel{\text{L } C_3H_8}\left(\frac{3\ \text{L } CO_2}{1\ \cancel{\text{L } C_3H_8}}\right) = 8.40\ \text{L } CO_2$$

A volume of 8.40 L of carbon dioxide would be produced at 25°C and 755 torr. However, we wish to know the volume of carbon dioxide at 315°C and 465 torr. We must simply correct the volume of 8.40 L to the final conditions stated in the problem.

SOLUTION

	t (°C)	T (K)	P (torr)	V (L)
Initial conditions	25	298	755	8.40
Final conditions	315	588	465	V_f

$$V_f = 8.40\ \text{L}\left(\frac{588\ \text{K}}{298\ \text{K}}\right)\left(\frac{755\ \text{torr}}{465\ \text{torr}}\right) = 26.9\ \text{L}$$

If we wish, we can combine the volume ratio from our first calculation with the correction for temperature and pressure to give one setup:

$$?\ \text{L CO}_2 = 2.80\ \text{L C}_3\text{H}_8\left(\frac{3\ \text{L CO}_2}{1\ \text{L C}_3\text{H}_8}\right)\left(\frac{588\ \text{K}}{298\ \text{K}}\right)\left(\frac{755\ \text{torr}}{465\ \text{torr}}\right) = 26.9\ \text{L CO}_2$$

We can also solve problems in which some of the reactants and products of interest are not gases. The following examples should illustrate the variety of problems that can be solved by combining the knowledge we gained in this chapter and in Chapter 10.

EXAMPLE 11.25

Calculate the volume of hydrogen, H_2, that is liberated at STP in the reaction of 5.00 g of aluminum, Al, with hydrochloric acid, HCl.

$$2\ \text{Al(s)} + 6\ \text{HCl(aq)} \rightarrow 2\ \text{AlCl}_3\text{(aq)} + 3\ \text{H}_2\text{(g)}$$

Symbolic statement

$$?\ \text{L H}_2\ \text{(at STP)} = 5.00\ \text{g Al}$$

Planning the solution

The problem asks for the volume of hydrogen produced, so hydrogen must be our unknown. As in our earlier stoichiometry problems, we can determine the moles of unknown as follows:

In this case, our known is 5.00 g Al. Thus, we can calculate the moles of hydrogen as follows:

g Al ⇨ **mol Al** ⇨ **mol H_2**

Once we have calculated the moles of hydrogen, we can find its volume at STP by multiplying by the molar volume of 22.4 L/mol.

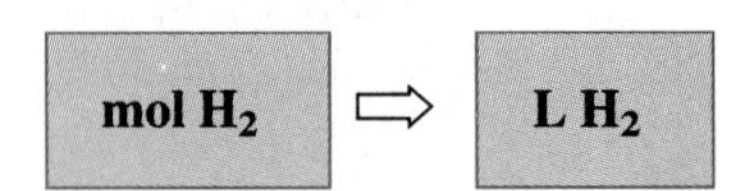

SOLUTION

First, we calculate the number of moles of hydrogen:

$$?\ \text{mol H}_2 = 5.00\ \text{g Al}\left(\frac{1\ \text{mol Al}}{27.0\ \text{g Al}}\right)\left(\frac{3\ \text{mol H}_2}{2\ \text{mol Al}}\right) = 0.278\ \text{mol H}_2$$

Then we use the molar volume of 22.4 L/mol to convert from moles to liters (at STP):

$$? \text{ L H}_2 = 0.278 \text{ mol H}_2 \left(\frac{22.4 \text{ L H}_2}{1 \text{ mol H}_2} \right) = 6.22 \text{ L H}_2$$

We could have combined both parts of this calculation into one setup as follows:

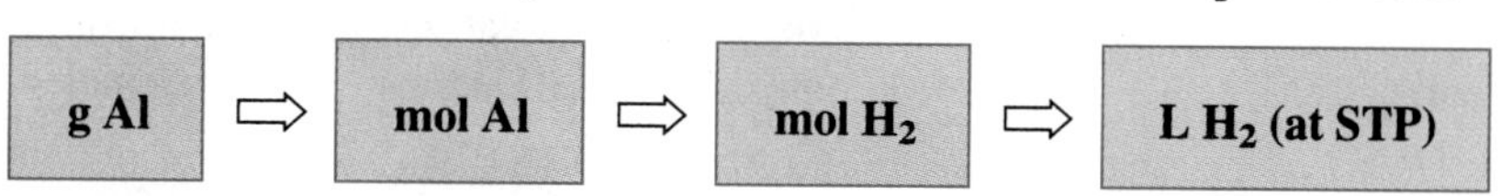

$$? \text{ L H}_2 = 5.00 \text{ g Al} \left(\frac{1 \text{ mol Al}}{27.0 \text{ g Al}} \right) \left(\frac{3 \text{ mol H}_2}{2 \text{ mol Al}} \right) \left(\frac{22.4 \text{ L H}_2}{1 \text{ mol H}_2} \right) = 6.22 \text{ L H}_2$$

EXAMPLE 11.26

What volume of sulfur dioxide, SO_2, at 215°C and 545 torr, is produced in the reaction of 7.43 g Fe_2S_3 (molar mass = 207.9 g/mol), according to the following reaction?

$$2 \text{ Fe}_2\text{S}_3(\text{s}) + 9 \text{ O}_2(\text{g}) \rightarrow 2 \text{ Fe}_2\text{O}_3(\text{s}) + 6 \text{ SO}_2(\text{g})$$

Symbolic statement

$$? \text{ L SO}_2 \text{ (at STP)} = 7.43 \text{ g Fe}_2\text{S}_3$$

Planning the solution

As in Example 11.25, we first calculate the volume of SO_2 produced at STP:

g Fe_2S_3 ⇨ **mol Fe_2S_3** ⇨ **mol SO_2** ⇨ **L SO_2 (at STP)**

Then we can correct the volume of SO_2 (at STP) to 215°C and 545 torr.

SOLUTION

We begin by calculating the volume of SO_2 at STP:

$$? \text{ L SO}_2 = 7.43 \text{ g Fe}_2\text{S}_3 \left(\frac{1 \text{ mol Fe}_2\text{S}_3}{207.9 \text{ g Fe}_2\text{S}_3} \right) \left(\frac{6 \text{ mol SO}_2}{2 \text{ mol Fe}_2\text{S}_3} \right) \left(\frac{22.4 \text{ L SO}_2}{1 \text{ mol SO}_2} \right)$$

$$= 2.40 \text{ L SO}_2 \text{ (at STP)}$$

Then we convert the volume to the conditions stated in the problem.

	t (°C)	*T* (K)	*P* (torr)	*V* (L)
Initial conditions	0	273	760	2.40
Final conditions	215	488	545	V_f

$$V_f = 2.40 \text{ L} \left(\frac{488 \text{ K}}{273 \text{ K}} \right) \left(\frac{760 \text{ torr}}{545 \text{ torr}} \right) = 5.98 \text{ L}$$

EXAMPLE 11.27

What mass of barium carbonate, $BaCO_3$, forms when 125 mL of carbon dioxide, CO_2, at 55°C and 745 torr, is bubbled through a solution of barium hydroxide, $Ba(OH)_2$, according to the following equation?

$$\text{Ba(OH)}_2(\text{aq}) + \text{CO}_2(\text{g}) \rightarrow \text{BaCO}_3(\text{s}) + \text{H}_2\text{O}(\ell)$$

Symbolic statement

$$? \text{ g } BaCO_3 = 125 \text{ mL } CO_2 \text{ (at 55°C, 745 torr)}$$

Planning the solution

In this problem, we are given the volume, temperature, and pressure of carbon dioxide. After correcting the volume to STP, we can determine the number of moles of carbon dioxide using the molar volume of 22.4 L/mol. The moles of carbon dioxide will then become our moles of known for the stoichiometric calculation of barium carbonate.

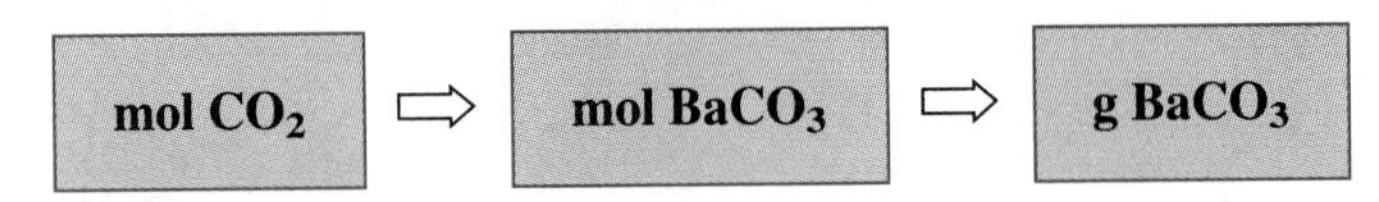

SOLUTION

We first calculate the number of moles of carbon dioxide:

	t (°C)	T (K)	P (torr)	V (L)
Initial conditions	55	328	745	0.125
Final conditions	0	273	760	V_{STP}

$$V_{STP} = 0.125 \text{ L}\left(\frac{273 \text{ K}}{328 \text{ K}}\right)\left(\frac{745 \text{ torr}}{760 \text{ torr}}\right) = 0.102 \text{ L}$$

$$? \text{ mol } CO_2 = 0.102 \text{ L}\left(\frac{1 \text{ mol } CO_2}{22.4 \text{ L}}\right) = 0.00455 \text{ mol } CO_2$$

Using the moles of carbon dioxide as our moles of known, we calculate the mass of barium carbonate via our usual stoichiometric methods:

$$? \text{ g } BaCO_3 = 0.00455 \text{ mol } CO_2\left(\frac{1 \text{ mol } BaCO_3}{1 \text{ mol } CO_2}\right)\left(\frac{197.3 \text{ g } BaCO_3}{1 \text{ mol } BaCO_3}\right)$$

$$= 0.898 \text{ g } BaCO_3$$

PROBLEM 11.20

Calculate the volume of carbon dioxide produced at STP upon combustion of 6.14 L of ethane, C_2H_6, at STP.

$$2\, C_2H_6(g) + 7\, O_2(g) \rightarrow 4\, CO_2(g) + 6\, H_2O(\ell)$$

PROBLEM 11.21

What volume of oxygen at STP is required in Problem 11.20?

PROBLEM 11.22

What volume of oxygen at STP is given off when 4.50 g of silver oxide are decomposed as follows?

$$2\, Ag_2O(s) \rightarrow 4\, Ag(s) + O_2(g)$$

PROBLEM 11.23

What volume of CO_2, at 255°C and 395 torr, is produced when 1.75 L of butane, C_4H_{10}, at 25°C and 765 torr, undergoes the following combustion reaction?

$$2\, C_4H_{10}(g) + 13\, O_2(g) \rightarrow 8\, CO_2(g) + 10\, H_2O(g)$$

PROBLEM 11.24

What mass of sulfur, S, is required to produce 1.35 L of sulfur dioxide, SO_2, at 185°C and 635 torr?

$$S(s) + O_2(g) \rightarrow SO_2(g)$$

PROBLEM 11.25

What volume of oxygen, O_2, at 23°C and 712 torr, is produced in the decomposition of 5.00 g of hydrogen peroxide, H_2O_2, according to the following reaction?

$$2\ H_2O_2(aq) \rightarrow 2\ H_2O(\ell) + O_2(g)$$

11.12 DETERMINING THE MOLAR MASS OF A GAS

OBJECTIVE 11.12 Given the mass of a sample of gas and its volume, temperature, and pressure, calculate its molar mass.

With the information we now have about gases, it is possible for us to go into the laboratory to find the molar mass of an unknown gaseous substance. If our goal is to determine the identity of the unknown substance, knowing its molar mass can help us to narrow the possibilities. Let us see how a molar mass can be determined.

Suppose we have a flask of known volume filled with a gaseous substance (Fig. 11-14). In addition, the flask is connected to a device for measuring the pressure inside. If we know the temperature of its contents, we can calculate the volume of the gas at STP. Using the molar volume of 22.4 L/mol at STP, we can then determine the number of moles present.

When we divide the number of grams of a substance by the number of moles it represents, we find the number of grams in 1 mole. The number of grams per mole, of course, is the molar mass:

$$\text{Molar mass} = \frac{\text{Grams of gas}}{\text{Moles of gas}} = \frac{\text{g}}{\text{mol}}$$

Thus, if we also know the mass of the gaseous substance in the flask, we can calculate its molar mass by dividing the mass of the gas by the number of moles already deter-

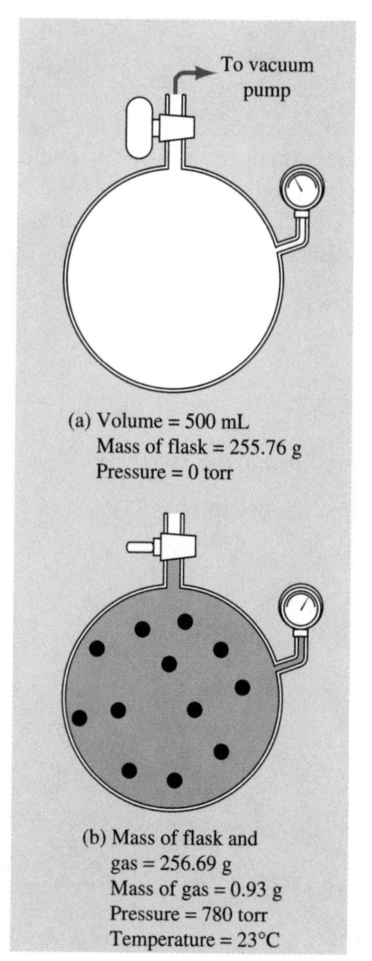

FIGURE 11-14 Determining the molar mass of a gaseous substance. (a) A flask of known volume is equipped with a gauge to measure gas pressure. The flask is evacuated, removing all gases, and the mass of the flask is determined. (b) The flask is filled with a gaseous substance and reweighed. The pressure is recorded.

mined. In order to determine the mass of the gas, it is necessary for us to evacuate the flask and weigh it prior to filling it with our gaseous substance. After filling the flask with the gaseous substance, we then reweigh it. The difference in mass between the evacuated flask and the filled flask represents the mass of the gaseous substance inside. We can use the following Summary to guide us through the subsequent examples.

SUMMARY Calculating the Molar Mass of an Unknown Gas

Step 1. Calculate the volume of the gas at STP.

Step 2. Calculate the number of moles this equals.

Step 3. Divide the mass of gas present by the number of moles it represents. The result is the molar mass.

*Alternatively, the number of moles of gas present (n) can be calculated using the Ideal Gas Law ($PV = nRT$).

EXAMPLE 11.28

(Refer to Fig. 11-14.) A 0.500-L flask is evacuated and weighed, and its mass is found to be 255.76 g. After being filled with a gas, the flask is reweighed, and its mass is determined to be 256.69 g. The pressure of the gas within the flask is 780.0 torr, and the temperature of its contents is 23°C. Calculate the molar mass of the gas.

Planning the solution

Since the molar mass represents the number of grams per mole, we divide the mass of gas by the number of moles it equals:

$$\text{Molar mass} = \frac{\text{Number of grams}}{\text{Number of moles}}$$

Let us carry out this calculation by following the steps outlined in the Summary.

SOLUTION

The first thing we need to do is to find the volume that the gas occupies at STP.

	t (°C)	T (K)	P (torr)	V (L)
Initial conditions	23	296	780.0	0.500
Final conditions	0	273	760.0	V_{STP}

$$V_{STP} = 0.500\ \text{L}\left(\frac{273\ \text{K}}{296\ \text{K}}\right)\left(\frac{780.0\ \text{torr}}{760.0\ \text{torr}}\right) = 0.473\ \text{L}$$

The volume at STP enables us to determine the number of moles present:

$$?\ \text{mol} = 0.473\ \text{L}\left(\frac{1\ \text{mol}}{22.4\ \text{L}}\right) = 0.0211\ \text{mol}$$

Next, we determine the mass of the gas by subtracting the mass of the evacuated flask from that of the gas-filled flask:

$$\text{Mass of gas} = 256.69 \text{ g} - 255.76 \text{ g} = 0.93 \text{ g}$$

Finally, we divide the mass of the gas by the number of moles it represents:

$$\text{Molar mass} = \frac{0.93 \text{ g}}{0.0211 \text{ mol}} = 44 \text{ g/mol}$$

EXAMPLE 11.29

A 1.00-L flask is evacuated and found to weigh 236.27 g. After being filled with a gas that has a molar mass that will be determined, the flask weighs 238.80 g. The unknown gas exerts a pressure of 740.0 torr at 27°C. Calculate the molar mass of the gas.

SOLUTION

First we must find the volume of the gas at STP.

	t (°C)	T (K)	P (torr)	V (L)
Initial conditions	27	300	740.0	1.00
Final conditions	0	273	760.0	V_{STP}

$$V_{STP} = 1.00 \text{ L}\left(\frac{273 \text{ K}}{300 \text{ K}}\right)\left(\frac{740.0 \text{ torr}}{760.0 \text{ torr}}\right) = 0.886 \text{ L}$$

Next, we convert this to moles:

$$? \text{ mol} = 0.886 \text{ L}\left(\frac{1 \text{ mol}}{22.4 \text{ L}}\right) = 0.0396 \text{ mol}$$

The mass of gas present is

$$238.80 \text{ g} - 236.27 \text{ g} = 2.53 \text{ g}$$

Dividing the mass by the number of moles gives the molar mass:

$$\text{Molar mass} = \frac{2.53 \text{ g}}{0.0396 \text{ mol}} = 63.9 \text{ g/mol}$$

PROBLEM 11.26

A 6.00-L flask is filled with 16.0 g of an unknown gas that exerts a pressure of 732 torr at 100°C. Calculate the molar mass of the gas.

PROBLEM 11.27

A 1.50-L sample of an unknown gas has a mass of 2.40 g. The gas exerts a pressure of 777 torr at –55°C. What is the molar mass of the gas?

PROBLEM 11.28 An unknown gas with a mass of 2.67 g occupies a volume of 0.500 L at 10°C. The pressure of the gas under these conditions is 0.900 atm. Determine the molar mass of the gas.

11.13 PARTIAL PRESSURE

OBJECTIVE 11.13A State Dalton's Law of Partial Pressures. Calculate the total pressure of a mixture of gases, given the partial pressure of each. Calculate the partial pressures of the components of a mixture of gases, given the total pressure and sufficient information to calculate the mole fraction of each.

We have said that all gases are affected in the same way by the various gas laws. You might wonder about the effect of having a mixture of gases. Mixtures of gases behave in exactly the same way as pure gases. Suppose that we have a container with a partition in the middle so that each side of the container has the same volume (Fig. 11-15). Now let us introduce oxygen on one side and nitrogen on the other side, so that each side has a pressure of 750 torr. Both sides have equal volumes of gases at the same temperature and pressure, so both sides must have equal numbers of molecules (Avogadro's Law). Furthermore, because both sides have the same pressure, the gases will mix when we remove the partition, but the pressure of the mixture will remain at 750 torr. Let us see why this is reasonable.

At the beginning of the experiment, oxygen is confined to the left side of the container and exerts a pressure of 750 torr. When the partition is removed, the volume to

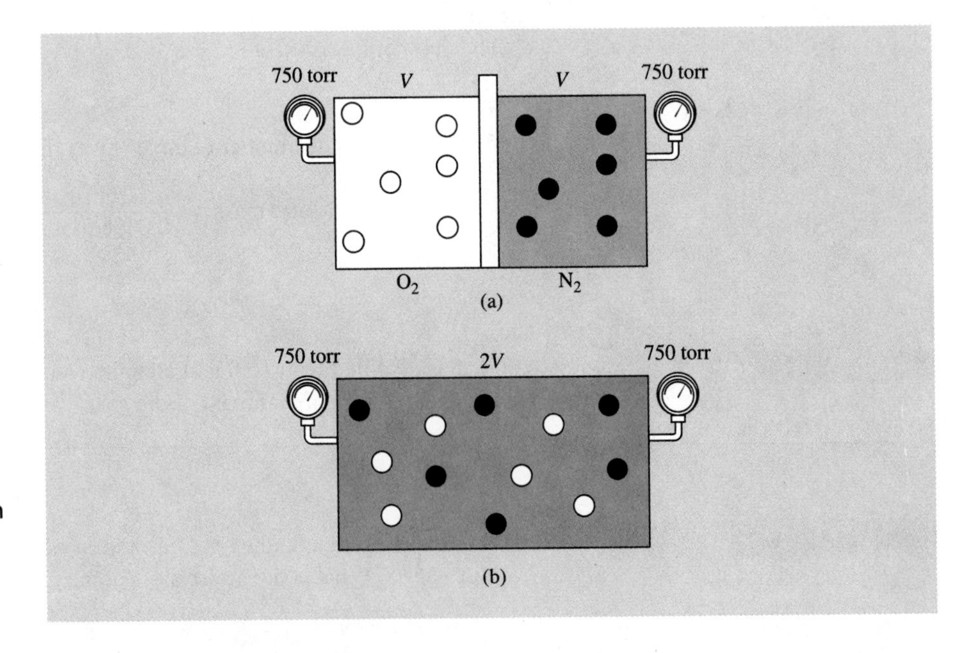

FIGURE 11-15 Partial pressure. (a) Oxygen and nitrogen in a container, separated by a partition. The volumes on the two sides are equal, and the pressure in each chamber is 750 torr. (b) After the partition is removed, the gases mix, and the resulting pressure is still 750 torr.

which the oxygen is confined doubles. Thus, oxygen makes half as many collisions with the walls of the container and must exert half of its original pressure: $\frac{1}{2}$(750 torr) = 375 torr. Similarly, the pressure caused by the nitrogen molecules must also be reduced to 375 torr at the end of the experiment. We call the pressure exerted by each of the gases a **partial pressure.** From this example, we can see that the total pressure of the mixture (750 torr) is equal to the sum of the partial pressures of the gases making up the mixture.

Partial pressure: The pressure that a component of a gaseous mixture would exert if it were alone in its vessel.

$$\begin{array}{ccccc} \text{Total pressure} & = & \text{Partial pressure } O_2 & + & \text{Partial pressure } N_2 \\ P_{\text{total}} & = & P_{O_2} & + & P_{N_2} \\ 750 \text{ torr} & = & 375 \text{ torr} & + & 375 \text{ torr} \end{array}$$

This behavior, which is typical of all mixtures of gases, was first noted by John Dalton. **Dalton's Law of Partial Pressures** states that the total pressure of a mixture of gases is equal to the sum of the partial pressures of the gases making up the mixture:

Dalton's Law of Partial Pressures: The total pressure of a mixture of gases is equal to the sum of the partial pressures.

$$P_{\text{total}} = P_1 + P_2 + \cdots + P_n$$

where $P_1, P_2, \ldots, P_n$ are the partial pressures of the various components of the mixture.

An additional point is also important. The partial pressure exerted by any component in a mixture of gases always equals the total pressure times the fraction of molecules (or moles) that gas represents in the mixture. In other words,

$$P_i = X_i P_{\text{total}}$$

where P_i is the partial pressure of substance i, and X_i is the fraction of molecules (or moles) of that substance. We generally refer to X_i as the **mole fraction** of substance i. The mole fraction of any component of a mixture can be calculated by dividing the number of moles of the component in question by the total number of moles of all substances in the mixture. Thus, if a mixture contains 2.00 mol H_2, 3.00 mol N_2, and 5.00 mol O_2, the mole fraction of the various components is calculated as follows:

Mole fraction: The number of moles of a component of a mixture, divided by the total number of moles in the mixture.

$$\text{Total number of moles} = 2.00 + 3.00 + 5.00 = 10.00$$

$$\text{Mole fraction of hydrogen:} \quad X_{H_2} = \frac{2.00}{10.00} = 0.200$$

$$\text{Mole fraction of nitrogen:} \quad X_{N_2} = \frac{3.00}{10.00} = 0.300$$

$$\text{Mole fraction of oxygen:} \quad X_{O_2} = \frac{5.00}{10.00} = 0.500$$

If this mixture of gases exerts a total pressure of 750.0 torr, we can determine the partial pressure of each by multiplying the mole fraction of each by the total pressure.

$$P_i = X_i P_{\text{total}}$$

$$\begin{array}{lll} \text{Hydrogen:} & P_{H_2} = (0.200)(750.0 \text{ torr}) & = 150 \text{ torr} \\ \text{Nitrogen:} & P_{N_2} = (0.300)(750.0 \text{ torr}) & = 225 \text{ torr} \\ \text{Oxygen:} & P_{O_2} = (0.500)(750.0 \text{ torr}) & = 375 \text{ torr} \\ \hline \text{Total pressure:} & P_{\text{total}} & = 750 \text{ torr} \end{array}$$

As you can see, the partial pressures *do* add up to the total pressure.

EXAMPLE 11.30

A mixture containing 11.0 g CO_2 and 12.0 g O_2 exerts a total pressure of 745 torr. Determine the partial pressure of each gas.

Planning the solution

The partial pressure of each gas is equal to its mole fraction times the total pressure: $P_i = X_i P_{total}$. Thus, we must first calculate the number of moles of each gas and its mole fraction. Multiplication of each mole fraction by the total pressure will give the partial pressure of each component.

SOLUTION

$$? \text{ mol } CO_2 = 11.0 \cancel{\text{g } CO_2}\left(\frac{1 \text{ mol } CO_2}{44.0 \cancel{\text{g } CO_2}}\right) = 0.250 \text{ mol } CO_2$$

$$? \text{ mol } O_2 = 12.0 \cancel{\text{g } O_2}\left(\frac{1 \text{ mol } O_2}{32.0 \cancel{\text{g } O_2}}\right) = 0.375 \text{ mol } O_2$$

$$\text{Total moles} = 0.250 \text{ mol} + 0.375 \text{ mol} = 0.625 \text{ mol}$$

$$P_{CO_2} = \left(\frac{0.250 \text{ mol}}{0.625 \text{ mol}}\right)745 \text{ torr} = 298 \text{ torr}$$

$$P_{O_2} = \left(\frac{0.375 \text{ mol}}{0.625 \text{ mol}}\right)745 \text{ torr} = 447 \text{ torr}$$

Checking the solution

As expected, the sum of the partial pressures equals the total pressure:

$$298 \text{ torr} + 447 \text{ torr} = 745 \text{ torr}$$

PROBLEM 11.29

The partial pressures of a mixture of four gases are determined to be $P_{O_2} = 135$ torr, $P_{N_2} = 624$ torr, $P_{CO_2} = 43$ torr, and $P_{CO} = 13$ torr. Calculate the total pressure of the mixture.

PROBLEM 11.30

A mixture containing 0.25 mol He, 0.33 mol Ar, and 0.42 mol Ne exerts a total pressure of 2.00 atm. What is the partial pressure of each gas?

PROBLEM 11.31

A container at 1.20 atm pressure contains three gases: 1.00 mol O_2, 2.00 mol H_2, and 7.00 mol N_2. What is the partial pressure of each gas?

PROBLEM 11.32

A mixture containing 2.80 g N_2 and 6.40 g O_2 exerts a pressure of 720 torr. Calculate the partial pressure of each of the gases in the mixture.

Collecting Gases Over Water

OBJECTIVE 11.13B Use a table of vapor pressures to determine the partial pressure of a gas collected over water at a given temperature. Given the volume of a gas collected over water, as well as its temperature and pressure, calculate the volume of dry gas at STP and the number of moles present.

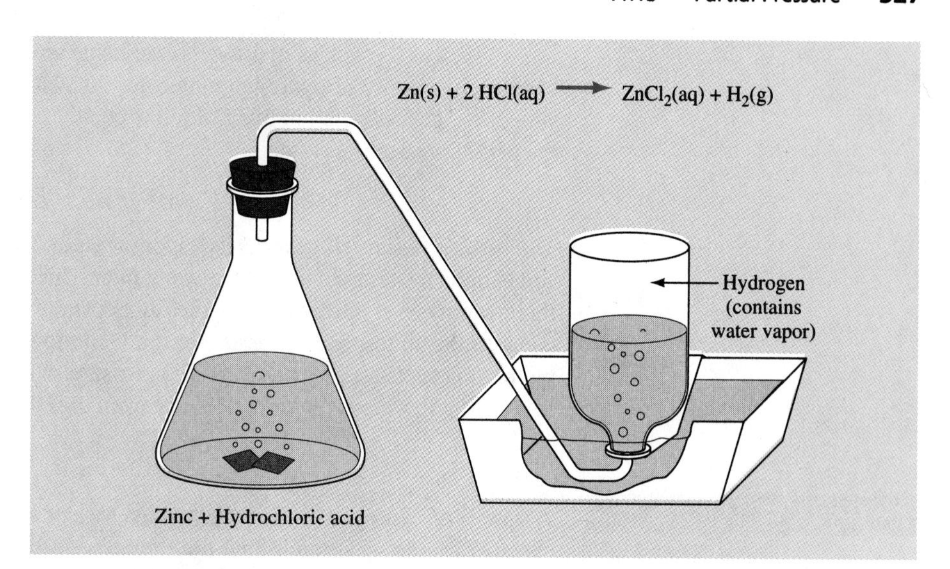

FIGURE 11-16 Collection of hydrogen over water. When zinc and hydrochloric acid are mixed, hydrogen is produced. The hydrogen gas collected in this fashion contains water vapor.

An important application of Dalton's Law of Partial Pressures is employed when gases are collected over water in the laboratory. For example, hydrogen is often generated in the laboratory by allowing hydrochloric acid to react with zinc, as shown in Fig. 11-16. The hydrogen produced passes through a tube and into a bottle, where its bubbles displace the water that previously filled the bottle. The gas that collects in the bottle is said to be "collected over water."

In the next chapter (Section 12.8), we will study a property of liquids known as *vapor pressure.* Molecules at the surface of a liquid have a tendency to escape from the liquid and enter the gaseous state. Once in the gaseous state, the molecules from the liquid behave like other gas molecules and exert a pressure. This pressure is the vapor pressure. The vapor pressure of water (and that of other liquids) varies with temperature and can be looked up in a table of vapor pressures. Table 11-5 is an abbreviated version of the table presented in Section 12.8. At 20°C, the vapor pressure of water is 17.5 torr.

TABLE 11-5 Selected vapor pressures of water

Temperature (°C)	Vapor pressure (torr)	Temperature (°C)	Vapor pressure (torr)
15	12.8	23	21.1
16	13.6	24	22.4
17	14.5	25	23.8
18	15.5	26	25.2
19	16.5	27	26.7
20	17.5	28	28.3
21	18.7	29	30.0
22	19.8	30	31.8

When a gas such as hydrogen is collected over water, as shown in Fig. 11-16, it is composed not only of hydrogen molecules but also of water molecules. The total pressure of the gas collected in this fashion is equal to the partial pressure of hydrogen plus the partial pressure of water:

$$P_{\text{total}} = P_{H_2} + P_{H_2O}$$

The partial pressure of water is its vapor pressure, and since the vapor pressure depends solely upon the temperature of the water over which the gas is collected, we can look up this quantity in a table of vapor pressures. The total pressure of the mixture is often equal to the atmospheric pressure in the laboratory and can be determined by reading the barometer there. Thus, the partial pressure of the hydrogen can be determined by subtracting the vapor pressure of water from the barometric pressure.

EXAMPLE 11.31

A sample of hydrogen gas is collected over water at 29°C on a day when the barometer reads 775.0 torr. Assuming that the pressure of the gas is the same as the barometric pressure, what is the partial pressure of the hydrogen?

Planning the solution

The total pressure of a gas collected over water equals the sum of the water vapor pressure plus the partial pressure of the gas being collected. Thus, we simply subtract the vapor pressure of water at 29°C from the barometric pressure.

SOLUTION

At 29°C, the vapor pressure of water is 30.0 torr.

$$P_{\text{total}} = P_{H_2} + P_{H_2O}$$

$$755.0 \text{ torr} = P_{H_2} + 30.0 \text{ torr}$$

$$755.0 \text{ torr} - 30.0 \text{ torr} = P_{H_2}$$

$$P_{H_2} = 725.0 \text{ torr}$$

We frequently refer to the partial pressure of hydrogen (or other gas collected in this fashion) as the pressure of the "dry" gas. This is the pressure that the gas would exert if we could remove the water molecules present. This can be useful in determining the moles of gas collected, as the following example illustrates.

EXAMPLE 11.32

A 2.00-L sample of carbon dioxide is collected over water at 24°C and at an atmospheric pressure of 762.0 torr. What volume of "dry" carbon dioxide is present at STP? How many moles of carbon dioxide are present?

Planning the solution

Since this gas was collected over water, the atmospheric pressure given in the problem represents the partial pressure of carbon dioxide plus the vapor pressure of water. To determine the volume of "dry" carbon dioxide at STP, we must use the partial pressure of carbon dioxide. When we subtract the vapor pressure of water from the atmospheric pressure, we are "drying" the gas. After that, we can calculate the volume at STP and the number of moles in the usual fashion.

SOLUTION

First, we must find the partial pressure of carbon dioxide:

$$P_{total} = P_{CO_2} + P_{H_2O}$$
$$762.0 \text{ torr} = P_{CO_2} + 22.4 \text{ torr}$$
$$762.0 \text{ torr} - 22.4 \text{ torr} = P_{CO_2}$$
$$739.6 \text{ torr} = P_{CO_2} \quad \leftarrow \textbf{Pressure of "dry" carbon dioxide}$$

Next, we can calculate the volume of dry carbon dioxide at STP. To do so, we must use the partial pressure of the carbon dioxide.

	t (°C)	T (K)	P (torr)	V (L)
Initial conditions	24	297	739.6	2.00
Final conditions	0	273	760.0	V_{STP}

$$V_{STP} = 2.00 \text{ L}\left(\frac{273 \cancel{\text{K}}}{297 \cancel{\text{K}}}\right)\left(\frac{739.6 \cancel{\text{torr}}}{760.0 \cancel{\text{torr}}}\right) = 1.79 \text{ L}$$

Finally, we use the molar volume at STP to determine the number of moles of dry carbon dioxide.

$$? \text{ mol} = 1.79 \cancel{\text{L}}\left(\frac{1 \text{ mol}}{22.4 \cancel{\text{L}}}\right) = 0.0799 \text{ mol}$$

PROBLEM 11.33 A sample of hydrogen is collected over water at 23°C and 770.0 torr. What is the pressure due to hydrogen?

PROBLEM 11.34 A 0.250-L sample of nitrogen is collected over water at 22°C and 758.8 torr. What volume would the dry nitrogen occupy at STP? How many moles are present?

CHAPTER SUMMARY

The behavior of gases can be explained by the **kinetic–molecular theory** of **ideal gases.** An ideal gas is composed of tiny particles that have negligible volumes and negligible intermolecular forces of attraction. The molecules are in constant linear motion, making *elastic* collisions with objects in their paths. Gases exert **pressures** that result from the impact of their molecules on the surfaces they strike. Gas pressure may be measured using a **Torricellian barometer.** In the absence of gaseous molecules, a **vacuum** exists, and this is accompanied by a zero pressure.

The kinetic behavior of gases is responsible for their ability to expand or contract to conform to the size and shape of the container in which they are placed. The velocities of the molecules depend on temperature: At higher temperatures the molecules move faster. The **absolute temperature** of a gas is a measure of the average kinetic energy of its molecules.

Three fundamental laws relating volume, temperature, and pressure characterize the behavior of gases:

1. **Boyle's Law:** The pressure of a gas kept at constant

temperature is inversely proportional to its volume.

2. **Gay–Lussac's Law:** The pressure of a gas kept at constant volume is directly proportional to the absolute temperature.
3. **Charles's Law:** The volume of a gas kept at constant pressure is directly proportional to the absolute temperature.

These three relationships can be summarized in the **Combined Gas Law,** which relates all three variables.

The behavior of gases is further characterized by **Avogadro's Law,** which states that equal volumes of gases at identical conditions of temperature and pressure contain the same number of moles. Thus, 1 mole of any gas occupies the same volume as 1 mole of any other gas if their temperatures and pressures are the same. We define an arbitrary set of such conditions, 0°C and 1 atm pressure, as **standard temperature and pressure (STP).** At STP, all gases have a **molar volume** of 22.4 L. Avogadro's Law may be combined with the Combined Gas Law to give the **Ideal Gas Law:** $PV = nRT$, which relates the volume, temperature, pressure, and number of moles of any sample of gas. The constant R is known as the **universal gas constant.** The **Law of Combining Volumes** is an extension of Avogadro's Law in its recognition that the ratio of volumes in a gaseous reaction is the same as the mole ratio from the balanced chemical equation.

Mixtures of gases behave as though each gas has a pressure independent of the others. **Dalton's Law of Partial Pressures** states that the total pressure of a gaseous mixture is equal to the sum of the **partial pressures** of its components. The partial pressure of each component is equal to the **mole fraction** of that component times the total pressure. A gas collected over water consists not only of the gas itself but also contains molecules of water vapor. The partial pressure of the gas can be determined by subtracting the vapor pressure of water from the total pressure of the gaseous mixture.

Key Terms

Review each of the following terms, which are discussed in this chapter and defined in the Glossary.

kinetic–molecular theory
ideal gas
pressure
Torricellian barometer
vacuum
Boyle's Law
absolute temperature
Gay–Lussac's Law
Charles's Law
standard temperature and pressure (STP)
Combined Gas Law
Avogadro's Law
molar volume
Ideal Gas Law
universal gas constant
Law of Combining Volumes
partial pressure
Dalton's Law of Partial Pressures
mole fraction

Additional Problems

SECTION 11.1

11.35 Answer each of the following questions about ideal gases.

(a) Describe the motion of gas molecules.
(b) Describe the strength of the intermolecular forces of attraction between gas molecules in an ideal gas.
(c) What is the effect of temperature on the velocity of gaseous molecules?
(d) Explain the meaning of an elastic collision.
(e) Under what two conditions does the Ideal Gas Law break down? Which of the assumptions are not true under those conditions?

SECTION 11.2

11.36 Answer each of the following questions about pressure.

(a) Define pressure.
(b) Explain how a gas produces pressure.
(c) Explain how a Torricellian barometer works.
(d) Describe how a mercury-filled U-tube can be used to measure pressure.
(e) Explain the physical meaning of a vacuum.

SECTION 11.3

11.37 What will be the final volume of a 43.0-mL sample of gas collected at 726 torr if its pressure is increased to 942 torr at constant temperature?

11.38 A small gas cylinder contains 1.50 L of helium at 12.0 atm. To what volume will this gas expand if its final pressure is 0.985 atm at constant temperature?

11.39 A bicyclist compresses 9.80 L of air at 755 torr into a tire that has a volume of 4.40 L. What is the final pressure of the air in the tire at constant temperature?

11.40 What is the final pressure of a 735-mL sample of gas at 780 torr (7.80×10^2 torr) after it is expanded to 1.00 L at constant temperature?

SECTION 11.4

11.41 Explain the physical meaning of absolute zero in terms of kinetic energy and motion.

SECTION 11.5

11.42 A fixed volume of gas at 25°C and 1.00 atm pressure is heated to 100°C. What is the final pressure of the gas?

11.43 A fixed volume of a gas at 37°C exerts a pressure of 760 torr (7.60×10^2 torr). The temperature of the gas is altered so that the final pressure of the gas is 600 torr (6.00×10^2 torr). What must be the new Celsius temperature?

SECTION 11.6

11.44 A balloon filled with 755 mL of helium at 20°C is taken outside, where it warms in the sun, reaching a temperature of 36°C. If the atmospheric pressure remains constant, to what volume does the balloon expand? (Assume that the balloon itself has a negligible effect on the pressure.)

11.45 A 2.65-L sample of gas at 22°C is cooled to –78°C at constant pressure. What is the final volume of the gas?

SECTION 11.7

11.46 State the specific conditions of standard temperature in both Celsius and Kelvin temperatures. State the specific conditions of standard pressure in atmospheres, in torr, and in pascals.

SECTION 11.8

11.47 A balloon holds 4.20 L of gas at 27°C and 740 torr (7.40×10^2 torr). A person equipped with a pressure suit carries the balloon into a high-altitude chamber, where the pressure is 0.105 atm and the temperature is –30°C. What is the volume of the balloon under these conditions? (Assume that the balloon itself has a negligible effect on the pressure.)

11.48 A 0.450-L sample of gas has a temperature of 35°C at a pressure of 1.25 atm. To what Celsius temperature must the gas be heated to bring the volume to 0.550 L if the pressure drops to 1.15 atm?

11.49 A balloon filled with 575 mL of oxygen at 1.05 atm and 25°C is taken underwater by a scuba diver, who descends to a depth where the pressure is 1.55 atm and the temperature is 15°C. What is the volume of the balloon at these underwater conditions? (Assume that the balloon itself has a negligible effect on the pressure.)

11.50 A 75.0-mL sample of oxygen is cooled from 28°C to –45°C at the same time that its volume is decreased to 55.0 mL. If the initial pressure of the sample is 535 torr, what is the final pressure?

SECTION 11.9

11.51 What is the volume of 1.00 mol of gas at 25°C and 1.00 atm?

11.52 How many moles are present in a 25.0-mL sample of nitrogen collected at STP?

11.53 Calculate the number of moles in 12.4 L of a gas having a temperature of 37°C and a pressure of 680 torr.

11.54 Determine the number of moles present in a 125-mL sample of hydrogen having a pressure of 735 torr and a temperature of 27°C.

11.55 Suppose that standard temperature and pressure were defined as 20°C and exactly 100 kPa. What would the molar volume be under these conditions?

SECTION 11.10

11.56 A 0.500-mole sample of argon gas is contained in a 3.00-L cylinder. What pressure does the gas exert at a temperature of 75°C?

11.57 A 15.0-g sample of sulfur dioxide, SO_2, fills a 5.00-L vessel at a temperature of 29°C. What is the pressure of the contents of the vessel?

11.58 A 325-mL flask contains 0.0123 mole of hydrogen at 744 torr. What is the temperature of the hydrogen?

11.59 A balloon is filled with 1.75 L of helium gas at 18°C and 755 torr. How many moles of helium fill the balloon?

SECTION 11.11

11.60 A 1.55-L sample of isobutylene, C_4H_8, is measured out at STP and allowed to undergo combustion according to the following equation:

$$C_4H_8(g) + 6\,O_2(g) \rightarrow 4\,CO_2(g) + 4\,H_2O(\ell)$$

(a) What volume of carbon dioxide, CO_2, is produced at STP?
(b) What volume of oxygen, O_2, at STP is required?

11.61 Ammonia, NH_3, may be produced according to the following reaction:

$$N_2(g) + 3\,H_2(g) \rightarrow 2\,NH_3(g)$$

What volume of hydrogen, H_2, at STP is required to produce 13.4 L of NH_3 at STP?

11.62 Aqueous barium hydroxide, $Ba(OH)_2(aq)$, solution may be used to test for carbon dioxide, CO_2. Bubbling carbon dioxide through the solution produces a precipitate of barium carbonate, $BaCO_3$:

$$Ba(OH)_2(aq) + CO_2(g) \rightarrow BaCO_3(s) + H_2O(\ell)$$

What mass of barium carbonate is precipitated by 576 mL of CO_2 at STP?

11.63 Aluminum, Al, reacts with hydrochloric acid, HCl(aq), to produce hydrogen gas, H_2:

$$2\,Al(s) + 6\,HCl(aq) \rightarrow 2\,AlCl_3(aq) + 3\,H_2(g)$$

What volume of hydrogen at STP is produced in this fashion by the reaction of 11.6 g of Al?

11.64 What volume of nitrogen dioxide, NO_2, at 325°C and 525 torr, is produced in the following combustion of 135 mL of gaseous hydrazine, N_2H_4, at 155°C and 646 torr?

$$N_2H_4(g) + 3\,O_2(g) \rightarrow 2\,NO_2(g) + 2\,H_2O(g)$$

11.65 How many milliliters of oxygen, O_2, at 20°C and 768 torr, are produced by the decomposition of 1.00 g of potassium chlorate, $KClO_3$, according to the following reaction?

$$2\,KClO_3(s) - 2\,KCl(s) + 3\,O_2(g)$$

11.66 How many grams of copper, Cu, are required to produce 175 mL of nitrogen monoxide, NO, at 35°C and 745 torr, by means of the following equation?

$$3\,Cu(s) + 8\,HNO_3(aq) \rightarrow 3\,Cu(NO_3)_2(aq) + 2\,NO(g) + 4\,H_2O(\ell)$$

SECTION 11.12

11.67 A 0.240-L flask contains a gas of unknown molar mass at 25°C and 1.00 atm. The mass of the flask with the gas is 90.88 g. When the flask is evacuated and weighed, its mass is found to be 90.24 g. Determine the molar mass of the gas.

11.68 A balloon with a mass of 1.05 g is filled with 0.600 L of gas at 30°C and 762 torr. The mass of the filled balloon is 4.79 g. Calculate the molar mass of the gas.

SECTION 11.13

11.69 A mixture of three noble gases has partial pressures as follows: P_{He} = 125 torr, P_{Ne} = 225 torr, P_{Ar} = 355 torr. What is the total pressure of the mixture?

11.70 A gaseous sample contains 0.250 mol O_2, 0.370 mol N_2, and 0.155 mol CO_2. If the sample exerts a total pressure of 755 torr, what is the partial pressure of each gas?

11.71 At sea level, approximately 20% of the molecules in a sample of air are oxygen molecules. Assuming an atmospheric pressure of 760 torr exists at sea level, what is the approximate partial pressure of oxygen? (Report your answer to two significant figures.)

11.72 A gaseous sample exerting a total pressure of 1.35 atm contains 5.00 g CO_2, 4.00 g CO, and 3.75 g Ne. What is the partial pressure of each gas?

11.73 A mixture containing 10.0 g each of Ne, Ar, and CO_2 exerts a total pressure of 750 torr (7.50×10^2 torr). What is the partial pressure of each gas?

11.74 Two 1.00-L flasks are connected by a tube fitted with a stopcock. One of the flasks contains oxygen, O_2, at a pressure of 608 torr. The other flask contains nitrogen, N_2, at a pressure of 732 torr. The stopcock is opened, allowing the gases in the two flasks to mix. What is the partial pressure of each gas? What is the final pressure in the system after mixing?

11.75 A sample of oxygen is collected over water at 21°C and 762.3 torr. What is the partial pressure of the oxygen?

11.76 A 275-mL sample of hydrogen is collected over water at 18°C on a day when the barometric pressure is 752.7 torr. Assuming the pressure of the collected sample is equal to the barometric pressure, what volume of dry hydrogen would be present at STP? How many moles of hydrogen is this?

GENERAL PROBLEMS

11.77 The following properties of gases are often listed in textbooks. Explain each of them in terms of the kinetic–molecular theory discussed in this chapter.

(a) Gases may be compressed or expanded.
(b) Gases have low densities.
(c) Gases uniformly fill the container in which they are placed.
(d) Gases mix uniformly.
(e) Gases exert a uniform pressure on all sides of their container.

11.78 State the relationship between each of the following pairs of variables. Name the law associated with each relationship.

(a) Pressure and volume (constant temperature and number of moles)
(b) Pressure and temperature (constant volume and number of moles)
(c) Volume and temperature (constant pressure and number of moles)
(d) Volume and number of moles (constant temperature and pressure).

11.79 Explain what happens to the size of a gas bubble and the rate at which it ascends if the bubble is released by a fish swimming at 50 meters below the surface of the water.

11.80 When a helium balloon is released from the earth's surface, it floats upward and expands until the balloon bursts. Explain why the balloon rises when it is released and why it expands as it floats to higher elevations.

11.81 A common frustration of many individuals is experienced when they blow up a flotation device for play in the water. Although the air pillow or ring is firm when first blown up, it gets soft when taken into the water. Explain the cause of this phenomenon, and suggest a remedy.

11.82 Aerosol cans may pose a hazard if handled improperly. Explain why each of the following actions represents a potential danger.

(a) Throwing an aerosol can into a fire.
(b) Putting an aerosol can into luggage transported in the checked luggage compartment of an airplane.

11.83 Sodium hydrogen carbonate, $NaHCO_3$, more commonly known as baking soda, can be used as an antacid for an upset stomach. The recommended dosage is generally 2 teaspoons in a glass of water. In its action, sodium hydrogen carbonate reacts with hydrochloric acid in the stomach, producing sodium chloride, water, and carbon dioxide. Assuming that 2 teaspoons of baking soda has a mass of 12.6 g, what volume of carbon dioxide is produced at 37°C (normal body temperature) and 1.00 atm pressure? (Does this suggest a reason why some antacids cause the user to burp?)

***11.84** The densities of gases are much lower than those of liquids and solids. They are generally reported in terms of grams per liter (g/L). Keeping in mind the fact that 1 mole of a substance equals its molecular mass in grams and that at STP its molar volume is 22.4 L, calculate the densities of the following gases at STP.

(a) Carbon dioxide, CO_2
(b) Helium, He
(c) Nitrogen, N_2
(d) Oxygen, O_2
(e) Use the results of calculations (a)–(d) to explain why a helium balloon rises in air.
(f) Neither carbon dioxide nor helium is capable of supporting combustion. However, (aside from cost) carbon dioxide is preferred as a fire extinguisher because of its high density. Explain why this property makes it more useful for fighting fires than helium.
(g) Calculate the density of carbon dioxide at 20°C and 1.25 atm.

***11.85** The following variation of the Ideal Gas Law is often used for solving molar mass problems: molar mass = gRT/PV. In this equation, g represents the number of grams of unknown.

(a) Explain how this equation is consistent with the method presented in Section 11.12.
(b) Solve Problem 11.68 using this equation.

***11.86** A 7.50-L flask contains a mixture of gases at a pressure of 740 torr and a temperature of 55°C. The partial pressure of oxygen, one of the components of the mixture, is 550 torr. How many moles of oxygen are present?

***11.87** A 1.00-kg block of dry ice (CO_2) is allowed to sublime in a room filled with pure nitrogen, N_2. Prior to sublimation, the pressure of the nitrogen is 745 torr. Following sublimation, the temperature of the room is 17°C. The dimensions of the room are 4.00 m × 3.00 m × 2.50 m. What is the partial pressure (in torr) of carbon dioxide after the entire block has sublimed?

***11.88** A 2.50-L flask contains a mixture of hydrogen and oxygen at 0°C. The partial pressure of hydrogen is 450 torr (4.50×10^2 torr); that of oxygen is 310 torr (3.10×10^2 torr). The contents of the flask are ignited, leading to the production of water.

(a) What is the mole fraction of hydrogen? Of oxygen?
(b) How many moles of hydrogen are present? How many moles of oxygen are present?
(c) Which is the limiting reagent?
(d) What mass of water is formed?

WRITING EXERCISES

11.89 One of your classmates has asked you to explain the concept of absolute temperature. You have agreed to put your explanation in writing. Use the effect of temperature changes on pressure to explain how absolute zero is determined. Then describe how pressure varies with absolute temperature. Finally, discuss the relationship between absolute temperature and kinetic energy, and explain why it is not possible to achieve a temperature less than absolute zero. Be sure to mention the reason that the pressure of a gas is zero when the absolute temperature reaches zero.

11.90 In your own words, explain how a gas exerts a pressure. Then explain how a Torricellian barometer works. Your explanation should include the reason that the mercury column is supported above the level of the mercury pool and why the column rises or falls as the external air pressure increases or decreases.

11.91 Explain the meanings of direct and inverse relationships. You may use Boyle's Law and Charles's Law to illustrate the difference.

11.92 Explain Boyle's Law. Your explanation should include a description of how a gas pressure depends upon collisions, as well as how the number of collisions is affected by changes in volume.

11.93 Without actually doing any calculations, explain how the number of moles in a gas sample can be calculated (using a method other than substitution into the Ideal Gas Law), given the volume, temperature, and pressure of the gas.

Properties of the Liquid and Solid States

OBJECTIVE 12.0 Describe the difference between intermolecular forces and intramolecular forces.

Perhaps you have wondered why some substances are solids at room temperature and others are liquids or gases. This is the same as wondering why substances have different melting and boiling points. A substance with a low boiling point (below 20°C) is a gas at room temperature, and a substance with a high melting point (greater than 20°C) is a solid.

Intramolecular forces: Forces *within* molecules, such as the forces found in a chemical bond.

Intermolecular forces: Forces *between* molecules, such as the forces that hold neighboring molecules together.

In Chapter 8 we examined the nature of chemical bonds, the forces that hold atoms together in molecules. These forces are referred to as **intramolecular forces** (*intra-* means "within"). By contrast, the forces that hold molecules together in the liquid or solid states are referred to as **intermolecular forces** (*inter-* means "between"). Figure 12-1 illustrates the difference between these terms.

Melting points and boiling points depend on the strength of intermolecular forces. When these forces are strong enough, the molecules or ions are held together at room temperature in a rigid network that is characteristic of the solid state. If the intermolecular forces are not strong enough to hold the molecules together rigidly but are still relatively strong, the substance may exist as a liquid at room temperature. However, when intermolecular forces are small, the kinetic energies of the molecules overcome any attractive forces, and the molecules exist in the gaseous state at room temperature. Recall that one of our assumptions about an ideal gas is that the intermolecular forces of attraction are negligible.

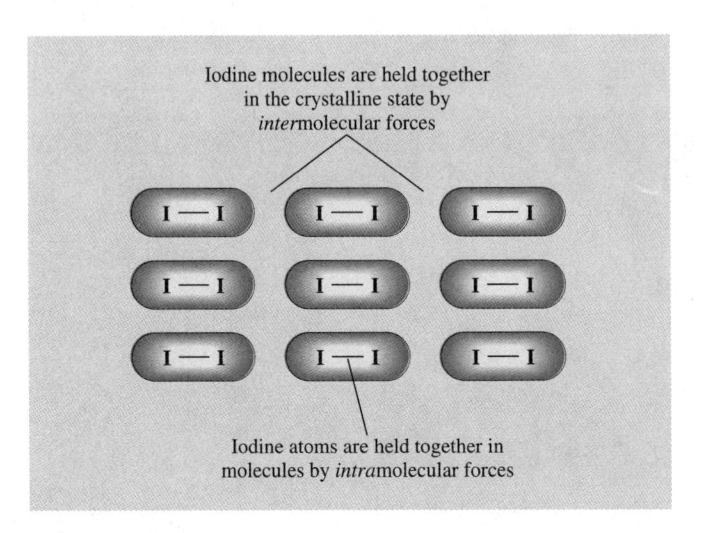

FIGURE 12-1 The difference between intermolecular and intramolecular forces.

12.1 PREDICTING MOLECULAR GEOMETRY: THE VSEPR APPROACH

OBJECTIVE 12.1 Use the VSEPR model to predict whether the geometry of a simple molecule of five atoms or less is linear, trigonal planar, tetrahedral, trigonal pyramidal, or angular.

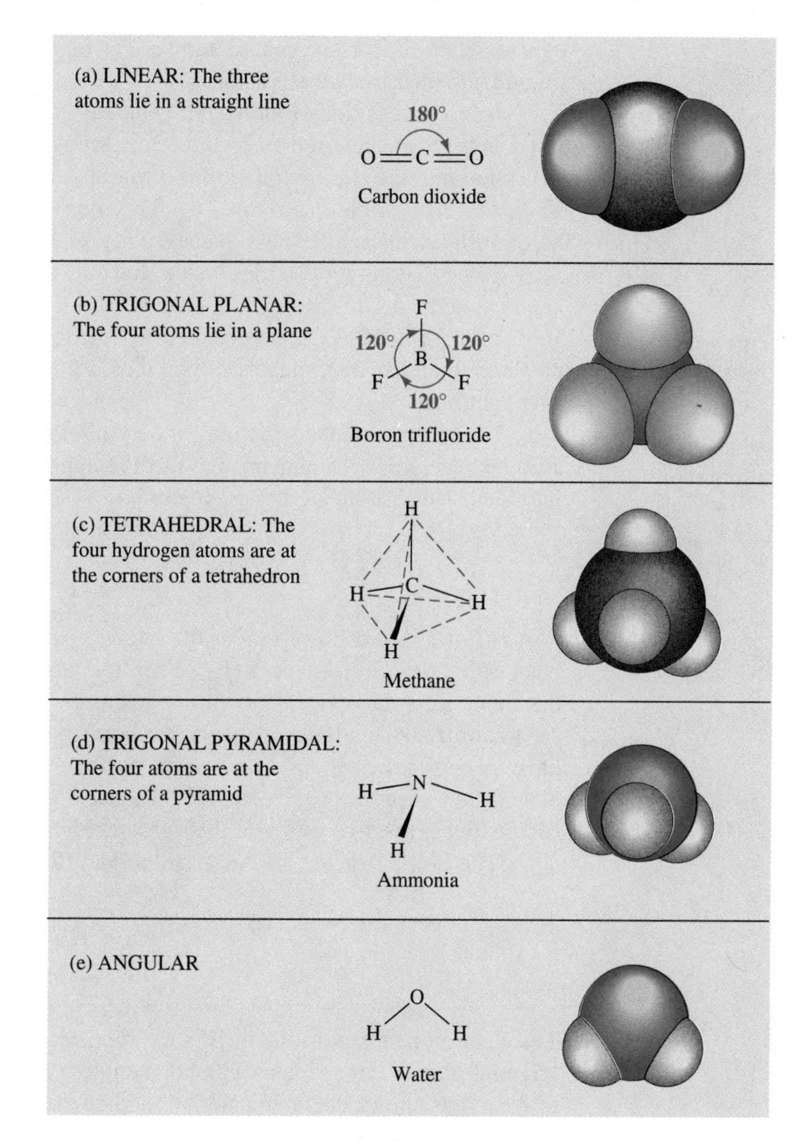

FIGURE 12-2 Shapes of molecules.

Before we examine the factors that affect intermolecular forces, it is helpful for us to understand the basic principles of molecular geometry. Molecules can have a rich variety of shapes, including *linear, trigonal planar, tetrahedral, trigonal pyramidal,* and *angular* geometries (Fig. 12-2). As a rule, the atoms in a molecule are arranged in a fashion that minimizes the repulsions between the various electron pairs. (Recall that like charges repel one another.) To simplify our discussion, we will consider only small molecules with a central atom surrounded by two, three, or four electron pairs. The ***V**alence **S**hell **E**lectron **P**air **R**epulsion (VSEPR) model* (generally pronounced "VES-per") predicts the geometry of a molecule by arranging the atoms and unshared pairs of electrons so that the bonding pairs and unshared pairs of electrons (the "valence"

VSEPR model: A model that predicts the geometry of a molecule with the pairs of valence electrons around each central atom as far from one another as possible.

electrons) attached to the central atom are as far from one another as possible, thereby minimizing their mutual repulsions.

Bond angle: The angle determined by three consecutive atoms.

In our description of molecular geometry, we often refer to bond angles. A **bond angle** is the angle determined by three consecutively bonded atoms. For example, both oxygen atoms and the central carbon atom in carbon dioxide lie along a straight line. Thus, the bond angle determined by the sequence O—C—O is 180° (Fig. 12-2a). In boron trifluoride, each F—B—F sequence makes a 120° bond angle (Fig. 12-2b).

Since diatomic molecules cannot have three consecutively bonded atoms, there are no bond angles in diatomic molecules. Thus, substances such as hydrogen fluoride, H—F, or chlorine, Cl—Cl, are necessarily **linear.** Some triatomic molecules (and other polyatomic molecules), such as carbon dioxide, are also linear if their atoms all fall on a straight line.

Linear: Molecular geometry in which atoms lie along a straight line.

Let us examine the structures of beryllium fluoride, BeF_2, boron trifluoride, BF_3, and methane, CH_4, to explain the VSEPR approach. In its gaseous state, beryllium fluoride has the following Lewis structure:

180°
:F̤̈—Be—F̤̈:

(We will use dashes to represent the bonded pairs of electrons.) There are only two pairs of bonded electrons in BeF_2. When the molecule assumes a linear arrangement, the bonds are as far away from one another as possible, and the electron-pair repulsions are minimized. This linear arrangement corresponds to a 180° bond angle. (You may have noted that beryllium does not satisfy the octet rule. That need not concern you. Not all substances obey the octet rule. The lack of this stabilizing feature is responsible for the highly reactive nature of this substance.)

The Lewis structure of boron trifluoride, BF_3, includes three bonded pairs:

:F̤̈:
| 120°
B
:F̤̈ F̤̈:

Trigonal planar: Planar molecular geometry in which a central atom is surrounded by atoms that occupy the corners of a triangle.

The electron-pair repulsions in BF_3 are minimized when the atoms are arranged in a **trigonal planar** arrangement with bond angles of 120° (Fig. 12-2b). This arrangement of the atoms allows the three pairs of bonding electrons to be located as far away from one another as possible. (Like BeF_2, boron trifluoride does not satisfy the octet rule, but that need not concern us.)

Finally, the Lewis structure of methane, CH_4, is the following:

H
|
H—C—H
|
H

Tetrahedral: Molecular geometry in which a central atom is surrounded by atoms that occupy the corners of a tetrahedron.

The **tetrahedral** arrangement (Fig. 12-2c) allows the four bonding pairs of electrons to be as far from one another as possible. The bond angles in methane are all 109.5°. (In the following three-dimensional representation, wedges are used to show hydrogens coming forward in front of the plane of the page: H►C; or going back behind the plane of the page: H◄C. The atoms connected by straight lines lie in the plane of the page.)

H
C
109.5° H H H

The geometry about a central atom is described concisely in the Summary.

SUMMARY Geometry Around a Central Atom

Number of electron pairs surrounding central atom	Geometry	Bond angle
2	Linear	180°
3	Trigonal planar	120°
4	Tetrahedral	109.5°

At this point, you might wonder about the geometries of ammonia and water, which are shown in Fig. 12-2. The nitrogen atom in ammonia has an unshared pair of electrons:

$$\mathrm{H{-}\ddot{N}{-}H}$$
$$\mathrm{|}$$
$$\mathrm{H}$$

The unshared pair of electrons on the nitrogen atom is treated like a bonded pair of electrons. Thus, there are four pairs of electrons around the nitrogen that repel one another: three bonded pairs and one unshared pair. (We frequently refer to an unshared pair of electrons as a *lone pair.*) The four pairs of electrons assume a tetrahedral orientation, just as in methane. The result is a tetrahedron, with hydrogen at three of the corners, and a lone pair pointing toward the fourth corner:

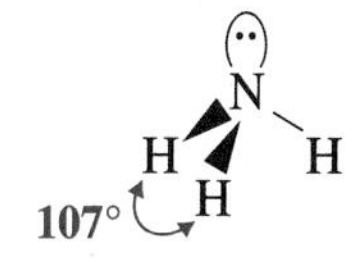

Trigonal pyramidal: Molecular geometry in which atoms occupy the corners of a pyramid with a triangular base.

If we consider the geometry of the atoms alone, the nitrogen and three hydrogens form a **trigonal pyramid,** as shown in Fig. 12-2d. The bond angles in ammonia are 107°—very close to the tetrahedral angle of 109.5°.

The geometry of a water molecule is determined in a fashion similar to that of ammonia. The two unshared pairs of electrons on oxygen are directed toward two positions of a tetrahedron, with the hydrogen atoms occupying the other two positions:

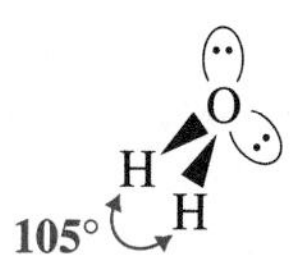

Angular: Molecular geometry in which three atoms lie in a nonlinear arrangement.

Considering the arrangement of the atoms alone, the oxygen and two hydrogens have the **angular** geometry depicted in Fig. 12-2e. The bond angle in water is 105°—again, very close to the tetrahedral angle of 109.5°. The slight distortions in ammonia and water from the tetrahedral angle of 109.5° are the result of the lone pairs of electrons. A lone pair of electrons exerts a greater repulsive force than a bonded pair of electrons, because a lone pair is not shared with a second atom. The greater repulsive force of the lone pair in ammonia pushes the bonding electron pairs slightly closer together, thereby decreasing the bond angles from 109.5° to 107°. In water, the presence of two lone pairs pushes the bonding electron pairs even closer together, reducing the bond angle to 105°.

To predict the geometry around a central atom, we simply look at a properly drawn Lewis structure and count the number of *atoms* plus *lone pairs* attached to the central atom in question. Each atom or lone pair is referred to as a "group." The number of groups determines the geometry as follows: 2 is linear, 3 is trigonal planar, and 4 is tetrahedral. For example, in carbon dioxide, CO_2, the central carbon atom is attached to two oxygen atoms; it has no unshared electron pairs. Since there are two groups attached to the central carbon, we predict a linear configuration for carbon dioxide:

$$\ddot{\underset{\cdot\cdot}{O}}=C=\ddot{\underset{\cdot\cdot}{O}}$$

Note that the presence of multiple bonds does not alter the total number of groups counted. The reason we do not count extra for multiple bonds is that all of the bonds in a double or triple bond point in the same direction, since they are connecting the same atoms. Therefore, the additional bonds do not affect the geometry.

Similarly, we predict a trigonal planar geometry about the central carbon atom in formaldehyde:

```
   :O:
    ‖
    C
   / \
  H   H
```

In this case the carbon is surrounded by three atoms and no lone pairs.

EXAMPLE 12.1

Predict the geometry of carbon tetrachloride, CCl_4. It has the following Lewis structure:

```
      ..
     :Cl:
  ..  |  ..
 :Cl—C—Cl:
  ..  |  ..
     :Cl:
      ..
```

Planning the solution

First we sum the number of atoms plus lone pairs surrounding the central carbon. If the sum is 2, the geometry is linear; if it is 3, the geometry is trigonal planar; if it is 4, the geometry is tetrahedral.

SOLUTION

There are four chlorine atoms and no unshared pairs of electrons attached to the carbon atom, so the total is 4. Thus, the geometry around the central carbon is tetrahedral.

EXAMPLE 12.2

Predict the geometry of phosphine, PH_3. It has the following Lewis structure:

SOLUTION

There are three hydrogen atoms plus one unshared pair of electrons, for a total of four groups. The four electron pairs are directed toward the corners of a tetrahedron:

If we consider only the geometry of the atoms, it is trigonal pyramidal, similar to that of ammonia.

EXAMPLE 12.3

Predict the geometry of phosgene, $COCl_2$. It has the following Lewis structure:

SOLUTION

The central carbon is attached to three groups: one oxygen and two chlorines, with no unshared pairs. It must have a trigonal planar geometry:

EXAMPLE 12.4

Predict the geometry of ozone, O_3. Its Lewis structure is as follows:

SOLUTION

The central oxygen is bonded to two oxygen atoms and has one lone pair. Thus, it contains three groups that must be arranged in a trigonal planar fashion. The two outside oxygen atoms occupy two of the trigonal planar positions, while the lone pair is pointed toward the third position.

If we consider only the geometry of the atoms, ozone has an angular shape with a bond angle of approximately 120°.

In this section we have considered only very simple molecules with a sole central atom. The geometries of more complex molecules may generally be described as combinations of the simple situations described here. Thus, the geometry of acetic acid includes one carbon atom with a tetrahedral arrangement, one carbon with a trigonal planar arrangement, and an oxygen with an angular arrangement.

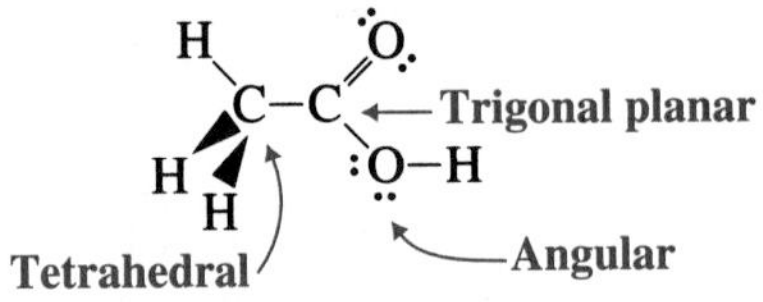

Trigonal planar

Tetrahedral

Angular

PROBLEM 12.1 Predict the geometry of each of the following molecules, given their Lewis structures. First determine the geometry *including* any unshared pairs of electrons around the central atom. Then give the geometry of the atoms alone (as in Examples 12.2–12.4).

(a) BF_3 **(b)** OCl_2 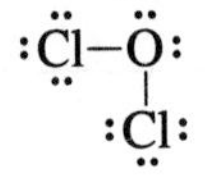 **(c)** NCl_3

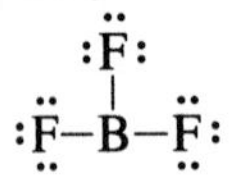

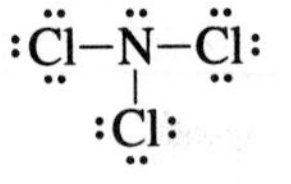

(d) CH_3Cl 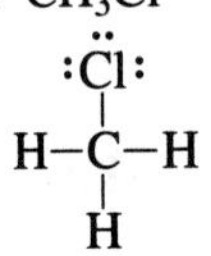 **(e)** CS_2

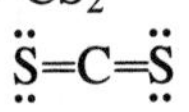

12.2 POLAR AND NONPOLAR MOLECULES

OBJECTIVE 12.2 Predict the polarity of simple molecules that have five or fewer atoms and one of the following geometries: linear, angular, trigonal planar, trigonal pyramidal, or tetrahedral.

Polar molecule: A molecule with an overall charge separation.

One of the factors that can create an attractive intermolecular force is the polarity of the substance. In Sections 8.9 and 8.10 we discussed electronegativity and introduced the concept of polar and nonpolar bonds. We saw that when the electronegativities of two bonded atoms differ, a polar bond results. Like bonds, molecules can be classified as polar and nonpolar. A **polar molecule** is one that has an overall charge separation, resulting in the molecule having a positive end and a negative end. For example, in a diatomic molecule such as HCl, the overall charge separation leaves Cl as the more negative end of the molecule and H as the more positive end of the molecule:

Positive end ⟷ **Negative end**
H–Cl

For a simple diatomic molecule such as hydrogen chloride, the charge separation is the same as that of the bond. However, the polarity of more complicated molecules depends not only on the dipoles of the component chemical bonds but also on the shape of the molecule.

Whereas molecules having only nonpolar bonds must be nonpolar, it is possible for molecules having polar bonds to be either polar or nonpolar. For example, both water and carbon dioxide have two polar bonds. However, water is polar and carbon dioxide is not. The reason for this behavior lies in the differing shapes of the two substances. Water has an angular geometry, whereas carbon dioxide is linear:

$$\overset{\delta^-}{O}(\text{-}H^{\delta^+})_2 \qquad \overset{\delta^-}{O}=\overset{\delta^+}{C}=\overset{\delta^-}{O}$$

(The symbols δ^+ and δ^- refer to the partial charges found in polar covalent bonds.)

As a result of its linear geometry, the dipoles in carbon dioxide point in opposite directions, canceling one another out. Compare this to two individuals of equal strength pulling on a box in opposite directions. Because they are pulling in opposite directions, the forces cancel one another out and the box will not move (Fig. 12-3a). Similarly, since there is no net dipole, carbon dioxide is a **nonpolar molecule.**

Nonpolar molecule: A molecule with no net charge separation.

O=C=O **Dipoles cancel**

On the other hand, the dipoles in water are at an angle to one another and do not completely cancel one another out. The net result of the individual dipoles is a net dipole through the center of the molecule.

Individual dipole → O ← **Individual dipole**
H H
Net dipole

Again, imagine two individuals of equal strength pulling on a box. This time, however, they are pulling on the box at an angle to one another. The box will move along the line that bisects the angle they make with the box (Fig. 12-3b). This is analogous to the effect of the individual bond dipoles on the net dipole in a water molecule. Thus, the orientation of the dipoles from individual bonds has a profound effect upon the overall polarity of a molecule.

Like the situation in carbon dioxide, a trigonal planar molecule consisting of a central atom surrounded by three atoms of the same element (such as BF_3) is nonpolar. Although it may not seem as obvious as in the carbon dioxide case, the overall effect of the three individual dipoles is to cancel one another out, leading to a net dipole of zero.

BF_3 **Dipoles cancel**

Similarly, a tetrahedral molecule consisting of a central atom surrounded by four atoms of the same element (such as CCl_4) is also nonpolar. Again, the orientations of the dipoles in mutually opposite directions cancel one another.

CCl_4 **Dipoles cancel**

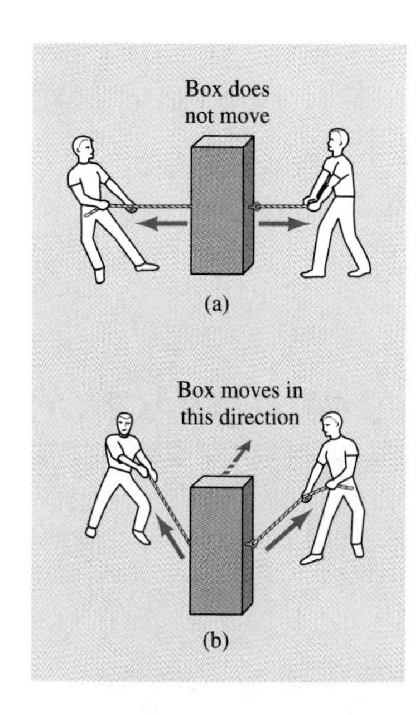

FIGURE 12-3 (a) When two individuals of equal strength pull a box in opposite directions, they cancel one another out, so the box does not move. (b) When they pull at an angle to one another, the box moves along the line that bisects the angle they make with the box.

However, if the individual dipoles in a molecule do not cancel one another completely (as in the case of water), a net dipole will exist. This may be true for linear, trigonal planar, or tetrahedral molecules, too. For example, consider carbonyl sulfide, COS, a linear molecule that is similar in structure to carbon dioxide: S=C=O. Whereas the carbon–oxygen bond is polar, the electronegativities of carbon and sulfur are similar, leading to little or no dipole for the carbon–sulfur bond. Thus, there is a net dipole.

$$\overset{\leftrightarrow}{\text{S=C=O}}$$

Similarly, formaldehyde (CH_2O) is trigonal planar. In this case, the carbon–hydrogen bonds are nonpolar and do not cancel out the dipole of the carbon–oxygen bond. Nor do the carbon–hydrogen bonds cancel the dipole of the carbon–chlorine bond in the tetrahedral molecule methyl chloride (CH_3Cl). Both formaldehyde and methyl chloride are polar.

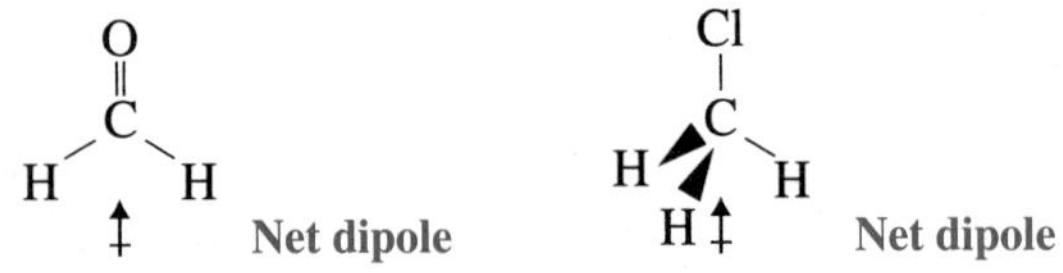

O, C, H, H — Net dipole — Trigonal planar

Cl, C, H, H, H — Net dipole — Tetrahedral

Likewise, the following molecules are all polar because, in each case, the orientation and magnitudes of the individual bond dipoles lead to a nonzero net dipole:

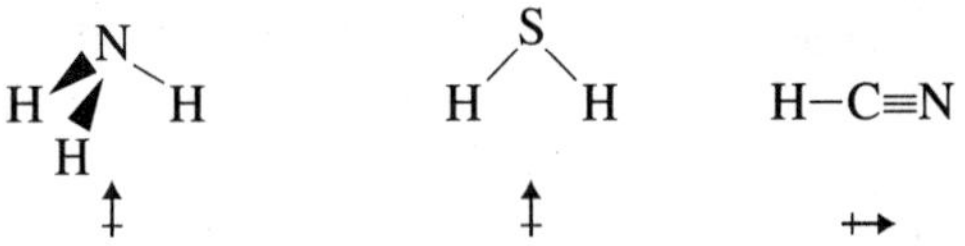

N, H, H, H — Trigonal pyramidal

S, H, H — Angular

H–C≡N — Linear

PROBLEM 12.2

Predict the polarity of each of the following molecules. (The geometry of each is supplied.)

(a) CCl_4 (tetrahedral) **(b)** CH_3F (tetrahedral) **(c)** H_2S (angular)
(d) CS_2 (linear) **(e)** NH_3 (trigonal pyramidal) **(f)** SiF_4 (tetrahedral)
(g) BF_3 (trigonal planar) **(h)** PF_3 (trigonal pyramidal)

12.3 INTERIONIC AND INTERMOLECULAR FORCES OF ATTRACTION

OBJECTIVE 12.3A Describe the origin of each of the following attractive forces: (a) ionic interactions, (b) hydrogen bonding, (c) dipole–dipole interactions, and (d) London forces. Classify substances according to their main type of attractive force.

OBJECTIVE 12.3B Identify substances capable of hydrogen bonding.

With our understanding of the polarities of molecules, we are now ready to examine the various interactions that hold individual molecules or ions together. We will consider four such interactions (which are arranged in order of decreasing strength):

1. Ionic interactions
2. Hydrogen bonding
3. Dipole–dipole interactions
4. London (or dispersion) forces

The strongest of these are the electrostatic forces that hold ionic compounds together. The others are intermolecular forces that hold groups of covalent molecules together.

Ionic Interactions

Ionic interactions: Electrostatic forces between ions.

Because of the full charges found on ions (as opposed to the partial charges of polar covalent bonds), ionic forces of attraction are quite high. The magnitude of these attractive forces is responsible for the very high melting points and boiling points of most ionic substances, which are held together by **ionic interactions.** Sodium chloride, NaCl, is an example of a substance exhibiting these interactions.

Hydrogen Bonding

Hydrogen bonding: Electrostatic attraction between an electropositive hydrogen atom from one molecule and an electronegative nitrogen, oxygen, or fluorine atom from a neighboring molecule.

Hydrogen bonding is a relatively strong interaction. It occurs when a hydrogen atom that is bonded to either a nitrogen, oxygen, or fluorine atom interacts with a nitrogen, oxygen, or fluorine atom on a neighboring molecule. For example, neighboring molecules of water can hydrogen-bond to one another as follows:

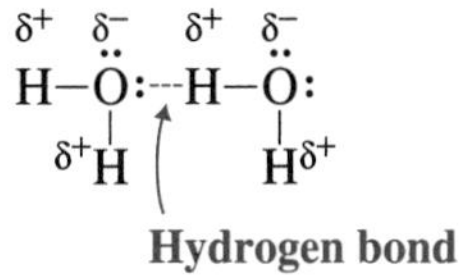

```
δ+  δ−     δ+  δ−
    ..          ..
H—O:---H—O:
    |           |
  δ+H   ↑      Hδ+
```

Hydrogen bond

We represent the hydrogen bond by a dashed line, as shown here. The hydrogen bonding in water is the result of a very strong attraction between the electropositive hydrogen of one molecule and an unshared pair of electrons on the oxygen of a neighboring molecule. When hydrogen bonding occurs, it is usually between a hydrogen atom from one molecule and an unshared pair of electrons on a nitrogen, oxygen, or fluorine atom in a neighboring molecule.

When both of the molecules involved in a hydrogen bond contain a hydrogen atom bonded to an electronegative atom (N, O, or F), the atoms of neighboring molecules can be arranged in a network, as illustrated by the following group of water molecules.

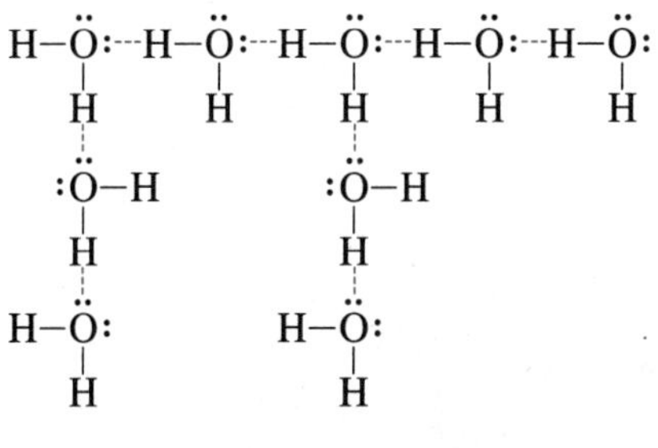

```
    ..      ..      ..      ..      ..
H—O:---H—O:---H—O:---H—O:---H—O:
    |       |       |       |       |
    H       H       H       H       H
    :               :
    ..              ..
   :O—H            :O—H
    |               |
    H               H
    :               :
    ..              ..
H—O:            H—O:
    |               |
    H               H
```

We observe similar hydrogen bonding in ammonia, $:NH_3$. In this case, the electronegative nitrogen provides the unshared pair of electrons, which attracts the partially positive hydrogen atom of a neighboring ammonia molecule.

```
    H       H        H
    | δ- δ+ | δ- δ+  |
  H—N:---H—N:---H—N:
    |       |        |
    H       H        H
```

For hydrogen bonding to occur, the hydrogen atom in the hydrogen bond must satisfy two conditions: (1) it must be bonded to either a nitrogen, oxygen, or fluorine atom in the compound it occurs in; and (2) it must make the hydrogen bond to a nitrogen, oxygen, or fluorine atom in the neighboring molecule. Consequently, hydrogen bonding may occur in either of two ways. In the examples just given, hydrogen bonding was shown between molecules of the same substance. In other words, water molecules were hydrogen-bonded only to water molecules; ammonia molecules were hydrogen-bonded only to ammonia molecules. For this type of hydrogen bonding to occur, the substance must have a hydrogen atom that is bonded to either a nitrogen, an oxygen, or a fluorine atom, as in the following examples:

```
   H  H                H
   |  |   ..           |  ..            ..
 H—C—C—O—H         H—C—N—H          H—F:
   |  |   ..           |  |             ..
   H  H                H  H
```

However, hydrogen bonding may also occur between molecules of different substances. For example, an ethyl ether molecule is not capable of hydrogen bonding with other molecules of itself:

```
   H  H     H  H
   |  |  .. |  |
 H—C—C—O—C—C—H
   |  |  .. |  |
   H  H     H  H
```

Although this molecule contains both hydrogen and oxygen, none of the hydrogen atoms is bonded to the oxygen. Because they are bonded to carbon, the hydrogen atoms in this molecule lack the electropositive character needed for hydrogen bonding. Only hydrogen atoms bonded to nitrogen, oxygen, and fluorine are capable of hydrogen bonding. However, the oxygen in ethyl ether may act as a receiver atom and engage in hydrogen bonding with a hydrogen atom from a molecule that can hydrogen-bond. For example, water molecules can hydrogen-bond to ethyl ether as follows:

```
          ..
         :O—H
          |
          H
   H  H   :  H  H
   |  |  ..  |  |
 H—C—C—O—C—C—H
   |  |  ..  |  |
   H  H   :  H  H
          H
          |
         :O—H
          ..
```

EXAMPLE 12.5

Which of the following molecules is capable of hydrogen bonding to other molecules of itself? Which is not capable of hydrogen bonding to itself but can receive the hydrogen from a neighboring molecule? Which is not capable of any involvement in hydrogen bonding?

(a) Ethyl alcohol **(b)** Methyl fluoride **(c)** Butane

```
    H H              H                  H H H H
    | |   ..         |  ..              | | | |
  H-C-C-O-H        H-C-F:             H-C-C-C-C-H
    | |   ..         |  ..              | | | |
    H H              H                  H H H H
```

SOLUTION

(a) Ethyl alcohol has a hydrogen atom bonded directly to oxygen. Thus, it is capable of hydrogen bonding to itself, as follows:

```
    H H         H H H
    | |   ..    | | |
  H-C-C-O-H---:O-C-C-H
    | |   ..  ..  | |
    H H           H H
```

(b) Methyl fluoride does not have any hydrogen atoms bonded directly to the fluorine atom present. Thus, it cannot hydrogen-bond to other molecules of itself. However, the fluorine on methyl fluoride can act as a receiver atom for hydrogen bonding with molecules that are capable of hydrogen bonding; for example, water:

```
    H
    |  ..     ..
  H-C-F:---H-O:
    |  ..     |
    H         H
```

(c) Butane lacks any nitrogen, oxygen, or fluorine; thus, it is not capable of being involved in hydrogen bonding at all.

Dipole–Dipole Interactions

Dipole–dipole interactions: The attractive forces between neighboring molecules that result from the alignment of dipoles.

Polar molecules are capable of aligning themselves with alternating charges, as shown in Fig. 12-4. This type of interaction between the positive and negative ends of polar molecules is called **dipole–dipole interaction.** Though this interaction might seem very similar to hydrogen bonding, it is considerably weaker. Hydrogen chloride (HCl) exhibits this type of interaction, as do other polar compounds, such as HBr, HI, CO, and PH_3.

London Forces

London forces: Relatively weak intermolecular forces of attraction that hold nonpolar molecules together.

London forces (also known as *dispersion forces*) are the weakest of the intermolecular interactions that hold substances together. Imagine a spherical, nonpolar molecule. The

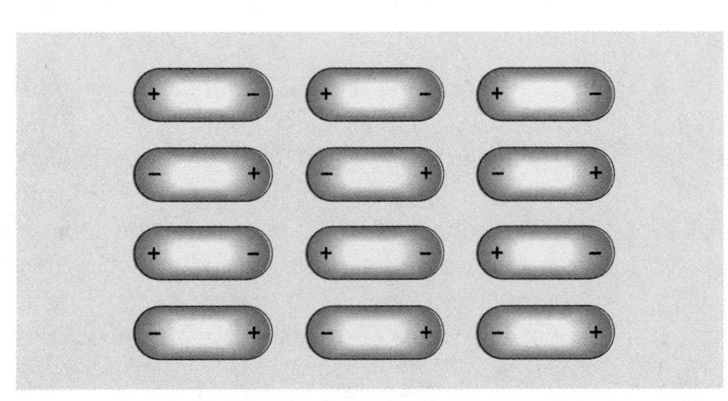

FIGURE 12-4 **Dipole–dipole interactions arise from the alignment of dipoles. The positive end of each molecule is adjacent to the negative end of neighboring molecules.**

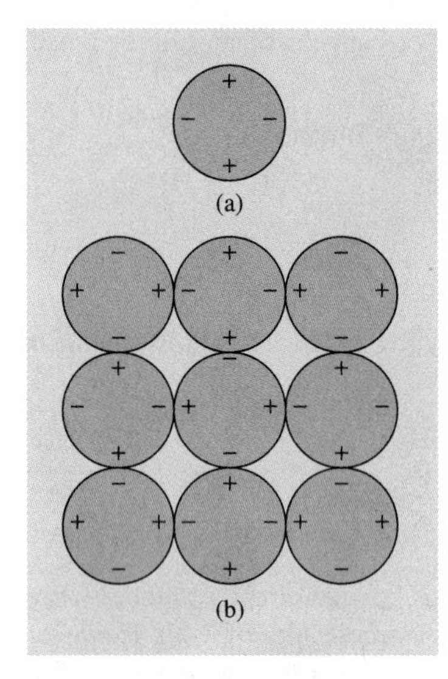

FIGURE 12-5 (a) London forces arise from momentary distortions of the electron distribution in molecules. (b) A distortion in one molecule induces similar distortions in neighboring molecules, thereby holding the molecules together.

electrons in the molecule are in constant motion, creating slight momentary distortions in the electron distribution. This, in turn, causes a slight electrical separation within the molecule (Fig. 12-5a). These distortions induce similar distortions in neighboring molecules, leading to weak attractive forces between the molecules (Fig. 12-5b). The charge separations created in this manner are not permanent but are continually changing their orientation.

While all molecules are capable of exhibiting London forces, these forces are most important for nonpolar molecules, because nonpolar molecules lack any of the other intermolecular forces we have discussed. Thus, nonpolar molecules are held together exclusively by London forces. Hydrocarbons (compounds composed only of carbon and hydrogen) are important examples of substances held together exclusively by London forces. For example, methane molecules (CH_4) are attracted to one another by London forces, as are molecules of propane (C_3H_8), butane (C_4H_{10}), and hexane (C_6H_{14}). The diatomic elements, such as Cl_2, Br_2, and I_2, are also nonpolar, and therefore experience London forces as their strongest intermolecular force of attraction. As we discuss in the next section, these forces become more significant with increasing molecular size.

The Summary Diagrams on pp. 349 and 350 give a synopsis of the various ionic and intermolecular forces we have been discussing, as well as a procedure for determining the strongest force present between molecules of a given substance.

EXAMPLE 12.6

Match each of the following compounds with the strongest type of interaction it is capable of making with other molecules of the same formula. (There is one compound for each type of interaction.)

(a) Br_2 — **1.** Ionic forces
(b) LiCl — **2.** Hydrogen bonding
(c) NO — **3.** Dipole–dipole interactions
(d) CH_3OH — **4.** London forces

Planning the solution

Since there is one compound for each type of interaction, we will go down the list of types of interactions and search for the substance that makes that type of interaction. As we locate the compound of each type, we will eliminate it as a possibility for the remaining types.

SOLUTION

1. Ionic forces: We must find an ionic compound. Lithium chloride, LiCl, is composed of a metal and a nonmetal and is therefore ionic.
2. Hydrogen bonding: Let us look for a compound with a hydrogen bonded to either nitrogen, oxygen, or fluorine. Only CH_3OH fits that description.
3. Dipole–dipole interactions: Of the remaining substances, we must find a compound that is polar. Since nitrogen and oxygen differ in electronegativity, the N—O bond must have a dipole.
4. London forces: Bromine, Br_2, is nonpolar and is held in aggregates by London forces.

SUMMARY DIAGRAM

Interionic and Intermolecular Forces

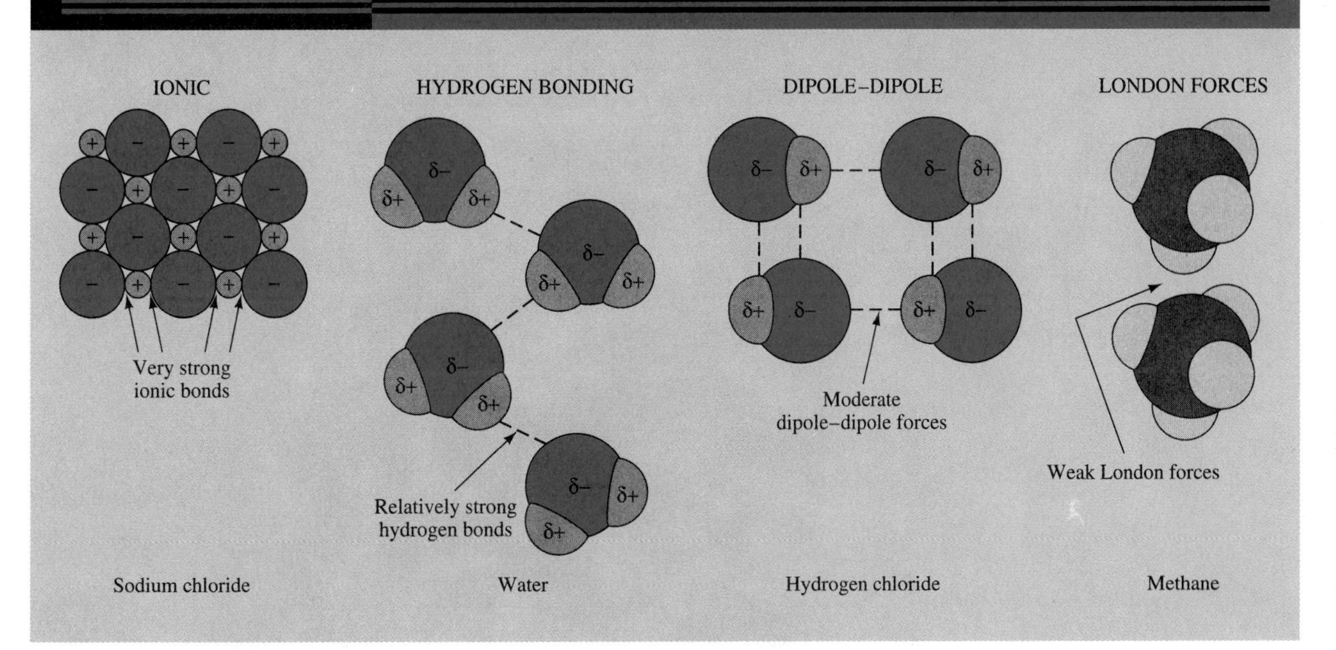

PROBLEM 12.3

Classify each of the following molecules as either (1) those capable of hydrogen bonding with other molecules of themselves; (2) those not capable of hydrogen bonding with themselves but capable of hydrogen bonding with an appropriate neighboring molecule of a different substance; or (3) those not capable of any involvement in hydrogen bonding.

(a) $\begin{array}{c} H \\ | \\ H-C-H \\ | \\ H \end{array}$ **(b)** $\begin{array}{c} H \quad\quad \\ | \quad\quad \\ H-C-\ddot{N}-H \\ |\quad\; | \\ H\quad H \end{array}$ **(c)** $\begin{array}{c} H \\ | \\ H-\underset{\cdot\cdot}{P}-H \end{array}$ **(d)** $\begin{array}{c} :\ddot{F}: \\ | \\ :\underset{\cdot\cdot}{\ddot{F}}-\underset{\cdot\cdot}{N}-\underset{\cdot\cdot}{\ddot{F}}: \end{array}$

(e) $\begin{array}{c} H \quad\quad \\ | \quad\quad \\ H-C-\underset{\cdot\cdot}{\ddot{O}}-H \\ | \quad\quad \\ H \quad\quad \end{array}$ **(f)** $\begin{array}{c} H \quad\quad H \\ | \quad\quad | \\ H-C-\underset{\cdot\cdot}{\ddot{O}}-C-H \\ | \quad\quad | \\ H \quad\quad H \end{array}$ **(g)** $H-\underset{\cdot\cdot}{\ddot{F}}:$

PROBLEM 12.4

Match each of the following compounds with the strongest type of interaction it is capable of making with other molecules of the same formula. (There is one compound for each type of interaction.)

(a) NH_3 **1.** Ionic forces
(b) I_2 **2.** Hydrogen bonding
(c) NaBr **3.** Dipole–dipole interactions
(d) CO **4.** London forces

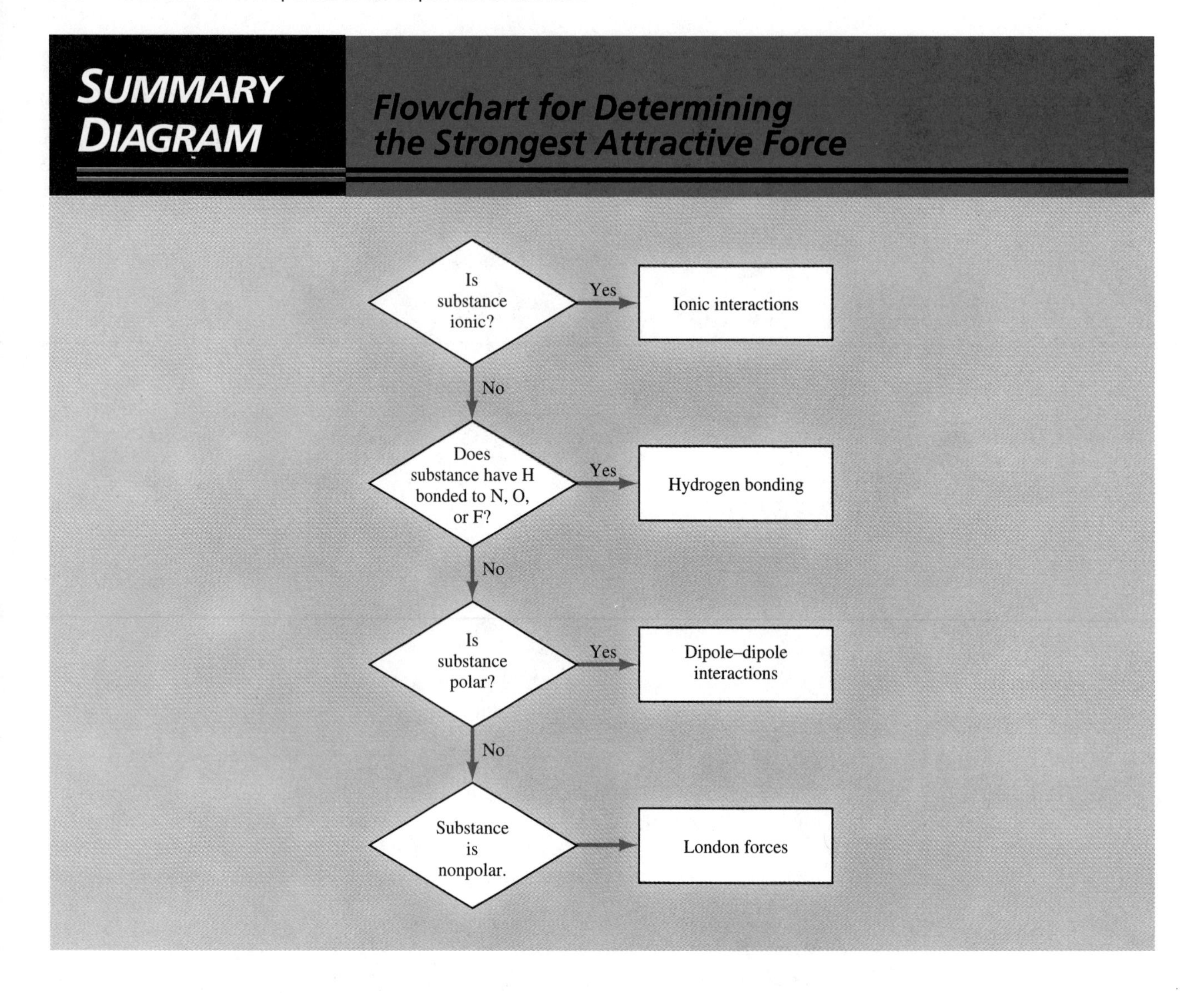

12.4 EFFECTS OF INTERIONIC AND INTERMOLECULAR FORCES ON MELTING POINTS AND BOILING POINTS

OBJECTIVE 12.4 Interpret the melting point and boiling point differences of substances in terms of their predominant attractive forces.

The various intermolecular and interionic forces we have just discussed contribute to the characteristic melting points and boiling points of substances. As we continue the discussion in this section, keep in mind that a substance may exhibit more than one of the four interactions discussed. For example, in addition to hydrogen bonding, dipole–dipole interactions also occur among water molecules. Nevertheless, the strongest interaction exhibited by a substance is the one that will have the most noticeable effect. Let us compare the melting points and boiling points of four compounds to see how different the relative strengths of these forces can be. Each compound exhibits one of the four forces as its strongest interaction.

1. *Ionic forces:* Sodium chloride is an ionic solid that exhibits the very strong ionic interactions characteristic of this class of substances. Like most ionic compounds, sodium chloride is a solid at room temperature and has a very high melting point (mp) and boiling point (bp):

   mp: 801°C   bp: 1413°C

2. *Hydrogen bonding:* Water has extensive hydrogen bonding, as discussed earlier. It is a liquid at room temperature.

   mp: 0°C   bp: 100°C

3. *Dipole–dipole interactions:* Hydrogen chloride is a polar covalent molecule, capable of dipole–dipole interactions. Hydrogen chloride is a gas at room temperature.

   mp: −115°C   bp: −85°C

4. *London forces:* Methane is a nonpolar compound whose strongest intermolecular interactions are London forces. Methane is a gas at room temperature.

   mp: −182°C   bp: −164°C

Note that the melting points and boiling points of the examples we have selected are highest for the ionic compound and lowest for the nonpolar substance. Furthermore, the melting point and boiling point of the ionic substance is so much higher than those of our other examples that it seems to belong in a different category (as is the case in distinguishing ionic from molecular substances). The melting point and boiling point of sodium chloride demonstrate the very high strength of the forces that exist between ions, as compared to those that exist between molecules.

Next, note that the boiling point and melting point of water are considerably higher than those of the two remaining examples. Again, this demonstrates that hydrogen bonding is a very significant intermolecular force (though clearly much weaker than ionic forces). Before we examine the relative strengths of dipole–dipole interactions and London forces, let us first consider some additional principles.

The effects of the various intermolecular forces on boiling points are cumulative. For example, both methane (CH_4, molecular mass = 16.0) and hexane (C_6H_{14}, molecular mass = 86.0) are nonpolar substances with London forces as their strongest (and only) intermolecular forces. However, hexane has a boiling point of 69°C, compared to −164°C for methane. The larger surface area of hexane allows an accumulation of the London forces that attract each molecule to its neighbor. (Think of these intermolecular forces as acting like Velcro. A long strip has much greater holding power than a small piece.) Since the surface areas of molecules generally increase with increasing molecular mass, the boiling points of substances that exhibit the same type of intermolecular force also tend to increase as their molecular masses increase. Thus, bromine (Br_2, molecular mass = 159.8) is a liquid at room temperature, even though it is a nonpolar substance. These principles are illustrated by the compounds listed in Table 12-1.

Now let us return to the question of the relative strength of dipole–dipole interactions. As we have just seen, in order to make any valid comparisons, we need to select examples of comparable molecular mass. Nitrogen, N_2, and carbon monoxide, CO, have virtually the same molecular mass. (The molecular mass is 28.0 for both if we round off to the nearest 0.1.) Nitrogen is nonpolar, whereas carbon monoxide has a modest dipole. Their boiling points are:

:N≡N: bp: −195.8C°

:C≡O: bp: −191.5C°

The rather modest dipole present in carbon monoxide leads to a very modest increase in boiling point. However, for compounds with larger net dipoles, the elevation is more noticeable. Phosphine (PH_3, molecular mass = 34.0) is a polar compound with a more substantial dipole than carbon monoxide. Let us compare its boiling point to that of silane (SiH_4, molecular mass = 32.1), a nonpolar compound:

SiH_4	PH_3
bp = −112°C	bp = −88°C

Although their molecular masses are not exactly the same, they are similar, and most of the 24°C elevation in boiling point for phosphine is the result of its dipole–dipole interactions.

In addition to the effects of ionic and intermolecular interactions, the melting points of substances are often dramatically affected by symmetry. As a general rule, the more symmetrical a substance, the more tightly its molecules can be packed together and the higher the melting point. Conversely, the more irregular the molecular shape of a substance, the less tightly its molecules can be packed together with other molecules and the lower its melting point. For example, methane (CH_4) is completely symmetrical about the central carbon atom and melts at −182°C. Propane (C_3H_8), which is less symmetrical, melts at −190°C, eight degrees lower than methane, despite having the higher molecular mass.

TABLE 12-1 The effect of molecular mass on boiling point

London forces[a]			Dipole–dipole[a]			Hydrogen bonding[a]		
Substance	Molecular mass	Boiling point	Substance	Molecular mass	Boiling point	Substance	Molecular mass	Boiling point
CH_4	16.0	−164°C	HCl	36.5	−85°C	CH_3OH	32.0	65°C
C_2H_6	30.0	−89°C	HBr	80.9	−67°C	C_2H_5OH	46.0	76°C
C_3H_8	44.0	−42°C	HI	127.9	−35°C	*n*-C_3H_7OH	60.0	97°C

[a]Strongest type of interaction for substances in this group.

$$\begin{array}{c} H \\ | \\ H-C-H \\ | \\ H \end{array} \qquad\qquad \begin{array}{c} H\ \ H\ \ H \\ |\ \ \ |\ \ \ | \\ H-C-C-C-H \\ |\ \ \ |\ \ \ | \\ H\ \ H\ \ H \end{array}$$

Methane
mp: −182°C
Molecular mass = 16.0

Propane
mp: −190°C
Molecular mass = 44.0

PROBLEM 12.5

Using the principles discussed in this section, arrange the following compounds in order of increasing boiling point: CH_4, CH_3OH, CH_3Cl.

PROBLEM 12.6

For each of the following pairs of compounds: (1) determine the most significant attractive force for both compounds, (2) predict which compound has the higher boiling point, and (3) explain your prediction.

(a) NaCl and $\begin{array}{c} H\ \ H \\ |\ \ \ | \\ H-C-C-Cl \\ |\ \ \ | \\ H\ \ H \end{array}$

(b) $\begin{array}{c} H\ \ \ \ \ \ \ H \\ |\ \ \ \ \ \ \ | \\ H-C-\ddot{O}-C-H \\ |\ \ \ \ \ \ \ | \\ H\ \ \ \ \ \ \ H \end{array}$ and $\begin{array}{c} H\ \ H\ \ H \\ |\ \ \ |\ \ \ | \\ H-C-C-C-H \\ |\ \ \ |\ \ \ | \\ H\ \ H\ \ H \end{array}$

(c) $\begin{array}{c} H\ \ H\ \ H \\ |\ \ \ |\ \ \ | \\ H-C-C-C-H \\ |\ \ \ |\ \ \ | \\ H\ \ H\ \ H \end{array}$ and $\begin{array}{c} H\ \ H\ \ H\ \ H \\ |\ \ \ |\ \ \ |\ \ \ | \\ H-C-C-C-C-H \\ |\ \ \ |\ \ \ |\ \ \ | \\ H\ \ H\ \ H\ \ H \end{array}$

(d) Br−Br and I−Cl

(e) $\begin{array}{c} H\ \ H\ \ H \\ |\ \ \ |\ \ \ | \\ H-C-C-C-\ddot{O}-H \\ |\ \ \ |\ \ \ | \\ H\ \ H\ \ H \end{array}$ and $\begin{array}{c} H\ \ H \\ |\ \ \ | \\ H-\ddot{O}-C-C-\ddot{O}-H \\ |\ \ \ | \\ H\ \ H \end{array}$

12.5 PHASE CHANGES AND ENERGY

OBJECTIVE 12.5A Interpret a warming curve by identifying what happens to a substance, at both the macroscopic level and the molecular level, along each segment of the curve. Identify the melting point, boiling point, heat of fusion, and heat of vaporization from the curve.

OBJECTIVE 12.5B Given the melting point and boiling point of a substance, determine its physical state at any temperature.

The strength of the intermolecular forces is only one of the factors that determine the physical state (solid, liquid, or gas) of a substance. The temperature of the substance is the other major factor. Temperature is a measure of the average kinetic energy (the energy of motion) of the particles of matter that make up the substance. For example, in a solid, the molecules that are held together in the crystal lattice vibrate about their fixed locations. Heating increases these vibrations. Similarly, when a liquid is heated, the increased kinetic energy causes the molecules to roll and tumble over one another more rapidly. Heating is a process that increases the ability of the molecules to escape from one another.

Warming curve: A diagram showing the temperature changes and changes of state that accompany the constant heating of a substance.

Under normal atmospheric pressure (1 atm), most substances are capable of existing in each of the three physical states—solid, liquid, and gas—depending on the temperature. For example, water can exist as a solid (ice), a liquid (water), or a gas (steam). A **warming curve** is a diagram that shows the various phase changes a substance goes through as it is heated. Let us consider what happens as a sample of water at –40°C is heated to 140°C (at 1 atm). We will refer to the warming curve of water (Fig. 12-6) for this discussion.

At –40°C, water exists in its solid form. In this state the molecules are close to one another, held in a rigid crystal lattice. As heat is absorbed, the molecules in the crystal lattice gain increasing kinetic energy and the solid begins to warm regularly (line segment *AB*). When the temperature of the ice reaches 0°C, the molecules of water in the crystal lattice possess enough kinetic energy to overcome the intermolecular forces holding them in place. As molecules break away from the lattice, the process of melting begins (point *B*). Thus, 0°C represents the melting point of ice.

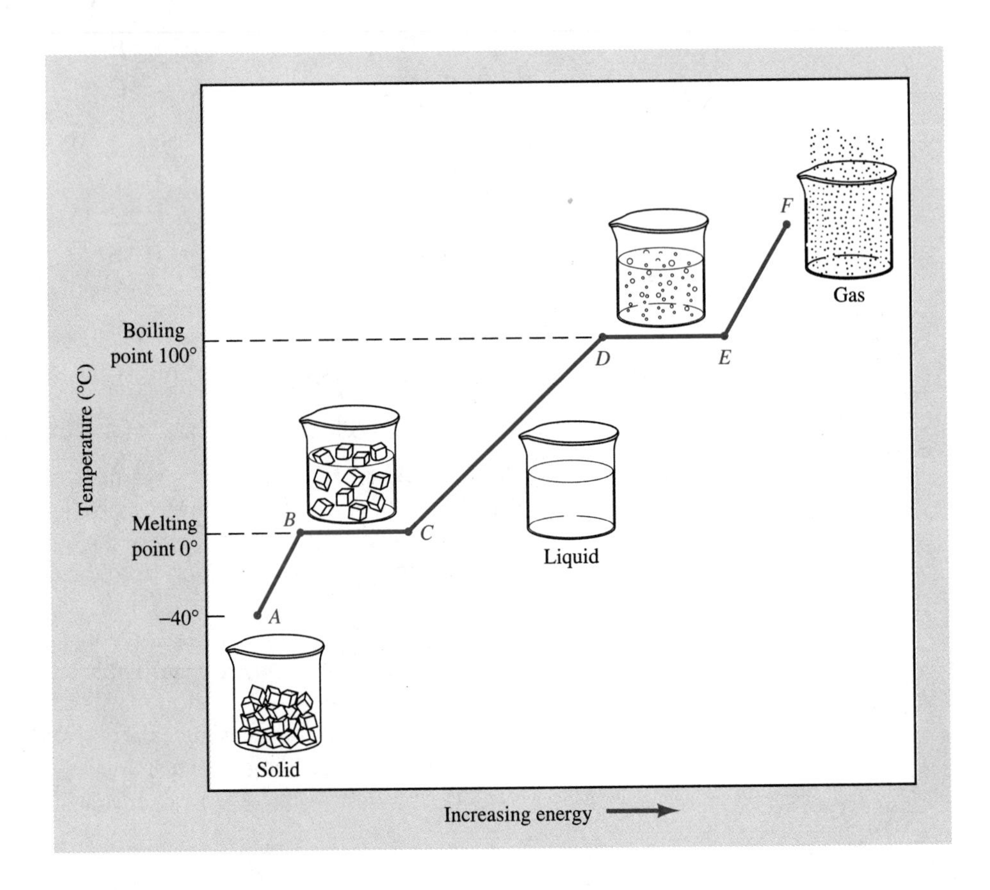

FIGURE 12-6 The warming curve for water.

Once the ice has reached 0°C, the heat being absorbed by the ice no longer results in an increase in temperature. Instead, the energy is now used to overcome the intermolecular forces holding the molecules together. The sample remains at 0°C until all of the ice has melted (line segment *BC*), at which point the sample is liquid water at 0°C (point *C*). The heat required to convert a solid at its melting point to a liquid at the same temperature is known as the **heat of fusion.** (The ability of ice to cool soft drinks results from the heat of fusion. When we put a few ice cubes in a glass of warm root beer, the root beer supplies the heat of fusion that melts the ice. As the root beer loses this heat, it cools off.)

Heat of fusion: The heat required to convert a solid to a liquid at its melting point.

Once all of the water is in the liquid state, continued heating raises the temperature again. The water temperature continues to rise until it reaches 100°C (line segment *CD*). At 100°C (point *D*) the molecules have sufficient kinetic energy to overcome the intermolecular forces holding them in the liquid state. Molecules break away from the surface of the liquid to form gaseous water (steam). Once again, the heat being applied does not change the temperature of the water but instead is used to effect a phase change—in this case, to convert liquid water at 100°C to gaseous water at 100°C (line segment *DE*). We refer to this temperature as the boiling point of water. The heat required to convert a liquid at its boiling point to a gas at the same temperature is known as the **heat of vaporization.** Once all of the liquid has been converted to a gas, continued heating raises the temperature of the gaseous sample (line segment *EF*).

Heat of vaporization: The heat required to convert a liquid to a gas at its boiling point.

The process described in Fig. 12-6 is completely reversible. Heat may be removed from steam, thereby cooling it to 100°C. At 100°C, the steam condenses to become liquid water. Further removal of heat results in cooling of the liquid to 0°C, at which point it begins to freeze. Once freezing is complete, further removal of heat causes the temperature of the ice to drop below 0°C.

PROBLEM 12.7

On the basis of the physical data given, identify each of the following substances as a solid, a liquid, or a gas at room temperature (20°C).

	mp (°C)	bp (°C)
(a) Ethyl alcohol	−11	78
(b) Potassium iodide	723	1420
(c) Butane	−135	−0.5
(d) Phosphorus trichloride	−112	76
(e) Benzene	5.5	80
(f) Bromine	−7	59
(g) Mercury	−39	357
(h) Naphthalene	80	218
(i) Hydrogen bromide	−88.5	−67
(j) Sulfur dioxide	−73	−10

PROBLEM 12.8

Classify each of the substances in Problem 12.7 as a solid, a liquid, or a gas at −50°C.

PROBLEM 12.9 Classify each of the substances in Problem 12.7 as a solid, a liquid, or a gas at −100°C; at +100°C.

12.6 SPECIFIC HEATS AND HEAT CAPACITY

OBJECTIVE 12.6 Define specific heat and molar heat capacity. Calculate the heat liberated or absorbed by a substance undergoing a temperature change, given the mass or number of moles, the specific heat or molar heat capacity, and the temperature change.

Specific heat: The heat required to raise the temperature of 1 gram of a substance by 1°C.

Molar heat capacity: The heat required to raise the temperature of 1 mole of a substance by 1°C.

The warming curve we have just examined shows that heat is required to raise the temperature of a substance that is not undergoing a phase change. Earlier we defined the calorie as the amount of heat required to raise the temperature of 1 gram of water by 1°C. Not all substances require the same amount of heat to raise the temperature of a gram by one degree. Consequently, we define the **specific heat** of a substance as the amount of energy required to raise the temperature of 1 gram by 1°C.

It is often desirable to express the heat required to raise 1 mole (rather than 1 gram) by 1°C. This quantity is known as the **molar heat capacity.** Because it requires 1.00 cal to raise the temperature of 1.00 g of water by one degree, it must require 18.0 cal to raise 1.00 mol (18.0 g) of water by one degree. Thus, the specific heat of water is 1.00 cal/g • °C; the molar heat capacity is 18.0 cal/mol • °C. A lowercase c symbolizes specific heat; a capital C is used for molar heat capacity. The SI units for these two quantities are J/g • °C and J/mol • °C, respectively. The specific heats and molar heat capacities of several substances are given in Table 12-2.

Note that the specific heat of water is quite high. A high specific heat means that a great deal of heat must be transferred in order to cause a given temperature change. Thus, a cup of hot coffee (mostly water) takes quite a few minutes to cool off. On the other hand, when an aluminum roasting pan is removed from the oven, the aluminum itself often can be handled very shortly after removal. Loss of a modest amount of heat from the aluminum pan results in a substantial drop in its temperature. (If you try this, be very careful! You will burn yourself if you handle the pan too soon or touch it too close to its contents.)

The relatively high specific heat of water plays an important role in our ability to maintain a constant body temperature. With water as the most abundant chemical in our

TABLE 12-2 Specific heats and molar heat capacities of several common substances

Substance	Specific heat, c $\frac{J}{g \cdot °C}$	$\frac{cal}{g \cdot °C}$	Molar heat capacity, C $\frac{J}{mol \cdot °C}$	$\frac{cal}{mol \cdot °C}$
Water	4.184	1.000	75.3	18.0
Aluminum	0.897	0.214	24.2	5.79
Copper	0.385	0.0920	24.5	5.85
Iron	0.449	0.107	25.1	5.99
Lead	0.129	0.0308	26.7	6.38

bodies (approximately two-thirds by mass), we are much like the cup of hot coffee. Sudden changes in the temperature of our surroundings (such as we experience when we go from a heated house into a winter's night) are met with resistance to rapid change. This gives our bodies time to respond by burning up some stored starch to produce heat (if we are getting colder) or to begin perspiring to cool off (if we are getting warmer).

Calculations Involving Specific Heat

We can calculate the heat, q, absorbed or liberated as an object changes temperature by using either of the following relationships:

$$q = m \cdot c \cdot \Delta T \qquad \text{where } m = \text{Mass in grams}$$
$$c = \text{Specific heat}$$
$$\Delta T = \text{Change in temperature}$$

or

$$q = n \cdot C \cdot \Delta T \qquad \text{where } n = \text{Number of moles}$$
$$C = \text{Molar heat capacity}$$
$$\Delta T = \text{Change in temperature}$$

The change in temperature, ΔT, equals the final temperature, T_f, minus the initial temperature, T_i:

$$\Delta T = T_f - T_i$$

It follows that, if the temperature is increasing, T_f is greater than T_i and ΔT is a positive value. Conversely, ΔT must be negative when the temperature decreases. Since heat must be absorbed when the temperature increases, it follows that calculation of a positive value of q indicates that heat is absorbed, whereas a negative value of q indicates a liberation of heat. For many applications of these equations, it is important to retain the sign of q. However, the examples and problems that follow ask us to calculate the *amount* of heat absorbed or liberated by various samples that are heated or allowed to cool. When asked in this fashion, it is acceptable to express our answers as positive quantities. To accomplish this, we will obtain the absolute value of ΔT by subtracting the lower temperature from the higher temperature, and then calculate the quantity of heat transferred as a positive quantity.

The specific heats and molar heat capacities in Table 12-2 can be used to solve the following examples and problems.

EXAMPLE 12.7

Calculate the heat liberated as 35.0 g of water cool from 98.0°C to 45.0°C.

Planning the solution

We simply substitute into the equation $q = m \cdot c \cdot \Delta T$.

SOLUTION

$$\Delta T = 98.0°\text{C} - 45.0°\text{C} = 53.0°\text{C}$$
$$q = m \cdot c \cdot \Delta T$$
$$= 35.0\ \cancel{\text{g}}\left(\frac{4.184\ \text{J}}{\cancel{\text{g}} \cdot \cancel{°\text{C}}}\right)53.0\cancel{°\text{C}}$$
$$= 7760\ \text{J} = 7.76\ \text{kJ}$$

EXAMPLE 12.8 Calculate the heat required to heat 145 g of aluminum from 25.0°C to 82.0°C.

SOLUTION

$$\Delta T = 82.0°C - 25.0°C = 57.0°C$$

$$q = m \cdot c \cdot \Delta T$$

$$= 145\ \cancel{g}\left(\frac{0.897\ J}{\cancel{g} \cdot \cancel{°C}}\right)57.0\cancel{°C}$$

$$= 7410\ J = 7.41\ kJ$$

EXAMPLE 12.9 Calculate the heat lost when 1.25 mol of copper cool from 55.0°C to 28.0°C.

SOLUTION

$$\Delta T = 55.0°C - 28.0°C = 27.0°C$$

$$q = n \cdot C \cdot \Delta T$$

$$= 1.25\ \cancel{mol}\left(\frac{24.5\ J}{\cancel{mol} \cdot \cancel{°C}}\right)27.0\cancel{°C}$$

$$= 827\ J$$

PROBLEM 12.10 Calculate the heat required to warm 55.0 g of water from 23.0°C to 76.0°C.

PROBLEM 12.11 How much heat is given off when a 155-g block of iron cools from 100.0°C to 22.0°C?

PROBLEM 12.12 How much heat must be supplied to 75.0 g of aluminum to raise its temperature from 15.0°C to 98.0°C?

12.7 HEAT OF FUSION AND HEAT OF VAPORIZATION

OBJECTIVE 12.7 Given the heat of fusion and the heat of vaporization of water, calculate the heat required to melt or vaporize a given mass of water.

In our discussion of phase changes, we saw that heat is required to melt a solid at its melting point. We referred to this as the heat of fusion. Similarly, the heat of vaporization represents the heat required to vaporize a liquid at its boiling point. These heats may be expressed in a variety of units, including kilojoules per gram and kilojoules per mole. We refer to a measurement in the latter units as the molar heat of fusion (or vaporization).

For water, the heat of fusion is

$$\frac{6.01\ kJ}{mol} = \frac{1.44\ kcal}{mol} \quad \text{or} \quad \frac{0.334\ kJ}{g} = \frac{79.8\ cal}{g}$$

The heat of vaporization is

$$\frac{40.7 \text{ kJ}}{\text{mol}} = \frac{9.72 \text{ kcal}}{\text{mol}} \quad \text{or} \quad \frac{2.26 \text{ kJ}}{\text{g}} = \frac{540 \text{ cal}}{\text{g}}$$

The heat of fusion and heat of vaporization of water are higher than those of most substances. The high heat of fusion is responsible for the cooling capability of ice. When an ice cube is placed in a glass of water at room temperature, the (liquid) water transfers heat that goes to melt the ice. As it loses this heat, the liquid water becomes cooler. The heat lost by the cooling liquid supplies the heat of fusion absorbed by the melting ice cube. Since a high heat of fusion means that a large amount of heat is required to melt the solid, a small piece of ice absorbs a relatively large amount of heat from the liquid in which it is placed. Consequently, it takes only two or three ice cubes to cool an entire glass of water.

The high heat of vaporization of water has important biological implications for us. When we perspire, water is vaporized as it evaporates from our skin. Because of the large heat of vaporization of water, the evaporation of a relatively small amount of water corresponds to the loss of a large amount of heat. This helps us to rid our bodies of excess internal heat through perspiration, thereby enabling us to maintain our normal body temperature.

Calculations Involving the Heat of Fusion and Heat of Vaporization of Water

We can calculate the heat required to melt a sample of ice by multiplying the appropriate form of the heat of fusion by the quantity of material being melted (Examples 12.10 and 12.11). Similarly, we determine the heat required to vaporize a sample of water at its boiling point by multiplying the heat of vaporization by the quantity of water (Example 12.12).

EXAMPLE 12.10

How much heat is required to melt 53.4 g of ice (at its melting point)?

Planning the solution

If we multiply the number of grams by the number of kilojoules per gram, we obtain the number of kilojoules. Used in this manner, the heat of fusion functions as a conversion factor from grams to kilojoules.

SOLUTION

$$q = 53.4\,\cancel{\text{g}}\left(\frac{0.334 \text{ kJ}}{\cancel{\text{g}}}\right) = 17.8 \text{ kJ}$$

EXAMPLE 12.11

How much heat must be removed to freeze 2.35 mol of water (at its melting point)?

SOLUTION

$$q = 2.35\,\cancel{\text{mol}}\left(\frac{6.01 \text{ kJ}}{\cancel{\text{mol}}}\right) = 14.1 \text{ kJ}$$

EXAMPLE 12.12

How much heat is required to vaporize 145 g of water (at its boiling point)?

SOLUTION

$$q = 145\,\cancel{\text{g}}\left(\frac{2.26 \text{ kJ}}{\cancel{\text{g}}}\right) = 328 \text{ kJ}$$

Other substances have their own characteristic heats of fusion and heats of vaporization. We will restrict ourselves here to problems involving water.

PROBLEM 12.13 Calculate the heat required to melt 17.4 g of ice (at 0°C).

PROBLEM 12.14 How much heat must be removed to freeze 7.82 mol of water (at 0°C)?

PROBLEM 12.15 How much heat must be supplied to completely vaporize 525 g of water (at 100°C)?

PROBLEM 12.16 Calculate the heat given off when 6.73 g of steam condense (at 100°C).

12.8 VAPOR PRESSURE

OBJECTIVE 12.8A Explain how a vapor pressure is created. State the major characteristics of an equilibrium system. Describe how the creation of a vapor pressure in a closed container conforms to the characteristics of a system at equilibrium.

OBJECTIVE 12.8B State how vapor pressure varies with temperature. Describe the vapor pressure condition that results in boiling. Distinguish between evaporation and boiling. Arrange a series of liquids in order of increasing normal boiling points, given their vapor pressures at a common temperature.

You have certainly witnessed the process of evaporation. For example, an uncovered glass of water set out in a room gradually evaporates or seems to disappear. In the process, molecules of water escape from the surface of the liquid and become gaseous water molecules. On the other hand, if the liquid is placed in a closed container (for example, a bottle with its cap screwed on), it will not evaporate. The molecules that escape from the surface of the liquid cannot diffuse into the atmosphere, because there is a cover preventing their escape. Nevertheless, some of the molecules at the surface of the liquid do escape to become vapor molecules above the surface of the liquid. The pressure exerted by these vapor molecules is known as the **vapor pressure.** You were introduced to the concept of vapor pressure in Section 11.13.

Vapor pressure: Pressure created by vapor molecules above the surface of the liquid from which they have escaped.

Evaporation: The escape of liquid molecules to the gaseous state at a temperature below the boiling point of the liquid.

Before we examine the concept of vapor pressure in detail, it will be helpful for us to define more precisely some of the terminology we will use. **Evaporation** occurs when molecules escape from the surface of an uncontained liquid that is below its

Vapor: Gaseous molecules of a substance that is normally a liquid at room temperature.

Vaporization: The process of forming a vapor.

boiling point. The gas formed in evaporation is referred to as a **vapor.** The term vapor is generally used to describe a gaseous material that is normally a liquid or solid at room temperature, such as water vapor in the air. The term *gas* is more correctly reserved for a substance that normally exists in the gaseous state at room temperature. **Vaporization** refers to the process of vapor formation.

Now let us return to our discussion of the vaporization of water. When we initially fill a glass with water and cover it (Fig. 12–7a), some of the molecules at the surface have enough energy to escape from the surface of the liquid to become vapor molecules in the space above the liquid. These molecules behave in the same manner as the molecules of any other gas, colliding with each other and with the walls of the glass. Some of the molecules collide with the surface of the liquid and return to the liquid state. When we first fill our glass and cover it, there are no water vapor molecules in the space above the liquid, so molecules only leave the liquid state. As time passes, however, the number of molecules in the space above the liquid surface increases, and so does the probability of a vapor molecule returning to the liquid state. Eventually, the number of molecules returning to the surface of the liquid equals the number leaving. Once this state is achieved, the vapor pressure remains constant.

Equilibrium: A dynamic condition in which the rates of opposing processes are equal.

The establishment of a constant vapor pressure illustrates a process that chemists refer to as **equilibrium.** In the establishment of a vapor pressure, molecules are constantly leaving the liquid state to enter the gaseous, or vapor, phase. At the same time, however, molecules of vapor are constantly returning to the liquid phase. During the initial stage of evaporation, the rate at which liquid molecules form vapor molecules is greater than that of the reverse process, because there are no vapor molecules. However, as the number of vapor molecules increases, the rate of the reverse process increases until the rates in both directions become equal. At that point vapor pressure equilibrium exists, and we will observe no further change in the vapor pressure. Equilibrium is characterized by the presence of two opposing processes that are occurring at equal rates.

We may express this establishment of a vapor pressure as follows:

$$H_2O(\ell) \rightleftharpoons H_2O(g)$$

The paired arrows, $\rightleftharpoons$, are used to show that the process occurs in both directions.

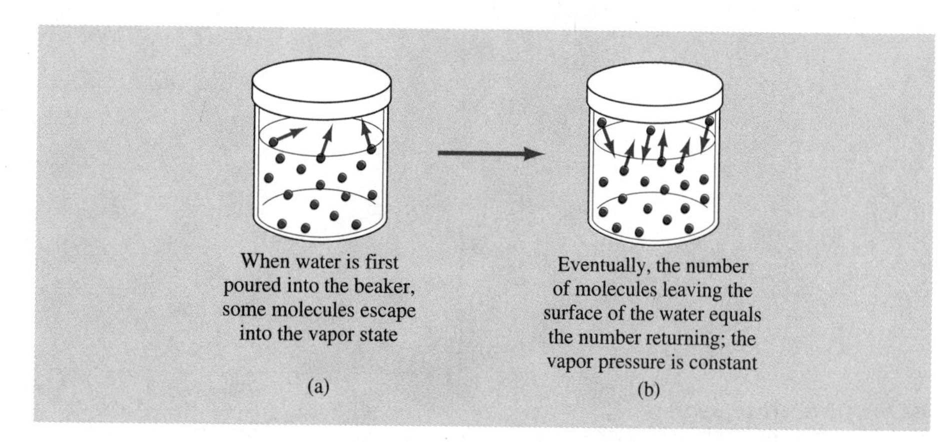

FIGURE 12-7 Evaporation of a liquid in a covered container.

It is important to remember that equilibrium is a dynamic process. It never stops. At the microscopic level (the level of the individual molecules), changes are constantly taking place. A molecule that is in the liquid state at one moment may be in the vapor state at a later time, and vice versa. However, at the macroscopic level (the larger level), we observe no further change in the vapor pressure once vapor pressure equilibrium is established. We discuss equilibrium in detail in Chapter 17. Meanwhile, the following three points summarize the characteristics of equilibrium.

1. Equilibrium is achieved when opposing processes take place at equal rates.
2. Equilibrium is a dynamic process—changes continue to occur at the microscopic (or molecular) level.
3. Equilibrium is characterized by constant macroscopic properties—at the gross level, no changes are observed.

As with gases, the temperature of a liquid is a measure of the average kinetic energy of its molecules. Consequently, as the temperature of a liquid sample is increased, the number of molecules with sufficient kinetic energy to escape from the surface of the liquid also increases. This behavior leads to an increase in the number and concentration of molecules in the vapor phase. *Increasing the temperature of a liquid always increases its vapor pressure* (Fig. 12-8). Furthermore, the vapor pressure of a pure substance depends *only* on the temperature of the liquid with which the vapor is in contact. It is independent of other factors, such as the volume or surface area of the liquid. Table 12-3 gives the vapor pressure of water over a wide range of temperatures. In it you can see that vapor pressure does increase with increasing temperature.

Closer inspection of Table 12-3 reveals that the vapor pressure of water at 100°C is 760.0 torr. It is no coincidence that the vapor pressure of water is equal to standard pressure at the temperature we usually think of as the boiling point. **Boiling** occurs when the vapor pressure of a liquid is equal to the atmospheric pressure acting on the surface of the liquid. Boiling differs from evaporation in that evaporation occurs only at the surface of the liquid, whereas boiling occurs within the body of the liquid. Thus, liquid molecules escaping to the gaseous state within the liquid form the bubbles we

Boiling: The process that occurs when the vapor pressure of a liquid equals the external pressure.

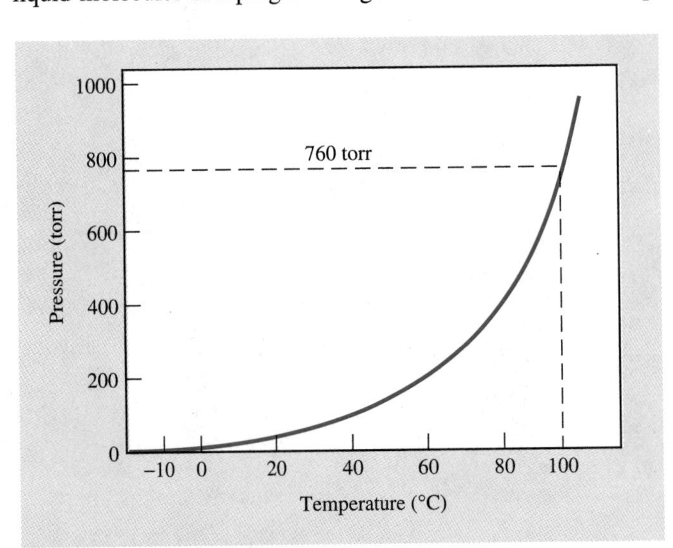

FIGURE 12-8 The vapor pressure of water increases with temperature.

TABLE 12-3 Vapor pressure of water at different temperatures

Temperature (°C)	Vapor pressure (torr)	Temperature (°C)	Vapor pressure (torr)
−15 (ice)	1.2	28	28.3
−10 (ice)	2.0	29	30.0
−5 (ice)	3.0	30	31.8
0	4.6	35	42.2
5	6.5	40	55.3
10	9.2	45	71.9
11	9.8	50	92.5
12	10.5	55	118.0
13	11.2	60	149.4
14	12.0	65	187.5
15	12.8	70	233.7
16	13.6	75	289.1
17	14.5	80	355.1
18	15.5	85	433.6
19	16.5	90	525.8
20	17.5	95	633.9
21	18.7	100	760.0
22	19.8	105	960.1
23	21.1	110	1074.6
24	22.4	120	1489.1
25	23.8	130	2026.2
26	25.2	140	2710.9
27	26.7		

generally associate with boiling. No such bubbles appear during evaporation, because evaporation is strictly a surface phenomenon.

We have said that, under ordinary circumstances, atmospheric pressure is about 760 torr. Since boiling occurs when the vapor pressure equals the atmospheric pressure, water will generally boil at a temperature close to 100°C. We refer to the boiling point of any liquid at standard pressure (1.000 atm) as its **normal boiling point (n-bp).** Thus, the normal boiling point of water is 100°C. However, the boiling point changes when the atmospheric pressure changes. Thus, if you go mountain climbing, the atmospheric pressure drops as you ascend to higher elevations. This decrease in atmospheric pressure is accompanied by a corresponding decrease in the boiling points of liquids, because a lower vapor pressure is required to equal the external pressure. The city of Denver, Colorado, is situated at an elevation of 1 mile above sea level. At this elevation the atmospheric pressure is about 630 torr. From Table 12-3 we can see that water generally boils at about 95°C in Denver.

Normal boiling point: The boiling point of a substance at standard pressure.

When you boil an egg, the speed with which the egg cooks depends on the temperature of the water rather than on the fact that the water is boiling. At higher temperatures, the rate at which an egg is cooked increases, whereas at lower temperatures, it cooks more slowly. Thus, people who go camping at elevations above about 8000 ft know that it takes a long time to cook a hard-boiled egg in an open pot. Although the water boils, the temperature at which it boils is too low to cook an egg in a convenient length of time. On the other hand, we frequently use a pressure cooker when we want to shorten the cooking time. A pressure cooker increases the pressure acting on its contents, so the water inside boils at a higher temperature than normal, decreasing the cooking time required.

Figure 12-9 shows how the vapor pressures of three common liquids (water, ethyl alcohol, and ethyl ether) vary with temperature. Note that the vapor pressure of each of the liquids increases with increasing temperature. Let us examine the vapor pressure of each of these substances at room temperature (20°C) to see how it is related to the normal boiling point.

Substance	Vapor pressure (20°C)	Normal boiling point
Water	17.5 torr	100°C
Ethyl alcohol	43.9 torr	78°C
Ethyl ether	442 torr	35°C

Note that water, with the lowest vapor pressure at room temperature, has the highest boiling point. The low vapor pressure of water reflects the relatively high intermolecular forces of attraction within the liquid. Thus, it is necessary to heat water to a higher temperature than either of the other two liquids in order to get it to boil (assuming equal external pressures). Conversely, ethyl ether has comparatively low intermolecular forces. Thus, ether molecules escape fairly easily from the surface of the liquid, leading to a substantial vapor pressure at room temperature. It requires only a slight elevation of temperature above 20°C to reach the normal boiling point of ethyl ether.

You may be aware that both boiling and evaporation are cooling processes. When water (or any other liquid) evaporates, it is the highest-energy molecules that escape from the surface of the liquid. The liquid sample left behind, having fewer high-energy

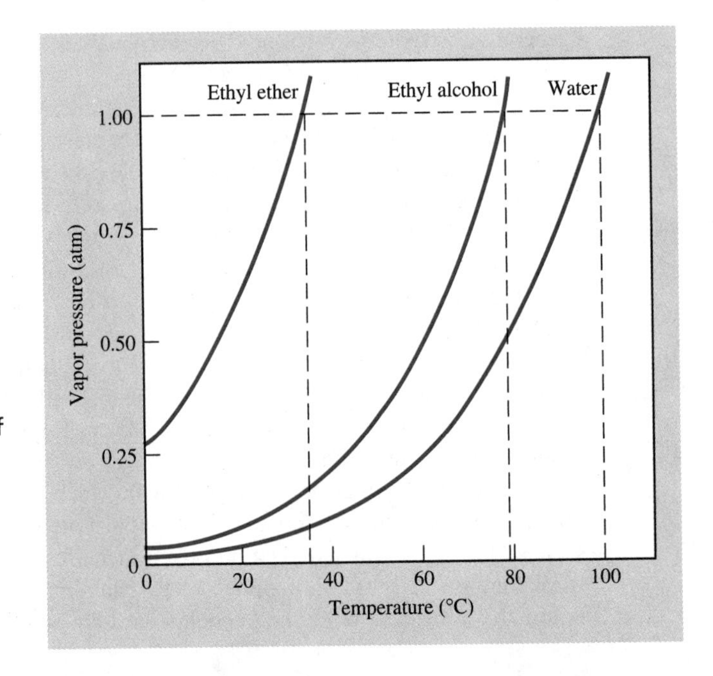

FIGURE 12-9 The variation of vapor pressure with temperature. The vapor pressure of each liquid increases with temperature. The normal boiling point of each is the temperature at which the vapor pressure is 1.00 atm.

molecules, has a lower temperature. Even on a warm day, you may experience a chill after stepping out of a swimming pool. The evaporation of water from the surface of your skin cools you. In our discussion of heats of vaporization, we mentioned the cooling that occurs when we perspire. The molecules that vaporize carry the heat of vaporization with them as they depart. When a liquid boils, heat is removed from the liquid as vapor bubbles leaving the liquid carry high-energy molecules with them. As a result, the temperature of a boiling liquid remains at its boiling point until all of the liquid has been vaporized. Turning up the flame under a boiling pot of water may increase the rate at which the liquid turns to vapor, but it will not increase the temperature of the liquid.

PROBLEM 12.17

The normal boiling points of three compounds (*A*, *B*, and *C*) are *A:* 53°C; *B:* 165°C; and *C:* 97°C.

(a) Which of these substances has the highest vapor pressure at room temperature?
(b) Which has the lowest vapor pressure at 115°C?

PROBLEM 12.18

Substances *D, E,* and *F* have the following vapor pressures at room temperature (20°C):

D: 135 torr *E:* 13 torr *F:* 960 torr

(a) Which of these substances has the highest boiling point?
(b) Which of these substances is a gas at room temperature (assuming the atmospheric pressure is 1 atm)?
(c) Do we have enough information to tell whether any of these substances is a solid at room temperature?

12.9 PROPERTIES OF THE LIQUID STATE

OBJECTIVE 12.9 Describe the following properties associated with liquids: viscosity, surface tension, capillary action, and miscibility.

Intermolecular forces of attraction play a role in a number of interesting physical properties associated with liquids.

Viscosity

Viscosity: The resistance of a liquid to flow.

The **viscosity** of a liquid represents its resistance to flow. A highly viscous liquid seems "thick" and pours slowly. The ratings given to various automobile motor oils are measures of their viscosity. A 40-weight oil is more viscous than a 30-weight oil. We can measure the viscosity of a liquid by determining how long it takes a given volume of the liquid to flow through a specific opening. For example, suppose we compare the time it takes to pour equal volumes of pancake syrup and water through the same funnel. The syrup is more viscous and takes longer. Viscosity is the result of intermolecular forces of attraction between liquid molecules. Viscosity decreases with increasing temperature. For example, when pancake syrup is heated, it pours more easily than unheated syrup. As the temperature of the liquid is increased, the molecules gain more kinetic energy and move more rapidly. Attractive forces between molecules are more easily overcome, so the molecules flow more freely over one another.

Surface Tension and Capillary Action

Surface tension: Property of liquids caused by the attraction of surface molecules to the interior of the liquid.

Capillary action: Phenomenon observed when a liquid climbs up a wick or rises in a glass tube.

Molecules in the center of a liquid sample are acted upon by intermolecular forces in all directions (Fig. 12-10). However, molecules at the surface of a liquid are drawn only toward the interior of the liquid, because there are no compensating molecules on the other side of the surface. These intermolecular forces create a **surface tension** that holds liquid droplets together.

Closely related to surface tension is the phenomenon of **capillary action,** which occurs when the molecules of a liquid are attracted to molecules of a different substance. We observe this phenomenon when water climbs up a paper towel or a wick. Capillary action is also responsible for the curved meniscus we observe when measuring volumes of aqueous solutions in the laboratory. Attractive forces between the water molecules and the glass surface of a burette or graduated cylinder draw the liquid up the sides, resulting in the characteristic curved shape seen in Fig. 12-11.

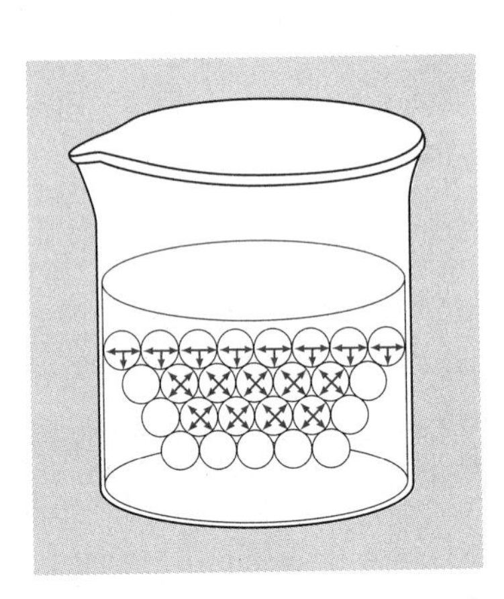

FIGURE 12-10 The creation of surface tension. Molecules in the center of a liquid are acted upon in all directions. Because molecules at the surface have no liquid molecules on the other side of the surface boundary, they are drawn toward the interior.

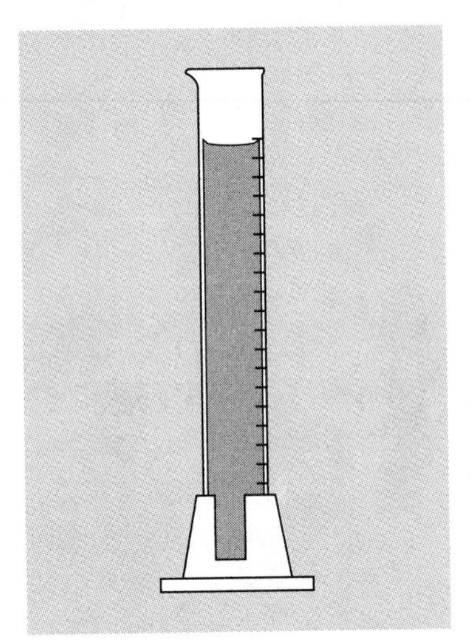

FIGURE 12-11 The meniscus of an aqueous solution has a curved shape caused by the intermolecular forces of attraction between water molecules and the glass container. These forces draw the water molecules up along the sides of the vessel.

Miscibility

Miscible: Refers to liquids that are capable of dissolving in one another.

Because of their ability to dissolve other substances, liquids are extremely useful as solvents. However, many combinations of liquids do not dissolve in one another; that is, they are not **miscible.** For example, gasoline and water do not dissolve in one another.

Immiscible: Refers to liquids that are incapable of dissolving in one another.

We refer to such "nonmixable" liquids as **immiscible.** In general, liquids are immiscible when one is a polar substance and one is nonpolar. In this case, gasoline is a mixture of nonpolar liquids, whereas water is a polar substance. Because a polar molecule experiences a greater attractive force for another polar molecule than for a nonpolar molecule, the polar molecules separate from the nonpolar molecules, resulting in the formation of two liquid layers. When immiscible liquids separate from one another, the less dense liquid always floats on top.

PROBLEM 12.19

Water (d = 1.00 g/mL) and chloroform (d = 1.48 g/mL) are immiscible. Which liquid will float on top?

12.10 PROPERTIES OF THE SOLID STATE

OBJECTIVE 12.10 Distinguish between an amorphous solid and a crystalline solid. List the four types of crystalline solids, and give an example of each.

Perhaps you have examined a granule of table salt under a magnifying lens. If you do so, you will notice that it has flat surfaces. The orderly nature of its surfaces is the result of an orderly arrangement of its ions. We refer to each granule of this type of substance as a *crystal.* Recalling Fig. 8-1, the ions in a sample of sodium fluoride are arranged in an orderly fashion, with a regular alternation of positive and negative ions. A similar arrangement of ions is present in a crystal of sodium chloride (table salt). We refer to the particular arrangement of ions as the **crystal lattice** structure.

Crystal lattice: A particular arrangement of the atoms, ions, or molecules in a crystal.

Crystalline solid: A solid with an orderly arrangement of atoms, ions, or molecules.

Amorphous solid: A solid that lacks an orderly arrangement of atoms, ions, or molecules.

Solids can be classified as **crystalline** or **amorphous.** Amorphous solids lack the well-defined arrangement that characterizes crystalline solids (Fig. 12-12, p. 368). Glass and rubber are examples of amorphous solids. The lack of internal order in these solids is reflected in their lack of sharp melting points. Whereas sodium fluoride melts sharply at 993°C, glass softens and melts over a wide temperature range.

Crystalline solids are generally divided into four categories: ionic crystals, molecular crystals, metallic crystals, and network solids. We have already examined sodium chloride, a typical ionic crystal. The same type of orderliness found in sodium chloride is also found in sucrose (table sugar), a molecular crystal, and is responsible for its crystalline appearance. In Chapter 8, we briefly examined the structure of metals and saw that a metallic solid is composed of a regular arrangement of metallic "kernels" surrounded by a sea of electrons (Fig. 8-8). This regular arrangement of the metal atoms allows us to classify metals as crystals, too. Finally, a network solid is one in which the atoms are bonded in a regular three-dimensional array. Diamond is a crystalline form of carbon in which each carbon atom is at the center of a tetrahedron, bonded to four other carbon atoms. The entire array extends regularly in three dimensions (Fig. 12-13).

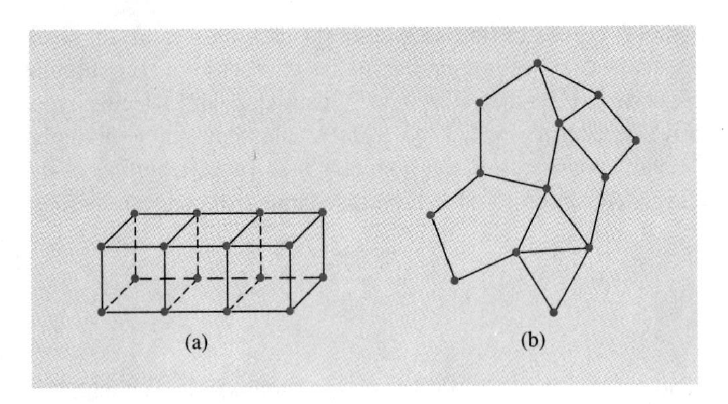

FIGURE 12-12 (a) A crystalline solid has a regular arrangement. (b) An amorphous solid lacks internal order.

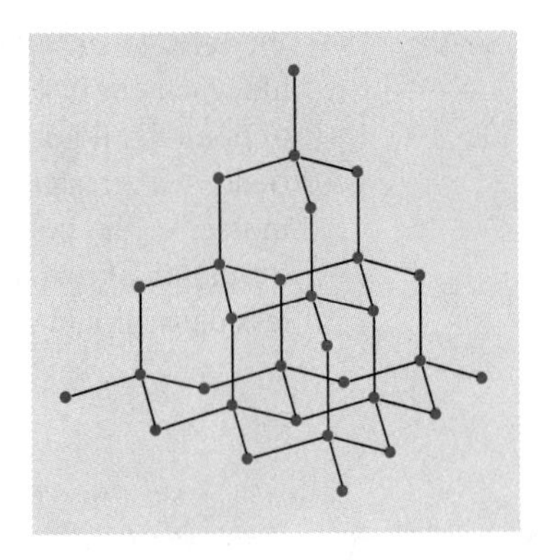

FIGURE 12-13 Diamond is a network solid. Each carbon atom is bonded to four other carbon atoms in a tetrahedral arrangement.

PROBLEM 12.20 Which of the four types of crystal would you expect to find in a snowflake? Explain your answer.

12.11 WATER

OBJECTIVE 12.11 Cite three unusual properties of water.

Let us conclude this chapter by examining several of the properties of water—one of the simplest molecules known, yet one of the most remarkable. Water is the most abundant compound in any living organism, and its unusually high specific heat helps protect living organisms from sudden changes in temperature. The heat of vaporization of water is also quite high (40.7 kJ/mol), providing additional protection by removing excess body heat through perspiration.

The high heats of vaporization and fusion of water also contribute to maintaining a relatively narrow range of temperatures on earth. Whereas the moon (which has no surface water) experiences a dramatic temperature variation between the light side and the dark side, the earth's surface has a much more uniform temperature profile. While this effect is partially caused by the earth's atmosphere (which traps radiant sunlight, maintaining the warmth we need for survival), both the evaporation and condensation, as well as the freezing and thawing of the earth's waters, help moderate the highs and lows of temperature over the planet's surface. This temperature stabilization is critical for the maintenance of life on this planet.

The density of water is another unusual property that has profound biological consequences. Most substances expand on heating, so we normally find that the density decreases with heating and increases with cooling. [*Remember:* Density and volume are inversely proportional: $d = m/V$.] As is true of most substances, the density of water increases as it is cooled to 4°C. At 4°C, however, water reaches a maximum density, and further cooling results in expansion of the liquid with a corresponding *decrease* in density. This expansion continues until water reaches its freezing point, at

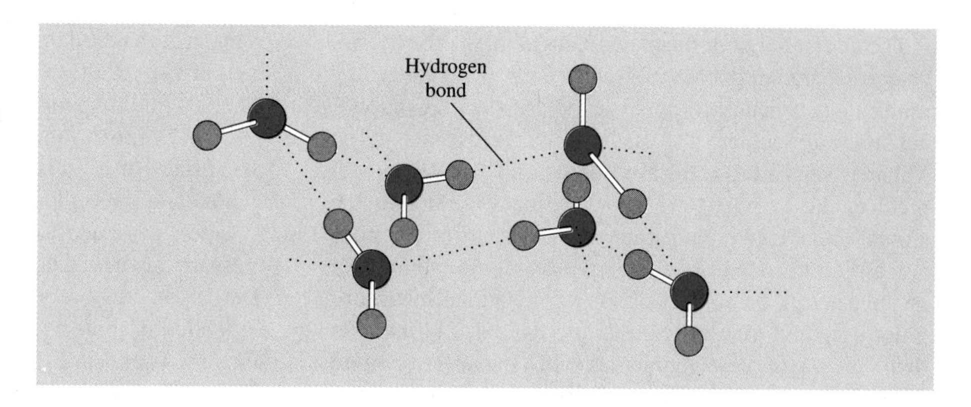

FIGURE 12-14 The structure of ice. Hydrogen bonds form between the hydrogen atoms of one water molecule and the oxygen atoms of other molecules.

which point it expands even further, producing a solid that is less dense than its liquid form. If you have ever forgotten a bottle of soda that you put in the freezer for quick cooling, you may have personally experienced the effects of the expansion of water at its freezing point! The beverage in a bottle allowed to freeze will expand, possibly cracking its container and leaving a mess.

Expansion of water comes about as the molecules orient themselves in the regular fashion shown in Fig. 12-14. This particular arrangement involves extensive hydrogen bonding between neighboring molecules, but it leaves considerable space between the molecules, causing the expansion just mentioned.

Because the density of water decreases as it freezes, ice has a lower density than the liquid in which it is immersed. Thus, ice always floats in liquid water. This may not seem to be of great importance, but its biological significance is profound. During the winter, lakes freeze at the surface, where the water comes in contact with the cold air above. This forms a layer of ice that not only floats but also insulates the water below, protecting the marine life in the lake. If ice were more dense than its liquid surroundings, it would sink to the bottom of the lake, permitting more water to freeze at the surface. In time the lake would be frozen solid, from the bottom up. The effect on the life forms that inhabit the lake would be disastrous.

Finally, the polar characteristics of water make it an excellent solvent for a large number of ionic substances, as well as for substances that hydrogen-bond. In the chapters that follow, we examine how water dissolves other substances, in addition to discussing a number of reactions that occur in aqueous solution.

PROBLEM 12.21 Explain why ice (solid water) is less dense than liquid water.

CHAPTER SUMMARY

The properties of matter in the solid and liquid states may be largely attributed to the **intermolecular forces** of attraction that exist between molecules. Whereas **intramolecular forces** are responsible for holding atoms together in molecules, intermolecular forces hold molecules together in aggregates, conferring many of their characteristic properties. In general, these forces arise from the charge characteristics of the molecules in question. We classify molecules as **polar** or **nonpolar,** depending on the overall charge separation in each molecule. Thus, the polarity of a molecule depends not only on the polarities of the individual bonds present, but also on the molecular geometry. The **Valence Shell Electron Pair Repulsion (VSEPR)** model predicts the geometry of a molecule by arranging the atoms so that the bonding pairs and lone pairs of electrons are as far from one another as possible. A number of geometries are commonly observed, including **linear, angular, trigonal planar, trigonal pyramidal,** and **tetrahedral.** Several of these geometries have characteristic **bond angles** associated with them.

Just as ionic compounds are held together as the result of electrostatic forces of attraction between ions, molecular substances are held together by forces that are generally electrostatic in nature. These intermolecular forces include **hydrogen bonding, dipole–dipole interactions,** and **London** (or **dispersion**) **forces.** If **ionic interactions** are included with intermolecular forces, the various forces may be arranged in order of decreasing strength as follows: ionic interactions > hydrogen bonding > dipole–dipole interactions > London forces. The boiling points and melting points of substances reflect the strengths of these various forces. In general, ionic compounds have much higher melting points and boiling points than covalent compounds. For substances that exhibit similar types of interaction, the boiling point generally increases with molecular mass.

Most substances can exist in any of the three physical states at atmospheric pressure, depending on the temperature. As a substance is heated, the average kinetic energy of its molecules increases. This results in an increase in the tendencies of the molecules to overcome the intermolecular forces of attraction between them. A **warming curve** is used to describe the temperature and phase changes a substance goes through as heat is supplied. Heat may be supplied to a solid, a liquid, or a gas, causing it to increase in temperature (so long as the material is not at its melting point or boiling point). The amount of heat required to raise the temperature of 1 gram of the material by 1°C is called that substance's **specific heat.** Alternatively, we refer to the **molar heat capacity,** the amount of heat required to raise the temperature of a mole of the substance by 1°C. Heat must also be supplied to a substance to cause a phase change. The **heat of fusion** is the heat required to melt a solid at its melting point. The **heat of vaporization** is the heat required to vaporize a liquid at its boiling point.

In **evaporation,** high-energy molecules escape from the surface of a liquid below its boiling point, creating a **vapor.** The molecules that **vaporize** in this fashion create a **vapor pressure** above the liquid surface. The vapor pressure always increases with increasing temperature. The vapor pressure at room temperature is a measure of the ability of molecules to escape from the liquid state and enter the vapor state. Molecules having low intermolecular forces of attraction have high vapor pressures, which correspond to low boiling points. **Boiling** occurs when the vapor pressure of a liquid equals the external pressure. The **normal boiling point** is the boiling point observed at standard pressure.

The vapor pressure that exists above the surface of a liquid in a closed container exhibits the characteristics of **equilibrium.** While some molecules are leaving the liquid state to become vapor molecules, other molecules of vapor are returning to the liquid state. An equilibrium state is achieved when opposing processes, such as this, take place at equal rates. Equilibrium is a dynamic process at the molecular level and is characterized by constant macroscopic properties.

Solids may be classified as either **crystalline** or **amorphous.** Crystalline solids are characterized by an orderly **crystal lattice.** Among the crystalline solids are ionic crystals, molecular crystals, metallic crystals, and network solids.

The liquid state is characterized by a number of interesting properties that result from intermolecular forces. These properties include **viscosity, surface tension,** and **capillary action.** Different liquids that are mutually soluble are said to be **miscible,** whereas liquids that do not dissolve completely in one another are **immiscible.**

Water is a rather remarkable liquid with many unusual properties. It hydrogen-bonds extensively to itself, leading to its unusually high boiling point for a substance of its molecular weight. Its polar properties and ability to hydrogen-bond make it an excellent solvent for a large number of substances. Water has a high specific heat, a high heat of fusion, a high heat of vaporization, and an unusual density curve.

KEY TERMS

Review each of the following terms, which are discussed in this chapter and defined in the Glossary.

intramolecular force
intermolecular force
linear
angular
trigonal planar
trigonal pyramidal
tetrahedral
Valence Shell Electron Pair Repulsion (VSEPR) model
bond angle
polar molecule
nonpolar molecule
ionic interaction
hydrogen bonding
dipole–dipole interaction
London forces
warming curve
heat of fusion
heat of vaporization
specific heat
molar heat capacity
vapor pressure
evaporation
vapor
vaporization
equilibrium
boiling
normal boiling point
viscosity
surface tension
capillary action
miscible
immiscible
crystal lattice
crystalline
amorphous

ADDITIONAL PROBLEMS

SECTION 12.1

12.22 Predict the geometry of each of the following molecules; the Lewis structures are given.

(a) BCl_3

```
      ..
     :Cl:
 ..   |   ..
:Cl — B — Cl:
 ..       ..
```

(b) SCl_2

```
 ..   ..
:Cl — S:
 ..   |
     :Cl:
      ..
```

(c) PF_3

```
 ..   ..   ..
:F —  P —  F:
 ..   |    ..
     :F:
      ..
```

(d) $CHCl_3$

```
      H
 ..   |   ..
:Cl — C — Cl:
 ..   |   ..
     :Cl:
      ..
```

(e) SO_2

```
 ..   ..   ..
:O —  S == O
 ..        ..
```

SECTION 12.2

12.23 Predict the polarity of each of the following molecules. (The geometry is supplied.) If the dipole moment is other than zero, state the direction of the net dipole.

(a) SiH_4 (tetrahedral) **(b)** CHF_3 (tetrahedral)
(c) SO_2 (angular, S in center) **(d)** CO_2 (linear)
(e) BCl_3 (trigonal planar) **(f)** PCl_3 (trigonal pyramidal)

SECTION 12.3

12.24 Match each of the following compounds with the strongest type of interaction it is capable of making. (There is one compound for each type of interaction.)

(a) HF **1.** Ionic forces
(b) CH_4 **2.** Hydrogen bonding
(c) KF **3.** Dipole–dipole interaction
(d) CH_3F **4.** London forces

12.25 Which of the following molecules are capable of hydrogen bonding with an identical molecule?

(a)

```
    H   H
    |   |
H — N — N — H
    ..  ..
```

(b)

```
    H   H
    |   |
H — C — C — H
    |   |
    H   H
```

(c)

```
    H   H   ..
    |   |
H — C — C — O — H
    |   |   ..
    H   H
```

(d)

```
    H   H   ..  H   H
    |   |       |   |
H — C — C — O — C — C — H
    |   |   ..  |   |
    H   H       H   H
```

(e)

```
     ..
H — Br:
     ..
```

SECTION 12.4

12.26 A substance has a melting point of 681°C and a boiling point of 1330°C. Which of the following compounds is it most likely to be? Explain your answer.
(a) O_2 **(b)** CH_3OH **(c)** CH_3F **(d)** KI

12.27 A substance has a melting point of –218°C and a boiling point of –183°C. Which of the following compounds is it most likely to be? Explain your answer.
(a) O_2 **(b)** CH_3OH **(c)** CH_3F **(d)** KI

12.28 Carbon monoxide (CO) and nitrogen (N_2) have the same molecular mass, yet carbon monoxide boils at a higher temperature than nitrogen. Explain why carbon monoxide has the higher boiling point.

12.29 Would you expect methane (CH_4, molecular mass = 16.0) or ammonia (NH_3, molecular mass = 17.0) to have a higher boiling point? Explain your answer. (Assume that the slight difference in molecular masses is not a significant factor in your answer.)

12.30 For each of the following pairs of compounds: (1) determine the most significant attractive force for both compounds, (2) predict which compound has the higher boiling point, and (3) explain your prediction.

```
         H
         |
(a)    H-C-F   and   LiF
         |
         H

         H  H  H  H  H
         |  |  |  |  |     ..
(b)    H-C--C--C--C--C--O-H      and
         |  |  |  |  |     ..
         H  H  H  H  H

          H  H  H  H
          |  |  |  |    ..
        H-C--C--C--C--O-H
          |  |  |  |    ..
          H  H  H  H

         H  H  H                   H          H
         |  |  |  ..               |    ..    |
(c)    H-C--C--C--N-H   and      H-C----N-----C-H
         |  |  |  |                |    |     |
         H  H  H  H                H    |     H
                                      H-C-H
                                        |
                                        H

(d) PH3  and  SiH4

         H  H  H  H
         |  |  |  |  ..
(e)    H-C--C--C--C--N-H      and
         |  |  |  |  |
         H  H  H  H  H

            H  H  H
         .. |  |  |  ..
        H-N-C--C--C--N-H
          | |  |  |  |
          H H  H  H  H
```

12.31 Many backpackers use butane stoves for cooking. (Butane is a gas that boils at 0.5°C.) A backpacker observes that she has difficulty lighting her stove on mornings when there is frost on the ground. Explain this observation.

12.32 On the basis of the physical data given, identify each of the following substances as a solid, a liquid, or a gas at room temperature (20°C).

	mp (°C)	bp (°C)
(a) Benzoic acid	122	249
(b) Methyl alcohol	–94	65
(c) Nitrogen	–210	–196
(d) Phenol	43	182
(e) Propane	–190	–42
(f) Phosphorus tribromide	–40	173
(g) *para*-Dichlorobenzene	53	174
(h) Phosgene	–118	8

12.33 Classify each of the substances in Problem 12.32 as a solid, a liquid, or a gas at 50°C; at –50°C.

12.34 Classify each of the substances in Problem 12.32 as a solid, a liquid, or a gas at –100°C; at +100°C.

Problems 12.36–12.38 and 12.42–12.45 require specific heats, heats of fusion, and heats of vaporization that are provided in Sections 12.6 and 12.7.

SECTION 12.6

12.35 When an aluminum tray is taken out of a hot oven, it cools rapidly. Explain why.

12.36 Calculate the heat given off when 73.0 g of water cool from 45.0°C to 22.0°C.

12.37 How much heat is required to heat 94.3 g of copper from 45.0°C to 76.0°C?

12.38 How much heat is given off when 75.0 g of lead cool from 100.0°C to 35.0°C?

12.39 The properties of *specific heat, heat of fusion,* and *heat of vaporization* are involved in each of the following statements. Match each statement with the property involved.

(a) As hot candle wax drips down the side of a candle, it solidifies.

(b) A cup of hot coffee cools when left to stand.

(c) If the flame under a pot of boiling water is maintained, the water eventually disappears.

SECTION 12.7

12.40 When an ice cube is placed in a glass of room-temperature water, a relatively small amount of ice is capable of cooling a fairly large volume of water.

(a) What physical property is responsible for the ability of ice to cool the liquid in which it is placed?

(b) Explain what takes place as ice acts to cool the liquid.

(c) Why is a small amount of ice able to cool a relatively large volume of liquid?

12.41 Water has a relatively high heat of vaporization. When we perspire, this property enables us to maintain our body temperature. Explain why.

12.42 Calculate the heat required to melt 3.17 mol of ice (at 0°C).
12.43 How much heat must be removed to freeze 7.82 g of water (at 0°C)?
12.44 Calculate the heat required to convert 8.50 mol of water at its boiling point to water vapor at the same temperature.
12.45 How much heat is given off when 76.0 g of steam condense at its boiling point?

SECTION 12.8

12.46 A constant vapor pressure exists above the liquid in a closed container. Describe how this phenomenon exhibits the characteristics of equilibrium.
12.47 Match each of the terms *boiling, evaporation, boiling point,* and *normal boiling point* with the statement below that best characterizes it.
(a) The temperature at which the vapor pressure of a liquid equals the external pressure.
(b) The temperature at which the vapor pressure of a liquid equals standard pressure.
(c) A process that occurs only at the surface of a liquid.
(d) A process that occurs within the entire liquid.
12.48 Explain how evaporation leads to cooling.
12.49 The normal boiling points of toluene, benzene, and acetone are 110°C, 80°C, and 56°C, respectively. Which of these substances has the lowest vapor pressure at room temperature?
12.50 Substances *H, I,* and *J* have the following vapor pressures at 78°C: *H,* 535 torr *I,* 1380 torr *J,* 760 torr
(a) Which of these substances has the lowest boiling point?
(b) What is the physical state of substance *I* at 78°C and standard pressure?
(c) What is the normal boiling point of substance *J?*

SECTION 12.9

12.51 Explain each of the following observations.
(a) Warm pancake syrup flows more easily than cold pancake syrup.
(b) When water is placed in a glass tube, it creeps up the walls of the glass, forming a meniscus.
(c) Liquids form droplets.

SECTION 12.10

12.52 Answer the following questions about solids.
(a) Distinguish between a crystalline and an amorphous solid.
(b) List the four types of crystalline solids and give an example of each.

GENERAL PROBLEMS

12.53 Distinguish between intermolecular forces and intramolecular forces.
12.54 Why would it be incorrect to consider ionic interactions as "intermolecular" forces?
12.55 Would you expect carbon tetrachloride, CCl_4, to be miscible with water? Would you expect methyl alcohol, CH_3OH, to be miscible with water? Explain your answers.
12.56 Ethyl ether, $CH_3CH_2OCH_2CH_3$, is slightly soluble in water (8 g/100 mL of water). Show how hydrogen bonding permits this substance to dissolve in water.
12.57 For each of the following molecules, write a Lewis structure, determine its geometry, and then classify it as polar or nonpolar.
(a) CF_4 **(b)** HCN **(c)** PF_3 **(d)** CH_2Cl_2 **(e)** SO_3
12.58 Arrange the elements Br_2, Cl_2, and F_2 in order of (a) increasing boiling point; (b) increasing vapor pressure; and (c) increasing molecular mass.
(d) Which of the intermolecular forces of attraction is the one responsible for holding molecules of these elements together in the liquid or solid states?
12.59 The *CRC Handbook of Chemistry and Physics* reports the solubility of fluoroethane (C_2H_5F) as *soluble* in water. The other halogen analogs (C_2H_5Cl, C_2H_5Br, and C_2H_5I) are all reported as only *slightly soluble.* Explain the enhanced solubility of fluoroethane in water.
12.60 When placed in a glass tube, water forms a meniscus that is concave (curves downward), whereas a mercury meniscus is convex (curves upward). The shape of a water meniscus may be explained by comparing the attraction of water molecules for the glass surface to the attractive forces that exist between the water molecules themselves. Elaborate on this comparison (complete the logic), thereby explaining the shape of a water meniscus, and then propose an explanation for the differing shape of a mercury meniscus.

***12.61** Two metal blocks (*A* and *B*) at 20°C have identical masses. Both are placed in identical insulated vessels containing identical masses of water at 80°C. The vessel of water containing metal *A* comes to thermal equilibrium at a higher final temperature than the vessel with metal *B*. Which metal has the greater specific heat? Explain your answer.

***12.62** *cis*-1,2-Dichloroethene and *trans*-1,2-dichloroethene are both planar molecules with the following structures:

Cl Cl
C=C
H H

cis-1,2-Dichloroethene

Cl H
C=C
H Cl

trans-1,2-Dichloroethene

The boiling points of these compounds are 48°C and 60°C. Predict which boiling point goes with each compound, and explain your selection.

***12.63** Mercury is a toxic metal that causes nervous system disorders. Although the vapor pressure of mercury is relatively low at normal temperatures (the vapor pressure is 2.5×10^{-6} atm at 25°C), it represents a high enough concentration to pose an unsafe environment. If a pool of mercury is allowed to stand in an unventilated room, vapor pressure equilibrium can be achieved. Suppose you work in an unventilated room with dimensions of 5.00 m by 4.00 m by 2.50 m. How many milliliters of liquid mercury would have to vaporize to achieve vapor pressure equilibrium at 25°C? The density of mercury is 13.6 g/mL. [*Hint:* Begin by calculating the number of moles of mercury vapor required to fill the room at the temperature and pressure stated in the problem.] If there are approximately 25 drops in a milliliter, how many drops of mercury would this volume equal?

***12.64** How many kilojoules are required to warm a 36-g cube of ice at −25°C to liquid water at +21°C? The average specific heat of ice is 2.0 J/g•°C over the temperature range from −25°C to 0°C. Refer to the text for additional values required to solve the problem. [*Caution:* What individual steps are required for the water sample to warm from −25°C to +21°C?] To what process is most of the heat supplied?

***12.65** A 20-g cube of ice at 0°C is dropped into 80 g of water at 70°C. What will be the final temperature of the water? To simplify your calculations, use the specific heat of 1.0 cal/g•°C for liquid water and 80 cal/g for the heat of fusion of ice. (Assume that all values stated in the problem are accurate to two significant figures. Finally, assume that no heat is lost to the surroundings.)

WRITING EXERCISES

12.66 Your father often cooks lasagne in a glass casserole dish, which he places in the oven on an aluminum cookie sheet. He has always been puzzled by the fact that when he removes the casserole (with the cookie sheet) from the oven, he can handle the cookie sheet with his bare hands long before he can grasp the glass casserole dish. Write a note to your father, explaining the concept of specific heat and why the cookie sheet cools off so much more rapidly than the glass dish.

12.67 In your own words, explain how you can tell whether a substance is a solid, a liquid, or a gas. Begin with a description of a typical warming curve, and conclude with a discussion of how the physical state of a substance can be determined at any temperature, given its melting point and boiling point.

12.68 One of your friends has some misconceptions about dry ice. She has observed that a puddle of water always remains after a block of dry ice "evaporates," and she thinks it is liquid carbon dioxide. Write her a note, explaining the process of sublimation and how it differs from the behavior of most substances under normal conditions. Then explain what the liquid is that she has observed and where it comes from.

12.69 In your own words, explain how the polarity of a molecule is determined by the geometry of the molecule as well as the polarities of its bonds.

12.70 One of your friends is an avid backpacker. He has noticed that all of his backpacking foods contain instructions to increase cooking time at higher elevations. Write a note to your friend, discussing the relationship between temperature and vapor pressure, and how the external pressure affects the boiling point of a liquid. Then explain how higher altitudes affect the boiling point of water and why this results in longer cooking times. Contrast this to the behavior observed in a pressure cooker, where the increased pressures result in higher boiling points and shorter cooking times.

SOLUTIONS

OBJECTIVE 13.0 Define *solution.* Explain what a liquid solution is. Give examples of two types of solutions other than liquid solutions.

Solution: Any homogeneous mixture.

When we carry out experiments in the laboratory, it is often helpful to dissolve the substances we are working with in a liquid. For example, we might dissolve sodium chloride in water. We refer to the resulting homogeneous mixture as a solution. A **solution** is any homogeneous mixture of two or more substances. Thus, a mixture of gases such as air is a solution. Similarly, metal alloys such as brass (which is a homogeneous mixture of copper and zinc) are solutions. In this chapter we consider only solutions prepared by dissolving a solid, a liquid, or a gas in a liquid. We refer to these as *liquid solutions.*

13.1 THE NATURE OF LIQUID SOLUTIONS

OBJECTIVE 13.1 For liquid solutions, define *solvent* and *solute.* Explain what happens to solute particles when a nonionic solute dissolves in a solvent. Contrast this behavior with that of ionic solutes. Explain why liquid solutions are widely used.

Solvent: The dissolving agent.

Solute: The substance dissolved.

Aqueous solution: A liquid solution in which the solvent is water.

When we work with liquid solutions, the liquid we use to dissolve the substance is known as the **solvent,** and the substance being dissolved is the **solute.** The solute can be either a solid, a liquid, or a gas. Most of the solutions we will consider are **aqueous solutions**—solutions in which the solute is dissolved in water. Many of the products we use in our homes are aqueous solutions. For example, vinegar is an aqueous solution of acetic acid. Similarly, many cosmetics, mouthwashes, and after-shave lotions are aqueous solutions.

When a solution is prepared by dissolving a solid in a liquid, the solid is the solute, and the liquid is the solvent. Thus, the solute in an aqueous sodium chloride solution is sodium chloride, and water is the solvent.

When referring to liquids dissolved in liquids, it is purely arbitrary which substance is considered the solute and which is the solvent. Under most circumstances, we call the substance that is in smaller quantity the solute, while the other is the solvent. However, that is not always the case. For example, ethyl alcohol solutions are frequently prepared by dissolving ethyl alcohol in water. The alcohol is generally considered the solute, even when it is found in the larger amount.

Solutions may also be prepared by dissolving a gas in a liquid. In this case, the gas is the solute, while the liquid is the solvent. Earlier we mentioned that hydrochloric acid, HCl(aq), is an aqueous solution of hydrogen chloride, HCl(g), dissolved in water.

When a solute is dissolved in a solvent, the molecules or ions that make up the solute separate from one another, as they become surrounded by molecules of the solvent. For example, when a molecular substance such as sucrose (table sugar, $C_{12}H_{22}O_{11}$) is dissolved in water, each sucrose molecule becomes surrounded by molecules of water (Fig. 13-1). When ionic substances dissolve in water, the various cations and anions of the solute become separated from one another. Thus, when sodium chloride is dissolved, each sodium ion becomes surrounded by water molecules, as does each chloride ion (Fig. 13-2). A molecule or ion that is surrounded by solvent is said to be *solvated;* if the solvent is water, it is *hydrated.*

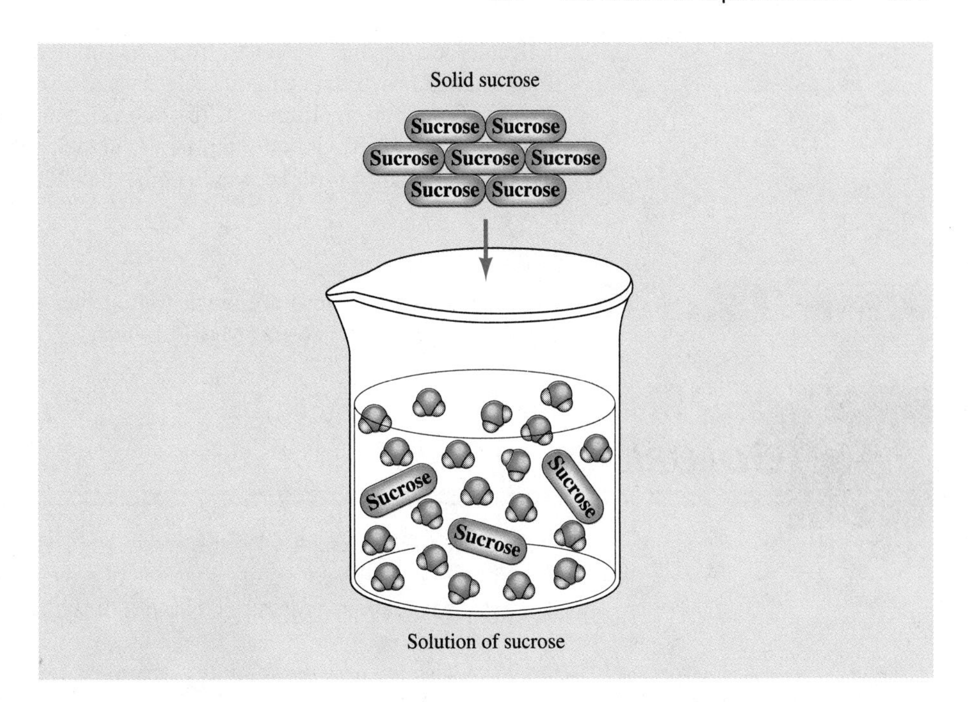

FIGURE 13-1 When sucrose dissolves in water, each molecule is surrounded by water molecules.

We use solutions for a number of reasons, including the ease with which they are dispensed. More important, however, is the excellent medium they provide for chemical reactions. When a chemical reaction occurs between two substances, it is necessary for molecules (or ions) to come in contact with one another. In a solution the solute

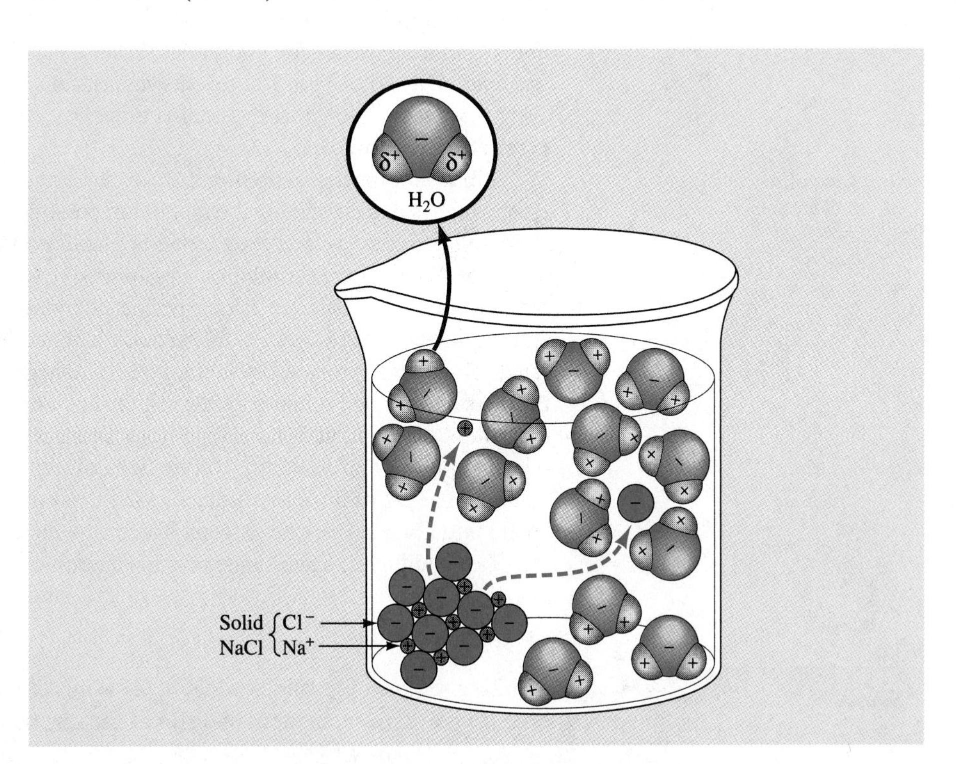

FIGURE 13-2 When sodium chloride dissolves, each ion becomes surrounded by water molecules.

molecules are not only separated from one another, but they are also in continuous motion. Thus, when we mix liquid solutions of two different substances together, it is possible for *all* the molecules of the two substances to collide—and react—with one another. (If we were to mix the substances in their solid states, only the molecules at the surfaces of the solid particles would come in contact with each other.)

PROBLEM 13.1

How many moles of ions are present altogether in a solution prepared by dissolving 1 mol of magnesium chloride, $MgCl_2$, in water?

13.2 SOLUBILITY

OBJECTIVE 13.2A Define *solubility.* Define *saturated solution.* Given the maximum mass of solute that will dissolve in a given mass of water, calculate the solubility in grams per 100 mL of water.

Solubility: The maximum possible concentration of a solute in a solvent.

Concentration: The quantity of solute dissolved in a given quantity of solvent or solution.

Saturated solution: A solution that has the maximum possible concentration of a solute.

Often we wish to know the solubility of a substance in a particular solvent. The **solubility** represents the maximum amount of solute that can be dissolved in a given volume of solvent. In other words, it is the maximum concentration of the solute in a particular solvent. By **concentration,** we mean the amount of solute dissolved in either a given quantity of solvent or a given quantity of solution as a whole. The solubility of a substance varies from one solvent to another. There are several ways to express solubility, but most often we give the mass of solute in grams per 100 mL of solvent. (The solubilities in water of several common substances are presented in Table 13-1.) A solubility reported as 25.0 g per 100 mL of water means that 25.0 g of solute will dissolve in 100 mL of water. Note that the final volume of solution prepared in this manner will be greater than 100 mL.

To determine the solubility of a substance, we must first prepare a **saturated solution,** which is one containing the maximum possible amount of dissolved solute. A saturated solution may be prepared by adding solute to the solvent until no further solute dissolves. We recognize a solution as saturated when undissolved solute remains in contact with the solution over a long period of time—for example, when a pile of solid solute remains at the bottom of the solution without dissolving further. Suppose the saturated solution was prepared by adding solid solute to 100 mL of solvent. One way to find the mass of dissolved solute is to filter off the undissolved solute and evaporate the solution to dryness. The residue is the solute from the saturated solution, and its mass represents the solubility of that solute per 100 mL of solvent. Alternatively, the mass of dissolved solute can be determined by subtracting the mass of undissolved solute from the mass of solute initially added to the solvent. The following example is a practical illustration of how the solubility of a substance may be determined in the laboratory.

EXAMPLE 13.1

A 9.21-g sample of a solid is added to 25.0 mL of water. The resulting mixture is covered (to prevent evaporation) and stirred for several hours to ensure that no more solid

TABLE 13-1 Solubilities of selected substances in water

Substance	Lower temperature g/100 mL	(°C)	Higher temperature g/100 mL	(°C)
Solids				
AgCl	0.000089	(10°C)	0.0021	(100°C)
$Ca(OH)_2$	0.185	(0°C)	0.077	(100°C)
$HgCl_2$	6.9	(20°C)	48	(100°C)
NaCl	35.7	(0°C)	39.1	(100°C)
KI	127.5	(0°C)	208	(100°C)
$C_{12}H_{22}O_{11}$ (sucrose)	179	(0°C)	487	(100°C)
Gases				
CO_2	0.348	(0°C)	0.145	(20°C)
NH_3	89.9	(0°C)	7.4	(100°C)
Liquids				
Acetic acid	∞ at all temperatures			
Ethyl alcohol	∞ at all temperatures			
Ethyl ether	8	(20°C)		

dissolves. At the end of the stirring period, there is still some undissolved solid in contact with the solution, indicating that the solution is saturated. The mixture is filtered, separating the undissolved solid from the solution. The solid in the filter is dried and found to weigh 3.26 g. Calculate the solubility of the solid in grams per 100 mL (g/1.00 × 10^2 mL) of water.

Planning the solution

Since there were 9.21 g of solid initially, and all but 3.26 g dissolved, there must be 5.95 g of solid dissolved in the 25.0 mL of water (9.21 g − 3.26 g = 5.95 g). We calculate the number of grams of solid per milliliter of water and then multiply our answer by 100 to get the number of grams per 100 mL of water.

SOLUTION

To calculate the number of grams per milliliter of water, we simply divide the number of grams by the number of milliliters:

$$\text{Solubility} = \frac{5.95\text{ g}}{25.0\text{ mL H}_2\text{O}} = 0.238\text{ g/mL H}_2\text{O}$$

Multiplying this by 100 gives us the number of grams that will dissolve in 100 mL H_2O (1.00 × 10^2 mL H_2O):

$$\text{Solubility} = 23.8\text{ g/100 mL H}_2\text{O} \qquad = \mathbf{23.8\text{ g/1.00} \times 10^2\text{ mL H}_2\text{O}}$$

Alternatively, since 5.95 g dissolves in 25.0 mL, we can write the following equivalency:

$$5.95\text{ g} = 25.0\text{ mL H}_2\text{O}$$

We may use this equivalency to derive a conversion factor for finding the number of grams that dissolve in 100 mL of water (1.00 × 10^2 mL H_2O):

$$1.00 \times 10^2\text{ mL H}_2\text{O}\left(\frac{5.95\text{ g}}{25.0\text{ mL H}_2\text{O}}\right) = 23.8\text{ g}$$

The solubility of the solid must be 23.8 g/100 mL H_2O.

PROBLEM 13.2

In an experiment similar to the one described in Example 13.1, a 32.6-g sample of a solid is added to 40.0 mL of water and stirred until no more solid dissolves. The presence of undissolved material indicates that the solution is saturated. After filtering and drying the undissolved solid, its mass is found to be 11.2 g. Calculate the solubility of the solid in grams per 100 mL (1.00×10^2 mL) of water.

Let us pause for a moment to examine the nature of a saturated solution more carefully. We have just learned to recognize a solution as saturated when the solution is in contact with undissolved solute and no more solute dissolves. When a solid is first added to a solvent, solute molecules (or ions) become separated from the surface of the solid. As more and more solute molecules accumulate in the liquid surrounding the solid, the reverse of dissolving may take place—a solute molecule may return to the crystal surface. We refer to this reverse process as *crystallization.* Initially, the only process that takes place is that of dissolving. However, as the concentration of solute molecules (or ions) in the solution increases, the rate of crystallization increases. A solution becomes saturated when the rate at which solute molecules returning to the solid phase equals the rate at which solute molecules are dissolving (Fig. 13.3). At this point there will be no further increase in the concentration of the solute.

$$\text{Solute (solid state)} \rightleftharpoons \text{Solute (dissolved)}$$

Note that this is a dynamic equilibrium process, just like the one we examined when we discussed vapor pressure (Section 12.8). Like all equilibrium processes, the creation of a saturated solution is characterized by opposing processes occurring at equal rates.

Under certain conditions, it is possible to prepare solutions that are described as *supersaturated.* These solutions have exceeded the maximum equilibrium concentration of the solute. The most common method for preparing a supersaturated solution makes use of the fact that most solids are more soluble at higher temperatures than at lower temperatures. To prepare a supersaturated solution, a saturated solution is prepared at an elevated temperature, the excess solid is filtered, and then the solution is allowed to cool. As the temperature drops, the concentration of the solution temporarily exceeds the solubility limit. Such solutions are usually short-lived. Addition of a "seed" crystal of solute or the spontaneous formation of such a crystal within the solution will cause the excess solute to crystallize out of solution, thereby reestablishing the equilibrium concentration characteristic of a saturated solution.

Handbooks frequently classify the solubility of a substance as very soluble, (*vs*), soluble (*s*), slightly soluble (*ss*), or insoluble (*i*). Although these classifications are not well defined, the following are useful approximations.

Very soluble:	At least 10 g/100 mL
Soluble:	At least 1 g/100 mL
Slightly soluble:	At least 0.1 g/100 mL
Insoluble:	Less than 0.1 g/100 mL

FIGURE 13-3 Characteristics of a saturated solution. When a solution is saturated, the rate at which molecules (or ions) dissolve equals the rate at which they return to the solid state.

Factors That Affect Solubility

OBJECTIVE 13.2B Explain what generalizations we can make about the variation of solubility with temperature for solids and gases in a liquid solvent. Define *miscible* and *immiscible.* State a generalization for the mutual solubilities of polar and nonpolar solutes and solvents.

A number of factors are involved in determining the extent of solubility of a given solute in a given solvent. Among these are temperature and intermolecular forces. The data in Table 13-1 help to illustrate several generalizations concerning the effect of temperature on solubility. In general, *the solubility of most solids in liquids increases as the temperature of the solution increases.* For example, the solubility of sucrose (table sugar), which is 179 g/100 mL at 0°C, increases to 487 g/100 mL at 100°C. Although the solubilities of *most* solids do increase with temperature, there are exceptions to this rule. For example, there is only a modest increase in the solubility of sodium chloride with increased temperature, while the solubility of calcium hydroxide *decreases* with increasing temperature.

Unlike solutions of solids in liquids, *the solubility of a gas in a liquid always decreases with increasing temperature.* For example, the solubility of carbon dioxide in water is 0.348 g/100 mL at 0°C. Warming this solution to room temperature (20°C) reduces its solubility to 0.145 g/100 mL. You probably have observed the collection of gas bubbles along the walls of a glass of water that has been allowed to warm up. These bubbles come from the escape of gases (such as carbon dioxide and oxygen) that had been dissolved in the water when it was colder.

Immiscible: Refers to two liquids that will not dissolve in one another.

Miscible: Refers to two liquids that dissolve in one another.

In order to discuss the solubilities of liquids in liquids, we need some additional terminology. As we have already mentioned, two liquids that will not dissolve in one another are said to be **immiscible.** We cited gasoline and water as an example of immiscible liquids. Because a mixture of immiscible liquids is not homogeneous, it is not classified as a solution. On the other hand, two liquids that *do* dissolve in one another, such as water and ethyl alcohol, are said to be **miscible,** and the homogeneous mixture formed *is* a solution. Liquids that are miscible in all proportions are said to be *infinitely* miscible. Table 13-1 expresses ethyl alcohol and acetic acid as infinitely soluble in water by using the symbol ∞. These liquids are miscible with water in all proportions. Ethyl ether, on the other hand, has a limited solubility in water; these liquids are miscible only when 8 g or less of ethyl ether are dissolved per 100 mL of water. Unlike the solubilities of solids or gases in liquids, the effect of temperature on the solubilities of liquids in liquids does not lead to any simple generalization.

The various intermolecular forces also play a major role in determining solubility. As a general rule, solutes are most soluble in solvents that exhibit similar types of intermolecular interactions. For example, sucrose ($C_{12}H_{22}O_{11}$, table sugar) is capable of extensive hydrogen bonding. This leads to its rather high solubility in water, another substance that hydrogen-bonds. By contrast, sucrose has a very limited solubility in carbon tetrachloride, a nonpolar liquid in which London forces are the primary intermolecular force of attraction. On the other hand, carbon tetrachloride is an excellent solvent for naphthalene, a nonpolar solid that is insoluble in water. Solvent and solute polarities are generally the most useful factors to examine in attempting to predict whether a given solute will dissolve in a given solvent. *Polar solutes are most likely to dissolve in polar solvents, whereas nonpolar solutes dissolve best in nonpolar solvents.* On the other hand, nonpolar solutes do not generally dissolve in polar solvents, nor do polar solutes dissolve in nonpolar solvents. These observations have led to the rule that *like dissolves like.*

These principles are utilized in our choice of solvents for various purposes. For example, the process of dry cleaning is not really dry. Nonpolar solvents are used to dissolve the nonpolar substances present in greases and oils, thereby removing them from the fabrics being cleaned. On the other hand, water is used to dissolve a rather wide range of solutes. Because of its ability to hydrogen-bond, water is an excellent solvent for many solutes that also hydrogen-bond. In addition, the polarity of water makes it an excellent solvent for many ionic solutes, such as sodium chloride and potassium iodide. (Not all ionic substances are soluble in water, however, as illustrated by silver chloride, whose very limited solubility is also indicated in Table 13-1.)

Factors That Affect the Rate at Which Substances Dissolve

> ***OBJECTIVE 13.2C*** State the three factors that determine the rate at which a solid dissolves, and describe how each factor affects the rate.

Three factors affect the rate at which substances dissolve: crystal size, stirring, and temperature. You have probably encountered all three of these in your personal experiences. If you have ever tried to dissolve sugar in iced tea, you know that table sugar dissolves somewhat slowly in a glass of iced tea unless you are using the very finely granulated sugar available in some restaurants. Earlier we saw that when a solid dissolves, solvent molecules interact with the solute molecules (or ions) at the surface of the solid, carrying them into solution. When a solid is finely granulated, there is more surface area available to interact with solvent and it dissolves more rapidly. The larger the crystal size, the greater the proportion of the substance located *within* the crystal, rather than at its surface, and the more slowly it dissolves.

You probably also know that stirring a glass of iced tea increases the rate at which the sugar dissolves. To understand why stirring speeds the rate of dissolving, let us consider what occurs if a solution is not stirred. As molecules of sugar enter the solution, solvated sugar molecules surround the crystal being dissolved. If nothing is done to carry the solvated sugar molecules away from the crystal, the reverse of dissolving (crystallization) can occur—molecules of sugar can return to the crystal. By stirring the mixture, solvated sugar molecules are moved away from the surface of the crystal, thereby allowing fresh solvent to approach and dissolve new solute molecules at the crystal surface.

Finally, let us consider the effect of temperature on the rate of dissolving. It is important not to confuse the *rate* of dissolving with the solubility—the *extent* to which the substance dissolves. We already have seen that the solubility of *most* substances increases with temperature. The rate at which a substance dissolves always increases with increasing temperature. When you add sugar to a cup of hot tea, it dissolves much more readily than it would in a cup of iced tea. With increased temperature comes an increase in the average kinetic energy of both the solvent and the solute. This increase in kinetic energy makes it easier for solute molecules (or ions) to break away from the solid surface and become solvated.

PROBLEM 13.3

Classify each of the *solids* in Table 13-1 as very soluble (*vs*), soluble (*s*), slightly soluble (*ss*), or insoluble (*i*), according to the guidelines given in this section. When more than one temperature is given, make the assignment at the lower temperature.

13.3 MOLARITY

OBJECTIVE 13.3 Define *molarity.* Describe a general procedure for preparing molar solutions. Given the mass of solute and the final volume of solution, calculate the molarity of a given solution.

Molarity: The number of moles of solute per liter of solution.

In the last section, we discussed the importance of knowing how much solute is in a given volume of a solution. We referred to this quantity as the concentration. One way we can express concentration is in terms of the number of grams of solute per liter of solution. For example, the concentration of a sodium chloride solution prepared by dissolving 58.5 g of NaCl in enough water to make a liter of solution may be described as 58.5 g/L. However, because we are generally more interested in knowing molar relationships, the most useful method for expressing concentration is in terms of the number of moles per liter. The **molarity (M)** of a solution is the number of moles of solute per liter of solution:

$$\text{Molarity} = \frac{\text{Moles of solute}}{\text{Liter of solution}} \quad \text{or} \quad \text{M} = \frac{\text{mol}}{\text{L soln}}$$

The symbol M stands for molarity. Because the solution just described contains 1 mol of NaCl (58.5 g NaCl = 1 mol NaCl) in 1 L of solution, it is a "one molar" sodium chloride solution: 1 M NaCl. Similarly, a solution prepared by dissolving 2 mol of NaCl in 1 L of solution would be two molar:

$$\frac{\text{2 mol NaCl}}{\text{L soln}} = \text{2 M NaCl}$$

Suppose we wish to prepare 1.00 L of a 2.00 M sodium chloride solution. We first measure out 2.00 mol of sodium chloride and then add sufficient water to make a total volume of 1.00 L. This gives us a 2.00 M solution of NaCl. If we had taken 1.00 L of water and then added 2.00 mol of sodium chloride, the final volume of the mixture would have been *greater* than 1.00 L because of the additional volume of the sodium chloride. The concentration of the solution would not have been 2.00 mol/L. Whenever we prepare a molar solution, we must measure out the solute first and then add enough water to achieve the desired final volume.

In preparing solutions, chemists often use a volumetric flask, as shown in Fig. 13-4. When a 1-L volumetric flask is filled to the line on the neck, it contains 1.000 L. Thus, to prepare a 2.00 M solution of sodium chloride, we first measure 2.00 mol of NaCl (117 g) into the volumetric flask and then add enough water to bring the level to the 1-L mark on the neck. After thorough mixing, we have 1.000 L of a 2.00 M NaCl solution.

To calculate the molarity of any solution, we need to know two things: the number of moles of solute and the total volume in liters of the final solution. We can then find the concentration by dividing the number of moles by the volume in liters:

$$\text{Molarity} = \frac{\text{Number of moles of solute}}{\text{Number of liters of solution}}$$

Let us find the molarity of a few solutions to illustrate the use of this formula.

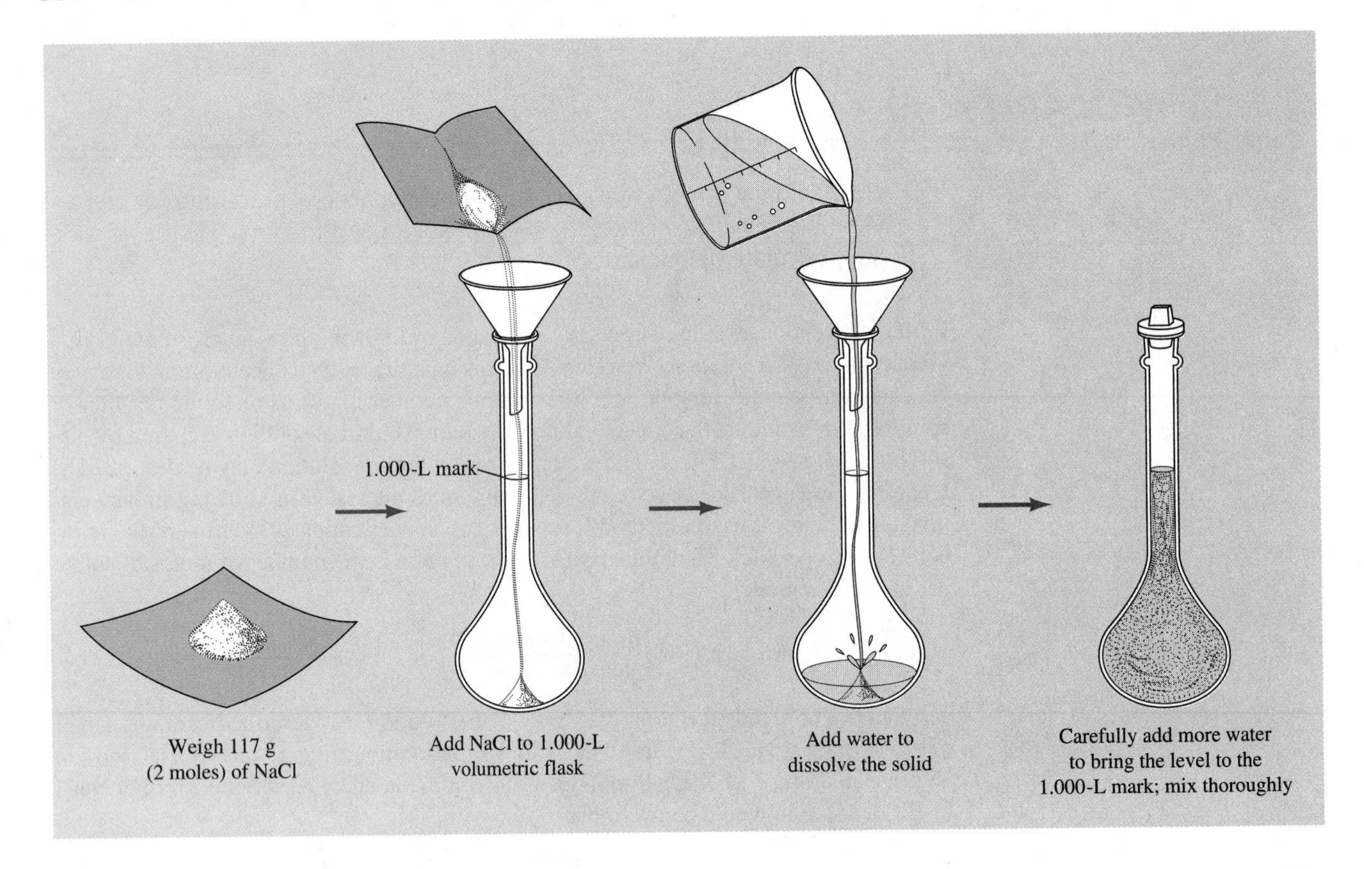

FIGURE 13-4 Preparing a 2.00 M NaCl solution.

(A reminder about significant figures: In the examples and problems presented in this text, we assume that all laboratory equipment used in the measurement of volume is accurate to at least the nearest milliliter. Thus, we will consider all of the zeros significant in each of the following: 100 mL, 250 mL, 500 mL, and 1000 mL.)

EXAMPLE 13.2

If 20.0 g of sodium hydroxide, NaOH, are dissolved in enough water to make a total volume of 2.00 L, what is the molar concentration of the solution?

Planning the solution

To find the molar concentration (which means the same thing as molarity), we must divide the number of moles of solute by the total volume of the solution in liters.

SOLUTION

First, let us find the number of moles in 20.0 g of NaOH:

$$40.0 \text{ g NaOH} = 1 \text{ mol NaOH}$$

$$? \text{ mol NaOH} = 20.0 \text{ g } \cancel{\text{NaOH}} \left(\frac{1 \text{ mol NaOH}}{40.0 \text{ g } \cancel{\text{NaOH}}} \right) = 0.500 \text{ mol NaOH}$$

Next, we divide the number of moles by the number of liters of solution:

$$\text{Molarity} = \frac{0.500 \text{ mol NaOH}}{2.00 \text{ L soln}} = 0.250 \text{ M NaOH}$$

EXAMPLE 13.3

What is the molar concentration of a solution prepared by dissolving 3.32 g KI in a total volume of 125 mL?

SOLUTION

As in the preceding example, we must first calculate the number of moles:

$$1 \text{ mol KI} = 166.0 \text{ g KI}$$

$$? \text{ mol KI} = 3.32 \cancel{\text{g KI}}\left(\frac{1 \text{ mol KI}}{166.0 \cancel{\text{g KI}}}\right) = 0.0200 \text{ mol KI}$$

However, before we can determine the molarity, we must convert the volume to liters of solution:

$$125 \text{ mL soln} = 0.125 \text{ L soln}$$

$$\text{Molarity} = \frac{0.0200 \text{ mol KI}}{0.125 \text{ L soln}} = 0.160 \text{ M KI}$$

EXAMPLE 13.4

What is the molar concentration of a solution prepared by dissolving 2.60 g $MgCl_2$ in a total volume of 50.0 mL of solution?

SOLUTION

$$1 \text{ mol MgCl}_2 = 95.3 \text{ g MgCl}_2$$

$$? \text{ mol MgCl}_2 = 2.60 \cancel{\text{g MgCl}_2}\left(\frac{1 \text{ mol MgCl}_2}{95.3 \cancel{\text{g MgCl}_2}}\right) = 0.0273 \text{ mol MgCl}_2$$

$$50.0 \text{ mL soln} = 0.0500 \text{ L soln}$$

$$\text{Molarity} = \frac{0.0273 \text{ mol MgCl}_2}{0.0500 \text{ L soln}} = 0.546 \text{ M MgCl}_2$$

PROBLEM 13.4

Calculate the molar concentration of each of the following solutions, which were prepared by dissolving the given amount of solute in enough water to achieve the final volume indicated.

(a) 444 g NaOH in 5.00L
(b) 4.04 g KNO_3 in 0.200 L
(c) 39.0 g NaBr in 500 mL (5.00×10^2 mL)
(d) 2.22 g $CaCl_2$ in 80.0 mL
(e) 0.385 g $MgBr_2$ in 5.00 mL
(f) 7.50 g KI in 40.0 mL
(g) 4.36 g Na_2SO_4 in 350 mL (3.50×10^2 mL)
(h) 12.3 g $Ca(NO_3)_2$ in 640 mL (6.40×10^2 mL)
(i) 2.65 g $CuSO_4$ in 45.0 mL
(j) 6.25 g NH_4Cl in 35.0 mL

13.4 CALCULATIONS INVOLVING MOLAR SOLUTIONS

Finding the Number of Moles in a Volume of Known Molarity

OBJECTIVE 13.4A Given the volume and molarity of a solution, calculate the number of moles present.

Much of our work in the laboratory is done with solutions that have already been prepared, and we often wish to know how many moles of solute are present in a given volume of such a solution. For example, suppose we wish to know the number of moles of sodium chloride contained in 2 L of a 4 M NaCl solution. A 4 M solution contains 4 mol per liter, so there must be 8 mol of sodium chloride in 2 L of solution. If we took 3 L of the same solution, we would have 12 mol of sodium chloride. Half a liter of this solution would contain 2 mol.

To find the number of moles of solute contained in each of these solutions, we multiplied the volume (in liters) by the molarity:

$$\text{Number of moles of solute} = \text{Volume} \times \text{Molarity}$$

$$? \text{ mol NaCl} = 2 \cancel{\text{L soln}}\left(\frac{4 \text{ mol NaCl}}{\cancel{\text{L soln}}}\right) = 8 \text{ mol NaCl}$$

$$? \text{ mol NaCl} = 3 \cancel{\text{L soln}}\left(\frac{4 \text{ mol NaCl}}{\cancel{\text{L soln}}}\right) = 12 \text{ mol NaCl}$$

$$? \text{ mol NaCl} = 0.5 \cancel{\text{L soln}}\left(\frac{4 \text{ mol NaCl}}{\cancel{\text{L soln}}}\right) = 2 \text{ mol NaCl}$$

You can see that we have used molarity (mol/L soln) as a conversion factor between liters of solution and moles of sodium chloride:

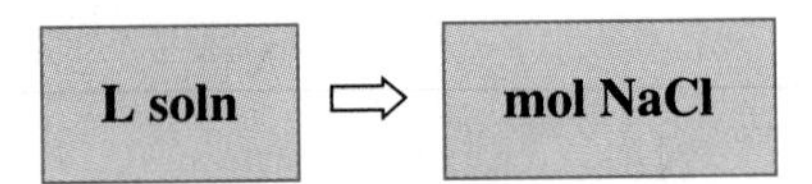

Let us do some sample calculations to illustrate this idea further.

EXAMPLE 13.5

How many moles of lithium chloride are contained in 3.00 L of 0.500 M LiCl?

Planning the solution

$$0.500 \text{ M LiCl} \quad \text{means} \quad \frac{0.500 \text{ mol LiCl}}{\text{L soln}}$$

We will use this as a conversion factor to convert liters of solution to moles of lithium chloride:

L soln ⇨ **mol LiCl**

SOLUTION

$$? \text{ mol LiCl} = 3.00 \cancel{\text{L soln}}\left(\frac{0.500 \text{ mol LiCl}}{\cancel{\text{L soln}}}\right) = 1.50 \text{ mol LiCl}$$

EXAMPLE 13.6

How many moles of magnesium bromide are contained in 500 mL of 4.00 M $MgBr_2$?

SOLUTION

First, we express our molarity in terms of units of moles per liter of solution:

$$4.00 \text{ M } MgBr_2 = \frac{4.00 \text{ mol } MgBr_2}{\text{L soln}}$$

Next, we must express the volume of solution in liters:

$$500 \text{ mL soln} = 0.500 \text{ L soln}$$

Finally, we use the molarity as a conversion factor to convert the liters of solution to moles of magnesium bromide:

L soln ⇨ mol $MgBr_2$

$$? \text{ mol } MgBr_2 = 0.500 \cancel{\text{L soln}} \left(\frac{4.00 \text{ mol } MgBr_2}{\cancel{\text{L soln}}} \right) = 2.00 \text{ mol } MgBr_2$$

EXAMPLE 13.7

How many moles of potassium iodide are contained in 75.0 mL of 0.650 M KI?

SOLUTION

$$0.650 \text{ M KI} = \frac{0.650 \text{ mol KI}}{\text{L soln}}$$

$$75.0 \text{ mL soln} = 0.0750 \text{ L soln}$$

$$? \text{ mol KI} = 0.0750 \cancel{\text{L soln}} \left(\frac{0.650 \text{ mol KI}}{\cancel{\text{L soln}}} \right) = 0.0488 \text{ mol KI}$$

PROBLEM 13.5

Calculate the number of moles in each of the following solutions.

(a) 3.00 L of 4.00 M KOH
(b) 1.50 L of 2.40 M $LiNO_3$
(c) 500 mL (5.00×10^2 mL) of 0.600 M $CaCl_2$
(d) 45.0 mL of 1.60 M $FeCl_3$
(e) 150 mL (1.50×10^2 mL) of 0.460 M $CuSO_4$
(f) 750 mL (7.50×10^2 mL) of 0.0440 M $MgBr_2$
(g) 3.50 mL of 3.00 M Na_2CrO_4
(h) 25.0 mL of 0.645 M $Ba(NO_3)_2$
(i) 2.0 mL of 12 M HCl
(j) 25 mL of 18 M H_2SO_4

Finding the Mass of Solute in a Volume of Known Molarity

OBJECTIVE 13.4B Given the volume and molarity of a solution, calculate the mass of solute present.

Suppose we wish to prepare 500 mL (5.00×10^2 mL) of 0.500 M NaCl solution. The number of moles present in the solution can be calculated using the method just discussed:

$$? \text{ mol NaCl} = 0.500 \cancel{\text{L soln}} \left(\frac{0.500 \text{ mol NaCl}}{\cancel{\text{L soln}}} \right) = 0.250 \text{ mol NaCl}$$

However, to prepare the desired solution, we need to know the mass of sodium chloride. We can easily calculate this by converting moles to grams:

$$? \text{ g NaCl} = 0.250 \cancel{\text{mol NaCl}} \left(\frac{58.5 \text{ g NaCl}}{1 \cancel{\text{mol NaCl}}} \right) = 14.6 \text{ g NaCl}$$

Thus, to prepare the solution, we dissolve 14.6 g of NaCl in enough water to make 500 mL (5.00×10^2 mL) of solution. The resulting solution is 0.500 M NaCl. Let us do a few more illustrative examples.

EXAMPLE 13.8

How do we prepare 250 mL (2.50×10^2 mL) of 0.600 M NaBr?

Planning the solution

To prepare this solution, we must find the mass of sodium bromide present in 250 mL of 0.600 M NaBr. From the volume and molarity, we can determine the number of moles of solute, which can be converted to grams of solute in the usual fashion:

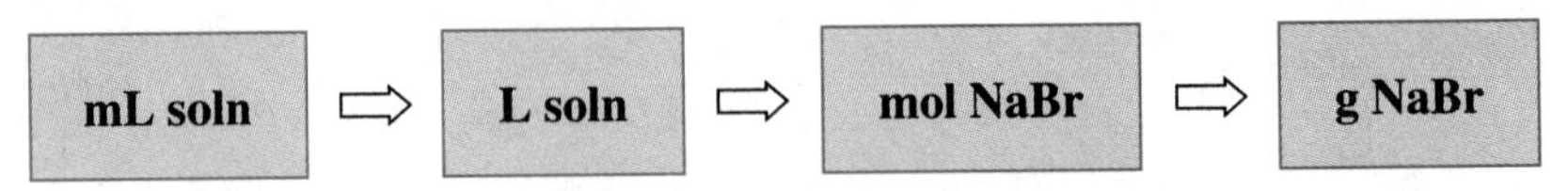

SOLUTION

First, we must convert 250 mL of solution to liters of solution:

$$250 \text{ mL soln} = 0.250 \text{ L soln}$$

Next, we find the number of moles of NaBr in the solution:

$$? \text{ mol NaBr} = 0.250 \cancel{\text{L soln}}\left(\frac{0.600 \text{ mol NaBr}}{\cancel{\text{L soln}}}\right) = 0.150 \text{ mol NaBr}$$

Then we convert this to grams of NaBr:

$$? \text{ g NaBr} = 0.150 \cancel{\text{mol NaBr}}\left(\frac{102.9 \text{ g NaBr}}{1 \cancel{\text{mol NaBr}}}\right) = 15.4 \text{ g NaBr}$$

To prepare the desired solution, we add enough water to 15.4 g of sodium bromide to make 250 mL (2.50×10^2 mL) of solution.

We could also solve this problem by combining all of these steps into one setup as follows:

$$? \text{ g NaBr} = 250 \cancel{\text{mL soln}}\left(\frac{1 \cancel{\text{L soln}}}{1000 \cancel{\text{mL soln}}}\right)\left(\frac{0.600 \cancel{\text{mol NaBr}}}{\cancel{\text{L soln}}}\right)\left(\frac{102.9 \text{ g NaBr}}{1 \cancel{\text{mol NaBr}}}\right)$$

$$= 15.4 \text{ g NaBr}$$

EXAMPLE 13.9

How do we prepare 80.0 mL of 1.50 M $NaNO_3$?

Planning the solution

We must find the mass of sodium nitrate to be dissolved to make 80.0 mL of solution. We proceed via the following strategy:

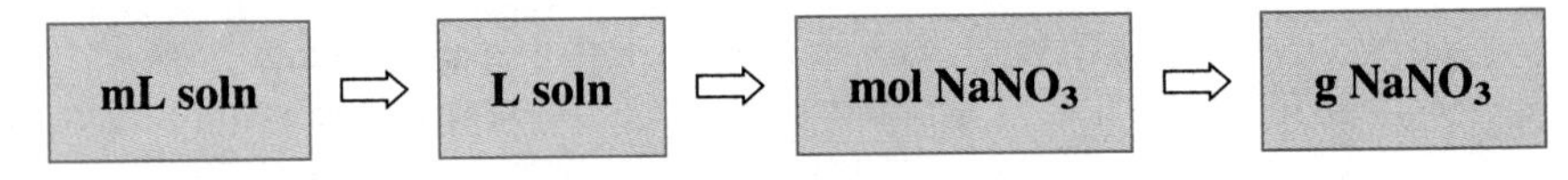

SOLUTION

First, we convert milliliters of solution to liters of solution:

$$80.0 \text{ mL soln} = 0.0800 \text{ L soln}$$

Next, we find the number of moles of $NaNO_3$ in 0.0800 L of 1.50 M $NaNO_3$:

$$? \text{ mol NaNO}_3 = 0.0800 \cancel{\text{L soln}}\left(\frac{1.50 \text{ mol NaNO}_3}{\cancel{\text{L soln}}}\right) = 0.120 \text{ mol NaNO}_3$$

Finally, we find the mass of 0.120 mol $NaNO_3$:

$$? \text{ g NaNO}_3 = 0.120 \cancel{\text{mol NaNO}_3}\left(\frac{85.0 \text{ g NaNO}_3}{1 \cancel{\text{mol NaNO}_3}}\right) = 10.2 \text{ g NaNO}_3$$

Thus, we dissolve 10.2 g of sodium nitrate in enough water to make 80.0 mL of solution.

As in Example 13.8, we could have saved time by combining our conversion factors as follows:

$$? \text{ g NaNO}_3 = 80.0\,\cancel{\text{mL soln}}\left(\frac{1\,\cancel{\text{L soln}}}{1000\,\cancel{\text{mL soln}}}\right)\left(\frac{1.50\,\cancel{\text{mol NaNO}_3}}{\cancel{\text{L soln}}}\right)\left(\frac{85.0\text{ g NaNO}_3}{1\,\cancel{\text{mol NaNO}_3}}\right)$$
$$= 10.2\text{ g NaNO}_3$$

PROBLEM 13.6

Calculate the mass of solute required to prepare each of the solutions described in parts (a)–(h) of Problem 13.5.

Finding the Volume Containing a Specified Number of Moles

OBJECTIVE 13.4C Given the molarity of a solution, calculate the volume containing a specified number of moles.

Occasionally, we know the molarity of a solution and wish to measure out a specified number of moles from it. This is often the case when we are measuring out a reactant for a chemical reaction. For example, suppose we wish to measure out 0.500 mol of HCl from a solution that is 2.50 M. In short, we need to know what volume of 2.50 M HCl contains 0.500 mol. To solve this problem, we again use the molarity as a conversion factor. However, this time we convert from moles to volume. Because a 2.50 M HCl solution contains 2.50 mol of HCl in 1.00 L of solution, the following ratio expresses the relationship between moles and volume *for this solution:*

$$2.50\text{ mol HCl} : 1.00\text{ L soln}$$

We can use this relationship to write a conversion factor for converting moles of the 2.50 M HCl solution to liters of solution:

$$? \text{ L soln} = 0.500\,\cancel{\text{mol HCl}}\left(\frac{1.00\text{ L soln}}{2.50\,\cancel{\text{mol HCl}}}\right) = 0.200\text{ L soln}$$

Note that we have used the reciprocal of the molarity (L soln/mol) as a conversion factor for converting moles to liters of solution. Earlier in this section we used molarity directly as a conversion factor for converting liters of solution to moles. The reciprocal of the molarity simply reverses the process.

EXAMPLE 13.10

What volume of 0.250 M KI contains 0.0750 mol KI?

Planning the solution

$$0.250\text{ M KI} \quad \text{means} \quad 0.250\text{ mol KI} : 1.00\text{ L soln}$$

Therefore,

$$\frac{1.00\text{ L soln}}{0.250\text{ mol KI}} = 1$$

We can use this conversion factor to determine the number of liters of solution that contains the desired number of moles of KI.

mol KI ⇨ **L soln**

SOLUTION

$$? \text{ L soln} = 0.0750\ \cancel{\text{mol KI}}\left(\frac{1.00 \text{ L soln}}{0.250\ \cancel{\text{mol KI}}}\right) = 0.300 \text{ L soln}$$

EXAMPLE 13.11 What volume of 18 M H_2SO_4 contains 0.25 mol H_2SO_4?

SOLUTION

$$18 \text{ M } H_2SO_4 \quad \text{means} \quad 18 \text{ mol } H_2SO_4 : 1.0 \text{ L soln}$$

Therefore,

$$\frac{1.0 \text{ L soln}}{18 \text{ mol } H_2SO_4} = 1$$

$$? \text{ L soln} = 0.25\ \cancel{\text{mol } H_2SO_4}\left(\frac{1.0 \text{ L soln}}{18\ \cancel{\text{mol } H_2SO_4}}\right) = 0.014 \text{ L soln}$$

If we wish the final answer to be expressed in milliliters, we simply carry out the additional conversion.

$$? \text{ mL soln} = 0.014\ \cancel{\text{L soln}}\left(\frac{1000 \text{ mL soln}}{1\ \cancel{\text{L soln}}}\right) = 14 \text{ mL soln}$$

The Summary Diagram provides a flowchart of the various relationships we have been using between quantities in solutions.

PROBLEM 13.7 Calculate the volume of solution that contains the indicated number of moles of solute.

(a) 0.50 mol from 6.0 M NaOH **(b)** 2.5 mol from 15 M NH_3
(c) 0.125 mol from 2.45 M KCl **(d)** 0.0575 mol from 0.785 M $H_2C_2O_4$
(e) 0.00350 mol from 0.120 M $CaCl_2$

13.5 DILUTION OF SOLUTIONS

OBJECTIVE 13.5 Given the molarity of a concentrated solution, calculate the volume required to prepare a specified volume and molarity of a diluted solution. Given the volume and molarity of a concentrated solution and the final volume to which it is diluted, calculate the molarity of the diluted solution.

Solutions are often prepared by *diluting* a more concentrated solution of the same substance. To dilute a concentrated aqueous solution, we simply mix it with water. Many acids are purchased from their manufacturers in concentrated form and then diluted to various concentrations for use in the laboratory. For example, concentrated hydrochloric

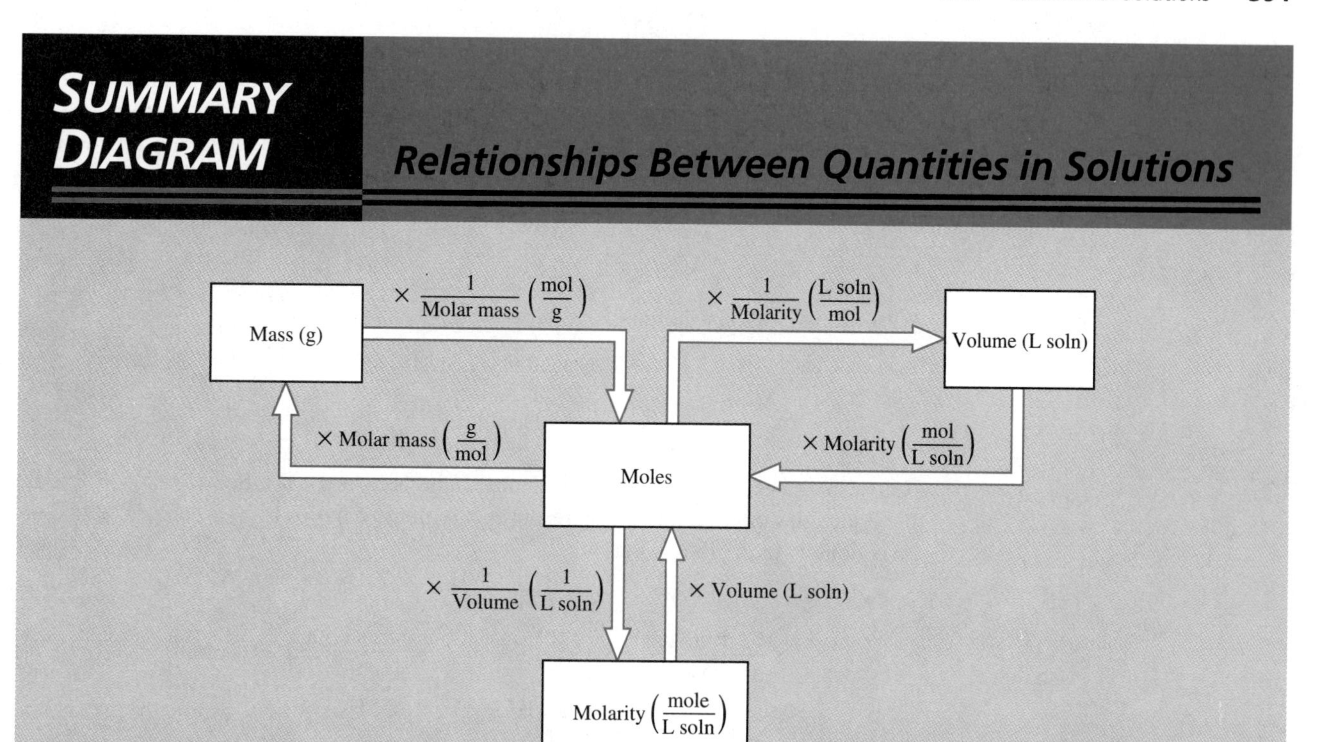

acid is 12 M HCl. If we wish to prepare 1.0 M HCl from this concentrated solution, we must dilute the concentrated acid to one-twelfth of its original concentration.

Suppose we wish to prepare a specific volume of the diluted solution, such as 3.0 L of 1.0 M HCl. We must calculate the volume of 12 M HCl that should be diluted to 3.0 L. One way to calculate this is to find the number of moles of hydrochloric acid that are present in the diluted solution. There are 3.0 mol of HCl in 3.0 L of 1.0 M HCl:

$$? \text{ mol HCl} = 3.0\ \text{L}\ \cancel{\text{soln}}\left(\frac{1.0 \text{ mol HCl}}{\text{L}\ \cancel{\text{soln}}}\right) = 3.0 \text{ mol HCl}$$

Since we obtain the 3.0 mol of HCl from the concentrated solution, we must calculate the volume of 12 M HCl (the concentrated solution) that contains 3.0 mol of HCl. To do so, we use the reciprocal of the molarity:

$$? \text{ L soln} = 3.0\ \cancel{\text{mol HCl}}\left(\frac{1.0 \text{ L soln}}{12\ \cancel{\text{mol HCl}}}\right) = 0.25 \text{ L soln}$$

Thus, if we take 0.25 L of 12 M HCl and add it to enough water to make 3.0 L, we will have the desired solution.

The approach we took to this problem makes sense, and it always gives us the correct result. However, there is a much easier way to perform calculations involving the dilution of solutions. We use the following formula, which we refer to as the *dilution formula:*

$$V_{\text{con}} \cdot M_{\text{con}} = V_{\text{dil}} \cdot M_{\text{dil}}$$

where V_{con} and M_{con} refer to the volume and molarity of the concentrated solution being diluted and V_{dil} and M_{dil} refer to the volume and molarity of the diluted solution we wish to prepare. For the preceding solution, we wanted to know what volume of 12 M HCl we needed to dilute in order to prepare 3.0 L of 1.0 M HCl. Our unknown in that example is the volume of concentrated solution, V_{con}. Our values are

$$M_{con} = 12\text{ M} \qquad V_{con} = ?$$
$$M_{dil} = 1.0\text{ M} \qquad V_{dil} = 3.0\text{ L soln}$$

Substituting into the dilution formula, we obtain

$$(V_{con})(12\text{ M}) = (3.0\text{ L soln})(1.0\text{ M})$$
$$V_{con} = 3.0\text{ L soln}\left(\frac{1.0\text{ M}}{12\text{ M}}\right) = 0.25\text{ L soln}$$

This is the same result we obtained before. The reason the dilution formula works is that both sides of the equation represent the number of moles of hydrochloric acid that are contained in the final solution.

$$V_{con} \cdot M_{con} = V_{dil} \cdot M_{dil}$$
$$\left(\frac{12\text{ mol HCl}}{\text{L soln}}\right)(0.25\text{ L soln}) = \left(\frac{1.0\text{ mol soln}}{\text{L soln}}\right)(3.0\text{ L soln})$$
$$3.0\text{ mol HCl} = 3.0\text{ mol HCl}$$

Let us work some additional examples.

EXAMPLE 13.12

Concentrated sulfuric acid (H_2SO_4) is 18 M H_2SO_4. How would you prepare 3.0 L of 0.30 M H_2SO_4?

Planning the solution

We simply substitute the appropriate quantities into the dilution formula.

SOLUTION

$$V_{con} \cdot M_{con} = V_{dil} \cdot M_{dil}$$
$$(V_{con})(18\text{ M}) = (3.0\text{ L soln})(0.30\text{ M})$$
$$V_{con} = 3.0\text{ L soln}\left(\frac{0.30\text{ M}}{18\text{ M}}\right) = 0.050\text{ L soln}$$

Add 0.050 L (50 mL) of 18 M H_2SO_4 to enough water to make 3.0 L of solution.

EXAMPLE 13.13

A physiologist has a solution of 2.00 M NaCl. How does she prepare 50.0 mL of 0.600 M NaCl?

Planning the solution

One of the convenient features of the dilution formula is that we can express our volumes directly in milliliters. The only restriction on this formula is that the units of volume must be the same on both sides of the equation.

SOLUTION

$$V_{con} \cdot M_{con} = V_{dil} \cdot M_{dil}$$
$$(V_{con})(2.00\text{ M}) = (50.0\text{ mL soln})(0.600\text{ M})$$

$$V_{con} = 50.0 \text{ mL soln}\left(\frac{0.600 \cancel{M}}{2.00 \cancel{M}}\right) = 15.0 \text{ mL soln}$$

The physiologist measures out 15.0 mL of 2.00 M NaCl and adds enough water to give a final volume of 50.0 mL.

EXAMPLE 13.14

A chemist dilutes 10.0 mL of 12 M HCl to a final volume of 250 mL (2.50×10^2 mL). What is the final concentration of the solution?

Planning the solution

In this example, we wish to find the molarity of the diluted solution, M_{dil}.

SOLUTION

$$V_{dil} \cdot M_{dil} = V_{con} \cdot M_{con}$$

$$(250 \text{ mL soln})M_{dil} = (10.0 \text{ mL soln})(12 \text{ M})$$

$$M_{dil} = \frac{10.0 \cancel{\text{mL soln}}}{250 \cancel{\text{mL soln}}}(12 \text{ M}) = 0.48 \text{ M}$$

EXAMPLE 13.15

If 25 mL of 6.0 M NaOH are diluted to a final volume of 1.00 L, what is the final molarity of the solution?

Planning the solution

In this problem, we must be careful to use the same units of volume on both sides of the equation. We will express 25 mL as 0.025 L.

SOLUTION

$$V_{dil} \cdot M_{dil} = V_{con} \cdot M_{con}$$

$$(1.00 \text{ L soln})M_{dil} = (0.025 \text{ L soln})6.0 \text{ M}$$

$$M_{dil} = \left(\frac{0.025 \cancel{\text{L soln}}}{1.00 \cancel{\text{L soln}}}\right)6.0 \text{ M} = 0.15 \text{ M}$$

PROBLEM 13.8

Write directions for preparing each of the following solutions from the concentrated solutions given.

(a) 500 mL (5.00×10^2 mL) of 3.0 M HCl from 12 M HCl
(b) 250 mL (2.50×10^2 mL) of 1.0 M H_2SO_4 from 18 M H_2SO_4
(c) 75 mL of 0.60 M NaOH from 3.0 M NaOH
(d) 150 mL (1.50×10^2 mL) of 0.10 M KI from 3.0 M KI
(e) 5.0 L of 0.25 M NH_3 from 15 M NH_3

PROBLEM 13.9

Calculate the final concentration when each of the following solutions is diluted to the final volume indicated.

(a) 75 mL of 18 M H_2SO_4 to 350 mL (3.50×10^2 mL)
(b) 25 mL of 2.4 M NaCl to 750 mL (7.50×10^2 mL)
(c) 10.0 mL of 1.25 M NH_3 to 60.0 mL
(d) 35.0 mL of 2.00 M $C_6H_{12}O_6$ to 0.500 L
(e) 5.00 mL of 6.00 M KOH to 1.25 L

13.6 PERCENT SOLUTIONS

OBJECTIVE 13.6A Define *percent solution by volume.* Given the volume of a liquid solute and the volume of the solution it is in, calculate its percent by volume. Calculate the volume of solute required to prepare a given volume of solution of a specified percent by volume

OBJECTIVE 13.6B Define *percent solution by mass.* Given the mass of solute and the total mass of the solution it is in, calculate its percent by mass. Calculate the mass of solute required to prepare a given mass of solution of a specified percent by mass. Given the mass of solute and the total mass of the solution it is in, calculate the concentration in parts per million.

For certain purposes it is useful to express the concentration of a solute in solution as a percentage of the total solution. Although concentrations expressed in this manner (*percent solutions*) are not as informative as molar concentrations, the compositions of these solutions are more easily calculated. No molar computations are involved, nor is it necessary to know the chemical formulas of the substances being mixed. There are several ways in which a percent solution can be expressed. We will discuss percent by volume and percent by mass. (A review of percentages is included in Appendix A.)

Percent solution by volume: The concentration of a solute as a percentage of the total volume.

The **percent solution by volume** expresses the concentration of a solute as a percentage of the total volume:

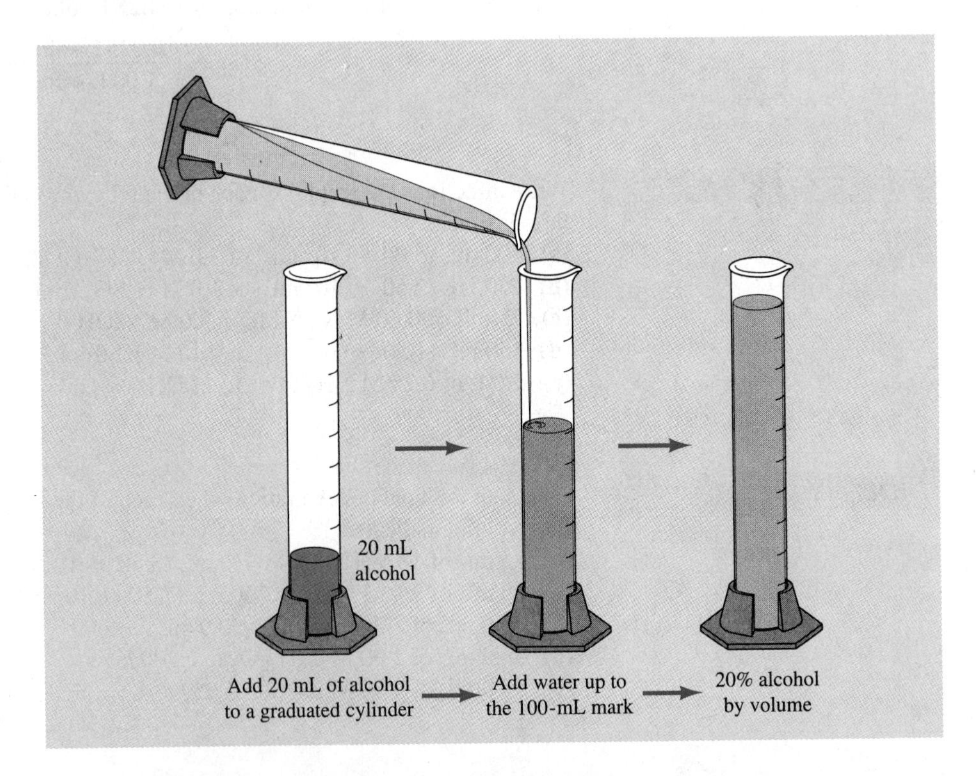

FIGURE 13-5 Preparing a volume percent solution (20% alcohol by volume).

$$\% \text{ by volume} = \frac{\text{Volume of solute}}{\text{Total volume}} \times 100\%$$

Volume percent solutions are limited to solutions made by dissolving one liquid in another. Thus, a 20% solution (*by volume*) of ethyl alcohol in water can be made by adding enough water to 20 mL of ethyl alcohol to make a total of 100 mL of solution (Fig. 13-5). On first thought, we might expect to simply mix 20 mL of alcohol and 80 mL of water. However, the volumes of different solutions are not necessarily additive. Thus, it is necessary to bring the final volume to 100 mL, in much the same way as we prepare molar solutions.

To determine the amount of solute needed to prepare a given solution, we simply take the appropriate percentage of the volume of solution we wish to prepare. For example, if we want to prepare 250 mL of a 14% solution (by volume) of isopropyl alcohol (rubbing alcohol), the required volume of solute is 14% of 250 mL. [*Remember:* To convert a percentage to its fractional equivalent, simply divide by 100; thus, 14% = 0.14.]

$$(0.14)(250 \text{ mL}) = 35 \text{ mL}$$

Thus, we take 35 mL of isopropyl alcohol (rubbing alcohol) and dilute it to 250 mL.

Percent solution by mass: The concentration of a solute as a percentage of the total mass.

The **percent solution by mass** expresses the concentration of a solute as a percentage of the total mass:

$$\% \text{ by mass} = \frac{\text{Mass of solute}}{\text{Total mass}} \times 100\%$$

$$= \frac{\text{Mass of solute}}{\text{Mass of solute} + \text{Mass of solvent}} \times 100\%$$

To prepare a 20% solution (*by mass*) of sodium chloride, we can add 20 g of sodium chloride to 80 g of water (Fig. 13-6, p. 396). Here we obtain the total mass of the solution by adding the mass of solute to the mass of solvent.

Calculations for percent solution by mass are done in the same way as those for percent volume, except that we work with the masses of solute and solvent. Thus, the mass of boric acid in 350 g of 4.0% boric acid solution (by mass) is simply 4.0% of 350 g:

$$(0.040)(350 \text{ g}) = 14 \text{ g}$$

To prepare this solution, we take 14 g of boric acid and add it to enough water (336 g) to make a total of 350 g.

EXAMPLE 13.16

How do you prepare 450 mL (4.50×10^2 mL) of 6.0% ethyl alcohol by volume?

SOLUTION

6.0% of 450 mL is

$$(0.060)(450 \text{ mL}) = 27 \text{ mL}$$

Add 27 mL of ethyl alcohol to enough water (approximately 423 mL) to make 450 mL.

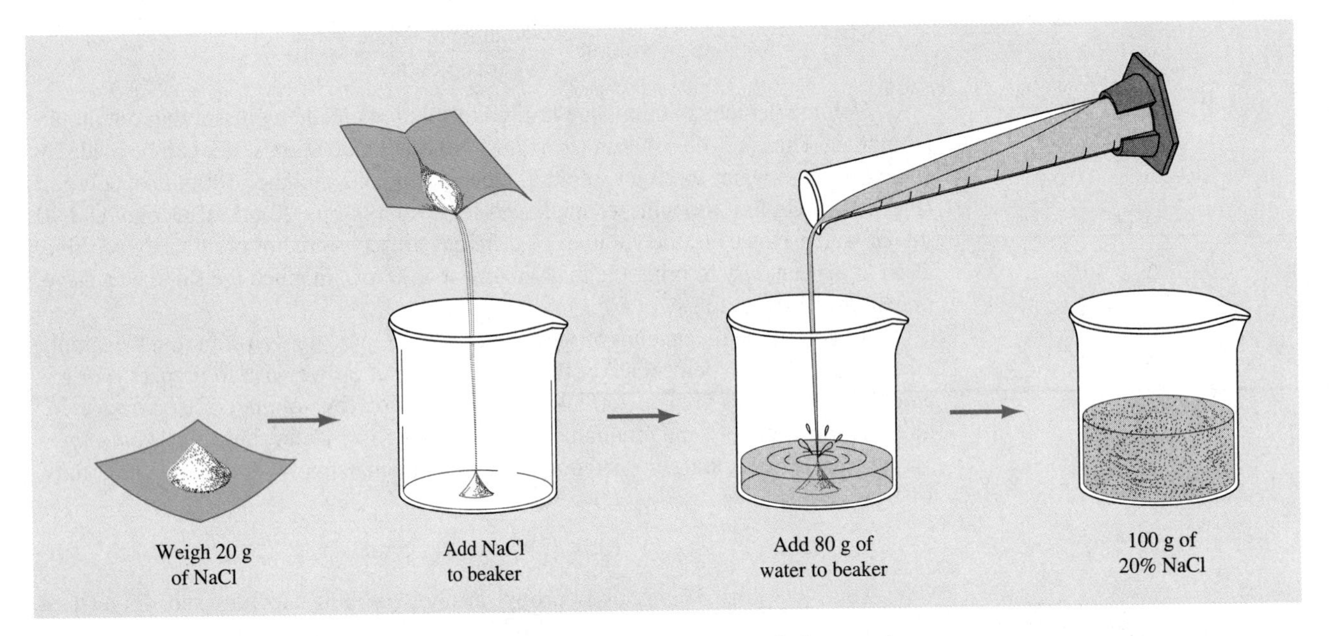

FIGURE 13-6 Preparing a mass percent solution (20% sodium chloride by mass).

EXAMPLE 13.17

How do you prepare 400 g (4.00×10^2 g) of 4.0% NaCl by mass?

SOLUTION

4.0% of 400 g is

$$(0.040)(400\text{ g}) = 16\text{ g}$$

Add 16 g of NaCl to 384 g of water.

PROBLEM 13.10

Calculate the percent concentration of each of the following solutions. For simplicity, assume the volumes to be additive in parts (a) and (b).

(a) 5.0 mL ethyl alcohol + 25.0 mL water (by volume)
(b) 15 mL acetic acid + 45 mL water (by volume)
(c) 3.0 g NaCl + 57.0 g H_2O (by mass)
(d) 14 g $NaHCO_3$ + 236 g H_2O (by mass)

PROBLEM 13.11

How would you prepare each of the following solutions?

(a) 500 mL (5.00×10^2 mL) of 7.0% acetic acid (by volume)
(b) 500 g (5.00×10^2 g) of 7.0% acetic acid (by mass)
(c) 250 mL (2.50×10^2 mL) of 12% ethyl alcohol (by volume)
(d) 750 g (7.50×10^2 g) of 6.0% H_3BO_3 (by mass)
(e) 25 g of 2.0% $NaHCO_3$ (by mass)

Parts per million: The number of grams of solute per million grams of solution.

Related to the percent composition is the concentration unit of **parts per million (ppm).** This unit is frequently used to describe the composition of natural bodies of water or the quality of air. In the case of natural bodies of water, the number of parts per million gives the number of grams of solute per million grams of solution. This is also equal to the number of milligrams of solute per million milligrams of solution. There are 1 million milligrams in a liter of water,

$$1 \text{ L} = 1000 \text{ mL} = 1000 \text{ g} = 1{,}000{,}000 \text{ mg}$$

So the number of parts per million is also equal to the number of milligrams of solute per liter of solution (mg/L). To give a qualitative idea of how small this concentration is, an eyedropper of alcohol (approximately 1 mL) added to a filled bathtub would produce roughly 1 ppm of alcohol in the bathwater.

You already know that a percentage represents the number of parts per hundred. Thus, to obtain a percentage, we multiply the number of grams of solute per gram of solution by 100:

$$\text{Percentage} = \frac{\text{Grams of solute}}{\text{Grams of solution}} \times 100\%$$

Similarly, to obtain the number of parts per million, we multiply the number of grams of solute per gram of solution by 1 million (10^6):

$$\text{Parts per million} = \frac{\text{Grams of solute}}{\text{Grams of solution}} \times 10^6 \text{ ppm}$$

Thus, if a 1.25-kg sample of river water is found to contain 23.7 mg of nitrate, we determine the parts per million by dividing the mass of nitrate (in grams) by the mass of river water (also in grams) and multiplying the result by 10^6:

$$\text{Parts per million} = \frac{0.0237 \text{ g}}{1250 \text{ g}} \times 10^6 \text{ ppm} = 19.0 \text{ ppm}$$

As a result of our increasing ability to detect the components of very dilute solutions, we often find the concentrations of water contaminants expressed in *parts per billion.* A part per billion (ppb) is one-thousandth the concentration of a part per million, and it represents the number of micrograms (10^{-6} g) per liter (μ g/L).

In describing air quality, determining the number of parts per million is somewhat more complicated than it is for water quality. Rather than work with masses, we compare the *volume* that a gas or vapor would occupy under a specified set of temperature and pressure conditions to the volume it actually occupies. Pollutants enter the air and waterways from a variety of sources, including not only industry but also private citizens, who drive automobiles, burn leaves, and so on. The release of chlorofluorocarbon gases into the atmosphere from both industrial and private sources is considered by most experts to be responsible for the increasing hole in the ozone layer.

Table 13-2 presents a typical set of maximum allowable pollution levels compiled from two government sources. The government agencies that monitor air and water quality try to ensure that the quality is never worse than these standards. When pollutants exceed these levels, the source of the pollution is determined, and action may be taken to stop the polluter. While businesses are strictly regulated with respect to the materials they discharge, private citizens are not. It is estimated that approximately 50% of the pollutants discharged into the environment are discharged by private citizens who pour household "chemicals" down their drains, spray hydrocarbon deodorants into the atmosphere, or simply throw their leftover household products in the garbage.

TABLE 13-2 Typical standards for water quality and air quality

Substance	Maximum contaminant level (ppm)
Water Quality	
Arsenic	0.05
Chromium	0.05
Lead	0.05
Mercury	0.002
Nitrate	45.0
Air Quality	
Ozone	0.12
Carbon monoxide	9
Nitrogen dioxide	0.05
Sulfur dioxide	0.03
Hydrocarbons	0.24

Concern over establishing and maintaining such minimum standards has increased as the effects of pollution on human health have become more apparent. For example, the symptoms of both lead poisoning and mercury poisoning are quite severe. Lead poisoning can lead to blood and nervous system disorders. Mercury poisoning can cause brain damage, leading to nervous system disorders that produce irritability, paralysis, and insanity. Mercury poisoning used to be quite common in the hat industry, where mercury compounds were used in the production of felt hats. Unaware of the toxicity of the chemicals they were using, workers frequently ingested these compounds inadvertently, through hand-to-mouth contact. As a result, the symptoms of mercury poisoning became known as "mad-hatter's syndrome."

Until fairly recently, mercury was used in a large number of industrial applications, and wastes were frequently discharged into natural waters. Then, in the 1950s, a tragic series of events in a small Japanese fishing village brought to public awareness the danger of indiscriminate mercury dumping. Mercury compounds used in the manufacture of plastics were being dumped into Minamata Bay. One of the serious concerns about the disposal of wastes into natural waters is the ability of marine life, such as fish, to concentrate contaminants in their bodies above the contaminant levels of the surrounding waters. While there are several ways that pollutants may become bioconcentrated, one of these is for contaminants to be absorbed by microscopic organisms, such as algae, and move up the food chain from smaller organisms to larger organisms. As a general rule, contaminant levels increase going up the food chain (though this is not always the case). In the Minamata Bay incident, fish taken from the bay for consumption had concentrated such dangerously high levels of mercury that many of the villagers who ate those fish became ill, and some died.

PROBLEM 13.12

A 15-kg sample of water is tested for arsenic and is found to contain 2.1 mg of this toxic element. How many parts per million does this represent? Does this level fall within the allowable limits given in Table 13-2?

13.7 SOLUTION STOICHIOMETRY

OBJECTIVE 13.7A Given a chemical equation and the volume and molarity of the limiting reagent, calculate the mass of any product formed.

OBJECTIVE 13.7B Given a chemical equation and the volume and molarity of one of the reactants, calculate the required volume of solution of another reactant, given its molarity.

OBJECTIVE 13.7C Given a chemical equation and the volume and molarity of the limiting reagent, calculate the volume of a gaseous product at a specified temperature and pressure.

In the stoichiometry problems presented in Chapter 10, we used the mass of a substance and its molar mass to determine the number of moles of known. Because much of our work in the laboratory is done with solutions, we often calculate the number of moles of known from the volume and molarity of a reacting solution. For example, suppose we wish to determine the mass of silver chloride produced when 25.0 mL of 0.100 M $AgNO_3$ are treated with an excess of sodium chloride solution:

$$AgNO_3(aq) + NaCl(aq) \rightarrow AgCl(s) + NaNO_3(aq)$$

As in the stoichiometry problems examined earlier, we have to determine the moles of known from the information given in the problem. Using the methods introduced earlier in this chapter, we can calculate the moles of silver nitrate from its volume and molarity:

$$? \text{ mol } AgNO_3 = 0.0250 \cancel{\text{L soln}} \left(\frac{0.100 \text{ mol } AgNO_3}{\cancel{\text{L soln}}} \right)$$
$$= 0.00250 \text{ mol } AgNO_3$$

The remainder of the problem is completed in the same way as the final steps of the mass–mass problems presented in Chapter 10:

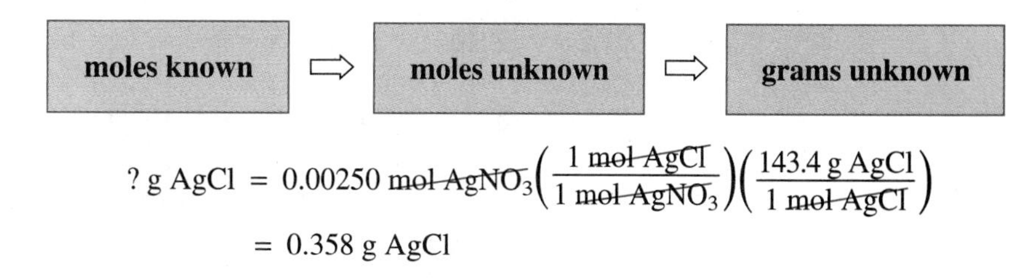

$$? \text{ g AgCl} = 0.00250 \cancel{\text{mol } AgNO_3} \left(\frac{1 \cancel{\text{mol AgCl}}}{1 \cancel{\text{mol } AgNO_3}} \right) \left(\frac{143.4 \text{ g AgCl}}{1 \cancel{\text{mol AgCl}}} \right)$$
$$= 0.358 \text{ g AgCl}$$

EXAMPLE 13.18

Calculate the mass of lead(II) iodide, PbI_2, that precipitates when 35.0 mL of 0.150 M KI are treated with excess lead(II) nitrate, $Pb(NO_3)_2$, according to the following chemical equation:

$$Pb(NO_3)_2(aq) + 2\,KI(aq) \rightarrow PbI_2(s) + 2\,KNO_3(aq)$$

Planning the solution

The volume and molarity of KI solution enables us to calculate the number of moles of KI available. This will become the moles of known for the problem. We can then complete the problem by using our usual strategy for solving stoichiometry problems:

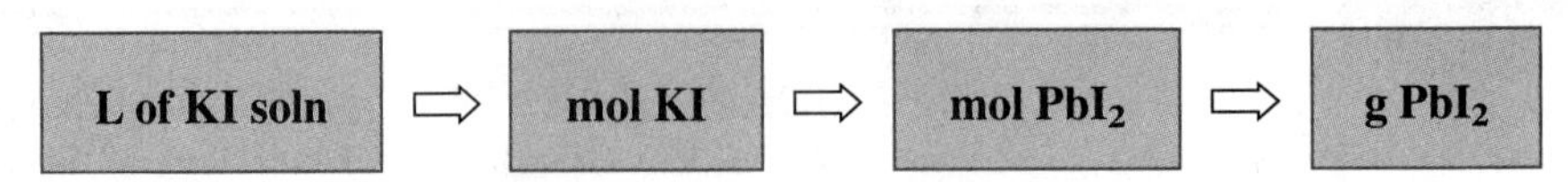

SOLUTION

Following the strategy just given, we first calculate the moles of potassium iodide:

$$? \text{ mol KI} = 0.0350 \text{ L soln}\left(\frac{0.150 \text{ mol KI}}{\text{L soln}}\right) = 0.00525 \text{ mol KI}$$

In this case, the mole ratio between moles of known (KI) and moles of unknown (PbI_2) is 2 : 1.

$$2 \text{ mol KI} : 1 \text{ mol PbI}_2$$

$$? \text{ g PbI}_2 = 0.00525 \text{ mol KI}\left(\frac{1 \text{ mol PbI}_2}{2 \text{ mol KI}}\right)\left(\frac{461.0 \text{ g PbI}_2}{1 \text{ mol PbI}_2}\right)$$
$$= 1.21 \text{ g PbI}_2$$

As usual, we can combine all of our calculations into one setup:

$$? \text{ g PbI}_2 = 0.0350 \text{ L soln}\left(\frac{0.150 \text{ mol KI}}{\text{L soln}}\right)\left(\frac{1 \text{ mol PbI}_2}{2 \text{ mol KI}}\right)\left(\frac{461.0 \text{ g PbI}_2}{1 \text{ mol PbI}_2}\right)$$
$$= 1.21 \text{ g PbI}_2$$

EXAMPLE 13.19

Calculate the volume of 0.120 M HCl required to react completely with 0.452 g Zn, according to the following equation:

$$\text{Zn(s)} + 2 \text{ HCl(aq)} \rightarrow \text{ZnCl}_2\text{(aq)} + \text{H}_2\text{(g)}$$

Planning the solution

In this problem we are given the mass of Zn (the known). Thus, we can find the number of moles of HCl (the unknown) required by using the following strategy:

This is the same strategy we employed in Chapter 10 for mass–mole problems. Once we know the number of moles of HCl required, we can use the molarity as a conversion factor to find the volume of 0.120 M HCl that contains this number of moles.

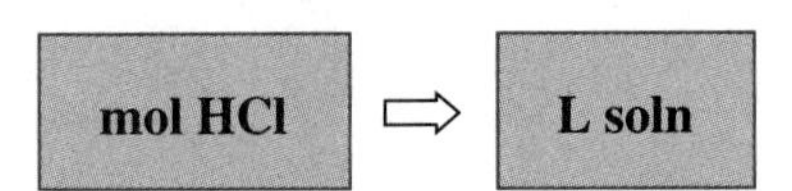

SOLUTION

First, we find the number of moles of HCl that are required:

$$? \text{ g Zn} = 0.452 \text{ g Zn}\left(\frac{1 \text{ mol Zn}}{65.4 \text{ g Zn}}\right)\left(\frac{2 \text{ mol HCl}}{1 \text{ mol Zn}}\right) = 0.0138 \text{ mol HCl}$$

Then we use the molarity as a conversion factor to determine the volume of 0.120 M HCl that contains 0.0138 mol HCl:

$$? \text{ L soln} = 0.0138 \text{ mol HCl} \left(\frac{1 \text{ L soln}}{0.120 \text{ mol HCl}} \right) = 0.115 \text{ L soln}$$

Thus, we need 0.115 L, or 115 mL, of 0.120 M HCl solution.

EXAMPLE 13.20

Calculate the volume of 0.180 M $AgNO_3$ required to react completely with 45.0 mL of 0.240 M K_2CrO_4 according to the following equation:

$$2\ AgNO_3(aq) + K_2CrO_4(aq) \rightarrow Ag_2CrO_4(s) + 2\ KNO_3(aq)$$

Planning the solution

In Example 13.18 we used the volume and molarity of a solution to calculate the number of moles of known in the first step of our strategy. In Example 13.19 we used the number of moles of unknown and its molarity to calculate the volume of unknown solution required in the last step of our strategy. In this problem, we will use both of these calculations, along with the mole ratio, to convert from moles of known to moles of unknown:

L K_2CrO_4 soln (vol known) ⇨ **mol K_2CrO_4 (mol known)** ⇨ **mol $AgNO_3$ (mol unknown)** ⇨ **L $AgNO_3$ soln (vol unknown)**

SOLUTION

First, we calculate the moles of K_2CrO_4:

$$? \text{ mol } K_2CrO_4 = 0.0450 \text{ L soln} \left(\frac{0.240 \text{ mol } K_2CrO_4}{\text{L soln}} \right) = 0.0108 \text{ mol } K_2CrO_4$$

Next, we use the mole ratio to find the moles of $AgNO_3$:

$$? \text{ mol } AgNO_3 = 0.0108 \text{ mol } K_2CrO_4 \left(\frac{2 \text{ mol } AgNO_3}{1 \text{ mol } K_2CrO_4} \right) = 0.0216 \text{ mol } AgNO_3$$

Finally, we use the molarity of $AgNO_3$ to find the volume of 0.180 M $AgNO_3$ that contains 0.0216 mol $AgNO_3$:

$$? \text{ L soln} = 0.0216 \text{ mol } AgNO_3 \left(\frac{1 \text{ L soln}}{0.180 \text{ mol } AgNO_3} \right) = 0.120 \text{ L soln}$$

Thus, we need 0.120 L of 0.180 M $AgNO_3$ solution.

In the following example, a gas is produced, so we must combine our knowledge of solutions, stoichiometry, and gas laws.

EXAMPLE 13.21

Sodium carbonate reacts with hydrochloric acid according to the following equation:

$$Na_2CO_3(aq) + 2\ HCl(aq) \rightarrow 2\ NaCl(aq) + H_2O(\ell) + CO_2(g)$$

What volume of carbon dioxide at STP is produced when 55.0 mL of 0.320 M Na_2CO_3 are treated with an excess of hydrochloric acid?

Planning the solution

In this case, our known is Na_2CO_3. We will calculate the number of moles of CO_2 as follows:

L Na_2CO_3 soln ⇨ **mol Na_2CO_3** ⇨ **mol CO_2**

Then we will apply the gas laws to find the volume of CO_2 at STP.

mol CO_2 ⇨ **L CO_2 (gas)**

SOLUTION

$$? \text{ mol } Na_2CO_3 = 0.0550 \text{ L soln}\left(\frac{0.320 \text{ mol } Na_2CO_3}{\text{L soln}}\right) = 0.0176 \text{ mol } Na_2CO_3$$

Our mole ratio gives the moles of carbon dioxide, the unknown:

$$? \text{ mol } CO_2 = 0.0176 \text{ mol } Na_2CO_3\left(\frac{1 \text{ mol } CO_2}{1 \text{ mol } Na_2CO_3}\right) = 0.0176 \text{ mol } CO_2$$

Finally, we calculate the volume of CO_2 at STP:

$$? \text{ L } CO_2 = 0.0176 \text{ mol } CO_2\left(\frac{22.4 \text{ L}}{1 \text{ mol } CO_2}\right) = 0.394 \text{ L}$$

This general method can also be applied when a problem asks for the volume of gas produced at a temperature and pressure other than STP. For example, if we are asked for the volume of CO_2 at 745 torr and 21°C (294 K), we simply correct the final volume of 0.394 L at STP to those conditions:

$$? \text{ L } CO_2 = 0.394 \text{ L}\left(\frac{760 \text{ torr}}{745 \text{ torr}}\right)\left(\frac{294 \text{ K}}{273 \text{ K}}\right) = 0.433 \text{ L}$$

Alternatively, we could have substituted 0.0176 mol, 745 torr, and 294 K into the Ideal Gas Law and calculated the same volume.

PROBLEM 13.13

When barium hydroxide is mixed with sulfuric acid, a precipitate of barium sulfate results.

$$Ba(OH)_2(aq) + H_2SO_4(aq) \rightarrow BaSO_4(s) + 2\, H_2O(\ell)$$

Calculate the mass of barium sulfate precipitated when 64.5 mL of 0.125 M H_2SO_4 are treated with excess $Ba(OH)_2$.

PROBLEM 13.14

Calculate the mass of silver sulfide that precipitates when 55.0 mL of 0.250 M $AgNO_3$ are treated with an excess of Na_2S according to the following equation.

$$2\, AgNO_3(aq) + Na_2S(aq) \rightarrow Ag_2S(s) + 2\, NaNO_3(aq)$$

PROBLEM 13.15

Calculate the volume of 0.275 M HNO_3 required to react completely with 3.55 g $CaCO_3$ according to the following equation.

$$CaCO_3(s) + 2\, HNO_3(aq) \rightarrow Ca(NO_3)_2(aq) + H_2O(\ell) + CO_2(g)$$

PROBLEM 13.16

Calculate the volume of 0.125 M $Pb(NO_3)_2$ required to react completely with 40.0 mL of 0.350 M Na_3PO_4 according to the following equation.

$$3\ Pb(NO_3)_2(aq) + 2\ Na_3PO_4(aq) \rightarrow Pb_3(PO_4)_2(s) + 6\ NaNO_3(aq)$$

PROBLEM 13.17

Magnesium metal reacts with hydrochloric acid to produce hydrogen gas.

$$Mg(s) + 2\ HCl(aq) \rightarrow MgCl_2(aq) + H_2(g)$$

(a) What volume of hydrogen at STP is produced when 16.0 mL of 0.150 M HCl react with an excess of magnesium metal?

(b) What volume of hydrogen at 1.15 atm and 35°C is produced using the same quantities of reactants as in part (a)?

13.8 TITRATION

OBJECTIVE 13.8 Describe the technique used to titrate one solution against another. Given the volume and molarity of one solution used in a titration, calculate the molarity of the other solution, given the volume of the second solution used and a chemical equation for their reaction with one another.

The concentration of one solution can be used to determine the concentration of another by comparing the volumes of each required in a known reaction. For example, hydrochloric acid and sodium hydroxide react according to the following equation:

$$HCl(aq) + NaOH(aq) \rightarrow NaCl(aq) + H_2O(\ell)$$

Suppose 50.0 mL of 0.100 M HCl is placed in a flask and just enough NaOH solution is added to consume all of the HCl. We can calculate the number of moles of NaOH that have been added by using the calculations we have been studying in this chapter.

L HCl soln ⇨ **mol HCl** ⇨ **mol NaOH**

$$?\ \text{mol NaOH} = 0.0500\ \cancel{\text{L soln}}\left(\frac{0.100\ \cancel{\text{mol HCl}}}{\cancel{\text{L soln}}}\right)\left(\frac{1\ \text{mol NaOH}}{1\ \cancel{\text{mol HCl}}}\right)$$

$$= 0.00500\ \text{mol NaOH}$$

Suppose the volume of the NaOH solution that is added is also measured and 25.0 mL is required. Since the 25.0-mL sample of solution contains 0.00500 mol NaOH, we can determine its concentration by dividing the number of moles of NaOH by its volume in liters:

$$\text{Molarity} = \frac{0.00500\ \text{mol NaOH}}{0.0250\ \text{L soln}} = 0.200\ \text{M NaOH}$$

Standardization: The determination of the concentration of a solution.

Titration: A technique used in the standardization of solutions.

We refer to the process of determining the precise concentration of a solution as **standardization.** The procedure for carrying out such an experiment is known as **titration** (Fig. 13-7). In the experiment we just described, the volume of each solution used can be determined accurately with a measuring device known as a *buret.* We also need some way of telling when the reactants have exactly consumed one another. For example, in the preceding experiment, at what point did the sodium hydroxide consume the last of the hydrochloric acid? This is known as the *endpoint.* An endpoint is often determined by adding an *indicator* to the reacting solution. In the titration just described, we sometimes use an indicator known as *phenolphthalein,* which is a substance that is colorless in an acid solution and violet in the presence of a base. (This color is often perceived as pink for solutions that are only slightly basic or are very dilute.) Initially, when phenolphthalein is added to the acid being titrated, it is colorless. However, when just enough sodium hydroxide has been added to consume the last of the hydrochloric acid, the very next drop of sodium hydroxide turns the solution from colorless to a faint pink. There are many variations of the procedures used in titration, including how the endpoint is determined. Regardless of how a titration is carried out, the calculations involve the principles of solution stoichiometry we have presented here.

FIGURE 13-7 Titration.

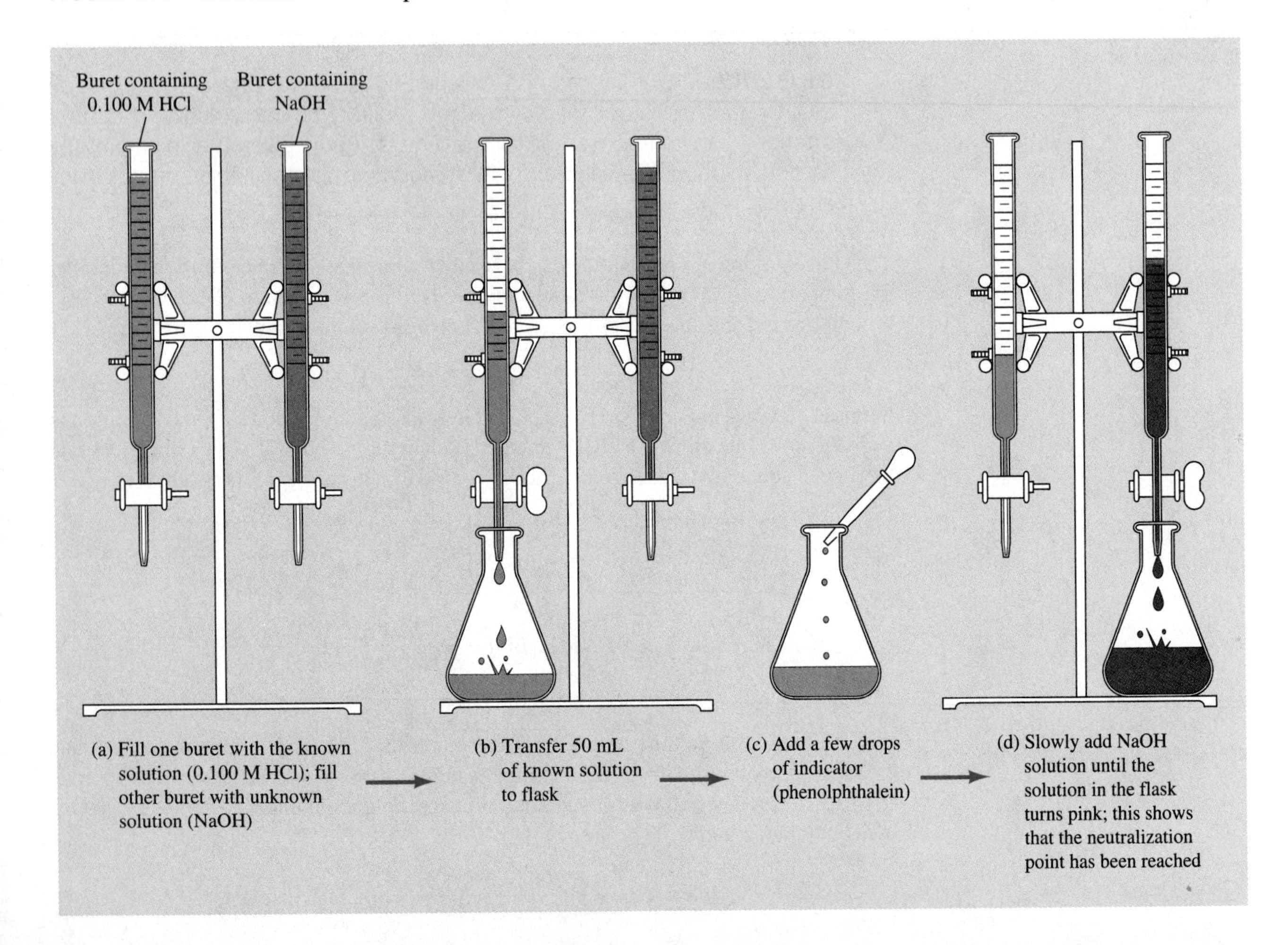

EXAMPLE 13.22

A 15.0-mL sample of NaCl solution is required to react completely with 25.0 mL of 0.210 M $AgNO_3$ according to the following equation:

$$AgNO_3(aq) + NaCl(aq) \rightarrow AgCl(s) + NaNO_3(aq)$$

Calculate the molarity of the sodium chloride solution.

Planning the solution

To calculate the molarity of a solution, we need to know the number of moles of solute and the volume in which it is dissolved. If we know the number of moles of NaCl present in the 15.0-mL sample used in the reaction, we can calculate the molarity. We determine the number of moles of NaCl as follows:

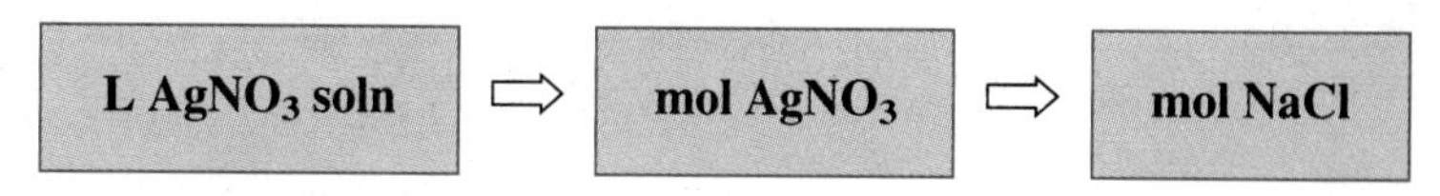

Then we divide the number of moles of NaCl by its volume in liters (15.0 mL = 0.0150 L).

SOLUTION

$$? \text{ mol NaCl} = 0.0250 \text{ L soln}\left(\frac{0.210 \text{ mol AgNO}_3}{\text{L soln}}\right)\left(\frac{1 \text{ mol NaCl}}{1 \text{ mol AgNO}_3}\right)$$

$$= 0.00525 \text{ mol NaCl}$$

$$\text{Molarity} = \frac{0.00525 \text{ mol NaCl}}{0.0150 \text{ L soln}} = 0.350 \text{ M NaCl}$$

EXAMPLE 13.23

A 35.0-mL sample of NaOH solution is required to consume all of the H_2SO_4 in 15.0 mL of 0.140 M H_2SO_4 according to the following equation:

$$H_2SO_4(aq) + 2\ NaOH(aq) \rightarrow Na_2SO_4(aq) + 2\ H_2O(\ell)$$

Calculate the molarity of the sodium hydroxide solution.

Planning the solution

We calculate the moles of NaOH using the following strategy:

L H_2SO_4 soln ⇨ **mol H_2SO_4** ⇨ **mol NaOH**

Then we divide the number of moles of NaOH by its volume in liters.

SOLUTION

$$? \text{ mol NaOH} = 0.0150 \text{ L soln}\left(\frac{0.140 \text{ mol H}_2\text{SO}_4}{\text{L soln}}\right)\left(\frac{2 \text{ mol NaOH}}{1 \text{ mol H}_2\text{SO}_4}\right)$$

$$= 0.00420 \text{ mol NaOH}$$

$$\text{Molarity} = \frac{0.00420 \text{ mol NaOH}}{0.0350 \text{ L soln}} = 0.120 \text{ M NaOH}$$

PROBLEM 13.18

A 25.0-mL sample of NaOH solution is required to react completely with 35.0 mL of 0.440 M HCl according to the following equation:

$$HCl(aq) + NaOH(aq) \rightarrow NaCl(aq) + H_2O(\ell)$$

What is the molarity of the NaOH solution?

PROBLEM 13.19

A 14.0-mL sample of H_3PO_4 solution reacts completely with 25.0 mL of 0.210 M KOH according to the following equation:

$$H_3PO_4(aq) + 3\ KOH(aq) \rightarrow K_3PO_4(aq) + 3\ H_2O(\ell)$$

Calculate the molarity of the H_3PO_4 solution.

13.9 COLLIGATIVE PROPERTIES OF SOLUTIONS

OBJECTIVE 13.9A List three colligative properties. Define *molality.* Calculate the boiling point of a solution, given the masses of nonionic solute and solvent, the boiling point of pure solvent, and the boiling point elevation constant. Calculate the freezing point of a solution, given the masses of nonionic solute and solvent, the melting point of pure solvent, and the cryoscopic constant.

Colligative properties: Properties of solutions that depend on the concentration of the dissolved particles.

Molality: The number of moles of solute per kilogram of solvent.

A number of properties associated with solutions are referred to as **colligative properties.** We will discuss three of these: boiling point elevation, freezing point depression, and osmotic pressure. These properties depend on the concentration of the solution in question. We use a concentration unit known as molality when working with colligative properties. The **molality** of a solution is the number of moles of solute per kilogram of solvent:

$$\text{Molality} = \frac{\text{Number of moles of solute}}{\text{Number of kilograms of solvent}}$$

The symbol m is used for molality and distinguishes it from molarity (M). As this definition implies, molal solutions are prepared differently from molar solutions. To prepare a 1.00 m solution of sucrose ($C_{12}H_{22}O_{11}$) in water, we weigh out 1.00 mol of sucrose (342 g) and add it to 1.00 kg (1000 g) of water. Note that this solution is not 1.00 M, because the final volume is greater than 1.00 L. (Although 1.00 kg of water does have a volume of 1.00 L, adding the sucrose results in a volume *greater* than 1.00 L.)

Boiling Point Elevation

Adding sucrose to water raises its boiling point. Sucrose is a *nonvolatile* solute (one that does not evaporate readily). When such a solute is dissolved in a solvent, the vapor pressure of the solution is reduced below that of the pure solvent at the same temperature. Boiling occurs when the vapor pressure of a liquid equals the external pressure. It follows that the temperature of such a solution must be elevated above the boiling point of the pure solvent before boiling can occur. In fact, the boiling point elevation depends on the molality of the solution. If water is the solvent, a 1.00 m solution of any nonvolatile *nonionic* substance boils at 100.52°C. This corresponds to a boiling point *elevation* of 0.52°C. The boiling point of a 2.00 m solution is elevated by twice this amount ($2 \times 0.52°C = 1.04°C$), leading to a boiling point of 101.04°C. The boiling point elevation, ΔT, is directly proportional to the molal concentration of the solution. That is,

$$\Delta T = K_b \cdot m$$

where K_b is a constant that tells the amount the boiling point is elevated for each 1.00 *m* concentration. For water, $K_b = 0.52°C/m$. Every pure liquid has its own value for K_b.

The boiling point elevation actually depends on the number of moles of *solute particles* per kilogram of solvent. Ionic substances separate into their component ions when they dissolve in water, so 1.00 mol of NaCl produces 1.00 mol of Na^+ and 1.00 mol of Cl^-. Consequently, a 1.00 *m* solution of NaCl in water boils at 101.04°C, the same temperature as a 2.00 *m* solution of a nonionic substance, such as sucrose. For simplicity, we will restrict our discussion to nonionic substances.

Freezing Point Depression

In a fashion similar to elevation of the boiling point, the freezing point of a liquid is lowered by the presence of nonvolatile solutes. If you have ever made homemade ice cream by hand, you may have used an ice–salt mixture to cool the ice cream being made. The addition of salt to the ice lowers the freezing point of the water, dropping the temperature below 0°C. The freezing point depression is proportional to the molality in the same manner as the boiling point elevation. That is,

$$\Delta T = K_f \cdot m$$

where K_f is known as the *cryoscopic constant* for the liquid, and ΔT represents the amount the freezing point is depressed. The cryscopic constant tells the amount the freezing point is depressed per 1.00 *m* concentration. For water, $K_f = 1.86°C/m$. This means that a 1.00 *m* solution of glucose in water freezes at a temperature 1.86°C below the freezing point of pure water, or (since water freezes at 0°C) at −1.86°C. A 2.00 *m* solution in water freezes at −3.72°C (−1.86°C × 2), and a 3.00 *m* solution freezes at −5.58°C (−1.86°C × 3).

The freezing point depression is used when antifreeze is added to the water in an automobile radiator. It helps prevent freezing by lowering the freezing point of the water. Antifreeze also helps prevent boiling over in the summer by raising the boiling point.

Table 13-3 lists the cryoscopic constants of a number of pure substances. Note that the cryoscopic constant is given as a positive quantity. Thus, when we calculate ΔT, we are determining the amount the freezing point will be depressed from the freezing point of the pure solvent. For example, pure acetic acid freezes at 16.6°C and has a cryoscopic constant of 3.9°C/*m*. A 1.00 *m* solution of a nonionic substance dissolved in acetic acid freezes 3.9°C below the freezing point of the pure solvent—that is, at

$$16.6°C - 3.9°C = 12.7°C$$

The freezing point of a 2.00 *m* solution is depressed by twice as much:

$$\Delta T = K_f \cdot m$$

$$= \left(\frac{3.9°C}{\cancel{m}}\right) 2.00 \cancel{m} = 7.8°C$$

Thus, it freezes at 16.6°C − 7.8°C = 8.8°C.

TABLE 13-3 Cryoscopic constants of selected solvents

Solvent	Normal melting point (°C)	Depression for a 1 *m* solution (°C)
Acetic acid	16.6	3.9
Benzene	5.5	4.9
t-Butyl alcohol	25.5	8.3
Camphor	17.6	40.0
Ethylene dibromide	10.0	11.8
Formic acid	8.4	2.7
Naphthalene	80.2	6.8
Nitrobenzene	5.7	7.0
Water	0.00	1.86

EXAMPLE 13.24

Calculate the freezing point depression and the freezing point of a 1.33 *m* solution of a nonionic solute in formic acid.

SOLUTION

First we must use the cryoscopic constant to determine the freezing point depression. The cryoscopic constant of formic acid is 2.7°C/*m*.

$$\Delta T = K_f \cdot m$$

$$= \left(\frac{2.7°\text{C}}{\cancel{m}}\right) 1.33\ \cancel{m} = 3.6°\text{C}$$

The freezing point is lowered by 3.6°C. The normal melting point of formic acid is 8.4°C, so the freezing point of the solution must be

$$8.4°\text{C} - 3.6°\text{C} = 4.8°\text{C}$$

PROBLEM 13.20

Calculate the freezing point depression and the freezing point of the following nonionic solutions.

(a) 1.20 *m* in water **(b)** 2.34 *m* in acetic acid **(c)** 0.670 *m* in camphor
(d) 0.23 *m* in benzene **(e)** 0.78 *m* in naphthalene

Osmotic Pressure

OBJECTIVE 13.9B Explain what a semipermeable membrane is. Describe the measurement of osmotic pressure. Give a generalization in terms of both concentration and osmotic pressure for the flow of solutions across a semipermeable membrane. Explain the terms *isotonic* and *physiological.*

If you have ever made pickles, you know that when a pickling cucumber is allowed to soak in a brine solution, it shrivels up. Water from inside the cucumber moves to the surrounding brine solution, which contains a very high concentration of salts. This

Osmotic pressure: A property of solutions that causes a net flow of solvent molecules across a semipermeable membrane.

Semipermeable membrane: A membrane that permits the flow of some molecules or ions, but not others, across its surface.

phenomenon is the result of osmotic pressure. **Osmotic pressure** arises when two solutions of different concentrations are separated by a barrier known as a **semipermeable membrane.** In the case of the cucumber, the peel serves this function.

A semipermeable membrane is a barrier that acts like a filter. It allows small particles through but prevents the passage of larger particles. For example, a typical semipermeable membrane might allow water molecules across, but it would prevent the passage of glucose molecules (Fig. 13-8). Suppose we have a U-tube (Fig. 13-9a) with a semipermeable membrane that allows solvent molecules to pass from one side to the other but does not allow solute particles across. Let us place an aqueous solution of glucose in one side of the tube and an equal height of water in the other. After a short period of time, the height of the liquid on the glucose side is higher than that of the water column (Fig. 13-9b). Solvent molecules (water) have traveled across the membrane from the pure water side to the glucose side. If we wish, we can apply an external pressure to the glucose solution (Fig. 13-9c), thereby "pushing" the columns back to equal heights. The pressure applied to equalize the heights represents the osmotic pressure of the glucose solution.

To further understand the phenomenon of osmotic pressure, imagine what would happen if we did not have a semipermeable membrane separating the two sides of the U-tube. There would be a natural tendency for the solutions to mix uniformly, causing the pure water to dilute the more concentrated solution. In other words, the natural course of events would be for the concentrations of the two solutions to equalize one another. With the semipermeable membrane in place, however, only solvent molecules can migrate. In an attempt to equalize the concentrations, water molecules must flow out of the pure water side, diluting the glucose solution. This creates an imbalance in the heights of the two columns, and the difference in heights is related to the osmotic pressure of the glucose solution (Fig. 13-9b).

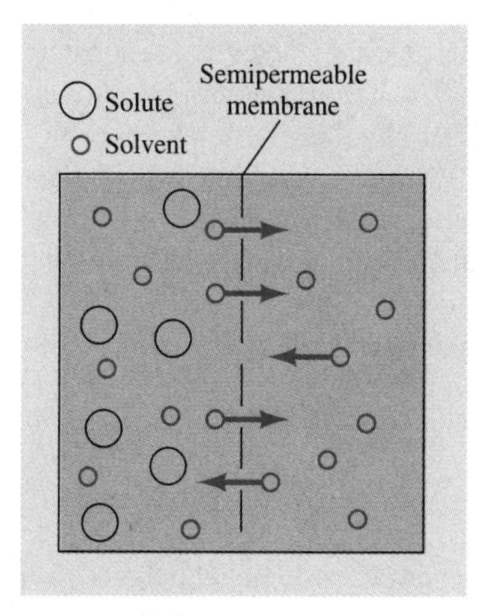

FIGURE 13-8 A semipermeable membrane allows solvent molecules to pass through, but solute molecules cannot cross from one side to the other.

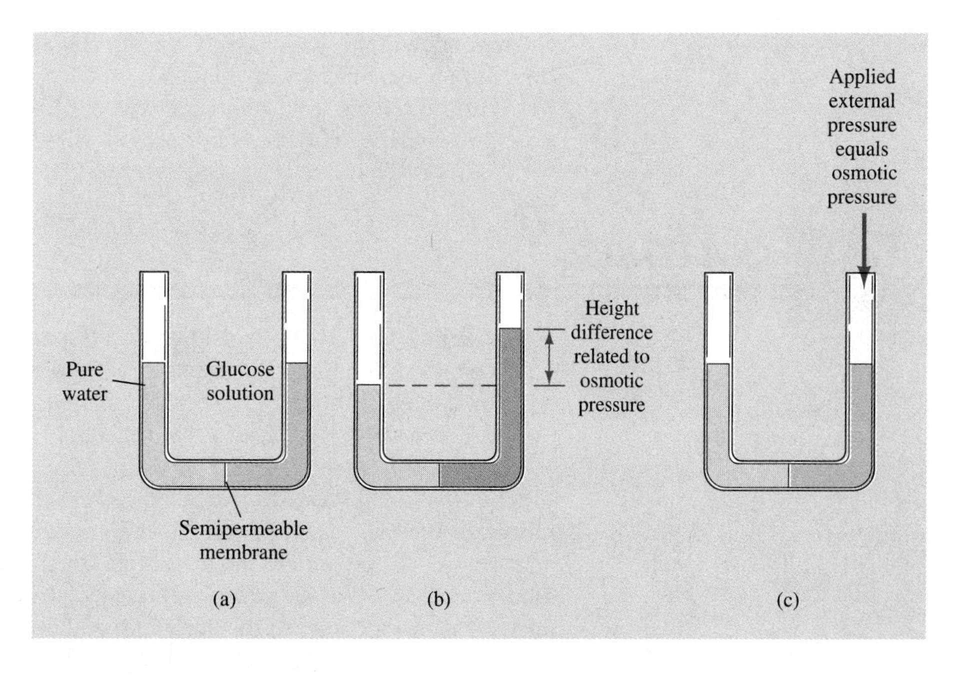

FIGURE 13-9 (a) A U-tube with arms separated by a semipermeable membrane is filled to equal heights with a glucose solution and water. Solvent molecules may cross in either direction, but solute molecules are confined to the glucose side. (b) Net movement of solvent molecules to the glucose side results in a difference in heights. The height difference is related to the osmotic pressure. (c) An external pressure may be applied to the glucose side, forcing the columns back to equal heights. The pressure required to accomplish this is the osmotic pressure.

When solvent molecules cross a semipermeable membrane, they flow from the more dilute side (lower osmotic pressure) to the more concentrated side (greater osmotic pressure). That is why the fluids (which are mostly water) migrate from the inside of a pickling cucumber to the surrounding brine solution. This principle also has considerable importance in biological systems. When blood cells are placed in pure water, water molecules cross the cell membrane (a semipermeable membrane) and enter the cell to dilute its contents. The result is that the cell enlarges and eventually bursts. On the other hand, when the cell is placed in a concentrated salt solution, water from the more dilute solution inside the cell leaves the cell, causing it to shrivel. In order to avoid either of these undesirable effects, it is necessary for the osmotic pressure of fluids injected into our veins to match the osmotic pressure of our blood. Solutions that have equal osmotic pressure are said to be **isotonic.** *Physiological solutions* of saline (0.9% NaCl) or glucose (5% $C_6H_{12}O_6$) are those that are isotonic with our cellular fluids.

Isotonic: Refers to solutions that have the same osmotic pressure.

The osmotic pressure of a solution is proportional to the number of moles of solute particles per kilogram of solution. A solute particle may be either a molecule or an ion. When 1.00 mol of NaCl dissolves in water, the solution contains 1.00 mol of Na^+ ions and 1.00 mol of Cl^- ions, or a total of 2.00 mol of solute particles. Physiologists use a concentration unit known as osmolality (osm) when working with the osmotic pressure of solutions. The **osmolality** of a solution is the number of moles of solute particles per kilogram of solvent. This is similar to molality, but for solutions of ionic substances, the osmolality differs from the molality. Since 1.00 mol of NaCl contains 2.00 mol of ions, a 1.00 m NaCl solution has an osmolality of 2.00 osm and exhibits the same osmotic pressure as a 2.00 m solution of a nonionic solute, such as glucose.

Osmolality: The number of moles of solute particles per kilogram of solvent.

PROBLEM 13.21

(a) Suppose the arms of the U-tube shown in Fig. 13-9 are filled to equal heights with 2.0% glucose on the left and 5.0% glucose on the right. In which direction will the water flow?

(b) A 0.9% sodium chloride solution is isotonic with a 5.0% glucose solution. If the U-tube is refilled to equal heights with 5.0% sodium chloride on the left and 5.0% glucose on the right, in which direction will the water flow?

13.10 COLLOIDS

OBJECTIVE 13.10 Explain what a dispersion is. Explain what property of a colloid differentiates it from other heterogeneous mixtures. Explain the meanings of the terms *dispersed phase* and *dispersing phase.* Give examples of three colloids and, for each, describe the dispersed and dispersing phases.

In Chapter 2 we discussed the difference between heterogeneous and homogeneous mixtures. A homogeneous mixture is uniform throughout; a heterogeneous mixture is not. We used a sugar–water mixture as an example of a homogeneous mixture and a sand–water mixture as an example of a heterogeneous mixture. Let us consider these mixtures more carefully. At the beginning of this chapter, we learned that when sugar dissolves in water, it separates into individual molecules, resulting in a uniform distribution of particles (sugar molecules). Thus, the particles in the mixture are quite small (the size of one molecule). By contrast, the particles in the sand–water mixture are quite large, and each grain is visible to the naked eye. (These particles are about 100,000 times the size of a molecule.)

Dispersion: A suspension of one substance in another.

Colloid: A heterogeneous mixture in which the phases do not separate out.

Dispersed phase: In a colloid, the phase that is suspended.

Dispersing phase: In a colloid, the phase suspending the dispersed phase.

A heterogeneous mixture such as the sand–water mixture may also be referred to as a **dispersion,** because the sand particles are not dissolved in the water but are simply dispersed in it. When the particles in a dispersion are large enough (about 1000 times the size of a molecule or larger), settling out of the particles occurs. However, it is possible to have a dispersion (a heterogeneous mixture) in which the particles are so much smaller than sand particles that they remain suspended rather than settling out. Such a dispersion is referred to as a colloid. **Colloids** are heterogeneous mixtures in which the particle size is such that separation of the phases does not occur.

When discussing a dispersion, we often refer to the **dispersed phase** and the **dispersing phase.** In a sand–water mixture, sand is the dispersed phase and water is the dispersing phase. It is possible for a dispersing phase to be a solid, a liquid, or a gas. The same is true of the dispersed phase. Since colloids are dispersions, it would appear that nine types of colloid are possible: a solid, liquid, or gas dispersed in a solid; a solid, liquid, or gas dispersed in a liquid; and a solid, liquid, or gas dispersed in a gas. However, any dispersion of one gas in another is of necessity a homogeneous mixture, so only eight types of colloids exist. Each has a special name, and they are summarized in Table 13-4, which also gives an example of each type.

It might appear that there is little distinction between a colloid and a homogeneous solution, because the particle size in a colloid is so small that the particles do not settle out. Indeed, when a colloidal suspension of a solid in a liquid is passed through a filter paper, both the dispersed solid and the dispersing liquid pass through. However, a

TABLE 13-4 Eight types of colloids

Dispersing substance	Dispersed substance	Special name	Example
Gas	Liquid	Liquid aerosol	Fog—finely divided water droplets are suspended in the air
Gas	Solid	Solid aerosol	Smoke—minute solid particles are suspended in air
Liquid	Gas	Foam	Whipped cream—air is suspended in cream (shaving creams are also colloids of this type)
Liquid	Liquid	Emulsion	Mayonnaise—oil and vinegar, two immiscible liquids, remain suspended in one another if beaten well with egg yolk
Liquid	Solid	Sol	Paints—these contain solids suspended in a liquid
Solid	Gas	Solid foam	Polystyrene—plastic foams such as this, frequently used for packaging or insulation, contain air suspended in a plastic
Solid	Liquid	Solid emulsion	Gelatins—desserts of this type contain liquids dispersed in solids
Solid	Solid	Solid sol	Stained glass—colored, solid particles are suspended in glass

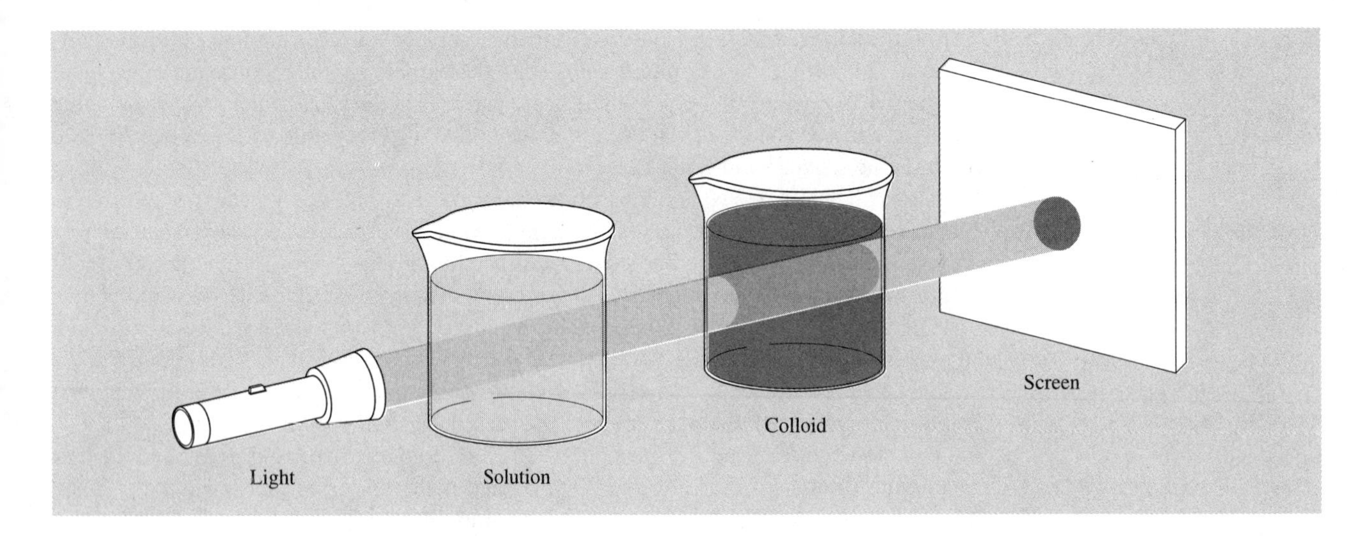

FIGURE 13-10 The Tyndall effect. Light passes through a solution with no noticeable effect. When a light beam passes through a colloid, the scattering of light by colloidal particles makes the path of the beam visible.

colloid exhibits behavior that is characteristic of the heterogeneous mixture that it is. Unlike the substances in a true solution, the substances in a colloidal suspension are insoluble in one another. Thus, whereas we can see through a solution with no interference, colloidal suspensions are frequently "milky" in appearance, resulting from light reflecting off the suspended particles. A colloidal suspension can be distinguished from a solution by shining a beam of light through it. The light beam passes through a solution with no noticeable effect, but the particles in a colloid scatter the light, enabling us to see the path of the beam (Fig. 13-10). We refer to the scattering of light in this manner as the *Tyndall effect.*

PROBLEM 13.22

Homogenized milk is a colloidal suspension. What two characteristics of milk enable you to identify it as such?

CHAPTER SUMMARY

A **solution** is any homogeneous mixture. The most commonly used solutions are *liquid solutions,* which are prepared by dissolving a substance, referred to as the **solute,** in a liquid, referred to as the **solvent.** If the solvent is water, we call the solution an **aqueous solution.**

The **concentration** of a solution is the amount of solute in a given quantity of solvent or solution as a whole. The maximum concentration possible for a solute (under specified conditions) is its **solubility.** A solution at the solubility limit is said to be **saturated.** Two liquids that are mutually soluble are said to be **miscible,** whereas **immiscible** liquids do not dissolve completely in one another.

The most useful unit for expressing the concentration of a solution is **molarity (M).** The molarity of a solution gives the number of moles of solute per liter of solution. Solutions can be prepared in either of two ways. We can measure out the mass of solute and dilute to the desired total volume, or we can dilute a more concentrated solution by adding solvent. When preparing solutions by dilution, we can use the dilution formula: $M_{con} \cdot V_{con} = M_{dil} \cdot V_{dil}$. This formula relates the volume and molarity of concentrated reagent to the molarity and volume of diluted solution.

Solution concentration can also be expressed in

terms of the **mass percent** or **volume percent** of the solution. Related to percentage concentration is the number of **parts per million (ppm),** which is used to describe water or air quality. For water quality, the number of parts per million gives the number of milligrams of solute per kilogram (or per million milligrams) of solution.

Since solutions are commonly used for carrying out reactions, the principles of stoichiometry are often applied to solutions. **Titration** is a technique that may be used for determining the concentration of an unknown solution. Determining the precise concentration of a solution is known as **standardization.**

Certain properties associated with solutions depend on the number of moles of solute particles in solution. These are known as **colligative properties,** and they depended on the **molality** of the solution—the number of moles of solute per kilogram of solvent. Among the colligative properties are boiling point elevation, freezing point depression, and osmotic pressure.

The presence of nonvolatile solute particles lowers the vapor pressure of a solvent. This in turn raises the boiling point. In a similar fashion, nonvolatile solutes lower the freezing point of a solvent.

The **osmotic pressure** of a solution expresses the tendency of solvent molecules to cross a **semipermeable membrane** from the more dilute side to the more concentrated side. Solutions that have matching osmotic pressures are said to be **isotonic.** The osmotic pressure of a solution is related to its **osmolality**—the number of moles of solute *particles* per kilogram of solvent.

A **colloid** exists when the phases of a heterogeneous **dispersion** remain suspended in one another. The particles of the substance being dispersed are much larger than those found in a true solution but small enough that they do not settle out. The substance being dispersed is known as the **dispersed phase;** the substance it is dispersed in is the **dispersing phase.** A colloidal suspension often has a milky appearance, which results from light being reflected off the dispersed particles. This phenomenon is referred to as the *Tyndall effect.*

KEY TERMS

Review each of the following terms, which are discussed in this chapter and defined in the Glossary.

solution
solvent
solute
aqueous solution
solubilty
concentration
saturated solution
immiscible
miscible
molarity
percent solution by volume
percent solution by mass
parts per million
standardization
titration
colligative properties
molality
osmotic pressure
semipermeable membrane
isotonic
osmolality
dispersion
colloid
dispersed phase
dispersing phase

ADDITIONAL PROBLEMS

SECTION 13.1

13.23 Answer each of the following questions about solutions.

(a) Define the term *solution.*
(b) Which of the following are solutions?
  (1) Oil and vinegar **(2)** Air
  (3) Bronze **(4)** Salt water
(c) What is a liquid solution?
(d) For a liquid solution, what is the solute and what is the solvent?
(e) What is an aqueous solution?
(f) Give two reasons why chemists use solutions.

SECTION 13.2

13.24 Answer each of the following questions about solubility.

(a) Define the term *solubility.*
(b) Does the solubility of a substance vary from one solvent to another?
(c) As a general rule, how does temperature affect the solubility of each of the following?
  (1) A solid in a liquid **(2)** A gas in a liquid
(d) What factor is the most helpful in predicting whether a given solute will dissolve in a particular solvent?

(e) Methyl alcohol is a polar liquid that hydrogen-bonds. Would you expect it to dissolve in water? Explain your answer.

(f) Pentane is a nonpolar liquid that interacts with neighboring molecules through London forces. Would you expect it to dissolve in water? Explain your answer.

(g) How do each of the following factors affect the rate at which a substance dissolves?

(i) Crystal size **(ii)** Stirring **(iii)** Temperature

13.25 Chemists frequently remove dissolved carbon dioxide from water by boiling the water before using it. Why does this technique work?

13.26 A saturated solution is prepared by mixing 9.85 g of a solid substance and 35.0 mL of water. The excess solid is then filtered and dried, and its mass is determined to be 2.73 g. Calculate the solubility of the substance in g/100 mL of water (g/1.00×10^2 mL H_2O).

SECTION 13.3

13.27 Define the term *molarity*. The mass of 0.500 mol of KI is 83.0 g. Describe the stepwise procedure you would use to prepare 1 L of a 0.500 M KI solution.

13.28 Calculate the molarity of each of the following solutions, which were prepared by dissolving the solute in enough water to achieve the final volume indicated.

(a) 135 g KCl in 2.50 L
(b) 403 g $MgSO_4$ in 5.25 L
(c) 5.00 g KBr in 0.500 L
(d) 0.554 g NH_3 in 15.0 mL
(e) 6.12 g $C_6H_{12}O_6$ in 50.0 L
(f) 1.24 g $AgNO_3$ in 125 mL

SECTION 13.4

13.29 Calculate the number of moles present in each of the following solutions.

(a) 2.50 L of 3.50 M NaCl
(b) 375 mL of 6.00 M HCl
(c) 55.0 mL of 3.00 M NaOH
(d) 23.6 mL of 0.125 M H_2SO_4

13.30 Calculate the mass of solute present in each of the solutions given in Problem 13.29.

13.31 Calculate the mass of solute needed to prepare each of the following solutions.

(a) 2.50 L of 0.250 M $(NH_4)_2CO_3$
(b) 125 mL of 2.50 M NH_4Cl
(c) 37.5 mL of 0.750 M KI
(d) 75.0 mL of 0.160 M $C_{12}H_{22}O_{11}$

13.32 Answer each of the following questions.

(a) What volume of 2.00 M KCl contains 0.500 mol KCl?
(b) What volume of 2.50 M NaI contains 0.655 mol NaI?
(c) What volume of 0.125 M H_2SO_4 contains 0.0435 mol H_2SO_4?
(d) What volume of 12 M HCl contains 0.50 mol HCl?
(e) What volume of 6.0 M NaOH contains 0.15 mol NaOH?
(f) What volume of 6.0 M NaOH would you use if you needed 2.0 g NaOH?

SECTION 13.5

13.33 How would you prepare each of the following dilute solutions from the concentrated solution indicated?

(a) 500 mL of 1.0 M HCl from 12 M HCl
(b) 250 mL of 2.0 M HNO_3 from 15 M HNO_3
(c) 150 mL of 0.60 M H_2SO_4 from 18 M H_2SO_4
(d) 750 mL of 0.25 M NaOH from 6.0 M NaOH

13.34 What is the final concentration when each of the following solutions is diluted as shown?

(a) 75 mL of 15 M HNO_3 diluted to 250 mL
(b) 25 mL of 18 M H_2SO_4 diluted to 125 mL
(c) 50.0 mL of 2.00 M NaCl diluted to 0.250 L
(d) 15 mL of 12 M HCl diluted to 0.50 L

SECTION 13.6

13.35 How would you prepare aqueous solutions of each of the following?

(a) 125 mL of 4.0% isopropyl alcohol (by volume)
(b) 500 mL of 5.0% acetic acid (by volume)
(c) 300 g of 5.0% glucose (by mass)
(d) 750 g of 0.90% sodium chloride (by mass)

13.36 A 750-mL sample of river water is found to have a total of 0.055 mg of mercury. How many parts per million is this? Would this water meet the minimum safety standards presented in Table 13-2?

13.37 What is the relationship between percent solution by mass and parts per million? What is the meaning of the term *parts per billion?*

SECTION 13.7

13.38 What mass of magnesium hydroxide precipitates from the reaction of 75.0 mL of 0.240 M $MgCl_2$ with excess sodium hydroxide?

$$MgCl_2(aq) + 2\,NaOH(aq) \rightarrow Mg(OH)_2(s) + 2\,NaCl(aq)$$

13.39 Barium hydroxide may be used to test for carbon dioxide. When carbon dioxide is bubbled through a barium hydroxide solution, barium carbonate precipitates:

$$Ba(OH)_2(aq) + CO_2(g) \rightarrow BaCO_3(s) + H_2O(\ell)$$

What mass of barium carbonate precipitates from 55.0 mL of 0.125 M $Ba(OH)_2$ when carbon dioxide is bubbled through the solution until all of the barium hydroxide has reacted?

13.40 Sodium carbonate reacts with hydrochloric acid according to the following equation:

$$2\,HCl(aq) + Na_2CO_3(aq) \rightarrow 2\,NaCl(aq) + H_2O(\ell) + CO_2(g)$$

(a) What volume of CO_2 is produced at STP when 75.0 mL of 0.125 M Na_2CO_3 is treated with an excess of HCl?

(b) What volume of CO_2 at 36°C and 715 torr is produced using the same quantities of reactants?

13.41 How many milliliters of 15 M HNO_3 is required to dissolve 2.5 g Cu according to the following equation?

$$3\,Cu(s) + 8\,HNO_3(aq) \rightarrow 3\,Cu(NO_3)_2(aq) + 2\,NO(g) + 4\,H_2O(\ell)$$

13.42 What volume of 0.180 M K_2CrO_4 is required to react completely with 12.0 mL of 0.375 M $AgNO_3$ according to the following equation?

$$2\,AgNO_3(aq) + K_2CrO_4(aq) \rightarrow Ag_2CrO_4(s) + 2\,KNO_3(aq)$$

SECTION 13.8

13.43 A 27.3-mL sample of KOH solution is required to consume all of the sulfuric acid in 15.2 mL of 0.240 M H_2SO_4, according to the following reaction:

$$H_2SO_4(aq) + 2\,KOH(aq) \rightarrow K_2SO_4(aq) + 2\,H_2O(\ell)$$

Calculate the molarity of the potassium hydroxide solution.

13.44 A 12.6-mL sample of $Pb(NO_3)_2$ solution is required to react completely with all of the potassium iodide in 32.8 mL of 0.225 M KI, according to the following equation:

$$Pb(NO_3)_2(aq) + 2\,KI(aq) \rightarrow PbI_2(s) + 2\,KNO_3(aq)$$

What is the molarity of the lead(II) nitrate solution?

SECTION 13.9

13.45 Use the data provided in Table 13-3 to determine the freezing point of a 2.50 *m* solution of a nonionic solute in each of the following solvents.

(a) Water **(b)** Formic acid

(c) Ethylene dibromide

13.46 A 5.0% glucose solution is isotonic with cellular fluids. Describe what occurs when a blood cell is placed in each of the following solutions.

(a) 1.0% glucose **(b)** 5.0% glucose **(c)** 10% glucose

13.47 The following questions refer to: (1) water, (2) a 0.12 *m* glucose solution in water, and (3) a 0.24 *m* glucose solution in water.

(a) Which has the highest boiling point?

(b) Which has the highest freezing point?

(c) Which has the highest osmotic pressure?

SECTION 13.10

13.48 Match each of the following colloids with the appropriate dispersed and dispersing phases. (There is one letter for each number, and vice versa.)

Colloid	Dispersed phase	/	Dispersing phase
(a) Fog	**(1)** Gas	in	Liquid
(b) Smoke	**(2)** Gas	in	Solid
(c) Whipped cream	**(3)** Liquid	in	Gas
(d) Paint	**(4)** Liquid	in	Liquid
(e) Mayonnaise	**(5)** Solid	in	Gas
(f) Styrofoam	**(6)** Solid	in	Liquid

GENERAL PROBLEMS

13.49 "Salting" icy highways is a common practice during the winter. What is the purpose of this practice, and why does it work?

13.50 As a liquid solution that contains a nonvolatile solute begins to freeze, the solid material that freezes out is pure solvent.

(a) Describe what happens to the freezing point of such a solution as the freezing progresses. Explain this behavior.

(b) Based on your answer to part (a), does it become easier or more difficult to freeze the water out of a salt water solution as freezing progresses?

13.51 Marketing personnel often use numbers to create the most favorable impression possible for their products, as illustrated below.

(a) The density of ethyl alcohol is 0.79 g/mL. Do you suppose that the percentage of alcohol reported on wine bottles is the percent by mass or the percent by volume? Explain your answer.

(b) Suppose you have discovered a pure liquid substance that stops the aging process. You have decided to market it as an aqueous solution. If the density of your miracle liquid is greater than the density of water, would it be more advantageous to label the contents of the bottle with the percent by mass or the percent by volume? Explain your answer.

13.52 In hospital work, intravenous solutions are often purchased from a manufacturer and prepared for use by a staff pharmacist. Morphine ($C_{15}H_{15}NO_3$) typically comes in a 20-mL vial that contains a solution of morphine having a concentration of 25 mg/mL. The staff pharmacist then dilutes the contents of the vial to meet the individual needs of a given patient.

(a) What is the molarity of the solution in the vial?

(b) What volume of solution having a concentration of 10 mg/mL can be prepared from one vial?

(c) What is the mass of the morphine in a 5.0-mL injection of the diluted solution?

(Assume that all values given in the problem are accurate to two significant figures.)

13.53 In advertising, phrases such as "98% fat free" can be misleading. For example, lowfat milk (which is 2% fat) is mostly water, with a small percentage of dissolved or suspended solids and liquids. Thus, the 2% fat is 2% of the entire mass of the milk, rather than 2% of the nutrients it contains. One popular nutrition advocate recommends a diet in which no more than 20% of the calories come from fat. There are 140 calories in an 8-oz (236-mL) serving of lowfat milk. If 1 g of fat is 9 calories, and the density of nonfat milk is 1.03 g/mL, what percent of the calories come from fat? Would lowfat milk fall within the 20% guideline?

13.54 A teaspoon of salt has a mass of approximately 8.0 g. If this quantity is added to a swimming pool that is 25 meters long and 15 meters wide, with an average depth of 2.0 meters, how many parts per million of salt have been added to the contents of the pool? Assume that the density of water in the pool is 1.00 g/mL. [*Reminder:* $1\ m^3 = 1000\ L$]

***13.55** The concentration of ions in water can be measured in moles of ion per liter of solution (molarity), grams per 100 mL of solution, or parts per million (ppm). For dilute solutions, a liter of solution has a mass of approximately 1 kg. What is the approximate molarity of lead ion in a water sample that is 5.0 ppm lead? (Assume that the equivalency between a liter of dilute solution and a kilogram is accurate to two significant figures.)

***13.56** When 35.0 mL of 0.300 M $Pb(NO_3)_2(aq)$ are mixed with 65.0 mL of 0.200 M KI(aq), a precipitate of $PbI_2(s)$ forms.

$$Pb(NO_3)_2(aq) + 2\ KI(aq) \rightarrow PbI_2(s) + 2\ KNO_3(aq)$$

(a) Which is the limiting reagent? [*Hint:* Begin by calculating the moles of each.]

(b) Calculate the mass of PbI_2 precipitated.

***13.57** In this chapter, we considered the freezing point depression of nonionic solutes only. The freezing point depression actually depends on the number of moles of *particles* per kilogram of solution. Because nonionic solutes do not dissociate, this is the same as the number of moles of solute. Using this additional information, answer the following questions.

(a) What is the freezing point of an aqueous 1.00 *m* NaCl solution?

(b) What is the freezing point of an aqueous 1.00 *m* $MgCl_2$ solution?

(c) Suppose an aqueous 1.00 *m* solution of a substance, AZ, is 50.0% ionized:

$$AZ \xrightarrow{50\%} A^+ + Z^-$$

What is the freezing point of this solution?

***13.58** Analytical instruments for determining the concentrations of metal ions are sensitive enough to measure *parts per trillion* (ppt). A commonly used instrument that is capable of detecting concentrations of cadmium in water as low as 3 ppt requires approximately 3 mL of sample. How many cadmium ions are in 3 mL of a water sample with a cadmium concentration of 3 ppt? [*Hint:* If one part per *billion* is equivalent to one *micro*gram per liter, how many grams per liter is one part per *trillion?*]

WRITING EXERCISES

13.59 Your mother's doctor has told her to soak her infected toe in a saturated solution of Epsom salts ($MgSO_4 \cdot 7H_2O$). Rather than waste a lot of excess Epsom salts, she has decided to determine its solubility so she can calculate the amount she needs for the foot bath she is using. She has asked you to explain how to do this. Write a note to your mother, describing the equipment she will need and outline the procedure she should follow. Your explanation should describe in words how to calculate the solubility from the data she gathers, and how to use that solubility to calculate the amount of Epsom salts she will need to use each time she prepares her foot bath. In the conclusion of your letter, explain how the solubilities of solids typically vary with temperature, and caution her about determining the solubility at the same temperature she plans to use for her foot bath.

13.60 In your own words, explain what osmotic pressure is. To illustrate osmotic pressure, explain what a physiological solution is and why solutions administered intravenously must be isotonic with cellular fluids. Your explanation should describe the effects of nonphysiological solutions on blood cells. Be sure to distinguish between the effects of solutions having greater osmotic pressures than cellular fluids and those with lower osmotic pressures than cellular fluids.

13.61 One of your friends has just read a newspaper article about air and water pollution contaminant levels in your local area. He realizes that he needs a better understanding of some of the terminology used, and he has asked you to explain the meanings of percentage (%), parts per million (ppm), and parts per billion (ppb). Write a note to your friend, explaining these three terms and how they are related to one another (how you can convert from one to another).

14 THE REACTIONS OF AQUEOUS SOLUTIONS

In Chapter 13 we examined some of the characteristics of solutions, focusing much of our attention on the most commonly used units of concentration. Over the next few chapters we will look at several types of reactions that occur in aqueous solution. We begin this chapter with an overview of reaction types; then we will concentrate most of our attention on double-replacement reactions, one of the types most commonly encountered in aqueous solution.

14.1 REACTION TYPES

OBJECTIVE 14.1 Classify chemical reactions as synthesis, decomposition, combustion, single-replacement, or double-replacement reactions.

Reactions are often classified according to the five following general types:

1. Synthesis reactions
2. Decomposition reactions
3. Combustion of organic compounds with oxygen
4. Single-replacement reactions
5. Double-replacement reactions

These reaction types are somewhat arbitrary, as other possible categories exist. Nevertheless, a brief examination of these general reaction types will help you organize your thinking about the different types of processes that may occur during chemical changes.

Synthesis Reactions

Synthesis reaction: Reaction in which two or more substances combine to form a single product.

Synthesis reactions are of the type

$$A + B \rightarrow C$$

in which two simpler substances combine to form a more complex product. For example,

$$C + O_2 \rightarrow CO_2$$
$$Mg + S \rightarrow MgS$$
$$CaO + CO_2 \rightarrow CaCO_3$$
$$C_2H_4 + Br_2 \rightarrow C_2H_4Br_2$$

Each reactant may be either an element or a compound, but the product *must* be a compound.

Decomposition Reactions

Decomposition reaction: Reaction in which a single reactant produces two or more products.

Decomposition reactions are the reverse of the synthesis type:

$$D \rightarrow E + F$$

Here, a more complex substance is broken down into simpler materials. For example,

$$2\,Ag_2O \rightarrow 4\,Ag + O_2$$
$$Mg(OH)_2 \rightarrow MgO + H_2O$$
$$2\,KClO_3 \rightarrow 2\,KCl + 3\,O_2$$

The simpler substances are not necessarily elements, as the last two examples illustrate.

Combustion of Organic Compounds With Oxygen

Combustion with oxygen: Type of reaction typical of hydrocarbons; when hydrocarbons undergo combustion, water and carbon dioxide form.

Reactions involving **combustion with oxygen** are the basis for most of our current energy production. Hydrocarbons represent a class of organic compounds made up exclusively of hydrogen and carbon. When these substances are burned completely in the presence of oxygen, carbon dioxide and water are formed. For example, methane (CH_4) and propane (C_3H_8) are fuels that may be burned in this fashion.

$$CH_4 + 2\,O_2 \rightarrow CO_2 + 2\,H_2O$$
$$C_3H_8 + 5\,O_2 \rightarrow 3\,CO_2 + 4\,H_2O$$

Heat, which may be used to serve our energy needs, is liberated in each of these exothermic reactions. Many of the examples and problems in Chapter 10 involved combustion reactions.

Occasionally, we encounter the incomplete combustion of hydrocarbons. When the amount of oxygen is insufficient to burn a hydrocarbon completely, carbon monoxide or carbon (soot) may be formed instead of carbon dioxide. For example,

$$2\,CH_4 + 3\,O_2 \rightarrow 2\,CO + 4\,H_2O$$
$$CH_4 + O_2 \rightarrow C + 2\,H_2O$$

Gasoline is a mixture of hydrocarbons. When improperly tuned, an automobile engine may mix insufficient oxygen with the hydrocarbon fuel. The engine will then emit large amounts of carbon monoxide and leave carbon deposits on the walls of the engine cylinders.

Carbohydrates are another class of organic compounds that react with oxygen to produce carbon dioxide and water. These organic substances, which include sugars, contain oxygen as well as carbon and hydrogen. Two commonly encountered examples are glucose, $C_6H_{12}O_6$, and sucrose, $C_{12}H_{22}O_{11}$. Our bodies decompose these substances and make use of the energy liberated in the following reactions:

$$C_6H_{12}O_6 + 6\,O_2 \rightarrow 6\,CO_2 + 6\,H_2O$$
$$C_{12}H_{22}O_{11} + 12\,O_2 \rightarrow 12\,CO_2 + 11\,H_2O$$

The production of glucose from water and carbon dioxide is endothermic. Plants harness solar energy to produce glucose during photosynthesis:

$$6\,CO_2 + 6\,H_2O + \text{Heat} \rightarrow C_6H_{12}O_6 + 6\,O_2$$

Then, when we eat fruits and vegetables (plants), we reverse the process and liberate this energy.

Single-Replacement Reactions

Single-replacement reaction: Reaction in which one element replaces another.

Single-replacement reactions are of the type

$$A{-}C + B \rightarrow B{-}C + A$$

The most important of these involve the replacement of one metal by another in an ionic compound. For example,

$$PbS + Mg \rightarrow MgS + Pb$$
$$CuSO_4 + Zn \rightarrow ZnSO_4 + Cu$$
$$2\,AgNO_3 + Cu \rightarrow Cu(NO_3)_2 + 2\,Ag$$

However, a single replacement may also involve a nonmetal replacing a nonmetal:

$$CaBr_2 + Cl_2 \rightarrow CaCl_2 + Br_2$$

or a metal replacing hydrogen:

$$2\,Na + 2\,H_2O \rightarrow 2\,NaOH + H_2$$
$$Mg + 2\,HCl \rightarrow MgCl_2 + H_2$$

All of these reactions also belong to a class known as oxidation-reduction reactions. These reactions are discussed in detail in Chapter 16.

Double-Replacement Reactions

Double-replacement reaction: Reaction in which two ionic substances exchange ions with one another.

Double-replacement reactions are also known as recombination or metathesis reactions. They are of the general type

$$A{-}C + B{-}D \rightarrow A{-}D + B{-}C$$

In a double-replacement reaction, the cation of each reactant combines with the anion of the other. In other words, the ions change partners. This category includes precipitation reactions—reactions in which an insoluble solid product separates from the reaction mixture, such as the following:

$$AgNO_3(aq) + NaCl(aq) \rightarrow AgCl(s) + NaNO_3(aq)$$
$$BaCl_2(aq) + K_2SO_4(aq) \rightarrow BaSO_4(s) + 2\,KCl(aq)$$

Also included are acid–base reactions:

$$HCl(aq) + NaOH(aq) \rightarrow NaCl(aq) + H_2O(\ell)$$
$$HNO_3(aq) + KOH(aq) \rightarrow KNO_3(aq) + H_2O(\ell)$$

Both of these types of double-replacement are carried out in aqueous solution. As we will discover later in this chapter, for a double-replacement reaction to occur, at least one of the products must be *nonionized* in water. In the next section we will discuss how to recognize a nonionized substance, as we examine the ionizing behavior of several types of solutes in water.

PROBLEM 14.1

Classify each of the following reactions according to the five categories discussed in this section.

(a) $2\,C_2H_6(g) + 7\,O_2(g) \rightarrow 4\,CO_2(g) + 6\,H_2O(\ell)$
(b) $KCl(aq) + AgNO_3(aq) \rightarrow AgCl(s) + KNO_3(aq)$
(c) $2\,HgO(s) \rightarrow 2\,Hg(\ell) + O_2(g)$
(d) $MgO(s) + CO_2(g) \rightarrow MgCO_3(s)$
(e) $Cu(NO_3)_2(aq) + Mg(s) \rightarrow Mg(NO_3)_2(aq) + Cu(s)$

PROBLEM 14.2

Write a balanced equation for each of the following double-replacement reactions.

(a) Aqueous lead(II) nitrate + Aqueous barium chloride
$\rightarrow$ Solid lead(II) chloride + Aqueous barium nitrate

(b) Aqueous sulfuric acid + Aqueous potassium hydroxide
$\rightarrow$ Aqueous potassium sulfate + Liquid water

14.2 CONDUCTIVITY OF IONIC SOLUTIONS: ELECTROLYTES

OBJECTIVE 14.2 Distinguish between the associated and dissociated forms of an ionizable substance. Distinguish between substances that are electrolytes, weak electrolytes, and nonelectrolytes.

In Chapter 13 you learned that when ionic substances dissolve in aqueous solution, the ions become separated from one another. For example, the sodium ions and chloride ions in a sodium chloride solution are separated from one another, with each ion surrounded by water molecules. We refer to the process of separation into ions as **dissociation.** When the ions of any substance become separated from one another, the substance is said to be **dissociated.**

Dissociation: The process of separation into ions.

Dissociated: Refers to the form of an ionizable substance in which the ions are separated.

On the other hand, some substances that are capable of producing ions remain largely nonionized. For example, water itself is capable of dissociating into hydrogen ions and hydroxide ions:

$$\underset{\text{Associated form}}{H_2O} \rightarrow \underset{\text{Dissociated form}}{H^+ + OH^-}$$

However, only two water molecules in every 1 billion actually dissociate in this fashion. We refer to the nonionized form of such a substance as the **associated** form.

Associated: Refers to the form of an ionizable substance in which the ions are together.

Before we examine the reactions of ionic substances, it is useful for us to have a method for examining the extent to which a substance produces ions in solution. The apparatus shown in Fig. 14-1 may be used to determine whether a given solution will conduct electricity. Solutions that contain a significant concentration of ions will conduct electricity, so we may use this apparatus to detect the presence of ions in solutions.

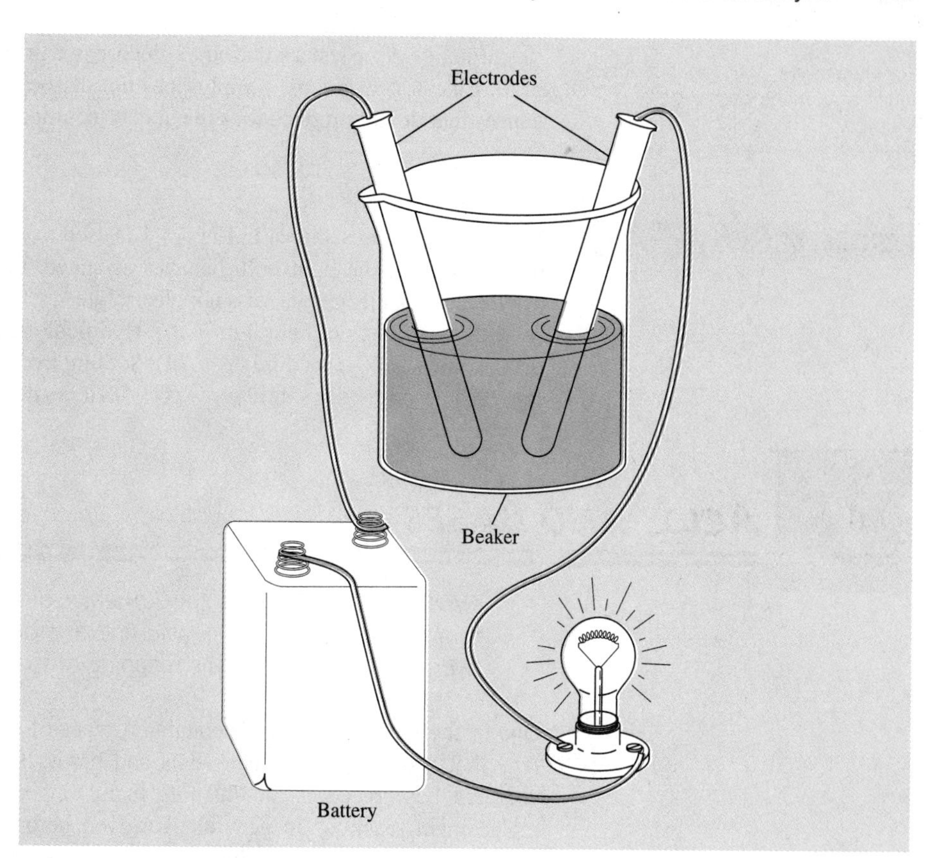

FIGURE 14-1 An apparatus used to determine whether a solution conducts electricity.

For example, if we place the electrodes of this apparatus in a beaker of pure water, we will find that the light does not light up, indicating that water is not a good conductor of electricity. However, if we dissolve some sodium chloride in the water, the bulb will glow brightly. We will also find that solutions of hydrochloric acid, sodium hydroxide, and potassium nitrate conduct electricity well. These are all ionic solutions. Any ionic solution will conduct electricity, provided there is a sufficient concentration of ions. Substances whose solutions conduct electricity are known as **electrolytes.**

Electrolyte: A substance whose aqueous solution conducts electricity.

At this point you may be wondering why it is dangerous to work with electricity when you are near a body of water or if you are wet. Although pure water itself is a nonconductor, the presence of ions from minerals dissolved in tap water or in natural waters permits the flow of electricity, thereby creating the potential for a serious electrical shock. Furthermore, the salts present in your skin also provide electrolytes, creating a hazard even in pure water.

If we test an aqueous solution of acetic acid, the light bulb on the conductivity apparatus lights, but only dimly. The fact that the solution conducts electricity very weakly indicates that acetic acid produces some ions in solution but that the concentration of ions present is small. Because acetic acid dissociates only slightly, it is classified as a weak acid. Substances whose solutions conduct electricity weakly are known as **weak electrolytes**. (Although pure water does produce some ions, it is such a weak electrolyte that it conducts too poorly to be observed in a typical apparatus of this type.)

Weak electrolyte: A substance whose aqueous solution conducts electricity weakly.

Nonelectrolyte: A substance whose aqueous solution does not conduct electricity.

Finally, if we test a solution of glucose, we find that it does not conduct electricity at all. Like water, glucose simply does not dissociate to any appreciable extent. Substances that do not produce ions in aqueous solution are known as **nonelectrolytes.**

PROBLEM 14.3

When the apparatus shown in Fig. 14-1 is used to test aqueous solutions of the following substances, the light bulb behaves as stated. Classify each substance as an electrolyte, a weak electrolyte, or a nonelectrolyte.

(a) Ethyl alcohol, does not light **(b)** Hydrochloric acid, glows brightly
(c) Formic acid, glows dimly **(d)** Sodium hydroxide, glows brightly
(e) Tartaric acid, glows dimly **(f)** Acetone, does not light

14.3 ACIDS AND BASES

OBJECTIVE 14.3 State the Arrhenius definition of an acid and a base. Distinguish between a weak acid and a strong acid. Write the formula of the hydronium ion. Explain the meaning of the term *alkaline*.

One of the most frequently encountered types of double-replacement reactions is the type that takes place between acids and bases. Because of their simplicity, we will use acid–base reactions to illustrate many important principles concerning double-replacement reactions in general. However, before we examine their reactions, it is helpful for us to learn a little bit about the general properties of acids and bases. You are probably already aware of the corrosive action of acids and bases. Many common household products that contain acids or bases carry warnings about using them properly. And the disastrous effects of acid rain on our environment have been widely reported.

Arrhenius acid: A substance that produces hydrogen ions in aqueous solution.

Arrhenius base: A substance that produces hydroxide ions in aqueous solution.

In Chapter 9, we described acids and bases as substances that contain either the hydrogen ion (in the case of acids) or the hydroxide ion (in the case of bases). We chose to delay any precise definitions of these substances until this chapter. As you will see in this and the next chapter, several different theories are used to describe acidic or basic characteristics. We begin in this chapter with the Arrhenius definitions of acids and bases: An **Arrhenius acid** is a substance that produces hydrogen ions in aqueous solution. An **Arrhenius base** is a substance that produces hydroxide ions in aqueous solution.

Acids

The acids we encounter most commonly in the laboratory are listed in Table 14-1. Acids are characterized by their sour taste, which is familiar to anyone who has tasted lemon juice or vinegar. The presence of acids may be detected via a number of dyes known as indicators. An indicator exhibits one color in the presence of an acid and a different color in the presence of a base. One of the most commonly used indicators is litmus. In the presence of an acid solution, blue litmus paper (a paper impregnated with the litmus dye) turns pink.

TABLE 14-1 Common acids

Hydrochloric acid	HCl	Hydrobromic acid	HBr
Sulfuric acid	H_2SO_4	Hydrofluoric acid	HF
Nitric acid	HNO_3	Hydriodic acid	HI
Phosphoric acid	H_3PO_4	Perchloric acid	$HClO_4$
Acetic acid	$HC_2H_3O_2$		

We can see that each of the acids listed in Table 14-1 is derived from one or more hydrogen ions, H^+, and one of the common anions. For example, hydrochloric acid is derived from hydrogen ion, H^+, and chloride ion, Cl^-, in a 1 : 1 ratio. Sulfuric acid is derived from hydrogen ion, H^+, and sulfate ion, SO_4^{2-}, in a 2 : 1 ratio. Similarly, phosphoric acid is derived from a 3 : 1 ratio of hydrogen ion, H^+, to phosphate ion, PO_4^{3-}. You may recall that a hydrogen ion, H^+, is a hydrogen atom that has lost its only electron. Because most hydrogen nuclei consist of a proton and no neutrons, we frequently refer to the hydrogen ions in acids as *protons.*

Weak acid: An acid that is only partially dissociated.

Strong acid: An acid that is completely dissociated.

The acids listed in Table 14-1 may be classified as strong acids or weak acids. A **weak acid** is one that is only partially dissociated. In the last section we saw that acetic acid is a weak electrolyte, indicating that it is only slightly dissociated. Acetic acid, hydrofluoric acid, and phosphoric acid are weak acids. On the other hand, a **strong acid** is one that is completely dissociated. The most common strong acids are nitric acid (HNO_3), sulfuric acid (H_2SO_4), hydrochloric acid (HCl), hydrobromic acid (HBr), hydriodic acid (HI), and perchloric acid ($HClO_4$).

In Chapter 9, we mentioned that hydrochloric acid is an aqueous solution of hydrogen chloride gas. Let us examine what takes place when HCl gas is bubbled into water to generate hydrochloric acid, HCl(aq):

$$HCl(g) \xrightarrow{H_2O} H^+(aq) + Cl^-(aq)$$

The hydrogen ions liberated in aqueous solution combine with water to form a hydrated species known as the **hydronium ion, H_3O^+:**

Hydronium ion: A hydrated hydrogen ion, H_3O^+.

$$\underset{\text{Hydrogen ion}}{H^+} + \underset{\text{Water}}{H_2O} \rightarrow \underset{\text{Hydronium ion}}{H_3O^+}$$

Thus, when gaseous hydrogen chloride is bubbled into water, the solution really contains hydronium ions and chloride ions:

$$HCl(g) + H_2O(\ell) \rightarrow H_3O^+(aq) + Cl^-(aq)$$

Similarly, a nitric acid solution contains hydronium ions and nitrate ions:

$$HNO_3(aq) + H_2O(\ell) \rightarrow H_3O^+(aq) + NO_3^-(aq)$$

Chemists use a wide variety of symbols when referring to aqueous hydrogen ions. These include $H^+(aq)$, H^+, $H_3O^+(aq)$, and H_3O^+. To simplify their symbolism, many chemists do not bother to write hydronium ions; it is understood that hydrogen ions are always hydrated in aqueous solution. We will use the simple H^+, except when there is some special need to emphasize the hydration of the hydrogen ions.

Bases

The bases most commonly encountered in the laboratory are shown in Table 14-2. Bases are caustic and may be found in many household products, such as oven cleaners, drain cleaners, and household ammonia (which usually contains soap). Like acids, bases affect the colors of indicators, but they have an opposite effect. For example, pink litmus paper turns blue in the presence of a base. Because most of the commonly used bases are compounds of one of the alkali or alkaline earth metals, the term *alkaline* is often used to refer to any solution that is "basic." Both terms have the same meaning.

Aqueous ammonia is also classified as a base, because ammonia gas produces hydroxide ions when it is dissolved in water:

$$NH_3(g) + H_2O(\ell) \rightarrow NH_4^+(aq) + OH^-(aq)$$

Depending on the concentration of ammonia, approximately 1% of the ammonia molecules that dissolve in water form hydroxide ions in this fashion. Because there is such a low concentration of hydroxide ions present, an aqueous ammonia solution conducts electricity weakly and is classified as a weak base.

TABLE 14-2 Common bases

Sodium hydroxide	$NaOH$
Potassium hydroxide	KOH
Calcium hydroxide	$Ca(OH)_2$
Barium hydroxide	$Ba(OH)_2$
Aqueous ammonia	$NH_3(aq)$

PROBLEM 14.4

Complete the following equation to describe what occurs when hydrogen bromide is bubbled into water.

$$HBr(g) + H_2O(\ell) \rightarrow ?$$

14.4 NEUTRALIZATION: AN ACID–BASE DOUBLE-REPLACEMENT REACTION

OBJECTIVE 14.4A Explain the meaning of neutralization. Given an acid and a base, write an equation for their complete neutralization. Write a balanced neutralization reaction to prepare any salt.

When hydrochloric acid is mixed with sodium hydroxide, a reaction takes place, producing water and sodium chloride:

$$HCl + NaOH \rightarrow H_2O + NaCl$$

As is the case with all double-replacement reactions, the cation of each reactant recombines with the anion of the other. In the case of an acid–base reaction, the products are always water and another compound known as a *salt.*

$$\text{Acid} + \text{Base} \rightarrow \text{Salt} + \text{Water}$$

Whenever an acid reacts with a base, the hydrogen ion from the acid combines with the hydroxide ion from the base to form water:

$$H^+ + OH^- \rightarrow H_2O$$

Because water is neither acidic nor basic (for example, it does not affect the color of litmus), we refer to it as a neutral substance. Thus, acids and bases are said to neutralize one another: They are chemical opposites that destroy (or "neutralize") each other's properties as acids and bases. The simple ionic reaction just demonstrated is the essence of every **neutralization** reaction.

Neutralization: A reaction in which an acid and a base destroy each other's acidic and basic properties.

Let us consider the neutralization of some other acids and bases. Nitric acid is neutralized by potassium hydroxide. This time the products are water and potassium nitrate.

$$HNO_3 + KOH \rightarrow H_2O + KNO_3$$

Suppose we carry out the reaction of sodium hydroxide with sulfuric acid. Unlike hydrochloric acid or nitric acid, sulfuric acid has two ionizable hydrogens. If we combine 1 mole of sulfuric acid and 1 mole of sodium hydroxide, the following reaction takes place:

$$H_2SO_4 + NaOH \rightarrow H_2O + NaHSO_4$$

Only one of the hydrogen ions from the sulfuric acid reacts with hydroxide ion. However, because the hydrogen sulfate ion, HSO_4^-, still has an ionizable hydrogen, adding a second mole of sodium hydroxide neutralizes this remaining hydrogen ion:

$$NaHSO_4 + NaOH \rightarrow H_2O + Na_2SO_4$$

These two reactions can be combined into a single reaction by using 2 moles of sodium hydroxide for each mole of sulfuric acid:

$$H_2SO_4 + 2\,NaOH \rightarrow 2\,H_2O + Na_2SO_4$$

This reaction represents the complete neutralization of sulfuric acid. We refer to sulfuric acid as a **diprotic acid,** because it has two hydrogen ions (or two protons) capable of reacting with base. Similarly, phosphoric acid, H_3PO_4, has three hydrogen ions and is called a **triprotic acid.** One mole of phosphoric acid requires 3 moles of sodium hydroxide for its complete neutralization:

Diprotic acid: An acid that has two hydrogen ions capable of reacting with a base.

Triprotic acid: An acid that has three hydrogen ions capable of reacting with a base.

$$H_3PO_4 + 3\,NaOH \rightarrow 3\,H_2O + Na_3PO_4$$

In like fashion, because calcium hydroxide, $Ca(OH)_2$, has two hydroxide ions, it requires two hydrogen ions for its complete neutralization. Thus, 2 moles of hydrochloric acid completely neutralize 1 mole of calcium hydroxide:

$$2\,HCl + Ca(OH)_2 \rightarrow 2\,H_2O + CaCl_2$$

As you can see, complete neutralization occurs when the number of hydrogen ions liberated by the acid equals the number of hydroxide ions liberated by the base. Thus, the complete neutralization of phosphoric acid, H_3PO_4, with barium hydroxide, $Ba(OH)_2$, requires a 2 : 3 ratio of phosphoric acid to barium hydroxide:

$$2\,H_3PO_4 + 3\,Ba(OH)_2 \rightarrow 6\,H_2O + Ba_3(PO_4)_2$$

PROBLEM 14.5

What ratio of oxalic acid, $H_2C_2O_4$, to calcium hydroxide, $Ca(OH)_2$, is required for the complete neutralization of each?

Salt: An ionic product of an acid–base reaction.

As we have already noted, whenever an acid and base neutralize one another, the products are water and a salt. A **salt** is an ionic compound made up of the cation of the base and the anion of the acid. Like all ionic compounds, salts are made up of the simplest ratio of ions that forms an uncharged compound.

The Summary reviews the reactions we have considered so far.

In each reaction, the cation of the acid (H^+) has combined with the anion of the base (OH^-) to form water. At the same time, the cation of the base has combined with the anion of the acid to form a salt. Earlier we stated that, for a double-replacement reaction to occur, at least one of the products must be nonionized. Since water exists primarily in its associated (or nonionized) form, neutralization reactions conform to this general principle.

We now have the skills we need to predict the products of any neutralization reaction and to write a balanced equation. We will consider only complete neutralizations.

EXAMPLE 14.1

Write the products of the following reaction and balance the equation.

$$HBr + KOH \rightarrow ?$$

SOLUTION

The reactants are an acid and a base, so the products must be water and a salt. The chemical formula of water is H_2O. The salt must be made up of potassium ion, K^+, and bromide ion, Br^-, combined in a 1:1 ratio. Thus, the completed equation is

$$HBr + KOH \rightarrow KBr + H_2O$$

Inspection shows the equation to be balanced as is.

SUMMARY Neutralization Reactions

Acid	+	Base	→	Salt	+	Water
HCl	+	$NaOH$	→	$NaCl$	+	H_2O
HNO_3	+	KOH	→	KNO_3	+	H_2O
H_2SO_4	+	$2\ NaOH$	→	Na_2SO_4	+	$2\ H_2O$
$2\ HCl$	+	$Ca(OH)_2$	→	$CaCl_2$	+	$2\ H_2O$
H_3PO_4	+	$3\ NaOH$	→	Na_3PO_4	+	$3\ H_2O$
$2\ H_3PO_4$	+	$3\ Ba(OH)_2$	→	$Ba_3(PO_4)_2$	+	$6\ H_2O$

EXAMPLE 14.2

Complete and balance a chemical equation for the complete neutralization of the following reactants.

$$HNO_3 + Ba(OH)_2 \rightarrow ?$$

SOLUTION

This time the products must be water and a salt made up of barium ion, Ba^{2+}, and nitrate ion, NO_3^-. Barium nitrate is made up of a 1 : 2 ratio of ions: $Ba(NO_3)_2$. Thus, the completed (unbalanced) equation is

$$HNO_3 + Ba(OH)_2 \rightarrow Ba(NO_3)_2 + H_2O \qquad \text{Unbalanced equation}$$

Adjusting the coefficients leads to the following balanced equation:

$$2\ HNO_3 + Ba(OH)_2 \rightarrow Ba(NO_3)_2 + 2\ H_2O \qquad \text{Balanced equation}$$

EXAMPLE 14.3

Write the products and balance the following equation.

$$H_2SO_4 + Al(OH)_3 \rightarrow ?$$

SOLUTION

The salt produced here must be aluminum sulfate, which is made up of aluminum ions, Al^{3+}, and sulfate ions, SO_4^{2-}. Its formula is $Al_2(SO_4)_3$. The complete unbalanced equation is

$$H_2SO_4 + Al(OH)_3 \rightarrow Al_2(SO_4)_3 + H_2O \qquad \text{Unbalanced equation}$$

Adjusting the coefficients leads to the final balanced equation:

$$3\ H_2SO_4 + 2\ Al(OH)_3 \rightarrow Al_2(SO_4)_3 + 6\ H_2O \qquad \text{Balanced equation}$$

Reversing our approach, we can determine which acid and base must be used to form a given salt. For example, to prepare sodium nitrate, $NaNO_3$, we must combine the base containing sodium ion (sodium hydroxide, NaOH) with the acid containing nitrate ion (nitric acid, HNO_3):

$$HNO_3 + NaOH \rightarrow NaNO_3 + H_2O$$

Similarly, if we wish to prepare potassium sulfate, K_2SO_4, we use the base containing potassium ion (potassium hydroxide, KOH) and the acid containing sulfate ion (sulfuric acid, H_2SO_4):

$$H_2SO_4 + 2\ KOH \rightarrow K_2SO_4 + 2\ H_2O$$

PROBLEM 14.6

Complete and balance each of the following equations for complete neutralization.

(a) $HCl + KOH \rightarrow ?$
(b) $HNO_3 + NaOH \rightarrow ?$
(c) $HBr + Ca(OH)_2 \rightarrow ?$
(d) $H_3PO_4 + KOH \rightarrow ?$
(e) $H_2SO_4 + Ca(OH)_2 \rightarrow ?$

PROBLEM 14.7

Write a balanced acid–base reaction to prepare each of the following salts.
(a) NaI **(b)** LiBr **(c)** $CaCl_2$ **(d)** Na_3PO_4 **(e)** BaI_2

Miscellaneous Salt-Producing Reactions

OBJECTIVE 14.4B Complete and balance the equation for the reaction of any metal oxide, carbonate, or hydrogen carbonate with an acid.

Salts are also produced in a number of other reactions. For example, the reactions of acids with metal oxides produce salts and water:

$$2\,HCl + MgO \rightarrow MgCl_2 + H_2O$$
$$6\,HNO_3 + Al_2O_3 \rightarrow 2\,Al(NO_3)_3 + 3\,H_2O$$

Salts also form when acids react with carbonates:

$$2\,HCl + Na_2CO_3 \rightarrow 2\,NaCl + H_2O + CO_2$$
$$H_2SO_4 + K_2CO_3 \rightarrow K_2SO_4 + H_2O + CO_2$$

In this case, the products are a salt, water, and carbon dioxide. While this reaction type may not look like a double-replacement, the products do result from the recombination of the ions. The cation of the carbonate combines with the anion of the acid to produce the salt, while the combination of hydrogen ions (from the acid) with the carbonate ion produces carbon dioxide and water, two nonionized substances.

$$2\,H^+ + CO_3^{2-} \rightarrow H_2O + CO_2$$

In a similar fashion, the reactions of acids with hydrogen carbonates form salts:

$$HCl + NaHCO_3 \rightarrow NaCl + H_2O + CO_2$$
$$HNO_3 + KHCO_3 \rightarrow KNO_3 + H_2O + CO_2$$

In this case, only one hydrogen ion is required to react with each hydrogen carbonate ion.

$$H^+ + HCO_3^- \rightarrow H_2O + CO_2$$

Finally, certain metals react with acids to produce salts:

$$Zn + 2\,HBr \rightarrow ZnBr_2 + H_2$$
$$Mg + 2\,HNO_3 \rightarrow Mg(NO_3)_2 + H_2$$

Although these reactions are examples of single-replacement reactions (which we will examine in Chapter 16), they are also included in our discussion of ionic reactions in Section 14.8.

PROBLEM 14.8

Complete and balance each of the following reactions.

(a) $HgO + HCl \rightarrow ?$ **(b)** $Na_2O + HI \rightarrow ?$
(c) $HBr + CaCO_3 \rightarrow ?$ **(d)** $HNO_3 + MgCO_3 \rightarrow ?$
(e) $HNO_3 + NaHCO_3 \rightarrow ?$ **(f)** $HI + LiHCO_3 \rightarrow ?$

14.5 THE FORMULAS OF IONS IN SOLUTION

OBJECTIVE 14.5 Given the formula of an ionic substance, write the formulas of the ions and state the ratios in which they are present.

In Section 14.4, we wrote equations for numerous neutralization reactions. Each of the examples we selected involved the reaction of a strong acid and a strong base. As we saw earlier, aqueous solutions of these substances contain the dissociated ions. When working with such solutions, we often find it useful to write a different form of equation—one that shows what happens to each of the ions during a reaction. We refer to these as *ionic equations,* and you will learn to write them shortly.

However, in order to write ionic equations, we need the ability to identify and write the formulas of the various ions present. For example, the ions present in a sodium bromide solution, $NaBr(aq)$, are sodium ion and bromide ion. The formulas of these ions should be familiar by now:

$$Na^{+}(aq) \text{ and } Br^{-}(aq) \text{ in a } 1:1 \text{ ratio}$$

(Because we are concerned with the states of reactants and products for the rest of the chapter, we have included the state symbols for these aqueous ions. We continue to do so throughout this chapter.)

Similarly, an aqueous magnesium bromide solution, $MgBr_2(aq)$, contains magnesium ion and bromide ion:

$$Mg^{2+}(aq) \text{ and } Br^{-}(aq) \text{ in a } 1:2 \text{ ratio}$$

Note that the formula of the bromide ion in $MgBr_2(aq)$ is exactly the same as in $NaBr(aq)$. However, in $MgBr_2(aq)$ the ions must be present in a 1 : 2 ratio in order to balance the charges. Similarly, an aqueous iron(III) bromide solution, $FeBr_3(aq)$, contains iron(III) ions and bromide ions—but this time in a 1 : 3 ratio:

$$Fe^{3+}(aq) \text{ and } Br^{-}(aq) \text{ in a } 1:3 \text{ ratio}$$

The point is that the formula for aqueous bromide ion is always $Br^{-}(aq)$. If we wish to indicate *two* bromide ions, we write $2\ Br^{-}(aq)$; for three bromide ions, we write $3\ Br^{-}(aq)$. For the moment, we just wish to identify the formulas of the ions present. If you are unsure about the formulas of any of the ions, now is a good time to review the list of ions found in the table inside the back cover. We present a few more examples here for basic review before continuing with the heart of this chapter.

EXAMPLE 14.4

What ions are present in an aqueous solution of calcium nitrate, $Ca(NO_3)_2(aq)$?

SOLUTION

$$Ca^{2+}(aq) \text{ and } NO_3^{-}(aq) \text{ in a } 1:2 \text{ ratio}$$

EXAMPLE 14.5

What ions are present in aqueous aluminum sulfate, $Al_2(SO_4)_3(aq)$?

SOLUTION

$$Al^{3+}(aq) \text{ and } SO_4^{2-}(aq) \text{ in a } 2:3 \text{ ratio}$$

EXAMPLE 14.6

What ions are present in aqueous potassium carbonate, $K_2CO_3(aq)$?

SOLUTION

$K^+(aq)$ and $CO_3^{2-}(aq)$ in a 2:1 ratio

PROBLEM 14.9

Write the formulas of the ions present in each of the following solutions, and then give the ratio in which they are found.

(a) $ZnCl_2(aq)$ **(b)** $Li_2SO_4(aq)$ **(c)** $Fe(NO_3)_3(aq)$
(d) $(NH_4)_2CO_3(aq)$ **(e)** $BaI_2(aq)$

14.6 TOTAL MOLECULAR, TOTAL IONIC, AND NET IONIC EQUATIONS

OBJECTIVE 14.6 Distinguish between total molecular, total ionic, and net ionic equations. Given a complete neutralization reaction between a strong acid and a strong base, write a total molecular equation, a total ionic equation, and a net ionic equation. Identify any spectator ions.

In Section 14.4 we looked at quite a number of acid–base reactions. We saw that every neutralization involves the reaction of a hydrogen ion with a hydroxide ion to form water.

$$H^+(aq) + OH^-(aq) \rightarrow H_2O(\ell)$$

Let us reexamine the reaction of hydrochloric acid with sodium hydroxide:

$$HCl(aq) + NaOH(aq) \rightarrow H_2O(\ell) + NaCl(aq)$$

Total molecular equation: An equation expressing the reactants and products in terms of their molecular formulas.

This type of chemical equation, which you have seen often throughout this text, is known as a **total molecular equation.** It is balanced and shows us what the reactants and products are. In a certain sense, however, it does not tell the entire story of what is happening during the reaction. When working with aqueous solutions of strong acids or ionic substances, we really have solutions of the materials in their dissociated (or ionized) forms. In other words, the hydrochloric acid solution, HCl(aq), is made up of hydrogen ions and chloride ions, and the sodium hydroxide solution, NaOH(aq), is composed of sodium ions and hydroxide ions. Thus, we could more accurately depict this reaction by rewriting the ionic reactants and products in their dissociated forms:

$$\underbrace{H^+(aq) + Cl^-(aq)}_{\textbf{Acid solution}} + \underbrace{Na^+(aq) + OH^-(aq)}_{\textbf{Base solution}} \rightarrow H_2O(\ell) + Na^+(aq) + Cl^-(aq)$$

Note that water is written in its molecular form because water is not appreciably ionized. This equation tells us that when a solution containing hydrogen ions and chloride ions is mixed with one containing sodium ions and hydroxide ions, water molecules are formed, and the sodium ions and chloride ions remain in solution (Fig. 14-2). Because we have shown everything that was present during the reaction, this equation is called a **total ionic equation.**

Total ionic equation: An ionic equation that expresses all of the ions present during the course of a reaction.

The sodium ions and the chloride ions appear in exactly the same form on both sides of the arrow. In other words, these ions are not really involved in the chemical

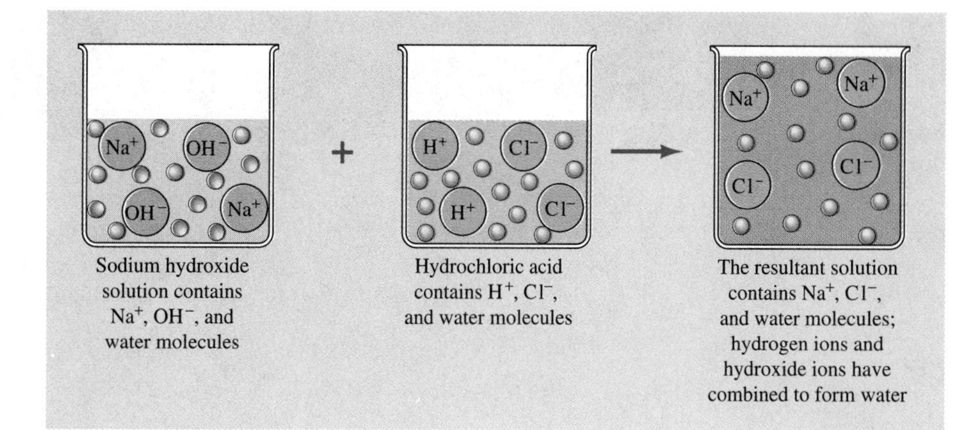

FIGURE 14-2 The reaction of hydrochloric acid with sodium hydroxide. Sodium ions and chloride ions do not react and are called spectator ions.

reaction that takes place, because they do not undergo any change during the reaction. We call these ions spectator ions. A **spectator ion** is one that is present in solution during the reaction but does not participate in the chemical reaction. We recognize the spectator ions by the fact that they appear in identical form on both sides of the arrow. Note that the hydrogen ions and hydroxide ions do *not* appear as such on the right side of the equation. They have combined to form water, thereby undergoing a chemical reaction.

Spectator ion: An ion that is present but does not undergo any chemical change during a reaction.

Let us eliminate the spectator ions from the total ionic equation, as follows:

$$H^+(aq) + \cancel{Cl^-(aq)} + \cancel{Na^+(aq)} + OH^-(aq) \rightarrow H_2O(\ell) + \cancel{Na^+(aq)} + \cancel{Cl^-(aq)}$$

$$H^+(aq) + OH^-(aq) \rightarrow H_2O(\ell)$$

We are left with an equation that involves only those substances that actually undergo chemical change. This equation is called a **net ionic equation.**

Net ionic equation: An ionic equation that shows only those species undergoing chemical change.

The net ionic equation represents the essence of the chemical reaction taking place and helps us focus on the similarities between two seemingly different reactions. For example, if we examine the reaction of nitric acid with potassium hydroxide, we find that the net ionic equation is the same as that for hydrochloric acid and sodium hydroxide. Only the spectator ions are different.

Total molecular equation:

$$HNO_3(aq) + KOH(aq) \rightarrow H_2O(\ell) + KNO_3(aq)$$

Total ionic equation:

$$H^+(aq) + NO_3^-(aq) + K^+(aq) + OH^-(aq) \rightarrow H_2O(\ell) + K^+(aq) + NO_3^-(aq)$$

Net ionic equation:

$$H^+(aq) + OH^-(aq) \rightarrow H_2O(\ell)$$

In each of these reactions, and in all neutralization reactions, the formation of water (a nonionized substance) provides the impetus for the double-replacement reaction to take place.

The Summary on the next page lists the three different ways of writing equations to represent the same reaction.

SUMMARY Three Types of Chemical Equations

1. The **total molecular equation** expresses the reaction in terms of the chemical formulas of the reactants and products.
2. The **total ionic equation** expresses any dissociated substances in terms of their ions.
3. The **net ionic equation** represents only those substances that actually undergo change.

EXAMPLE 14.7

Balance the following total molecular equation, and write a total ionic equation and a net ionic equation.

$$HBr(aq) + Ca(OH)_2(aq) \rightarrow H_2O(\ell) + CaBr_2(aq)$$

SOLUTION

First, we must balance the equation:

$$2\ HBr(aq) + Ca(OH)_2(aq) \rightarrow 2\ H_2O(\ell) + CaBr_2(aq)$$

To write the total ionic equation, we express each ionized substance in terms of all the ions generated:

$$2\ H^+(aq) + 2\ Br^-(aq) + Ca^{2+}(aq) + 2\ OH^-(aq) \rightarrow 2\ H_2O(\ell) + Ca^{2+}(aq) + 2\ Br^-(aq)$$

This example differs slightly from the previous cases in that here we have coefficients other than 1. Thus, 2 mol HBr(aq) ionize to form 2 mol $H^+(aq)$ and 2 mol $Br^-(aq)$; and 1 mol $Ca(OH)_2(aq)$ ionizes to form 1 mol $Ca^{2+}(aq)$ and 2 mol $OH^-(aq)$. Similarly, 1 mol $CaBr_2(aq)$ exists as 1 mol $Ca^{2+}(aq)$ and 2 mol $Br^-(aq)$.

Examining the total ionic equation reveals that $Ca^{2+}(aq)$ and $Br^-(aq)$ are the spectator ions. Eliminating these ions leads to the following multiple of the net ionic equation:

$$2\ H^+(aq) + 2\ OH^-(aq) \rightarrow 2\ H_2O(\ell)$$

Simplifying the coefficients to the simplest whole-number ratio provides the correct form of the net ionic equation:

$$H^+(aq) + OH^-(aq) \rightarrow H_2O(\ell)$$

In this section, we have considered only reactions of strong acids (100% dissociated). Neutralization reactions between acids and bases are only one type of ionic reaction. The next section introduces ionic reactions that produce precipitates. Later in the chapter we deal with reactions that involve weak acids.

PROBLEM 14.10

Complete and balance each of the following neutralization reactions, and then write a total ionic equation and a net ionic equation for each.

(a) $HClO_4(aq) + KOH(aq) \rightarrow ?$ **(b)** $HBr(aq) + Ba(OH)_2(aq) \rightarrow ?$
(c) $HI(aq) + Ca(OH)_2(aq) \rightarrow ?$

14.7 PRECIPITATION REACTIONS: ANOTHER DOUBLE-REPLACEMENT

OBJECTIVE 14.7 Given two ionic substances that undergo a precipitation reaction, write a balanced total molecular equation for their reaction and identify the precipitate from a solubility table. Then write total ionic and net ionic equations.

Precipitate: An insoluble solid product.

Frequently, when ionic solutions are mixed together, one (or more) of the products is an insoluble salt that separates from solution. We refer to such an insoluble product as a **precipitate** and to the process of forming such a product as *precipitation.* A precipitate is always an uncharged compound, and it can often be separated from the soluble products by filtration.

Precipitation occurs any time we bring together two ions that form an insoluble compound. Table 14-3 gives the solubilities of a wide range of ionic compounds. For example, silver chloride is an insoluble solid. When we mix aqueous solutions of silver nitrate and sodium chloride, the silver ions from the silver nitrate combine with the chloride ions from the sodium chloride, forming silver chloride:

$$Ag^+(aq) + Cl^-(aq) \rightarrow AgCl(s)$$

Precipitate

The state symbol (s) tells us that the silver chloride formed has precipitated from solution in its solid state. We can write total molecular, total ionic, and net ionic equations for this reaction. The preceding equation is the net ionic equation.

TABLE 14-3 Solubilities of selected salts[a]

	Anions									
Cations	NO_3^-	SO_4^{2-}	OH^-	Cl^-	Br^-	I^-	S^{2-}	$C_2H_3O_2^-$	CO_3^{2-}	PO_4^{3-}
Na^+	S	S	S	S	S	S	S	S	S	S
K^+	S	S	S	S	S	S	S	S	S	S
NH_4^+	S	S	S	S	S	S	S	S	S	S
Ag^+	S	ss	–	I	I	I	I	ss	I	I
Al^{3+}	S	S	I	S	S	S	d	S	–	I
Ba^{2+}	S	I	S	S	S	S	d	S	I	I
Ca^{2+}	S	I	ss	S	S	S	ss	S	I	I
Co^{2+}	S	S	I	S	S	S	I	S	I	I
Cu^{2+}	S	S	I	S	S	–	I	S	–	I
Hg^{2+}	S	d	I	S	S	S	I	S	–	I
Mg^{2+}	S	S	I	S	S	S	d	S	I	I
Pb^{2+}	S	I	I	I	I	I	I	S	I	I
Sr^{2+}	S	I	S	S	S	S	I	S	I	I
Zn^{2+}	S	S	I	S	S	S	I	S	I	I

[a]S = soluble, ss = slightly soluble, I = insoluble, d = decomposes in water

Examining the total molecular equation reveals that this reaction is a double-replacement reaction:

$$AgNO_3(aq) + NaCl(aq) \rightarrow AgCl(s) + NaNO_3(aq)$$

Silver chloride forms from the combination of silver ions and chloride ions; the sodium ions and nitrate ions recombine as a soluble by-product. If you will recall, when an ionic substance dissolves, it dissociates into ions. When two ions form a precipitate, the process of dissociation is reversed; in other words, the precipitate that forms is in its associated (or nonionized) form. Thus, the impetus for the double-replacement to occur is the formation of the precipitate. Let us develop a systematic method for analyzing such precipitation reactions.

EXAMPLE 14.8

Complete and balance the following equation. Then write a total ionic equation and a net ionic equation for the reaction.

$$AgNO_3(aq) + NaCl(aq) \rightarrow ?$$

SOLUTION

To complete the products of a precipitation reaction, we begin by listing the ions present in each of the solutions being mixed:

Solution 1: Ag^+ ⟶ NO_3^-

Solution 2: Na^+ ⟶ Cl^-

The reaction involves a recombination of the ions, so the products must be silver chloride and sodium nitrate. The formula of each is written as the simplest whole-number ratio of the recombined ions. Next, we refer to Table 14-3 to determine the solubilities of these products. We find that silver chloride is insoluble and sodium nitrate is soluble:

$$AgNO_3(aq) + NaCl(aq) \rightarrow AgCl(s) + NaNO_3(aq)$$

Inspection of this total molecular equation shows it to be balanced as is.

To write the total ionic equation, we express any dissociated substances in terms of the ions in solution:

$$\underbrace{Ag^+(aq) + NO_3^-(aq)}_{\textbf{Solution 1}} + \underbrace{Na^+(aq) + Cl^-(aq)}_{\textbf{Solution 2}} \rightarrow AgCl(s) + Na^+(aq) + NO_3^-(aq)$$

Note that the precipitate is written in its associated form rather than in its dissociated form. At the end of the reaction, the silver ions and chloride ions are no longer dissolved in solution but instead have formed large aggregates of alternating cations and anions, characteristic of all ionic solids. These solid particles separate from the solution, and we observe them in the form of a precipitate.

Examining the total ionic equation reveals that sodium ion and nitrate ion are spectator ions. Eliminating them gives the net ionic equation for the precipitation of AgCl(s):

$$Ag^+(aq) + Cl^-(aq) \rightarrow AgCl(s)$$

Let us consider a slightly more complicated example.

EXAMPLE 14.9

When aqueous solutions of lead(II) nitrate, $Pb(NO_3)_2(aq)$, and potassium iodide, $KI(aq)$, are mixed, a precipitate forms. Write a balanced total molecular equation for the reaction. Then write a total ionic equation and a net ionic equation.

$$Pb(NO_3)_2(aq) + KI(aq) \rightarrow ?$$

SOLUTION

To write the products, we first identify the ions present:

Solution 1: Pb^{2+} NO_3^-

Solution 2: K^+ I^-

The precipitate must be either lead(II) iodide or potassium nitrate (or both). Table 14-3 reveals that lead(II) iodide is insoluble and that potassium nitrate is a soluble salt:

$$Pb(NO_3)_2(aq) + KI(aq) \rightarrow PbI_2(s) + KNO_3(aq) \qquad \textbf{Unbalanced molecular}$$

Adjusting the coefficients leads to a balanced total molecular equation:

$$Pb(NO_3)_2(aq) + 2\,KI(aq) \rightarrow PbI_2(s) + 2\,KNO_3(aq) \qquad \textbf{Balanced molecular}$$

Next, we express the ionized substances in terms of the ions present:

$$\underbrace{Pb^{2+}(aq) + 2\,NO_3^-(aq)}_{\textbf{Solution 1}} + \underbrace{2\,K^+(aq) + 2\,I^-(aq)}_{\textbf{Solution 2}} \rightarrow PbI_2(s) + 2\,K^+(aq) + 2\,NO_3^-(aq) \qquad \textbf{Total ionic}$$

Finally, eliminating the spectator ions (K^+ and NO_3^-) yields the net ionic equation:

$$Pb^{2+}(aq) + 2\,I^-(aq) \rightarrow PbI_2(s)$$

EXAMPLE 14.10

Aqueous solutions of magnesium sulfate, $MgSO_4(aq)$, and barium hydroxide, $Ba(OH)_2(aq)$, react together to form a precipitate. Identify the precipitate and write total molecular, total ionic, and net ionic equations.

$$MgSO_4(aq) + Ba(OH)_2(aq) \rightarrow ?$$

SOLUTION

The ions in solution are

Solution 1: Mg^{2+} SO_4^{2-}

Solution 2: Ba^{2+} OH^-

The recombination products must be magnesium hydroxide, $Mg(OH)_2$, and barium sulfate, $BaSO_4$. Either or both of these may be precipitates. Examination of Table 14-3 reveals that *both* are insoluble salts. Completing and balancing the equation leads to the following set of equations.

Total molecular equation:

$$MgSO_4(aq) + Ba(OH)_2(aq) \rightarrow BaSO_4(s) + Mg(OH)_2(s)$$

Because neither of the products is soluble, both are written in the associated form in the total ionic equation.

Total ionic equation:

$$\underbrace{Mg^{2+}(aq) + SO_4^{2-}(aq)}_{\textbf{Solution 1}} + \underbrace{Ba^{2+}(aq) + 2\,OH^-(aq)}_{\textbf{Solution 2}} \rightarrow BaSO_4(s) + Mg(OH)_2(s)$$

There are no spectator ions. Every ion was involved in a chemical change. Thus, the net ionic equation is the same as the total ionic equation.

The information given in Table 14-3 can be organized into several rules that are equally helpful in enabling us to predict the solubilities of ionic compounds. For instance, we can look at Table 14-3 and see that all common salts containing sodium, potassium, ammonium, or nitrate ion are soluble. Several generalizations about the solubilities of common salts are given in the Summary.

SUMMARY Solubility Rules

1. *Nitrates.* All nitrate salts are soluble.
2. *Alkali metals.* The salts of lithium, sodium, potassium, rubidium, and cesium are generally very soluble.
3. *Ammonium salts.* Almost all ammonium salts are soluble.
4. *Sulfates.* The sulfates of most common elements are soluble, with the exception of calcium, strontium, barium, and lead(II) ions. Silver sulfate is slightly soluble.
5. *Hydroxides.* Most hydroxides are insoluble except those of the alkali metals and barium; calcium hydroxide is slightly soluble.
6. *Halides.* Chloride, bromide, and iodide salts are generally soluble except those of silver, lead(II), and mercury(I) ions.
7. *Sulfides.* Most sulfides are insoluble, except those of the alkali metals and ammonium ion.
8. *Acetates.* All acetates are soluble except silver acetate, which is slightly soluble.
9. *Silver salts.* All silver salts are insoluble except silver nitrate, $AgNO_3$; silver nitrite, $AgNO_2$; and silver perchlorate, $AgClO_4$. Silver acetate, $AgC_2H_3O_2$, and silver sulfate, Ag_2SO_4, are slightly soluble.
10. *Carbonates.* All carbonates are insoluble except those of ammonium, sodium, potassium, and the other alkali metals.
11. *Phosphates.* Phosphates are insoluble except those of ammonium, sodium, potassium, and the other alkali metals.

PROBLEM 14.11

Complete and balance each of the following recombination reactions, using Table 14-3 (or the Summary of Solubility Rules) to identify any precipitates in the total molecular equation. Then write a total ionic equation and a net ionic equation for each.

(a) $AgNO_3(aq) + NH_4Cl(aq) \rightarrow ?$ **(b)** $AgNO_3(aq) + Na_2CO_3(aq) \rightarrow ?$
(c) $Sr(NO_3)_2(aq) + K_2SO_4(aq) \rightarrow ?$

14.8 OTHER TYPES OF IONIC REACTIONS

OBJECTIVE 14.8 Given a complete or partial molecular equation involving an insoluble substance, a gaseous substance, or a weak acid, complete and balance the molecular equation, and then write a total ionic equation and a net ionic equation.

In addition to acid–base and precipitation reactions, a rather wide variety of reaction types exist that we might wish to express in ionic form. These include reactions in which (a) a solid, (b) a gas, or (c) a weak acid is either produced or consumed. In each case, ionic equations express dissociated substances as ions in solution and nonionized substances in molecular form.

In our examination of precipitation reactions, we always wrote the insoluble product in its associated solid form, because that is the actual nature of the precipitate. The state symbol (s) told us that the precipitate was not actually in solution but instead existed in a separate solid phase. Similarly, when we wrote ionic equations for neutralization reactions, we expressed the water produced in its associated form. This was because only a small fraction of water molecules are actually dissociated. The state symbol (ℓ) told us that water was formed in its liquid state. We have not yet written ionic equations that involve gases. When gases escape from (or are bubbled into) aqueous solution, they exist in molecular form. Thus, in ionic equations for reactions that involve gases (such as hydrogen, oxygen, or carbon dioxide), these substances are written in their usual molecular form, along with the state symbol (g).

Reactions Involving Insoluble Reactants or Gaseous Substances

In the examples we have studied, the precipitates formed were products. However, we may also encounter insoluble reactants. For example, calcium carbonate is an insoluble solid that dissolves upon reaction with nitric acid. The products formed include carbon dioxide, water, and an aqueous solution of calcium nitrate.

Total molecular equation:

$$CaCO_3(s) + 2\,HNO_3(aq) \rightarrow Ca(NO_3)_2(aq) + H_2O(\ell) + CO_2(g)$$

To write a total ionic equation, we proceed as before, writing each reactant and product in terms of the ions actually present in solution.

Total ionic equation:

$$CaCO_3(s) + 2\,H^+(aq) + 2\,NO_3^-(aq) \rightarrow Ca^{2+}(aq) + 2\,NO_3^-(aq) + H_2O(\ell) + CO_2(g)$$

Because calcium carbonate is an insoluble solid, it is treated just like the precipitates we examined earlier. Similarly, nitric acid, aqueous calcium nitrate, and water are all written as in our earlier neutralization reactions. The only other substance is carbon dioxide gas. Carbon dioxide is a nonionic molecular substance, so it is always written in molecular form.

Examining the total ionic equation reveals that nitrate ions are the only spectator ions. Eliminating them results in the following net ionic equation.

Net ionic equation:

$$CaCO_3(s) + 2\,H^+(aq) \rightarrow Ca^{2+}(aq) + H_2O(\ell) + CO_2(g)$$

Insoluble salts are not the only type of insoluble reactants we are likely to encounter. Zinc metal reacts with hydrochloric acid in the following single-replacement reaction, producing aqueous zinc chloride and hydrogen gas.

Total molecular equation:

$$Zn(s) + 2\,HCl(aq) \rightarrow ZnCl_2(aq) + H_2(g)$$

The metallic zinc is an elemental substance and, as such, is uncharged. (In other words, it is *not* an ion.) Therefore, it appears in the ionic equation exactly the same as in the molecular equation. On the product side, however, zinc exists in its ionic form (Zn^{2+}) as part of the aqueous zinc chloride.

Total ionic equation:

$$Zn(s) + 2\,H^+(aq) + 2\,Cl^-(aq) \rightarrow Zn^{2+}(aq) + 2\,Cl^-(aq) + H_2(g)$$

Chloride is the only spectator ion. Eliminating the chloride ions leads to the net ionic equation.

Net ionic equation:

$$Zn(s) + 2\,H^+(aq) \rightarrow Zn^{2+}(aq) + H_2(g)$$

Reactions Involving Weak Acids

Earlier in the chapter we examined the conductivity of weak acids. We observed that these substances are only slightly dissociated, as reflected by their very modest conductivity. For example, a 0.10 M acetic acid solution is about 1% ionized, leaving most of the acetic acid in the associated molecular form. Consequently, acetic acid and other weak acids are written in the associated form in ionic equations. (This is the same as our treatment of water, which also dissociates only slightly.) Consequently, the neutralization of sodium hydroxide with acetic acid leads to a slightly different set of equations from those we observed earlier for strong acids.

Total molecular equation:

$$HC_2H_3O_2(aq) + NaOH(aq) \rightarrow NaC_2H_3O_2(aq) + H_2O(\ell)$$

Total ionic equation:

$$HC_2H_3O_2(aq) + Na^+(aq) + OH^-(aq) \rightarrow Na^+(aq) + C_2H_3O_2^-(aq) + H_2O(\ell)$$

Net ionic equation:

$$HC_2H_3O_2(aq) + OH^-(aq) \rightarrow C_2H_3O_2^-(aq) + H_2O(\ell)$$

The reaction of acetic acid with hydroxide ion produces acetate ion and water.

Also of interest are reactions in which a weak acid is produced. Consider the reaction of sodium acetate with hydrochloric acid.

Total molecular equation:

$$NaC_2H_3O_2(aq) + HCl(aq) \rightarrow NaCl(aq) + HC_2H_3O_2(aq)$$

Total ionic equation:

$$Na^+(aq) + C_2H_3O_2^-(aq) + H^+(aq) + Cl^-(aq) \rightarrow Na^+(aq) + Cl^-(aq) + HC_2H_3O_2(aq)$$

Net ionic equation:

$$H^+(aq) + C_2H_3O_2^-(aq) \rightarrow HC_2H_3O_2(aq)$$

Note that the net ionic equation represents the reverse of the ionization of acetic acid. Considering the fact that acetic acid is only slightly dissociated, it is not surprising to find that the component ions combine when they are placed in solution together. The Summary Diagram on p. 442 provides a review of how to express substances in ionic equations.

PROBLEM 14.12

Balance the following molecular equations, each of which produces a gas. Then write a total ionic equation and a net ionic equation for each.

(a) $K_2CO_3(aq) + HClO_4(aq) \rightarrow KClO_4(aq) + H_2O(\ell) + CO_2(g)$
(b) $Mg(s) + HNO_3(aq) \rightarrow Mg(NO_3)_2(aq) + H_2(g)$
(c) $Al(s) + HCl(aq) \rightarrow AlCl_3(aq) + H_2(g)$
(d) $BaCO_3(s) + HNO_3(aq) \rightarrow Ba(NO_3)_2(aq) + H_2O(\ell) + CO_2(g)$

PROBLEM 14.13

The following single-replacement reactions are balanced. Write a total ionic equation and a net ionic equation for each.

(a) $Cu(NO_3)_2(aq) + Mg(s) \rightarrow Mg(NO_3)_2(aq) + Cu(s)$
(b) $2\ AgNO_3(aq) + Zn(s) \rightarrow Zn(NO_3)_2(aq) + 2\ Ag(s)$
(c) $3\ Pb(NO_3)_2(aq) + 2\ Al(s) \rightarrow 2\ Al(NO_3)_3(aq) + 3\ Pb(s)$

PROBLEM 14.14

Complete and balance the following reactions, each of which produces a weak acid. Then write a total ionic equation and a net ionic equation for each.

(a) $NaNO_2(aq) + HCl(aq) \rightarrow ?$ **(b)** $KF(aq) + HNO_3(aq) \rightarrow ?$
(c) $KCN(aq) + HCl(aq) \rightarrow ?$

14.9 PREDICTING DOUBLE-REPLACEMENT REACTIONS

OBJECTIVE 14.9 Predict whether a double-replacement reaction will occur between a given pair of reactants, and if it will, write a balanced equation.

Suppose we were to mix aqueous solutions of sodium iodide and potassium nitrate together:

$$NaI(aq) + KNO_3(aq) \rightarrow ?$$

SUMMARY DIAGRAM

Checklist for Writing Formulas in Ionic Equations

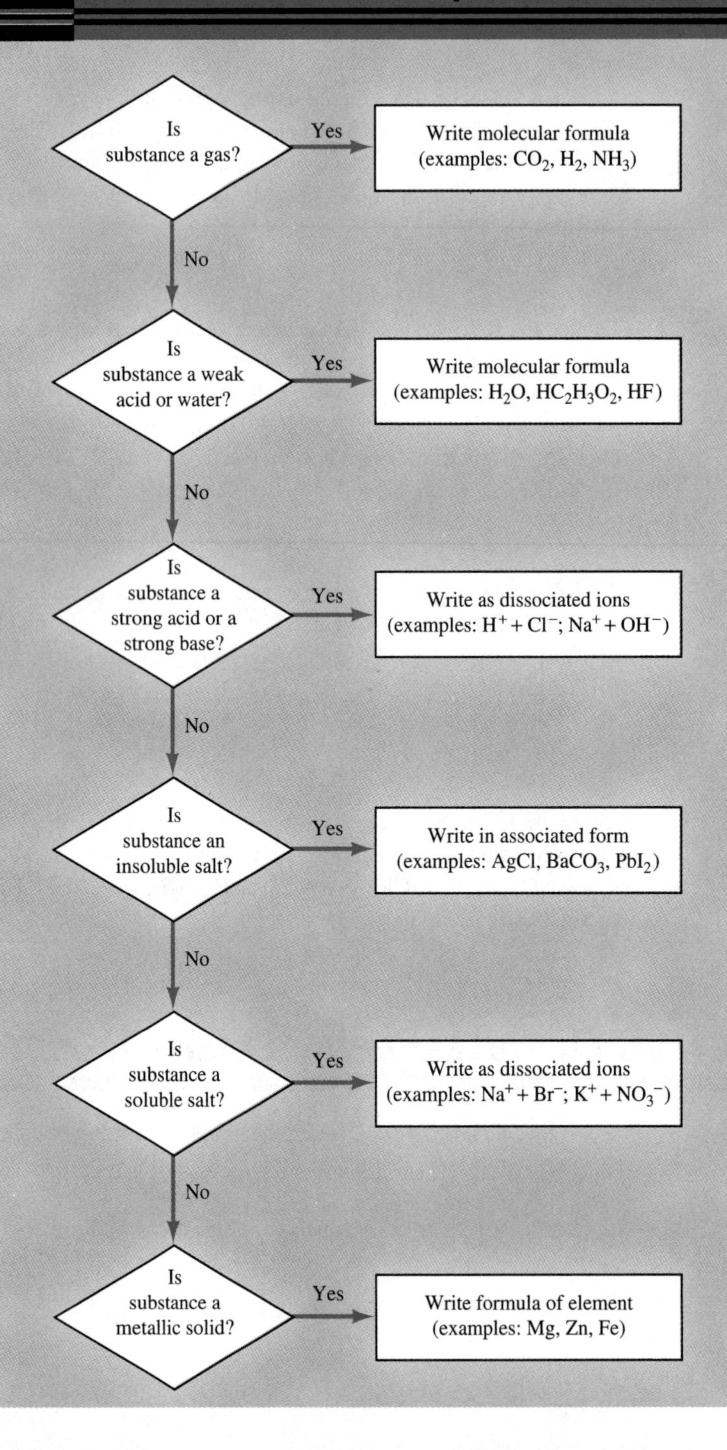

If a double-replacement reaction were to occur, the recombination products would be sodium nitrate, $NaNO_3$, and potassium iodide, KI. We have stated that for a recombination reaction to occur, one of the products must be nonionized. In this case, both of the possible products are ionic compounds that are soluble in water, so no reaction takes place. Let us see what result we obtain if we attempt to write ionic equations for this potential reaction.

Total molecular equation:

$$NaI(aq) + KNO_3(aq) \xrightarrow{?} NaNO_3(aq) + KI(aq)$$

Total ionic equation:

$$Na^+(aq) + I^-(aq) + K^+(aq) + NO_3^-(aq) \longrightarrow Na^+(aq) + NO_3^-(aq) + K^+(aq) + I^-(aq)$$

Every ion that appears on the reactant side of the total ionic equation also appears on the product side. Every ion is a spectator. There is no net ionic equation. In fact, there is no reaction! When we mix the two solutions (NaI and KNO_3), we simply obtain a mixture of the four ions: Na^+, I^-, K^+, and NO_3^-. We obtain a meaningful net ionic equation only when a nonionized product forms.

We have considered several of the possible types of double-replacement reactions that lead to a nonionized product. *If any of the products is a gas, a precipitate, a weak acid, or water, a reaction will occur.* Thus, both of the following mixtures lead to a reaction:

$$HI(aq) + KOH(aq) \longrightarrow ?$$

$$HI(aq) + NaF(aq) \longrightarrow ?$$

In the first reaction, the combination of H^+ and OH^- forms water:

$$H^+(aq) + OH^-(aq) \longrightarrow H_2O(\ell)$$

In the second reaction, the combination of H^+ and F^- forms hydrofluoric acid, a weak acid:

$$H^+(aq) + F^-(aq) \longrightarrow HF(aq)$$

We will examine weak acids in greater detail in Chapter 15. The weak acids we have mentioned in this chapter include acetic acid ($HC_2H_3O_2$), hydrofluoric acid (HF), and phosphoric acid (H_3PO_4).

We have also seen that carbonates and acids react with one another. Thus, the following mixture will undergo a reaction:

$$HBr(aq) + K_2CO_3(aq) \longrightarrow ?$$

In this case, the combination of hydrogen ions with carbonate ion produces nonionized products:

$$2\,H^+(aq) + CO_3^{2-}(aq) \longrightarrow CO_2(g) + H_2O(\ell)$$

In a similar fashion, when ammonium salts are mixed with bases, ammonia is released:

$$NH_4^+(aq) + OH^-(aq) \longrightarrow NH_3(g) + H_2O(\ell)$$

In our earlier description of aqueous ammonia, we observed that about 1% of the ammonia molecules combine with water to form ammonium ion and hydroxide ion. On the other hand, when ammonium ion and hydroxide ion are combined, the reverse of that process produces ammonia and water. (The reversibility of such reactions is discussed in Chapters 15 and 17.) Thus, mixing aqueous solutions of ammonium nitrate and potassium hydroxide results in the following reaction:

$$NH_4NO_3(aq) + KOH(aq) \rightarrow KNO_3(aq) + NH_3(g) + H_2O(\ell)$$

The formation of ammonia and water (two nonionized substances) provides the impetus for this reaction to take place.

PROBLEM 14.15

For each of the following mixtures, decide whether a reaction will occur, and if it will, complete and balance the equation. (Use Table 14-3 or the Summary of Solubility Rules to determine solubilities.)

(a) $KI(aq) + Pb(NO_3)_2(aq) \rightarrow ?$
(b) $HClO_4(aq) + Ca(OH)_2(aq) \rightarrow ?$
(c) $NaBr(aq) + Al_2(SO_4)_3(aq) \rightarrow ?$
(d) $HBr(aq) + KC_2H_3O_2(aq) \rightarrow ?$
(e) $HI(aq) + Li_2CO_3(aq) \rightarrow ?$
(f) $NH_4Cl(aq) + NaOH(aq) \rightarrow ?$

CHAPTER SUMMARY

Ionic substances in aqueous solution may separate into component ions. The process of separation is known as **dissociation.** The nonionized form of a substance capable of dissociation is referred to as the **associated** form. The ionized form is the **dissociated** form. Substances that produce ions in solution are **electrolytes.** Substances that dissociate only partially are called **weak electrolytes.** Those that do not dissociate at all are called **nonelectrolytes.**

The most common description of acids and bases is the Arrhenius description. An **Arrhenius acid** is a substance that produces hydrogen ions in solution; an **Arrhenius base** produces hydroxide ions in solution. When an acid dissociates in water, the hydrogen ions that are liberated combine with water to form **hydronium ions.** Acids may be classified according to their strengths. A **strong acid** is one that ionizes completely; a **weak acid** is one that is only partially dissociated into its component ions. Acids also may be classified according to the number of ionizable hydrogens per molecule; for example, as **diprotic** or **triprotic** acids.

Acids and bases react to **neutralize** one another, thereby destroying each other's acid and base properties. In the process, water and a salt are formed. The water results from combination of the hydrogen ion from the acid and the hydroxide ion from the base. The **salt** is an ionic compound made up of the cation from the base and the anion from the acid.

A *precipitation* reaction occurs when two ions that form an insoluble solid product are brought together in solution. The insoluble solid, known as a **precipitate,** is not appreciably dissociated.

Reactions are often classified according to their types as **synthesis, decomposition, combustion with oxygen, single-replacement,** and **double-replacement** reactions. In a double-replacement reaction, the ions of the two reactants exchange partners. The incentive for a double-replacement reaction to occur is the formation of a nonionized or slightly dissociated product. A weak acid, water, a precipitate, or a gas all constitute such a product. Neutralization reactions and precipitation reactions are examples of double-replacement reactions.

We can describe chemical reactions by writing three types of equations:

1. The **total molecular equation** expresses the reaction in terms of the chemical formulas of the substances involved.
2. The **total ionic equation** expresses any dissociated ionic substances in terms of their ions. It may contain **spectator ions,** which do not participate in the chemical change.
3. The **net ionic equation** represents only those substances that undergo change.

KEY TERMS

Review each of the following terms, which are discussed in this chapter and defined in the Glossary.

synthesis reaction
decomposition reaction
combustion with oxygen
single-replacement reaction
double-replacement reaction
dissociation
dissociated
associated
electrolyte
weak electrolyte
nonelectrolyte
Arrhenius acid
Arrhenius base
weak acid
strong acid
hydronium ion
neutralization
diprotic acid
triprotic acid
salt
total molecular equation
total ionic equation
spectator ion
net ionic equation
precipitate

ADDITIONAL PROBLEMS

SECTION 14.1

14.16 Classify each of the following reactions according to the five categories discussed in this section.
(a) $Mg + Cu(NO_3)_2 \rightarrow Cu + Mg(NO_3)_2$
(b) $S + O_2 \rightarrow SO_2$
(c) $H_3PO_4 + 3\,KOH \rightarrow K_3PO_4 + 3\,H_2O$
(d) $C_5H_{12} + 8\,O_2 \rightarrow 5\,CO_2 + 6\,H_2O$
(e) $Ca(OH)_2 \rightarrow CaO + H_2O$

14.17 Write a balanced equation for each of the following double-replacement reactions.
(a) Aqueous hydrobromic acid plus aqueous potassium hydroxide forms aqueous potassium bromide plus liquid water.
(b) Aqueous silver nitrate plus aqueous ammonium chromate forms solid silver chromate plus aqueous ammonium nitrate.
(c) Aqueous sulfuric acid plus aqueous barium hydroxide forms solid barium sulfate plus liquid water.

SECTION 14.2

14.18 Water can be decomposed into its elements by passing an electrical current through water to which some sodium sulfate has been added.
(a) Would electricity pass through pure water?
(b) What is the purpose of the sodium sulfate?

SECTION 14.3

14.19 What is the difference between a strong acid and a weak acid? Suppose you were given two substances and told that one is a strong acid and one is a weak acid. What simple test could you use to distinguish between the two? Describe exactly what you would do and what you would observe in each case.

SECTION 14.4

14.20 Explain the meaning of the following terms as they are used to describe acids and bases.
(a) Proton **(b)** Diprotic **(c)** Triprotic
(d) Alkaline **(e)** Hydronium ion

14.21 Complete and balance each of the following equations for complete neutralization.
(a) $HI + NaOH \rightarrow$ **(b)** $H_2SO_4 + KOH \rightarrow$
(c) $HNO_3 + Ca(OH)_2 \rightarrow$ **(d)** $H_3PO_4 + NaOH \rightarrow$

14.22 Write a balanced acid–base reaction to prepare each of the following salts.
(a) KBr **(b)** $Ba(NO_3)_2$ **(c)** $MgSO_4$ **(d)** K_3PO_4

14.23 Write a balanced equation between a carbonate and an acid to produce the following salts.
(a) KBr **(b)** NH_4Cl

SECTION 14.5

14.24 Write the formulas of the ions present in each of the following solutions, and then give the ratio in which they are found.
(a) $CaI_2(aq)$ **(b)** $(NH_4)_2CrO_4(aq)$
(c) $Mg(NO_3)_2(aq)$ **(d)** $Al_2(SO_4)_3(aq)$

SECTION 14.6

14.25 Complete and balance each of the following acid–base reactions, and then write a total ionic equation and a net ionic equation for each.
(a) $HI(aq) + NaOH(aq) \rightarrow ?$
(b) $HClO_4(aq) + Ca(OH)_2(aq) \rightarrow ?$

SECTION 14.7

14.26 Answer each of the following questions about the reaction that takes place when aqueous solutions of sodium chloride and silver nitrate are mixed.

(a) What are the ionic species present in an aqueous sodium chloride solution?
(b) What are the ionic species present in an aqueous silver nitrate solution?
(c) When these two solutions are mixed, which of the ions react? What happens to these ions?
(d) Which of the ions are spectator ions? What happens to these ions?

14.27 Complete and balance each of the following recombination reactions, using Table 14-3 to identify any precipitates in the total molecular equation. Then write a total ionic equation and a net ionic equation for each.
(a) $Pb(NO_3)_2(aq) + Na_2SO_4(aq) \rightarrow ?$
(b) $AgNO_3(aq) + MgBr_2(aq) \rightarrow ?$
(c) $MgCl_2(aq) + Na_2CO_3(aq) \rightarrow ?$

SECTION 14.8

14.28 Each of the following reactions is balanced. Write a total ionic equation and a net ionic equation for each. For parts (g) and (h), remember that HNO_2 and HCN are weak acids.
(a) $2\ HNO_3(aq) + (NH_4)_2CO_3(aq) \rightarrow 2\ NH_4NO_3(aq) + H_2O(\ell) + CO_2(g)$
(b) $Zn(s) + 2\ HCl(aq) \rightarrow ZnCl_2(aq) + H_2(g)$
(c) $CaCO_3(s) + 2\ HI(aq) \rightarrow CaI_2(aq) + H_2O(\ell) + CO_2(g)$
(d) $2\ Al(s) + 3\ Cu(NO_3)_2(aq) \rightarrow 2\ Al(NO_3)_3(aq) + 3\ Cu(s)$
(e) $2\ Na(s) + 2\ H_2O(\ell) \rightarrow 2\ NaOH(aq) + H_2(g)$
(f) $3\ Cu(s) + 8\ HNO_3(aq) \rightarrow 3\ Cu(NO_3)_2(aq) + 2\ NO(g) + 4\ H_2O(\ell)$
(g) $HI(aq) + KNO_2(aq) \rightarrow KI(aq) + HNO_2(aq)$
(h) $HCN(aq) + NaOH(aq) \rightarrow NaCN(aq) + H_2O(\ell)$

14.29 Calcium carbonate is often found as the insoluble scale left behind on pots and pans from hard water. Household vinegar (aqueous acetic acid) can be used to dissolve it. Write a balanced total molecular equation, a total ionic equation, and a net ionic equation for this reaction of a weak acid with an insoluble salt.

SECTION 14.9

14.30 For each of the following mixtures, decide whether a reaction will occur, and if it will, complete and balance the equation. (Use Table 14-3 or the Summary of solubility rules to determine solubilities.)
(a) $BaCl_2(aq) + K_2SO_4(aq) \rightarrow ?$
(b) $CaI_2(aq) + Mg(NO_3)_2(aq) \rightarrow ?$
(c) $HI(aq) + KOH(aq) \rightarrow ?$
(d) $HCl(aq) + NaCN(aq) \rightarrow ?$
(e) $H_2SO_4(aq) + Na_2CO_3(aq) \rightarrow ?$

GENERAL PROBLEMS

14.31 Suppose the beaker in Fig. 14-1 is emptied and a small amount of 0.1 M NaCl is added, covering just enough of the electrodes to cause the bulb to glow brightly. What will happen to the intensity of the light as distilled water is added slowly?

14.32 Some substances can act both as an acid and as a base. A substance possessing this property is said to be *amphoteric*. The hydrogen carbonate ion performs this function in the blood, where it can neutralize small amounts of acid or base. Write a net ionic equation for the reaction of hydrogen carbonate ion with a hydrogen ion, and a net ionic equation for its reaction with a hydroxide ion.

14.33 Sodium hydrogen carbonate is often used to clean up acid spills, both in the laboratory and on the highway. What is the role of this reagent, and how many grams would be required to neutralize a spill of 125 mL of 0.100 M HCl?

***14.34** Sulfuric acid can be neutralized by adding aqueous barium hydroxide:

$$H_2SO_4(aq) + Ba(OH)_2(aq) \rightarrow BaSO_4(s) + 2\ H_2O(\ell)$$

We can monitor the progress of this reaction by carrying it out in a beaker, using a conductivity apparatus such as the one shown in Fig. 14-1. When the apparatus is first placed in the beaker of sulfuric acid, the light glows brightly. Then, as barium hydroxide is added dropwise, the light eventually goes out. However, if the addition of barium hydroxide is continued beyond this point, the light will again glow brightly.
(a) Why does the light glow brightly at first?
(b) At what point does the light go out?
(c) Why does the light go out at this point?
(d) Why does the light go on again when barium hydroxide is added beyond this point?

14.35 Write equations to describe what occurs when the following acids and bases are mixed in the proportions indicated.

(a) 1 mol $H_2C_2O_4$ + 1 mol NaOH → ?
(b) 1 mol $NaHC_2O_4$ + 1 mol NaOH → ?
(c) 1 mol H_3PO_4 + 1 mol KOH → ?
(d) 1 mol H_3PO_4 + 2 mol KOH → ?
(e) 1 mol KH_2PO_4 + 1 mol KOH → ?

***14.36** Although sulfuric acid is considered a strong acid, only the first hydrogen ion is completely dissociated:

$$H_2SO_4(aq) \xrightarrow{100\%} H^+(aq) + HSO_4^-(aq)$$

The second hydrogen ion is only partially dissociated, like the ions of weak acids. Write a total ionic equation and a net ionic equation to correspond to each of the following molecular equations.

(a) $H_2SO_4(aq) + NaOH(aq) \rightarrow NaHSO_4(aq) + H_2O(\ell)$
(b) $NaHSO_4(aq) + NaOH(aq) \rightarrow Na_2SO_4(aq) + H_2O(\ell)$
(c) $H_2SO_4(aq) + 2\ NaOH(aq) \rightarrow Na_2SO_4(aq) + 2\ H_2O(\ell)$

***14.37** Use the general knowledge you have acquired in this chapter to write total molecular, total ionic, and net ionic equations (including state symbols) for each of the following reactions.

(a) A gas is evolved when a piece of aluminum is dropped into a hydrobromic acid solution.
(b) When a sodium sulfide solution is added to hydrochloric acid, a foul odor (characteristic of rotten eggs) is observed.
(c) Mixing aqueous solutions of calcium iodide and ammonium phosphate results in a precipitate.
(d) A sharp odor is observed when concentrated solutions of ammonium nitrate and sodium hydroxide are mixed.
(e) A lead(II) acetate solution reacts with hydriodic acid.

***14.38** When concentrated hydrochloric acid and concentrated aqueous ammonia solutions are placed in open beakers in a fume hood, a cloud of very fine white solid particles is observed over the beakers. Write a total molecular equation, with state symbols, to describe this reaction.

WRITING EXERCISES

14.39 Your nephew is going to summer camp in a location where there are lots of afternoon thunder and lightning storms. You recently read an article about a child being electrocuted while swimming in a mountain lake, and you are concerned about your nephew's safety. He is very interested in science, so you thought you would send him a letter discussing electrolytes and nonelectrolytes, and explaining why it is dangerous to be in a lake during a lightning storm, despite the very low conductivity of pure water. Write this letter.

14.40 In your own words, describe a total molecular equation, a total ionic equation, and a net ionic equation. Explain how to convert a total molecular equation to a total ionic equation, and then convert the total ionic equation to a net ionic equation. Be sure to explain the meaning of spectator ions and their role in a chemical reaction.

14.41 In your own words, explain what acids and bases are, and give several examples of each. Discuss the general properties of these substances, and explain how they are able to neutralize one another when they are mixed in the proper proportions.

15 WORKING WITH ACIDS AND BASES

In Chapter 14, you were introduced to acids and bases as we examined their neutralization reactions. In this chapter we examine some of the major theories used to describe acid–base behavior, as well as some of the calculations used by practicing chemists when working with acids and bases. We begin by examining the concentrations of ions in solutions of strong acids and strong bases.

15.1 Determining the Hydrogen Ion Concentration of a Strong Acid or the Hydroxide Ion Concentration of a Strong Base

OBJECTIVE 15.1 Given the molarity of a strong monoprotic acid, determine the hydrogen ion concentration; given the molarity of an alkali metal hydroxide, determine the hydroxide ion concentration.

When working with solutions of acids, we often wish to know the concentrations of the hydrogen ions present. For example, suppose we wish to know the hydrogen ion concentration of a 0.10 M HCl solution. Hydrochloric acid is a strong acid. Consequently, the hydrogen ions and chloride ions dissociate completely. Since each HCl molecule produces one H^+ ion, there is 0.10 mol of H^+ in 0.10 mol of HCl. In a solution containing 0.10 mol of HCl per liter, there is 0.10 mol of H^+ per liter. In other words, the hydrogen ion concentration is 0.10 M H^+. We frequently use brackets, [ ], when we wish to indicate molar concentrations. Thus, we express this hydrogen ion concentration as follows:

$$[H^+] = 0.10 \text{ M}$$

The hydrogen ion concentrations of other strong monoprotic acids are obtained in the same fashion. For example, the hydrogen ion concentration of 0.25 M HI is $[H^+] = 0.25$ M. However, dissociation of the second and subsequent protons in polyprotic acids, such as H_2SO_4, is not complete. Consequently, the hydrogen ion concentrations of these acids are not so readily obtained. Nor are the hydrogen ion concentrations of weak acids.

The hydroxide ion concentrations of strong bases, such as metallic hydroxides, are similarly determined. For example, a 0.35 M NaOH solution has a hydroxide ion concentration of $[OH^-] = 0.35$ M. To determine the hydroxide ion concentration of a 0.010 M $Ca(OH)_2$ solution, we must keep in mind that for each $Ca(OH)_2$, there are two OH^- ions. Consequently, there is 0.020 mol of OH^- in 0.010 mol of $Ca(OH)_2$; or a solution that contains 0.010 mol of $Ca(OH)_2$ per liter must have a hydroxide ion concentration of $[OH^-] = 0.020$ M. The hydroxide ion concentrations of weak bases, such as aqueous ammonia, are not so readily obtained. To simplify our discussion in this chapter, we will restrict our calculations to strong monoprotic acids and the alkali metal hydroxides.

PROBLEM 15.1

Find the concentration of the ion indicated in each of the following solutions.

(a) 0.30 M KOH(aq); $[OH^-]$ = ______
(b) 0.50 M HBr(aq); $[H^+]$ = ______
(c) 0.65 M NaOH(aq); $[OH^-]$ = ______
(d) 0.73 M HNO_3(aq); $[H^+]$ = ______

15.2 THE DISSOCIATION OF WATER

OBJECTIVE 15.2 Write a mathematical expression that relates the hydrogen ion concentration to the hydroxide ion concentration in aqueous solution. Given the concentration of either one, calculate the other.

In Chapter 14, you learned that water dissociates to a very slight extent into hydrogen ions and hydroxide ions:

$$H_2O \rightarrow H^+ + OH^-$$

However, you also learned that when hydrogen ions and hydroxide ions are placed together, they combine to form water:

$$H^+ + OH^- \rightarrow H_2O$$

These processes illustrate the concept that a reaction may be reversible. We can combine the two processes using paired arrows, as follows:

$$H_2O \rightleftharpoons H^+ + OH^-$$

This is the same type of arrow we used when we discussed the fundamentals of vapor pressure equilibrium in Section 12.8. Like the establishment of a vapor pressure, the dissociation of water is an equilibrium process. In the last chapter, you learned that most of the molecules in a sample of water exist in the associated state. Nevertheless, the process of dissociation is a dynamic process. Associated molecules are constantly undergoing dissociation, while at the same time, hydrogen ions are combining with hydroxide ions to form molecules of water. When the reaction is in a state of equilibrium, the rate at which molecules of water are dissociating equals the rate at which ions are combining to form water molecules.

We describe a solution as being *neutral* when the hydrogen ion concentration equals the hydroxide concentration, or $[H^+] = [OH^-]$. An acidic solution is one in which the hydrogen ion concentration exceeds the hydroxide ion concentration, or $[H^+] > [OH^-]$. And a basic solution is one in which the hydroxide ion concentration is greater than the hydrogen ion concentration, or $[H^+] < [OH^-]$.

Because water produces the same number of hydrogen ions as hydroxide ions, it is a neutral substance. The concentrations of hydrogen ion, $[H^+]$, and hydroxide ion, $[OH^-]$, in pure water are both 1.0×10^{-7} M.

$$[H^+] = 1.0 \times 10^{-7}\,M \qquad [OH^-] = 1.0 \times 10^{-7}\,M$$

The dissociation of water is such that when the system is at equilibrium, multiplying the hydrogen ion concentration by the hydroxide ion concentration gives a constant with a numerical value of 1.0×10^{-14}. This constant is given a special symbol, $\boldsymbol{K_w}$.

K_w: A constant relating the hydrogen ion concentration and the hydroxide ion concentration of aqueous solutions: 1.0×10^{-14}.

$$K_w = [H^+][OH^-] = 1.0 \times 10^{-14}$$

Let us verify this relationship for pure water, where the concentrations of hydrogen and hydroxide ions are both 1.0×10^{-7} M:

$$[H^+][OH^-] = [1.0 \times 10^{-7}][1.0 \times 10^{-7}] = 1.0 \times 10^{-14}$$

When acid is added to water, the hydrogen ion concentration increases. However, some of the added hydrogen ions combine with dissociated hydroxide ions already present to form water molecules via the association reaction:

$$\underset{\leftarrow \text{ Association}}{\overset{\text{Dissociation } \rightarrow}{H_2O \rightleftharpoons H^+ + OH^-}}$$

Thus, the increase in hydrogen ion concentration is accompanied by a corresponding decrease in the hydroxide ion concentration. When equilibrium is reestablished, the hydrogen ion concentration times the hydroxide ion concentration once again equals 1.0×10^{-14}. For example, when the hydrogen ion concentration, $[H^+]$, is increased to 1.0×10^{-5} M, the hydroxide ion concentration, $[OH^-]$, decreases to 1.0×10^{-9} M:

$$[H^+][OH^-] = \underset{\text{Concentration increases}}{[1.0 \times 10^{-5}]}\underset{\text{Concentration decreases}}{[1.0 \times 10^{-9}]} = 1.0 \times 10^{-14}$$

On the other hand, adding base to water causes the hydroxide ion concentration to increase. This is accompanied by a corresponding decrease in the hydrogen ion concentration. For example, an increase in the hydroxide ion concentration to 1.0×10^{-4} M is accompanied by a corresponding decrease in the hydrogen ion concentration to 1.0×10^{-10} M:

$$[H^+][OH^-] = \underset{\text{Concentration decreases}}{[1.0 \times 10^{-10}]}\underset{\text{Concentration increases}}{[1.0 \times 10^{-4}]} = 1.0 \times 10^{-14}$$

No matter what the conditions, if we multiply the hydrogen ion concentration by the hydroxide ion concentration, the product always equals 1.0×10^{-14}. Table 15-1 (p. 452) gives this information for a number of hydrogen ion and hydroxide ion concentrations.

EXAMPLE 15.1

The hydrogen ion concentration in a solution is $[H^+] = 1.0 \times 10^{-3}$ M. Calculate the hydroxide ion concentration, $[OH^-]$.

Planning the solution

We substitute the values of K_w and $[H^+]$ into the following relationship:

SOLUTION

$$K_w = [H^+][OH^-]$$

$$1.0 \times 10^{-14} = [1.0 \times 10^{-3}][OH^-]$$

$$\frac{1.0 \times 10^{-14}}{1.0 \times 10^{-3}} = [OH^-]$$

$$1.0 \times 10^{-11} = [OH^-]$$

The brackets denote moles per liter, so the answer with appropriate units is

$$1.0 \times 10^{-11}\ \text{M} = [OH^-]$$

TABLE 15-1 Selected hydrogen ion and hydroxide ion concentrations

Hydrogen ion concentration (moles/liter)		Hydroxide ion concentration (moles/liter)	
Decimal	Power of 10	Power of 10	Decimal
1.0	10^{0}	10^{-14}	0.00000000000001
0.1	10^{-1}	10^{-13}	0.0000000000001
0.01	10^{-2}	10^{-12}	0.000000000001
0.001	10^{-3}	10^{-11}	0.00000000001
0.0001	10^{-4}	10^{-10}	0.0000000001
0.00001	10^{-5}	10^{-9}	0.000000001
0.000001	10^{-6}	10^{-8}	0.00000001
0.0000001	10^{-7}	10^{-7}	0.0000001
0.00000001	10^{-8}	10^{-6}	0.000001
0.000000001	10^{-9}	10^{-5}	0.00001
0.0000000001	10^{-10}	10^{-4}	0.0001
0.00000000001	10^{-11}	10^{-3}	0.001
0.000000000001	10^{-12}	10^{-2}	0.01
0.0000000000001	10^{-13}	10^{-1}	0.1
0.00000000000001	10^{-14}	10^{0}	1.0

EXAMPLE 15.2

The hydroxide ion concentration in a solution is $[OH^-] = 4.0 \times 10^{-6}$ M. Calculate the hydrogen ion concentration, $[H^+]$.

SOLUTION

$$K_w = [H^+][OH^-]$$

$$1.0 \times 10^{-14} = [H^+][4.0 \times 10^{-6}]$$

$$\frac{1.0 \times 10^{-14}}{4.0 \times 10^{-6}} = [H^+]$$

$$0.25 \times 10^{-8} = [H^+]$$

Expressing this number in scientific notation gives

$$2.5 \times 10^{-9}\text{ M} = [H^+]$$

PROBLEM 15.2

Calculate the missing hydrogen ion or hydroxide ion concentration for each of the following aqueous solutions.

(a) $[H^+] = 1.0 \times 10^{-10}$ M; $[OH^-] =$ ________

(b) $[H^+] =$ ________; $[OH^-] = 1.0 \times 10^{-6}$ M

(c) $[H^+] = 2.0 \times 10^{-5}$ M; $[OH^-] =$ ________

(d) $[H^+] =$ ________; $[OH^-] = 6.5 \times 10^{-4}$ M

15.3 pH

OBJECTIVE 15.3 Define pH and pOH. Given either one, calculate the other. Given enough information to calculate the hydrogen ion concentration or the hydroxide ion concentration, determine the pH and pOH of a solution. Given the pH, calculate the hydrogen ion concentration; given the pOH, calculate the hydroxide ion concentration.

pH: $-\log[H^+]$

In Section 15.2 you saw that the concentrations of H^+ and OH^- cover a broad range of values. Rather than work with negative exponents (as would generally be the case), we use a scale called the pH scale as a shorthand method for expressing the acidity or basicity of a solution. The **pH** of any solution is equal to the negative logarithm of the hydrogen ion concentration. That is,

$$pH = -\log[H^+]$$

Before we apply this definition, let us point out that a logarithm is nothing more than an exponent. For example, the logarithm of 10^3 is 3. The logarithm of 10^{-4} is −4. Since you will need to round off logarithms to the correct number of significant figures, let us examine the logarithm of the following number, expressed in scientific notation, to see which digits in a logarithm are significant figures:

$$\log(2.3 \times 10^4) = 4.36$$

The digit 4 in 4.36 comes from the exponential term, 10^4. The digits 36 following the decimal point come from the term 2.3. This is true for all logarithms. The portion of the logarithm derived from the exponential term *precedes* the decimal point, and the portion of the logarithm derived from the term preceding the exponential term *follows* the decimal point. Since the number of significant figures in a scientific number is determined only by the term preceding the exponential term, it follows that *only the digits following the decimal point in the logarithm are significant figures.* (This is true, regardless of whether the number was initially expressed in scientific notation or not.) Thus, a logarithm of 4.36 has two significant figures, the same as 2.3×10^4. We use the following rule to round off logarithms to the correct number of significant figures: *To express a logarithm to the correct number of significant figures, round off the logarithm so that the number of digits following the decimal point equals the number of significant figures in the number for which the logarithm is being calculated.*

By this rule, the logarithm of 1.0×10^{-3} (0.0010) is −3.00 (two significant figures). Similarly, the logarithm of 4.5×10^{-12} is −11.35 (also two significant figures). For simplicity, the problems in this section have been designed to be rounded off to the nearest 0.01, the usual limit of most pH meters.

Applying the definition of pH, we find that a solution with a hydrogen ion concentration of 1.0×10^{-3} M H^+ has a pH of 3.00:

$$pH = -\log[1.0 \times 10^{-3}] = -(-3.00) = 3.00$$

In the last section you saw that the concentration of hydrogen ions in pure water is $[H^+] = 1.0 \times 10^{-7}$ M. Applying the definition of pH to this concentration reveals that the pH of water is 7.00:

$$pH = -\log[H^+] = -\log[1.0 \times 10^{-7}] = -(-7.00) = 7.00$$

Thus a pH of 7.00 represents a neutral pH. Acidic solutions have pH values less than 7.00, whereas the pH values of basic solutions are greater than 7.00.

pOH: $-\log[OH^-]$

Closely related to the pH is the **pOH:**

$$pOH = -\log[OH^-]$$

Because the hydroxide concentration in water is also 1.0×10^{-7} M, the pOH of water is 7.00:

$$pOH = -\log[OH^-] = -\log[1.0 \times 10^{-7}] = -(-7.00) = 7.00$$

The pH and pOH are related to one another as follows:

$$pH + pOH = 14.00$$

For example, the values of the hydrogen ion and hydroxide ion concentrations for pure water are

$$[H^+] = 1.0 \times 10^{-7} \quad \text{leading to pH} = 7.00$$
$$[OH^-] = 1.0 \times 10^{-7} \quad \text{leading to pOH} = 7.00$$
$$pH + pOH = 7.00 + 7.00 = 14.00$$

The pOH of a solution that has a hydroxide ion concentration of 1.0×10^{-2} M must be 2.00:

$$pOH = -\log[1.0 \times 10^{-2}] = -(-2.00) = 2.00$$

The sum of the pH and the pOH always equals 14.00, so the pH of this solution must be 12.00:

$$pH + pOH = 14.00$$
$$pH + 2.00 = 14.00$$
$$pH = 14.00 - 2.00 = 12.00$$

In practice, the pH of an acid solution is calculated directly from the hydrogen ion concentration, $[H^+]$. However, the pH of a basic solution is generally obtained by first calculating the pOH from the hydroxide ion concentration, $[OH^-]$, followed by subtraction of the pOH from 14.00. (Because the pH is so readily computed from the pOH, the acidity or basicity of solutions is generally reported only in terms of pH.)

EXAMPLE 15.3

Calculate the pH of a solution that contains 7.3 g of HCl in 20.0 L of solution.

Planning the solution

We calculate the molarity of the solution the way we did in Chapter 13. This will enable us to determine the hydrogen ion concentration, $[H^+]$, and then the pH.

SOLUTION

First, we calculate the molarity as follows:

$$? \text{ mol HCl} = 7.3 \text{ g HCl}\left(\frac{1 \text{ mol HCl}}{36.5 \text{ g HCl}}\right) = 0.20 \text{ mol HCl}$$

$$\text{Molarity} = \frac{0.20 \text{ mol HCl}}{20.0 \text{ L}} = 0.010 \text{ M HCl} = 1.0 \times 10^{-2} \text{ M HCl}$$

Because each mole of HCl contains one mole of hydrogen ion, H^+, and because HCl completely dissociates (it is a strong acid), the hydrogen ion concentration must also be 1.0×10^{-2} M H^+.

$$pH = -\log[H^+] = -\log[1.0 \times 10^{-2}] = -(-2.00) = 2.00$$

The pH of this acidic solution is 2.00.

EXAMPLE 15.4 Calculate the pH of a solution that contains 1.2 g of NaOH in 30.0 L of solution.

Planning the solution For this basic solution, we calculate the pOH first and then determine the pH by subtracting from 14.00.

SOLUTION We begin by determining the molarity:

$$? \text{ mol NaOH} = 1.2 \text{ g NaOH}\left(\frac{1 \text{ mol NaOH}}{40.0 \text{ g NaOH}}\right) = 0.030 \text{ mol NaOH}$$

$$\text{Molarity} = \frac{0.030 \text{ mol NaOH}}{30.0 \text{ L}} = 0.0010 \text{ M NaOH} = 1.0 \times 10^{-3} \text{ M NaOH}$$

There is one mole of hydroxide ion per mole of sodium hydroxide, so the hydroxide concentration is also 1.0×10^{-3} M.

$$\text{pOH} = -\log[\text{OH}^-] = -\log[1.0 \times 10^{-3}] = -(-3.00) = 3.00$$

Finally, we calculate the pH by subtracting the pOH from 14.00:

$$\text{pH} = 14.00 - \text{pOH} = 14.00 - 3.00 = 11.00$$

The pH of this basic solution is 11.00.

In the preceding examples, we calculated pH values from H^+ and OH^- concentrations that were exact powers of 10. As a result, all of the pH values we determined were integers. In fact, pH values are not limited to such whole numbers but cover a continuous range of values, as illustrated by the following examples. (We assume that your calculator has a LOG function. To determine a logarithm with your calculator, follow the directions that came with it or refer to Appendix B, Section B.6.)

EXAMPLE 15.5 Calculate the pH of a solution that has a hydrogen ion concentration, $[H^+]$, of 0.054 M.

Planning the solution As in the previous examples, we simply substitute the hydrogen ion concentration into the definition of pH:

$$\text{pH} = -\log[\text{H}^+] = -\log[0.054]$$

SOLUTION To calculate this pH, we need to take the logarithm of 0.054, a number that is not an exact power of 10. You may enter the number into your calculator either in decimal form or in scientific notation.

$$0.054 = 5.4 \times 10^{-2}$$

$$\log(0.054) = \log(5.4 \times 10^{-2}) = -1.27$$

$$\text{pH} = -\log[\text{H}^+] = -(-1.27) = 1.27$$

EXAMPLE 15.6 Calculate the pH of a KOH solution that has a hydroxide concentration, $[OH^-]$, of 0.0029 M.

Planning the solution We calculate the pOH and subtract from 14.00.

SOLUTION

$$\log(0.0029) = -2.54$$
$$pOH = -\log[OH^-] = -(-2.54) = 2.54$$
$$pH = 14.00 - pOH = 14.00 - 2.54 = 11.46$$

EXAMPLE 15.7 Calculate the pH of a solution that contains 0.72 g of NaOH in 50.0 L of total solution.

SOLUTION First we calculate the molarity of the NaOH solution:

$$? \text{ mol NaOH} = 0.72 \not{\text{g NaOH}}\left(\frac{1 \text{ mol NaOH}}{40.0 \not{\text{g NaOH}}}\right) = 0.018 \text{ mol NaOH}$$

$$\text{Molarity} = \frac{0.018 \text{ mol NaOH}}{50.0 \text{ L}} = 0.00036 \text{ M NaOH} = 3.6 \times 10^{-4} \text{ M NaOH}$$

Since there is one mole of hydroxide ion per mole of sodium hydroxide, the hydroxide ion concentration, $[OH^-]$, also must be 3.6×10^{-4}. Using the hydroxide ion concentration, we find the pOH and, from that, the pH:

$$[OH^-] = 3.6 \times 10^{-4}$$
$$\log(3.6 \times 10^{-4}) = -3.44$$
$$pOH = -\log[OH^-] = -(-3.44) = 3.44$$
$$pH = 14.00 - pOH = 14.00 - 3.44 = 10.56$$

From the results of these examples, we can see that acidic solutions have a pH of less than 7.00, whereas the pH of basic solutions is greater than 7.00. Those solutions with a pH of 7.00 are neutral. Figure 15-1 shows how acid and base strength varies with pH.

Many of the biological processes carried out within our bodies are extremely sensitive to pH. For example, the stomach has a pH in the range of 1–2. If we were to neutralize our stomach acid completely, the digestive processes that take place in the stomach could not be carried out properly.

In addition to calculating the pH from the hydrogen ion concentration, it is useful to be able to reverse the process and determine the hydrogen ion concentration, $[H^+]$, from the pH. To do so requires the use of antilogs. An antilog is readily obtained by using the INVERSE LOG function on your calculator. If you do not know how to take an antilog on your calculator, refer to the directions that came with the calculator or see Appendix B, Section B.6.

When we take an antilog, we are asking the question, "What number has this as its logarithm?" For example, if the logarithm in question is –5.00, we want to know the number that has a logarithm of –5.00. The number in question is 1.0×10^{-5}. The procedure for determining the number of significant figures of an antilog is the reverse of the procedure we discussed when obtaining a log. Since the number –5.00 is a logarithm, only the digits to the right of the decimal are significant. Thus, there are two significant figures. Consequently, the antilog must have two significant figures, as is the case for 1.0×10^{-5}.

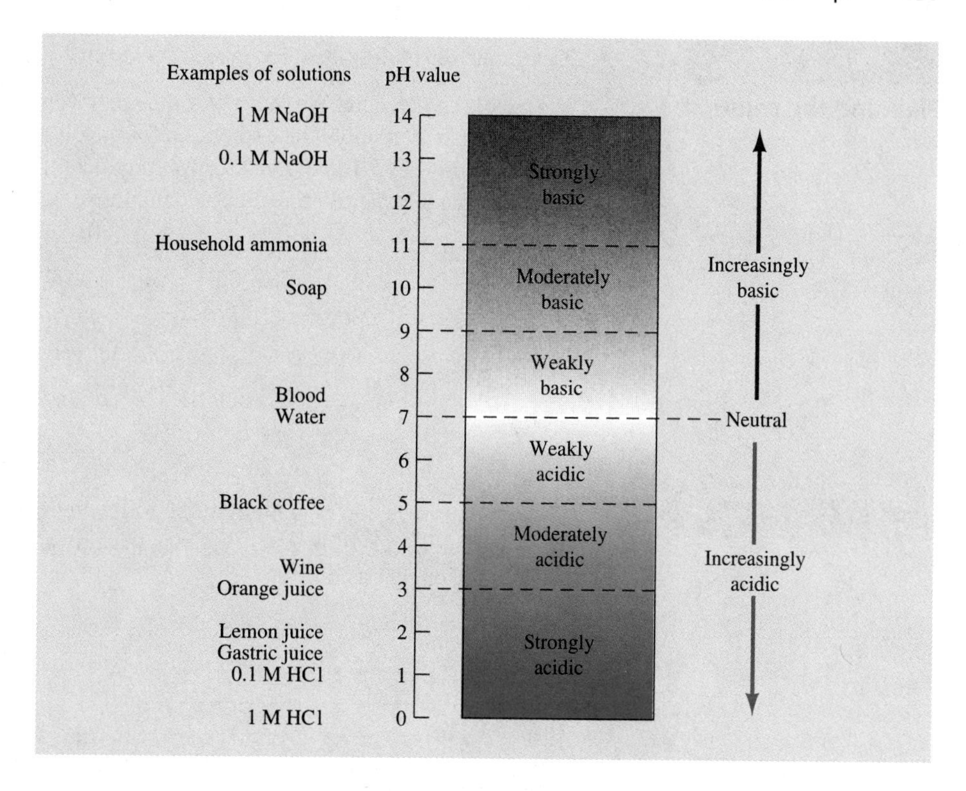

FIGURE 15-1 The pH scale.

To determine the hydrogen ion concentration, $[H^+]$, from the pH, we simply reverse the procedure we use to obtain a pH. Thus, if the pH of a solution is 5.00, we find $[H^+]$ as follows:

$$\text{pH} = -\log[H^+] = 5.00$$
$$\log[H^+] = -5.00$$

Since the logarithm of $[H^+]$ is –5.00, the quantity $[H^+]$ is equal to the antilog of –5.00:

$$[H^+] = \text{antilog}(-5.00) = 1.0 \times 10^{-5}\text{ M}$$

The hydroxide ion concentration, $[OH^-]$, is obtained from the pOH in exactly the same fashion as we determine the hydrogen ion concentration from the pH.

EXAMPLE 15.8

Determine the hydrogen ion concentration, $[H^+]$, of a solution with a pH of 3.32.

Planning the solution

Since the pH is the negative log of the hydrogen ion concentration, the hydrogen ion concentration must be the antilog of –3.32.

SOLUTION

$$\text{pH} = -\log[H^+] = 3.32$$
$$\log[H^+] = -3.32$$
$$[H^+] = \text{antilog}(-3.32) = 4.8 \times 10^{-4}\text{ M}$$

EXAMPLE 15.9

Determine the hydroxide ion concentration, $[OH^-]$, of a solution with a pH of 5.23.

Planning the solution

We have a choice here. We may determine the hydrogen ion concentration (as in Example 15.8) and then apply the expression for the dissociation of water: $K_w = [H^+][OH^-]$ (as in Examples 15.1 and 15.2); or we may determine the pOH from the pH and then apply the same general procedure we employed in Example 15.8. The latter method is simpler.

SOLUTION

$$\text{pOH} = 14.00 - \text{pH} = 14.00 - 5.23 = 8.77$$
$$\text{pOH} = -\log[OH^-] = 8.77$$
$$\log[OH^-] = -8.77$$
$$[OH^-] = \text{antilog}(-8.77) = 1.7 \times 10^{-9}\ M$$

PROBLEM 15.3

Give pOH values corresponding to the following pH values. Then use the information provided in Fig. 15-1 to classify each as strongly acidic, weakly acidic, neutral, weakly basic, or strongly basic.

	pH	pOH	Classification
(a)	7.00	______	__________
(b)	7.50	______	__________
(c)	5.80	______	__________
(d)	11.70	______	__________
(e)	2.60	______	__________

PROBLEM 15.4

Calculate the pH and the pOH of each of the following solutions. Parts (a)–(e) are integers; parts (f)–(j) are not.

(a) 0.00010 M HCl **(b)** 0.00010 M NaOH **(c)** 0.010 M HNO_3
(d) 1.0×10^{-5} M KOH **(e)** 1.0 M HCl **(f)** 0.068 M HCl
(g) 0.00031 M HNO_3 **(h)** 0.000058 M HBr **(i)** 0.022 M NaOH
(j) 0.0074 M KOH

PROBLEM 15.5

Calculate the pH of each of the following solutions, which contain the mass of solute indicated in the final total volume. Parts (a)–(d) are integers; parts (e)–(h) are not. [Assume the volumes in parts (a) and (b) are accurate to the nearest liter.]

(a) 0.63 g HNO_3 in 1000 L **(b)** 4.0 g NaOH in 100 L
(c) 0.080 g NaOH in 0.200 L **(d)** 73 g HCl in 20.0 L
(e) 0.051 g HCl in 15 L **(f)** 0.10 g NaOH in 5.0 L
(g) 0.084 g KOH in 250 mL **(h)** 0.59 g HBr in 4.0 L

PROBLEM 15.6

Calculate the hydrogen ion concentration of each of the following solutions. Parts (a)–(c) are integers; parts (d)–(f) are not.

(a) pH = 4.00 **(b)** pH = 11.00 **(c)** pOH = 2.00
(d) pH = 8.44 **(e)** pH = 2.88 **(f)** pOH = 10.33

15.4 THEORIES OF ACIDS AND BASES

OBJECTIVE 15.4A State the Arrhenius definition of an acid and a base. Given an Arrhenius base that lacks the hydroxide ion in its formula, write an equation to show how it produces hydroxide ions in aqueous solution.

Thus far we have restricted ourselves to the Arrhenius definition of acids and bases, which enabled us to identify most acids and bases by the presence of hydrogen ions or hydroxide ions. In addition to the Arrhenius model, we also find it useful to recognize the Brønsted–Lowry and Lewis theories of acids and bases.

Arrhenius Acids and Bases

Arrhenius acid: A substance that produces hydrogen ions in aqueous solution.

Arrhenius base: A substance that produces hydroxide ions in aqueous solution.

In Section 14.3 we defined **Arrhenius acids** and **Arrhenius bases** as substances that produce hydrogen ions and hydroxide ions, respectively, in aqueous solutions. For example, hydrochloric acid (HCl) and nitric acid (HNO_3) produce hydrogen ions in solution and therefore are Arrhenius acids. Similarly, metal hydroxides such as sodium hydroxide (NaOH) and potassium hydroxide (KOH) are Arrhenius bases. Most Arrhenius acids and bases can be identified by the presence of ionizable hydrogens or hydroxides in their formulas. All of the acids and bases we have described so far fit the Arrhenius model.

However, the presence of the word *produce* in the Arrhenius definitions causes some chemists to include some additional substances we might not think of as Arrhenius acids and bases. We have already noted that ammonia produces hydroxide ions when it is dissolved in water:

$$NH_3 + H_2O \rightleftharpoons NH_4 + OH^-$$

As in the dissociation of water, this process is readily reversible. Hence, we have used paired arrows to indicate that this is an equilibrium process. We noted earlier that the concentration of hydroxide ions (at equilibrium) is approximately 1% of the ammonia concentration. This small concentration of hydroxide ions adds to those hydroxide ions present from the dissociation of water, resulting in a basic solution.

In a fashion similar to that of ammonia, the salts of weak acids (such as sodium acetate) produce hydroxide ions when dissolved in water. In the case of sodium acetate, hydroxide ions are produced as the following reaction takes place between water and the acetate ions present:

$$C_2H_3O_2^- + H_2O \rightleftharpoons HC_2H_3O_2 + OH^-$$

In contrast, when carbon dioxide is dissolved in water, hydrogen ions are formed:

$$CO_2 + H_2O \rightleftharpoons H^+ + HCO_3^-$$

Thus, carbon dioxide may be considered an Arrhenius acid. You may be aware that carbonated beverages are acidic in nature due to the carbon dioxide present. Excess carbon dioxide in the blood causes a condition known as *acidosis.* Individuals with respiratory diseases, such as emphysema, often suffer from this condition, because of the inability of their lungs to release carbon dioxide from the bloodstream.

PROBLEM 15.7

Write equations to show how each of the following anions can react with water to produce hydroxide ions.
(a) ClO^- (as in KClO) **(b)** NO_2^- (as in $NaNO_2$) **(c)** $C_7H_5O_2^-$ (as in $NaC_7H_5O_2$)

Brønsted–Lowry Acids and Bases

OBJECTIVE 15.4B State the Brønsted–Lowry definition of an acid and a base. Explain the relationship between a conjugate acid and its conjugate base. Identify conjugate acids and their conjugate bases.

We have previously seen that when hydrogen chloride is dissolved in water, hydronium ions are formed:

$$HCl + H_2O \rightleftharpoons H_3O^+ + Cl^-$$

In this reaction, hydrogen chloride donates a hydrogen ion to water. Turning this around, we may think of water as accepting the hydrogen ion. Reflecting the fact that we often refer to a hydrogen ion as a proton (as in the terms *diprotic* and *triprotic*), the Brønsted–Lowry theory defines acids and bases in terms of their ability to donate or accept protons: A **Brønsted–Lowry acid** is a proton donor. A **Brønsted–Lowry base** is a proton acceptor. In the preceding example, hydrogen chloride is the Brønsted–Lowry acid, and water serves as the Brønsted–Lowry base. Note that water, a substance we have considered neutral until now, may act as a base in the Brønsted–Lowry sense.

Brønsted–Lowry acid: A proton donor.

Brønsted–Lowry base: A proton acceptor.

The Brønsted–Lowry theory is consistent with the Arrhenius definitions we have already learned. The simple reaction of a hydrogen ion with a hydroxide ion (the essence of an Arrhenius neutralization reaction) may be viewed as a Brønsted–Lowry acid–base reaction. However, we must recall that an aqueous hydrogen ion really exists as a hydronium ion:

$$H_3O^+ + OH^- \rightleftharpoons H_2O + H_2O$$

The hydronium ion is the proton donor; the hydroxide ion is the proton acceptor.

Similarly, when gaseous ammonia is dissolved in water, a proton is donated from water to ammonia:

$$NH_3 + H_2O \rightleftharpoons NH_4^+ + OH^-$$

In this case, water is the Brønsted–Lowry acid and ammonia is the Brønsted–Lowry base. In the previous example, water acted as the base; in this case, it acts as the acid. We can define a Brønsted–Lowry acid or base only in terms of the particular reaction we are discussing—never in an isolated context.

The following examples should further illustrate these ideas.

Acid	Base	⇌	Acid	Base
H_2SO_4	$+ OH^-$	⇌	H_2O	$+ HSO_4^-$
H_2O	$+ NH_3$	⇌	NH_4^+	$+ OH^-$
H_2O	$+ C_2H_3O_2^-$	⇌	$HC_2H_3O_2$	$+ OH^-$
HCl	$+ H_2O$	⇌	H_3O^+	$+ Cl^-$

Note that if we reverse the direction of each of the preceding reactions, we can again describe a Brønsted–Lowry acid and base. The acid on each side is related to the base on the opposite side, differing only by a proton (hydrogen ion). Thus, sulfuric acid, H_2SO_4, is related to the hydrogen sulfate ion, HSO_4^-. Hydroxide ion, OH^-, is related to water, H_2O. In each case, we can describe two species that bear the relationship of a **conjugate acid** and its **conjugate base.** A conjugate acid and base differ only by a hydrogen ion. Symbolically, we may represent this as HB/B^- (or as HB^+/B). So far we have used the following conjugate-acid/conjugate-base pairs:

Conjugate-acid/conjugate-base pair: Two species that differ only by a hydrogen ion.

Conjugate acid / Conjugate base
H_3O^+ / H_2O
H_2O / OH^-
H_2SO_4 / HSO_4^-
$HC_2H_3O_2$ / $C_2H_3O_2^-$
NH_4^+ / NH_3

PROBLEM 15.8

Fill in the missing conjugate acid or base for each of the following pairs.

	Conjugate acid / Conjugate base		Conjugate acid / Conjugate base
(a)	HBr / ______	**(b)**	______ / PH_3
(c)	$H_2C_2O_4$ / ______	**(d)**	H_2S / ______
(e)	HS^- / ______	**(f)**	______ / $C_2O_4^{2-}$
(g)	______ / $HC_2O_4^-$		

PROBLEM 15.9

(a) What is the conjugate acid of HSO_4^-?
(b) What is the conjugate base of HSO_4^-?

PROBLEM 15.10

Identify the Brønsted–Lowry acid and base on the reactant (left) side of each of the following equations.

(a) $HCN + OH^- \rightleftharpoons H_2O + CN^-$
(b) $NH_3 + H_2SO_4 \rightleftharpoons NH_4^+ + HSO_4^-$
(c) $H_2O + HNO_3 \rightleftharpoons H_3O^+ + NO_3^-$
(d) $H_2O + F^- \rightleftharpoons HF + OH^-$
(e) $HPO_4^{2-} + HSO_4^- \rightleftharpoons H_2PO_4^- + SO_4^{2-}$

Lewis Acids and Bases

OBJECTIVE 15.4C State the Lewis definition of an acid and a base. Given a chemical reaction showing the electron pairs, identify the Lewis acid and base.

In Chapter 8 we examined the Lewis electron-dot structures of molecules. The importance of electronic structure to acid–base theory is expressed in the Lewis theory of

Lewis acid: An electron-pair acceptor.

Lewis base: An electron-pair donor.

acids and bases, which defines these substances in terms of the behavior of their unshared pairs of electrons. A **Lewis base** is an electron-pair donor. A **Lewis acid** is an electron-pair acceptor. Thus, in the reaction of a hydrogen ion with a hydroxide ion, one of the lone pairs on the oxygen is donated to the hydrogen ion:

$$H^+ + :\underset{\cdot\cdot}{\overset{\cdot\cdot}{O}}:H^- \rightleftharpoons H:\underset{\cdot\cdot}{\overset{\cdot\cdot}{O}}:H$$

In our discussion of Brønsted–Lowry acids and bases, we saw that ammonia is a base. From the Lewis standpoint, we might consider this the reaction of ammonia with a hydrogen ion:

$$\underset{\text{Electron-pair acceptor}}{H^+} + \underset{\text{Electron-pair donor}}{\begin{array}{c} H \\ :\underset{\cdot\cdot}{\overset{\cdot\cdot}{N}}:H \\ H \end{array}} \rightleftharpoons \left[\begin{array}{c} H \\ H:\underset{\cdot\cdot}{\overset{\cdot\cdot}{N}}:H \\ H \end{array}\right]^+$$

Ammonia acts as the Lewis base, donating an electron pair. The hydrogen ion is the Lewis acid; it accepts the electron pair. You can see that the Lewis theory, like the Brønsted–Lowry theory, is consistent with the ideas we have already developed about acids and bases. However, the Lewis theory includes some reactions we would not ordinarily think of as acid–base reactions. For example, the following reaction of silver ion with ammonia is a Lewis acid–base reaction:

$$\underset{\text{Electron-pair acceptor}}{Ag^+} + \underset{\text{Electron-pair donor}}{2\,:NH_3} \rightleftharpoons [H_3N:Ag:NH_3]^+$$

In this reaction, the silver ion is the Lewis acid and ammonia is the Lewis base.

In a similar fashion, aluminum chloride combines with chloride to form the $AlCl_4^-$ ion:

$$\underset{\text{Electron-pair acceptor}}{\begin{array}{c} :\overset{\cdot\cdot}{Cl}: \\ :\underset{\cdot\cdot}{\overset{\cdot\cdot}{Cl}}:\overset{\cdot\cdot}{Al} \\ :\underset{\cdot\cdot}{Cl}: \end{array}} + \underset{\text{Electron-pair donor}}{:\underset{\cdot\cdot}{\overset{\cdot\cdot}{Cl}}:^-} \rightleftharpoons \left[\begin{array}{c} :\overset{\cdot\cdot}{Cl}: \\ :\underset{\cdot\cdot}{\overset{\cdot\cdot}{Cl}}:\underset{\cdot\cdot}{\overset{\cdot\cdot}{Al}}:\underset{\cdot\cdot}{\overset{\cdot\cdot}{Cl}}: \\ :\underset{\cdot\cdot}{Cl}: \end{array}\right]^-$$

Here, aluminum chloride is the Lewis acid and chloride ion is the Lewis base. The Lewis theory of acids and bases enables us to include many nonaqueous reactions that do not fit either the Arrhenius or the Brønsted–Lowry definition.

The Summary reviews the Arrhenius, Brønsted–Lowry, and Lewis theories.

PROBLEM 15.11

Identify the Lewis acid and base for each of the following reactions.

(a) $H^+ + :PH_3 \rightarrow PH_4^+$

(b) $Cu^+ + 2\,:NH_3 \rightarrow [H_3N:Cu:NH_3]^+$

(c) $H_3N: + BF_3 \rightarrow H_3N:BF_3$

(d) $\begin{array}{c} :\overset{\cdot\cdot}{Cl}: \\ :\underset{\cdot\cdot}{\overset{\cdot\cdot}{Cl}}:\overset{\cdot\cdot}{Fe} \\ :\underset{\cdot\cdot}{Cl}: \end{array} + :\underset{\cdot\cdot}{\overset{\cdot\cdot}{Cl}}:^- \rightarrow \left[\begin{array}{c} :\overset{\cdot\cdot}{Cl}: \\ :\underset{\cdot\cdot}{\overset{\cdot\cdot}{Cl}}:\underset{\cdot\cdot}{\overset{\cdot\cdot}{Fe}}:\underset{\cdot\cdot}{\overset{\cdot\cdot}{Cl}}: \\ :\underset{\cdot\cdot}{Cl}: \end{array}\right]^-$

SUMMARY Acid–Base Theories

Name of theory	Acid	Base
Arrhenius	Produces hydrogen ions in aqueous solution	Produces hydroxide ions in aqueous solution
Brønsted–Lowry	Proton donor	Proton acceptor
Lewis	Electron-pair acceptor	Electron-pair donor

15.5 WEAK ACIDS

OBJECTIVE 15.5 Distinguish between a solution of a weak acid and a solution that is weakly acidic. Given a table of relative Brønsted–Lowry acid strengths, predict whether the reaction between a given Brønsted–Lowry acid and base will proceed to any appreciable extent.

In Section 14.3 we defined weak acids as those that do not dissociate completely when dissolved in water. This is in contrast to strong acids, which are completely dissociated. Do not confuse a weak acid with a *solution* that is weakly acidic. For example, a 10^{-5} M HCl solution has a pH of 5, so it is classified as weakly acidic. Nevertheless, because hydrochloric acid is completely dissociated, it is classified as a strong acid, regardless of its concentration. On the other hand, acetic acid, $HC_2H_3O_2$, the main component of vinegar (other than water), is a weak acid. Acetic acid can dissociate into a hydrogen ion, H^+, and an acetate ion, $C_2H_3O_2^-$. (The three hydrogen atoms in the acetate ion do not dissociate.) If we make up a solution of 0.1 M $HC_2H_3O_2$, we find that the hydrogen ion concentration is only 0.001 M. If acetic acid ionized completely, the concentration of hydrogen ions would be 0.1 M, corresponding to a pH of 1. In other words, we would have a strongly acidic solution. Instead, we find the solution to be only moderately acidic, having a pH of about 3. This means that only one molecule out of about every 100 is ionized.

$$\underset{\substack{0.1\text{ M}\\ \textbf{Associated}}}{HC_2H_3O_2} \rightleftharpoons \underbrace{\underset{0.001\text{ M}}{H^+} + \underset{0.001\text{ M}}{C_2H_3O_2^-}}_{\textbf{Dissociated}}$$

As in the case of water, we refer to the two forms of acetic acid as the associated form and the dissociated form; and like water, the two forms exist in equilibrium with one another. Molecules that exist in the neutral molecular form (on the left side of the arrows) are associated. Molecules that have ionized (on the right) are dissociated.

Recalling our discussion of Brønsted–Lowry acids and bases, the associated form of acetic acid on the left ($HC_2H_3O_2$) is the *conjugate acid* of the acetate anion on the right ($C_2H_3O_2^-$), while the acetate ion is the *conjugate base* of acetic acid.

There are many weak acids, and chemists compile tables of their relative strengths. The more dissociated an acid is, the stronger its acid strength. Table 15-2 shows the percent dissociation of 0.10 M solutions of several strong acids and several weak acids.

In our discussion of Arrhenius bases, we noted that the acetate ion is a base:

$$C_2H_3O_2^- + H_2O \rightleftharpoons HC_2H_3O_2 + OH^-$$

In fact, all of the conjugate bases (the anions to the right of the arrows) in Table 15-2 can be considered Arrhenius bases. Furthermore, *the stronger a conjugate acid, the weaker its conjugate base; and the weaker a conjugate acid, the stronger its conjugate base.* Let us consider this statement more carefully. The hydrogen ion in a strong acid is readily separated from its anion (the conjugate base). It follows that the anion has only a very weak attraction for the hydrogen ion. In other words, it is a very weak base. On the other hand, most of the hydrogen ions in a weak acid solution remain associated with their anions. This indicates that there is a much stronger attraction between the hydrogen ion of a weak acid and its anion, or that the anion is a stronger base than is the case for a strong acid. Thus, cyanide ion (CN^-), which is the conjugate base of the weakest acid (HCN) & in Table 15-2, is the strongest base in the table; whereas chloride ion (Cl^-), which is the conjugate base of a strong acid (HCl), is such a weak base that it does not produce enough hydroxide ions to affect the pH of water. Thus, an aqueous solution of sodium chloride is neutral. We generally do not consider the conjugate bases of the strong acids (those that are 100% dissociated) to be bases in the Arrhenius sense. The other conjugate bases in Table 15-2 are all considered weak bases, with cyanide ion (the strongest base shown) being comparable to aqueous ammonia in its basicity.

TABLE 15-2 Percent dissociation of some acids

Name		Associated form		Dissociated form		Percent dissociation (0.10 M solution)
Perchloric acid	↑	$HClO_4$	$\rightleftharpoons$	$H^+ + ClO_4^-$		100
Hydriodic acid		HI	$\rightleftharpoons$	$H^+ + I^-$		100
Hydrobromic acid		HBr	$\rightleftharpoons$	$H^+ + Br^-$		100
Sulfuric acid		H_2SO_4	$\rightleftharpoons$	$H^+ + HSO_4^-$		100
Nitric acid		HNO_3	$\rightleftharpoons$	$H^+ + NO_3^-$		100
Hydrochloric acid	Increasing	HCl	$\rightleftharpoons$	$H^+ + Cl^-$	Increasing	100
Phosphoric acid	conjugate	H_3PO_4	$\rightleftharpoons$	$H^+ + H_2PO_4^-$	conjugate	24
Hydrofluoric acid	acid	HF	$\rightleftharpoons$	$H^+ + F^-$	base	8.0
Nitrous acid	strength	HNO_2	$\rightleftharpoons$	$H^+ + NO_2^-$	strength	6.5
Formic acid		$HCHO_2$	$\rightleftharpoons$	$H^+ + CHO_2^-$		4.5
Benzoic acid		$HC_7H_5O_2$	$\rightleftharpoons$	$H^+ + C_7H_5O_2^-$		2.5
Acetic acid		$HC_2H_3O_2$	$\rightleftharpoons$	$H^+ + C_2H_3O_2^-$		1.3
Hypochlorous acid		HClO	$\rightleftharpoons$	$H^+ + ClO^-$		0.056
Hydrocyanic acid		HCN	$\rightleftharpoons$	$H^+ + CN^-$	↓	0.0063

Whenever a Brønsted–Lowry acid and base are brought together, the product formed from the Brønsted–Lowry acid is its conjugate base, and the product formed from the Brønsted–Lowry base is its conjugate acid. For example, in the following reaction, the conjugate base of HCl is Cl^-, and the conjugate acid of CN^- is HCN:

HCl	+	CN^-	→	Cl^-	+	HCN
Conjugate acid 1		Conjugate base 2		Conjugate base 1		Conjugate acid 2

When an acid and a base are mixed together, the extent of reaction depends on the relative strengths of the acids and bases in both the reactants and products. The preceding reaction forms products as written. However, the reverse of that reaction forms a negligible amount of product because chloride ion is too weak a base to remove the hydrogen ion from hydrocyanic acid.

$$Cl^- + HCN \rightarrow \text{No appreciable reaction}$$

A reaction takes place when the conjugate acid on the reactant side is stronger than the conjugate acid on the product side. When this is the case, the conjugate base on the reactant side is also stronger than the conjugate base on the product side. Thus, *the stronger conjugate acid and conjugate base react to form the weaker conjugate acid and conjugate base.* If a reaction would have to produce a stronger conjugate acid and conjugate base from a weaker conjugate acid and conjugate base, no appreciable reaction will occur. (The actual amount of product formed depends on the relative strengths of acids and bases. We will only attempt to predict whether an appreciable amount of reaction is likely to occur.)

Imagine that, in the preceding reaction, a chloride ion and a cyanide ion were having a tug-of-war for a hydrogen ion. Since cyanide ion is the stronger base, it will end up with the hydrogen ion. Thus, when HCl and CN^- are mixed, CN^- takes H^+ away from Cl^-. However, when Cl^- and HCN are mixed, Cl^- is too weak to remove H^+ from CN^-. The following examples further illustrate this principle.

EXAMPLE 15.10

Predict whether the following reaction produces any appreciable amount of product.

$$HBr + C_2H_3O_2^- \xrightarrow{?} Br^- + HC_2H_3O_2$$

Planning the solution

We examine Table 15-2 to determine whether the stronger acid is HBr or $HC_2H_3O_2$. Alternatively, we could determine whether Br^- or $C_2H_3O_2^-$ is the stronger base.

SOLUTION

Since HBr (a reactant) is a stronger acid than $HC_2H_3O_2$ (a product), this reaction is proceeding from a stronger acid to a weaker acid and will therefore take place as written.

EXAMPLE 15.11

Predict whether the following reaction produces any appreciable amount of product.

$$HClO + F^- \xrightarrow{?} ClO^- + HF$$

SOLUTION

Since ClO^- (a product) is a stronger base than F^- (a reactant), this reaction would have to proceed from a weaker base to a stronger base and will therefore not occur to any appreciable extent.

PROBLEM 15.12

Predict whether each of the following reactions produces any appreciable amount of product.

(a) $HF + I^- \xrightarrow{?} F^- + HI$ **(b)** $HNO_3 + NO_2^- \xrightarrow{?} NO_3^- + HNO_2$

15.6 BUFFERS

OBJECTIVE 15.6 Describe what a buffer is and how it is prepared. Explain why a buffer works.

Buffer: A chemical system resistant to changes in pH.

Human blood has a pH of 7.4. When the pH of the blood varies too far from this value, death results. The body's organs also operate at specific pH values. The various parts of the body maintain their constant pH through the use of buffers. A **buffer** is a system that is resistant to large changes in pH caused by added acid or base.

A buffer is composed of a Brønsted–Lowry conjugate-acid/conjugate-base pair in which neither substance is a strong acid or a strong base. In the laboratory, a buffer is often prepared by mixing a weak acid and its conjugate base in comparable concentrations. For example, a solution made up of 0.10 M $HC_2H_3O_2$ and 0.10 M $NaC_2H_3O_2$ functions as a buffer having a pH of 4.74. Modest amounts of acid or base have little effect on the pH of this solution. To understand how this buffer works, let us examine the various species in solution.

	$HC_2H_3O_2$ ⇌	H^+	$+ C_2H_3O_2^-$
Approximate concentration	0.10 M	10^{-5} M	0.10 M

(When $NaC_2H_3O_2$ dissolves, it dissociates completely into Na^+ and $C_2H_3O_2^-$. Since the Na^+ ions act as spectator ions, we have ignored them. The $C_2H_3O_2^-$ ions serve as the conjugate base of the buffer.) In our discussion of weak acids, we noted that their conjugate bases are weak bases. Thus, if we add H^+ to this buffer, it combines with free acetate ion, $C_2H_3O_2^-$ (the conjugate base), thereby forming associated acetic acid, $HC_2H_3O_2$ (the conjugate acid):

$$H^+ + C_2H_3O_2^- \rightleftharpoons HC_2H_3O_2$$

On the other hand, if we add OH^-, it is neutralized by the associated acetic acid present:

$$HC_2H_3O_2 + OH^- \rightleftharpoons C_2H_3O_2^- + H_2O$$

Because the buffer solution contains both the conjugate acid and its conjugate base, adding small amounts of acid or base has little effect on the final hydrogen ion concentration.

In a similar fashion, buffers can also be prepared by combining a weak base and its conjugate acid. For example, a solution that is 0.10 M NH_4Cl and 0.10 M NH_3 functions as a buffer with a pH of 9.26:

	NH_4^+ ⇌	H^+	$+ NH_3$
Approximate concentration	0.10 M	10^{-9} M	0.10 M

(In this case, the ammonium ion from the ammonium chloride acts as the conjugate acid of ammonia, the conjugate base. The chloride ion is a spectator ion that has no effect on the pH.)

We cannot prepare a buffer by combining a strong acid with its conjugate base (or a strong base with its conjugate acid), however. As we have seen, the conjugate base of a strong acid is itself such a weak base that it is not capable of neutralizing acid that might be added to it. Thus, a solution of hydrochloric acid and sodium chloride has no buffering power.

One of the important buffers used by the body is the hydrogen phosphate buffer, which is composed of dihydrogen phosphate ion and monohydrogen phosphate ion:

$$H_2PO_4^- \rightleftharpoons H^+ + HPO_4^{2-}$$

The dihydrogen phosphate ion is the conjugate acid, and the monohydrogen phosphate ion is the conjugate base. This buffer helps maintain a constant pH in the body's cells.

PROBLEM 15.13

Which of the following conjugate-acid/conjugate-base pairs could *not* be used as a buffer? [*Hint:* Use Table 15-2 to see which are derived from strong acids.]
(a) HCN/CN^- **(b)** HNO_3/NO_3^- **(c)** HNO_2/NO_2^- **(d)** HBr/Br^-

15.7 EQUIVALENTS OF ACIDS AND BASES

In a neutralization reaction, one hydrogen ion combines with one hydroxide ion to form a molecule of water:

$$H^+ + OH^- \rightarrow H_2O$$

For example, in the reaction of hydrochloric acid with sodium hydroxide, the hydrogen ion from the hydrochloric acid combines with the hydroxide ion from sodium hydroxide to form water:

$$HCl + NaOH \rightarrow H_2O + NaCl$$

HCl	NaOH
1 mol HCl can deliver 1 mol H^+	1 mol NaOH can deliver 1 mol OH^-

Because sulfuric acid contains two hydrogen ions per molecule, complete neutralization of a mole of sulfuric acid requires two moles of sodium hydroxide:

$$H_2SO_4 + 2\,NaOH \rightarrow 2\,H_2O + Na_2SO_4$$

H_2SO_4	2 NaOH
1 mol H_2SO_4 can deliver 2 mol H^+	2 mol NaOH can deliver 2 mol OH^-

Each hydroxide combines with one of the hydrogen ions.

Similarly, the complete neutralization of a mole of phosphoric acid requires three moles of sodium hydroxide:

$$H_3PO_4 + 3\,NaOH \rightarrow 3\,H_2O + Na_3PO_4$$

H_3PO_4	$3\,NaOH$
1 mol H_3PO_4 can deliver 3 mol H^+	3 mol NaOH can deliver 3 mol OH^-

In each neutralization, the number of moles of hydrogen ion equals the number of moles of hydroxide ion.

Defining an Equivalent

OBJECTIVE 15.7A Define the term *equivalent.* Determine the number of equivalents in one mole of an acid or a base from its formula.

Because we work so frequently with acids and bases, it is convenient to define some new terms that relate the number of moles of hydrogen ion and the number of moles of hydroxide ion, regardless of the particular acids or bases used. In order to simplify our discussion, we will restrict ourselves to complete neutralization reactions.

Equivalent: The quantity of acid or base that delivers 1 mol of hydrogen ions or hydroxide ions.

An **equivalent (equiv)** is defined as the quantity of acid or base that delivers 1 mol of hydrogen ion or 1 mol of hydroxide ion. Thus, a mole of hydrochloric acid, HCl, contains 1 equiv, because 1 mol of the acid can deliver 1 mol of hydrogen ion. In a complete neutralization, 1 mol of sulfuric acid, H_2SO_4, can deliver 2 mol of hydrogen ions. Thus, a mole of sulfuric acid contains 2 equiv. Similarly, because 1 mol of phosphoric acid, H_3PO_4, can deliver 3 mol of hydrogen ions, there are 3 equiv per mole of phosphoric acid. A similar principle holds for bases.

$$1\text{ mol HCl} = 1\text{ equiv HCl} \qquad 1\text{ mol NaOH} = 1\text{ equiv NaOH}$$

$$1\text{ mol }H_2SO_4 = 2\text{ equiv }H_2SO_4 \qquad 1\text{ mol }Ca(OH)_2 = 2\text{ equiv }Ca(OH)_2$$

$$1\text{ mol }H_3PO_4 = 3\text{ equiv }H_3PO_4 \qquad 1\text{ mol }Al(OH)_3 = 3\text{ equiv }Al(OH)_3$$

No matter what acid or base is used in a neutralization reaction, the number of equivalents of acid equals the number of equivalents of base. The same statement cannot necessarily be made for moles of acid and base.

$$HCl + NaOH \rightarrow H_2O + NaCl$$

HCl	NaOH
1 mol	1 mol
1 equiv	1 equiv

$$H_2SO_4 + 2\,NaOH \rightarrow 2\,H_2O + Na_2SO_4$$

H_2SO_4	$2\,NaOH$
1 mol	2 mol
2 equiv	2 equiv

$$2\,H_3PO_4 + 3\,Ba(OH)_2 \rightarrow 6\,H_2O + Ba_3(PO_4)_2$$

$2\,H_3PO_4$	$3\,Ba(OH)_2$
2 mol	3 mol
6 equiv	6 equiv

PROBLEM 15.14

How many equivalents are in one mole of each of the following compounds?
(a) H_3PO_4 **(b)** HBr **(c)** $Al(OH)_3$ **(d)** LiOH **(e)** $H_2C_2O_4$ **(f)** $Mg(OH)_2$

Converting Between Moles and Equivalents

OBJECTIVE 15.7B Given the formula of an acid or base, convert between moles and equivalents.

It is quite easy to determine the number of equivalents in a sample if we know the number of moles and the chemical formula of the substance. For example, 0.400 mol H_2SO_4 is equal to 0.800 equiv H_2SO_4, because there are 2 equiv in each mole of sulfuric acid:

$$1 \text{ mol } H_2SO_4 = 2 \text{ equiv } H_2SO_4$$

Using this as a conversion factor, we find that

$$? \text{ equiv } H_2SO_4 = 0.400 \cancel{\text{mol } H_2SO_4}\left(\frac{2 \text{ equiv } H_2SO_4}{1 \cancel{\text{mol } H_2SO_4}}\right) = 0.800 \text{ equiv } H_2SO_4$$

Likewise, 0.450 mol H_3PO_4 contains 1.35 equiv:

$$1 \text{ mol } H_3PO_4 = 3 \text{ equiv } H_3PO_4$$

$$? \text{ equiv } H_3PO_4 = 0.450 \cancel{\text{mol } H_3PO_4}\left(\frac{3 \text{ equiv } H_3PO_4}{1 \cancel{\text{mol } H_3PO_4}}\right) = 1.35 \text{ equiv } H_3PO_4$$

In a similar fashion, we can convert the number of equivalents to moles. For example, to find the number of moles in 0.340 equiv $Ca(OH)_2$, we proceed as follows:

$$1 \text{ mol } Ca(OH)_2 = 2 \text{ equiv } Ca(OH)_2$$

$$? \text{ mol } Ca(OH)_2 = 0.340 \cancel{\text{equiv } Ca(OH)_2}\left(\frac{1 \text{ mol } Ca(OH)_2}{2 \cancel{\text{equiv } Ca(OH)_2}}\right) = 0.170 \text{ mol } Ca(OH)_2$$

PROBLEM 15.15

Calculate the number of equivalents in each of the following.
(a) 0.250 mol HCl **(b)** 0.550 mol $Ba(OH)_2$ **(c)** 2.50 mol H_3PO_4
(d) 0.0140 mol H_2SO_4 **(e)** 0.630 mol $Al(OH)_3$

PROBLEM 15.16

Calculate the number of moles in each of the following.
(a) 0.150 equiv H_3PO_4 **(b)** 0.650 equiv HBr **(c)** 1.55 equiv $Ca(OH)_2$
(d) 3.25 equiv KOH **(e)** 0.0750 equiv $H_2C_2O_4$

Calculating Equivalents from Masses

OBJECTIVE 15.7C Calculate the number of equivalents present in a given mass of acid or base.

We can also determine the number of equivalents in a given mass of an acid or a base. For example, perhaps we are interested in knowing how many equivalents are in 3.90 g $Al(OH)_3$. We can use dimensional analysis to carry out this conversion.

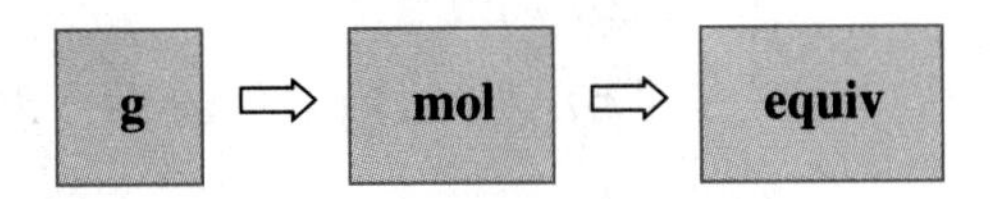

The following examples demonstrate the technique.

EXAMPLE 15.12

Calculate the number of equivalents in 3.90 g $Al(OH)_3$.

Planning the solution

We use the following strategy:

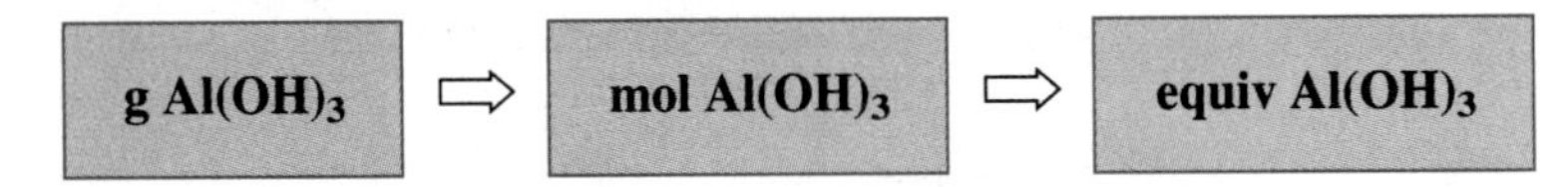

SOLUTION

First, we convert from grams to moles:

$$? \text{ mol } Al(OH)_3 = 3.90 \text{ g } Al(OH)_3 \left(\frac{1 \text{ mol } Al(OH)_3}{78.0 \text{ g } Al(OH)_3} \right)$$
$$= 0.0500 \text{ mol } Al(OH)_3$$

Next, we convert from moles to equivalents:

$$? \text{ equiv } Al(OH)_3 = 0.0500 \text{ mol } Al(OH)_3 \left(\frac{3 \text{ equiv } Al(OH)_3}{1 \text{ mol } Al(OH)_3} \right)$$
$$= 0.150 \text{ equiv } Al(OH)_3$$

Of course, we can simplify the process by combining our conversion factors, as shown in the next example.

EXAMPLE 15.13

How many equivalents are in 1.35 g $H_2C_2O_4$?

SOLUTION

g $H_2C_2O_4$ ⇨ mol $H_2C_2O_4$ ⇨ equiv $H_2C_2O_4$

$$? \text{ equiv } H_2C_2O_4 = 1.35 \text{ g } H_2C_2O_4 \left(\frac{1 \text{ mol } H_2C_2O_4}{90.0 \text{ g } H_2C_2O_4} \right) \left(\frac{2 \text{ equiv } H_2C_2O_4}{1 \text{ mol } H_2C_2O_4} \right)$$
$$= 0.0300 \text{ equiv } H_2C_2O_4$$

PROBLEM 15.17

Calculate the number of equivalents in each of the following.
(a) 3.65 g HCl **(b)** 0.0200 g NaOH **(c)** 7.41 g $Ca(OH)_2$ **(d)** 0.162 g HBr
(e) 21.0 g $H_2C_2O_4$ **(f)** 2.94 g H_2SO_4 **(g)** 0.392 g H_3PO_4

15.8 NORMALITY

OBJECTIVE 15.8A Define the term *normality.* Calculate the normality of a solution, given the mass of solute and the total volume of solution.

Normality: The number of equivalents per liter.

In our work with acids and bases, it is often more useful to know the number of equivalents than the number of moles. And since so much of our work is done with solutions, it is helpful to define a new concentration unit. **Normality** is the number of equivalents per liter:

$$\text{Normality} = \frac{\text{Number of equivalents}}{\text{Liter of solution}}$$

or

$$\text{N} = \frac{\text{equiv}}{\text{L soln}}$$

We use the symbol N to stand for normality.

Calculating Normality

To find the normality of any solution, we merely divide the number of equivalents of acid or base by the volume (in liters) in which it is dissolved. Thus, if 0.100 equiv HCl is dissolved in 200 mL of solution, the normality is 0.500 N.

$$\text{Normality} = \frac{0.100 \text{ equiv HCl}}{0.200 \text{ L soln}}$$

$$= \frac{0.500 \text{ equiv HCl}}{\text{L soln}} = 0.500 \text{ N HCl}$$

Note the similarity between normality and molarity. Instead of expressing the number of moles per liter, the normality gives the number of *equivalents* per liter. Thus, if we know the mass of acid or base and the volume of solution it is dissolved in, we can calculate the normality by much the same means as we use to calculate molarity. First we determine the number of equivalents present, and then we divide by the volume in liters.

EXAMPLE 15.14

What is the normality of a solution containing 5.00 g H_3PO_4 in 350 mL (3.50×10^2 mL) of solution?

SOLUTION

First, we determine the number of equivalents of H_3PO_4, using the same strategy as in Examples 15.12 and 15.13:

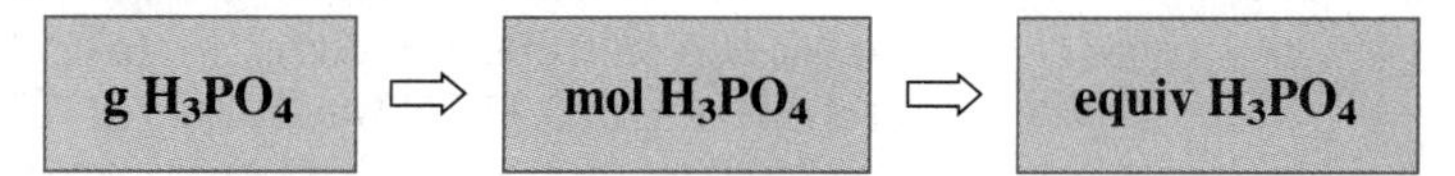

$$? \text{ equiv } H_3PO_4 = 5.00 \cancel{\text{g } H_3PO_4}\left(\frac{1 \cancel{\text{mol } H_3PO_4}}{98.0 \cancel{\text{g } H_3PO_4}}\right)\left(\frac{3 \text{ equiv } H_3PO_4}{1 \cancel{\text{mol } H_3PO_4}}\right)$$

$$= 0.153 \text{ equiv } H_3PO_4$$

Next, we divide the number of equivalents by the volume in liters (350 mL = 0.350 L).

$$\text{Normality} = \frac{0.153 \text{ equiv } H_3PO_4}{0.350 \text{ L soln}}$$

$$= 0.437 \text{ N } H_3PO_4$$

PROBLEM 15.18

Calculate the normality of each of the following solutions.
(a) 1.00 g NaOH in 50.0 mL **(b)** 0.500 g H_2SO_4 in 250 mL
(c) 2.00 g H_3PO_4 in 150 mL **(d)** 0.350 g $Mg(OH)_2$ in 2.50 L
(e) 9.87 g HBr in 1.80 L

Calculating Equivalents from Normality and Volume

OBJECTIVE 15.8B Given the volume and normality of a solution, calculate the number of equivalents present.

In Section 13.4 we discussed how to calculate the number of moles contained in a given volume of a solution with known molarity. For example, to calculate the number of moles in 400 mL of 1.20 M NaCl, we multiply the volume (in liters) by the molarity:

$$? \text{ mol NaCl} = 0.400 \cancel{\text{L soln}}\left(\frac{1.20 \text{ mol NaCl}}{\cancel{\text{L soln}}}\right) = 0.480 \text{ mol NaCl}$$

In a similar fashion, we can determine the number of equivalents contained in a given volume of a solution with known normality. In this case, multiplying the volume (in liters) by the normality yields the number of equivalents:

$$V_{(L)} \times N = \text{Number of equivalents}$$

Thus, 0.300 L of 0.250 N HCl contains 0.0750 equiv HCl:

$$? \text{ equiv HCl} = 0.300 \cancel{\text{L soln}}\left(\frac{0.250 \text{ equiv HCl}}{\cancel{\text{L soln}}}\right) = 0.0750 \text{ equiv HCl}$$

Similarly, 55.0 mL of 1.50 N NaOH contain 0.0825 equiv NaOH:

$$? \text{ equiv NaOH} = 0.0550 \cancel{\text{L soln}}\left(\frac{1.50 \text{ equiv NaOH}}{\cancel{\text{L soln}}}\right) = 0.0825 \text{ equiv NaOH}$$

PROBLEM 15.19

Calculate the number of equivalents in each of the following.
(a) 2.00 L of 0.300 N H_2SO_4 **(b)** 500 mL of 1.20 N HCl
(c) 350 mL of 0.750 N NaOH **(d)** 175 mL of 0.180 N H_3PO_4
(e) 50.0 mL of 0.125 N $H_2C_2O_4$

Using Normality in Titration

OBJECTIVE 15.8C Given the normality and volume of an acid or base used in titration, as well as the volume of an unknown acid or base solution being determined, calculate the normality of the unknown solution.

In Section 15.7 we noted that the equivalents of acid and base required to neutralize one another are the same. This fact may be used to simplify titration calculations that involve normality. For example, suppose 25.0 mL of NaOH solution are required to neutralize 50.0 mL of 0.100 N HCl. Because neutralization requires the same number of equivalents of acid and base, the neutralization point is reached when an equal number of equivalents of acid and base have been added. Since the volume of sodium hydroxide used is only half that of the hydrochloric acid, we can conclude that the sodium hydroxide solution must be twice as concentrated as the hydrochloric acid. In other words, the sodium hydroxide solution must be 0.200 N NaOH.

In this example the numbers were so simple that we were able to determine the sodium hydroxide concentration quite readily. However, when we have more complicated numbers to work with, it is convenient to use the following relationship between the concentrations and volumes of acid and base used:

$$V_a \cdot N_a = V_b \cdot N_b$$

where N_a and N_b are the normality of the acid and base, and V_a and V_b are the volumes of each used in the titration. This relationship holds because the left side of the equation represents the equivalents of acid used and the right side of the equation represents the equivalents of base. At the neutralization point, these must be equal. (This formula works only for equivalents; a similar relationship would *not* hold for moles of all acids and bases.)

For the previous example, the normality of the acid, N_a, is 0.100 N. The normality of the base, N_b, is the unknown we wish to determine. The volume of the acid used is 50.0 mL; the volume of the base used is 25.0 mL. (As with the dilution formula, the units of volume may be expressed in either liters or milliliters. The only restriction is that we must use the same units on both sides of the equation.)

$$V_a \cdot N_a = V_b \cdot N_b$$

$$(50.0\ \text{mL})(0.100\ \text{N}) = (25.0\ \text{mL})N_b$$

$$\frac{(50.0\ \cancel{\text{mL}})(0.100\ \text{N})}{(25.0\ \cancel{\text{mL}})} = N_b = 0.200\ \text{N}$$

Titration may be used to find the concentration of either an unknown base or an unknown acid, as the following example illustrates.

EXAMPLE 15.15

If 50.0 mL of H_2SO_4 of unknown concentration exactly neutralizes 35.0 mL of 0.600 N NaOH solution, what is the normality of the acid?

Planning the solution

We wish to determine the normality of the acid, N_a. We simply substitute the appropriate quantities into the relationship $V_a \cdot N_a = V_b \cdot N_b$.

SOLUTION

$$V_a \cdot N_a = V_b \cdot N_b$$

$$(50.0\text{ mL})N_a = (35.0\text{ mL})(0.600\text{ N})$$

$$N_a = \frac{(35.0\text{ mL})(0.600\text{ N})}{(50.0\text{ mL})} = 0.420\text{ N}$$

PROBLEM 15.20 If 45.0 mL of 0.200 N NaOH neutralizes 55.0 mL of HCl, what is the normality of the unknown acid?

PROBLEM 15.21 If 25.0 mL of 0.450 N H_2SO_4 neutralizes 35.0 mL of KOH, what is the normality of the unknown base?

PROBLEM 15.22 If 23.14 mL of 0.1200 N HCl neutralizes 21.20 mL of NaOH, what is the normality of the unknown base?

PROBLEM 15.23 If 40.00 mL of 0.1000 N NaOH neutralizes 52.00 mL of H_2SO_4, what is the normality of the sulfuric acid?

PROBLEM 15.24 A 42.4-mL portion of calcium hydroxide solution is found to neutralize 50.9 mL of 0.0400 N H_2SO_4. What is the normality of the calcium hydroxide solution?

Chapter Summary

Water is a slightly dissociated substance with hydrogen ion and hydroxide ion concentrations that are both equal to 1.0×10^{-7} M. When acid is added to water, the hydrogen ion concentration exceeds that of the hydroxide ion. When base is added to water, the reverse happens. However, regardless of the conditions employed, the hydrogen ion concentration times the hydroxide ion concentration is always $K_w = 1.0 \times 10^{-14}$.

The molar concentrations of the ions in a solution of a strong electrolyte can be determined by multiplying the molarity of the solution by the number of moles of each ion produced per mole of solute. The hydrogen ion concentrations of strong monoprotic acids can be determined in similar fashion, as can the hydroxide ion concentrations of strong bases.

We can express the acidity of a solution by its **pH.** The pH is defined as follows: pH $= -\log[H^+]$. Related to the pH is the **pOH** $= -\log[OH^-]$. The pH and pOH are related such that pH + pOH = 14. The pH of an acidic solution is less than 7, whereas that of a basic solution is greater than 7. Pure water has a pH of 7.

Acids and bases may be described according to three major theories: (a) the **Arrhenius** theory, (b) the **Brønsted–Lowry** theory, and (c) the **Lewis** theory. The Arrhenius theory defines an acid as a substance that produces hydrogen ions in solution; an Arrhenius base produces hydroxide ions in solution. The Brønsted–Lowry theory defines an acid as a proton (or hydrogen ion) donor and a base as a proton acceptor. The Lewis theory defines an acid as an electron–pair acceptor and a base as an electron–pair donor. As a consequence of the Brønsted–Lowry definitions, we describe a **conjugate-acid/**

conjugate-base relationship such that a conjugate acid and its conjugate base differ by a hydrogen ion.

A weak acid is one that is only partially dissociated. The associated form of a weak acid exists in equilibrium with its dissociated ions. The conjugate base of a Brønsted–Lowry acid is itself a Brønsted–Lowry base. The stronger the conjugate acid, the weaker its conjugate base, and vice versa. The products of a Brønsted–Lowry acid–base reaction are themselves a Brønsted–Lowry acid and base. For a Brønsted–Lowry acid–base reaction to produce an appreciable amount of product, the reaction must proceed from a stronger acid and base to form a weaker acid and base.

A **buffer** is a chemical system that resists changes in pH when acid or base is added. A buffer may be prepared by mixing a weak acid and its conjugate base in comparable concentrations. When acid is added to a buffer, the conjugate base neutralizes it. When base is added, the conjugate acid acts as the neutralizing agent.

An **equivalent** is the quantity of acid or base that contains one mole of hydrogen ions or one mole of hydroxide ions. In a neutralization reaction, the number of equivalents of acid and base that react are equal. The concentration of an acid or base solution may be expressed in terms of its **normality,** the number of equivalents per liter. The normality of an unknown acid or base can be determined by titrating it in a neutralization reaction with an acid or base of known normality. At the neutralization point, the following relationship can be used to calculate the normality of the unknown acid or base:

$$V_a \cdot N_a = V_b \cdot N_b$$

KEY TERMS

Review the following terms, which are discussed in this chapter and defined in the Glossary.

K_w	Arrhenius base	conjugate base	buffer
pH	Brønsted–Lowry acid	Lewis acid	equivalent
pOH	Brønsted–Lowry base	Lewis base	normality
Arrhenius acid	conjugate acid		

ADDITIONAL PROBLEMS

SECTION 15.1

15.25 For each of the following solutions, determine the concentration of the ion indicated.

(a) 0.21 M HCl(aq) $[H^+]$ = ______

(b) 0.16 M KOH(aq) $[OH^-]$ = ______

(c) 3.7×10^{-3} M $HClO_4$ $[H^+]$ = ______

(d) 5.2×10^{-4} M NaOH $[OH^-]$ = ______

SECTION 15.2

15.26 Calculate the missing hydrogen ion concentration or hydroxide ion concentration for each of the following aqueous solutions.

(a) $[H^+] = 1.0 \times 10^{-2}$ M $[OH^-]$ = __________

(b) $[H^+]$ = __________ $[OH^-] = 1.0 \times 10^{-11}$ M

(c) $[H^+] = 3.5 \times 10^{-8}$ M $[OH^-]$ = __________

(d) $[H^+]$ = __________ $[OH^-] = 8.5 \times 10^{-1}$ M

SECTION 15.3

15.27 Use the guidelines presented in Fig. 15-1 to characterize solutions with each of the following pH values as either (1) strongly acidic, (2) weakly acidic, (3) weakly basic, or (4) strongly basic.

(a) 11.30 **(b)** 6.20 **(c)** 1.70 **(d)** 8.00

15.28 Give the pOH value that corresponds to each of the following pH values.

pH	pOH
(a) 4.50	______
(b) 7.00	______
(c) 9.10	______
(d) 0.00	______

15.29 Fill in the pH and pOH values for each of the following aqueous solutions.

Concentration	pH	pOH
(a) 0.0010 M HNO_3	______	______
(b) 0.000010 M KOH	______	______
(c) 1.0×10^{-4} M HCl	______	______
(d) 1.0 M NaOH	______	______

15.30 Calculate the pH of each of the following solutions. (Answers are integers.)
(a) 0.63 g HNO_3 in 1.0 L **(b)** 2.8 g KOH in 500 mL
(c) 0.73 g HCl in 20.0 L **(d)** 0.10 g NaOH in 25 L

15.31 Calculate the pH of each of the following solutions. (Answers are nonintegers.)
(a) 0.016 M HCl **(b)** 0.00037 M NaOH
(c) 0.24 M HNO_3 **(d)** 6.6×10^{-6} M KOH

15.32 Calculate the pH of each of the following solutions, which contain the mass of solute indicated in the final total volume. (Answers are nonintegers.)
(a) 2.5 g HBr in 50.0 L **(b)** 0.076 g NaOH in 25 L
(c) 0.22 g HNO_3 in 750 mL **(d)** 4.2 mg KOH in 5.0 L

15.33 Fill in the following table.

pH	pOH	[H⁺]	[OH⁻]
3.27	______	______	______
______	5.60	______	______
______	______	5.8×10^{-6} M	______
______	______	______	7.2×10^{-5} M

SECTION 15.4

15.34 Match each of the following terms with its meaning.

______	**(a)** Arrhenius acid	**(1)** Electron-pair acceptor
______	**(b)** Arrhenius base	**(2)** Electron-pair donor
______	**(c)** Brønsted–Lowry acid	**(3)** Produces hydrogen ions
______	**(d)** Brønsted–Lowry base	**(4)** Produces hydroxide ions
______	**(e)** Lewis acid	**(5)** Proton acceptor
______	**(f)** Lewis base	**(6)** Proton donor

15.35 Write equations to show how each of the following anions can react with water to produce hydroxide ions.
(a) CN^- (as in KCN)
(b) F^- (as in NaF)
(c) $C_2H_3O_2^-$ (as in $NaC_2H_3O_2$)
(d) CO_3^{2-} (as in Na_2CO_3)

15.36 Identify the Brønsted–Lowry acid and base on the reactant side of each of the following equations.
(a) $HBr + C_2H_3O_2^- \rightarrow HC_2H_3O_2 + Br^-$
(b) $H_2O + NO_2^- \rightarrow HNO_2 + OH^-$
(c) $H_2O + HClO_4 \rightarrow H_3O^+ + ClO_4^-$
(d) $HSO_4^- + OH^- \rightarrow H_2O + SO_4^{2-}$

15.37 Fill in the missing conjugate acid or base for each of the following pairs.

	Conjugate acid / Conjugate base		Conjugate acid / Conjugate base
(a)	HNO_3 / ______	**(b)**	______ / NH_3
(c)	H_2SO_4 / ______	**(d)**	HCO_3^- / ______
(e)	______ / HPO_4^{2-}	**(f)**	______ / HCO_3^-

15.38 Answer the following questions about the $HC_2O_4^-$ ion:
(a) What is the conjugate acid of $HC_2O_4^-$?
(b) What is the conjugate base of $HC_2O_4^-$?

15.39 Identify the Lewis acid and base on the reactant side of each of the following equations:
(a) $H_3N{:} + AlCl_3 \rightarrow H_3N{:}AlCl_3$
(b) $Ag^+ + 2{:}NH_3 \rightarrow [H_3N{:}Ag{:}NH_3]^+$
(c) $Cu^+ + 2{:}CN^- \rightarrow [NC{:}Cu{:}CN]^-$
(d) $Pd^{2+} + 4{:}\ddot{\underset{..}{Cl}}{:}^- \rightarrow [PdCl_4]^{2-}$ (Lewis structure: :Cl: above and below, :Cl:Pd:Cl: across)

SECTION 15.5

15.40 Answer the following about weak acids:
(a) What is a weak acid?
(b) How does a weak acid differ from a strong acid?
(c) Explain the difference between a solution that is weakly acidic and a solution of a weak acid.

15.41 Use Table 15-2 to arrange the following sets of solutions in order of increasing pH. [*Caution:* Which has the higher pH, a solution that is more strongly acidic or one that is more weakly acidic?]
(a) 0.10 M HF, 0.10 M $HC_2H_3O_2$, 0.10 M $HC_7H_5O_2$
(b) 0.10 M HCN, 0.10 M HCl, 0.10 M HClO
(c) 0.010 M HNO_2, 0.10 M HNO_2, 0.10 M HNO_3

15.42 Use Table 15-2 to predict whether each of the following reactions will produce an appreciable amount of product.
(a) $HCN + C_2H_3O_2^- \xrightarrow{?} HC_2H_3O_2 + CN^-$
(b) $HClO_4 + ClO^- \xrightarrow{?} HClO + ClO_4^-$
(c) $H_2SO_4 + H_2PO_4^- \xrightarrow{?} H_3PO_4 + HSO_4^-$

SECTION 15.6

15.43 Which of the following conjugate-acid/conjugate-base pairs can *not* be used as a buffer? [*Hint:* Use Table 15-2 to see which are derived from strong acids.]
(a) $HClO/ClO^-$ **(b)** H_2SO_4/HSO_4^-
(c) HI/I^- **(d)** HF/F^-

SECTION 15.7

15.44 How many equivalents are in one mole of each of the following?
(a) HI (b) H_2SO_3 (c) KOH
(d) $Ba(OH)_2$ (e) $H_4P_2O_7$

15.45 Calculate the number of equivalents in each of the following.
(a) 0.255 mol $HClO_4$ (b) 0.450 mol H_3PO_4
(c) 0.125 mol $Ca(OH)_2$ (d) 0.735 mol $H_2C_2O_4$

15.46 Calculate the number of moles in each of the following.
(a) 0.275 equiv HI (b) 0.834 equiv H_3PO_4
(c) 0.510 equiv $Al(OH)_3$ (d) 2.44 equiv H_2SO_4

15.47 Calculate the number of equivalents in each of the following.
(a) 4.32 g HNO_3 (b) 1.00 g $H_2C_2O_4$
(c) 0.935 g H_3PO_4 (d) 0.0352 g $Mg(OH)_2$

SECTION 15.8

15.48 Calculate the normality of each of the following solutions.
(a) 5.00 g KOH in 0.150 L
(b) 0.525 g H_2SO_4 in 75.0 mL
(c) 0.115 g $Ca(OH)_2$ in 0.250 L
(d) 7.55 g H_3PO_4 in 1.50 L

15.49 Calculate the number of equivalents in each of the following solutions.
(a) 0.250 L of 1.24 N H_3PO_4
(b) 40.0 mL of 6.00 N NaOH
(c) 375 mL of 0.350 N HCl
(d) 5.0 mL of 12 N H_3PO_4
(e) 75.0 mL of 0.500 N $H_2C_2O_4$

15.50 A 35.0-mL quantity of NaOH is required to neutralize 20.0 mL of 0.215 N HCl. Calculate the normality of the sodium hydroxide solution.

15.51 A 22.0-mL sample of HCl exactly neutralizes 36.5 mL of 0.355 N KOH. What is the normality of the hydrochloric acid solution?

15.52 A 12.5-mL sample of 0.344 N NaOH is found to neutralize 17.4 mL of H_2SO_4. Calculate the normality of the sulfuric acid solution.

GENERAL PROBLEMS

15.53 What would be the pH of a neutral solution if the value of K_w was 1.0×10^{-16}?

15.54 Use the percent dissociation given in Table 15-2 to calculate the pH of 0.10 M HCN.

15.55 Blood is carried to and from the lining of the lungs, where gas exchange occurs. Would you expect the pH of the blood to be higher before it enters the lungs or after it leaves? Explain your answer.

***15.56** A bottle of sodium hydroxide is standardized and placed in the laboratory. Over a period of several months, the apparent concentration of sodium hydroxide drops. Account for this observation.

15.57 In Chapter 14, you were introduced to phenolphthalein and litmus, two acid–base indicators. An acid–base indicator changes colors over a pH range that is characteristic of the indicator. The accompanying table gives the colors exhibited by a number of indicators over a wide pH range. Use the information in the table to answer the following questions.
(a) What is the pH of a solution that turns orange in the presence of Methyl Red?
(b) What is the pH of a solution that turns orange in the presence of Methyl Orange?
(c) What is the pH of a solution that turns yellow in the presence of either Thymol Blue, Methyl Red, or Bromthymol Blue?
(d) Is it true that a solution that turns red in the presence of Methyl Orange is acidic? Explain your answer.
(e) Is it true that a solution that turns yellow in the presence of Methyl Orange is basic? Explain your answer.
(f) Is it possible for a single solution to turn all of the indicators in the table yellow? Explain your answer.

	pH										
Indicator	1	2	3	4	5	6	7	8	9	11	13
Bromcresol Green	y	y	y	y-g	b-g	b	b	b	b	b	b
Bromthymol Blue	y	y	y	y	y	y	g	b	b	b	b
Methyl Orange	r	r	r	o	y	y	y	y	y	y	y
Methyl Red	r	r	r	r	o	y	y	y	y	y	y
Thymol Blue	r	o	y	y	y	y	y	g	b	b	b

Key: y = yellow, g = green, b = blue, r = red, o = orange.

15.58 The normality of a solution can be calculated from its molarity by using the relationship between moles and equivalents to convert the units of "moles per liter" to "equivalents per liter." For example,

$$0.125\text{ M HCl} = \left(\frac{0.125\text{ }\cancel{\text{mol HCl}}}{\text{L}}\right)\left(\frac{1\text{ equiv HCl}}{1\text{ }\cancel{\text{mol HCl}}}\right) = \frac{0.125\text{ equiv HCl}}{\text{L}} = 0.125\text{ N HCl}$$

The molarity can be calculated from the normality in a similar fashion. Use this method to carry out the following conversions.

(a) 0.240 M H_2SO_4 = ___?___ N H_2SO_4
(b) 0.015 M $Ca(OH)_2$ = ___?___ N $Ca(OH)_2$
(c) ___?___ M HNO_3 = 0.350 N HNO_3
(d) ___?___ M H_3PO_4 = 0.420 N H_3PO_4

15.59 The concentrations of ions in solutions of soluble salts can be determined in the same fashion as that described in Section 15.1. Determine the ionic concentrations of the ions in each of the following solutions. Part (a) is done for you.

(a) 0.25 M $Cu(NO_3)_2$
$[Cu^{2+}]$ = 0.25 M $[NO_3^-]$ = 0.50 M

(b) 0.35 M $MgBr_2$
$[Mg^{2+}]$ = ______ $[Br^-]$ = ______

(c) 0.12 M K_3PO_4
$[K^+]$ = ______ $[PO_4^{3-}]$ = ______

(d) 0.22 M $Al_2(SO_4)_3$
$[Al^{3+}]$ = ______ $[SO_4^{2-}]$ = ______

***15.60** In Sections 15.7 and 15.8, we restricted our discussion of equivalents and normality to complete neutralizations. When a partial neutralization is carried out on a polyprotic acid, the number of equivalents per mole is equal to the number of ionizable hydrogens neutralized. Suppose a titration is carried out using the following partial neutralization of phosphoric acid:

$$H_3PO_4 + 2\text{ KOH} \rightarrow K_2HPO_4 + 2\text{ }H_2O$$

(a) For this reaction, what is the number of equivalents per mole of phosphoric acid?
(b) What is the normality of a solution that contains 5.78 g of H_3PO_4 in 0.500 L of solution?
(c) If 25.4 mL of this solution neutralizes 16.8 mL of KOH solution via the partial neutralization above, what is the normality of the KOH solution?

WRITING EXERCISES

15.61 You are out to lunch with some friends, when the subject of pH-balanced cosmetic products comes up in conversation. It turns out that you are the only one at the table who has studied any chemistry, and suddenly everyone is counting on you to explain pH. In your own words, explain what the pH scale is, what it is used for, and give several examples of familiar substances or products (such as coffee or shampoo) and their typical pH values. Be sure to explain why someone would be concerned with the pH of the cosmetic products they use.

15.62 A friend in your chemistry class has asked you to explain equivalents and normality to her. She is not interested in a lot of mathematics. She just needs some help understanding why chemists define these terms, what they mean, and how they are used. In your own words, write an explanation covering these points, and explain how these concepts can be used to simplify certain acid–base calculations.

15.63 Compare and contrast the Arrhenius, Brønsted–Lowry, and Lewis definitions of an acid and a base. Your discussion should be descriptive, explaining each definition and the differences among the three theories without resorting to any chemical equations. You may, however, use the formulas of substances to illustrate your ideas.

OXIDATION–REDUCTION

In this chapter we consider a class of reactions known as oxidation–reduction reactions. These reactions involve the transfer of electrons from one substance to another. In general, when such a transfer occurs, one substance loses electrons and another gains them. In fact, many of the reactions we have previously encountered are oxidation–reduction reactions, although we were not specific about the details of their electron transfer. For example, the combustion of hydrocarbons with oxygen is an example of oxidation–reduction. Let us begin by defining the terms we use to describe these processes.

16.1 Oxidation and Reduction

OBJECTIVE 16.1 Define the terms *oxidation* and *reduction.* Explain why these processes must accompany one another. Explain what a half-reaction is. Explain what oxidizing and reducing agents are.

Oxidation: A loss of electrons.

Reduction: A gain of electrons.

We define **oxidation** as a loss of electrons and **reduction** as a gain of electrons. In an oxidation–reduction reaction, the substance that loses electrons is said to be oxidized, while the substance that gains those electrons is reduced. Because the electrons lost in oxidation must go somewhere, it should not be surprising that oxidation is always accompanied by reduction, and vice versa. The term *oxidation–reduction* emphasizes the interdependence of these two processes. The reaction of metallic zinc with hydrochloric acid belongs to this class of reaction.

Total molecular equation: $Zn + 2\,HCl \rightarrow ZnCl_2 + H_2$

Total ionic equation: $Zn + 2\,H^+ + 2\,Cl^- \rightarrow Zn^{2+} + 2\,Cl^- + H_2$

Net ionic equation: $Zn + 2\,H^+ \rightarrow Zn^{2+} + H_2$

We have omitted state symbols from these equations in order to simplify our examination of the electron-transfer process. In general, we will continue this practice throughout the chapter, except when we want to emphasize some detail relative to the physical state of the substance.

Let us examine the net ionic equation more carefully, considering separately the chemical transformation of each species. Zinc is undergoing a change from its neutral (or "uncharged") elemental state (on the left-hand side) to an ion bearing a +2 charge (on the right-hand side). To accomplish this, zinc must *lose* two electrons:

$$\underset{\begin{pmatrix}30\,p^+\\30\,e^-\end{pmatrix}}{Zn} \rightarrow \underset{\begin{pmatrix}30\,p^+\\28\,e^-\end{pmatrix}}{Zn^{2+}} + \underset{(2\,e^-)}{2\,e^-} \qquad \textbf{Oxidation}$$

On the other hand, two hydrogen ions (on the left-hand side) are *gaining* a pair of electrons to form molecular hydrogen, H_2:

$$\underset{2\begin{pmatrix}1\,p^+\\0\,e^-\end{pmatrix}}{2\,H^+} + \underset{(2\,e^-)}{2\,e^-} \rightarrow \underset{2\begin{pmatrix}1\,p^+\\1\,e^-\end{pmatrix}}{H_2} \qquad \textbf{Reduction}$$

Note that the number of electrons lost in oxidation is equal to the number of electrons gained in reduction:

$$\begin{array}{rcll} Zn & \rightarrow & Zn^{2+} + 2\,e^- & \textbf{Oxidation} \\ 2\,H^+ + 2\,e^- & \rightarrow & H_2 & \textbf{Reduction} \\ \hline 2\,H^+ + Zn & \rightarrow & Zn^{2+} + H_2 & \end{array}$$

If this were not so, the reaction would involve the creation or destruction of electrons, and that is not possible.

One way to keep from confusing oxidation and reduction is to remember that the substance undergoing reduction is accompanied by a reduction in charge (from +1 to 0 in the case of hydrogen). Thus, the charges on species undergoing oxidation and reduction occur in the following directions:

Oxidation (loss of electrons) →

−7 −6 −5 −4 −3 −2 −1 0 +1 +2 +3 +4 +5 +6 +7

← **Reduction (gain of electrons)**

Let us consider another simple oxidation–reduction reaction, the reaction of silver nitrate, $AgNO_3$, with metallic copper, Cu. When a piece of copper wire is placed in a silver nitrate solution, metallic silver, Ag, forms, accompanied by the production of copper(II) nitrate, $Cu(NO_3)_2$. This reaction is a single-replacement reaction. The net ionic equation shows that the reaction that takes place is between metallic copper and silver ion.

Total molecular equation: $2\,AgNO_3 + Cu \rightarrow 2\,Ag + Cu(NO_3)_2$
Total ionic equation: $2\,Ag^+ + 2\,NO_3^- + Cu \rightarrow 2\,Ag + Cu^{2+} + 2\,NO_3^-$
Net ionic equation: $2\,Ag^+ + Cu \rightarrow 2\,Ag + Cu^{2+}$

Once again, we can see that one of the elements is losing electrons while the other is gaining them:

$$\begin{array}{rcll} Cu & \rightarrow & Cu^{2+} + 2\,e^- & \textbf{Oxidation} \\ 2\,Ag^+ + 2\,e^- & \rightarrow & 2\,Ag & \textbf{Reduction} \\ \hline 2\,Ag^+ + Cu & \rightarrow & 2\,Ag + Cu^{2+} & \end{array}$$

Furthermore, because each silver ion, Ag^+, requires only one electron to become neutral silver metal, Ag, it is necessary to take two silver ions for every copper atom. This way the number of electrons gained in reduction equals the number of electrons lost in oxidation. This principle will guide us through the sections on balancing oxidation–reduction reactions.

Half-reaction or **half-cell:** Either the oxidation or the reduction portion of an oxidation–reduction reaction.

In each of the preceding examples, we were able to separate the equation into two halves: the oxidation reaction and the reduction reaction. We refer to each of these as a **half-reaction.** (Sometimes the term **half-cell** is used instead.) Every oxidation–reduction reaction is made up of two half-reactions: one for oxidation and one for reduction.

Returning to the reaction of zinc with hydrogen ion, we can define two more terms, the oxidizing agent and the reducing agent:

$$\underset{\text{Reducing agent}}{Zn} + \underset{\text{Oxidizing agent}}{2\,H^+} \rightarrow Zn^{2+} + H_2$$

During the course of this reaction, zinc is oxidized; its electrons are taken away by the hydrogen ions. Hydrogen ions remove the electrons, so we may think of them as the agents causing the oxidation. Thus, we refer to hydrogen ion as the oxidizing agent in this reaction. On the other hand, we might wish to concentrate on the reduction of hydrogen ions to molecular hydrogen. The metallic zinc donates the electrons that bring about the reduction, so we refer to zinc as the reducing agent. The **oxidizing agent** is the substance that causes oxidation. The **reducing agent** causes reduction to occur. It follows that the oxidizing agent is always the substance undergoing reduction, and the reducing agent is the substance undergoing oxidation. For the reaction of silver ion with metallic copper,

Oxidizing agent: A substance that causes oxidation.

Reducing agent: A substance that causes reduction.

$$\underset{\text{Oxidizing agent}}{2\,Ag^+} + \underset{\text{Reducing agent}}{Cu} \rightarrow 2\,Ag + Cu^{2+}$$

Silver ion is the oxidizing agent and elemental copper is the reducing agent.

PROBLEM 16.1

Identify each of the following half-reactions as oxidation or reduction.

(a) $Ag \rightarrow Ag^+ + e^-$ **(b)** $S + 2\,e^- \rightarrow S^{2-}$
(c) $Fe^{2+} \rightarrow Fe^{3+} + e^-$ **(d)** $Br_2 + 2\,e^- \rightarrow 2\,Br^-$

PROBLEM 16.2

Define the following terms.

(a) Oxidation **(b)** Reduction **(c)** Half-reaction
(d) Oxidizing agent **(e)** Reducing agent

PROBLEM 16.3

Identify the oxidizing agent and the reducing agent in each of the following reactions. [*Remember:* The oxidizing agent is the substance that is reduced, while the reducing agent is the substance that is oxidized.]

(a) $Mg + Cu^{2+} \rightarrow Mg^{2+} + Cu$ **(b)** $Br_2 + 2\,I^- \rightarrow I_2 + 2\,Br^-$
(c) $2\,Na + Cl_2 \rightarrow 2\,Na^+ + 2\,Cl^-$

16.2 OXIDATION NUMBERS

OBJECTIVE 16.2 Define the term *oxidation number.* Assign an oxidation number to each atom in an element, compound, or ion.

Oxidation number: A charge assigned to an atom; used to keep track of electrons during oxidation–reduction reactions.

You may recall from our discussion of nomenclature (Sections 9.4 and 9.6) that the **oxidation number** for any metallic ion is simply the charge on that ion. Thus, iron(II) oxide, FeO, is so-named because it contains iron(II) ions: Fe^{2+}. On the other hand, iron(III) oxide, Fe_2O_3, contains iron(III) ions: Fe^{3+}. In our discussion of nomenclature, we gave primary attention to the oxidation numbers of monatomic metal ions, such as copper(II), Cu^{2+}. However, chemists do not restrict their assignment of oxidation numbers to such metallic ions. Instead, we may assign oxidation numbers to each element in any compound or ion. In this section we shall present the rules that enable us to make these assignments.

As you will see, oxidation numbers are very helpful in determining whether a given chemical reaction is an oxidation–reduction reaction, and in balancing certain chemical equations that are difficult to balance using the techniques of Section 10.2. The function of assigning oxidation numbers is simply one of bookkeeping. You can keep track of electrons gained and lost—and after all, the transfer of electrons from one element to another is what oxidation–reduction is all about.

With these remarks in mind, let us examine the set of rules given in the Summary for assigning oxidation numbers. *The list of rules is arranged so that each rule takes precedence over the rules that follow it.* There is a guiding principle that governs the assignment of all oxidation numbers. We shall call this principle the zeroth rule (Rule 0).

SUMMARY Important Rules for Assigning Oxidation Numbers

0. *The sum of the oxidation numbers of the atoms in any molecule or ion must equal the charge on the molecule or ion.* This means that, for an uncharged molecular substance such as water, H_2O, the sum of the oxidation numbers must equal zero. By contrast, because the charge on a sulfate ion, SO_4^{2-}, is –2, the sum of its oxidation numbers must be –2. In view of the fact that an oxidation number is just the charge on each atom in a molecule or ion, this principle makes considerable sense.
1. *All neutral elemental substances are assigned an oxidation number of zero.* Because these are uncharged elements, the zeroth rule dictates that each atom must bear a charge of zero. For example, in molecular hydrogen, H_2, each hydrogen atom has an oxidation number of zero. Similarly, each atom in molecular oxygen, O_2, and molecular iodine, I_2, is assigned an oxidation number of zero, as are metallic zinc, Zn; copper, Cu; and other monatomic elements.
2. *All monatomic ions have an oxidation number equal to their charge.* Thus, the oxidation number of copper(II), Cu^{2+}, is +2. Likewise, the oxidation number of chloride ion, Cl^-, is –1. We used this rule when we assigned oxidation numbers in the chapter on nomenclature.
3. *Hydrogen is assigned an oxidation number of +1.* Thus, hydrogen is assigned a +1 oxidation state in water, H_2O; hydrochloric acid, HCl; and sulfuric acid, H_2SO_4. The exceptions to this rule include molecular hydrogen, H_2, which is assigned an oxidation state of zero. Metallic hydrides, such as sodium hydride, NaH, are another exception. In these compounds hydrogen is treated as a hydride ion, H^-, and is assigned a –1 charge.
4. *Oxygen is assigned an oxidation number of –2.* Thus, in water, H_2O, each hydrogen is assigned a +1 oxidation state and the oxygen is assigned –2. Note that the sum of the oxidation numbers is zero, as the zeroth rule dictates:

$$\begin{array}{c} +1 - 2 + 1 = 0 \\ \text{H—O—H} \end{array}$$

Molecular oxygen is an exception to this rule, because it is a neutral elemental substance (Rule 1). Another exception is hydrogen peroxide, H_2O_2. Rule 3 (which takes precedence over Rule 4) requires that hydrogen have an oxidation number of +1. If the oxidation numbers are to add up to zero (the zeroth rule), each oxygen must be assigned a –1 charge.

$$\begin{array}{c} +1 - 1 - 1 + 1 = 0 \\ \text{H–O–O–H} \end{array}$$

These rules are usually sufficient to assign oxidation numbers to most compounds. However, the following rules are often helpful in assigning oxidation numbers to certain ionic compounds:

SUMMARY Further Rules for Assigning Oxidation Numbers

5. *In ionic compounds, alkali metals (group IA or 1 of the periodic table) are assigned an oxidation number of +1.*
6. *In ionic compounds, the alkaline earth metals (group IIA or 2 of the periodic table) are assigned an oxidation number of +2.*
7. *Halogens (group VIIA or 17 of the periodic table) are often assigned an oxidation number of −1 when they are combined with metals and not with oxygen.* (Example 16.3 illustrates why this rule does not hold for compounds or polyatomic ions containing oxygen.)

In assigning oxidation numbers, it is often helpful to write the oxidation number of *each* element above the proper atom in the formula and to tabulate sums below the element. The following examples illustrate a good method for assigning oxidation numbers.

EXAMPLE 16.1

Assign an oxidation number to each element in H_2S.

Planning the solution

This is a neutral molecule, so the sum of the oxidation numbers must be zero (Rule 0).

SOLUTION

The first element we can assign is hydrogen, which has an oxidation number of +1 (Rule 3).

Oxidation number:	+1 ?
	H_2S
Sum of oxidation numbers:	+2 +? = 0

Because there are two hydrogen atoms and each has a charge of +1, the two hydrogen atoms carry a +2 charge. For the charges to add up to zero, sulfur must have an oxidation number of −2.

Oxidation number:	+1 −2
	H_2S
Sum of oxidation numbers:	+2 −2 = 0

EXAMPLE 16.2

Assign an oxidation number to magnesium, Mg.

SOLUTION

This is a neutral elemental substance; the oxidation number is zero (Rule 1).

Oxidation number:	0
	Mg
Sum of oxidation numbers:	0 = 0

EXAMPLE 16.3

Assign an oxidation number to each element in iodate ion, IO_3^-.

SOLUTION

The sum of the oxidation numbers is −1 (Rule 0). Oxygen has an oxidation number of −2 (Rule 4). There are three oxygen atoms, for a total negative charge of −6.

Oxidation number:	? −2
	IO_3^-
Sum of oxidation numbers:	? −6 = −1

Thus, iodine must have an oxidation number of +5.

Oxidation number:	+5 −2
	IO_3^-
Sum of oxidation numbers:	+5 −6 = −1

This example demonstrates why the oxidation number of a halogen is not −1 when oxygen is present (Rule 7).

EXAMPLE 16.4

Assign an oxidation number to each element in dichromate ion, $Cr_2O_7^{2-}$.

SOLUTION

The sum of the oxidation numbers is −2 (Rule 0). Each oxygen has an oxidation number of −2 (Rule 4), so the seven oxygen atoms have a charge of −14.

Oxidation number:	? −2
	$Cr_2O_7^{2-}$
Sum of oxidation numbers:	? −14 = −2

Because the two chromium atoms must have a total charge of +12, the oxidation number of each chromium atom is +6.

Oxidation number:	+6 −2
	$Cr_2O_7^{2-}$
Sum of oxidation numbers:	+12−14 = −2

EXAMPLE 16.5

Assign an oxidation number to each element in $KMnO_4$.

SOLUTION

The sum of the oxidation numbers is zero (Rule 0). Oxygen has an oxidation number of −2 (Rule 4).

Oxidation number:	? ? −2
	$KMnO_4$
Sum of oxidation numbers:	? ? −8 = 0

Potassium has an oxidation number of +1 (Rule 5).

Oxidation number:	+1 ? −2
	$KMnO_4$
Sum of oxidation numbers:	+1 ? −8 = 0

Then manganese must have an oxidation number of +7.

Oxidation number:	+1 +7 −2
	$KMnO_4$
Sum of oxidation numbers:	+1 +7 −8 = 0

EXAMPLE 16.6

Assign an oxidation number to each element of $Cu(NO_3)_2$.

SOLUTION

When we deal with compounds containing polyatomic ions, we must treat the compound as separate ions. In this example, $Cu(NO_3)_2$ is made up of a copper ion and two nitrate ions:

$$Cu^{?} + 2\,NO_3^-$$

Because the compound is neutral and two nitrate ions carry a total charge of −2, copper must be in the +2 state:

$$Cu^{2+} + 2\,NO_3^- = Cu(NO_3)_2$$

Assigning the oxidation numbers to the nitrate ion, we find that oxygen is −2 and nitrogen is +5.

$$\begin{array}{rl} \textbf{Oxidation number:} & \overset{+5\ -2}{NO_3^-} \\ \textbf{Sum of oxidation numbers:} & +5\ -6 = -1 \end{array}$$

Finally, assembling the oxidation numbers determined for the individual ions allows us to summarize the compound as a whole.

$$\textbf{Oxidation number:}\quad \overset{+2\ +5-2}{Cu(NO_3)_2}$$

The Summary Diagram presents a simplified flowchart for assigning oxidation numbers.

PROBLEM 16.4

Assign an oxidation number to each element in the following molecules or ions.
(a) H_2 **(b)** Al^{3+} **(c)** H_3PO_4 **(d)** NO **(e)** N_2O_5
(f) HNO_2 **(g)** CrO_4^{2-} **(h)** MnO_2 **(i)** MnO_4^- **(j)** $C_2O_4^{2-}$

PROBLEM 16.5

Assign an oxidation number to each element in the following.
(a) F_2 **(b)** CO_2 **(c)** BrO_2^- **(d)** KI **(e)** $CuBr_2$
(f) $LiNO_3$ **(g)** Na_2SO_4 **(h)** $Al_2(SO_4)_3$

16.3 IDENTIFYING OXIDATION–REDUCTION REACTIONS

OBJECTIVE 16.3 Assign oxidation numbers to determine whether an unbalanced or balanced equation represents an oxidation–reduction reaction.

Occasionally we are called upon to balance oxidation–reduction equations that are difficult to balance using the methods described in Section 10.2. For example, it would probably take some lucky guessing to balance the following equation using those techniques. (Try it!)

$$Cu + HNO_3 \rightarrow Cu(NO_3)_2 + NO + H_2O$$

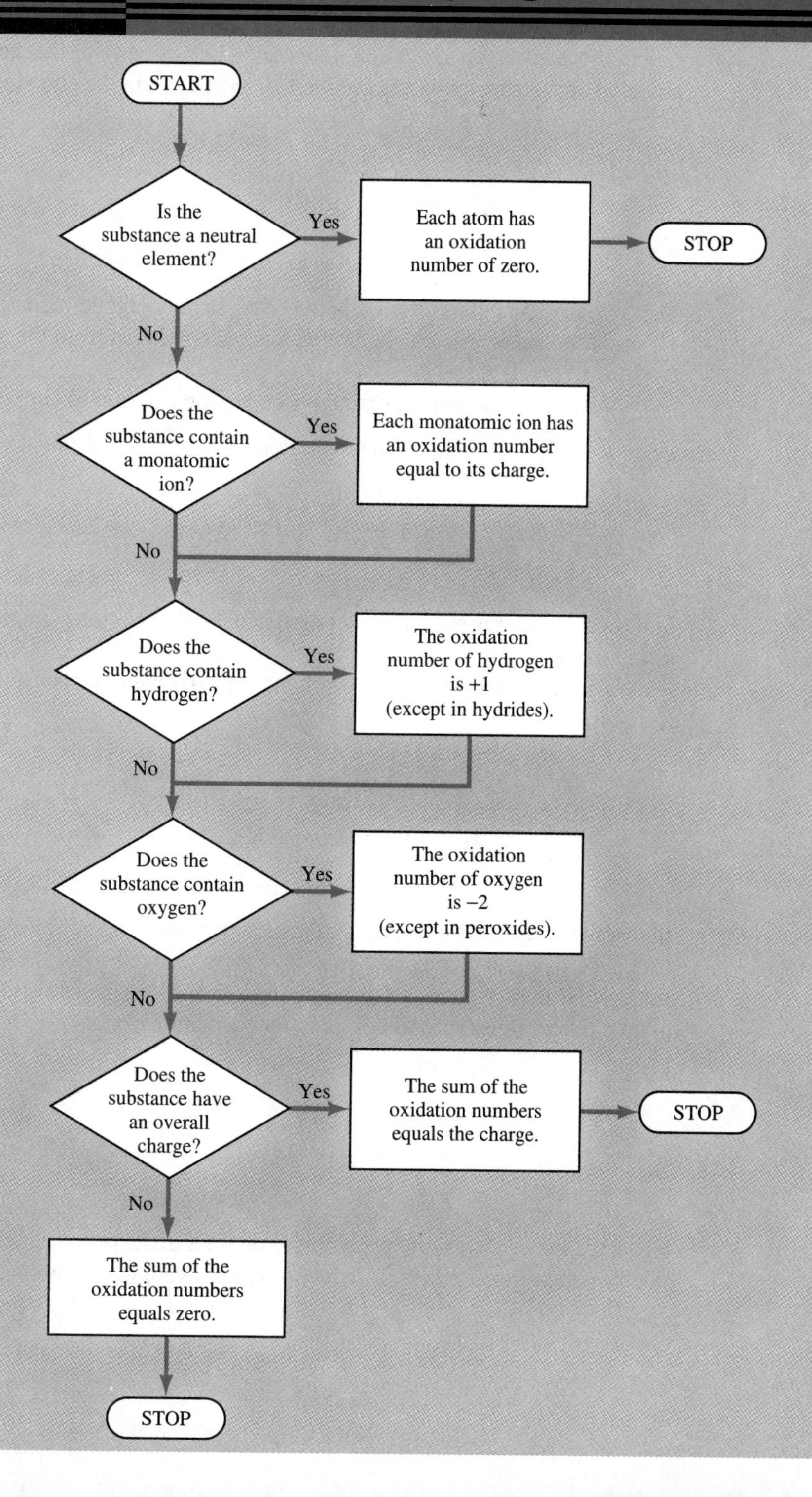
SUMMARY DIAGRAM
Scheme for Assigning Oxidation Numbers
START
Is the substance a neutral element?
Yes
Each atom has an oxidation number of zero.
STOP
No
Does the substance contain a monatomic ion?
Yes
Each monatomic ion has an oxidation number equal to its charge.
No
Does the substance contain hydrogen?
Yes
The oxidation number of hydrogen is +1 (except in hydrides).
No
Does the substance contain oxygen?
Yes
The oxidation number of oxygen is −2 (except in peroxides).
No
Does the substance have an overall charge?
Yes
The sum of the oxidation numbers equals the charge.
STOP
No
The sum of the oxidation numbers equals zero.
STOP

The next section introduces a special technique for balancing oxidation–reduction equations such as this one. First, however, we need to learn how to determine when an unbalanced equation actually involves oxidation–reduction.

In the opening section of this chapter, we identified the reaction of zinc with hydrochloric acid as an oxidation–reduction. Let us rewrite the unbalanced molecular equation and assign oxidation numbers to each atom in the equation:

Each Zn atom loses 2 e⁻ — **Oxidation**

$$\overset{0}{Zn} + \overset{+1-1}{HCl} \rightarrow \overset{+2\,-1}{ZnCl_2} + \overset{0}{H_2} \qquad \textbf{Unbalanced equation}$$

Each H atom gains 1 e⁻ — **Reduction**

Inspection of the equation reveals that zinc has undergone oxidation from the 0 to the +2 oxidation state, whereas hydrogen has been reduced from the +1 to the 0 oxidation state.

We also examined the oxidation–reduction reaction of copper with silver nitrate:

Each Cu atom loses 2 e⁻ — **Oxidation**

$$\overset{+1}{Ag}NO_3 + \overset{0}{Cu} \rightarrow \overset{0}{Ag} + \overset{+2}{Cu}(NO_3)_2 \qquad \textbf{Unbalanced equation}$$

Each Ag atom gains 1 e⁻ — **Reduction**

In this case, silver is reduced from the +1 to the 0 oxidation state, whereas copper is oxidized from the 0 to the +2 oxidation state.

The combustion of hydrocarbons with oxygen is also an oxidation–reduction reaction:

Each C atom loses 8 e⁻ — **Oxidation**

$$\overset{-4}{C}H_4 + \overset{0}{O_2} \rightarrow \overset{+4-2}{CO_2} + H_2\overset{-2}{O} \qquad \textbf{Unbalanced equation}$$

Each O atom gains 2 e⁻ — **Reduction**

In this particular example (the combustion of methane), carbon is oxidized from the −4 to the +4 oxidation state, whereas oxygen is reduced from the 0 to the −2 oxidation state. Note that oxygen is the oxidizing agent. As a general rule, molecular oxygen often oxidizes other substances, as in the formation of rust, Fe_2O_3:

Each Fe atom loses 3 e⁻ — **Oxidation**

$$\overset{0}{Fe} + \overset{0}{O_2} \rightarrow \overset{+3\,-2}{Fe_2O_3} \qquad \textbf{Unbalanced equation}$$

Each O atom gains 2 e⁻ — **Reduction**

We may contrast the preceding oxidation–reduction reactions with the following double-replacement reactions, which involve a simple recombination of ions.

$$\overset{+1+5-2}{AgNO_3(aq)} + \overset{+1\,-1}{NaCl(aq)} \rightarrow \overset{+1\,-1}{AgCl(s)} + \overset{+1+5-2}{NaNO_3(aq)}$$

$$\overset{+1-1}{HCl(aq)} + \overset{+1-2+1}{NaOH(aq)} \rightarrow \overset{+1\,-1}{NaCl(aq)} + \overset{+1-2}{H_2O(\ell)}$$

Assignment of oxidation numbers reveals that each element retains its original oxidation state throughout the course of its particular reaction. In other words, neither oxidation nor reduction takes place in either example. A reaction occurs in each case, but it does not involve a transfer of electrons. Let us determine whether each of the unbalanced equations in the following examples involves oxidation–reduction.

EXAMPLE 16.7

Assign an oxidation number to each element in the following unbalanced equation, and determine whether it represents an oxidation–reduction reaction.

Planning the solution

We will assign an oxidation number to each element on both sides of the equation. If any of the elements change oxidation state, it is an oxidation–reduction reaction.

SOLUTION

$$\overset{+1+5-2}{AgNO_3} + \overset{0}{Zn} \rightarrow \overset{+2\ +5-2}{Zn(NO_3)_2} + \overset{0}{Ag} \qquad \textbf{Unbalanced equation}$$

Silver is being reduced from the +1 to the 0 oxidation state, whereas zinc is being oxidized from the 0 to the +2 oxidation state. Thus, this is an oxidation–reduction reaction.

EXAMPLE 16.8

Determine whether the following reaction involves oxidation–reduction.

SOLUTION

$$\overset{+1+7\ -2}{HMnO_4} + \overset{+1+3-2}{H_2C_2O_4} \rightarrow \overset{+4\ -2}{MnO_2} + \overset{+4-2}{CO_2} + \overset{+1-2}{H_2O} \qquad \textbf{Unbalanced equation}$$

Inspection of the oxidation states shows that manganese is changing from a +7 to a +4 oxidation state (a reduction because the oxidation number is being reduced), whereas carbon is changing from a +3 to a +4 oxidation state (oxidation). This is an oxidation–reduction reaction.

EXAMPLE 16.9

Is the following an oxidation–reduction reaction?

SOLUTION

$$\overset{+2\ -1}{BaCl_2(aq)} + \overset{+1+6-2}{H_2SO_4(aq)} \rightarrow \overset{+2+6-2}{BaSO_4(s)} + \overset{+1-1}{HCl(aq)} \qquad \textbf{Unbalanced equation}$$

No oxidation states change during this precipitation reaction. It is not an oxidation–reduction reaction.

PROBLEM 16.6

Determine which of the following unbalanced equations involve oxidation–reduction.

(a) $Zn + H_2SO_4 \rightarrow ZnSO_4 + H_2$ **(b)** $KI + Pb(NO_3)_2 \rightarrow PbI_2 + KNO_3$

(c) $Mn + HCl \rightarrow MnCl_2 + H_2$ **(d)** $HgO + HNO_3 \rightarrow Hg(NO_3)_2 + H_2O$

PROBLEM 16.7

Assign an oxidation number to each atom in the following reaction of copper with nitric acid:

$$Cu + HNO_3 \rightarrow Cu(NO_3)_2 + NO + H_2O \qquad \textbf{Unbalanced equation}$$

Which element is undergoing oxidation? Which element is undergoing reduction? What do you notice about nitrogen? (We will discuss this reaction in the next section.)

16.4 USING OXIDATION NUMBERS TO BALANCE EQUATIONS

OBJECTIVE 16.4 Use the oxidation-number method to balance oxidation–reduction equations.

There are two general methods for balancing oxidation–reduction equations: (1) the oxidation-number method and (2) the ion–electron method. Both methods utilize the same chemical principles, and either can be used successfully to balance oxidation–reduction equations. The first method is presented here, and the second method is presented in the next section. You are urged to try them both. Then, if one method seems easier than the other, feel free to adopt it.

Both methods are based on a fundamental principle of all oxidation–reduction reactions. When the equation is finally balanced, *the number of electrons lost by the substance being oxidized must equal the number of electrons gained by the substance being reduced.* Because all oxidation–reduction reactions involve a transfer of electrons from one substance to another, that principle holds for every oxidation–reduction reaction.

There are two general ways in which an unbalanced oxidation–reduction equation can be presented:

1. The unbalanced equation may express the reactants and products in terms of their molecular formulas, as in a total molecular equation:

$$Zn + HCl \rightarrow ZnCl_2 + H_2 \qquad \textbf{Unbalanced equation}$$

2. The unbalanced equation may be an unbalanced net ionic equation, which eliminates any spectator ions:

$$Zn + H^+ \rightarrow Zn^{2+} + H_2 \qquad \textbf{Unbalanced equation}$$

We shall balance equations of the first type by the *oxidation-number method* and equations of the second type by the *ion–electron method.*

To balance an equation using the oxidation-number method, we begin by assigning oxidation numbers to the elements. Once we have identified the species undergoing oxidation and reduction, we can proceed with balancing the equation. We shall use the following example to illustrate:

$$C + H_2SO_4 \rightarrow CO_2 + SO_2 + H_2O \qquad \textbf{Unbalanced equation}$$

Assignment of oxidation numbers reveals that carbon and sulfur are changing oxidation states during the course of this reaction:

Loss of 4 e⁻ — **Oxidation**

$$\overset{0}{C} + H_2\overset{+6}{S}O_4 \rightarrow \overset{+4}{C}O_2 + \overset{+4}{S}O_2 + H_2O$$

Gain of 2 e⁻ — **Reduction**

Because carbon is changing from a 0 to a +4 oxidation state, each carbon atom must lose four electrons. Similarly, in changing from a +6 to a +4 oxidation state, each sulfur atom must gain two electrons. If the number of electrons gained by sulfur is to equal the number lost by carbon, there need to be *two* sulfur atoms for each carbon atom. We transfer this information to the unbalanced equation by assigning a coefficient of 1 to each substance containing carbon and a coefficient of 2 to each substance containing sulfur on both sides of the equation:

Loss of 4 e⁻ — **Oxidation**

$$\overset{0}{1\,C} + 2\,H_2\overset{+6}{S}O_4 \rightarrow 1\,\overset{+4}{C}O_2 + 2\,\overset{+4}{S}O_2 + ?\,H_2O$$

2(Gain of 2 e⁻) = Gain of 4 e⁻ — **Reduction**

Having balanced the electrons being transferred, we can balance the water by the usual method of counting atoms. There are four hydrogen atoms on the left, so we need two water molecules on the right:

$$C + 2\,H_2SO_4 \rightarrow CO_2 + 2\,SO_2 + 2\,H_2O$$

Checking the final result, we find that every element is balanced, including the oxygen, which has eight atoms on each side.

In balancing this example, we followed several steps, which we list in the Summary.

EXAMPLE 16.10

Balance the following equation, using the oxidation-number method.

Planning the solution

We follow the steps outlined in the Summary.

SOLUTION

$$Na_2\overset{-2}{S} + HCl + H\overset{+5}{N}O_3 \rightarrow NaCl + \overset{0}{S} + \overset{+2}{N}O + H_2O$$

Assignment of oxidation numbers (Step 1) reveals that sulfur is changing from the −2 to the 0 oxidation state, whereas nitrogen is changing from +5 to the +2 oxidation state:

SUMMARY: The Oxidation-Number Method for Balancing Equations

Step 1. Assign oxidation numbers to each element, and determine which substances are changing their oxidation state.

Step 2. Determine the number of electrons lost by each atom undergoing oxidation and the number of electrons gained by each atom being reduced.

Step 3. Balance the number of electrons lost in oxidation with those gained in reduction by adjusting the coefficients of the substances being oxidized and reduced.

Step 4. Balance the remaining substances by the usual method of counting atoms.

Step 5. Check the final equation to be sure each element is balanced.

Loss of 2 e⁻ — Oxidation

$$\overset{-2}{Na_2S} + HCl + \overset{+5}{HNO_3} \rightarrow NaCl + \overset{0}{S} + \overset{+2}{NO} + H_2O$$

Gain of 3 e⁻ — Reduction

In order to balance the electrons, it is necessary for us to take three atoms of sulfur for every two atoms of nitrogen (Step 3):

3(Loss of 2 e⁻) = Loss of 6 e⁻ — Oxidation

$$3\ \overset{-2}{Na_2S} + ?\ HCl + 2\ \overset{+5}{HNO_3} \rightarrow ?\ NaCl + 3\ \overset{0}{S} + 2\ \overset{+2}{NO} + ?\ H_2O$$

2(Gain of 3 e⁻) = Gain of 6 e⁻ — Reduction

We can now balance the remainder of the substances by the usual method of counting atoms. There are six atoms of sodium on the left, so we must balance the sodium with 6 NaCl on the right. This results in six atoms of chlorine on the right, requiring 6 HCl on the left. Now there are a total of eight hydrogen atoms on the left. To balance this requires 4 H_2O on the right (Step 4).

$$3\ Na_2S + 6\ HCl + 2\ HNO_3 \rightarrow 6\ NaCl + 3\ S + 2\ NO + 4\ H_2O$$

A final check of the elements reveals that oxygen is also balanced, with six atoms on each side (Step 5).

EXAMPLE 16.11

Balance the following equation by the oxidation-number method.

$$\overset{+5}{HNO_3} + \overset{-1}{HI} \rightarrow \overset{+2}{NO} + \overset{0}{I_2} + H_2O$$

SOLUTION

Assignment of oxidation numbers (Step 1) reveals that nitrogen and iodine are involved in the oxidation–reduction:

Each I *atom* loses 1 e⁻ — Oxidation

$$\overset{+5}{HNO_3} + \overset{-1}{HI} \rightarrow \overset{+2}{NO} + \overset{0}{I_2} + H_2O$$

Each N atom gains 3 e⁻ — Reduction

However, we must be careful when we balance iodine! On the right-hand side of the equation, iodine appears as I_2. This means that we must take iodine atoms in pairs if we are to balance the equation successfully (Step 2). We temporarily place a coefficient of 2 in front of HI. Each *pair* of iodine atoms loses *two* electrons:

Loss of 2 e⁻ — Oxidation

$$\overset{+5}{HNO_3} + 2\ \overset{-1}{HI} \rightarrow \overset{+2}{NO} + \overset{0}{I_2} + H_2O$$

Gain of 3 e⁻ — Reduction

Now to balance the electrons, we must take two nitrogen atoms for every three iodine *molecules* (I_2). This requires a coefficient of 6 in front of HI (Step 3):

3(Loss of 2 e⁻) = Loss of 6 e⁻ — **Oxidation**

$$\overset{+5}{2\ HNO_3} + 6\ H\overset{-1}{I} \rightarrow 2\ \overset{+2}{N}O + 3\ \overset{0}{I_2} + ?\ H_2O$$

2(Gain of 3 e⁻) = Gain of 6 e⁻ — **Reduction**

Finally, to balance the hydrogen atoms (eight atoms on each side), we must assign a coefficient of 4 to H_2O:

$$2\ HNO_3 + 6\ HI \rightarrow 2\ NO + 3\ I_2 + 4\ H_2O$$

Checking the final equation reveals that the oxygen is correctly balanced, with six atoms on each side.

EXAMPLE 16.12

Balance the following equation by the oxidation-number method.

SOLUTION

$$Cu + HNO_3 \rightarrow Cu(NO_3)_2 + NO + H_2O$$

Assignment of oxidation numbers (Step 1) reveals that copper and nitrogen are involved in the oxidation–reduction:

Loss of 2 e⁻ — **Oxidation**

$$\overset{0}{Cu} + H\overset{+5}{N}O_3 \rightarrow \overset{+2}{Cu}(\overset{+5}{N}O_3)_2 + \overset{+2}{N}O + H_2O$$

Gain of 3 e⁻ — **Reduction**

However, this particular example involves an additional pitfall if we do not balance it carefully. Nitrogen appears in two different oxidation states on the right. The nitrogen on the left is found exclusively in the +5 state. On the right, however, some of it is in the +5 state and some is in the +2 state. What is happening is very simple. On the left, nitrogen is found exclusively as nitrate ion. Some of the nitrate ions present on the left are not involved in the oxidation–reduction reaction but act as spectator ions instead. These spectator ions are found on the right as part of the copper(II) nitrate. The remaining nitrate ions on the left undergo reduction, being converted to nitrogen monoxide. To balance this oxidation–reduction equation, we begin by balancing the nitrogen involved in the oxidation–reduction with copper. After balancing the electron transfer, we can change the coefficient of the nitric acid to include the additional nitrate ions that are needed as spectators.

Initial balancing of the electron transfer results in the following partially balanced equation:

3(Loss of 2 e⁻) = Loss of 6 e⁻ — **Oxidation**

$$3\ \overset{0}{Cu} + 2\ H\overset{+5}{N}O_3 \rightarrow 3\ \overset{+2}{Cu}(NO_3)_2 + 2\ \overset{+2}{N}O + ?\ H_2O$$

2(Gain of 3 e⁻) = Gain of 6 e⁻ — **Reduction**

The two molecules of nitric acid on the left provide the nitrogen that is changing oxidation state (thereby forming the two molecules of nitrogen monoxide on the right.) However, we also need an additional six molecules of nitric acid to balance the six nitrate ions that are present in the 3 $Cu(NO_3)_2$. Therefore, we change the coefficient of nitric acid from 2 to 8:

$$3\ Cu + 8\ HNO_3 \rightarrow 3\ Cu(NO_3)_2 + 2\ NO + ?\ H_2O$$

Then we finish balancing the equation in the usual fashion.

$$3\ Cu + 8\ HNO_3 \rightarrow 3\ Cu(NO_3)_2 + 2\ NO + 4\ H_2O$$

PROBLEM 16.8

Balance the following equations, using the oxidation-number technique. In each case, identify the elements undergoing oxidation and reduction.

(a) $Sn + HNO_3 \rightarrow SnO_2 + NO_2 + H_2O$
(b) $HBr + H_2SO_4 \rightarrow SO_2 + Br_2 + H_2O$
(c) $HNO_3 + HCl \rightarrow NO + Cl_2 + H_2O$
(d) $Na_2Cr_2O_7 + FeCl_2 + HCl \rightarrow CrCl_3 + NaCl + FeCl_3 + H_2O$
(e) $I_2 + Cl_2 + H_2O \rightarrow HIO_3 + HCl$
(f) $Zn + HNO_3 \rightarrow Zn(NO_3)_2 + NO_2 + H_2O$

16.5 USING THE ION–ELECTRON METHOD TO BALANCE EQUATIONS

OBJECTIVE 16.5 Use the ion–electron method to balance oxidation–reduction equations.

Often, an unbalanced oxidation–reduction equation is presented as its unbalanced net ionic equation (no spectator ions are present). The ion–electron method is particularly useful for balancing such equations, and you may find it easier than the oxidation-number method. (However, the oxidation-number method will work equally well if you prefer that method.)

To balance equations by the ion–electron method, we use the set of rules listed in the Summary. We state the rules first and then illustrate their meaning in the examples that follow.

EXAMPLE 16.13

Balance the following equation.

$$Cu + H^+ + NO_3^- \rightarrow Cu^{2+} + NO + H_2O$$

Planning the solution

We follow the steps outlined in the Summary.

SOLUTION

Step 1. Identify the main elements. Assignment of oxidation numbers reveals that copper is changing from a 0 to a +2 oxidation state, whereas nitrogen is changing from a +5 to a +2 oxidation state.

SUMMARY The Ion–Electron Method for Balancing Equations*

Step 1. Identify the main elements undergoing oxidation and reduction.

Step 2. For each main element, begin a skeleton half-reaction by writing the molecule or ion that contains the element on both sides of the equation.

Step 3. Balance each half-reaction as follows:

(a) Balance the main element.

(b) Balance any unbalanced oxygen by adding water, H_2O, to the appropriate side of the equation.

(c) Balance any unbalanced hydrogen by adding hydrogen ions, H^+, to the appropriate side of the equation.

(d) Balance the charges by adding electrons, e^-, to the appropriate side of the equation. When this step is completed, the half-reaction should be balanced.

Step 4. Multiply each half-reaction by an appropriate coefficient so that the number of electrons lost in oxidation equals the number of electrons gained in reduction.

Step 5. Add the two half-reactions together.

Step 6. Eliminate anything that appears in identical form on both sides of the equation.

Step 7. Check the equation to be sure that the charges balance on both sides.

*These rules apply to reactions carried out in acidic or neutral solution. For simplicity, we will restrict ourselves to such problems and examples in this chapter.

Step 2. *Begin a skeleton half-reaction for each main element:*

$$Cu \rightarrow Cu^{2+}$$
$$NO_3^- \rightarrow NO$$

We have written each species as it appears on both sides of the equation, and we are now ready to balance each half-reaction. In the ion–electron method, it is not necessary for us to analyze the oxidation numbers of individual atoms. Instead, we work with the entire species undergoing oxidation or reduction.

Step 3. *Balance one of the half-reactions:*

$$NO_3^- \rightarrow NO$$

(a) *Balance the main element.* The main element is balanced as is. (There is one nitrogen on each side of the equation.)

(b) *Add H_2O to balance O.* We add two water molecules to the right-hand side, thereby balancing the oxygen:

$$NO_3^- \rightarrow NO + 2\,H_2O$$

(c) *Add H^+ to balance H.* Next we add 4 H^+ to the left-hand side of the equation to balance the hydrogen:

$$4\,H^+ + NO_3^- \rightarrow NO + 2\,H_2O$$

(d) *Add e^- to balance the charge.* At this point the atoms are balanced. However, the total charge on the left is +3 (+4 − 1), whereas the charge on the right side is 0 (0 + 0). To balance the charge, we have to add 3 e^- to the left:

$$3\,e^- + 4\,H^+ + NO_3^- \rightarrow NO + 2\,H_2O$$

When you carry out this step, remember that the charge on an electron is *minus* 1. Thus, we always add electrons to the side that is more positive. In this case the left side (with a charge of +3) is more positive than the right side (with a charge of 0), so we added electrons to the left. The difference between the two charges tells us how many electrons to add.

Everything we did in Steps 3 (a)–(d) simply comprised a technique for arriving at a balanced half-reaction for the reduction of nitrogen. In the process we found that three electrons must be gained in the reduction. The resulting half-reaction is balanced with respect to the atoms present and with respect to charge. Recalling Section 16.4, we need to know how many electrons are involved in each half-reaction in order to balance the electron transfer of the entire reaction. We must now balance the other half-reaction.

Step 3. *Balance the second half-reaction:*

$$Cu \rightarrow Cu^{2+}$$

(a) *Balance the main element.* The main element is balanced as is. (There is one copper atom on each side.)

(b) *Add H_2O to balance O.* There are no oxygen atoms on either side, so the oxygen is already balanced.

(c) *Add H^+ to balance H.* There is no hydrogen on either side, so the hydrogen is balanced.

(d) *Add e^- to balance the charge.* To balance the charge, we have to add 2 e^- to the right-hand side (the more positive side).

$$Cu \rightarrow Cu^{2+} + 2\,e^-$$

Step 4. *Multiply the half-reactions by a ratio that balances the electrons.* We can see that the first half-reaction involved a gain of 3 e^-, whereas the second half-reaction involved a loss of 2 e^-. If the loss of electrons is to equal the gain of electrons, we must multiply the nitrogen-containing half-reaction by 2 and the copper-containing half-reaction by 3:

$$2(3\,e^- + 4\,H^+ + NO_3^- \rightarrow NO + 2\,H_2O) \qquad \textbf{Oxidation}$$

$$3(Cu \rightarrow Cu^{2+} + 2\,e^-) \qquad \textbf{Reduction}$$

Step 5. *Add up the two half-reactions.* After we multiply the half-reactions, adding them gives the following result:

$$\begin{array}{r} 6\,e^- + 8\,H^+ + 2\,NO_3^- \rightarrow 2\,NO + 4\,H_2O \\ 3\,Cu \rightarrow 3\,Cu^{2+} + 6\,e^- \\ \hline 6\,e^- + 8\,H^+ + 2\,NO_3^- + 3\,Cu \rightarrow 2\,NO + 4\,H_2O + 3\,Cu^{2+} + 6\,e^- \end{array}$$

Step 6. *Eliminate anything that appears on both sides of the equation.* Finally, we eliminate the electrons from both sides, leaving the following balanced equation:

$$3\,Cu + 8\,H^+ + 2\,NO_3^- \rightarrow 3\,Cu^{2+} + 2\,NO + 4\,H_2O$$

Step 7. Check the charge balance. A check of the charges reveals that the total charge on the left is the same as that on the right:

$$3(0) + 8(+1) + 2(-1) = +6 = 3(+2) + 2(0) + 4(0)$$

The equation is balanced with respect to charge.

The last step in Example 16.13 is quite important, because it is possible to have an equation that is balanced with respect to atoms but not with respect to charge. For example, the following equation might at first glance appear to be balanced:

$$Ag^+ + Cu \rightarrow Ag + Cu^{2+}$$

However, the charge on the left is +1 and that on the right is +2. If this equation were correctly balanced, an electron would have to be destroyed during the course of the reaction. That, of course, is not possible. Closer examination of this equation reveals the following half-reactions:

$$Ag^+ + e^- \rightarrow Ag$$
$$Cu \rightarrow Cu^{2+} + 2\,e^-$$

To balance the electrons, we must multiply the first half-reaction by 2:

$$2\,Ag^+ + 2\,e^- \rightarrow 2\,Ag$$
$$Cu \rightarrow Cu^{2+} + 2\,e^-$$
$$\overline{2\,Ag^+ + Cu \rightarrow 2\,Ag + Cu^{2+}}$$

EXAMPLE 16.14

Balance the following oxidation–reduction reaction by the ion–electron method.

$$Sn^{2+} + IO_3^- + H^+ \rightarrow Sn^{4+} + I^- + H_2O$$

SOLUTION

Analysis of the oxidation numbers reveals the main elements to be tin and iodine (Steps 1 and 2):

$$Sn^{2+} \rightarrow Sn^{4+}$$
$$IO_3^- \rightarrow I^-$$

Next, we balance each half-reaction (Step 3). For the IO_3^-/I^- half-reaction:

(a) Balance the main element:	$IO_3^- \rightarrow I^-$	**Already balanced**
(b) Add H_2O:	$IO_3^- \rightarrow I^- + 3\,H_2O$	
(c) Add H^+:	$6\,H^+ + IO_3^- \rightarrow I^- + 3\,H_2O$	
(d) Add e^-:	$6\,e^- + 6\,H^+ + IO_3^- \rightarrow I^- + 3\,H_2O$	

For the Sn^{2+}/Sn^{4+} half-reaction:

(a) Balance the main element:	$Sn^{2+} \rightarrow Sn^{4+}$	**Already balanced**
(b) Add H_2O:	$Sn^{2+} \rightarrow Sn^{4+}$	**No O present**
(c) Add H^+:	$Sn^{2+} \rightarrow Sn^{4+}$	**No H present**
(d) Add e^-:	$Sn^{2+} \rightarrow Sn^{4+} + 2\,e^-$	

Balancing the electrons (Step 4) leads to the final balanced equation (Steps 5 and 6):

$$3(Sn^{2+} \rightarrow Sn^{4+} + 2\,e^-)$$
$$1(6\,e^- + IO_3^- + 6\,H^+ \rightarrow I^- + 3\,H_2O)$$
$$3\,Sn^{2+} + IO_3^- + 6\,H^+ \rightarrow 3\,Sn^{4+} + I^- + 3\,H_2O$$

Checking the charges (Step 7) reveals that the equation is balanced with respect to charge:

$$3(+2) + 1(-1) + 6(+1) = +11 = 3(+4) + 1(-1) + 3(0)$$

EXAMPLE 16.15

Balance the following equation.

$$Cr_2O_7^{2-} + AsO_3^{3-} + H^+ \rightarrow Cr^{3+} + AsO_4^{3-} + H_2O$$

SOLUTION

The main elements are chromium and arsenic (Steps 1 and 2):

$$Cr_2O_7^{2-} \rightarrow Cr^{3+}$$
$$AsO_3^{3-} \rightarrow AsO_4^{3-}$$

Next, we balance these half-reactions. For the AsO_3^{3-}/AsO_4^{3-} half-reaction:

(a) Balance the main element: $AsO_3^{3-} \rightarrow AsO_4^{3-}$ **Already balanced**

(b) Add H_2O: $H_2O + AsO_3^{3-} \rightarrow AsO_4^{3-}$

(c) Add H^+: $H_2O + AsO_3^{3-} \rightarrow AsO_4^{3-} + 2\,H^+$

(d) Add e^-: $H_2O + AsO_3^{3-} \rightarrow AsO_4^{3-} + 2\,H^+ + 2\,e^-$

For the $Cr_2O_7^{2-}/Cr^{3+}$ half-reaction:

(a) Balance the main element: $Cr_2O_7^{2-} \rightarrow 2\,Cr^{3+}$

(b) Add H_2O: $Cr_2O_7^{2-} \rightarrow 2\,Cr^{3+} + 7\,H_2O$

(c) Add H^+: $14\,H^+ + Cr_2O_7^{2-} \rightarrow 2\,Cr^{3+} + 7\,H_2O$

(d) Add e^-: $6\,e^- + 14\,H^+ + Cr_2O_7^{2-} \rightarrow 2\,Cr^{3+} + 7\,H_2O$

Note that for the chromium half-reaction, it was necessary to place a coefficient of 2 in front of Cr^{3+}. *Failure to balance the main element is one of the most common causes of error in balancing oxidation–reduction equations!*

Having balanced both half-reactions, we multiply the arsenic-containing half-reaction by 3 (Step 4) and add the two half-reactions (Step 5):

$$1(6\,e^- + Cr_2O_7^{2-} + 14\,H^+ \rightarrow 2\,Cr^{3+} + 7\,H_2O)$$
$$3(AsO_3^{3-} + H_2O \rightarrow AsO_4^{3-} + 2\,H^+ + 2\,e^-)$$
$$6\,e^- + Cr_2O_7^{2-} + 14\,H^+ + 3\,AsO_3^{3-} + 3\,H_2O \rightarrow 2\,Cr^{3+} + 3\,AsO_4^{3-} + 6\,H^+ + 7\,H_2O + 6\,e^-$$

Finally, we can eliminate 6 e^- from both sides, as well as the 3 H_2O and 6 H^+ that appear on both sides of the equation (Step 6). We have the final balanced equation:

$$Cr_2O_7^{2-} + 8\,H^+ + 3\,AsO_3^{3-} \rightarrow 2\,Cr^{3+} + 3\,AsO_4^{3-} + 4\,H_2O$$

A check of the charges shows that the equation is balanced with respect to charge (Step 7):

$$1(-2) + 8(+1) + 3(-3) = -3 = 2(+3) + 3(-3) + 4(0)$$

We could have balanced any of these examples using the oxidation-number method; for instance, in Example 16.14:

Loss of 2 e⁻ — Oxidation

$$\overset{+5}{}\; Sn^{2+} + IO_3^- + H^+ \rightarrow Sn^{4+} + I^- + H_2O$$

Gain of 6 e⁻ — Reduction

Assignment of oxidation numbers shows that tin is losing 2 e^- and that iodine is gaining 6 e^-. To balance the electron transfer, we need three atoms of tin for each iodine atom:

3(Loss of 2 e⁻) = Loss of 6 e⁻ — Oxidation

$$\overset{+5}{}\; 3\,Sn^{2+} + IO_3^- + ?\,H^+ \rightarrow 3\,Sn^{4+} + I^- + ?\,H_2O$$

Gain of 6 e⁻ — Reduction

Finally, to complete the equation requires 3 H_2O to balance the oxygen, followed by 6 H^+ to balance the hydrogen:

$$3\,Sn^{2+} + IO_3^- + 6\,H^+ \rightarrow 3\,Sn^{4+} + I^- + 3\,H_2O$$

To balance molecular equations using the ion–electron method, it is necessary first to express unbalanced equations in ionic form and then to eliminate any spectator ions, giving an unbalanced net ionic equation. Thus, in Example 16.10 (Section 16.4), we would use the ion–electron methods as follows:

Total molecular equation:

$$Na_2S + HCl + HNO_3 \rightarrow NaCl + S + NO + H_2O$$

Total ionic equation (unbalanced):

$$Na^+ + S^{2-} + H^+ + Cl^- + NO_3^- \rightarrow Na^+ + Cl^- + S + NO + H_2O$$

Net ionic equation (unbalanced):

$$S^{2-} + H^+ + NO_3^- \rightarrow S + NO + H_2O$$

Balancing this equation by the ion–electron method leads to the following balanced net ionic equation:

$$3\,S^{2-} + 8\,H^+ + 2\,NO_3^- \rightarrow 3\,S + 2\,NO + 4\,H_2O$$

We can translate this information back to the unbalanced molecular equation by assigning coefficients of 3 to Na_2S and 2 to HNO_3 on the left. On the right we assign 3 to S, 2 to NO, and 4 to H_2O:

$$3\,Na_2S + ?\,HCl + 2\,HNO_3 \rightarrow ?\,NaCl + 3\,S + 2\,NO + 4\,H_2O$$

With these assignments in place, we balance the six sodium atoms on the left with 6 NaCl on the right. Doing so necessitates 6 HCl on the left to balance the chlorine:

$$3\,Na_2S + 6\,HCl + 2\,HNO_3 \rightarrow 6\,NaCl + 3\,S + 2\,NO + 4\,H_2O$$

Note that the 6 HCl + 2 HNO_3 in the balanced molecular equation account for the 8 H^+ from the net ionic equation.

PROBLEM 16.9

Complete and balance each of the following half-reactions, using the ion–electron method. Identify each as an oxidation or a reduction.

(a) $Sn^{2+} \rightarrow Sn^{4+}$ **(b)** $MnO_4^- \rightarrow Mn^{2+}$ **(c)** $C_2O_4^{2-} \rightarrow CO_2$
(d) $I^- \rightarrow I_2$ **(e)** $Cr_2O_7^{2-} \rightarrow Cr^{3+}$

PROBLEM 16.10

Use the ion–electron method to balance each of the following oxidation–reduction reactions. In each case, identify the oxidation half-reaction and the reduction half-reaction.

(a) $Cu + NO_3^- + H^+ \rightarrow Cu^{2+} + NO + H_2O$
(b) $NO_3^- + S^{2-} + H^+ \rightarrow NO + SO_4^{2-} + H_2O$
(c) $MnO_4^- + NO + H^+ \rightarrow Mn^{2+} + NO_3^- + H_2O$
(d) $Ag + Cr_2O_7^{2-} + H^+ \rightarrow Ag^+ + Cr^{3+} + H_2O$
(e) $H_2O_2 + I^- + H^+ \rightarrow I_2 + H_2O$
(f) $NO_3^- + Cl^- + H^+ \rightarrow NO + Cl_2 + H_2O$

16.6 OXIDIZING AND REDUCING AGENTS

OBJECTIVE 16.6 Explain how the oxidizing and reducing agent strengths of the reactants and products determine the spontaneous direction of an oxidation–reduction reaction. Predict the spontaneous direction of any oxidation–reduction reaction that can be constructed from half-reactions in Table 16-1.

If we immerse a piece of zinc metal in a solution containing copper(II) ions, a reaction takes place, and electrons are transferred from zinc to the copper(II) ions:

$$Zn + Cu^{2+} \rightarrow Zn^{2+} + Cu \qquad \textbf{Spontaneous direction}$$

Analysis of the products would show that the reaction proceeds essentially to completion.

On the other hand, if we were to try to carry out the reverse of this reaction by placing a copper wire in a solution of zinc ions, we would find that no appreciable reaction takes place.

$$Zn^{2+} + Cu \rightarrow \text{No reaction} \qquad \textbf{Nonspontaneous direction}$$

We might wonder what factors are responsible for the first reaction occurring rather than the second. We can answer this question in a variety of ways. If we examine the energetics of the reaction, we find that the reaction of zinc with copper(II) ion releases energy and therefore occurs spontaneously (Fig. 16-1). The reverse process requires energy and therefore does not occur spontaneously.

However, we can also explain the favored direction of this reaction by examining the reactants and products in terms of their strengths as oxidizing and reducing agents. Remember that an oxidizing agent is a substance that causes oxidation to occur by taking electrons away from the substance being oxidized. In so doing, the oxidizing agent is itself reduced. Thus, in the previous reaction (spontaneous direction), copper(II) ion

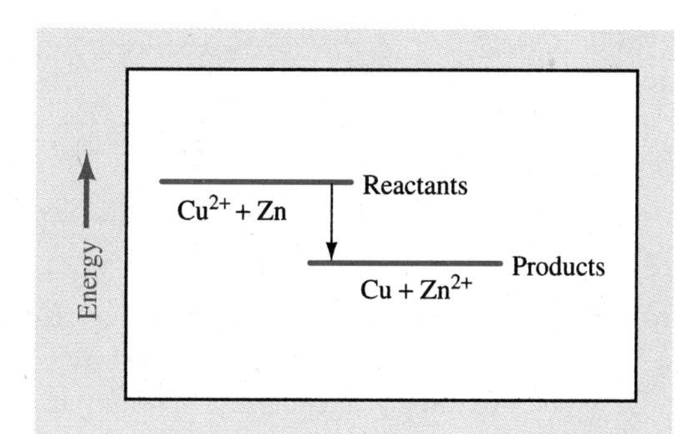

FIGURE 16-1 **Energetics of the oxidation of zinc by copper(II) ion. The spontaneous direction is $Cu^{2+} + Zn \rightarrow Cu + Zn^{2+}$. The reverse reaction is nonspontaneous.**

is the oxidizing agent: It removes electrons from metallic zinc and is itself reduced. On the other hand, we may describe the substance that donates the electrons as the reducing agent. Zinc metal is the reducing agent, as it reduces Cu^{2+} to Cu.

$$\underset{\text{Reducing agent}}{Zn} + \underset{\text{Oxidizing agent}}{Cu^{2+}} \rightarrow Zn^{2+} + Cu$$

To determine the spontaneous direction of an oxidation–reduction reaction, we use tables similar to Table 16-1, which lists a number of simple half-reactions, all written in the direction of reduction from left to right. Reversing the direction of any of the half-reactions in Table 16-1 results in an oxidation half-reaction. (We use paired arrows to indicate the reversibility of these half-reactions.) Thus, the reaction of zinc with cop

TABLE 16-1 Properties of selected half-reactions

		Oxidizing agent	+	Electrons	⇌	Reducing agent	
Potassium	Increasing strength of oxidizing agents (↓)	K^+	+	e^-	⇌	K	Increasing strength of reducing agents (↑)
Barium		Ba^{2+}	+	$2\,e^-$	⇌	Ba	
Calcium		Ca^{2+}	+	$2\,e^-$	⇌	Ca	
Sodium		Na^+	+	e^-	⇌	Na	
Magnesium		Mg^{2+}	+	$2\,e^-$	⇌	Mg	
Aluminum		Al^{3+}	+	$3\,e^-$	⇌	Al	
Manganese		Mn^{2+}	+	$2\,e^-$	⇌	Mn	
Zinc		Zn^{2+}	+	$2\,e^-$	⇌	Zn	
Iron		Fe^{2+}	+	$2\,e^-$	⇌	Fe	
Cadmium		Cd^{2+}	+	$2\,e^-$	⇌	Cd	
Tin		Sn^{2+}	+	$2\,e^-$	⇌	Sn	
Lead		Pb^{2+}	+	$2\,e^-$	⇌	Pb	
Hydrogen		$2\,H^+$	+	$2\,e^-$	⇌	H_2	
Copper		Cu^{2+}	+	$2\,e^-$	⇌	Cu	
Mercury		Hg_2^{2+}	+	$2\,e^-$	⇌	2 Hg	
Silver		Ag^+	+	e^-	⇌	Ag	
Gold		Au^{3+}	+	$3\,e^-$	⇌	Au	

per(II) ion is made up of two of the half-reactions in Table 16-1, with the zinc half-reaction being reversed.

$$\begin{array}{lr} Zn \rightarrow Zn^{2+} + 2\,e^- & \text{Oxidation} \\ Cu^{2+} + 2\,e^- \rightarrow Cu & \text{Reduction} \\ \hline Cu^{2+} + Zn \rightarrow Cu + Zn^{2+} & \end{array}$$

Notice that each of the substances on the left-hand side of the arrow in Table 16-1 is an oxidizing agent capable of gaining electrons during an oxidation–reduction reaction. Each such oxidizing agent has a counterpart on the right-hand side of the arrow that is capable of giving up electrons and thereby acting as a reducing agent. As a result of this relationship between an oxidizing agent and its corresponding reducing agent, all oxidation–reduction reactions have an oxidizing agent and a reducing agent on each side of the arrow. For example, consider again the reaction of metallic zinc with copper(II) ions:

$$\underset{\text{Reducing agent 1}}{Zn} + \underset{\text{Oxidizing agent 2}}{Cu^{2+}} \rightleftharpoons \underset{\text{Oxidizing agent 1}}{Zn^{2+}} + \underset{\text{Reducing agent 2}}{Cu}$$

If the reaction were to proceed in the opposite (nonspontaneous) direction, zinc ion would act as the oxidizing agent, and metallic copper would be the reducing agent.

Every oxidizing or reducing agent has a certain strength to oxidize or reduce another substance. The stronger a given oxidizing agent, the greater its ability to remove electrons from another material. The stronger a reducing agent, the greater its ability to donate electrons to another substance. The oxidizing and reducing agents in the half-reactions in Table 16-1 are listed in order of their strength. Note that the reducing agents increase in strength from bottom to top, while the oxidizing agents increase in strength from top to bottom.

Before proceeding, let us consider why the strengths of oxidizing agents on the left should increase with decreasing strength of their corresponding reducing agents. If a particular reducing agent is quite strong, such as potassium, that reducing agent readily gives up its electrons to other substances. It follows that the corresponding oxidizing agent formed in the process must have little tendency to reacquire those electrons. In other words, the corresponding oxidizing agent is a weak oxidizing agent.

$$\underset{\text{Very strong reducing agent}}{K} \rightleftharpoons \underset{\text{Very weak oxidizing agent}}{K^+} + e^-$$

On the other hand, a substance that is a very strong oxidizing agent, such as gold(III) ion, Au^{3+}, has a very strong tendency to gain electrons. It follows that, once it has acquired electrons to form metallic gold, the corresponding reducing agent (Au) has very little tendency to give up those electrons. In other words, the corresponding reducing agent is a weak reducing agent.

$$\underset{\text{Very strong oxidizing agent}}{Au^{3+}} + 3\,e^- \rightleftharpoons \underset{\text{Very weak reducing agent}}{Au}$$

Table 16-1 tells us that copper(II) ion is a stronger oxidizing agent than zinc ion. At the same time, metallic zinc is a stronger reducing agent than metallic copper. In this reaction, the stronger oxidizing agent reacted with the stronger reducing agent to form the weaker oxidizing agent and the weaker reducing agent.

Zn	+	Cu^{2+}	$\rightarrow$	Zn^{2+}	+	Cu
Stronger reducing agent		Stronger oxidizing agent		Weaker oxidizing agent		Weaker reducing agent

In any oxidation–reduction reaction, the spontaneous direction will always be from the stronger oxidizing and reducing agent to the weaker oxidizing and reducing agent. Thus, metallic zinc reacts with copper(II) ion, but the reverse process (between metallic copper and zinc ion) does not occur spontaneously.

You might wonder whether the stronger oxidizing agent and the stronger reducing agent can ever be on opposite sides of the equation. Examination of Table 16-1 reveals that this can never happen. Because the strengths of oxidizing and reducing agents increase in opposite directions (top-to-bottom for the oxidizing agents and bottom-to-top for the reducing agents), when we turn one of the half-reactions around to create the oxidation half-cell, the stronger oxidizing and reducing agents end up on the same side of the arrow.

EXAMPLE 16.16

Does lead(II) ion react with metallic zinc?

$$Pb^{2+} + Zn \rightarrow ?$$

Planning the solution

Before we can tell whether a reaction occurs, we must complete the chemical equation as though a reaction does occur. We begin by looking for Pb^{2+} in Table 16-1. The Pb^{2+} ion is an oxidizing agent. If a reaction occurs, it must form Pb metal, its corresponding reducing agent. Similarly, Zn is a reducing agent, so it would form Zn^{2+} ion, its corresponding oxidizing agent. Thus, if lead(II) ion reacts with zinc metal, the following reaction must take place:

$$Pb^{2+} + Zn \rightarrow Pb + Zn^{2+}$$

If the stronger oxidizing and reducing agents are on the reactant side, the reaction will occur. If the stronger oxidizing and reducing agents are on the product side, the reaction will not proceed.

SOLUTION

Referring to Table 16-1, we see that metallic zinc is a stronger reducing agent than metallic lead and that lead(II) ion is a stronger oxidizing agent than zinc ion:

Pb^{2+}	+	Zn	$\rightarrow$	Pb	+	Zn^{2+}
Stronger oxidizing agent		Stronger reducing agent		Weaker reducing agent		Weaker oxidizing agent

Consequently, this reaction proceeds as written.

EXAMPLE 16.17

Does metallic tin react with a solution containing aluminum ions?

$$Sn + Al^{3+} \rightarrow ?$$

Planning the solution

Metallic tin, Sn, is a reducing agent; its corresponding oxidizing agent is Sn^{2+} ion. Aluminum ion, Al^{3+}, is an oxidizing agent; its corresponding reducing agent is aluminum metal, Al. If this reaction takes place, the unbalanced equation is

$$Sn + Al^{3+} \rightarrow Sn^{2+} + Al$$

and the balanced equation has to be

$$3\,Sn + 2\,Al^{3+} \rightarrow 3\,Sn^{2+} + 2\,Al$$

SOLUTION

Table 16-1 shows that tin metal is a weaker reducing agent than aluminum metal and that tin(II) ion is a stronger oxidizing agent than aluminum ion:

$3\,Sn$	+ $2\,Al^{3+}$ $\not\rightarrow$	$3\,Sn^{2+}$ +	$2\,Al$
Weaker reducing agent	Weaker oxidizing agent	Stronger oxidizing agent	Stronger reducing agent

Thus, this reaction does not proceed from left to right as written.

EXAMPLE 16.18

Does metallic copper react with zinc?

$$Cu + Zn \rightarrow ?$$

SOLUTION

This question is tricky! Both copper and zinc metal are reducing agents. An oxidation–reduction does not occur between two reducing agents.

EXAMPLE 16.19

Does silver ion react with aluminum ion?

$$Ag^{+} + Al^{3+} \rightarrow ?$$

SOLUTION

No. In this example, both substances are oxidizing agents. An oxidation–reduction reaction requires one oxidizing agent and one reducing agent.

Activity series or **electromotive series:** A series of metals arranged in order of decreasing reactivity with hydrogen ions.

Chemists frequently refer to the arrangement of the metals in Table 16-1 as either an **activity series** or an **electromotive series.** The metals are arranged in order of their ability to reduce hydrogen ions (or to replace them in a single-replacement reaction). Let us examine the location of hydrogen in Table 16-1:

$$K > Ba > Ca > Na > Mg > Al > Mn > Zn > Fe > Cd > Sn > Pb > \mathbf{H_2} > Cu > Hg > Ag > Au$$

Because all the metals above hydrogen reduce hydrogen ion, each of them undergoes reaction in the presence of acid. For example,

$$2\,K + 2\,HCl \rightarrow 2\,KCl + H_2 \qquad \textbf{K metal reduces H}^{+}$$

$$Mg + 2\,HCl \rightarrow MgCl_2 + H_2 \qquad \textbf{Mg metal reduces H}^{+}$$

The most active of these metals (the strongest reducing agents) are so reactive toward hydrogen ion that they react even with water (which has a very low concentration of hydrogen ions, $[H^{+}] = 10^{-7}$ M).

$$2\,Na + 2\,H_2O \rightarrow 2\,NaOH + H_2$$

$$Ba + 2\,H_2O \rightarrow Ba(OH)_2 + H_2$$

The metals located below hydrogen do not reduce hydrogen ion, so they do not undergo reactions of the type just described.

$$Cu + HCl \rightarrow \text{No reaction}$$
$$Ag + HCl \rightarrow \text{No reaction}$$
$$Au + HCl \rightarrow \text{No reaction}$$

In fact, this class of metals includes many of our precious metals, such as silver and gold, which are resistant to reaction with most acids. On the other hand, the reverse of each of these reactions does take place, as in the reduction of silver ion by hydrogen gas.

$$2\,Ag^+ + H_2 \rightarrow 2\,Ag + 2\,H^+$$

PROBLEM 16.11

Identify the oxidizing and reducing agents in each of the following reactions. Then use Table 16-1 to predict whether the reaction occurs in the direction written.

(a) $Ba + Cd^{2+} \rightarrow Ba^{2+} + Cd$ **(b)** $3\,Cu + 2\,Al^{3+} \rightarrow 3\,Cu^{2+} + 2\,Al$
(c) $Mg + 2\,H^+ \rightarrow Mg^{2+} + H_2$ **(d)** $2\,Au + 6\,H^+ \rightarrow 2\,Au^{3+} + 3\,H_2$
(e) $Pb + 2\,H^+ \rightarrow Pb^{2+} + H_2$

16.7 ELECTROCHEMICAL CELLS

OBJECTIVE 16.7 Explain the operation of a voltaic cell and of an electrolytic cell. Identify the anode and the cathode of each. Explain the function of the porous partition. State the overall type of energy change that accompanies each of these types of cells.

Energy Changes

The spontaneous direction of oxidation–reduction reactions represents a downhill process, whereas the nonspontaneous direction is uphill. Let us draw an analogy between this behavior and that of water, which spontaneously flows downhill (Fig. 16-2).

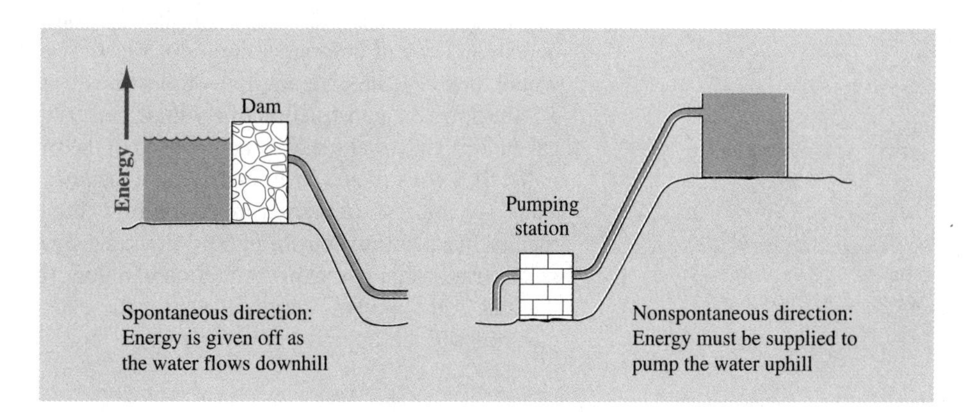

FIGURE 16-2 Spontaneous and nonspontaneous processes.

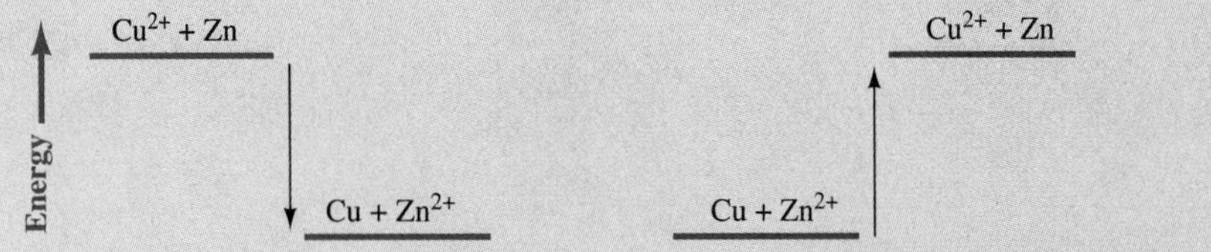

FIGURE 16-3 Electrochemical cells can operate in both the spontaneous direction and the nonspontaneous direction.

Hydroelectric power plants are designed to harness the excess energy given off as water flows downhill, from a higher to a lower potential energy. On the other hand, we often find it desirable to pump water uphill, in the nonspontaneous direction, which is readily accomplished by using water pumps. However, we must supply energy to the pumps in order to raise the water from the lower to the higher potential energy.

It is possible to harness the energy given off when an oxidation–reduction reaction is run in its spontaneous direction. Or we can drive the reaction in its nonspontaneous direction by supplying an external source of energy. Either process may be accomplished through a device known as an **electrochemical cell.** A flashlight battery is one type of electrochemical cell.

An electrochemical cell that is designed to operate in the spontaneous direction is referred to as a **voltaic cell.** An electrochemical cell that is designed to operate in the nonspontaneous direction is called an **electrolytic cell** (Fig. 16-3).

Electrochemical cell: An arrangement of two half-reactions that allows the transferred electrons to pass through an external circuit.

Voltaic cell: An electrochemical cell that operates in the spontaneous direction.

Electrolytic cell: An electrochemical cell that is driven in the nonspontaneous direction.

Voltaic Cells

One of the reactions we examined in the last section was that of zinc metal with copper(II) ions:

$$Zn + Cu^{2+} \rightarrow Zn^{2+} + Cu$$

We saw that electrons were transferred from the zinc metal to the copper(II) ions. If we place a zinc strip into a solution containing copper(II) ions, the actual transfer of electrons occurs at the surface of the zinc strip where it comes into contact with the copper(II) ions (Fig. 16-4a). As a result, copper metal is deposited on the surface of the zinc strip.

Suppose we could arrange for the electrons to travel from the zinc strip through an external circuit before coming into contact with the solution of copper(II) ions. We would then have an external flow of electrons—and thus a source of electricity. Figure 16-4b shows the construction of a voltaic cell that may be used to produce such electrical energy. Note that the cell is composed of the two half-cells in question and that each half-cell is constructed by immersing the metal strip in a solution containing its cation. Thus, Cu metal is immersed in a solution containing Cu^{2+} ions, and Zn metal is immersed in a solution containing Zn^{2+} ions. A **porous partition** separating the two half-cells permits the internal net migration of anions (NO_3^-) from the copper(II) solution to the zinc solution. This compensates for the external migration of negatively charged electrons (in the opposite direction) from the zinc strip to the solution containing the

Porous partition: A barrier that separates the half-cells in an electrochemical cell.

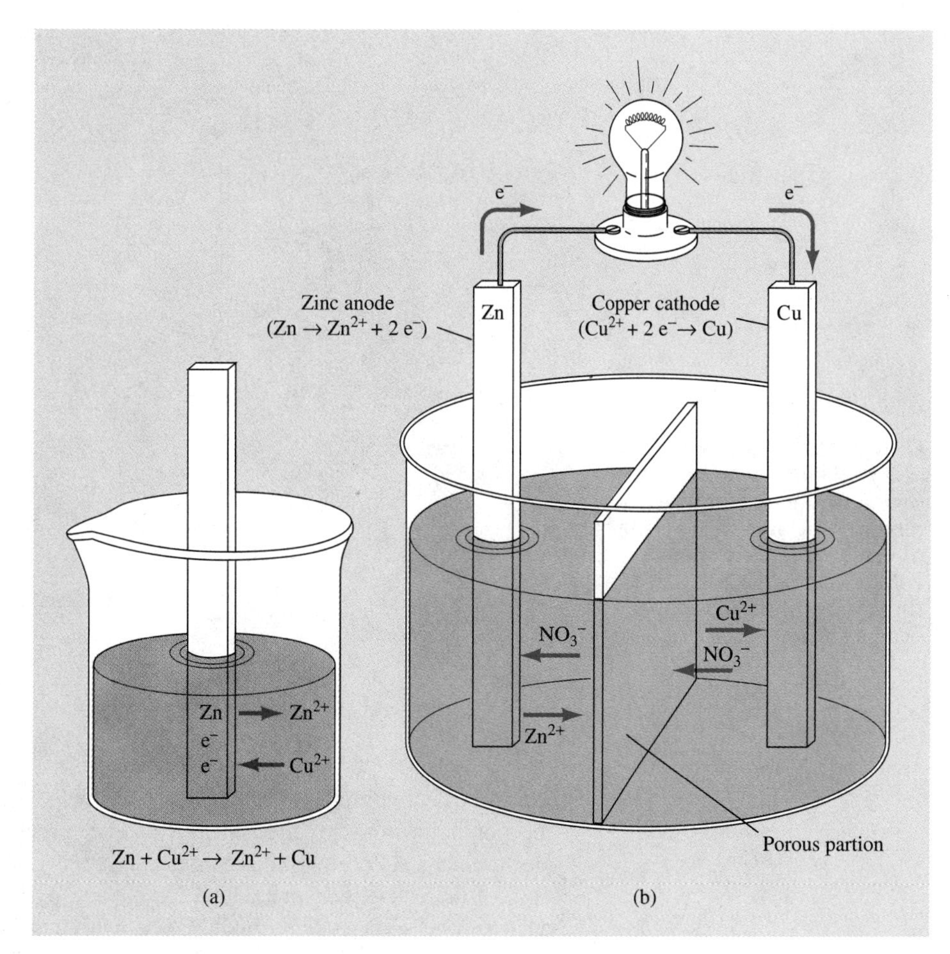

FIGURE 16-4 (a) When a zinc strip is placed in a solution containing copper(II) ions, reaction takes place at the metal surface. (b) Construction of a voltaic cell. Oxidation occurs at the anode. Reduction occurs at the cathode. The porous partition permits ions to migrate between solutions to maintain a charge balance. (Anions migrate toward the anode and cations migrate toward the cathode.) Electrons travel through the external circuit.

Electrode: A conductor that carries electrons to or from a half-cell.

Anode: An electrode at which oxidation occurs.

Cathode: An electrode at which reduction occurs.

copper(II) ions, and it keeps the electrical charges balanced in each solution. The strip in each solution is referred to as an **electrode.** The electrode at which oxidation takes place is known as the **anode,** and the electrode at which reduction occurs is called the **cathode.**

Examining Fig. 16-4 reveals that *an*ions in solution migrate toward the *an*ode, whereas *cat*ions migrate toward the *cat*hode.

Electrolytic Cells

Often, we wish to reverse the spontaneous direction of an oxidation–reduction reaction—for instance, to recharge a battery or to decompose a substance into its elements. For example, the formation of water from its elements proceeds via the following energy-releasing oxidation–reduction reaction:

$$\overset{0}{2\,H_2} + \overset{0}{O_2} \rightarrow \overset{+1-2}{2\,H_2O}$$

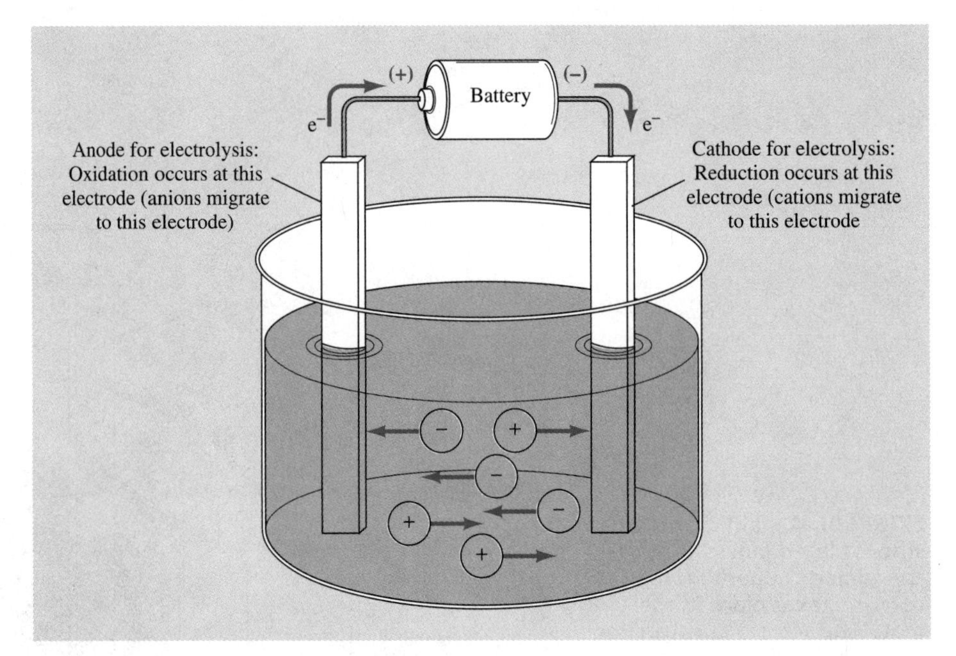

FIGURE 16-5 **An electrolytic cell is an electrochemical cell driven in the nonspontaneous direction by an external source of electrical energy. Like a voltaic cell, oxidation occurs at the anode while reduction occurs at the cathode.**

Suppose we wish to reverse this reaction and decompose water into its elements:

$$2\,H_2O \rightarrow 2\,H_2 + O_2$$

Electrolysis: Operation of an electrochemical cell in its nonspontaneous direction.

This may be accomplished by using an external source of electrical energy to drive the reaction in the nonspontaneous direction. The process is known as **electrolysis,** and an electrochemical cell that utilizes electrolysis is referred to as an electrolytic cell (Fig. 16-5). Like voltaic cells, oxidation in an electrolytic cell occurs at the anode while reduction takes place at the cathode. Electrolysis may be used to decompose water into its elements as shown in Fig. 16-6. The following half-reactions are involved in the decomposition of water.

Cathode:	$2(2\,H^+ + 2\,e^- \rightarrow H_2)$
Anode:	$2\,H_2O \rightarrow O_2 + 4\,H^+ + 4\,e^-$
Overall reaction:	$2\,H_2O \rightarrow 2\,H_2 + O_2$

Water is a nonconductor, so a noninterfering electrolyte (usually sodium sulfate) must be added to allow the conduction of electricity. The volume of hydrogen produced in this electrolysis is twice that of the oxygen, in accordance with the stoichiometry of the process.

The electrolytic technique is also commonly employed to electroplate metallic surfaces. For example, jewelry is frequently produced by silver-plating less expensive metals. To accomplish this, an electrolytic cell is constructed as shown in Fig. 16-7 (p. 510). The object to be plated is used as the cathode and is dipped into a solution containing silver ions. A silver strip is used as the anode. As electrons flow through the circuit, silver ions are reduced to silver metal, plating out on the surface of the cathode. At the same time, the oxidation of silver metal at the anode restores silver ions to the solution as fast as they are used up.

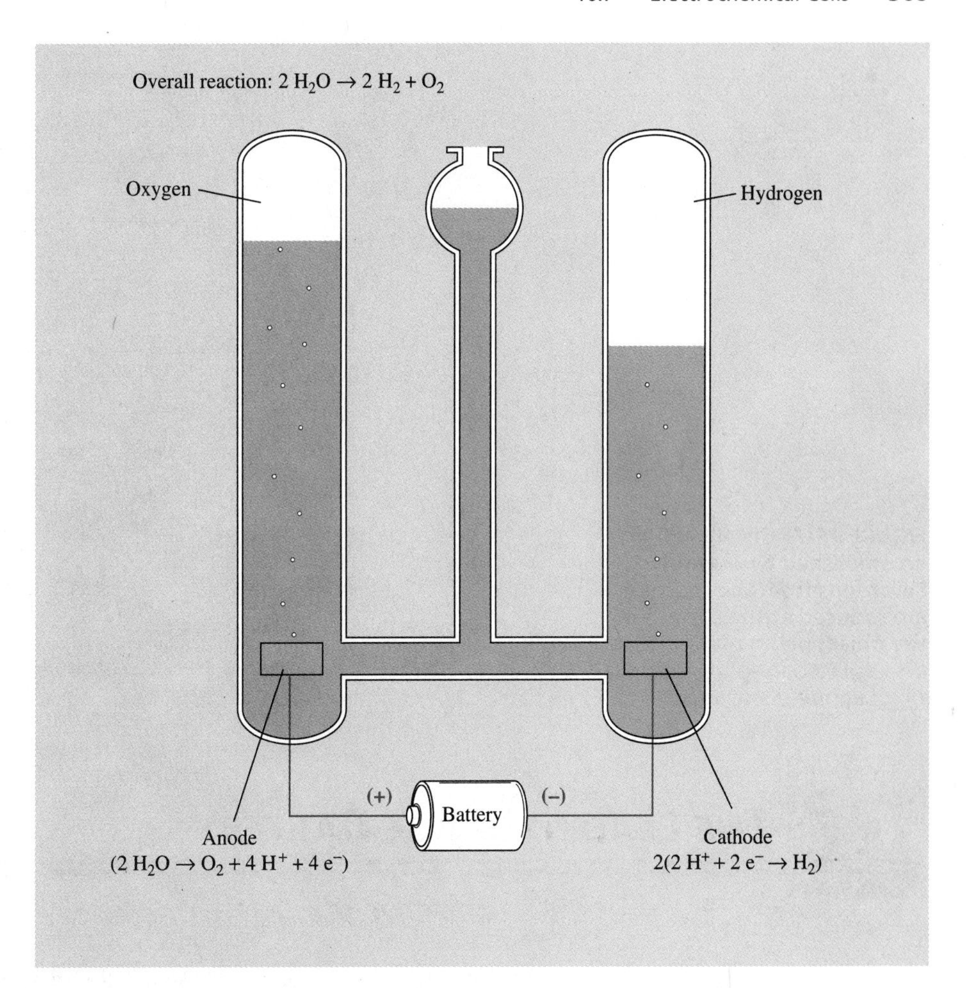

FIGURE 16-6 The electrolysis of water. The volume of hydrogen produced is twice that of oxygen.

PROBLEM 16.12

Answer the following questions about electrochemical cells.

(a) What is an electrochemical cell?
(b) What is the difference between a voltaic cell and an electrolytic cell?
(c) At which electrode does oxidation occur?
(d) At which electrode does reduction occur?
(e) What is the purpose of the porous partition?
(f) In the direction of which electrode do anions migrate across the porous partition?

PROBLEM 16.13

Suppose the spoon being silver-plated in Fig. 16-7 (p. 510) is made of copper, and the battery is mistakenly connected backwards (so that electrons flow from right to left).

(a) Write the half-reaction that takes place at the surface of the spoon.
(b) Is the spoon the anode or the cathode?
(c) What becomes of the spoon?

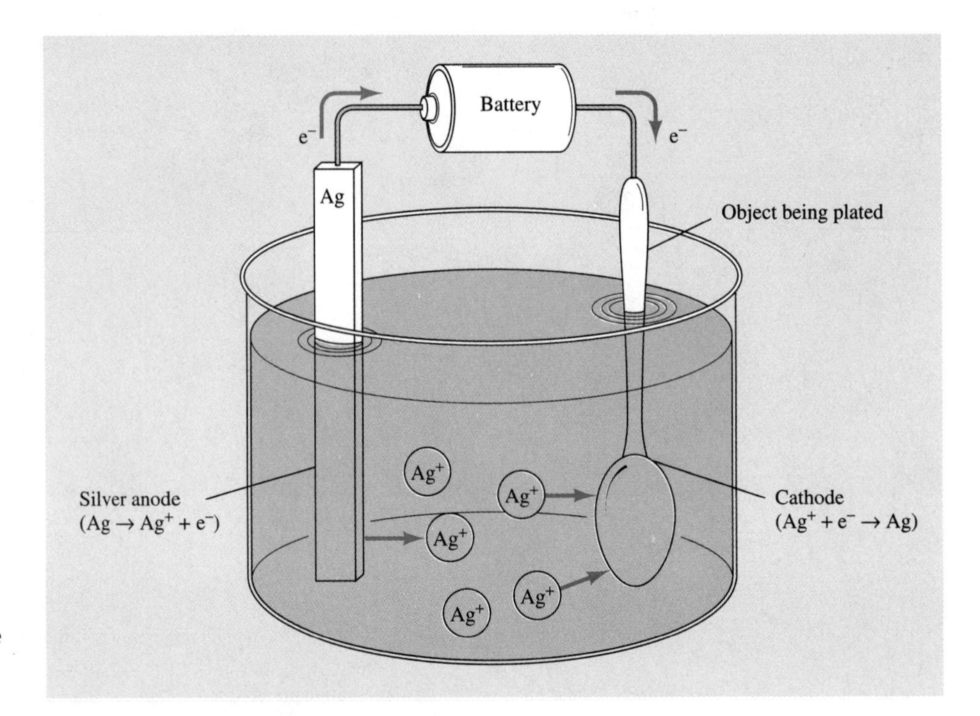

FIGURE 16-7 Silver-plating is accomplished by electrolysis. Silver ions from the solution are reduced at the surface of the object being plated. These are replaced in solution as the silver anode dissolves.

16.8 THE LEAD STORAGE BATTERY

OBJECTIVE 16.8 Explain the operation of a lead storage battery.

One of the devices our society relies on is the 12-volt lead storage battery used to start automobiles. This battery is constructed by connecting a series of voltaic cells, each capable of delivering 2 volts. In fact, the term *battery* is derived from the military description of a series (or "battery") of cannons lined up together. The following reaction is used in each voltaic cell:

$$Pb(s) + PbO_2(s) + 2\,H_2SO_4(aq) \rightarrow 2\,PbSO_4(s) + 2\,H_2O(\ell)$$

The Pb(s) and PbO_2(s) used are solids that are immersed in the sulfuric acid solution and act as the electrodes. Because they are not in physical contact with one another, no reaction occurs until they are connected via the external circuit. This differs slightly from the voltaic cells we examined earlier, in which it was necessary to keep the half-cells physically separated from one another.

When the electrodes are connected to complete the circuit, the reaction proceeds, and metallic lead, Pb, is oxidized to lead(II) ion:

$$\textbf{Anode} \qquad Pb \rightarrow Pb^{2+} + 2\,e^-$$

The electrons that are given off travel through the circuit to the lead(IV) oxide, PbO_2, electrode, where reduction of lead(IV) takes place:

$$\textbf{Cathode} \qquad Pb^{4+} + 2\,e^- \rightarrow Pb^{2+}$$

The lead(II) produced at each electrode precipitates as lead(II) sulfate, which coats the surface of each electrode. The overall equation also shows that sulfuric acid is one of the reactants, and it must be present for the reaction to take place. A schematic representation of the cell is shown in Fig. 16-8. The electrical energy produced in this fashion is used to power the starter motor of a car and to run the various accessories when the engine is not in operation.

Several aspects of the operation of the lead storage battery are interesting. As the chemical equation shows, sulfuric acid is one of the reactants required for the reaction. Thus, it is important to check the water in your battery periodically to make sure there is enough sulfuric acid covering the electrodes for the chemical reaction to take place. Often, heat causes evaporation, leaving the electrodes partially exposed. If a battery seems to be working poorly, the attendant may also check whether the sulfuric acid concentration is high enough. This is done with a hydrometer, which checks the density of the sulfuric acid solution. If the concentration of sulfuric acid drops, so does the density of the solution inside each cell.

The chemical equation also reveals that, each time the battery is used, the electrodes disintegrate slightly, producing a coating of $PbSO_4(s)$. If we were to use a car battery without ever reversing the reaction, the battery would last only a short time. The reactants would be consumed quickly, and the battery would go dead. In fact, that is what happens when we inadvertently leave our lights on after turning the engine off; we return to a car that does not start.

Once the engine has been started, however, the preceding reaction is reversed as the automobile's alternator (or generator) produces electrical energy, "recharging" the battery. In essence, the voltaic cells of the battery are turned into electrolytic cells. Of course, the electrical energy produced for recharging comes ultimately from the

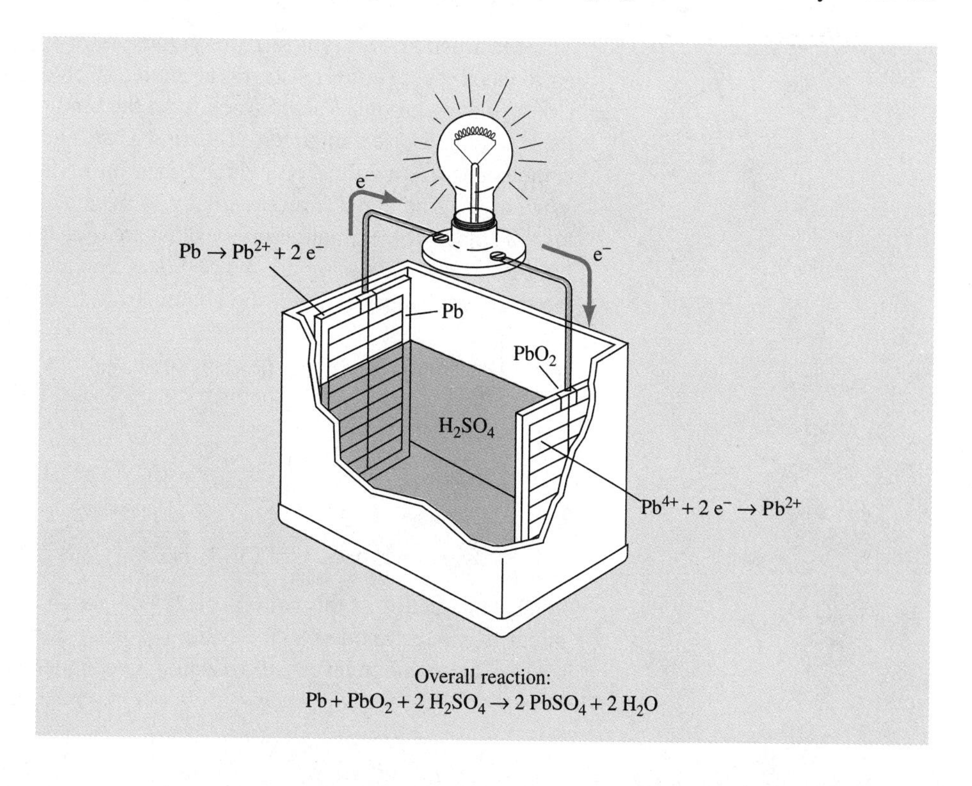

FIGURE 16-8 The lead storage battery.

gasoline (fuel) being consumed as the engine runs. Thus, we are using fossil fuel energy to produce electrical energy, which in turn "recharges" the battery after each start. In the recharging process, the lead(II) sulfate deposited on each electrode is returned to its original state, regenerating the electrodes.

Even with the best of care, every automobile battery eventually goes dead, and attempts to recharge such a battery result in failure. The reason for this has to do with the quality of the electrodes, which deteriorate over time. Lead(II) sulfate produced in the operation of the battery may be shaken loose by road vibrations, falling to the bottom of the battery case, where it is no longer available to participate in the electrolytic reaction. The loosening of internal parts may also result in the electrodes shorting out as they come into physical contact with one another. Eventually, the quality of the electrodes becomes so poor that the electrical output of the battery is inadequate to start the car. It is time to buy a new battery.

Other Batteries

Before concluding our discussion of batteries, there are two other types of batteries worth mentioning: the zinc–carbon (drycell) battery and the nickel–cadmium (nicad) battery. Both types of batteries are used to power a wide range of portable devices such as flashlights, electronic games, and AM/FM/cassette players.

The zinc–carbon battery (Fig. 16-9) utilizes the following half-reactions:

Anode: $$Zn \rightarrow Zn^{2+} + 2\,e^-$$

Cathode: $$2\,NH_4^+ + 2\,MnO_2 + 2\,e^- \rightarrow 2\,NH_3 + Mn_2O_3 + H_2O$$

Overall reaction: $$Zn + 2\,NH_4^+ + 2\,MnO_2 \rightarrow Zn^{2+} + 2\,NH_3 + Mn_2O_3 + H_2O$$

The construction of a typical battery consists of a central graphite (carbon) electrode surrounded by a moist paste of ammonium chloride (NH_4Cl), manganese dioxide (MnO_2), and carbon. (The chloride from the ammonium chloride is a spectator ion in the reaction.) These components function as the cathode of the battery. Surrounding the cathode paste is a zinc shell, which functions as the anode. As the reaction proceeds, some of the zinc metal from the casing forms zinc ions, which diffuse into the paste to take the place of the ammonium ions that are used up. The top of the carbon rod and the bottom of the zinc casing are the points of connection generally used to complete the circuit. As a battery is used, the chemicals are consumed, and eventually the battery goes dead.

The outer appearance of a nickel–cadmium battery is very similar to that of a zinc–carbon battery, but its chemistry is different. The half-reactions that drive the battery are the following:

Anode: $$Cd + 2\,OH^- \rightarrow Cd(OH)_2 + 2\,e^-$$

Cathode: $$NiO_2 + 2\,H_2O + 2\,e^- \rightarrow Ni(OH)_2 + 2\,OH^-$$

Overall reaction: $$Cd + NiO_2 + 2\,H_2O \rightarrow Cd(OH)_2 + Ni(OH)_2$$

In the construction of this battery, nickel(IV) oxide, NiO_2, forms the cathode while cadmium forms the anode. As the battery is used, the products adhere to the electrodes, making it possible to reverse the reaction. Consequently, nicad batteries are rechargeable.

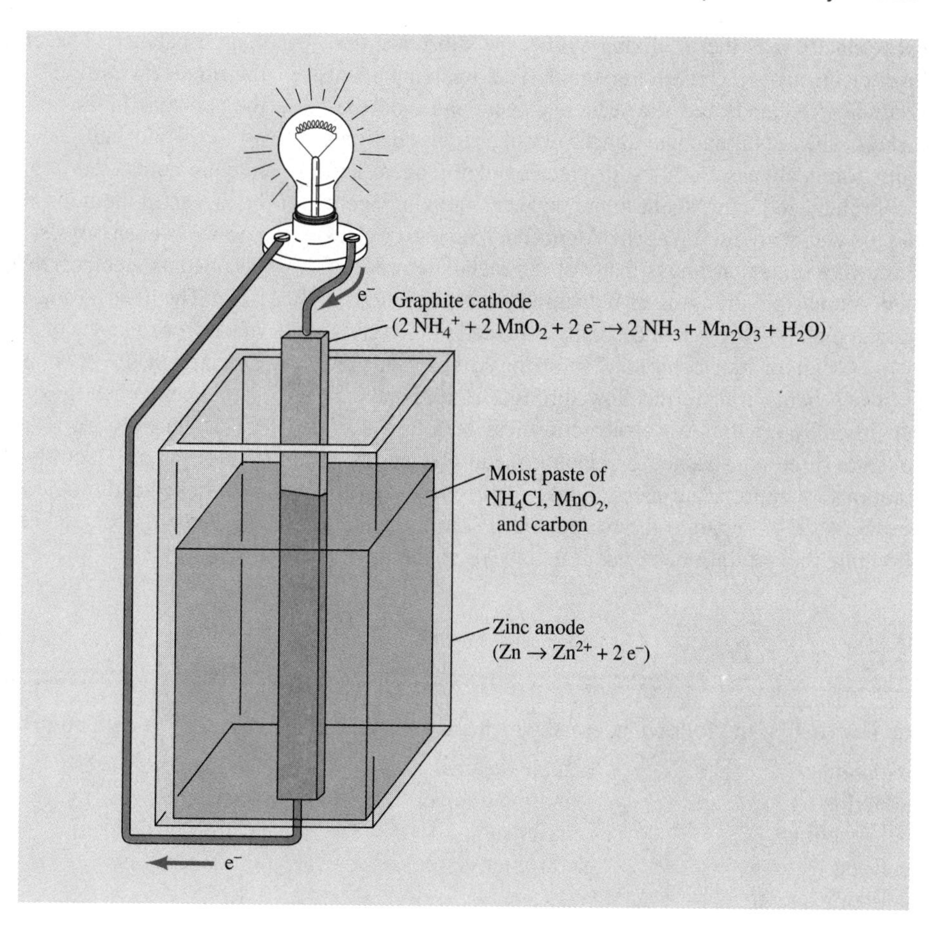

FIGURE 16-9 The zinc–carbon battery.

PROBLEM 16.14

Write a chemical equation for the reaction that takes place when a lead storage battery is recharged. When the battery is being recharged, is it acting as a voltaic cell or as an electrolytic cell?

CHAPTER SUMMARY

Many reactions involve the transfer of electrons from one substance to another. We refer to such reactions as **oxidation–reduction** reactions. The substance that loses the electrons is said to be **oxidized,** whereas the substance gaining the electrons is **reduced.** The individual processes of oxidation and reduction are referred to as either **half-reactions** or **half-cells.** For each oxidation–reduction reaction, the number of electrons lost in oxidation must equal the number of electrons gained in reduction.

Several rules exist for assigning **oxidation numbers** (or charges) to each element in a substance. The sum of the oxidation numbers must equal the total charge on the species being assigned. All monatomic ions are assigned the charge on the ion. All elements have the oxidation number 0. In compounds, hydrogen is usually assigned +1 and oxygen −2. Other elements are assigned with these principles in mind.

In every oxidation–reduction, the substance that

gains electrons is the **oxidizing agent;** the substance that loses electrons is the **reducing agent.** The oxidizing agent is always reduced, and the reducing agent must be oxidized. The spontaneous direction of an oxidation–reduction is always from the stronger oxidizing agent and the stronger reducing agent to the weaker oxidizing agent and the weaker reducing agent. An **activity series** (or **electromotive series**) arranges many of the metals in order of their decreasing strengths as reducing agents. Although it is not a metal, hydrogen is generally included in the series.

A pair of half-cells may be arranged such that the electrons being transferred flow through an external circuit, thereby creating an **electrochemical cell.** The half-cells are often constructed by placing metal **electrodes** in solutions of appropriate electrolytes. A **porous partition** may be used to separate the two half-cell solutions while allowing the migration of enough ions to maintain the charge balance. The electrode where oxidation occurs is the **anode,** while reduction takes place at the **cathode.** If the cell runs in the spontaneous direction, it is referred to as a **voltaic cell.** Alternatively, an electrochemical cell may be connected to an external source of electricity and be driven in the nonspontaneous direction. This process is known as **electrolysis,** and a cell that employs electrolysis is called an **electrolytic cell.**

The lead storage battery, used in automobiles, is made from a series of cells connected together. The battery operates in the spontaneous direction when it is used to start a car. When the battery is "recharged" by the alternator or generator, it is operating in the nonspontaneous direction as an electrolytic cell. The zinc–carbon (drycell) and nickel–cadmium (nicad) batteries are other commonly used electrochemical cells.

KEY TERMS

Review each of the following terms, which are discussed in this chapter and defined in the Glossary.

oxidation
reduction
half-reaction
half-cell
oxidizing agent
reducing agent
oxidation number
activity series
electromotive series
electrochemical cell
voltaic cell
electrolytic cell
porous partition
electrode
anode
cathode
electrolysis

ADDITIONAL PROBLEMS

SECTION 16.1

16.15 Identify each of the following changes as oxidation or reduction. In each case, determine the number of electrons that must be gained or lost, and add them to the appropriate side of the equation.

(a) $Na \rightarrow Na^+$ **(b)** $Se \rightarrow Se^{2-}$
(c) $Co^{2+} \rightarrow Co^{3+}$ **(d)** $Cu^{2+} \rightarrow Cu^+$
(e) $Cl_2 \rightarrow 2\,Cl^-$

16.16 Answer the following questions about oxidation and reduction:

(a) Why must oxidation always be accompanied by reduction?
(b) Why is the oxidizing agent reduced?
(c) Why is the reducing agent oxidized?

16.17 Identify the oxidizing agent and the reducing agent in each of the following reactions.

(a) $Mg + 2\,H^+ \rightarrow Mg^{2+} + H_2$
(b) $Cl_2 + 2\,Br^- \rightarrow 2\,Cl^- + Br_2$
(c) $2\,K + I_2 \rightarrow 2\,K^+ + 2\,I^-$

SECTION 16.2

16.18 Assign an oxidation number to each element in the following compounds.

(a) NO_2 **(b)** P_2O_5 **(c)** H_2SO_3 **(d)** H_3PO_2
(e) Tl_2O **(f)** $HClO$ **(g)** $ZrCl_3$ **(h)** $KMnO_4$

16.19 Assign an oxidation number to each element in the following.

(a) CrO_3 **(b)** HIO_4 **(c)** K_2Se **(d)** $Cr_2O_7^{2-}$
(e) UF_6 **(f)** $KHCO_3$ **(g)** $Al_2(CO_3)_3$ **(h)** $S_2O_3^{2-}$

SECTION 16.3

16.20 Determine which of the following *unbalanced* equations involve oxidation–reduction.

(a) $Al + HCl \rightarrow AlCl_3 + H_2$
(b) $H_2SO_4 + Ba(OH)_2 \rightarrow BaSO_4 + H_2O$
(c) $AgNO_3 + K_2CrO_4 \rightarrow Ag_2CrO_4 + KNO_3$
(d) $NaCl + MnO_2 + H_2SO_4 \rightarrow MnSO_4 + Na_2SO_4 + H_2O + Cl_2$

SECTION 16.4

16.21 Balance the following equations, using the oxidation-number technique. In each case, identify the elements undergoing oxidation and reduction.
(a) $HNO_3 + CuS \rightarrow CuSO_4 + NO + H_2O$
(b) $Sb + HNO_3 \rightarrow Sb_2O_5 + NO + H_2O$
(c) $AuCl_3 + KI \rightarrow AuCl + I_2 + KCl$
(d) $KMnO_4 + HCl \rightarrow MnCl_2 + Cl_2 + H_2O + KCl$
(e) $Bi_2S_3 + HNO_3 \rightarrow Bi(NO_3)_3 + NO + S + H_2O$

SECTION 16.5

16.22 Complete and balance each of the following half-reactions, using the ion–electron method. Identify each as an oxidation or a reduction.
(a) $Cu^+ \rightarrow Cu^{2+}$ (b) $NO_3^- \rightarrow NO$
(c) $S_2O_3^{2-} \rightarrow S_4O_6^{2-}$ (d) $Br^- \rightarrow Br_2$
(e) $H_2O_2 \rightarrow O_2$

16.23 Use the ion–electron method to balance each of the following oxidation–reduction reactions. In each case, identify the oxidation and reduction half-reactions.
(a) $Sn^{2+} + MnO_4^- + H^+ \rightarrow Sn^{4+} + Mn^{2+} + H_2O$
(b) $I^- + SO_4^{2-} + H^+ \rightarrow I_2 + S^{2-} + H_2O$
(c) $Cr_2O_7^{2-} + C_2O_4^{2-} + H^+ \rightarrow Cr^{3+} + CO_2 + H_2O$
(d) $Br^- + SO_4^{2-} + H^+ \rightarrow Br_2 + SO_2 + H_2O$
(e) $H_2O_2 + NO_3^- + H^+ \rightarrow O_2 + NO + H_2O$

SECTION 16.6

16.24 For each of the following reactions, use Table 16-1 to complete the equation. Identify the oxidizing agent and the reducing agent, and predict whether the reaction will occur spontaneously in the direction written.
(a) $Ba + Fe^{2+} \rightarrow ?$ (b) $Cu + Cd^{2+} \rightarrow ?$
(c) $Mg + Al^{3+} \rightarrow ?$ (d) $Au + Cu^{2+} \rightarrow ?$
(e) $K + Zn^{2+} \rightarrow ?$ (f) $H_2 + Hg_2^{2+} \rightarrow ?$

SECTION 16.7

16.25 A voltaic cell may be constructed as shown in Fig. 16-10, using the Pb/Pb^{2+} and Mg/Mg^{2+} half-reactions. Answer the following questions about this cell.
(a) Which is the stronger oxidizing agent, Pb^{2+} or Mg^{2+}? (Refer to Table 16-1.)

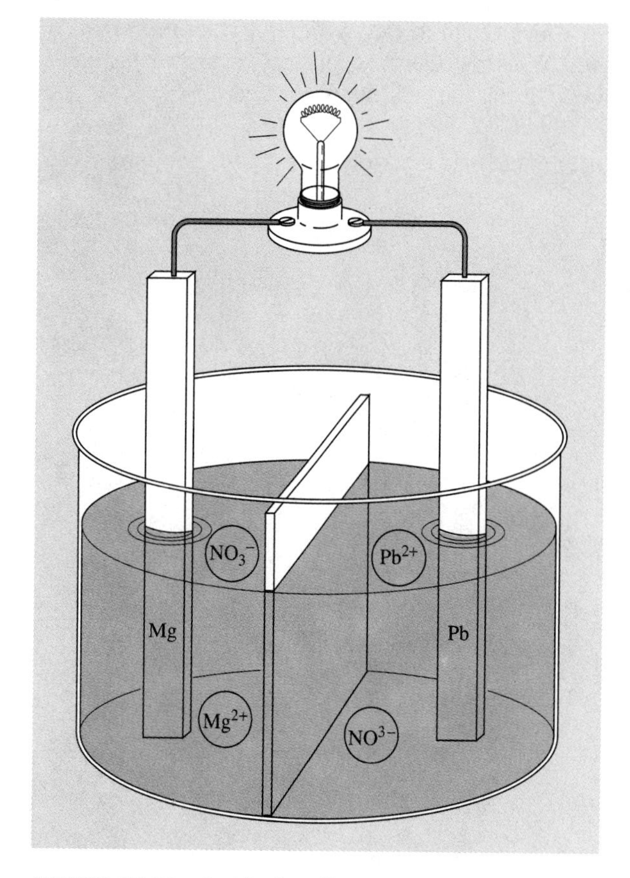

FIGURE 16-10 A voltaic cell.

(b) Write a chemical equation for the overall reaction that takes place during the operation of this cell.
(c) In which direction (to the left or to the right) do electrons flow through the external circuit?
(d) In which direction (to the left or to the right) do nitrate ions flow across the porous partition?
(e) At which electrode (Pb or Mg) does oxidation occur?
(f) At which electrode (Pb or Mg) does reduction occur?
(g) Which electrode is the anode? Which electrode is the cathode?

16.26 Figure 16-11 (p. 516) shows an apparatus for the electrolysis of molten sodium chloride. When the battery is connected as shown, sodium metal and chlorine gas are produced. Refer to this diagram as you answer the following questions.
(a) At which electrode (left or right) is sodium metal deposited? Write the half-reaction that occurs at this electrode. Is this the anode or the cathode?

(b) At which electrode (left or right) is chlorine gas given off? Write the half-reaction that occurs at this electrode. Is this the anode or the cathode?

FIGURE 16-11 The electrolysis of molten sodium chloride.

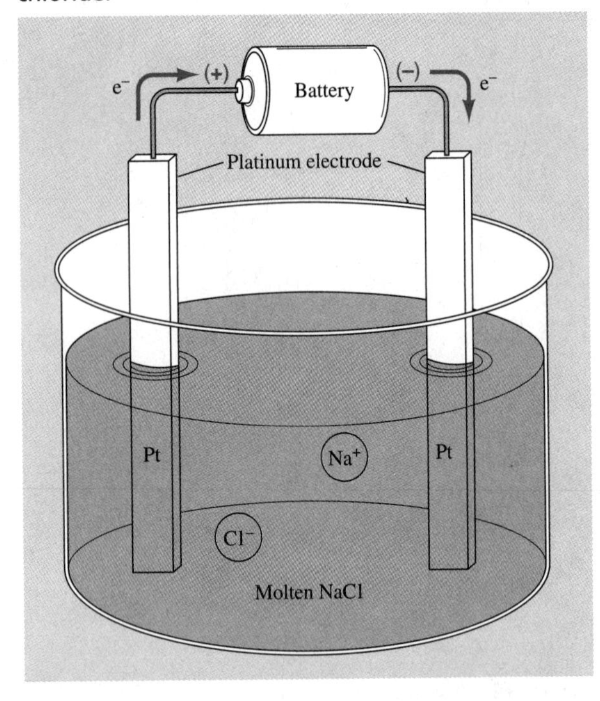

SECTION 16.8

16.27 The lead storage battery utilizes the following chemical reaction:

$$Pb(s) + PbO_2(s) + 2\ H_2SO_4(aq) \rightarrow 2\ PbSO_4(s) + 2\ H_2O(\ell)$$

Refer to Fig. 16-8 as you answer the following questions.

(a) Explain how the battery works.

(b) Explain why it is necessary to check the "water" in the battery periodically.

(c) Explain what happens when a lead storage battery is recharged and why it cannot be recharged indefinitely.

16.28 Refer to Fig. 16-9 to answer the following questions about the zinc–carbon battery.

(a) What half-reaction takes place at the cathode?

(b) What half-reaction takes place at the anode?

(c) Use these two half-reactions to explain how the battery works.

(d) What is the function of the graphite (carbon) rod?

(e) How is the charge balance of the moist paste maintained as ammonium ions are consumed?

(f) What is the function of the chloride ions from the ammonium chloride?

(g) In what important way does a zinc–carbon battery differ from a nickel–cadmium battery?

GENERAL PROBLEMS

16.29 Answer the following questions about oxidation numbers and chemical nomenclature:

(a) Assign an oxidation number to sulfur in sulfate, SO_4^{2-}, and in sulfite, SO_3^{2-}.

(b) Assign an oxidation number to nitrogen in nitrate, NO_3^-, and in nitrite, NO_2^-.

(c) Assign an oxidation number to chlorine in chlorate, ClO_3^-, and in chlorite, ClO_2^-.

(d) What relationship exists between oxidation number and the suffixes *-ate* and *-ite?*

(e) Would you expect phosphorus to have a higher oxidation number in phosphorous acid or phosphoric acid?

16.30 Balance each of the following oxidation–reduction equations, using the method you prefer.

(a) $HNO_3 + I_2 \rightarrow NO_2 + HIO_3 + H_2O$

(b) $Cr_2O_7^{2-} + SO_2 + H^+ \rightarrow Cr^{3+} + SO_4^{2-} + H_2O$

(c) $MnO_2 + HCl \rightarrow MnCl_2 + Cl_2 + H_2O$

(d) $C_2O_4^{2-} + MnO_4^- + H^+ \rightarrow CO_2 + Mn^{2+} + H_2O$

(e) $CuO + NH_3 \rightarrow Cu + N_2 + H_2O$

(f) $IO_4^- + I^- + H^+ \rightarrow I_2 + H_2O$

(g) $KIO \rightarrow KIO_3 + KI$

(h) $C_3H_8 + O_2 \rightarrow CO_2 + H_2O$

16.31 Answer each of the following questions.

(a) Would a lead vessel make a suitable container for hydrochloric acid? Explain your answer.

(b) Would a copper vessel be suitable for holding a solution of hydrochloric acid? Explain.

(c) Would you expect the copper(II) ions in CuO to be reduced by hydrogen gas, H_2? Explain your answer.

(d) Would you expect the magnesium ions in MgO to be reduced by hydrogen gas? Explain.

***16.32** The following half-reactions represent the reduction of four of the halogens to their corresponding halide ion:

$$Br_2 + 2\,e^- \rightleftharpoons 2\,Br^-$$
$$Cl_2 + 2\,e^- \rightleftharpoons 2\,Cl^-$$
$$F_2 + 2\,e^- \rightleftharpoons 2\,F^-$$
$$I_2 + 2\,e^- \rightleftharpoons 2\,I^-$$

Use the following information to arrange these halogens into a table like Table 16-1:

Iodine, I_2, does not react with chloride ion.
Bromine, Br_2, reacts with iodide ion but not with chloride ion.
Fluorine, F_2, reacts with chloride ion.

16.33 People with silver fillings in their teeth often experience an uncomfortable sensation when they chew on a piece of aluminum foil. Their pain is the result of an electrical potential created as silver cations in a filling are reduced by the aluminum metal. Write a balanced oxidation–reduction equation for the reaction associated with this process. Is the aluminum foil the cathode or anode in the electrochemical cell created?

***16.34** When chlorine gas is added to water to kill bacteria, hydrochloric acid and hypochlorous acid are formed.
(a) Write a balanced equation for the reaction.
(b) Use the ion–electron method to write a balanced half-reaction for the oxidation. Identify the reducing agent. [*Remember:* The oxidizing agent is the substance reduced.]
(c) Use the ion–electron method to write a balanced half-reaction for the reduction. Identify the oxidizing agent. [*Remember:* The reducing agent is the substance oxidized.]
(d) Show how these two half-reactions add up to the balanced equation you wrote in part (a).
(e) What do you notice about the oxidizing and reducing agents?

***16.35** Corrosion is a phenomenon that is often observed when metals are buried beneath the earth's surface. Underground storage tanks (such as those used to store gasoline and other petroleum products) often leak as a result of the corrosion of steel. The process, which involves water, oxygen, and electrolytes in the ground, results in the oxidation (and accompanying rust formation) of iron in the steel. In order to protect tanks and pipes against corrosion, a rod made of magnesium (or some other readily oxidized metal) is buried in the ground and wired to the tank or pipe to be protected. Over time this sacrificial rod is oxidized but not the steel tank or pipe. Explain the principles upon which this process of *cathodic protection* works.

***16.36** An *ampere* is a measure of electrical current. A total of 6.24×10^{18} electrons passes through a circuit carrying a 1.00-ampere current in 1.00 second. Suppose an electrolytic cell with a Cu/Cu^{2+} cathode is exposed to a 10.0-ampere current for a period of 10.0 minutes. What mass of copper metal will plate out on the cathode? [*Remember:* 1.00 ampere • second = 6.24×10^{18} electrons]

***16.37** The rules presented in this chapter for determining oxidation numbers allow assignment of most chemical species. However, a more precise rule calls for both electrons in a bond to be attributed to the more electronegative element in the pair. The charge (or oxidation number) on each atom is then determined by comparing the number of electrons on each atom to the number it normally has as an uncharged atom. For example, since oxygen is more electronegative than hydrogen, both electrons in each bond in water are assigned to oxygen:

H) :Ö: (H

This gives each hydrogen a +1 charge and the oxygen a −2 charge. When both atoms forming a bond have identical electronegativities, the electrons in the bond are divided equally between the two atoms. Thus, the electrons in hydrogen peroxide are assigned as follows:

H) :Ö:|:Ö: (H

In this case, each oxygen has a −1 charge. Use this method and the electronegativities presented in Table 8-3 to determine the oxidation number of each atom in the following molecules and ions.

(a) :C̈l–F̈:

(b) [H–Al–H, with H above and H below Al]⁻

(c) :F̈–Ö–F̈:

(d) H–C̈l–Ö:

(e) H–F̈–Ö:

WRITING EXERCISES

16.38 Your uncle has always enjoyed collecting metal objects. When he heard that you know how the process of electroplating works, he decided that it would be fun to silver-plate some of the artifacts in his collection. Consequently, he has asked you for directions on how to do this. He has already purchased a bottle of silver nitrate (to make up the silver cation solution), some silver rods (that can be used as electrodes), and several batteries (to drive the electrolysis). Write a note to your uncle, explaining how an electrolytic cell works, and write specific directions that he can use to set up an electrolytic cell for silver-plating the objects in his collection.

16.39 In your own words, explain how to assign oxidation numbers to the elements in any atom, molecule, or ion.

16.40 Without using any diagrams, explain what a voltaic cell is and how it works. Your explanation should include directions for setting one up. Be sure to give the names of the two types of electrodes and describe the processes that occur at each. Then indicate the direction in which electrons flow in the external circuit and the directions in which ions flow within the cell.

16.41 One of your friends is taking a biology course that requires some knowledge of oxidation–reduction. You have agreed to provide your friend with a brief written explanation of this topic. Your discussion should include the definitions of oxidation and reduction and the definitions of oxidizing and reducing agents. In addition, you should explain what a half-reaction is and why oxidation is always accompanied by reduction.

REACTION RATES AND EQUILIBRIUM

17

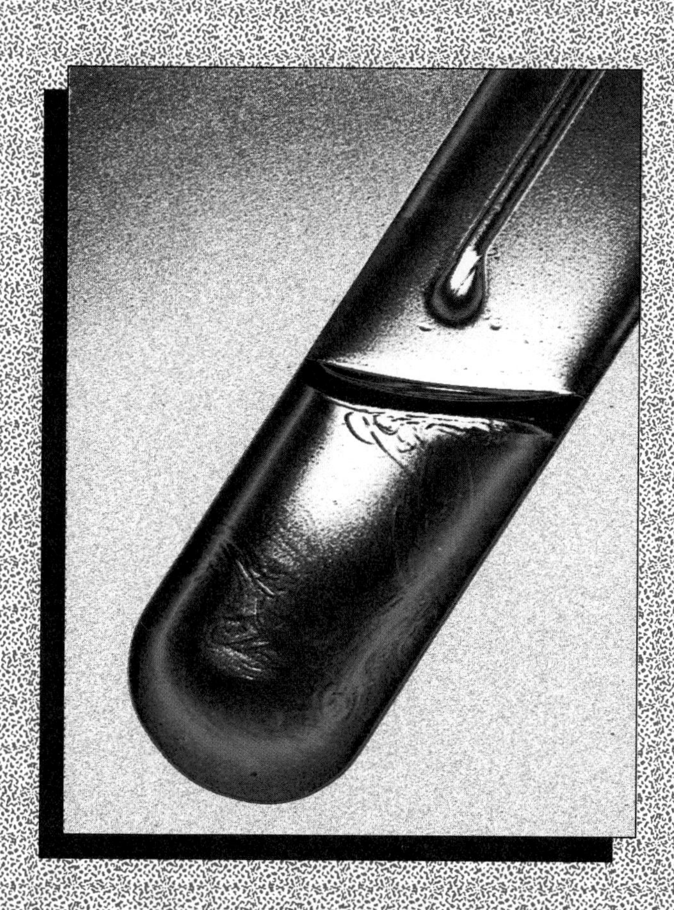

In Chapter 10 we examined the fundamentals of chemical reactions, focusing on the mass relationships (stoichiometry) of such processes. In this chapter, we consider the rates of chemical reactions and the various factors involved in the conversion of a set of reactants to products. For example, when we light a gas burner (such as those found on gas stoves), the combustion of one or more hydrocarbons (the components of natural gas) takes place. We have already seen that this type of reaction liberates energy in the form of heat:

$$CH_4 + 2\,O_2 \rightarrow CO_2 + 2\,H_2O + \text{Heat}$$

If you have ever lit a gas burner manually, you might wonder why it is necessary to strike a match in order to get this reaction started. Alternatively, you might be surprised to learn that certain reactions that require several hours to take place in a test tube are complete in a matter of microseconds inside a living cell.

Reaction rate: The speed with which a chemical reaction occurs.

Chemical kinetics: The study of the rates of reaction.

The answers to both of these puzzles require an understanding of **reaction rates,** the speed with which chemical reactions take place. The study of reaction rates belongs to a subject area called **chemical kinetics.** Just as the kinetic energy of an object depends on its velocity, the term *kinetic* here refers to the speed with which a reaction takes place. As we shall see, reaction rates are affected by a variety of factors, including concentration, temperature, and the presence of chemical agents known as catalysts.

17.1 THE REACTION PROFILE

OBJECTIVE 17.1 Explain what a reaction profile is. Identify the reaction coordinate, the activation energy, and the transition state on a reaction profile, and explain the significance of each.

In order for two molecules to react, it is necessary for them to collide with one another. Because a chemical reaction involves a "rearrangement" of atoms, the reacting substances must come in contact at some point if they are to exchange atoms with one another.

Let us consider the reaction of carbon monoxide, CO, and nitrogen dioxide, NO_2, to form carbon dioxide, CO_2, and nitrogen monoxide, NO (Fig. 17-1):

$$CO(g) + NO_2(g) \rightarrow CO_2(g) + NO(g)$$

For this reaction to occur, a carbon monoxide molecule must collide with a molecule of nitrogen dioxide. Furthermore, because bonds must be broken during the conversion from reactants to products, some minimum amount of energy is required for the reaction to occur. This energy is supplied by the impact of the colliding molecules. If the molecules do not collide with sufficient energy, they merely bounce off one another and no reaction takes place.

Reaction profile: A diagram showing the energy changes that occur as a reaction progresses.

Reaction coordinate: The horizontal coordinate of a reaction profile.

Activation energy: The minimum amount of energy required for a reaction to occur.

Figure 17-2 shows the progress of this reaction in an energy diagram known as a **reaction profile.** The vertical axis represents the energy changes that occur as the reaction proceeds. The horizontal axis, known as the **reaction coordinate,** is an arbitrary scale that represents the progress of the reaction as it proceeds from reactants to products. The energy required to go from the reactant state to the high point in the diagram is referred to as the activation energy (E_{act}). The **activation energy** represents the

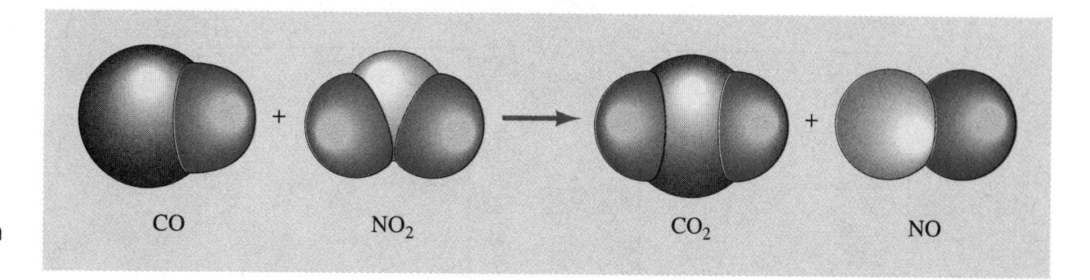

FIGURE 17-1 The reaction of carbon monoxide and nitrogen dioxide.

Transition state or **activated complex:** The arrangement of atoms at the highest point on a reaction profile.

amount of energy with which the reactant molecules must collide if they are to form products. We refer to the arrangement of atoms at this high point as either the **activated complex** or the **transition state;** this represents the point of transition from reactants to products. Molecules that collide with insufficient energy to overcome this energy barrier merely bounce off one another unreacted (Fig. 17-3a). However, molecules that collide with sufficient energy to get over the energy barrier form products (Fig. 17-3b).

We can draw an analogy between this reaction profile and driving over a mountain pass. Imagine that we are on a mountain road at an elevation of 8000 ft. Several miles away is a town at an elevation of 7000 ft. Although the overall trip is downhill, let us suppose that the road to town goes over a mountain pass that reaches 9000 ft. We must first go uphill to cross the pass before we can descend into the town below. If we do not have enough gas in our car (that is, if we do not have enough energy), we will not be able to reach the top of the pass and get to town. On the other hand, once we reach the top of the pass, we can coast the rest of the way down, even if we run out of gas.

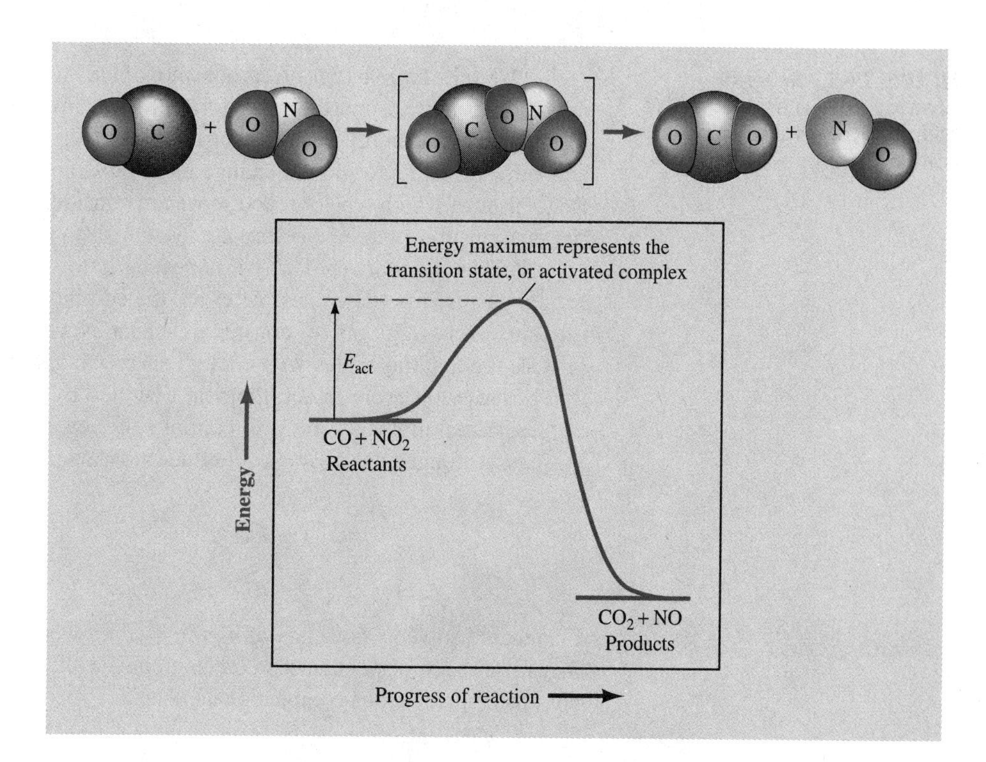

FIGURE 17-2 A reaction profile for $CO + NO_2 \rightarrow CO_2 + NO$

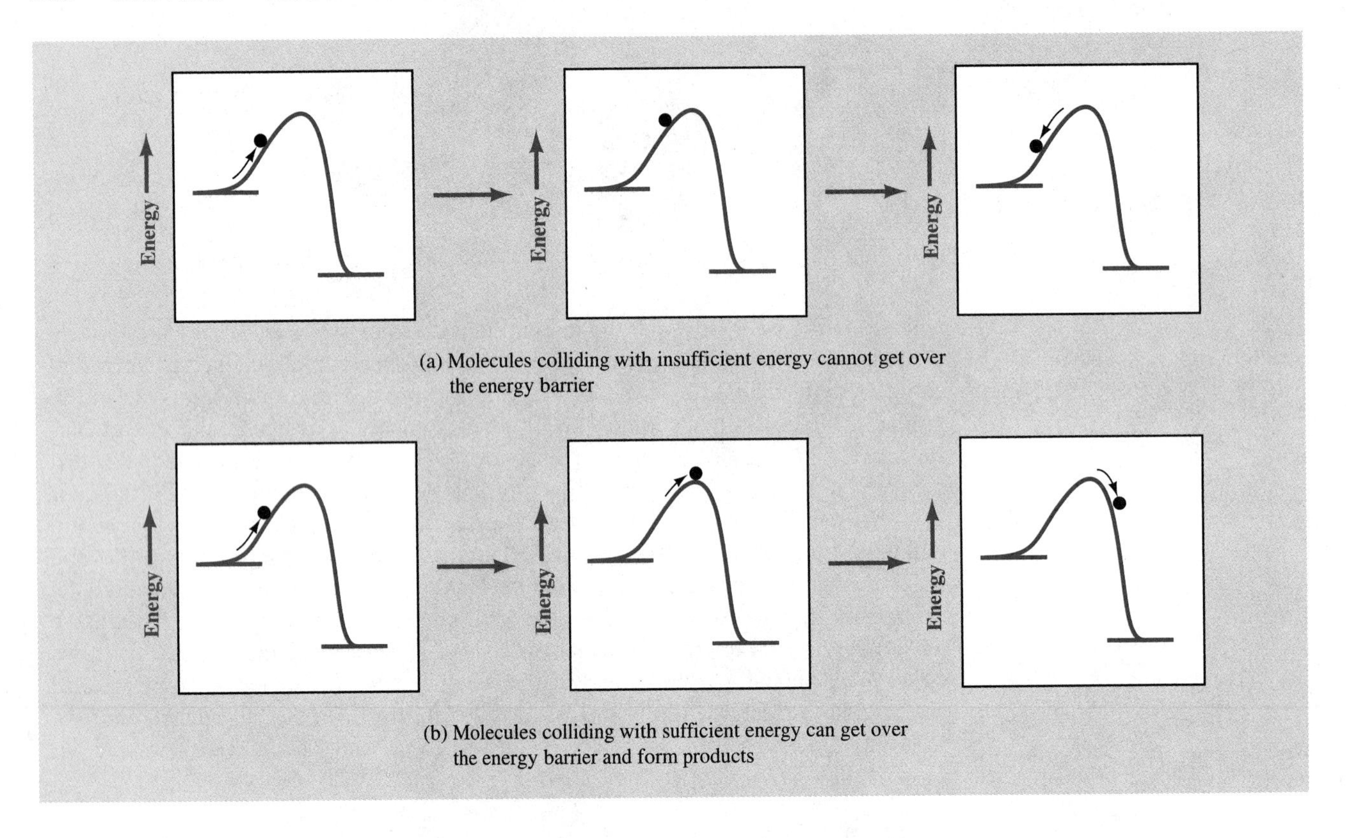

FIGURE 17-3 Reaction progress for (a) an ineffective collision; (b) a successful collision.

Overall, the reaction of carbon monoxide with nitrogen dioxide involves a net release of energy, primarily in the form of heat (Fig. 17-4a). On the other hand, many reactions absorb energy from the surroundings (Fig. 17-4b). It is also possible for a reaction to show little or no net change in going from reactants to products (Fig. 17-4c). Regardless of which type of reaction we are considering, there is always an energy barrier that must be overcome before the reactants can go on to products.

Earlier, we wondered why it is necessary to strike a match in order to ignite the natural gas used in our stoves. After all, the combustion of hydrocarbons is an energy-releasing process, giving off considerable heat. Nevertheless, we must strike a match to provide the first molecules with enough energy to get over the energy barrier. Once the first of these molecules react, the heat liberated by their reaction provides the energy that additional molecules need to continue the reaction. Thus, once the stove has been lit, it is no longer necessary to continue holding the burning match to keep the fire going.

PROBLEM 17.1

Flammable solvents have an *autoignition point,* which is the temperature above which they may undergo spontaneous combustion. Use the principles just discussed to explain how such spontaneous combustion can occur.

17.2 REACTION RATES

OBJECTIVE 17.2 Explain how the concentration, temperature, and orientation of reacting molecules affect the rate of a reaction.

In Section 17.1, we saw that a reaction will take place if the colliding molecules have enough energy to overcome the activation energy of the reaction. However, we did not discuss the *rate* of a reaction. The rate of a reaction is a measure of how quickly the reactant molecules are able to cross the energy barrier and form product molecules. In fact, different reactions proceed at different rates—some very rapidly and some very slowly. Generally, there are three factors that influence the rate of a reaction:

1. How often do molecules collide?
2. How many of the collisions have sufficient energy to get over the energy barrier?
3. How many of these collisions occur with an orientation that allows products to form?

Frequency of Collision—Concentration

The greater the number of collisions between molecules, the greater the chance that a successful collision will take place and a reaction will occur. The most important factor influencing the frequency of collision is concentration. The greater the concentration of reacting molecules, the greater the chance that they will collide with one another. In general, then, *increased concentration leads to an increased reaction rate.*

Activation Energy—Temperature

Temperature is a critical factor in determining the number of molecules with sufficient energy to get over the activation energy barrier. Figure 17-5 shows the distribution of kinetic energies in a reaction sample at two different temperatures. The sample at the

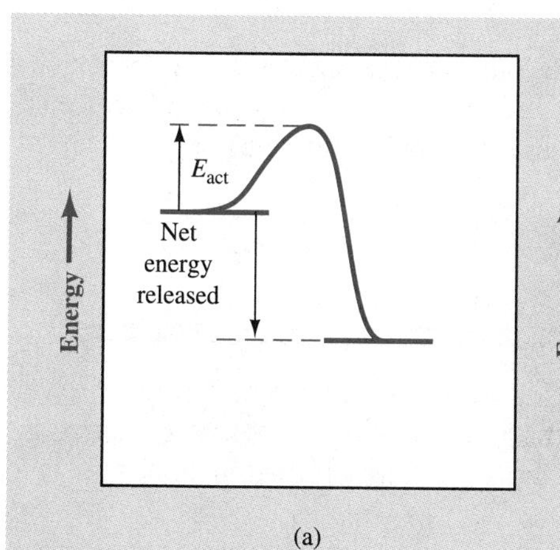

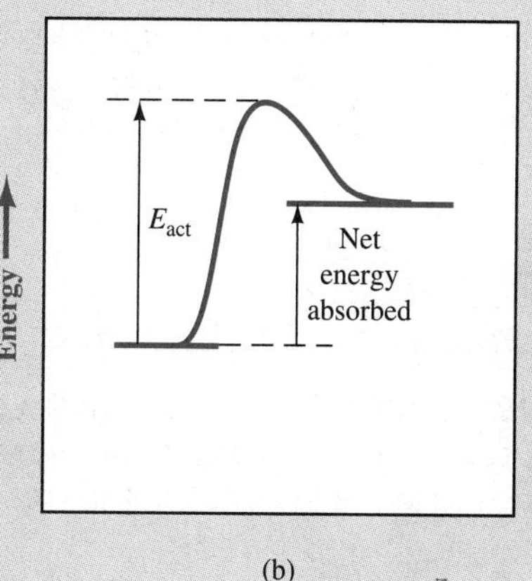

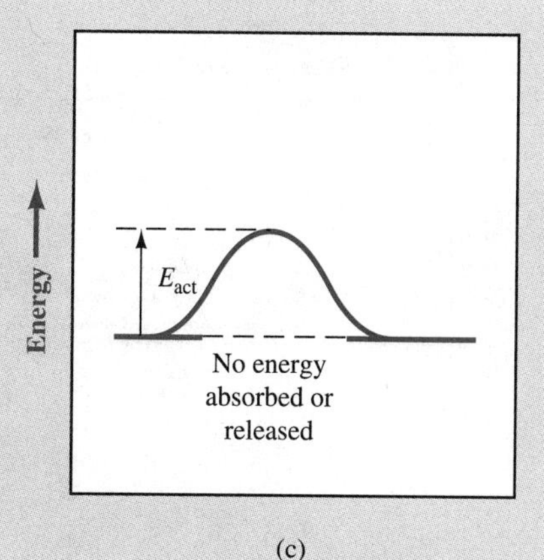

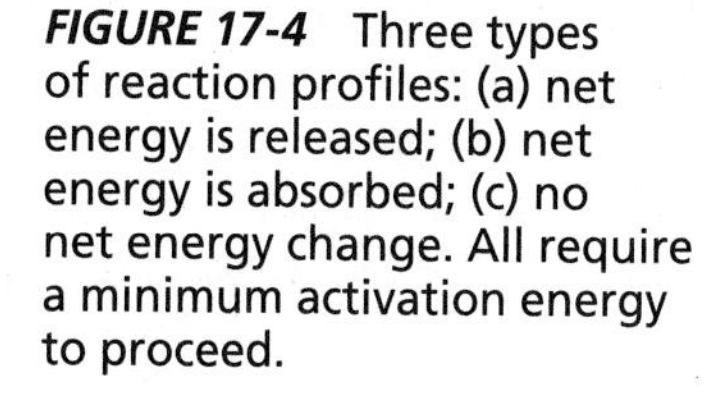
FIGURE 17-4 Three types of reaction profiles: (a) net energy is released; (b) net energy is absorbed; (c) no net energy change. All require a minimum activation energy to proceed.

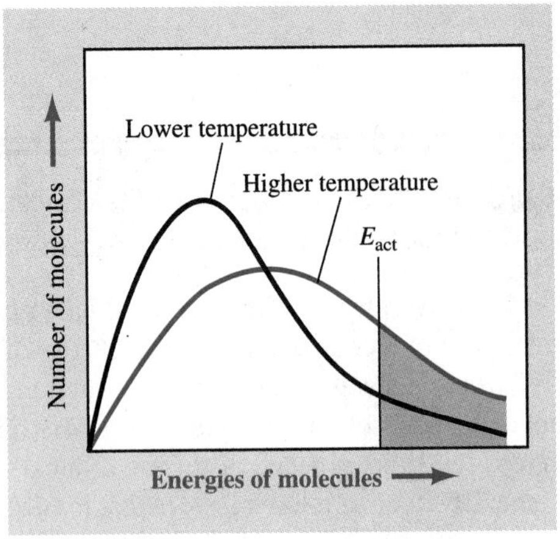

FIGURE 17-5 The distributions of energies at two temperatures show that at higher temperatures, a greater number of molecules have sufficient energy to get over an energy barrier.

higher temperature has more high-energy molecules and fewer low-energy molecules than the sample at the lower temperature. Suppose the line labeled E_{act} in Fig. 17-5 represents the activation energy for a reaction. There will be more molecules having sufficient activation energy at the higher temperature. Thus, *an increase in temperature always leads to an increase in reaction rate.*

Orientation

In order for a reaction to take place, the molecules that collide must have a favorable orientation. For example, when carbon monoxide and nitrogen dioxide collide, a favorable geometry of atoms for the activated complex might be the arrangement shown in Fig. 17-6a. If molecules should collide with one another as shown in Fig. 17-6b, the molecules would bounce away from one another unreacted. Thus, only a fraction of the collisions that occur actually lead to product formation, even if the molecules collide with an energy greater than the activation energy.

PROBLEM 17.2

Increasing the temperature not only affects the number of molecules with sufficient energy to overcome the energy barrier, but it also increases the frequency of collision. How is this accomplished, and what effect does it have on the reaction rate?

17.3 THE RATE EXPRESSION

OBJECTIVE 17.3 Explain what a rate expression and a rate constant are. Given the rate expression for a reaction, predict the effect that a change in the concentration of one of the reactants will have on the reaction rate.

Rate expression: A relationship between the rate of a reaction and the concentrations of the reactants.

The rate of a reaction can be described by a mathematical expression known as a **rate expression.** Rate expressions must be determined experimentally, and they vary

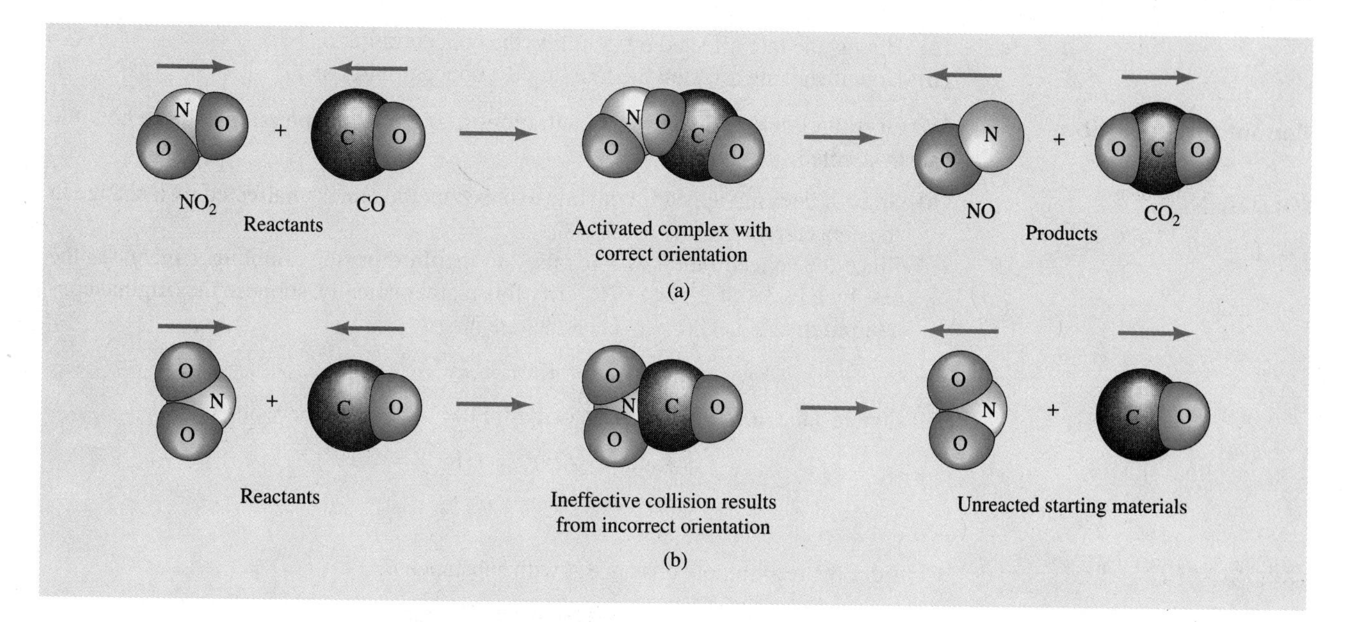

FIGURE 17-6 The molecules that collide in the transition state must have the correct orientation. (a) When the molecules that collide have the correct orientation, products can result. (b) Incorrect orientation results in an ineffective collision.

considerably in complexity. Let us consider a relatively simple example, the rate expression for the reaction of hydrogen and iodine.

$$H_2(g) + I_2(g) \rightarrow 2\ HI(g)$$

If this reaction is carried out at an elevated temperature so that iodine is in the gaseous state, the rate expression is

$$\text{Rate} = k[H_2][I_2]$$

This expression relates the rate of the reaction to the concentrations of the reactants (hydrogen and iodine). The proportionality constant, k, is referred to as the *rate constant.* The brackets around H_2 and I_2 indicate that we are referring to the molar concentrations (moles per liter) of these substances. We can see that when the concentration of hydrogen is doubled, the rate is also doubled. This makes sense, because we would expect twice as many collisions to occur between hydrogen and iodine. Doubling the concentration of iodine likewise doubles the rate of this reaction. If we were to double both, the reaction rate would be quadrupled.

The rate expression for the reaction of hydrogen with iodine is quite logical and might seem to be predictable from the balanced equation. However, rate expressions do not necessarily conform to the balanced chemical equation, as the following example illustrates.

EXAMPLE 17.1

The following reaction of substance X with substance Y obeys the accompanying rate expression:

$$X + Y \rightarrow \text{Products}$$
$$\text{Rate} = k[Y]^3$$

(a) How is the rate affected by doubling the concentration of *X*?
(b) How is the rate affected by doubling the concentration of *Y*?

Planning the solution

We substitute each of the new concentrations into the rate expression and see how the rate is affected.

SOLUTION

(a) Since *X* does not appear in the rate expression, the rate is unaffected by a change in the concentration of *X*.
(b) Since the concentration of *Y* is raised to the third power, doubling *Y* increases the rate by a factor of 2^3, or 8. To verify this mathematically, suppose the original concentration of *Y* is $[Y] = y$. Then the rate is

$$\text{Rate} = ky^3$$

If *Y* doubles, the new concentration of *Y* is $[Y] = 2y$, and the new rate is

$$\text{Rate} = k(2y)^3 = k(8y^3) = 8(ky^3)$$

PROBLEM 17-3

Consider the reaction of substance *A* with substance *B*:

$$A + B \rightarrow \text{Products}$$

(a) Suppose the rate expression for the reaction of *A* with *B* is

$$\text{Rate} = k[A][B]$$

How is the rate of the reaction affected by doubling the concentration of *A*? By doubling the concentration of *B*?

(b) Suppose the rate expression for the reaction of *A* with *B* is

$$\text{Rate} = k[A][B]^2$$

How is this rate affected by doubling $[A]$? By doubling $[B]$?

(c) Suppose the rate expression for the reaction of *A* with *B* is

$$\text{Rate} = k[A]$$

How is this rate affected by doubling $[A]$? By doubling $[B]$?

17.4 CATALYSTS

OBJECTIVE 17.4 Explain the function of a catalyst. Explain how a catalyst works in terms of the reaction profile. Explain what an enzyme is.

Catalyst: A substance that increases the rate of a chemical reaction without being consumed.

Reaction rates may be affected by factors other than concentration and temperature. A **catalyst** is a substance that causes a reaction to proceed at a more rapid rate but is not itself consumed by the reaction. For example, the decomposition of potassium chlorate, $KClO_3$, proceeds sluggishly, even on heating. However, the addition of manganese dioxide, MnO_2, causes the reaction to proceed smoothly even at moderate temperatures:

$$2\ KClO_3 \xrightarrow{MnO_2} 2\ KCl + 3\ O_2$$

The catalyst is not used up in the reaction, so it is not shown as a reactant or product but is written above the arrow instead.

A catalyst performs its function by *lowering the activation energy* of the reaction. It does not affect the energies of the reactants or products. Instead, a catalyst is a chemical that provides reactants with an alternative "route" with a lower activation energy than the uncatalyzed pathway. Figure 17-7 shows the catalyzed pathway and the uncatalyzed pathway of the same reaction. Lowering the activation energy increases the number of reacting molecules with sufficient energy to overcome the energy barrier, thereby leading to an increased reaction rate.

Enzyme: A biological catalyst.

Enzymes are biological catalysts that enable the chemical reactions in living organisms to proceed enormously faster than they would in a test tube. For example, enzymes in the mouth cause starches to be broken down into simpler substances, even before reaching our stomachs. The same reaction occurs much more sluggishly in the absence of enzymes. These remarkable catalysts are essential to the process we call life. In spite of their complexity and the important tasks they perform, they operate in basically the same manner as all other catalysts.

PROBLEM 17.4

Why is a catalyst written above the arrow in a chemical equation, rather than with the reactants?

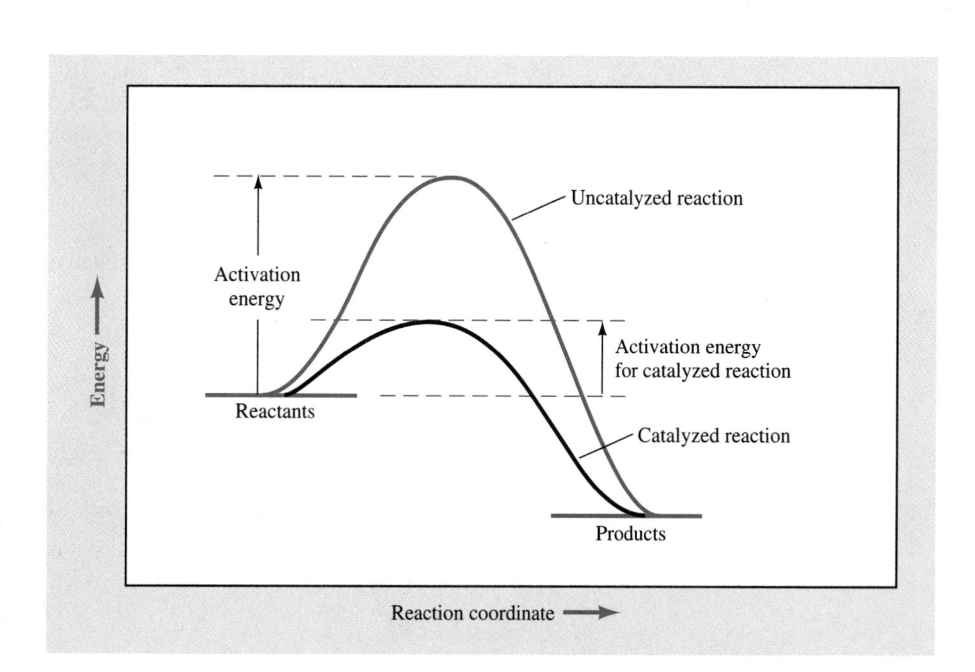

FIGURE 17-7 Energy profiles of catalyzed and uncatalyzed pathways.

17.5 EQUILIBRIUM

OBJECTIVE 17.5 Explain what an equilibrium mixture is. Characterize the equilibrium state in terms of the rates of opposing processes. Characterize equilibrium in terms of its macroscopic and microscopic properties.

In our discussion of vapor pressure (Section 12.8), you were introduced to the general concepts of equilibrium. We expanded on those concepts slightly when we discussed the dissociation of water and of weak acids (Sections 15.2 and 15.5), and the reversibility of oxidation–reduction half-reactions (Section 16.6). With the information you now have concerning reaction rates, we are ready to embark on a more comprehensive examination of equilibrium.

Let us consider the reaction of molecular hydrogen with iodine to form hydrogen iodide:

$$H_2(g) + I_2(g) \rightarrow 2\,HI(g)$$

If the reaction is carried out at 520°C, all of the reactants and products, including iodine, exist in the gaseous state. In its gaseous state, iodine is a dark purple vapor, and we can detect its presence by this characteristic color. Furthermore, we can detect any changes in the iodine concentration by observing changes in the intensity of its color. A darkening of the purple color indicates an increase in the concentration of iodine, whereas a lightening of the color indicates a decrease in its concentration. (Because hydrogen and hydrogen iodide are both colorless gases, their presence does not impair our ability to monitor the iodine concentration in this manner.)

Let us carry out the reaction in a transparent 1.00-L container so that we can observe the iodine concentration. If we mix 0.100 mol of hydrogen and 0.100 mol of iodine together, the initial color (due to iodine) is a deep purple. As time passes, we observe a gradual decrease in the intensity of the iodine color, indicating that iodine is reacting with the hydrogen, resulting in hydrogen iodide formation. After a while we observe no further decrease in the intensity of color. However, the persistence of some purple color indicates that iodine is still present, meaning that the reaction has not gone to completion (Fig. 17-8a). An experimental determination of the actual concentrations would give the following results:

	$H_2(g)$	+ $I_2(g)$	$\rightarrow$ $2\,HI(g)$
Initial concentration	0.100 M	0.100 M	0.000 M
Final concentration	0.020 M	0.020 M	0.160 M

Before we attempt to explain why the reaction has not gone to completion, let us carry out a second experiment. In a 1.00-L container identical to the one we just used, let us measure out 0.200 mol of hydrogen iodide, the amount of hydrogen iodide we expected to form in our first experiment. (This container will also be maintained at 520°C, the temperature we used in the first experiment.) If the reverse of our first reaction were to occur, hydrogen and iodine would be produced from the hydrogen iodide:

$$2\,HI(g) \rightarrow H_2(g) + I_2(g)$$

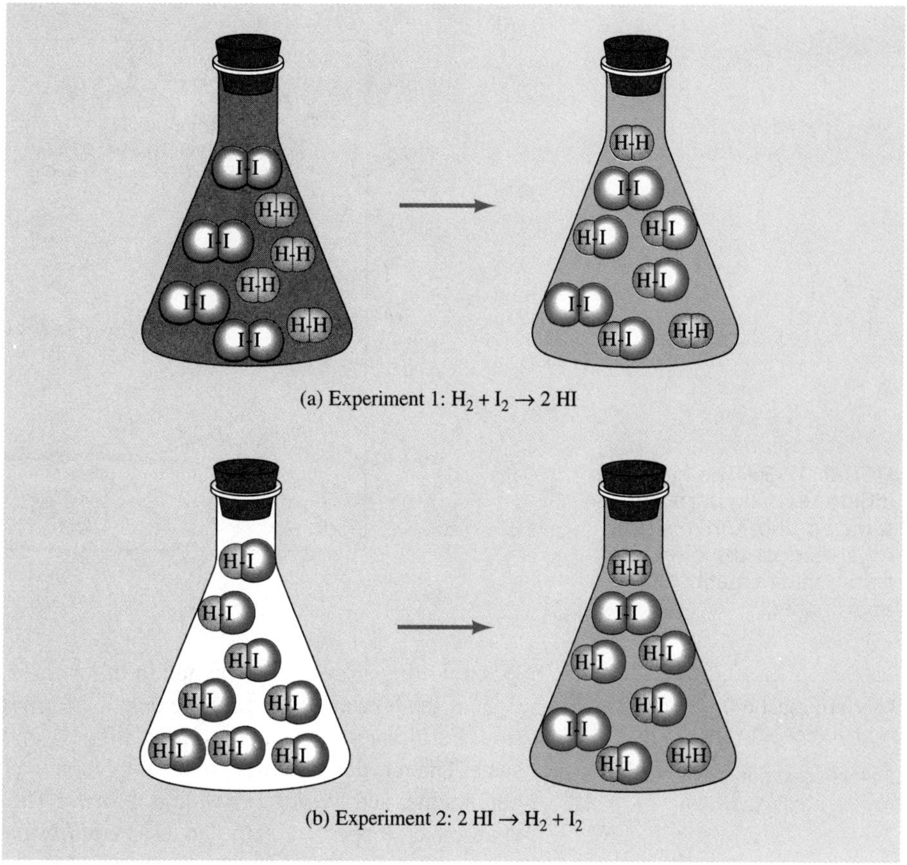

(a) Experiment 1: $H_2 + I_2 \rightarrow 2\ HI$

(b) Experiment 2: $2\ HI \rightarrow H_2 + I_2$

FIGURE 17-8 (a) When hydrogen and iodine react, the purple color of iodine fades. (b) Hydrogen iodide reacts with itself to produce hydrogen and iodine. The same point of equilibrium is reached from either direction.

Indeed, as time passes, we begin to notice the appearance of the purple color characteristic of iodine. The hydrogen iodide must be reacting with itself to form iodine (and hydrogen). The purple color of the iodine continues to intensify, until at some point we observe no further increase in the concentration of iodine. The measured concentrations would be as follows:

	$2\ HI(g)$	$\rightarrow$	$H_2(g)$	$+$	$I_2(g)$
Initial concentration	0.200 M		0.000 M		0.000 M
Final concentration	0.160 M		0.020 M		0.020 M

We find that the intensity of the iodine color in the second experiment is identical to that in the first experiment (Fig. 17-8b). Figure 17-9 (p. 530) shows the results of these two experiments graphically.

We can draw several conclusions from these experiments. First, the reaction of hydrogen and iodine to form hydrogen iodide is a readily reversible reaction. We might better combine the two chemical equations we have written to show this reversibility:

$$H_2(g) + I_2(g) \rightleftharpoons 2\ HI(g)$$

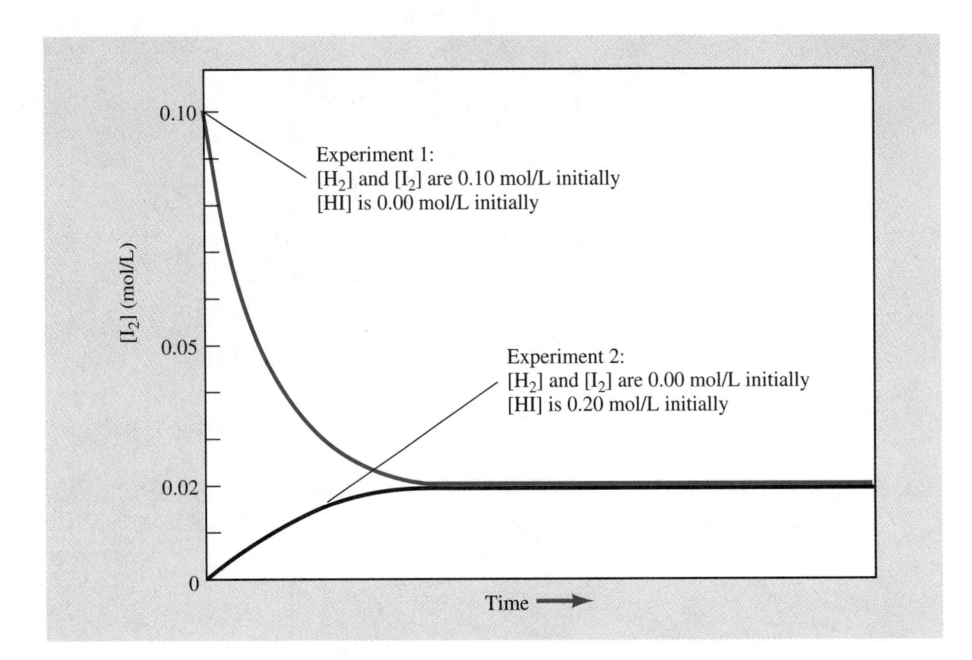

FIGURE 17-9 The hydrogen iodide reaction produces the same equilibrium mixture, regardless of the direction from which equilibrium is approached.

Forward reaction: The reaction written from left to right.

Reverse reaction: The reaction written from right to left.

When writing a reversible reaction in this fashion, we refer to the reaction from left to right as the **forward reaction;** the reaction from right to left is called the **reverse reaction.** Furthermore, whether we start with H_2 and I_2 or with HI, the end result is the same. The reaction does not go to completion in either direction, so a mixture of hydrogen, iodine, and hydrogen iodide is formed. The mixture that is present when all net change has stopped is referred to as an *equilibrium mixture.*

Let us consider why we observe this behavior. In our first experiment, we placed hydrogen molecules and iodine molecules together in a flask. As molecules of hydrogen collided with molecules of iodine, some of the collisions occurred with enough energy to form hydrogen iodide (the forward reaction, Fig. 17-10). Consequently, iodine was consumed by its reaction with hydrogen, and the intensity of the iodine color gradually diminished. At the same time, the number of molecules of hydrogen iodide (HI) was increasing. These molecules could collide with each other to form hydrogen and iodine (the reverse reaction). As more and more molecules of hydrogen iodide were formed, the probability increased that two molecules of hydrogen iodide would collide with sufficient energy to undergo the reverse reaction, thereby forming hydrogen and iodine again. At some point the rate at which molecules of hydrogen iodide were being used up exactly equaled the rate at which they were being formed. In other words, the rates of the forward reaction and the reverse reaction were equal, and we observed no further change in the iodine concentration.

Rate of forward reaction = Rate of hydrogen iodide formation

Rate of reverse reaction = Rate of hydrogen iodide consumption

Equilibrium: A dynamic condition in which the rates of opposing processes are equal.

Equilibrium is achieved when two opposing processes proceed at equal rates.

Rate of forward reaction = Rate of reverse reaction

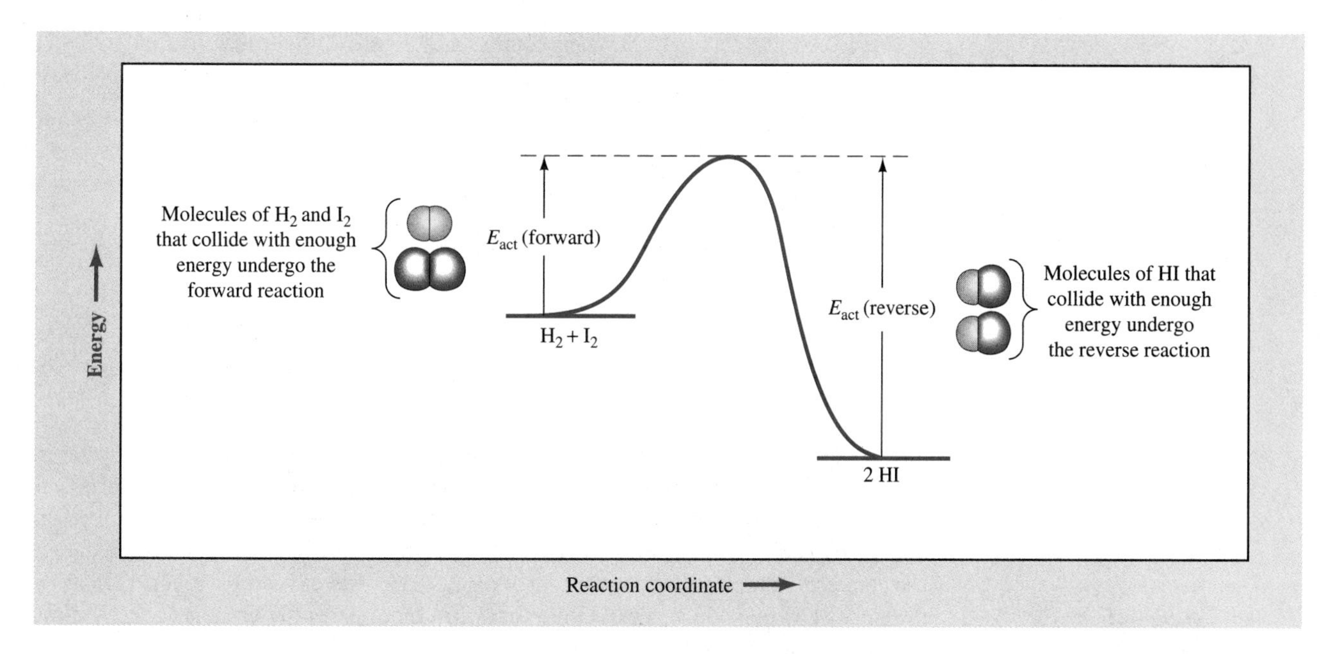

FIGURE 17-10 A reaction profile for the hydrogen iodide equilibrium. At equilibrium, the rates of the forward and reverse reactions are equal.

Equilibrium is a dynamic process. Individual molecules of hydrogen, iodine, and hydrogen iodide continue to undergo reaction with one another. However, because the opposing processes proceed at equal rates, we observe no net change in the system. Equilibrium is characterized by constant macroscopic properties (such as iodine intensity). At the microscopic level, however, changes continue to occur.

When hydrogen iodide was placed in a similar container in the second experiment, pairs of hydrogen iodide molecules collided with one another to form hydrogen and iodine (Fig. 17-10). As more and more hydrogen and iodine molecules appeared, their probability of colliding and undergoing the opposite reaction (formation of hydrogen iodide) increased. Once again, a point was reached at which the rates of the opposing processes were equal. Regardless of the starting direction, equilibrium was reached.

PROBLEM 17.5

Suppose we introduce 0.100 mol of HI and 0.100 mol of I_2 into a 1.00-L flask under the same conditions as those just discussed and note the intensity of the iodine color. If we allow a period of time to elapse, will the intensity of the iodine color increase, decrease, or stay the same? Explain your answer.

17.6 THE EQUILIBRIUM CONSTANT

OBJECTIVE 17.6 Given a chemical equation for a reaction at equilibrium, write an equilibrium expression.

As described in Section 17.5, when equimolar amounts of hydrogen and iodine react, they form an equilibrium mixture containing hydrogen, iodine, and hydrogen iodide:

	$H_2(g)$ +	$I_2(g)$ ⇌	$2\ HI(g)$
Initial concentration	0.100 M	0.100 M	0.000 M
Equilibrium concentration	0.020 M	0.020 M	0.160 M

By contrast, when hydrogen and chlorine react, virtually all of the reactants are consumed in the reaction:

	$H_2(g)$ +	$Cl_2(g)$ ⇌	$2\ HCl(g)$
Initial concentration	0.100 M	0.100 M	0.000 M
Equilibrium concentration	0.000 M	0.000 M	0.200 M

You might wonder why some reactions seem to proceed to completion, whereas other reactions reach equilibrium with only part of the starting materials actually consumed. The reason for the difference in behavior is that the point at which equilibrium is reached is different for every reaction, depending on the reaction in question and on the conditions employed. For every reaction, however, we can write a mathematical expression known as the **equilibrium expression** that relates the various concentrations of reactants and products at equilibrium.

Equilibrium expression: A relationship between the concentrations of reactants and products at equilibrium.

For the reaction $A + B \rightleftharpoons C + D$, the equilibrium expression is

$$K_{eq} = \frac{[C][D]}{[A][B]}$$

Equilibrium constant: Numerical constant that relates the mutual concentrations of reactants and products at equilibrium.

In this expression, K_{eq} is a mathematical constant known as the **equilibrium constant.** Brackets are used to designate the concentrations of A, B, C, and D in moles per liter.

For any reaction $nA + mB \rightleftharpoons xC + yD$, the equilibrium expression is written

$$K_{eq} = \frac{[C]^x[D]^y}{[A]^n[B]^m} \quad \begin{array}{l}\leftarrow \textbf{Products} \\ \leftarrow \textbf{Reactants}\end{array}$$

where K_{eq} is the equilibrium constant and n, m, x, and y are the coefficients of each substance in the balanced chemical equation. Note that the concentrations of the products appear in the numerator, whereas the concentrations of the reactants are in the denominator. The coefficient preceding each reactant and product appears in the equilibrium expression as an exponent. For the hydrogen iodide equilibrium,

$$H_2(g) + I_2(g) \rightleftharpoons 2\ HI(g)$$

the equilibrium expression is

$$K_{eq} = \frac{[HI]^2}{[H_2][I_2]}$$

This means that, no matter what initial amounts of hydrogen, iodine, and hydrogen iodide we mix together, when equilibrium is established, the square of the hydrogen iodide concentration divided by the mathematical product of the hydrogen concentration and the iodine concentration equals a constant.

As we have stated, each reaction has its own unique equilibrium constant, and each one must be determined experimentally. Furthermore, the equilibrium constant for a given reaction varies with temperature.

Before concluding this section, let us emphasize an important feature of the equilibrium expression. The equilibrium expression *always* places the products in the numerator and the reactants in the denominator. Occasionally we wish to consider the reverse direction of a reaction. For example, we might wish to examine the decomposition of hydrogen iodide:

$$2\ HI(g) \rightleftharpoons H_2(g) + I_2(g)$$

When we reverse the direction of a reaction, we also reverse the reactants and products. Because the new equilibrium expression is just the reciprocal of the original one, the numerical value of the new equilibrium constant (K'_{eq}) is the reciprocal of the value K_{eq} for the old (reverse) direction:

$$K'_{eq} = \frac{[H_2][I_2]}{[HI]^2} = \frac{1}{K_{eq}}$$

It is important, for the sake of accuracy, to write both the reaction and the equilibrium expression together.

The format for writing equilibrium expressions is different when the reactants and products are not all in homogeneous contact with one another. We will simplify our discussion here by restricting ourselves to gaseous equilibria, which are always homogeneous.

Let us write the equilibrium expressions for a few more reactions to illustrate these principles.

EXAMPLE 17.2

Ammonia may exist in equilibrium with its elements. Write an equilibrium expression for its formation from nitrogen and hydrogen.

$$N_2(g) + 3\ H_2(g) \rightleftharpoons 2\ NH_3(g)$$

Planning the solution

The equilibrium expression gives the concentrations of the products in the numerator and the concentrations of the reactants in the denominator. The concentration of each reactant and product is raised to a power equal to its coefficient in the balanced equation. The terms in the numerator and denominator are multiplied and divided—never added or subtracted.

SOLUTION

$$K_{eq} = \frac{[NH_3]^2}{[N_2][H_2]^3}$$

EXAMPLE 17.3

Write an equilibrium expression for the following reaction.

$$CH_4(g) + 2\ O_2(g) \rightleftharpoons CO_2(g) + 2\ H_2O(g)$$

SOLUTION

$$K_{eq} = \frac{[CO_2][H_2O]^2}{[CH_4][O_2]^2}$$

EXAMPLE 17.4

Write an equilibrium expression for the following decomposition of dinitrogen tetroxide.

$$N_2O_4(g) \rightleftharpoons 2\,NO_2(g)$$

SOLUTION

$$K_{eq} = \frac{[NO_2]^2}{[N_2O_4]}$$

PROBLEM 17.6

Write an equilibrium expression for each of the following equilibria. (Note that all reactants and products are in the gaseous state.)

(a) $2\,SO_2(g) + O_2(g) \rightleftharpoons 2\,SO_3(g)$
(b) $N_2(g) + 3\,Cl_2(g) \rightleftharpoons 2\,NCl_3(g)$
(c) $2\,NO(g) + Br_2(g) \rightleftharpoons 2\,NOBr(g)$
(d) $CO(g) + H_2O(g) \rightleftharpoons CO_2(g) + H_2(g)$
(e) $CO(g) + 3\,H_2(g) \rightleftharpoons CH_4(g) + H_2O(g)$
(f) $PCl_5(g) \rightleftharpoons PCl_3(g) + Cl_2(g)$
(g) $3\,O_2(g) \rightleftharpoons 2\,O_3(g)$
(h) $CO(g) + Cl_2(g) \rightleftharpoons COCl_2(g)$
(i) $H_2(g) + F_2(g) \rightleftharpoons 2\,HF(g)$

17.7 Calculating the Equilibrium Constant

OBJECTIVE 17.7 Given a reaction at equilibrium and a set of equilibrium conditions, calculate the equilibrium constant. Explain the relationship between the magnitude of the equilibrium constant and the relative quantities of the reactants and products.

Each equilibrium has its own unique equilibrium constant, which must be determined experimentally. Furthermore, the value of K_{eq} varies with temperature. To evaluate K_{eq} at a given temperature, we must obtain a set of concentrations for each of the reactants and products at equilibrium. For example, suppose the indicated equilibrium concentrations are observed for N_2O_4 and NO_2 in the following equilibrium:

	$N_2O_4(g)$	$\rightleftharpoons$ $2\,NO_2(g)$
Equilibrium concentrations	0.40 M	0.60 M

We substitute these values into our equilibrium expression to obtain K_{eq}:

$$K_{eq} = \frac{[NO_2]^2}{[N_2O_4]} = \frac{(0.60)^2}{(0.40)} = 0.90$$

As another example, suppose the following equilibrium concentrations are observed for the formation of ammonia:

	$N_2(g)$	$+\ 3\,H_2(g)$	$\rightleftharpoons$ $2\,NH_3(g)$
Equilibrium concentrations	0.040 M	0.500 M	0.050 M

Again, simple substitution into the equilibrium expression gives the equilibrium constant:

$$K_{eq} = \frac{[NH_3]^2}{[N_2][H_2]^3} = \frac{(0.050)^2}{(0.040)(0.500)^3} = 0.50$$

EXAMPLE 17.5

Phosphorus pentachloride, PCl_5, decomposes into phosphorus trichloride, PCl_3, and chlorine, Cl_2, according to the following equilibrium:

$$PCl_5(g) \rightleftharpoons PCl_3(g) + Cl_2(g)$$

When 0.300 mol of $PCl_5(g)$ is introduced into a 1.00-L vessel, the equilibrium concentrations are found to be

$$[PCl_5] = 0.200\ M \qquad [PCl_3] = 0.100\ M \qquad [Cl_2] = 0.100\ M$$

Evaluate K_{eq}.

SOLUTION

To evaluate K_{eq}, we must first write an equilibrium expression:

$$K_{eq} = \frac{[PCl_3][Cl_2]}{[PCl_5]}$$

Substituting the equilibrium concentrations into this expression gives the value of K_{eq}:

$$K_{eq} = \frac{(0.100)(0.100)}{(0.200)} = 0.0500$$

EXAMPLE 17.6

A 2.00-mol sample of NOCl is introduced into a 1.00-L flask and allowed to come to equilibrium as follows:

$$2\ NOCl(g) \rightleftharpoons 2\ NO(g) + Cl_2(g)$$

The equilibrium concentrations are found to be

$$[NOCl] = 0.20\ M \qquad [NO] = 1.80\ M \qquad [Cl_2] = 0.90\ M$$

Calculate the equilibrium constant, K_{eq}.

SOLUTION

Again, we must first write an equilibrium expression:

$$K_{eq} = \frac{[NO]^2[Cl_2]}{[NOCl]^2}$$

Substituting the equilibrium concentrations into the equilibrium expression gives the value of K_{eq}:

$$K_{eq} = \frac{(1.80)^2(0.90)}{(0.20)^2} = 73$$

PROBLEM 17.7

Calculate the value of K_{eq} for each of the following equilibria. In each case, the equilibrium concentrations are given.

(a) $N_2(g) + O_2(g) \rightleftharpoons 2\,NO(g)$
$[N_2] = 1.00\text{ M}$ $[O_2] = 0.10\text{ M}$ $[NO] = 0.10\text{ M}$

(b) $2\,ICl(g) \rightleftharpoons I_2(g) + Cl_2(g)$
$[ICl] = 0.67\text{ M}$ $[I_2] = 0.25\text{ M}$ $[Cl_2] = 0.20\text{ M}$

(c) $PCl_5(g) \rightleftharpoons PCl_3(g) + Cl_2(g)$
$[PCl_5] = 0.40\text{ M}$ $[PCl_3] = 0.20\text{ M}$ $[Cl_2] = 0.10\text{ M}$

Earlier, we pointed out that the reaction of hydrogen with chlorine goes virtually to completion, whereas the reaction of hydrogen with iodine does not. The difference in behavior of these two reactions reflects the large difference between the values of their equilibrium constants under comparable conditions.

The equilibrium expression places the concentrations of the products in the numerator and the concentrations of the reactants in the denominator. This means that an equilibrium that favors products has a large value of K_{eq}, because there is a much greater concentration of products than of reactants. An equilibrium that favors reactants does not proceed to any appreciable extent and has a small value of K_{eq}. Those equilibria that have comparable concentrations of reactants and products have moderate values of K_{eq} (close to 1).

The equilibrium expressions for the formation of hydrogen iodide and hydrogen chloride are quite similar:

Hydrogen iodide formation: $H_2(g) + I_2(g) \rightleftharpoons 2\,HI(g)$ $\quad K_{eq} = \dfrac{[HI]^2}{[H_2][I_2]}$

Hydrogen chloride formation: $H_2(g) + Cl_2(g) \rightleftharpoons 2\,HCl(g)$ $\quad K_{eq} = \dfrac{[HCl]^2}{[H_2][Cl_2]}$

However, the value of K_{eq} is larger ($>10^{10}$) for the hydrogen chloride reaction. This means that, when equilibrium is achieved, virtually all of the substances present are products (HCl). On the other hand, the value of K_{eq} for the hydrogen iodide equilibrium is close to 1, so comparable amounts of reactants and products exist at equilibrium. Figure 17-11 shows the relationship between the magnitude of K_{eq} and the relative quantities of reactants and products for three types of equilibria: (a) where the value of K_{eq} is large, (b) where K_{eq} is moderate, and (c) where K_{eq} is small.

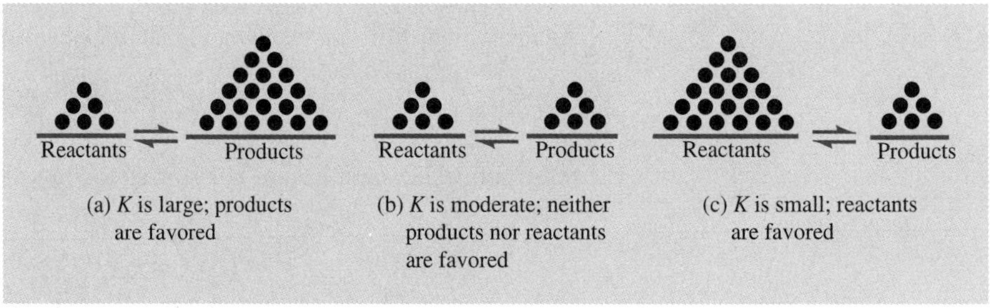

Reactants ⇌ Products
(a) K is large; products are favored
Reactants ⇌ Products
(b) K is moderate; neither products nor reactants are favored
Reactants ⇌ Products
(c) K is small; reactants are favored

FIGURE 17-11 Relationship between the magnitude of K_{eq} and the relative quantities of reactants and products in equilibrium.

PROBLEM 17.8

At 25°C, the equilibrium constant for the formation of NO(g) from $N_2(g)$ and $O_2(g)$ is $K_{eq} = 1 \times 10^{-30}$.

$$N_2(g) + O_2(g) \rightleftharpoons 2\,NO(g)$$

Suppose 1 mol of nitrogen and 1 mol of oxygen are introduced into a 1.00-L flask, and the contents of the flask are allowed to come to equilibrium. Will the equilibrium mixture be primarily products, will it be primarily reactants, or will there be comparable quantities of each? Explain your answer.

17.8 LE CHATELIER'S PRINCIPLE

OBJECTIVE 17.8 State Le Chatelier's Principle. Apply Le Chatelier's Principle to changes in concentration, pressure, or temperature of a gaseous system at equilibrium.

Let us return to the reaction of hydrogen and iodine and see what happens when we change some of the conditions of the equilibrium mixture.

	$H_2(g)$	+	$I_2(g)$	$\rightleftharpoons$	$2\ HI(g)$
Equilibrium mixture	0.020 M		0.020 M		0.160 M

Suppose, for example, we introduce an additional 0.034 mol of hydrogen into the equilibrium mixture, thereby increasing its concentration. The purple color of iodine that had achieved a constant intensity begins to fade again, indicating that iodine is reacting with the additional hydrogen to produce more hydrogen iodide. After a period of time, the iodine color becomes constant once more, indicating that equilibrium has been reestablished.

The introduction of additional hydrogen into the reaction mixture has resulted in more product formation. We say that the equilibrium has "shifted to the right." We might say that the addition of hydrogen to the original equilibrium mixture (which we will refer to as the "system") placed a *stress* on the system. Suddenly there was extra hydrogen present, available to collide with iodine molecules. The system responded to this sudden increase in hydrogen concentration by removing some of it through reaction with the unreacted iodine that was still present. This phenomenon is quite general, and it is expressed in **Le Chatelier's Principle:** When a stress is placed on a reaction mixture at equilibrium, the system responds in such a way as to relieve the stress. By increasing the hydrogen concentration, we placed a stress on the equilibrium mixture. The reaction relieved the stress by removing some of the additional hydrogen through reaction with the iodine that was present.

Le Chatelier's Principle: When a stress is applied to a system at equilibrium, the system responds in such a way as to relieve the stress.

Le Chatelier's Principle makes good sense. When additional hydrogen was placed in the mixture, there was a sudden increase in the probability that hydrogen molecules would collide with iodine molecules to form hydrogen iodide. There was no corresponding increase in the probability that two hydrogen iodide molecules would react together to form hydrogen and iodine. In other words, the rate of the forward reaction temporarily exceeded the rate of the reverse reaction. Consequently, more of the iodine reacted to form hydrogen iodide. As additional molecules of hydrogen iodide were formed, however, the rate of the reverse reaction increased until the rates became equal in both directions and a new equilibrium was established.

If we were to analyze the composition of the new equilibrium mixture, we would find that the new concentrations of hydrogen, iodine, and hydrogen iodide still fit our equilibrium expression.

	$H_2(g)$ +	$I_2(g)$ ⇌	$2\ HI(g)$
Old equilibrium	0.020 M	0.020 M	0.160 M
New equilibrium	0.045 M	0.011 M	0.178 M

Old conditions: $$K_{eq} = \frac{(0.160)^2}{(0.020)(0.020)} = 64$$

New conditions: $$K_{eq} = \frac{(0.178)^2}{(0.045)(0.011)} = 64$$

The result of the stress on the system (increased hydrogen concentration) and the response to the stress (decreased iodine concentration and increased hydrogen iodide concentration) is such that substituting all three new concentrations into the equilibrium expression gives the same value of K_{eq} as that found for the original set of conditions.

A stress is also applied to a system when the concentration of one of the reactants or products is decreased. Suppose that, in the original equilibrium mixture, we were to remove the hydrogen iodide as it formed. This would place a stress on the system by diminishing the hydrogen iodide concentration. The system would respond by producing more hydrogen iodide. In other words, the equilibrium would again shift to the right. Let us look at a few more examples.

EXAMPLE 17.7

In which direction will the following gaseous equilibrium shift when the concentration of N_2O_4 is increased?

$$N_2O_4(g) \rightleftharpoons 2\ NO_2(g)$$

SOLUTION

The equilibrium will shift to the right, thereby relieving the excess N_2O_4.

EXAMPLE 17.8

In which direction will the following equilibrium shift if the ammonia concentration is decreased?

$$3\ H_2(g) + N_2(g) \rightleftharpoons 2\ NH_3(g)$$

SOLUTION

The equilibrium will shift to the right, thereby producing more ammonia.

EXAMPLE 17.9

In which direction will the following equilibrium shift when the oxygen concentration is increased?

$$2\ NOCl(g) \rightleftharpoons N_2(g) + O_2(g) + Cl_2(g)$$

SOLUTION

The equilibrium will shift to the left, thereby using up the added oxygen.

PROBLEM 17.9

In which direction will each of the following gaseous equilibria shift when the indicated stress is applied?

(a) $2\ SO_2(g) + O_2(g) \rightleftharpoons 2\ SO_3(g)$ — Increased $[O_2]$
(b) $N_2(g) + O_2(g) \rightleftharpoons 2\ NO(g)$ — Decreased $[N_2]$
(c) $CO(g) + Cl_2(g) \rightleftharpoons COCl_2(g)$ — Increased $[COCl_2]$

(d) $CO(g) + 3\,H_2(g) \rightleftharpoons CH_4(g) + H_2O(g)$ Decreased $[H_2O]$
(e) $CO(g) + H_2O(g) \rightleftharpoons CO_2(g) + H_2(g)$ Increased $[CO_2]$

The Effect of Pressure on Equilibrium

Le Chatelier's Principle applies to stresses other than simple concentration changes. Altering the pressure of a gaseous system imposes another type of stress. For example, we can increase the pressure in the N_2O_4/NO_2 equilibrium by decreasing the volume to which the gaseous mixture is confined. The change in pressure places a stress on the system. Examination of the chemical equation reveals that there are two moles of gaseous products but only one mole of gaseous reactants:

$$\underset{\textbf{1 mole}}{N_2O_4(g)} \rightleftharpoons \underset{\textbf{2 moles}}{2\,NO(g)}$$

If some of the product is converted to reactant, the total number of moles in the system is reduced, resulting in a decrease in pressure. By shifting to the left, this system can relieve some of the stress caused by the increase in pressure.

On the other hand, shifting the hydrogen iodide equilibrium would have no effect on the number of moles present, because the total number of moles of gas is the same on both sides of the equation:

$$\underbrace{H_2(g) + I_2(g)}_{\textbf{2 moles}} \rightleftharpoons \underset{\textbf{2 moles}}{2\,HI(g)}$$

Shifting this equilibrium would not alter the pressure, so changes in pressure have no effect on this equilibrium.

Our discussion of equilibrium thus far has been restricted to gaseous equilibria. Changes in pressure affect only those equilibria in which one or more of the reactants and products are gases. The discussion presented here concerning the effect of pressure on equilibrium holds only for equilibria in which all of the reactants and products are gases.

PROBLEM 17.10

For each of the following gaseous equilibria, predict the direction in which the equilibrium would shift if the pressure were altered as indicated.

(a) $PCl_3(g) + Cl_2(g) \rightleftharpoons PCl_5(g)$ Increased pressure
(b) $CO_2(g) + H_2(g) \rightleftharpoons CO(g) + H_2O(g)$ Decreased pressure
(c) $N_2(g) + 3\,H_2(g) \rightleftharpoons 2\,NH_3(g)$ Decreased pressure
(d) $2\,NOCl(g) \rightleftharpoons N_2(g) + O_2(g) + Cl_2(g)$ Increased pressure

The Effect of Temperature on Equilibrium

Earlier we stated that the equilibrium constant can vary with temperature. We can predict the direction of that effect on the equilibrium if we know whether the reaction is exothermic or endothermic. For example, the formation of ammonia is exothermic. We may write the chemical equation to include the heat term as follows:

$$N_2(g) + 3\,H_2(g) \rightleftharpoons 2\,NH_3(g) + 92.4\text{ kJ}$$

Applying Le Chatelier's Principle to this problem, we may treat the heat term like any of the reactants or products. Raising the temperature adds heat to the system, so this equilibrium shifts to the left in order to absorb the heat. This corresponds to a decrease in the value of K_{eq} as the temperature of the system is raised.

On the other hand, raising the temperature of an endothermic reaction such as

$$N_2(g) + O_2(g) + 180.8\ kJ \rightleftharpoons 2\ NO(g)$$

causes the equilibrium to shift to the right, because that is the direction needed to absorb the added heat. Thus, increasing the temperature of this system causes the nitrogen monoxide concentration to increase, corresponding to an increase in K_{eq}.

The Summary reviews Le Chatelier's Principle for changes in concentration, pressure, and temperature.

SUMMARY

Effects of Changing Concentration, Pressure, or Temperature for a Reaction at Equilibrium (Reactants ⇌ Products)

Applied stress	Direction of equilibrium shift
Concentration of reactants	
Increase	→ Products
Decrease	Reactants ←
Concentration of products	
Increase	Reactants ←
Decrease	→ Products
***Pressure: increase**	
Moles of reactants = Moles of products	No change
Moles of reactants < Moles of products	Reactants ←
Moles of reactants > Moles of products	→ Products
***Pressure: decrease**	
Moles of reactants = Moles of products	No change
Moles of reactants < Moles of products	→ Products
Moles of reactants > Moles of products	Reactants ←
Temperature: increase	
Exothermic	Reactants ←
Endothermic	→ Products
Temperature: decrease	
Exothermic	→ Products
Endothermic	Reactants ←

*True only when all reactants and products are gases.

PROBLEM 17.11

In which direction will the following equilibrium be shifted by the given stresses?

$$H_2(g) + Cl_2(g) \rightleftharpoons 2\,HCl(g) + 184.6\text{ kJ}$$

(a) Increased $[H_2]$ **(b)** Decreased pressure **(c)** Increased temperature

PROBLEM 17.12

In which direction will the following equilibrium be shifted by the given stresses?

$$N_2(g) + O_2(g) + Cl_2(g) + 106\text{ kJ} \rightleftharpoons 2\,NOCl(g)$$

(a) Decreased $[O_2]$
(b) Increased $[Cl_2]$
(c) Increased pressure
(d) Decreased temperature

17.9 EQUILIBRIUM AND WEAK ACIDS

OBJECTIVE 17.9 Write an equilibrium expression for the dissociation of a weak acid. Calculate the equilibrium constant for a weak acid, given a set of equilibrium concentrations or the percent dissociation of a solution of known molarity.

We have seen that weak acids dissociate only partially. The dissociation of a weak acid is a readily reversible equilibrium process. The following equation represents the dissociation of acetic acid:

$$HC_2H_3O_2(aq) \rightleftharpoons H^+(aq) + C_2H_3O_2^-(aq)$$

Like other equilibrium processes, this reaction has an equilibrium expression and an equilibrium constant:

$$K_a = \frac{[H^+][C_2H_3O_2^-]}{[HC_2H_3O_2]}$$

We generally use the symbol K_a when referring to the equilibrium constants of weak acids. For acetic acid, $K_a = 1.8 \times 10^{-5}$.

The formula HB is often used to refer to weak acids in general. Here, B^- represents the conjugate base of the weak acid. The equilibrium of the acid is always written in the direction that produces dissociation:

$$HB(aq) \rightleftharpoons H^+(aq) + B^-(aq)$$

where

$$K_a = \frac{[H^+][B^-]}{[HB]}$$

We can calculate the K_a value of a weak acid when we know the concentrations of the associated acid, HB, the hydrogen ion, H^+, and the conjugate base, B^-. These may be obtained by measuring the pH value of a solution of known concentration. For example, the pH of a 0.25 M solution of hydrocyanic acid, HCN, is 5.00. This

TABLE 17-1

	$HCN(aq)$	$\rightleftharpoons$	$H^+(aq)$	+	$CN^-(aq)$
Initial concentration before dissociation	0.25 M		0 M		0 M
Moles per liter that dissociate	-1.0×10^{-5} M		$+1.0 \times 10^{-5}$ M		$+1.0 \times 10^{-5}$ M
Equilibrium concentration after dissociation	0.25 M		1.0×10^{-5} M		1.0×10^{-5} M

corresponds to a hydrogen ion concentration of $[H^+] = 1.0 \times 10^{-5}$ M (pH = $-\log[1.0 \times 10^{-5}] = 5.00$). Because each hydrogen ion present must be accompanied by a cyanide ion, the hydrogen ion and cyanide ion concentrations must be the same: $[CN^-] = 1.0 \times 10^{-5}$. Finally, the hydrogen cyanide concentration, [HCN], equals the initial concentration of 0.25 M less the amount of acid that dissociates. Although the dissociation of hydrocyanic acid uses up some of the associated molecules, the amount dissociated (1.0×10^{-5} M) is so much smaller than 0.25 M that this dissociation does not significantly affect the concentration of associated acid. We may compare this to the removal of a grain of sand from a beach. The loss is negligible compared to the large number of grains present.

Substituting the values in Table 17-1 into the equilibrium expression gives us the value of K_a:

$$K_a = \frac{[H^+][CN^-]}{[HCN]} = \frac{(1.0 \times 10^{-5})(1.0 \times 10^{-5})}{(0.25)} = 4.0 \times 10^{-10}$$

EXAMPLE 17.10

Use the following set of equilibrium concentrations for 2.0 M $HC_2H_3O_2(aq)$ to calculate K_a.

$[H^+] = 0.0060$ M $\qquad [C_2H_3O_2^-] = 0.0060$ M $\qquad [HC_2H_3O_2] = 2.0$ M

Planning the solution

We must first write an equilibrium expression for the dissociation of the weak acid and then substitute the equilibrium concentrations into the equilibrium expression.

SOLUTION

The dissociation of the weak acid is

$$HC_2H_3O_2(aq) \rightleftharpoons H^+(aq) + C_2H_3O_2^-(aq)$$

Substituting the given values into the equilibrium expression gives the value of K_a.

$$K_a = \frac{[H^+][C_2H_3O_2^-]}{[HC_2H_3O_2]} = \frac{(0.0060)(0.0060)}{(2.0)} = 1.8 \times 10^{-5}$$

EXAMPLE 17.11

A 1.00 M solution of a weak acid, HB(aq), is 20% (= 0.20) dissociated. Calculate the value of K_a.

SOLUTION

If 20% of the acid is dissociated, the concentration of H^+ and B^- must be 0.20 M (0.20×1.00 M). Furthermore, that leaves 0.80 M of the acid in its associated form.

	$HB(aq) \rightleftharpoons$	$H^+(aq)$ +	$B^-(aq)$
Initial concentration:	1.00 M	0 M	0 M
Dissociated acid:	−0.20 M	+0.20 M	+0.20 M
Equilibrium concentration:	0.80 M	0.20 M	0.20 M

$$K_a = \frac{[H^+][B^-]}{[HB]} = \frac{(0.20)(0.20)}{(0.80)} = 0.050 = 5.0 \times 10^{-2}$$

PROBLEM 17.13

The equilibrium concentrations in a 1.00 M HF(aq) solution are

$$[HF] = 0.97 \text{ M} \qquad [H^+] = 0.026 \text{ M} \qquad [F^-] = 0.026 \text{ M}$$

Calculate K_a for the dissociation $HF(aq) \rightleftharpoons H^+(aq) + F^-(aq)$.

PROBLEM 17.14

The equilibrium concentration in a 0.10 M HClO(aq) solution are

$$[HClO] = 0.10 \text{ M} \qquad [H^+] = 5.6 \times 10^{-4} \text{ M} \qquad [ClO^-] = 5.6 \times 10^{-4} \text{ M}$$

Calculate K_a for the dissociation $HClO(aq) \rightleftharpoons H^+(aq) + ClO^-(aq)$

PROBLEM 17.15

A 0.10 M solution of a weak acid, HZ(aq), is 50% dissociated. Calculate K_a. What is the pH of this solution?

17.10 SOLUBILITY EQUILIBRIA

OBJECTIVE 17.10 Describe the characteristics of solubility equilibria. Write an equilibrium expression for an ionic solid in equilibrium with a solution of its ions. Calculate the solubility product of an ionic substance, given its solubility.

Earlier, we classified insoluble substances as those having solubilities of less than 0.1 g per 100 mL of water. In fact, all ionic substances have some limited solubility in water, though it may be quite low.

Let us describe what happens when lead(II) iodide dissolves. This is a salt we would classify as slightly soluble. As we have seen, when ionic salts dissolve, the ions become separated from one another.

$$PbI_2(s) \rightleftharpoons Pb^{2+}(aq) + 2\,I^-(aq)$$

This process is an equilibrium process and we shall describe it in terms of the same factors we examined in our study of equilibrium. Note that the reverse reaction represents the net ionic equation for the precipitation of lead(II) iodide.

If we place a sample of lead(II) iodide in a beaker of water, it will begin to dissolve. As this happens, the number of lead(II) ions and iodide ions in solution increases.

Along with this increase in their concentration, the probability that ions will return to the solid state also increases. Eventually, the rate at which ions leave the solid state to enter solution equals the rate at which they return to the solid state. When this point is achieved, solubility equilibrium exists. We may use the principles of equilibrium to arrive at an equilibrium expression for the dissociation process:

Dissociation $PbI_2(s) \rightleftharpoons Pb^{2+}(aq) + 2\,I^-(aq)$

Equilibrium expression: $K_{sp} = [Pb^{2+}][I^-]^2$

Solubility product: An equilibrium constant for a slightly soluble salt.

where K_{sp} is a constant called the **solubility product,** the brackets, [ ], indicate molar concentrations, and exponents come from the coefficients in the ionic equation.

The solubility equilibrium of an ionic compound is always written in the direction of dissociation, and the equilibrium expression is arrived at in the same fashion as that of other equilibria. But there is one difference: the associated solid is not included in the equilibrium expression. The solubility product, K_{sp}, represents the mathematical product of the concentrations of ions in solution. The values of several solubility products are given in Table 17-2.

We can explain the absence of $PbI_2(s)$ in the preceding equilibrium expression by recognizing that the undissociated solid is not actually in solution. Instead, it is simply in contact with the solution. Solubility equilibrium is achieved when the rate at which ions enter the solution equals the rate at which they leave. The rate at which ions enter solution depends on the surface area of solid in contact with the solution. If we place more solid in the solution, there is a greater surface area, so the rate at which solid dissolves increases. However, this increased surface area is also available for contact with ions returning to the solid state. Thus, the rate at which solid reforms also increases, so there is no net increase in the actual concentration of solute dissolved. Once the solution is saturated, adding more solid does not increase the solubility of the substance. The position of equilibrium depends only on the concentration of ions in solution.

The solubility product can be determined from the solubility of a substance, as shown in the following examples.

TABLE 17-2 Selected solubility products, K_{sp}

Bromides	$AgBr$	5×10^{-13}	Fluorides	BaF_2	2.4×10^{-5}
	$PbBr_2$	4.6×10^{-6}		CaF_2	1.7×10^{-10}
				PbF_2	4×10^{-8}
Carbonates	$BaCO_3$	1.6×10^{-9}			
	$CaCO_3$	4.8×10^{-9}	Iodides	AgI	8.5×10^{-17}
	$MgCO_3$	3×10^{-5}		PbI_2	8.3×10^{-9}
	$PbCO_3$	1.5×10^{-13}			
			Sulfates	$BaSO_4$	1.5×10^{-9}
Chlorides	$AgCl$	2.8×10^{-10}		$CaSO_4$	2.4×10^{-5}
	$PbCl_2$	1.6×10^{-5}		$PbSO_4$	1.3×10^{-8}
Chromates	Ag_2CrO_4	1.9×10^{-12}	Sulfides	Ag_2S	1×10^{-49}
	$BaCrO_4$	1.2×10^{-10}		CuS	8×10^{-36}
	$PbCrO_4$	2×10^{-16}		PbS	1.3×10^{-28}
				ZnS	1.6×10^{-23}

EXAMPLE 17.12

The solubility of zinc sulfide is 4.0×10^{-12} M. (It's not very soluble!) Calculate the solubility product.

Planning the solution

We must determine the concentrations of the ions in solution and then substitute them into the equilibrium expression. Multiplication of the values substituted into the equilibrium expression will give the solubility product.

SOLUTION

We begin by determining the molar concentration of the ions, using the same procedure we employed in Section 15.1. The concentration of zinc sulfide is

$$[ZnS] = 4.0 \times 10^{-12}\ M$$

Since each mole of ZnS contains 1 mol of Zn^{2+} ion and 1 mol of S^{2-} ion, the molar concentrations of the ions must be

$$[Zn^{2+}] = 4.0 \times 10^{-12}\ M \qquad \text{and} \qquad [S^{2-}] = 4.0 \times 10^{-12}\ M$$

The equation for the dissociation of zinc sulfide is

$$ZnS(s) \rightleftharpoons Zn^{2+}(aq) + S^{2-}(aq)$$

This leads to the following equilibrium expression:

$$K_{sp} = [Zn^{2+}][S^{2-}]$$

Substituting the Zn^{2+} and S^{2-} concentrations gives

$$K_{sp} = [4.0 \times 10^{-12}][4.0 \times 10^{-12}] = 16 \times 10^{-24}$$
$$= 1.6 \times 10^{-23}$$

EXAMPLE 17.13

The solubility of lead(II) chloride, $PbCl_2$, is 1.6×10^{-2} M. Calculate the value of K_{sp}.

SOLUTION

The dissociation of lead(II) chloride is

$$PbCl_2(s) \rightleftharpoons Pb^{2+}(aq) + 2\ Cl^-(aq)$$

Since 1 mol of $PbCl_2$ contains 1 mol of Pb^{2+} ion and 2 mol of Cl^- ion, the concentrations of these ions in a 1.6×10^{-2} M $PbCl_2$ solution must be

$$[Pb^{2+}] = 1.6 \times 10^{-2}\ M \qquad \text{and} \qquad [Cl^-] = 3.2 \times 10^{-2}\ M$$

The solubility product relates these concentrations as follows:

$$K_{sp} = [Pb^{2+}][Cl^-]^2$$
$$= [1.6 \times 10^{-2}][3.2 \times 10^{-2}]^2 = 16 \times 10^{-6}$$
$$= 1.6 \times 10^{-5}$$

PROBLEM 17.16

Use the following solubilities to calculate K_{sp}. Check your answers with the values provided in Table 17-2.

(a) $[AgBr] = 7.1 \times 10^{-7}$ M **(b)** $[BaCO_3] = 4.0 \times 10^{-5}$ M
(c) $[CaF_2] = 3.5 \times 10^{-4}$ M

CHAPTER SUMMARY

Chemical kinetics is the study of **reaction rates** and the factors affecting them. Reactions proceed according to a **reaction profile,** which is a graphic description of the energy changes that take place in the transition from reactants to products. Energy changes are represented in the reaction profile along the vertical axis, whereas the horizontal axis, known as the **reaction coordinate,** represents the progress of the reaction. Reacting molecules must pass through a **transition state** or **activated complex,** which represents an energy maximum on the reaction profile. The energy required to go from the reactant state to this energy maximum is referred to as the **activation energy.**

The rate of a reaction depends on three factors: the frequency of collision of the reacting molecules, the fraction of these collisions with enough energy to surmount the energy barrier, and the fraction of these collisions that have the proper orientation to undergo reaction. The first of these factors is most influenced by the concentrations of the reacting molecules, and the second is most affected by temperature. Thus, increased concentrations or temperatures usually lead to an increase in reaction rate. The rates of reactions may also increase in the presence of **catalysts** or **enzymes.** These chemical agents lower the activation energy of the reaction, thereby increasing the fraction of molecules having sufficient energy to undergo reaction. The catalysts themselves are not consumed in the reaction.

The rates of reactions are related to the concentration of reactants in a mathematical formulation known as the **rate expression,** which has the general form

$$\text{Rate} = k[A]^x[B]^y$$

The rate expression for a given reaction must be experimentally determined.

The progress of chemical reactions is also governed by **chemical equilibrium.** Equilibrium is a dynamic phenomenon characterized by constant macroscopic properties despite changes at the microscopic level. Equilibrium exists when opposing processes occur at equal rates in both directions. Thus, chemical equilibrium is achieved when the rate of the **forward reaction** equals the rate of the **reverse reaction.** At equilibrium, the concentrations of the reactants and products are related to one another such that the generalized chemical reaction

$$mA + nB \rightleftharpoons xC + yD$$

leads to the following **equilibrium expression:**

$$K_{eq} = \frac{[C]^x[D]^y}{[A]^m[B]^n}$$

where K_{eq} is known as the **equilibrium constant.**

Chemical equilibrium may further be described by **Le Chatelier's Principle,** which states that a system at equilibrium responds to a stress placed on the system by relieving that stress. Le Chatelier's Principle can be used to predict the effect brought about by changes in concentration, pressure, or temperature.

The dissociation of weak acids can be described in terms of the concepts of equilibrium:

$$HB \rightleftharpoons H^+ + B^-$$

The concentrations of the conjugate acid, HB, the conjugate base, B^-, and hydrogen ion, H^+, are related by the following equilibrium expression:

$$K_a = \frac{[H^+][B^-]}{[HB]}$$

where K_a is an equilibrium constant for the dissociation of the acid.

At equilibrium, the rate at which a solid substance dissolves equals the rate at which it recrystallizes. When such a solubility equilibrium is achieved, the product of the ionic concentrations equals a constant known as the **solubility product.**

$$A_nB_m(s) \rightleftharpoons nA(aq) + mB(aq) \qquad K_{sp} = [A]^n[B]^m$$

KEY TERMS

Review each of the following terms, which are discussed in this chapter and defined in the Glossary.

reaction rate
chemical kinetics
reaction profile
reaction coordinate
activation energy
activated complex
transition state
rate expression
catalyst
enzyme
forward reaction
reverse reaction
equilibrium
equilibrium expression
equilibrium constant
Le Chatelier's Principle
solubility product

ADDITIONAL PROBLEMS

SECTION 17.1

17.17 Explain the meaning of each of the following terms.

(a) Reaction profile **(b)** Reaction coordinate
(c) Activation energy **(d)** Transition state

SECTION 17.2

17.18 Explain the role of each of the following in determining the rate of a chemical reaction.

(a) Concentration of the reactants
(b) Temperature
(c) Orientation of collisions

SECTION 17.3

17.19 The reaction of hydrogen iodide with itself to form hydrogen and iodine has the following rate expression:

$$2\,HI(g) \rightarrow H_2(g) + I_2(g)$$
$$\text{Rate} = k[HI]^2$$

What effect will each of the following have on the rate?

(a) Doubling the HI concentration
(b) Tripling the HI concentration
(c) Dividing the HI concentration in half

SECTION 17.4

17.20 Answer the following questions concerning catalysts:

(a) Explain how a catalyst works in terms of its effect on the activation energy.
(b) What is an enzyme?

SECTION 17.5

17.21 Answer the following questions about equilibrium.

(a) Compare the rates of the forward and reverse reactions for a chemical reaction that is at equilibrium.
(b) How does a system at equilibrium vary with respect to its macroscopic properties?
(c) How does a system at equilibrium vary with respect to its microscopic properties?

SECTION 17.6

17.22 Write an equilibrium expression for each of the following reactions.

(a) $2\,IBr(g) \rightleftharpoons I_2(g) + Br_2(g)$
(b) $2\,H_2(g) + O_2(g) \rightleftharpoons 2\,H_2O(g)$
(c) $2\,CO_2(g) \rightleftharpoons 2\,CO(g) + O_2(g)$
(d) $N_2(g) + O_2(g) \rightleftharpoons 2\,NO(g)$
(e) $2\,N_2O_5(g) \rightleftharpoons 4\,NO_2(g) + O_2(g)$
(f) $C_3H_8(g) + 5\,O_2(g) \rightleftharpoons 3\,CO_2(g) + 4\,H_2O(g)$

SECTION 17.7

17.23 Use the equilibrium concentrations provided to calculate the value of K_{eq} for each of the following gaseous equilibria.

(a) $CO(g) + 2\,H_2(g) \rightleftharpoons CH_3OH(g)$
$[CO] = 2.00\text{ M}$ $[H_2] = 1.00\text{ M}$
$[CH_3OH] = 1.00\text{ M}$
(b) $2\,CO_2(g) \rightleftharpoons 2\,CO(g) + O_2(g)$
$[CO_2] = 0.30\text{ M}$ $[CO] = 0.046\text{ M}$
$[O_2] = 0.023\text{ M}$
(c) $CO(g) + H_2O(g) \rightleftharpoons CO_2(g) + H_2(g)$
$[CO] = 3.7\text{ M}$ $[H_2O] = 3.7\text{ M}$
$[CO_2] = 2.3\text{ M}$ $[H_2] = 2.3\text{ M}$

17.24 Explain how the magnitude of K_{eq} is related to the tendency of a reaction to go to completion.

SECTION 17.8

17.25 State Le Chatelier's Principle. In which direction would the equilibrium

$$N_2(g) + 3\,H_2(g) \rightleftharpoons 2\,NH_3(g) + 92\text{ kJ}$$

be affected by each of the following conditions?

(a) Decreased temperature **(b)** Increased pressure
(c) Increased $[H_2]$ **(d)** Decreased $[N_2]$
(e) Increased $[NH_3]$

17.26 In which direction would the equilibrium

$$CH_4(g) + 2\,O_2(g) \rightleftharpoons CO_2(g) + 2\,H_2O(g) + 943\text{ kJ}$$

be affected by each of the following conditions?

(a) Increased temperature **(b)** Increased pressure
(c) Increased $[O_2]$ **(d)** Increased $[CO_2]$

17.27 In which direction would the equilibrium

$$2\,H_2O(g) + 484\text{ kJ} \rightleftharpoons 2\,H_2(g) + O_2(g)$$

be affected by each of the following conditions?

(a) Increased temperature **(b)** Decreased pressure
(c) Increased $[O_2]$ **(d)** Decreased $[H_2O]$

17.28 In which direction would each of the equilibria described in Problem 17.22 be affected by a decrease in pressure?

SECTION 17.9

17.29 The equilibrium concentrations in a 0.100 M $HNO_2(aq)$ solution are

$[HNO_2] = 0.094$ M $[H^+] = 0.0065$ M

$[NO_2^-] = 0.0065$ M

Calculate K_a for the dissociation

$$HNO_2(aq) \rightleftharpoons H^+(aq) + NO_2^-(aq)$$

17.30 The equilibrium concentrations in a 0.200 M $HC_7H_5O_2(aq)$ solution are

$[HC_7H_5O_2] = 0.196$ M $[H^+] = 0.0036$ M

$[C_7H_5O_2^-] = 0.0036$ M

Calculate K_a for the dissociation

$$HC_7H_5O_2(aq) \rightleftharpoons H^+(aq) + C_7H_5O_2^-(aq)$$

17.31 A 1.0 M solution of a weak acid, HW(aq), is 25% dissociated. Calculate K_a. What is the pH of this solution?

17.32 A 0.0140 M solution of a weak acid, HF(aq), is 20% dissociated. Use this information to calculate K_a.

SECTION 17.10

17.33 Use the following solubilities to calculate K_{sp}. Check your answers with the values provided in Table 17-2.

(a) $[PbCO_3] = 3.9 \times 10^{-7}$ M
(b) $[Ag_2CrO_4] = 7.8 \times 10^{-5}$ M
(c) $[AgI] = 9.2 \times 10^{-9}$ M

GENERAL PROBLEMS

17.34 Does the value of K_{eq} for an exothermic reaction increase or decrease with increasing temperature? Explain your answer.

17.35 The creation of a vapor pressure above the surface of water contained in a closed jar is an example of an equilibrium process. Describe the dynamic nature of this process, and explain how it conforms to the characteristics of equilibrium.

***17.36** When the ions of an insoluble substance are brought together, precipitation may take place. The solubility product, K_{sp}, represents the maximum mutual concentration of ions that may coexist in the same solution without precipitation occurring. Use Table 17-2 to predict whether or not a precipitate will form when the following concentrations of ions are brought together.

(a) $[Ag^+] = 5.0 \times 10^{-3}$ M; $[Br^-] = 3.5 \times 10^{-5}$ M
(b) $[Ba^{2+}] = 1.0 \times 10^{-3}$ M; $[F^-] = 6.0 \times 10^{-4}$ M
(c) $[Pb^{2+}] = 3.5 \times 10^{-3}$ M; $[Br^-] = 1.0 \times 10^{-2}$ M
(d) $[Cu^{2+}] = 1.5 \times 10^{-4}$ M; $[S^{2-}] = 2.0 \times 10^{-5}$ M

Problems 17.37 and 17.38 involve the calculation of equilibrium concentrations. Equilibrium concentrations are often obtained by determining the equilibrium concentration of one of the reactants or products after a reaction mixture has come to equilibrium. Once the equilibrium concentrations are known, K_{eq} may be calculated readily. Problems 17.37 and 17.38 will lead you through this technique.

***17.37** Phosphorus pentachloride, PCl_5, decomposes into phosphorus trichloride, PCl_3, and chlorine, Cl_2, according to the following equilibrium:

$$PCl_5(g) \rightleftharpoons PCl_3(g) + Cl_2(g)$$

When 0.30 mol of PCl_5 is introduced into a 1.00-L vessel and allowed to come to equilibrium, the equilibrium concentration of PCl_5 is found to be 0.20 mol/L.

(a) What is the initial concentration of PCl_5 *before* any reaction takes place?
(b) What are the initial concentrations of PCl_3 and Cl_2 *before* any reaction takes place?
(c) If the equilibrium concentration of PCl_5 is 0.20 mol/L, how many moles per liter of PCl_5 have been used up?
(d) On the basis of your answer to part (c), determine how many moles per liter of PCl_3 form.
(e) What is the equilibrium concentration of PCl_3?
(f) Repeat parts (d) and (e) for Cl_2.
(g) Use the equilibrium concentrations of PCl_5, PCl_3, and Cl_2 to calculate the value of K_{eq}.

***17.38** A 1.00-L flask is filled with 4.0 mol of CO(g) and 5.0 mol of $H_2O(g)$. The contents are allowed to come to equilibrium according to the following reaction:

$$CO(g) + H_2O(g) \rightleftharpoons CO_2(g) + H_2(g)$$

When equilibrium is achieved, the hydrogen concentration is 3.0 mol/L. Use the technique described in Problem 17.37 to calculate K_{eq}.

***17.39** In their discussions of chemical kinetics, chemists often refer to the *order* of a reaction, which is determined from the rate expression by summing the exponents of all of the concentration terms. Thus, either of the following rate expressions would represent *second-order* reactions:

$$\text{Rate} = k[A]^2 \qquad \text{or} \qquad \text{Rate} = k[A][B]$$

Similarly, both of the following would represent *third-order* reactions:

$$\text{Rate} = k[A]^2[B] \qquad \text{or} \qquad \text{Rate} = [A][B][C]$$

Use these principles to determine the order of the reactions described by each of the following rate expressions.

(a) Rate $= k[A]$ **(b)** Rate $= k[A]^3$
(c) Rate $= k[A]^2[B]^3$ **(d)** Rate $= k[A][B]^{1/2}$
(e) Rate $= k[A]^2/[B]$ **(f)** Rate $= k$

***17.40** Account for the following observations using the principles of equilibrium behavior as the basis for your explanations.

(a) A saturated solution of potassium iodide is prepared by adding solid potassium iodide to a stirred beaker of water until no further solid dissolves. Saturation is assured by the fact that undissolved solid potassium iodide remains in contact with the saturated solution. At this point, several crystals of radioactive potassium iodide are added to the beaker. The radioactivity is in the iodide ion, rather than the potassium ion. Several hours later, the solution is decanted from the solid and found to contain radioactive iodide. Explain how the radioactive iodide was able to enter the saturated solution.

(b) A saturated solution of potassium iodide is prepared as in part (a). However, prior to adding radioactive iodide, several milliliters of lead(II) nitrate solution are added to the beaker, precipitating lead(II) iodide. The lead(II) iodide is free of any radioactive iodide ions. At this point, several large crystals of radioactive potassium iodide are added to the beaker as in part (a). Several hours later, the crystals (which are large enough to grasp with tweezers) are removed from the beaker and the contents of the beaker are filtered. Radioactive iodide is found in the precipitate. Explain how the radioactive iodide got from the potassium iodide crystals into the lead(II) iodide precipitate. (You may assume that none of the original potassium iodide crystals are mixed in with the precipitate tested.)

***17.41** The ionization constant of a weak acid, K_a, may be determined by measuring the pH of a solution of known concentration. Consider the weak acid, HA:

$$HA \rightleftharpoons H^+ + A^-$$

The hydrogen ion concentration, $[H^+]$, may be calculated from the pH. Since each weak acid molecule that dissociates produces one hydrogen ion and one conjugate base ion, the conjugate base concentration must equal the hydrogen ion concentration, $[A^-] = [H^+]$. Finally, since each weak acid molecule that dissociates produces one hydrogen ion, the weak acid concentration at equilibrium may be obtained by subtracting the hydrogen ion concentration from the initial concentration of the weak acid solution:

$$[HA]_{eq} = [HA]_{initial} - [H^+]$$

In most cases, this subtraction is negligible, and the weak acid concentration at equilibrium is equal to the concentration of the solution prepared. Calculate the ionization constants of the acids described below from the data given. (The dissociations of the acids are given in parentheses.)

(a) The pH of 1.0 M $HC_2H_3O_2$ is 2.37 ($HC_2H_3O_2 \rightleftharpoons H^+ + C_2H_3O_2^-$)

(b) The pH of 0.10 M HClO is 4.23 ($HClO \rightleftharpoons H^+ + ClO^-$)

WRITING EXERCISES

17.42 Your father has always been intrigued by the fact that it is necessary to strike a match to light a propane torch, yet the flame sustains itself once it is lit. Without using any diagrams, write him a note in which you discuss reaction rates, activation energies, and energy profiles in order to explain this phenomenon. In your note, compare the energetics of this type of reaction to those of a drive over a mountain pass to get to a lower elevation on the other side of the pass.

17.43 Explain how each of the three major factors of concentration, temperature, and orientation of collisions affects the rate of a reaction.

17.44 Your niece is very curious about science, and she wonders why water evaporates out of an open glass but not out of a closed container. You have decided to provide her with a written explanation that will also act as a mini-lesson in the basics of equilibrium. Write her a note that explains how a vapor pressure is created and how it reaches equilibrium in a closed container. Then use that discussion as the basis for explaining the fundamental characteristics of all equilibrium processes. (You may wish to refer to Section 12.8 to review your understanding of vapor pressure equilibrium.)

USEFUL MATHEMATICAL SKILLS FOR SCIENTIFIC WORK

APPENDIX A

Many of the calculations described in this appendix may be carried out rapidly with a calculator. Appendix B describes the use of a calculator, but it does not attempt to explain the mathematical foundation underlying each operation. This appendix explains the mathematical basis for the various types of calculations encountered in this text.

A.1 ROUNDING OFF NUMBERS

Rounding off numbers is particularly important in scientific work where we must obey rules concerning significant figures (Section 3.3). When you are using a calculator to carry out your computations, you must round off the answer displayed on the calculator to conform to the rules for significant figures.

Before listing the rules for rounding off, we should outline the *rationale* behind the rules. The following is a number line from 0 to 10 on which several points have been located:

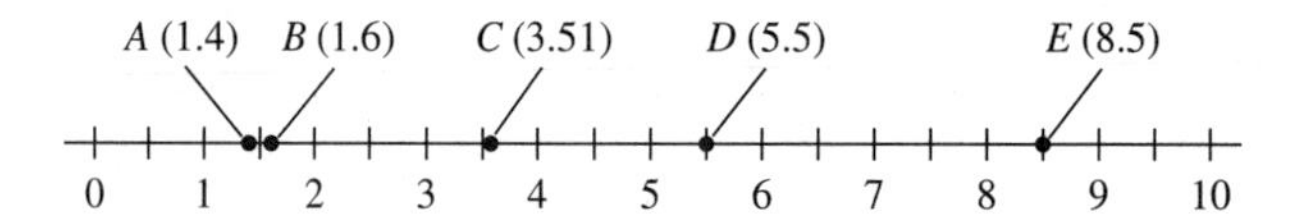

Suppose that, for each of the points along the line, we wish to find the closest integer (whole number). Point *A*, at 1.4, is closer to 1 than to 2. On the other hand, point *B*, at 1.6, is closer to 2 than to 1. When we round off a number to the nearest integer, we simply state which whole number along a number line is closest to the number being rounded off. Thus, we would round off 1.4 to 1 (Rule 1 following) and 1.6 to 2 (Rule 2 following).

In fact, when deciding how to round off a number, we must decide whether it is more than halfway to the next number. If it is, we round the number up. For example, 3.5 is the halfway point between 3 and 4. Point *C*, located at 3.51, is more than halfway to 4. Thus, we must round it up (Rule 3a following). On the other hand, what should we do with numbers such as 5.5 (point *D*) and 8.5 (point *E*), which are *exactly* halfway between two numbers? If we were always to round off such numbers upward, we would introduce statistical inaccuracy whenever we worked with a large sample of numbers that required rounding off. We would add a similar bias if we always rounded off to the lower number. However, if we round off half of our numbers upward and half of our numbers downward, our errors will average out. To accomplish this, we always round off numbers to an *even* digit. Thus, we round off 5.5 upward to 6 (Rule 3b) and 8.5 downward to 8 (Rule 3c). These rules, which we have just explained using integers, apply when we round off numbers to *any* decimal place. The rules for rounding are stated in the Summary.

SUMMARY Rounding Off Numbers

Rule 1. When the first digit of those to be dropped is less than 5, leave the preceding digit unchanged. For example, 1.7463 rounded off to the nearest tenth is 1.7.

Rule 2. When the first digit of those to be dropped is greater than 5, raise the preceding digit by 1. For example, 21.612 rounded off to the nearest unit is 22.

Rule 3. When the first digit of those to be dropped is 5:

(a) *Raise the preceding digit if any nonzero digits follow the 5.* For example, 13.653 rounded off to the nearest tenth is 13.7.

(b) *Raise the preceding digit if it is odd and no nonzero digits follow the 5.* For example, 4.750 rounded off to the nearest tenth is 4.8.

(c) *Leave the preceding digit unchanged if it is even and no nonzero digits follow the 5.* For example, 4.850 rounded off to the nearest tenth is 4.8.

In carrying out these rules, it is important that we base our rounding off strictly on the digit immediately following the decimal place to which we are rounding off. In other words, *do not* round off sequentially from right to left. Suppose, for example, that we wish to round off 7.46 to the units place. Because a 4 immediately follows the 7, we must drop all of the digits to the right of the 7, giving a rounded-off value of 7. We *do not* round off 7.46 to 7.5, and then round off 7.5 to 8.

Correct	~~Incorrect~~
7.46 → 7	~~7.46 → 7.5 → 8~~

EXAMPLE A.1

Round off each of the following numbers to the indicated decimal place.

	Rounded Off to Nearest	Before Rounding Off		After Rounding Off	Rule Applied
(a)	0.1	57.431	→	57.4	Rule 1: The digit immediately following the 4 in the tenths column is less than 5, so we must drop all of the digits following the 4.
(b)	0.01	3.4763	→	3.48	Rule 2: The digit immediately following the 7 in the hundredths column is greater than 5, so we must increase the 7 in the hundredths column.

	Rounded Off to Nearest	Before Rounding Off		After Rounding Off	Rule Applied
(c)	1	8.52	→	9	Rule 3a: The digit following the 8 in the units column is 5, so Rule 3 applies. However, a nonzero digit follows the 5, so the 8 must be increased.
(d)	0.1	6.75	→	6.8	Rule 3b: A 5 follows the 7 in the tenths column, so Rule 3 applies. In this example, no nonzero digits follow the 5, so we must choose between Rule 3b and Rule 3c. Because the 7 is odd, we raise it to 8.
(e)	0.0001	0.03285	→	0.0328	Rule 3c: A 5 follows the 8 in the column to which we are rounding off, so Rule 3 applies. No nonzero digits follow the 5, so we must choose between Rule 3b and Rule 3c. Because the 8 is even, we leave it unchanged.
(f)	0.1	3.99	→	4.0	Rule 1: To increase the 9 in the tenths column, add 0.1 to 3.9, raising it to 4.0.
(g)	0.01	7.896	→	7.90	Rule 1: To increase the 9 in the hundredths column, add 0.01 to 7.89, raising it to 7.90.

PROBLEM A.1

Round off each of the following numbers to the indicated decimal place.

(a) 0.1	71.838	**(b)** 0.001	6.4374	**(c)** 0.01	0.83500
(d) 1	417.53	**(e)** 0.01	0.265001	**(f)** 0.1	13.85
(g) 10	452	**(h)** 0.0001	1.00095	**(i)** 0.1	7.550
(j) 0.1	14.547	**(k)** 0.01	7.665	**(l)** 0.001	9.9995

A.2 EXPRESSING FRACTIONS IN DECIMAL NOTATION

In scientific work, fractions are generally expressed in their equivalent decimal notation. Thus, we would express $\frac{1}{8}$ as 0.125. To convert a fraction to a decimal, simply divide the numerator (the number on top) by the denominator (the number on the bottom), using your calculator. For example,

$$\tfrac{1}{2} = 0.5$$
$$\tfrac{4}{25} = 0.16$$

For certain fractions, you might need to round off the answer displayed on your calculator. The number of significant digits to which a fraction is rounded off depends on the particular measuring device used. Each of the following has been rounded off to the nearest one-thousandth.

$$\tfrac{1}{3} = 0.3333333 \qquad \text{Rounded off to } 0.333$$
$$\tfrac{13}{19} = 0.6842105 \qquad \text{Rounded off to } 0.684$$

PROBLEM A.2

Express each of the following fractions as a decimal rounded off to the nearest one-thousandth.

(a) $\frac{3}{8}$ **(b)** $\frac{5}{16}$ **(c)** $\frac{11}{17}$ **(d)** $\frac{3}{5}$

A.3 Scientific Notation

The Meaning of Exponents

Exponents are used when we wish to show that a number is multiplied by itself, usually in a repeated fashion. For example, each of the following multiplications shown on the left is expressed on the right using an exponent:

$$\begin{aligned}(2)(2)(2) &= 2^3\\(5)(5)(5)(5) &= 5^4\\(10)(10) &= 10^2\\8 &= 8^1\end{aligned}$$

Exponents

The exponent, written as a superscript in the upper right, is often referred to as a *power.* Thus, in the first example, we refer to 2^3 as "two to the third power." Similarly, 5^4 is "five to the fourth power." Note that any number raised to the first power (such as 8^1) is just equal to itself. As you will see, exponents are especially useful when we work with very large or very small numbers.

PROBLEM A.3

Express each of the following using an exponent.

(a) $(3)(3)(3)(3)(3) = 3^?$ **(b)** $(6)(6)(6)(6)(6)(6)(6) = 6^?$

(c) $(7)(7)(7) = 7^?$ **(d)** $(10) = 10^?$

Rules About Exponents

The following rules about exponents are applied when we work with numbers expressed in scientific notation.

$$(x^a)(x^b) = x^{a+b}$$

$$\frac{1}{x^a} = x^{-a} \quad \text{or} \quad \frac{1}{x^{-a}} = x^a$$

EXAMPLE A.2

(a) $(2^2)(2^3) = 2^{2+3} = 2^5$ **(b)** $(10^5)(10^{-2}) = 10^{5-2} = 10^3$

(c) $\frac{1}{3^2} = 3^{-2}$ **(d)** $\frac{1}{10^4} = 10^{-4}$

(e) $\frac{1}{10^{-4}} = 10^4$ **(f)** $(5^{-6})(5^4) = 5^{-2} = \frac{1}{5^2}$

PROBLEM A.4

Evaluate each of the following.

(a) $(4^5)(4^3) = 4^?$ **(b)** $(7^{-2})(7^9)(7^{-3}) = 7^?$ **(c)** $\frac{3^4}{3^7} = 3^?$

Powers of 10

The following numbers can all be expressed as powers of 10:

$$\begin{aligned}
1{,}000{,}000 &= 10^6 \\
100{,}000 &= 10^5 \\
10{,}000 &= 10^4 \\
1{,}000 &= 10^3 \\
100 &= 10^2 \\
10 &= 10^1 \\
1 &= 10^0 \\
0.1 &= 10^{-1} = \frac{1}{10} \\
0.01 &= 10^{-2} = \frac{1}{100} \\
0.001 &= 10^{-3} = \frac{1}{1{,}000} \\
0.0001 &= 10^{-4} = \frac{1}{10{,}000} \\
0.00001 &= 10^{-5} = \frac{1}{100{,}000} \\
0.000001 &= 10^{-6} = \frac{1}{1{,}000{,}000} \\
&\vdots \\
&\text{etc.}
\end{aligned}$$

When we multiply (or divide) a number by a power of 10, we determine the final answer by moving the decimal point in the number being multiplied. For example,

$$2.12 \times 100 = 212$$

The decimal point has been moved two places to the right.

$$45.1 \times \frac{1}{10{,}000} = 0.00451$$

The decimal point has been moved four places to the left.

We can easily determine how far and in which direction to move the decimal point by expressing the power of 10 in exponential form. Thus,

$$2.12 \times 100 = 2.12 \times 10^2 = 212$$

The exponent +2 tells us to move the decimal point two places to the right.

$$45.1 \times \frac{1}{10{,}000} = 45.1 \times 10^{-4} = 0.00451$$

The exponent –4 tells us to move the decimal point four places to the left.

To help determine how far and in which direction to move the decimal point, imagine a number line across the page:

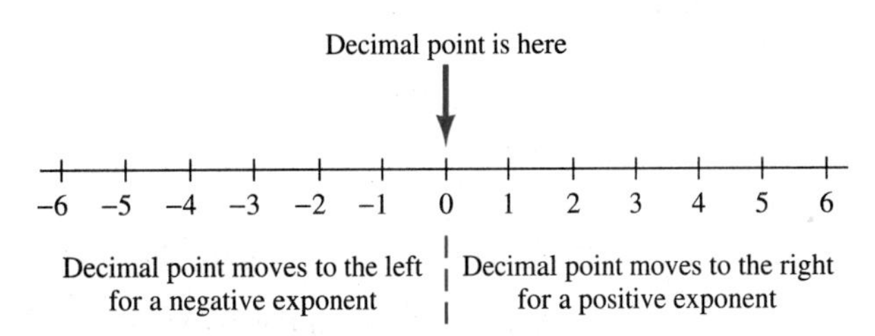

Expressing Numbers in Scientific Notation

Frequently, scientists find it useful to express numbers in a form known as *scientific notation,* whereby the number is rewritten as a number between 1 and 10 times an appropriate power of 10. The following examples illustrate this idea.

EXAMPLE A.3

(a) $300 = 3 \times 100 = 3 \times 10^2$

(b) $4200 = 4.2 \times 1000 = 4.2 \times 10^3$

(c) $0.064 = 6.4 \times \frac{1}{100} = 6.4 \times 10^{-2}$

(d) $5{,}600{,}000 = 5.6 \times 1{,}000{,}000 = 5.6 \times 10^6$

(e) $0.00000479 = 4.79 \times \frac{1}{1{,}000{,}000} = 4.79 \times 10^{-6}$

To express a number in scientific notation, we simply reverse the process used for multiplying by a power of 10. For example, to express 4200 in scientific notation, we move the decimal point three places to the left:

$$4200. \quad \text{becomes} \quad 4.2 \times 10^{?}$$

Because the number 4200 is greater than 1, its power of 10 must be positive:

$$4200 = 4.2 \times 10^3$$

On the other hand, to express 0.00000479 in scientific notation, we must move the decimal point six places to the right:

$$0.00000479 \quad \text{becomes} \quad 4.79 \times 10^{?}$$

Because this number is less than 1, its power of 10 is negative:

$$0.00000479 = 4.79 \times 10^{-6}$$

PROBLEM A.5

Express each of the following numbers as its decimal equivalent. (The first one is done for you.)

(a) $3.75 \times 10^2 = 375$ **(b)** 4.21×10^4 **(c)** 7.39×10^{-3}
(d) 8.32×10^{-1} **(e)** 5.93×10^6 **(f)** 5.93×10^{-6}

PROBLEM A.6

Express each of the following numbers in scientific notation.

(a) 47,600 **(b)** 541 **(c)** 0.0943
(d) 0.000315 **(e)** 2,190,000 **(f)** 0.00000219

Multiplication and Division Using Scientific Notation

Scientific notation is especially convenient when we are multiplying or dividing very large or very small numbers. Much of the ease of working with scientific numbers comes from the fact that, when we multiply or divide numbers, it does not matter in what order we multiply or divide them. For example, when 2, 3, and 8 are multiplied together, the product is always 48, regardless of the order in which the numbers are taken.

$$(8)(3)(2) = (8)(2)(3) = (3)(2)(8) = (3)(8)(2) = (2)(3)(8) = (2)(8)(3) = 48$$

We refer to this as the *commutative* property of multiplication and division. Thus, if we are multiplying two (or more) numbers expressed in scientific notation, we may rearrange the order of multiplication so that all of the powers of 10 are grouped together. For example,

$$(2.5 \times 10^5)(3.0 \times 10^2) = (2.5)(3.0) \times (10^5)(10^2)$$

We can complete the multiplication by first evaluating 2.5 times 3.0 and then evaluating 10^5 times 10^2. Because $(x^a)(x^b) = x^{a+b}$, we can readily multiply the powers of 10 by summing their exponents.

$$(2.5)(3.0) \times (10^5)(10^2) = 7.5 \times 10^{5+2} = 7.5 \times 10^7$$

EXAMPLE A.4

Multiply or divide the following numbers.

(a)
$$\begin{aligned}(0.32)(25{,}000) &= (3.2 \times 10^{-1})(2.5 \times 10^4)\\ &= (3.2)(2.5) \times (10^{-1} \times 10^4)\\ &= 8.0 \times 10^3\end{aligned}$$

(b)
$$\frac{0.00045}{1800} = \frac{4.5 \times 10^{-4}}{1.8 \times 10^3} = \frac{4.5}{1.8} \times 10^{-4} \times 10^{-3}$$
$$= 2.5 \times 10^{-7}$$

(c) $(0.000075)(240) = (7.5 \times 10^{-5})(2.4 \times 10^{2})$

$$= (7.5)(2.4) \times (10^{-5} \times 10^{2})$$
$$= 18 \times 10^{-3}$$

Because 18 is not a number between 1 and 10, this answer is not expressed in scientific notation. However, we can easily complete the calculation by moving the decimal point in 18 one place to the left:

$$18 \times 10^{-3} = 1.8 \times 10^{1} \times 10^{-3} = 1.8 \times 10^{-2}$$

Moving the decimal point one place to the left required us to raise the power of 10 by +1.

(d) $\dfrac{(0.05)(8000)}{(0.0002)} = \dfrac{(5 \times 10^{-2})(8 \times 10^{3})}{2 \times 10^{-4}}$

$$= \frac{(5)(8)}{2} \times \frac{(10^{-2})(10^{3})}{10^{-4}} = \frac{(5)(8)}{2} \times (10^{-2})(10^{3})(10^{4})$$
$$= 20 \times 10^{5} = 2 \times 10^{1} \times 10^{5}$$
$$= 2 \times 10^{6}$$

When multiplying and dividing exponential numbers, it is important that we express our final answer according to the rules discussed in Section 3.3 on significant figures. Remember that we ignore the power of 10 in determining the number of significant figures in an exponential number.

PROBLEM A.7

Carry out the following multiplications and divisions, using scientific notation.

(a) $(4{,}000{,}000)(0.0002)$ **(b)** $\dfrac{2200}{0.0055}$ **(c)** $\dfrac{(436)(0.00539)}{(0.0620)}$ **(d)** $\dfrac{(0.000318)(21.7)}{4830}$

Addition and Subtraction Using Scientific Notation

Extra care must be taken when adding or subtracting numbers that are expressed in scientific notation. For example, if we wish to add

$$(1.6 \times 10^{-4}) + (3.28 \times 10^{-3}) = ?$$

we must keep in mind that the numbers being added are as follows:

$$\begin{array}{r} 0.00016 \\ +0.00328 \\ \hline 0.00344 \end{array} = 3.44 \times 10^{-3}$$

We may *not* add 1.6 to 3.28 directly. However, rather than change both numbers to decimal notation as we just did, it is generally simpler to express both numbers to the same power of 10. For example,

$$1.6 \times 10^{-4} = 0.16 \times 10^{-3}$$

Now we can add the two numbers:

$$\begin{array}{r} 0.16 \times 10^{-3} \\ +3.28 \times 10^{-3} \\ \hline 3.44 \times 10^{-3} \end{array}$$

Note that the powers of 10 are *not* added. The reason stems from the *distributive law* of algebra:

$$(a)(c) + (b)(c) = (a + b)(c)$$

For example,

$$\begin{aligned} (3)(2) + (4)(2) &= (3 + 4)(2) \\ 6 + 8 &= (7)(2) = 14 \end{aligned}$$

As applied to scientific notation,

$$\begin{aligned} (0.16 \times 10^{-3}) + (3.28 \times 10^{-3}) &= (0.16 + 3.28) \times 10^{-3} \\ &= 3.44 \times 10^{-3} \end{aligned}$$

As in multiplication and division, we must observe the rules for significant figures when expressing the final answer.

EXAMPLE A.5

Add the following:

$$(1.47 \times 10^{-2}) + (2.83 \times 10^{-3}) = ?$$

SOLUTION

First we match the powers of 10:

$$2.83 \times 10^{-3} = 0.283 \times 10^{-2}$$

Then we add the numbers:

$$\begin{array}{r} 1.47 \times 10^{-2} \\ +0.283 \times 10^{-2} \\ \hline +1.753 \times 10^{-2} \end{array} \rightarrow 1.75 \times 10^{-2}$$

The final answer must be rounded off to 1.75×10^{-2}, in accordance with the rules for significant figures.

EXAMPLE A.6

Subtract the following:

$$(1.27 \times 10^{-2}) - (4.16 \times 10^{-4}) = ?$$

SOLUTION

First we match the powers of 10:

$$4.16 \times 10^{-4} = 0.0416 \times 10^{-2}$$

Then we subtract these numbers in exactly the same way that we added in the last example.

$$\begin{aligned} 1.27 \times 10^{-2} &= 1.27 \quad \times 10^{-2} \\ -4.16 \times 10^{-4} &= \underline{-0.0416 \times 10^{-2}} \\ & \quad 1.2284 \times 10^{-2} \rightarrow 1.23 \times 10^{-2} \end{aligned}$$

PROBLEM A.8

Carry out the following calculations, expressing the final answer in scientific notation.
(a) $(3.5 \times 10^{-5}) + (5.13 \times 10^{-4})$ **(b)** $(8.02 \times 10^{-5}) - (7 \times 10^{-7})$
(c) $(2.04 \times 10^{-2}) + (3.4 \times 10^{-3}) + (7 \times 10^{-4})$

A.4 PERCENTAGES

Taking a Percentage

A percentage may be described in either of the following two ways:

1. *A percentage is the number of parts per hundred.* For example, 25% means 25 per 100, $\frac{25}{100}$, or 0.25.
2. *A percentage is a fraction multiplied by 100.* To convert a fraction to a percentage, just multiply by 100. For the example just cited, $\frac{25}{100} \times 100\% = 25\%$. To reverse the process, we can convert a percentage to a fraction by dividing by 100. This corresponds to moving the decimal point two places to the left. For example, 16% = 0.16.

Thus, if a class of 100 students has 56 women and 44 men, we can express the composition of the class as 56% women and 44% men. Similarly, if a class of 25 students has 14 women and 11 men, we calculate the percentage of women by finding the fraction of women and multiplying by 100%.

$$\text{Fraction of women:} \quad \frac{14}{25} = 0.56$$

$$\text{Percentage of women:} \quad \frac{14}{25} \times 100\% = 0.56 \times 100\% = 56\%$$

Reversing the procedure, if we know that 25% of a class of 16 students are men, we determine the actual number of men as follows: The fraction of men is 0.25, or $\frac{25}{100}$. Then,

$$0.25 \times 16 = 4$$

(In interpreting percentage problems, it is helpful to remember that the word *of* tells us to multiply. Thus, to obtain 25% of 16, we had to multiply 0.25 times 16.)

Using dimensional analysis, we interpret 25% as follows:

$$\frac{25 \text{ men}}{100 \text{ students}}$$

$$? \text{ men} = 16 \cancel{\text{students}}\left(\frac{25 \text{ men}}{100 \cancel{\text{students}}}\right) = 4 \text{ men}$$

This technique may be extended to chemical samples as follows: If a salt water solution is 12% salt, how much salt is in 55 g of solution? The fraction of salt is 0.12, or $\frac{12}{100}$. Then,

$$0.12 \times 55 \text{ g} = 6.6 \text{ g}$$

Using dimensional analysis, this is interpreted as follows:

$$\frac{12 \text{ g salt}}{100 \text{ g solution}}$$

$$? \text{ g salt} = 55 \cancel{\text{g solution}}\left(\frac{12 \text{ g salt}}{100 \cancel{\text{g solution}}}\right) = 6.6 \text{ g salt}$$

PROBLEM A.9

(a) A solution contains 16 g of sugar in a total of 225 g of solution. What is the percentage of sugar in the solution?

(b) A salt solution is 18% salt. How many grams of salt are there in 250 g of solution?

Percentage Error

In our laboratory work, we often wish to compare an experimentally obtained result with a known accepted value. This comparison is reported as a percentage error. (The vertical lines, | |, tell us to take the absolute value of the quantity inside. Thus, $|3 - 5| = 2$.)

$$\frac{|\text{Accepted} - \text{Experimental}|}{\text{Accepted}} \times 100\% = \% \text{ error}$$

The percentage error represents the fraction of the accepted value by which our experimentally determined value is "off," expressed as a percentage. For example, if a compound is known to contain 25 g of carbon and we measure 26 g of carbon, we are in error by 1 g of carbon per 25 g of carbon. Our fraction of error is $\frac{1}{25}$. This corresponds to a percentage of 4%.

$$\frac{|25 - 26|}{25} \times 100\% = \frac{|-1|}{25} \times 100\% = \frac{1}{25} \times 100\% = 4\%$$

PROBLEM A.10

(a) The molar mass of an unknown gas is experimentally determined to be 42.8 g/mol. The accepted value for the substance is known to be 44.0 g/mol. Calculate the percentage error.

(b) The accepted heat of formation of magnesium oxide is 602 kJ/mol. An experimental determination gives a result of 588 kJ/mol. What is the percentage error?

A.5 GRAPHS

In scientific work, we frequently take experimental measurements over a range of values. Presenting such results graphically often gives the clearest picture of the relationships between the quantities measured. For example, we might obtain the following data if we were to measure the effect of temperature on the pressure of a cylinder of gas.

Temperature	Pressure
100°C	1038 torr
20°C	816 torr
0°C	760 torr
−78°C	543 torr
−196°C	214 torr

We can display the data graphically by plotting the temperature along one axis and the pressure along the other, as shown in Fig. A-1. As a general rule, it is purely arbitrary which variable is plotted along which axis. However, for certain applications, a preferred orientation may exist. (The problems accompanying this section will specify which variable to place along each axis.)

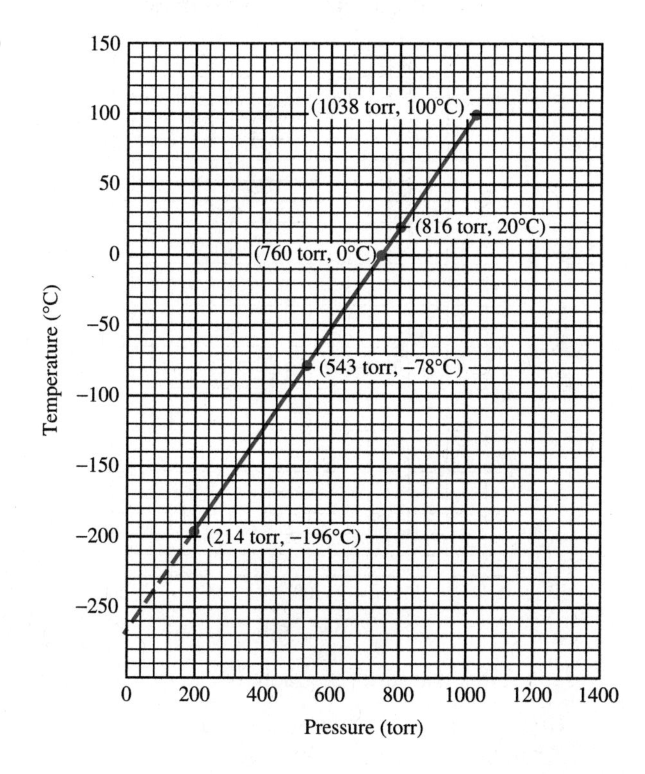

FIGURE A-1 Graph of temperature versus pressure for an ideal gas at constant volume.

Each temperature and its corresponding pressure determine a data point located at the appropriate intersection on the graph. The points are usually connected with a smooth curve. In this example, the points determine a straight line. Often a graph is extended *beyond* the range for which experimental values have been determined. This procedure is known as *extrapolation.* The extrapolated portion of the graph in Fig. A-1 is shown by the dashed portion of the line.

Experimental data do not necessarily fall along a straight line. For example, the following set of data might represent the progress over time of uniformly cooling a sample of molten naphthalene.

Temperature (°C)	Time (sec)	Temperature (°C)	Time (sec)
90	0	80	270
88	30	80	300
86	60	80	330
84	90	79	360
82	120	78	390
81	150	76	420
80	180	74	450
80	210	72	480
80	240		

In this case, a smooth curve is drawn through the points, as shown in Fig. A-2.

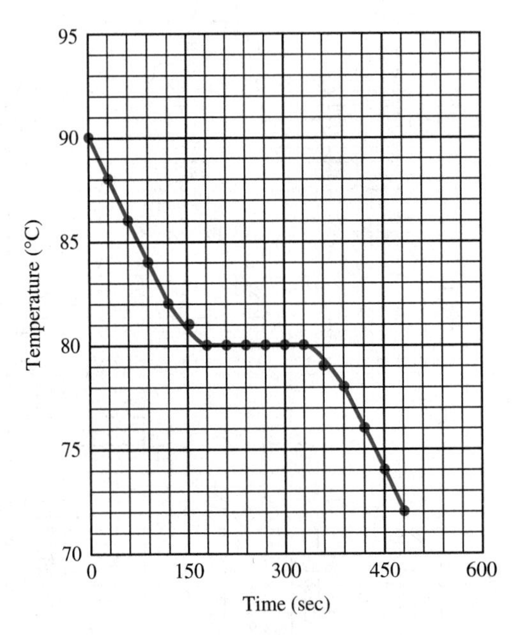

FIGURE A-2 Cooling curve of naphthalene.

Note that for each graph, we do the following.

1. Give the graph a title.
2. Outline the two coordinate axes being used.
3. Label each axis, and indicate the unit of measurement. For example, in the second graph the vertical axis is labeled "Temperature," and we have indicated that

temperature is being measured in degrees Celsius (°C). If we are not specific about units, someone else who looks at the graph cannot be certain whether the temperature is measured in °C, °F, or K. Similarly, the horizontal coordinate is labeled "Time," and we have indicated that the unit of measurement is seconds (sec).

4. For each axis, subdivide the units equally, as shown. How much each square represents is usually chosen so as to spread the data points out and give an informative visual display.

In the two preceding examples, the points fell nicely along a straight line or curve. However, that is not always the case, as shown in the following example. When faced with such irregularity, we *do not* attempt to connect the dots. Instead we draw the *best* straight line or curve, using the points as our guide. The following set of points might be obtained by a driver keeping track of mileage (see Fig. A-3).

Time (hr)	Distance (miles)	
0.0	0	**Start**
1.0	50	
2.0	95	
3.0	145	
4.0	210	
5.0	250	

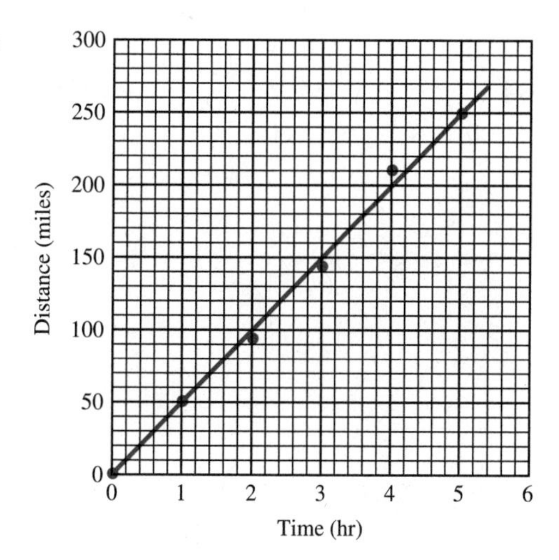

FIGURE A-3 Graph of distance versus time.

PROBLEM A.11

A balloon containing a gas sample is heated, and its volume is measured at various temperatures. Graph the following data, plotting temperature along the vertical axis and volume along the horizontal axis. Extrapolate the graph to zero volume.

Volume (mL)	Temperature (°C)
866	200
683	100
500	0
316	–100
134	–200

PROBLEM A.12

A pH curve may be obtained by measuring the pH of an acid solution as base is added. Plot the following data, with the pH of the solution along the vertical axis and the volume of the base along the horizontal axis. Draw a smooth curve through the points.

Volume of base (mL)	pH	Volume of base (mL)	pH
0.00	3.61	15.10	6.00
1.00	3.70	15.20	6.30
2.00	3.78	15.30	6.50
3.00	3.85	15.40	7.50
4.00	3.94	15.50	9.20
5.00	4.01	15.60	9.40
6.00	4.10	15.70	9.65
7.00	4.17	15.80	9.80
8.00	4.26	15.90	9.90
9.00	4.35	16.00	9.95
10.00	4.45	16.50	10.33
11.00	4.56	17.00	10.60
12.00	4.70	17.50	10.80
13.00	4.86	18.00	10.90
14.00	5.12	19.00	11.09
14.50	5.35	20.00	11.20
15.00	5.80		

CALCULATORS

APPENDIX B

The ready availability of inexpensive hand-held calculators makes them a valuable aid for solving the problems in this text. You should have at hand a calculator that is at least capable of multiplying, dividing, adding, and subtracting. It will be helpful if your calculator can also take logarithms and roots. Most inexpensive calculators operate on the basis of the Algebraic Operating System (AOS), which we describe here. Your calculator may instead use the Reverse Polish Notation (RPN), which is somewhat different. To determine the type of calculator you have, check the keys. If your calculator lacks an ENTER key, you probably have an AOS calculator, and the directions provided here will help you learn the basic mathematical operations required for the problems you will encounter in this text. To become proficient at using the full range of your calculator's capabilities, you should consult the operating manual that was included when you bought it. (Refer to the manual for *all* instructions if you own an RPN calculator.)

B.1 SIGNIFICANT FIGURES

Most calculators are not designed to round numbers to any particular number of significant digits. As a general rule, the answer displayed on a calculator ends either at the last nonzero digit or when the display itself runs out of room to place digits. Thus, you will have to assign the correct number of significant digits to the final form of any calculation. For example,

You enter	Calculator displays		Should be expressed as
18.0 ÷ 4.00 =	4.5	→	4.50
3.76 ÷ 5.91 =	0.63620981	→	0.636

B.2 THE FUNCTION KEYS

Depending on the complexity of your calculator, it may offer a wide range of mathematical functions, each served by a different key. We will restrict ourselves to the following:

Key	Function	Key	Function
+	Plus	log	Logarithm
−	Minus	$\sqrt{x}$	Square root
×	Times	INV	Inverse
÷	Divided by	y^x	"y" to the "x" power
=	Equals	EE	Exponential
+/−	Change sign	$1/x$	Reciprocal

To enter numbers, press the numeral keys exactly as though you were typing the number on a typewriter.

B.3 ADDITION, SUBTRACTION, MULTIPLICATION, AND DIVISION

The operation of an AOS calculator is relatively simple to learn, because the order of operation is exactly the same as a verbal statement of the problem being solved. For example, to carry out the following addition, we press the keys on the calculator in the sequence indicated.

Addition: $6 + 2 = ?$

1. Type the number 6.
2. Press the operation [+] .
3. Type the number 2.
4. Press the operation [=] .
5. Read the answer 8 on the digital display.

Subtraction, multiplication, and division are carried out in essentially the same fashion as addition.

Subtraction: $6 - 2 = ?$

1. Type 6.
2. Press [−] .
3. Type 2.
4. Press [=] .
5. Read 4 on the display.

Multiplication: $6 \times 2 = ?$

1. Type 6.
2. Press [×] .
3. Type 2.
4. Press [=] .
5. Read 12 on the display.

Division: $6 \div 2 = ?$

1. Type 6.
2. Press [÷] .
3. Type 2.
4. Press [=] .
5. Read 3 on the display.

B.4 SUCCESSIVE CALCULATIONS

To carry out successive calculations, it is not necessary to start over after each computation. For example, here is how we carry out the following calculation.

$12.30 + 0.65 - 4.73 = ?$

1. Type 12.30.
2. Press [+] .
3. Type 0.65.
4. Press [−] .
5. Type 4.73.
6. Press [=] .
7. Read 8.22 on the display.

B.5 EXPONENTIAL NUMBERS

To enter numbers in scientific notation, such as 6.02×10^{23}, use the following set of operations.

1. Type 6.02.
2. Press EE .
3. Type 23.

The display will read 6.02 23, which is interpreted as 6.02×10^{23}.

On the other hand, if you wish to enter a number with a negative power of 10, such as 6.63×10^{-34}, use the "change sign" button as follows:

1. Type 6.63.
2. Press EE .
3. Type 34.
4. Press +/− .

The display will read 6.63 −34, which is interpreted as 6.63×10^{-34}.

To carry out calculations with exponential numbers, enter the numbers as just shown, and carry out operations exactly as shown in the section on addition, subtraction, multiplication, and division. For example, suppose we want to carry out the following:

$$(4.5 \times 10^4)(8.0 \times 10^{-11}) = ?$$

1. Type 4.5.
2. Press EE .
3. Type 4.
4. Press × .
5. Type 8.0.
6. Press EE .
7. Type 11.
8. Press +/− .
9. Press = .

Read the answer 3.6 −6, which corresponds to 3.6×10^{-6}.

When exponential numbers are added, the answer will be expressed in scientific notation.

B.6 LOGARITHMS AND ANTILOGARITHMS

To obtain the logarithm of a number, such as log 0.064, simply type the number and press the logarithm key.

1. Type 0.064.
2. Press [log].
3. Read –1.19.

In accordance with the rules for significant figures, the logarithm (in this case, –1.19) has been rounded off to have the same number of digits to the right of the decimal point as the number of significant figures in the number for which the logarithm is being obtained (in this case 0.064, which has two significant figures).

The logarithms of exponential numbers, such as $\log(2.5 \times 10^{-5})$, are obtained in essentially the same fashion as just shown.

1. Type 2.5.
2. Press [EE].
3. Type 5.
4. Press [+/–].
5. Press [log].
6. Read –4.60.

To take an antilog, such as antilog(–2.49), type the number and then press [INV] followed by [log].

1. Type 2.49.
2. Press [+/–].
3. Press [INV].
4. Press [log].
5. Read 0.0032 or 3.2 –03.

We may instead enter the negative number by beginning as though we were subtracting it from zero (replace Steps 1 and 2 above by Steps 1, 2, and 3 in the following):

1. Press [–].
2. Type 2.49.
3. Press [=].
4. Press [INV].
5. Press [log].
6. Read 0.0032 or 3.2 –03.

B.7 Roots

If your calculator has a square root function, take the square root of 14.76 as follows:

1. Type 14.76.
2. Press [$\sqrt{x}$].
3. Read 3.84 on display.

For roots other than the square root, the easiest procedure is to consider the problem the inverse of raising a number to a power. We can combine the [INV] function with the [y^x] function. For example, let us take the cube root of 35.2, that is, $\sqrt[3]{35.2}$:

1. Type 35.2.
2. Press [INV].
3. Press [y^x].
4. Type 3.
5. Press [=].
6. Read 3.28 on display.

Similarly, let us take the fifth root of 89.7, that is, $\sqrt[5]{89.7}$.

1. Type 89.7.
2. Press [INV].
3. Press [y^x].
4. Type 5.
5. Press [=].
6. Read 2.46 on display.

Useful Conversion Factors — Appendix C

Conversion factors

Length:	1 m = 1000 mm = 100 cm
	2.54 cm = 1 in.
	1.00 km = 0.621 mile
Mass:	1 kg = 1000 g
	1 g = 1000 mg
	454 g = 1.00 lb
	1.00 kg = 2.20 lb
Volume:	1 L = 1000 mL
	946 mL = 1.00 qt
	3.785 L = 1.000 gal
Energy:	4.184 J = 1 cal
Temperature:	$T_F = 1.8T_C + 32$
	$T_K = T_C + 273$
Gases:	1 mol = 22.4 L (at STP)
Avogadro's number:	6.02×10^{23}
Universal gas constant:	$R = \frac{0.0821 \text{ L} \cdot \text{atm}}{\text{mol} \cdot \text{K}}$

Important SI prefixes

Greater than 1:

kilo (k) means 10^3 or 1000

mega (M) means 10^6 or 1,000,000

Less than 1:

deci (d) means 10^{-1} = 0.1

centi (c) means 10^{-2} = 0.01

milli (m) means 10^{-3} = 0.001

micro (μ) means 10^{-6} = 0.000001

nano (n) means 10^{-9} = 0.000000001

pico (p) means 10^{-12} = 0.000000000001

APPENDIX D ELECTRONIC CONFIGURATIONS OF THE ELEMENTS

Element	Atomic Number	Electronic Structure
H	1	$1s^1$
He	2	$1s^2$
Li	3	[He]$2s^1$
Be	4	[He]$2s^2$
B	5	[He]$2s^22p^1$
C	6	[He]$2s^22p^2$
N	7	[He]$2s^22p^3$
O	8	[He]$2s^22p^4$
F	9	[He]$2s^22p^5$
Ne	10	[He]$2s^22p^6$
Na	11	[Ne]$3s^1$
Mg	12	[Ne]$3s^2$
Al	13	[Ne]$3s^23p^1$
Si	14	[Ne]$3s^23p^2$
P	15	[Ne]$3s^23p^3$
S	16	[Ne]$3s^23p^4$
Cl	17	[Ne]$3s^23p^5$
Ar	18	[Ne]$3s^23p^6$
K	19	[Ar]$4s^1$
Ca	20	[Ar]$4s^2$
Sc	21	[Ar]$3d^14s^2$
Ti	22	[Ar]$3d^24s^2$
V	23	[Ar]$3d^34s^2$
Cr	24	[Ar]$3d^54s^1$
Mn	25	[Ar]$3d^54s^2$
Fe	26	[Ar]$3d^64s^2$
Co	27	[Ar]$3d^74s^2$
Ni	28	[Ar]$3d^84s^2$
Cu	29	[Ar]$3d^{10}4s^1$
Zn	30	[Ar]$3d^{10}4s^2$
Ga	31	[Ar]$3d^{10}4s^24p^1$
Ge	32	[Ar]$3d^{10}4s^24p^2$
As	33	[Ar]$3d^{10}4s^24p^3$
Se	34	[Ar]$3d^{10}4s^24p^4$
Br	35	[Ar]$3d^{10}4s^24p^5$
Kr	36	[Ar]$3d^{10}4s^24p^6$
Rb	37	[Kr]$5s^1$
Sr	38	[Kr]$5s^2$
Y	39	[Kr]$4d^15s^2$
Zr	40	[Kr]$4d^25s^2$
Nb	41	[Kr]$4d^45s^1$
Mo	42	[Kr]$4d^55s^1$
Tc	43	[Kr]$4d^65s^1$
Ru	44	[Kr]$4d^75s^1$
Rh	45	[Kr]$4d^85s^1$
Pd	46	[Kr]$4d^{10}$
Ag	47	[Kr]$4d^{10}5s^1$
Cd	48	[Kr]$4d^{10}5s^2$
In	49	[Kr]$4d^{10}5s^25p^1$
Sn	50	[Kr]$4d^{10}5s^25p^2$
Sb	51	[Kr]$4d^{10}5s^25p^3$
Te	52	[Kr]$4d^{10}5s^25p^4$
I	53	[Kr]$4d^{10}5s^25p^5$
Xe	54	[Kr]$4d^{10}5s^25p^6$
Cs	55	[Xe]$6s^1$
Ba	56	[Xe]$6s^2$
La	57	[Xe]$5d^16s^2$
Ce	58	[Xe]$4f^26s^2$
Pr	59	[Xe]$4f^36s^2$
Nd	60	[Xe]$4f^46s^2$
Pm	61	[Xe]$4f^56s^2$
Sm	62	[Xe]$4f^66s^2$
Eu	63	[Xe]$4f^76s^2$
Gd	64	[Xe]$4f^75d^16s^2$
Tb	65	[Xe]$4f^96s^2$
Dy	66	[Xe]$4f^{10}6s^2$
Ho	67	[Xe]$4f^{11}6s^2$
Er	68	[Xe]$4f^{12}6s^2$
Tm	69	[Xe]$4f^{13}6s^2$
Yb	70	[Xe]$4f^{14}6s^2$
Lu	71	[Xe]$4f^{14}5d^16s^2$
Hf	72	[Xe]$4f^{14}5d^26s^2$
Ta	73	[Xe]$4f^{14}5d^36s^2$
W	74	[Xe]$4f^{14}5d^46s^2$
Re	75	[Xe]$4f^{14}5d^56s^2$
Os	76	[Xe]$4f^{14}5d^66s^2$
Ir	77	[Xe]$4f^{14}5d^76s^2$
Pt	78	[Xe]$4f^{14}5d^96s^1$
Au	79	[Xe]$4f^{14}5d^{10}6s^1$
Hg	80	[Xe]$4f^{14}5d^{10}6s^2$
Tl	81	[Xe]$4f^{14}5d^{10}6s^26p^1$
Pb	82	[Xe]$4f^{14}5d^{10}6s^26p^2$
Bi	83	[Xe]$4f^{14}5d^{10}6s^26p^3$
Po	84	[Xe]$4f^{14}5d^{10}6s^26p^4$
At	85	[Xe]$4f^{14}5d^{10}6s^26p^5$
Rn	86	[Xe]$4f^{14}5d^{10}6s^26p^6$
Fr	87	[Rn]$7s^1$
Ra	88	[Rn]$7s^2$
Ac	89	[Rn]$6d^17s^2$
Th	90	[Rn]$6d^27s^2$
Pa	91	[Rn]$5f^26d^17s^2$
U	92	[Rn]$5f^36d^17s^2$
Np	93	[Rn]$5f^46d^17s^2$
Pu	94	[Rn]$5f^67s^2$
Am	95	[Rn]$5f^77s^2$
Cm	96	[Rn]$5f^76d^17s^2$
Bk	97	[Rn]$5f^97s^2$
Cf	98	[Rn]$5f^{10}7s^2$
Es	99	[Rn]$5f^{11}7s^2$
Fm	100	[Rn]$5f^{12}7s^2$
Md	101	[Rn]$5f^{13}7s^2$
No	102	[Rn]$5f^{14}7s^2$
Lr	103	[Rn]$5f^{14}6d^17s^2$
Unq	104	[Rn]$5f^{14}6d^27s^2$
Unp	105	[Rn]$5f^{14}6d^37s^2$
Unh	106	[Rn]$5f^{14}6d^47s^2$
Uns	107	[Rn]$5f^{14}6d^57s^2$
Uno	108	[Rn]$5f^{14}6d^67s^2$
Une	109	[Rn]$5f^{14}6d^7s^2$

Solubilities of Selected Salts[a]

Cations	Anions NO_3^-	SO_4^{2-}	OH^-	Cl^-	Br^-	I^-	S^{2-}	$C_2H_3O_2^-$	CO_3^{2-}	PO_4^{3-}
Na^+	S	S	S	S	S	S	S	S	S	S
K^+	S	S	S	S	S	S	S	S	S	S
NH_4^+	S	S	S	S	S	S	S	S	S	S
Ag^+	S	ss	–	I	I	I	I	ss	I	I
Al^{3+}	S	S	I	S	S	S	d	S	–	I
Ba^{2+}	S	I	S	S	S	S	d	S	I	I
Ca^{2+}	S	I	ss	S	S	S	ss	S	I	I
Co^{2+}	S	S	I	S	S	S	I	S	I	I
Cu^{2+}	S	S	I	S	S	–	I	S	–	I
Hg^{2+}	S	d	I	S	S	S	I	S	–	I
Mg^{2+}	S	S	I	S	S	S	d	S	I	I
Pb^{2+}	S	I	I	I	I	I	I	S	I	I
Sr^{2+}	S	I	S	S	S	S	I	S	I	I
Zn^{2+}	S	S	I	S	S	S	I	S	I	I

[a]S = soluble, ss = slightly soluble, I = insoluble, d = decomposes in water

GLOSSARY

absolute temperature Temperature expressed in the Kelvin temperature scale. See also *Kelvin temperature scale.*

accuracy The degree to which a measured value agrees with the correct value.

acid Substance capable of liberating hydrogen ions (H^+) in a chemical reaction (simple working definition). See also *Arrhenius acid, Brønsted–Lowry acid,* and *Lewis acid.*

actinide series Series of 14 metals following actinium (atomic numbers 90–103).

activated complex The arrangement of atoms at the transition state of a reaction coordinate.

activation energy The minimum energy that reacting molecules must possess in order to react.

activity series A series of metals arranged in order of decreasing ability to react with hydrogen ion, H^+.

addition reaction Characteristic reaction of organic compounds containing carbon–carbon multiple bonds. During an addition reaction, the organic substance in question combines with another substance, thereby eliminating a multiple bond.

ADP Adenosine diphosphate; a mononucleotide formed as a by-product during the cleavage of ATP to release energy; consists of an adenine base, a ribose sugar, and a diphosphate group.

alcohol Organic substance that has a hydroxyl group bonded to an alkyl group, R—OH.

aldehyde Organic substance that has an aldehyde group, $R-\overset{\overset{\displaystyle O}{\|}}{C}-H$.

alkali metal Any of the elements in the first column of the periodic table (group IA or 1), excluding hydrogen.

alkaline earth metal Any of the elements in the second column of the periodic table (group IIA or 2).

alkane Hydrocarbon having no multiple bonds. The general formula of noncyclic alkanes is C_nH_{2n+2}. See also *hydrocarbon.*

alkene Hydrocarbon containing a carbon–carbon double bond.

alkyl group Hydrocarbon group derived from an alkane by removal of a hydrogen atom. The free bond that results is bonded to other atoms or groups of atoms. The symbol R— is used to represent alkyl groups.

alkyl halide See *halogenated alkane.*

alkyne Hydrocarbon containing a carbon–carbon triple bond.

alpha radiation Emission of alpha particles given off during a radioactive process. An alpha particle is identical in mass and charge to a helium nucleus, ${}^4_2He$ or ${}^4_2\alpha$.

amide Functional group derived from a carboxylic acid and either ammonia or an amine, $R-\overset{\overset{\displaystyle O}{\|}}{C}-NH_2$ (also $RCONHR'$ or $RCONR'_2$).

amine Organic substance having an amino group, $R-NH_2$ (also R_2NH, R_3N).

amino acid A compound that contains an amino group and a carboxylic acid group. Biologically important polypeptides and proteins are composed of polymers of *α-amino acids,* which have the following general structure: $H_2N-\underset{\underset{\displaystyle R}{|}}{CH}-COOH$.

amorphous Refers to a solid that lacks an orderly arrangement of atoms, ions, or molecules.

analytical chemistry Branch of chemistry concerned with finding out either what is present in a given sample (qualitative analysis) or how much of some component is present (quantitative analysis).

angular Geometry in which three atoms lie in a nonlinear or bent arrangement. May describe a triatomic molecule, such as water.

anhydrous Without water.

anion A negatively charged ion.

anode In an electrochemical cell, the electrode at which oxidation occurs.

aqueous solution A liquid solution in which the solvent is water.

aromatic hydrocarbon Unsaturated cyclic hydrocarbon, for which the properties of the carbon–carbon bonds differ from those of alkenes and alkynes. The most noteworthy of these is benzene, C_6H_6.

Arrhenius acid A substance that produces hydrogen ions, H^+, when dissolved in water.

Arrhenius base A substance that produces hydroxide ions, OH^-, when dissolved in water.

associated Refers to the nonionized or nondissociated form of a substance capable of dissociation.

-ate Suffix used with oxyanions. Denotes one more oxygen than the similar oxyanion named with an *-ite* ending. For example, nitrate NO_3^-, has one more oxygen than nitrite, NO_2^-.

atom Building block of matter composed of protons, neutrons, and electrons. The protons and neutrons are found in a central nucleus; the electrons are found in a larger volume surrounding the nucleus.

atomic mass Relative mass of a naturally occurring sample of any element compared to a pure sample of carbon-12, which is assigned an atomic mass of exactly 12 u.

atomic mass unit (u) One-twelfth the mass of one atom of carbon-12.

atomic number The number of protons in the nucleus of an element.

atomic weight See *atomic mass.*

ATP Adenosine triphosphate; the substance used to supply energy during cellular activity; consists of an adenine base, a ribose sugar, and a triphosphate group. Energy is released when the triphosphate bond is broken.

Aufbau principle Principle that electrons fill the available orbitals in an atom starting with the lowest available energy level and working up.

Avogadro's Law Equal volumes of gases (either the same or different) under equal conditions of temperature and pressure contain equal numbers of molecules. It follows that 1 mole of any gas at STP occupies the same volume as 1 mole of any other gas (22.4 L).

Avogadro's number (6.02×10^{23}) The number of particles in 1 mole. One mole of any molecular substance contains Avogadro's number of molecules.

balanced equation See *chemical equation.*

barometer See *Torricellian barometer.*

base Substance capable of liberating hydroxide ions (OH^-) in a chemical reaction (simple working definition). See also *Arrhenius base, Brønsted–Lowry base,* and *Lewis base.*

base pair A pair of complementary bases; in DNA, either adenine and thymine or guanine and cytosine; in RNA, either adenine and uracil or guanine and cytosine.

beta radiation Emission of beta particles given off during a radioactive process. A beta particle is identical in mass and charge to an electron, ${}_{-1}^{0}e$ or ${}_{-1}^{0}\beta$.

binary compound Compound composed of two different elements.

biochemistry Study of molecules and chemical processes important to the functioning of life.

boiling The process that occurs when the vapor pressure of a liquid equals the external pressure; characterized by the formation of bubbles within the liquid.

boiling point Temperature at which the vapor pressure of a liquid is equal to the external pressure. At this temperature, vapor bubbles form within the liquid and rise to the surface. See *normal boiling point.*

bond angle Angle determined by any three consecutive atoms bonded together in a molecule.

bond energy The energy required to break the bond between two atoms.

Boyle's Law At constant temperature, the pressure exerted by a gas is inversely proportional to its volume: $PV = k$.

breeder reactor Nuclear fission reactor in which nonfissionable isotopes capture stray neutrons to produce new fissionable isotopes, which may be used as fuel.

Brønsted–Lowry acid A proton (or hydrogen ion) donor.

Brønsted–Lowry base A proton (or hydrogen ion) acceptor.

buffer Chemical system that is resistant to changes in pH. Composed of the conjugate acid and conjugate base of a weak acid.

calorie The quantity of heat required to raise the temperature of 1 g of water by 1°C (1 cal = 4.184 J).

capillary action Phenomenon observed when liquid molecules are attracted to another substance; for example, when a liquid climbs up a wick or rises in a glass tube.

carbohydrate A polyhydroxy aldehyde or ketone.

carbonyl group The organic group, $-\overset{O}{\overset{\|}{C}}-$; a component of many functional groups.

carboxylic acid Organic substance containing the carboxylic acid group, $R-\overset{O}{\overset{\|}{C}}-OH$.

catalyst Substance that increases the rate of a chemical reaction without being consumed in the reaction.

cathode In an electrochemical cell, the electrode at which reduction occurs.

cation A positively charged ion.

Celsius temperature scale Temperature scale based on the freezing point (0°C) and the boiling point (100°C) of water.

centigrade temperature scale See *Celsius temperature scale.*

chain reaction Any reaction that sustains itself once started, such as nuclear fission.

chalcogen Any of the elements in group VIA (16) of the periodic table.

Charles's Law At constant pressure, the volume of a gas is proportional to its Kelvin temperature: $V = kT$.

chemical change A change in the fixed composition of one or more substances. During such a change, starting materials known as reactants are converted into new substances known as products.

chemical energy Energy that is either liberated or absorbed during a chemical change.

chemical equation Symbolic representation of a chemical change in which the starting materials (reactants) and final materials (products) are written on opposite sides of an arrow that points from the reactants to the products.

chemical family The elements in the same vertical column of the periodic table. (*Exception:* Hydrogen, which appears in the same column as the alkali metals, is considered a family of its own.)

chemical formula Symbolic representation of the composition of a substance expressing the elements present. In the case of molecular substances, the number of atoms of each element per molecule is given. For nonmolecular substances, the simplest ratio of atoms or ions is generally given.
chemical kinetics The study of the rates of reactions.
chemical property A property associated with a change in the fixed composition of some substance.
chemical reaction See *chemical change.*
chemistry Study of the composition of matter and the changes it undergoes.
codon A sequence of three mononucleotide bases that codes for a specific amino acid during protein synthesis.
coenzyme A substance that must be present for an enzyme to function; many vitamins are coenzymes.
colligative properties Properties of solutions that depend on the concentration of dissolved particles. These include boiling point elevation, freezing point depression, and osmotic pressure.
colloid Heterogeneous mixture in which the phases present do not separate out.
Combined Gas Law Relationship describing the mutual effects of changes in the pressure, volume, and temperature of a fixed sample of gas : $PV/T = k$.
combustion with oxygen Type of reaction that is characteristic of the reaction of hydrocarbons with oxygen.
common name Name traditionally associated with a substance; not based on any systematic method of nomenclature.
compound A pure substance composed of more than one element. Compounds can be decomposed into simpler substances.
concentration The quantity of solute dissolved in a given quantity of solvent or solution.
conjugate acid The chemical species obtained when a Brønsted–Lowry base accepts a proton. A conjugate acid differs from its corresponding conjugate base by H^+.
conjugate base The chemical species obtained when a Brønsted–Lowry acid donates a proton. A conjugate base differs from its conjugate acid by H^+.
conservation of mass See *Law of Conservation of Mass.*
continuous spectrum A continuous band of light waves such as the colors in the rainbow. A continuous spectrum contains radiations of all frequencies within the range of the spectrum. See also *discontinuous spectrum.*
controlled experiment An experiment in which only one variable (or factor) is changed at a time.
conversion factor A mathematical relationship that enables the conversion of a quantity from one unit to another.
core symbol An abbreviation that may be used to represent the electronic configuration of a particular noble gas.
covalent bond A bond that arises from the sharing of a pair of electrons between two atoms. A covalent bond forms when two orbitals on two atoms overlap to share a pair of electrons between them.
critical mass The minimum mass of fissionable material needed to sustain a chain reaction.
cryoscopic constant Physical constant that relates the freezing point depression of a solvent to the molality of a solution prepared in that solvent.
crystal lattice The particular arrangement of molecules or ions in a solid substance.
crystalline Refers to a solid with an orderly arrangement of atoms, ions, or molecules.

Dalton's Law of Partial Pressures The total pressure of a mixture of gases is equal to the sum of the pressures that each of the gases would exert if confined to the same volume in the absence of the other gases present.
decay Process in which a nucleus emits radiation to form a more stable (and usually smaller) nucleus.
decomposition reaction Reaction type in which a single reactant produces two or more products.
density A proportion that relates the mass of a substance to its volume. The density of a substance is equal to its mass per unit volume.
deuteron A hydrogen-2 nucleus, ${}^2_1H$.
diatomic Having two atoms. Usually used to refer to molecules composed of two atoms, such as molecular oxygen, O_2, or hydrogen, H_2.
dimensional analysis Mathematical technique for solving problems by converting a quantity from one unit of measure to another.
dipeptide A molecule composed of two amino acids connected via an amide (or peptide) bond.
dipole Electrical separation of charges within a chemical bond or a molecule as a whole.
dipole–dipole interaction The alignment of dipoles in a collection of polar molecules, such that the negative end of each molecule lies next to the positive end of its neighbor, thereby creating an attractive force between neighboring molecules.
diprotic acid An acid possessing two ionizable hydrogens per molecule, such as sulfuric acid, H_2SO_4.
disaccharide A sugar composed of two monosaccharides. See also *monosaccharide.*
discontinuous spectrum A band of light waves that contains only certain frequencies within the range, such as those found in the hydrogen spectrum. See also *continuous spectrum.*
disintegration series A series of nuclear decays by which a radioactive isotope eventually produces a stable nucleus; for example, the uranium-238 series.
dispersed phase In a colloid, the phase that is considered to be suspended in the dispersing phase. For example, water is the dispersed phase in fog, a suspension of water droplets in air.
dispersing phase In a colloid, the phase in which the dispersed phase is suspended. For example, air is the dispersing phase in fog, a suspension of water droplets in air.
dispersion A suspension of one substance in another.
dispersion forces See *London forces.*

dissociated Refers to the ionized form of a substance that is capable of separation into ions.
dissociation The process of separation into ions, or of ionization.
DNA Deoxyribonucleic acid; genetic material that directs cellular activity.
double bond Type of covalent bonding in which two pairs of electrons are shared between the same two atoms. Each shared pair of electrons represents one covalent bond.
double helix The double-stranded structure of DNA in which the two strands intertwine in a helical arrangement around one another.
double-replacement reaction Reaction in which two ionic substances exchange ions (or "change partners") with one another.
drug A substance taken to cure or relieve the symptoms of disease.

effective charge The charge experienced by an outer-shell electron, after subtracting the electrons of the last noble gas core from the nuclear charge.
Einstein equation Relationship between mass and its energy equivalent, $E = mc^2$. The conversion of small amounts of mass produces enormous amounts of energy.
elastic collision A collision in which there is no loss of energy. Collisions between ideal gas molecules are elastic.
electrochemical cell An arrangement of two half-reactions such that the electrons transferred in the oxidation–reduction must pass through an external circuit from one electrode to another.
electrode A conductor that carries electrons to or from a half-cell in an electrochemical cell.
electrolysis The operation of an electrochemical cell in its nonspontaneous direction, as in an electrolytic cell.
electrolyte A substance that, when dissolved in water, permits the aqueous solution to conduct electricity.
electrolytic cell An electrochemical cell that is driven in the nonspontaneous direction by an external source of electrical energy.
electromotive series See *activity series.*
electron Fundamental particle found outside of the nucleus and having an electrical charge of −1 and a mass of 0.0005486 u.
electron affinity The energy change that accompanies the gain of an electron by a gaseous atom of an element.
electron-dot structure See *Lewis electron-dot structure.*
electron transfer The transfer of electrons from one chemical species to another. Occurs during oxidation–reduction reactions.
electronegativity A measure of the tendency of an atom to attract electrons in a bonded pair.
electronic configuration A shorthand description of the locations of the various electrons in their respective orbitals.
electrostatic force Force that arises from the electrical charges on like or unlike charged particles. Unlike charges have an attractive electrostatic force; like charges have a repulsive electrostatic force.
electrovalent bond See *ionic bond.*
element Substance that cannot be decomposed into simpler substances by ordinary means. An element is composed of only one kind of atom.
empirical formula Simplest whole-number ratio of atoms in a given substance.
endothermic reaction A reaction that absorbs heat from the surroundings.
energy The ability to do useful work. Energy is required to apply a force through a distance.
English system System of units still used in the United States.
enthalpy of reaction Heat liberated or absorbed during a chemical change.
enzyme A biological catalyst that accelerates chemical reactions in living organisms.
equilibrium Dynamic condition in which the rates of opposing processes are equal. Equilibrium is characterized by constant macroscopic properties despite continuing change at the microscopic level.
equilibrium constant Numerical constant expressing the equilibrium relationship for a given equilibrium reaction, usually abbreviated K_{eq}.
equilibrium expression Mathematical expression relating the concentrations of the reactants and products of a chemical system at equilibrium.
equilibrium mixture A mixture of reactants and products that conforms to the equilibrium expression and equilibrium constant for a chemical system at equilibrium.
equilibrium state A state in which the rates of two opposing processes have become equal.
equivalent The quantity of an acid or base that can deliver one mole of hydrogen ions or hydroxide ions.
ester Functional group derived from a carboxylic acid and an alcohol, $R{-}\overset{\overset{\displaystyle O}{\|}}{C}{-}OR'$.
ether Functional group in which an oxygen connects two hydrocarbon groups, R—O—R′.
evaporation Process whereby the molecules of a liquid substance leave the liquid state and enter the gaseous state at a temperature below that of the boiling point of the liquid.
exact number A number that contains no uncertainty. For example, one dozen is exactly 12.
excited state An electronic state other than the lowest-energy, or ground-state, configuration. See also *ground state.*
exothermic reaction A reaction in which heat is given off.

Fahrenheit temperature scale Temperature scale in the English system.
fat A triester derived from glycerol and three fatty acids; a liquid fat is referred to as an *oil.*
fission See *nuclear fission.*
fissionable isotope An isotope that can undergo fission.

formula mass The sum of the atomic masses in a chemical formula.
formula weight See *formula mass.*
forward reaction The reaction written from left to right.
fractional distillation Process by which the components in a mixture of liquids are separated in order of increasing boiling points.
frequency Quantity used to describe wavelike characteristics. Tells the number of cycles per unit time.
functional group Group of atoms in an organic compound that give it characteristic chemical and physical properties.
fundamental particle Subatomic particle from which atoms are composed.
fusion See *nuclear fusion.*

galvanic cell See *voltaic cell.*
gamma radiation High-energy electromagnetic radiations (nonvisible light waves) associated with most nuclear transformations; γ.
gas One of the physical states of matter. The particles of matter in a gas are far away from one another and are in constant motion.
Gay–Lussac's Law At constant volume, the pressure of a gas is proportional to its Kelvin temperature: $P = kT$.
Geiger–Müller counter Instrument used in the detection of ionizing radiation produced by nuclear decay.
gene A segment of DNA that codes for an enzyme or polypeptide.
ground state Electronic configuration in which the electrons are in the lowest possible energy state.
group Vertical column of the periodic table. See also *chemical family.*

half-cell See *half-reaction.*
half-life Period of time required for exactly one-half of a sample of a radioactive isotope to decay.
half-reaction In an oxidation–reduction reaction, either the oxidation portion or the reduction portion.
halogen Any of the elements in group VIIA (17) of the periodic table.
halogenated alkane Alkane in which one hydrogen atom (or more) has been replaced by a halogen.
heat Form of energy associated with molecular motion.
heat capacity See *molar heat capacity.*
heat of fusion Heat required to convert a solid to a liquid at its melting point.
heat of reaction The heat liberated or absorbed during a given reaction.
heat of vaporization Heat required to convert a liquid to a vapor at its boiling point.
heterogeneous matter Matter that is nonuniform.
homogeneous matter Matter that is uniform throughout.
hormone A regulatory chemical produced by an endocrine gland.
Hund's rule Principle that, within a given subshell, electrons half-fill each of the orbitals before pairing.
hydrated salt A salt possessing a fixed number of moles of water per mole of the salt.
hydration energy Energy liberated when solute molecules or ions become surrounded by water molecules.
hydrocarbon An organic compound composed exclusively of carbon and hydrogen.
hydrogen bonding Formation of loose electrostatic attractions between an electropositive hydrogen atom from one molecule and an electronegative atom of a neighboring molecule. The hydrogen atom in question must be bonded to N, O, or F in its parent compound, and the hydrogen bond it forms must be to N, O, or F in the neighboring molecule.
hydrometer Instrument used to determine the density of a liquid.
hydronium ion (H_3O^+) The hydrated species that forms when hydrogen ions are dissolved in water.
hypo- Prefix used in the naming of oxyanions. Denotes one less oxygen than the ion of the same name without the prefix. For example, hypochlorite, ClO^-, has one less oxygen than chlorite, ClO_2^-.
hypothesis An educated guess used to explain an observation.

-ic Suffix used in nomenclature. Denotes the higher of two possible oxidation states.
ideal gas A gas that obeys the Ideal Gas Law.
Ideal Gas Law Mathematical expression that relates the pressure, volume, temperature, and number of moles of a gas: $PV = nRT$.
immiscible Refers to liquids that are incapable of dissolving in one another, such as oil and vinegar.
indicator A chemical substance that indicates the endpoint of a chemical reaction by changing colors when the reaction is complete. Acid–base indicators display different colors in acid and base.
inorganic chemistry Branch of chemistry concerned with substances other than those classified as organic substances—for example, minerals and water.
intermolecular force An attractive or repulsive force that exists between molecules.
International System (SI) Currently accepted system of units adopted by the international scientific community.
intramolecular force An attractive or repulsive force that exists within a molecule, such as a chemical bond.
ion An atom or group of atoms that bears a charge other than zero.
ionic bond Chemical bond that results from the electrostatic forces of attraction between two oppositely charged ions.
ionic interaction Electrostatic interaction between ions.
ionization energy Energy required to remove an electron from an atom of an element in the gaseous state.
ionizing radiation High-energy radiation that causes ionization of atoms with which it interacts. Radiations emitted by radioisotopes are ionizing radiations.
isoelectronic Refers to two or more chemical species that have the same electronic configuration.

isomer One of two or more compounds having the same chemical formula but different structures, and therefore different physical and chemical properties.

isotonic solutions Solutions that have the same osmotic pressure.

isotope One of two or more atoms that have the same number of protons (the same atomic number or element) but differ in their number of neutrons (different mass numbers). Often used to designate a particular nuclide.

-ite Suffix used with oxyanions. Denotes one less oxygen than the similar oxyanion named with an *-ate* ending. For example, sulfite, SO_3^{2-}, has one less oxygen than sulfate, SO_4^{2-}.

IUPAC system Systematic method for naming substances, agreed to by the International Union of Pure and Applied Chemistry.

K_w Constant relating the hydrogen ion concentration and the hydroxide ion concentration of aqueous solutions: $K_w = 1.0 \times 10^{-14}$.

Kelvin temperature scale Temperature scale based on extrapolation of the temperature–pressure relationship of gases to zero pressure.

ketone Organic molecule having a carbonyl group bonded to two hydrocarbon groups, $R{-}\underset{\underset{O}{\|}}{C}{-}R'$.

kinetic energy Energy of an object associated with its motion: $KE = \frac{1}{2}mv^2$.

kinetic–molecular theory Theory describing matter in terms of particles in constant motion.

lanthanide series Series of 14 metals following lanthanum (atomic numbers 58–71).

law A set of observed regularities expressed in a concise verbal or mathematical statement.

Law of Combining Volumes The volumes of reacting gases and their products are in simple, whole-number ratios.

Law of Conservation of Energy Energy can be neither created nor destroyed.

Law of Conservation of Mass Matter can be neither created nor destroyed during a chemical change.

Law of Constant Composition A pure substance has a fixed and definite composition.

Le Chatelier's Principle Principle governing the behavior of systems at equilibrium: When a stress is applied to a system at equilibrium, the system responds in such a way as to relieve the stress.

Lewis acid An electron-pair acceptor.

Lewis base An electron-pair donor.

Lewis electron-dot structure Representation of the valence electrons in molecules showing the location of such electrons and the resultant chemical bonds.

limiting reagent In a chemical reaction, the reactant that is completely consumed and thereby limits the extent to which all other reactants are consumed.

linear Refers to the geometry of atoms that lie along a straight line.

lipid A biomolecule that is soluble in nonpolar solvents.

liquid One of the physical states of matter. The particles of matter in a liquid lie close together but have the freedom to flow over and around one another.

liquid solution A solution in which the dispersing phase is a liquid.

liquid–vapor phase equilibrium Equilibrium process in which liquid particles enter the vapor state at the same rate that vapor particles return to the liquid state.

London forces Relatively weak intermolecular forces of attraction created by momentary distortions in the electronic distributions of the molecules.

mass A measure of the quantity of matter.

mass defect The loss of mass that accompanies an energy-releasing nuclear transformation. Corresponds to an energy equivalent via the Einstein equation, $E = mc^2$. See also *Einstein equation.*

mass number The sum of the protons and neutrons in the nucleus of an atom.

matter Anything that has mass and occupies space.

measured number A number that has been determined through measurement and therefore has some uncertainty associated with it.

melting point Temperature at which a substance in the solid state may be converted to its liquid state at the same temperature. More precisely, the temperature at which a solid and a liquid can exist in equilibrium. Identical to the freezing point.

metabolism The biochemical processes of a living organism or cell.

metal An element that is shiny, malleable (ductile), and capable of conducting electricity and heat. Metals generally lose electrons to form positive ions and are located on the left-hand side of the periodic table.

metallic bond Bonding between metal atoms that occurs through a network of overlapping outer orbitals, which share their "sea" of electrons.

metalloid See *semimetal.*

miscible Refers to liquids that are capable of dissolving in one another.

model A concrete mental image used to describe observations explained in laws and theories.

molality (*m*) Unit of concentration. The number of moles of solute per kilogram of solvent.

molar heat capacity Quantity of heat required to raise the temperature of 1 mole of a substance by 1°C.

molar mass Mass of 1 mole of a substance; numerically equal to the atomic or molecular mass in grams.

molar volume The volume occupied by 1 mole of any gas. At STP, 1 mole occupies 22.4 L.

molarity (M) Unit of concentration. The number of moles of solute per liter of solution.

mole Avogadro's number of particles, 6.02×10^{23}. The mass of 1 mole of a molecular (or atomic) substance equals its molecular (or atomic) mass taken in grams.

mole fraction The number of moles represented by a component of a mixture, divided by the total number of moles of all components of the mixture.

mole ratio Ratio of the coefficients of two substances in a balanced chemical equation.

molecular formula Chemical formula that tells how many atoms of each element are present in each molecule. The term *molecular formula* is often used to refer to the formula derived from the simplest whole-number ratio of atoms or ions in nonmolecular substances.

molecular mass The sum of the atomic masses of the atoms in a molecular formula. It represents either the mass of one molecule of the substance in atomic mass units or the mass of 1 mole of the substance in grams.

molecular weight See *molecular mass.*

molecule Smallest unit of a pure substance that can exist independently and exhibit all of the properties of the substance.

monatomic Having one atom. Refers to a chemical substance composed of only one atom per molecule or ion.

monatomic ion An ion containing only one atom.

monomer Basic unit repeated in a polymer. See also *polymer.*

mononucleotide A building-block unit of DNA and RNA; composed of a ribose or deoxyribose sugar, a phosphate group, and a nucleotide base (adenine, cytosine, guanine, thymine, or uracil).

monosaccharide A sugar composed of a single unit; often referred to as a *simple sugar.*

net ionic equation An ionic equation that expresses only those ions or molecules that actually undergo change during a chemical reaction.

neutralization Process by which an acid and a base react with one another to counteract one another's acid or base properties.

neutron Nuclear particle having a mass of 1.0086649 u and a charge of zero.

neutron emission A stream of neutrons given off during some nuclear reactions.

noble gas Any of the elements belonging to group VIIIA (18) of the periodic table.

nonelectrolyte Substance that, when dissolved in water or in its molten state, does not conduct electricity.

nonmetal An element that lacks the properties of a metal; usually nonlustrous and lacking significant electrical conductivity; found on the upper right-hand side of the periodic table.

nonpolar bond See *nonpolar covalent bond.*

nonpolar covalent bond Chemical bond that is formed between atoms of identical or similar electronegativities and lacks a significant charge separation.

nonpolar molecule Molecule whose net dipole is approximately zero.

normal boiling point The boiling point of a substance at standard pressure (1 atm).

normality (N) Unit of concentration. The number of equivalents of solute per liter of solution.

nuclear fission The process by which a nucleus fragments into smaller nuclei and other nuclear particles.

nuclear fusion The process by which small nuclei are brought together ("fused") into larger nuclei.

nuclear symbol Symbol of the form $^{A}_{Z}X$, in which A represents the mass number, Z represents the atomic number, and X is the chemical symbol for the element.

nucleon Either a proton or a neutron.

nucleus Central portion of an atom that is made up of protons and neutrons; it contains almost the entire mass of an atom.

nuclide A specific nucleus for which the numbers of protons and neutrons are known.

observation An event that has been witnessed.

octet rule Rule used to predict the sharing of electrons in chemical bonding. When we draw Lewis structures for elements in the first three rows, each element must be surrounded by eight electrons, except hydrogen, which must have two electrons.

oil Liquid fat. See also *fat.*

orbital An energy state describing a region of space within which an electron may be found. No more than two electrons can occupy the same orbital at the same time.

organic chemistry Branch of chemistry concerned with compounds that are composed primarily of carbon and hydrogen.

osmolality The number of moles of solute particles per kilogram of solvent.

osmotic pressure Colligative property of solutions that causes the net migration of solvent molecules across a semipermeable membrane from the more dilute solution (lower osmotic pressure) to the more concentrated solution (greater osmotic pressure).

-ous Suffix used in nomenclature. Denotes the lower of two possible oxidation states.

oxidation Loss of electrons.

oxidation number The assigned charge on an atom (also known as oxidation state).

oxidizing agent In an oxidation–reduction reaction, the substance that causes oxidation in another substance. The oxidizing agent is reduced.

oxyanion Anion composed of two elements, one of which is oxygen.

partial pressure Pressure that a gas in a mixture would exert if the other gases were removed and the conditions of volume and temperature remained the same. See also *Dalton's Law of Partial Pressures.*

parts per million (ppm) Unit of concentration. Generally, the number of grams of solute per million grams of solution. For dilute aqueous solutions this corresponds to the number of milligrams of solute per liter.

peptide bond An amide bond that connects two amino acids.

per- Prefix used in the nomenclature of oxyanions. Denotes one more oxygen than the anion of similar name. For example, perchlorate, ClO_4^-, has one more oxygen than chlorate, ClO_3^-.

percent ionization The extent to which a substance is dissociated, expressed as a percentage of the substance present prior to dissociation.

percent solution by mass Unit of concentration expressing the mass of solute as a percentage of the total mass.

percent solution by volume Unit of concentration expressing the volume of solute as a percentage of the total volume.

percentage composition A description of a compound in terms of the percentage of the total mass that each element represents.

percentage yield The actual yield of a reaction, expressed as a percentage of the theoretical yield. See also *theoretical yield.*

period A horizontal row of the periodic table. One cycle through the recurring, or "periodic," properties of the elements.

Periodic Law Recurring patterns of physical and chemical properties exist when the elements are arranged in order of increasing atomic number. These patterns are the result of recurring patterns of electronic configuration.

periodic table Table of the elements in which the elements are arranged according to similarities in their properties.

petroleum Hydrocarbon mixture found beneath the surface of the earth that serves as our most widely used energy source. It is also a source of raw materials for many products.

pH Measure of the acidic or basic strength of a solution; $pH = -\log[H^+]$.

phase The physical state of a substance.

phase change A change in the physical state of a substance, such as occurs during melting or boiling.

phenol Any organic substance having a hydroxyl group, $-OH$, bonded to a benzene ring. Phenol is also the name of the parent compound for this class of substances.

photon A single "bundle" of light energy.

physical change Change in a substance that does not involve a change in its fixed composition, such as a phase change.

physical chemistry Branch of chemistry concerned with the physical properties of chemicals and the physics of chemical processes.

physical property Property of a substance that can be observed in the absence of a change in its fixed composition; examples include color and melting point.

physical state The solid, liquid, or gaseous state of matter.

Planck's constant Physical constant that relates the frequency and the energy of a photon of light.

pOH $-\log[OH^-]$; related to the pH as follows: $pOH = 14 - pH$.

polar covalent bond Covalent bond with a significant charge separation.

polar molecule Molecule with a net dipole other than zero.

polyatomic Having more than one atom. Usually used to refer to molecules or ions.

polyatomic ion An ion composed of more than one atom.

polymer Large molecule made up of repeating units of a smaller substance. See also *monomer.*

polynucleotide DNA or RNA; a polymer composed of mononucleotides.

polypeptide A molecule composed of many amino acids connected through peptide (amide) bonds. A polypeptide of at least 100 amino acid units is considered a protein.

polysaccharide A sugar composed of many monosaccharides. See also *monosaccharide.*

porous partition A material used to separate the half-cells in an electrochemical cell. The porous partition allows ions to migrate from one half-cell to the other in order to maintain electrical neutrality in each half-cell.

positron emission A stream of positron particles given off during some nuclear processes. A positron has the same mass as an electron but a charge of +1.

potential energy Energy due to the position of an object in a force field. For example, the potential energy of a weight suspended above the ground depends upon its height above the ground and the strength of the gravitational field.

precipitate An insoluble solid product that separates from the solution in which it is formed.

precipitation The formation of a precipitate.

precision The agreement between two or more measured quantities.

pressure Force per unit area. Gas pressure is related to the number of collisions that gaseous molecules make per unit area during a given time.

primary structure The amino acid sequence of a protein or polypeptide.

principal energy level One of the discrete electronic energy levels, n.

product A substance formed during a chemical reaction. See also *reactant.*

protein A long-chain polymer composed of amino acid units; generally a polypeptide of at least 100 amino acid units.

proton Nuclear particle having a mass of 1.0072765 u and a +1 charge. When we are referring to acids, proton means the same as hydrogen ion, H^+.

pure substance A substance having a fixed and definite composition.

quaternary structure The packing arrangement of multiple protein units.

radioactivity The spontaneous decay of a nucleus, accompanied by radiation. May be classified as natural radioactivity for naturally occurring nuclides, or induced radioactivity for artificial radionuclides. The most common forms of radiation are alpha, beta, and gamma radiation.

radioisotope A radioactive isotope. See *radionuclide.*

radionuclide A radioactive nuclide. See *radioisotope.*

rare earth metal Any of the metals for which an *f* sublevel is the last filled sublevel.

rate constant Constant used to relate the rate of a reaction to the concentrations of the reactants.

rate expression Mathematical expression that relates the rate of a reaction to the rate constant and the concentrations of the reactants.

reactant A starting material that undergoes a change during a chemical reaction. See also *product.*

reaction coordinate Horizontal coordinate of a reaction profile that represents the progress of a reaction from reactants to products.

reaction profile Diagram showing the energy changes that occur as a reaction progresses from reactants to products.

reaction rate The change with time in the concentration of a reactant or product in a chemical reaction.

reducing agent In an oxidation–reduction reaction, the substance that causes reduction in another substance. The reducing agent is oxidized.

reduction Gain of electrons.

relative mass The mass of an object, as compared to some standard.

replication The process by which DNA is reproduced; each strand forms its complementary strand.

representative element Any of the elements for which an *s* or *p* sublevel is the last filled sublevel.

reverse reaction The reaction written from right to left.

RNA Ribonucleic acid; assists DNA in directing cellular activity. Messenger RNA is the form of RNA involved in producing proteins or polypeptides from DNA gene segments.

salt Ionic substance that may be formed from an acid–base reaction.

saturated For hydrocarbons, the lack of carbon–carbon double or triple bonds. See also *saturated solution.*

saturated solution A solution having the maximum possible concentration of a solute.

scientific method Approximate set of steps used in the pursuit of scientific inquiry. Requires the testing of hypotheses through experimentation.

secondary structure The presence of an α–helix or a β–pleated sheet in protein or polypeptide chains.

semimetal An element that has properties intermediate between those of metals and nonmetals; generally solids that are nonlustrous, brittle, and conduct electricity. They are located between the metals and nonmetals on the periodic table and include silicon and germanium, two semiconductors.

semipermeable membrane A membrane that permits the selective flow of some molecules or ions, but not others, across its surface. For example, a semipermeable membrane may allow the flow of solvent molecules but not that of solute particles.

shell One of the principal electronic energy levels.

SI See *International System.*

significant digit See *significant figure.*

significant figure In a measured number, any of the digits that are known exactly plus the last digit, which has been estimated.

single-replacement reaction Reaction in which one element replaces another element. Single replacements involve oxidation–reduction.

solid One of the physical states of matter. The particles of matter in a solid lie next to one another and are held in a rigid arrangement.

solubility The maximum possible concentration of a solute in a solvent.

solubility product Equilibrium expression that characterizes the maximum mutual solution concentrations of the ions of an ionic solid. Also refers to the equilibrium constant, K_{sp}.

solute The component of a liquid solution that is dissolved in the solvent.

solution Any homogeneous mixture.

solvent The component of a liquid solution that dissolves the solute.

specific heat Quantity of heat required to raise the temperature of 1 gram of a substance by 1°C.

spectator ion An ion that is present during an ionic reaction but does not take part in the chemical transformation.

standard temperature and pressure (STP) Arbitrary conditions of temperature and pressure that are used as a reference point in the discussion of gases. 0°C (273 K) and 1 atm pressure (760 torr).

standardization Determination of the concentration of a solution.

state symbol A symbol used to specify the physical state of a chemical species. The most commonly used state symbols are (s), (ℓ), (g), and (aq).

stoichiometry Quantitative mass relationships between reactants and products in a chemical reaction.

STP See *standard temperature and pressure.*

strong acid An acid that is completely dissociated in aqueous solution.

sublevel One of the subdivisions of a principal energy level; usually *s, p, d,* or *f.*

sublimation Phase change in which a substance passes directly from the solid state to the gaseous state without passing through the liquid state.

sublimation point For substances that undergo sublimation, the temperature at which sublimation occurs.

subshell An electronic energy sublevel. See *sublevel.*

surface tension Attractive forces at the surface of a liquid that draw the liquid molecules toward the center of the liquid.

synthesis reaction Reaction type in which two (or more) substances combine to form a single product.

systematic name Derived name; based on a specific set of rules for naming substances.

temperature A measure of the average kinetic energy of a collection of molecules.

tertiary structure The three-dimensional arrangement of a protein chain. The chain is held in its tertiary structure by ionic and polar interactions (including hydrogen bonding), hydrophobic interactions, and disulfide linkages.

tetrahedral Refers to the geometry of a molecule or group of atoms in which a central atom is surrounded by four atoms that occupy the corners of a tetrahedron.

theoretical yield The amount of product expected in a reaction, as determined by a stoichiometric calculation based on the limiting reagent.

theory Explanation for an observation or series of observations that is substantiated by a considerable body of evidence.

titration Technique used in quantitative analysis in which a precisely measured volume of a solution is reacted with a precisely measured quantity of another reactant until stoichiometric equivalence is achieved, as evidenced by an indicator. Commonly used for neutralization reactions.

Torricellian barometer Device used to measure atmospheric pressure.

total ionic equation Ionic equation that expresses all of the ions present in solution during the course of a reaction.

total molecular equation Chemical equation expressing the reactants and products in terms of their molecular formulas.

transcription The formation of RNA from DNA.

transition metal Any of the metals for which a *d* sublevel is the last filled sublevel.

transition state Activated complex. Arrangement of atoms at the energy maximum of the reaction profile. Represents the transition from reactants to products.

translation The formation of an enzyme or polypeptide from messenger RNA.

transmutation The nuclear conversion of one element into another.

transuranium element Any element having an atomic number greater than 92; those elements beyond uranium in the periodic table.

triatomic Having three atoms. Usually used to refer to molecules composed of three atoms, such as carbon dioxide, CO_2.

trigonal planar Refers to the geometry of a molecule or group of atoms in which a central atom is surrounded by three atoms that occupy the corners of a triangle.

trigonal pyramidal Refers to the geometry of a molecule in which the atoms are located at the corners of a trigonal pyramid.

triple bond Type of covalent bonding in which three pairs of electrons are shared between the same two atoms. Each shared pair of electrons represents one covalent bond.

triprotic acid An acid possessing three ionizable hydrogens per molecule, such as phosphoric acid, H_3PO_4.

Tyndall effect The scattering of light passed through a dispersion, created by reflection of the light off the suspended particles.

unbalanced equation A chemical equation that shows the formulas of the reactants and products, but for which the coefficients have not been adjusted to balance the atoms.

uncertainty The degree to which the estimated digit in a measured number is in doubt.

universal gas constant Physical constant that relates the quantities of volume, temperature, pressure, and moles in the Ideal Gas Law.

unsaturated For hydrocarbons, a substance possessing a carbon–carbon double or triple bond.

vacuum The absence of gaseous molecules.

valence electrons Those electrons beyond the last noble gas core. The valence electrons are the ones involved in the bonding of an atom to other atoms.

Valence Shell Electron Pair Repulsion (VSEPR) model Model that predicts the geometry of a molecule with the pairs of valence electrons around each central atom as far from one another as possible.

Van der Waal's forces Intermolecular forces of attraction that hold molecules together in the condensed physical states. These include hydrogen bonding, dipole–dipole interaction, and London forces.

vapor Gaseous molecules of a substance that is normally a liquid at room temperature.

vapor pressure The pressure exerted by vapor molecules in equilibrium with the surface of the liquid they have escaped from.

vaporization The process of forming a vapor.

velocity of a wave The velocity with which a point on a wave travels.

velocity of light The velocity of a light wave, $c = 3.00 \times 10^8$ m/sec.

viscosity A measure of the resistance of a solution to flow.

vitamin A substance required in the diet for proper health; often functions as a coenzyme.

voltaic cell An electrochemical cell that operates in the spontaneous direction and is capable of producing electrical energy. Also known as a *galvanic cell.*

warming curve Curve that shows the temperature and physical changes that accompany the constant heating of a sample.

wavelength The distance between the crests of two consecutive waves.

weak acid An acid that only partially dissociates in aqueous solution.

weak electrolyte A substance that, when dissolved in water, conducts electricity weakly.

weight Force with which gravity attracts an object.

word equation A chemical equation that gives the names, rather than formulas, of the reactants and products of a chemical reaction.

zwitterion An internal salt; a substance in which the cation and anion are both part of the same molecule. Amino acids form zwitterions.

SOLUTIONS TO PROBLEMS

Chapter 1

1.1 The following general problems and their solutions have pervaded the news for over 20 years: **(a)** The problem of worldwide food shortage can be eased by means of improved agricultural techniques that use chemicals to promote faster growth of crops and by means of gene-splicing techniques that produce heartier crops. **(b)** The energy shortage can be eased by means of the production of new fuels such as alcohol or shale oil or by means of the development of fusion energy. **(c)** Air and water pollution can be eased through better control of the release of contaminants and through an increased knowledge of how to remove such contaminants from the environment.

In recent years, the following major news stories have appeared: the Chernobyl disaster, the discovery of toxic radon gas levels in certain locations, the development of high-temperature superconductors, the Exxon Valdez oil spill, the depletion of the ozone layer, and the report of a possible cold-fusion reaction.

1.2 **(a)** 2 **(b)** 5 **(c)** 1 **(d)** 3 **(e)** 4

1.3 The progress of science depends upon experimentation that is based on earlier findings. By reporting experimental findings in the literature, investigators make their results available to the entire scientific community for further research. The effective use of penicillin as an antibiotic required its synthesis in the laboratory. Although Fleming never synthesized penicillin himself, by reporting his discoveries in the journals, he enabled Florey and Chain to continue his work.

Chapter 2

2.1 A solid has a definite shape and volume and tends to resist any change in its shape. A liquid flows freely, having a definite volume, but its shape changes to fit the shape of its container. A gas has no definite shape or volume but expands or contracts to fill exactly the container it is in. Thus, gases can be compressed easily. The solid and liquid states are known as the "condensed" phases.

2.2 **(a)** Homogeneous **(b)** Heterogeneous
(c) Homogeneous **(d)** Heterogeneous
(e) Heterogeneous **(f)** Heterogeneous
(g) Heterogeneous **(h)** Homogeneous
(i) Heterogeneous

2.3 Homogeneous matter is uniform throughout. Heterogeneous matter is not uniform.

2.4 A pure substance has a fixed and definite composition. A homogeneous mixture may be prepared in different proportions.

2.5 A compound can be decomposed into simpler substances. An element cannot be decomposed into simpler substances by ordinary means.

2.6 A: compound; B: insufficient information; C: compound; D: element; E: insufficient information

2.7 **(a)** A molecule is the smallest unit of a pure substance that possesses all of the characteristic properties of the substance. An atom is a building block from which molecules are made.

(b) An element is composed of only one kind of atom. A compound is composed of two or more kinds of atoms.

(c) Diatomic substances have two atoms per molecule. Monatomic substances have one atom per molecule. Oxygen and hydrogen are diatomic elements. Helium and argon are monatomic elements. Carbon monoxide is a diatomic compound. It is not possible to have a monatomic compound, since a compound must contain atoms of at least two different elements.

2.8 **(a)** HCl **(b)** N_2O_5 **(c)** C_6H_6 **(d)** H_2SO_4
(e) $NaCl$ **(f)** Ne **(g)** $MgBr_2$ **(h)** Al_2S_3

2.9 **(a)** 4 **(b)** 12 **(c)** 2 **(d)** 9

2.10 When a substance undergoes a phase change, such as melting or boiling, the fixed composition of the substance does not change.

2.11 A chemical change involves a change in the fixed composition of the reactants, as new substances with different composition are formed. A physical change does not involve any change in the composition of the substance.

2.12 **(a)** Physical property **(b)** Physical property **(c)** Physical property **(d)** Chemical property **(e)** Physical property **(f)** Chemical property **(g)** Physical property

Chapter 3

3.1 **(a)** Exact **(b)** Measured **(c)** Exact **(d)** Measured

3.2 **(a)** 3 **(b)** 2 **(c)** 2 (The last zero holds the decimal.)
(d) 4 **(e)** 3 **(f)** 3
(g) 2 (The leading zeros hold the decimal.)
(h) 3 (The leading zeros hold the decimal, but the last zero must be measured and is therefore significant.)
(i) 4 **(j)** 3
(k) 8 (Since the last zero is significant, so are all those in between.)
(l) 4 **(m)** 2 **(n)** 3

3.3 **(a)** 17.6 **(b)** 0.00627 **(c)** 36.0 **(d)** 36
(e) 40 (4×10^1) **(f)** 7.0×10^{-6}

3.4 4 significant figures; 6.100×10^4

3.5 **(a)** 43.6 (remains as is)
(b) 45 (rounded off from 45.28)
(c) 12.7 (rounded off from 12.70)
(d) 0.8 (rounded off from 0.843)
(e) 1840 (rounded off from 1838.113)

3.6 **(a)** 19.1 **(b)** 3.88 **(c)** 2.248 **(d)** 1100
(e) 2.0 **(f)** 5

3.7 **(a)** $(11.62)(0.55) = 6.4$ **(b)** $\dfrac{57.80}{99.0} = 0.584$

3.8 **(a)** $3.0\text{ ft}\left(\dfrac{12\text{ in.}}{1\text{ ft}}\right) = 36\text{ in.}$

(b) $54\text{ in.}\left(\dfrac{1\text{ ft}}{12\text{ in.}}\right) = 4.5\text{ ft}$

(c) $2.0\text{ yd}\left(\dfrac{3\text{ ft}}{1\text{ yd}}\right)\left(\dfrac{12\text{ in.}}{1\text{ ft}}\right) = 72\text{ in.}$

3.9 1 m = 100 cm; 1 L = 1000 mL

3.10 **(a)** 1 kg = 1000 g
(b) 1 mm = 10^{-3} m or 1000 mm = 1 m
(c) 1 cL = 10^{-2} L or 100 cL = 1 L
(d) 1 kJ = 1000 J
(e) 1 μsec = 10^{-6} sec or 10^6 μsec = 1 sec
(f) 1 mg = 10^{-3} g or 1000 mg = 1 g
(g) 1 MW = 10^6 W
(h) 1 μL = 10^{-6} L or 10^6 μL = 1 L

3.11 **(a)** Exact **(b)** Measured **(c)** Exact
(d) Exact **(e)** Measured

3.12 **(a)** $42.6\text{ cm}\left(\dfrac{1\text{ m}}{100\text{ cm}}\right) = 0.426\text{ m}$

(b) $1.5\text{ km}\left(\dfrac{1000\text{ m}}{1\text{ km}}\right) = 1500\text{ m}\ (1.5 \times 10^3\text{ m})$

(c) $502\text{ m}\left(\dfrac{1\text{ km}}{1000\text{ m}}\right) = 0.502\text{ km}$

(d) $0.372\text{ m}\left(\dfrac{100\text{ cm}}{1\text{ m}}\right) = 37.2\text{ cm}$

(e) $943\text{ mm}\left(\dfrac{1\text{ m}}{1000\text{ mm}}\right) = 0.943\text{ m}$

(f) $10.0\text{ cm}\left(\dfrac{1\text{ in.}}{2.54\text{ cm}}\right) = 3.94\text{ in.}$

(g) $3.86\text{ in.}\left(\dfrac{2.54\text{ cm}}{1\text{ in.}}\right) = 9.80\text{ cm}$

(h) $15.0\text{ m}\left(\dfrac{100\text{ cm}}{1\text{ m}}\right) = 1500\text{ cm}\ (1.50 \times 10^3\text{ cm})$

3.13 **(a)** km → m → mm

$$3.23\text{ km}\left(\frac{1000\text{ m}}{1\text{ km}}\right)\left(\frac{1000\text{ mm}}{1\text{ m}}\right) = 3{,}230{,}000\text{ mm}\ (3.23 \times 10^6\text{ mm})$$

(b) km → m → cm

$$1.5\text{ km}\left(\frac{1000\text{ m}}{1\text{ km}}\right)\left(\frac{100\text{ cm}}{1\text{ m}}\right) = 150{,}000\text{ cm}\ (1.5 \times 10^5\text{ cm})$$

(c) ft → in. → cm

$$1.40\text{ ft}\left(\frac{12\text{ in.}}{1\text{ ft}}\right)\left(\frac{2.54\text{ cm}}{1\text{ in.}}\right) = 42.7\text{ cm}$$

(d) cm → in. → ft

$$125\text{ cm}\left(\frac{1\text{ in.}}{2.54\text{ cm}}\right)\left(\frac{1\text{ ft}}{12\text{ in.}}\right) = 4.10\text{ ft}$$

(e) yd → ft → in. → cm

$$3.20\text{ yd}\left(\frac{3\text{ ft}}{1\text{ yd}}\right)\left(\frac{12\text{ in.}}{1\text{ ft}}\right)\left(\frac{2.54\text{ cm}}{1\text{ in.}}\right) = 293\text{ cm}$$

(f) ft → in. → cm → m

$$10.0\text{ ft}\left(\frac{12\text{ in.}}{1\text{ ft}}\right)\left(\frac{2.54\text{ cm}}{1\text{ in.}}\right)\left(\frac{1\text{ m}}{100\text{ cm}}\right) = 3.05\text{ m}$$

3.14 **(a)** m → cm → in.

$$1.00\text{ m}\left(\frac{100\text{ cm}}{1\text{ m}}\right)\left(\frac{1\text{ in.}}{2.54\text{ cm}}\right) = 39.4\text{ in.}$$

(b) km → m → cm → in. → ft → mile

$$1.00\text{ km}\left(\frac{1000\text{ m}}{1\text{ km}}\right)\left(\frac{100\text{ cm}}{1\text{ m}}\right)\left(\frac{1\text{ in.}}{2.54\text{ cm}}\right) \times \left(\frac{1\text{ ft}}{12\text{ in.}}\right)\left(\frac{1\text{ mile}}{5280\text{ ft}}\right) = 0.621\text{ mile}$$

3.15 (a) kg → g → lb

$$1.00\ \text{kg}\left(\frac{1000\ \text{g}}{1\ \text{kg}}\right)\left(\frac{1\ \text{lb}}{454\ \text{g}}\right) = 2.20\ \text{lb}$$

3.16 (a) $1235\ \text{g}\left(\frac{1\ \text{kg}}{1000\ \text{g}}\right) = 1.235\ \text{kg}$

(b) $3.45\ \text{g}\left(\frac{1000\ \text{mg}}{1\ \text{g}}\right) = 3450\ \text{mg}\ (3.45 \times 10^3\ \text{mg})$

(c) $598\ \text{mg}\left(\frac{1\ \text{g}}{1000\ \text{mg}}\right) = 0.598\ \text{g}$

(d) $3.00\ \text{lb}\left(\frac{454\ \text{g}}{1\ \text{lb}}\right) = 1360\ \text{g}\ (1.36 \times 10^3\ \text{g})$

(e) $7.00\ \text{kg}\left(\frac{1000\ \text{g}}{1\ \text{kg}}\right)\left(\frac{1\ \text{lb}}{454\ \text{g}}\right) = 15.4\ \text{lb}$

(f) $3.50\ \text{lb}\left(\frac{454\ \text{g}}{1\ \text{lb}}\right)\left(\frac{1\ \text{kg}}{1000\ \text{g}}\right) = 1.59\ \text{kg}$

(g) $0.620\ \text{lb}\left(\frac{454\ \text{g}}{1\ \text{lb}}\right)\left(\frac{1000\ \text{mg}}{1\ \text{g}}\right)$

$= 281{,}000\ \text{mg}\ (2.81 \times 10^5\ \text{mg})$

3.17 (a) $525\ \text{mL}\left(\frac{1\ \text{L}}{1000\ \text{mL}}\right) = 0.525\ \text{L}$

(b) $4.95\ \text{L}\left(\frac{1000\ \text{mL}}{1\ \text{L}}\right) = 4950\ \text{mL}\ (4.95 \times 10^3\ \text{mL})$

(c) $1.32\ \text{qt}\left(\frac{946\ \text{mL}}{1\ \text{qt}}\right) = 1250\ \text{mL}\ (1.25 \times 10^3\ \text{mL})$

(d) $2.25\ \text{qt}\left(\frac{946\ \text{mL}}{1\ \text{qt}}\right)\left(\frac{1\ \text{L}}{1000\ \text{mL}}\right) = 2.13\ \text{L}$

(e) $10.8\ \text{gal}\left(\frac{4\ \text{qt}}{1\ \text{gal}}\right)\left(\frac{946\ \text{mL}}{1\ \text{qt}}\right)\left(\frac{1\ \text{L}}{1000\ \text{mL}}\right) = 40.9\ \text{L}$

(f) $45.3\ \text{L}\left(\frac{1000\ \text{mL}}{1\ \text{L}}\right)\left(\frac{1\text{qt}}{946\ \text{mL}}\right)\left(\frac{1\ \text{gal}}{4\ \text{qt}}\right) = 12.0\ \text{gal}$

3.18 (a) $V = (12.0\ \text{cm})(8.00\ \text{cm})(9.50\ \text{cm}) = 912\ \text{cm}^3$

(b) $912\ \text{cm}^3\left(\frac{1\ \text{mL}}{1\ \text{cm}^3}\right) = 912\ \text{mL}$

(c) $912\ \text{mL}\left(\frac{1\ \text{L}}{1000\ \text{mL}}\right) = 0.912\ \text{L}$

3.19 (a) $\left(\frac{4.0\ \text{g}}{2.5\ \text{mL}}\right) = 1.6\ \text{g/mL}$

(b) $\left(\frac{34.0\ \text{g}}{2.50\ \text{mL}}\right) = 13.6\ \text{g/mL}$

(c) $\left(\frac{59.2\ \text{g}}{75.0\ \text{mL}}\right) = 0.789\ \text{g/mL}$

(d) $\left(\frac{15.6\ \text{g}}{5.00\ \text{mL}}\right) = 3.12\ \text{g/mL}$

(e) $\left(\frac{8.69\ \text{g}}{11.0\ \text{mL}}\right) = 0.790\ \text{g/mL}$

3.20 (a) $11.5\ \text{mL}\left(\frac{0.85\ \text{g}}{1\ \text{mL}}\right) = 9.8\ \text{g}$

(b) $12\ \text{g}\left(\frac{1\ \text{mL}}{1.59\ \text{g}}\right) = 7.5\ \text{mL}$

(c) $65.0\ \text{mL}\left(\frac{0.879\ \text{g}}{1\ \text{mL}}\right) = 57.1\ \text{g}$

(d) $27.0\ \text{g}\left(\frac{1\ \text{mL}}{1.48\ \text{g}}\right) = 18.2\ \text{mL}$

(e) $1.00\ \text{lb}\left(\frac{454\ \text{g}}{1\ \text{lb}}\right)\left(\frac{1\ \text{mL}}{13.6\ \text{g}}\right) = 33.4\ \text{mL}$

(f) $245\ \text{mL}\left(\frac{0.789\ \text{g}}{1\ \text{mL}}\right) = 193\ \text{g}$

3.21 (a) $\text{sp gr} = \frac{\text{Density of bromine}}{\text{Density of water}} = \frac{3.12\ \text{g/mL}}{1.00\ \text{g/mL}} = 3.12$

(b) sp gr × Density of water = Density of bromine

$3.12 \times 62.4\ \text{lb/ft}^3 = 195\ \text{lb/ft}^3$

(c) $\text{sp gr} = \frac{\text{Density of carbon tetrachloride}}{\text{Density of water}}$

$= \frac{99.5\ \text{lb/ft}^3}{62.4\ \text{lb/ft}^3} = 1.59$

(d) sp gr × Density of water = Density of carbon tetrachloride

$1.59 \times 1.00\ \text{g/mL} = 1.59\ \text{g/mL}$

3.22 (a) 293 K (b) 418 K (c) 195 K (d) 50 K (e) 25°C (f) 304°C (g) −173°C (h) −250°C

3.23 (a) $87 = 1.8T_C + 32;\ T_C = 31°C$
(b) $68 = 1.8T_C + 32;\ T_C = 20°C$
(c) $-40 = 1.8T_C + 32;\ T_C = -40°C$
(d) $0 = 1.8T_C + 32;\ T_C = -18°C$
(e) $-48 = 1.8T_C + 32;\ T_C = -44°C$
(f) $-104 = 1.8T_C + 32;\ T_C = -76°C$
(g) $932 = 1.8T_C + 32;\ T_C = 500°C$
(h) $400 = 1.8T_C + 32;\ T_C = 204°C$

3.24 (a) 304 K (b) 293 K (c) 233 K (d) 255 K (e) 229 K (f) 197 K (g) 773 K (h) 477 K

3.25 (a) $T_F = 1.8(95) + 32;\ T_F = 203°F$
(b) $T_F = 1.8(60) + 32;\ T_F = 140°F$
(c) $T_F = 1.8(20) + 32;\ T_F = 68°F$
(d) $T_F = 1.8(-40) + 32;\ T_F = -40°F$
(e) $T_F = 1.8(-70) + 32;\ T_F = -94°F$
(f) $T_F = 1.8(37) + 32;\ T_F = 99°F$

3.26 (a) $59 = 1.8T_C + 32;\ T_C = 15°C;\ T_K = 288\ K$
(b) $-13 = 1.8T_C + 32;\ T_C = -25°C;\ T_K = 248\ K$
(c) $257 = 1.8T_C + 32;\ T_C = 125°C;\ T_K = 398\ K$

3.27 (a) $25.0 \text{ cal}\left(\frac{4.184 \text{ J}}{1 \text{ cal}}\right) = 105 \text{ J}$

(b) $0.575 \text{ cal}\left(\frac{4.184 \text{ J}}{1 \text{ cal}}\right) = 2.41 \text{ J}$

(c) $1.43 \text{ J}\left(\frac{1 \text{ cal}}{4.184 \text{ J}}\right) = 0.342 \text{ cal}$

(d) $1.75 \text{ kcal}\left(\frac{4.184 \text{ kJ}}{1 \text{ kcal}}\right) = 7.32 \text{ kJ}$

(e) $325 \text{ cal}\left(\frac{4.184 \text{ J}}{1 \text{ cal}}\right)\left(\frac{1 \text{ kJ}}{1000 \text{ J}}\right) = 1.36 \text{ kJ}$

3.28 temperature, heat, heat, temperature

3.29 (a) $4.50 \text{ yd}^2\left(\frac{3 \text{ ft}}{1 \text{ yd}}\right)^2 = 4.5 \text{ yd}^2\left(\frac{3^2 \text{ ft}^2}{1^2 \text{ yd}^2}\right) = 40.5 \text{ ft}^2$

(b) $3.50 \text{ m}^3\left(\frac{10 \text{ dm}}{1 \text{ m}}\right)^3 = 3.50 \text{ m}^3\left(\frac{10^3 \text{ dm}^3}{1^3 \text{ m}^3}\right)$

$= 3.50 \times 10^3 \text{ dm}^3$

(c) $3.50 \text{ m}^3\left(\frac{100 \text{ cm}}{1 \text{ m}}\right)^3\left(\frac{1 \text{ mL}}{1 \text{ cm}^3}\right)\left(\frac{1 \text{ L}}{1000 \text{ mL}}\right)$

$= 3.50 \text{ m}^3\left(\frac{100^3 \text{ cm}^3}{1^3 \text{ m}^3}\right)\left(\frac{1 \text{ mL}}{1 \text{ cm}^3}\right)\left(\frac{1 \text{ L}}{1000 \text{ mL}}\right)$

$= 3.50 \times 10^3 \text{ L}$

$1 \text{ L} = 1 \text{ dm}^3$

(d) $7.25 \text{ ft}^3\left(\frac{12 \text{ in.}}{1 \text{ ft}}\right)^3\left(\frac{2.54 \text{ cm}}{1 \text{ in.}}\right)^3\left(\frac{1 \text{ mL}}{1 \text{ cm}^3}\right)\left(\frac{1 \text{ mL}}{1000 \text{ mL}}\right)$

$= 7.25 \text{ ft}^3\left(\frac{12^3 \text{ in.}^3}{1^3 \text{ ft}^3}\right)\left(\frac{2.54^3 \text{ cm}^3}{1^3 \text{ in.}^3}\right)\left(\frac{1 \text{ mL}}{1 \text{ cm}^3}\right)\left(\frac{1 \text{ mL}}{1000 \text{ mL}}\right)$

$= 205 \text{ L}$

3.30 $1.00 \text{ ft}^3\left(\frac{12 \text{ in.}}{1 \text{ ft}}\right)^3\left(\frac{2.54 \text{ cm}}{1 \text{ in.}}\right)^3\left(\frac{1 \text{ mL}}{1 \text{ cm}^3}\right)\left(\frac{1 \text{ qt}}{946 \text{ mL}}\right) \times \left(\frac{1 \text{ gal}}{4 \text{ qt}}\right)$

$= 1.00 \text{ ft}^3\left(\frac{12^3 \text{ in.}^3}{1^3 \text{ ft}^3}\right)\left(\frac{2.54^3 \text{ cm}^3}{1^3 \text{ in.}^3}\right)\left(\frac{1 \text{ mL}}{1 \text{ cm}^3}\right)\left(\frac{1 \text{ qt}}{946 \text{ mL}}\right)\left(\frac{1 \text{ gal}}{4 \text{ qt}}\right)$

$= 7.48 \text{ gal}$

3.31 $\frac{0.00133 \text{ g}}{\text{cm}^3}\left(\frac{1 \text{ kg}}{1000 \text{ g}}\right)\left(\frac{100 \text{ cm}}{1 \text{ m}}\right)^3$

$= \frac{0.00133 \text{ g}}{\text{cm}^3}\left(\frac{1 \text{ kg}}{1000 \text{ g}}\right)\left(\frac{100^3 \text{ cm}^3}{1^3 \text{ m}^3}\right) = 1.33 \text{ kg/m}^3$

3.32 $\frac{60.0 \text{ miles}}{\text{hr}}\left(\frac{1 \text{ hr}}{60 \text{ min}}\right)\left(\frac{1 \text{ min}}{60 \text{ sec}}\right)\left(\frac{5280 \text{ ft}}{1 \text{ mile}}\right)\left(\frac{12 \text{ in.}}{1 \text{ ft}}\right) \times \left(\frac{2.54 \text{ cm}}{1 \text{ in.}}\right)\left(\frac{1 \text{ m}}{100 \text{ cm}}\right) = 26.8 \text{ m/sec}$

3.33 $\frac{0.163 \text{ lb}}{\text{in.}^3}\left(\frac{1 \text{ in.}}{2.54 \text{ cm}}\right)^3\left(\frac{454 \text{ g}}{1 \text{ lb}}\right)$

$= \frac{0.163 \text{ lb}}{\text{in.}^3}\left(\frac{1^3 \text{ in.}^3}{2.54^3 \text{ cm}^3}\right)\left(\frac{454 \text{ g}}{1 \text{ lb}}\right) = 4.52 \text{ g/cm}^3$

Chapter 4

4.1 (a) C_2H_5 (b) CCl_4 (c) BH_3 (d) CH_2O (e) NH_2

4.2 (a) 83.8 (b) 85.5 (c) 238.0 (d) 102.9 (e) 195.1 (f) 79.9 (g) 40.1 (h) 27.0 (i) 126.9 (j) 31.0

4.3 (a) $2(23.0) + 12.0 + 3(16.0) = 106.0$
(b) $209.0 + 3(35.5) = 315.5$
(c) $39.1 + 54.9 + 4(16.0) = 158.0$
(d) $23.0 + 1.0 + 12.0 + 3(16.0) = 84.0$
(e) $2(12.0) + 5(1.0) + 35.5 = 64.5$
(f) $40.1 + 2(16.0) + 2(1.0) = 74.1$
(g) $2(27.0) + 3(32.1) + 12(16.0) = 342.3$

4.4 (a) $1 \times \text{Mn} = 1 \times 54.9 = 54.9$
$2 \times \text{O} = 2 \times 16.0 = 32.0$
$MnO_2 = 86.9$

$\%\text{Mn} = \frac{54.9}{86.9} \times 100\% = 63.2\%$

$\%\text{O} = \frac{32.0}{86.9} \times 100\% = 36.8\%$

(b) $3 \times \text{C} = 3 \times 12.0 = 36.0$
$8 \times \text{F} = 8 \times 19.0 = 152.0$
$C_3F_8 = 188.0$

$\%\text{C} = \frac{36.0}{188.0} \times 100\% = 19.1\%$

$\%\text{F} = \frac{152.0}{188.0} \times 100\% = 80.9\%$

(c) $1 \times \text{Cu} = 1 \times 63.5 = 63.5$
$2 \times \text{Br} = 2 \times 79.9 = 159.8$
$CuBr_2 = 223.3$

$\%\text{Cu} = \frac{63.5}{223.3} \times 100\% = 28.4\%$

$\%\text{Br} = \frac{159.8}{223.3} \times 100\% = 71.6\%$

(d) $3 \times \text{H} = 3 \times 1.0 = 3.0$
$1 \times \text{P} = 1 \times 31.0 = 31.0$
$4 \times \text{O} = 4 \times 16.0 = 64.0$
$H_3PO_4 = 98.0$

$$\%H = \frac{3.0}{98.0} \times 100\% = 3.1\%$$

$$\%P = \frac{31.0}{98.0} \times 100\% = 31.6\%$$

$$\%O = \frac{64.0}{98.0} \times 100\% = 65.3\%$$

(e)
$$\begin{array}{l} 1 \times C = 1 \times 12.0 = 12.0 \\ 2 \times H = 2 \times 1.0 = 2.0 \\ 1 \times Br = 1 \times 79.9 = 79.9 \\ \underline{1 \times F = 1 \times 19.0 = 19.0} \\ CH_2BrF = 112.9 \end{array}$$

$$\%C = \frac{12.0}{112.9} \times 100\% = 10.6\%$$

$$\%H = \frac{2.0}{112.9} \times 100\% = 1.8\%$$

$$\%Br = \frac{79.9}{112.9} \times 100\% = 70.8\%$$

$$\%F = \frac{19.0}{112.9} \times 100\% = 16.8\%$$

(f)
$$\begin{array}{l} 1 \times Zn = 1 \times 65.4 = 65.4 \\ 2 \times N = 2 \times 14.0 = 28.0 \\ \underline{6 \times O = 6 \times 16.0 = 96.0} \\ Zn(NO_3)_2 = 189.4 \end{array}$$

$$\%Zn = \frac{65.4}{189.4} \times 100\% = 34.5\%$$

$$\%N = \frac{28.0}{189.4} \times 100\% = 14.8\%$$

$$\%O = \frac{96.0}{189.4} \times 100\% = 50.7\%$$

(g)
$$\begin{array}{l} 1 \times Mg = 1 \times 24.3 = 24.3 \\ 2 \times C = 2 \times 12.0 = 24.0 \\ \underline{2 \times N = 2 \times 14.0 = 28.0} \\ Mg(CN)_2 = 76.3 \end{array}$$

$$\%Mg = \frac{24.3}{76.3} \times 100\% = 31.8\%$$

$$\%C = \frac{24.0}{76.3} \times 100\% = 31.5\%$$

$$\%N = \frac{28.0}{76.3} \times 100\% = 36.7\%$$

4.5 **(a)**
$$\begin{array}{l} 1 \times Na = 1 \times 23.0 = 23.0 \\ \underline{1 \times Cl = 1 \times 35.5 = 35.5} \\ NaCl = 58.5 \end{array}$$

$$3.50\text{ g NaCl}\left(\frac{23.0\text{ g Na}}{58.5\text{ g NaCl}}\right) = 1.38\text{ g Na}$$

(b)
$$\begin{array}{l} 1 \times Ag = 1 \times 107.9 = 107.9 \\ 1 \times N = 1 \times 14.0 = 14.0 \\ \underline{3 \times O = 3 \times 16.0 = 48.0} \\ AgNO_3 = 169.9 \end{array}$$

$$5.65\text{ g AgNO}_3\left(\frac{107.9\text{ g Ag}}{169.9\text{ g AgNO}_3}\right) = 3.59\text{ g Ag}$$

(c)
$$\begin{array}{l} 2 \times N = 2 \times 14.0 = 28.0 \\ 4 \times H = 4 \times 1.0 = 4.0 \\ \underline{3 \times O = 3 \times 16.0 = 48.0} \\ NH_4NO_3 = 80.0 \end{array}$$

$$454\text{ g NH}_4\text{NO}_3\left(\frac{28.0\text{ g N}}{80.0\text{ g NH}_4\text{NO}_3}\right) = 159\text{ g N}$$

(d)
$$\begin{array}{l} 1 \times Fe = 1 \times 55.8 = 55.8 \\ 1 \times S = 1 \times 32.1 = 32.1 \\ \underline{4 \times O = 4 \times 16.0 = 64.0} \\ FeSO_4 = 151.9 \end{array}$$

$$125\text{ mg FeSO}_4\left(\frac{55.8\text{ mg Fe}}{151.9\text{ mg FeSO}_4}\right) = 45.9\text{ mg Fe}$$

4.6 **(a)** $6.00\text{ g C}\left(\frac{1\text{ mol C}}{12.0\text{ g C}}\right) = 0.500\text{ mol C}$

(b) $4.19\text{ g Kr}\left(\frac{1\text{ mol Kr}}{83.8\text{ g Kr}}\right) = 0.0500\text{ mol Kr}$

(c) $47.6\text{ g U}\left(\frac{1\text{ mol U}}{238.0\text{ g U}}\right) = 0.200\text{ mol U}$

(d) $8.37\text{ g Fe}\left(\frac{1\text{ mol Fe}}{55.8\text{ g Fe}}\right) = 0.150\text{ mol Fe}$

(e) $8.82\text{ g Ca}\left(\frac{1\text{ mol Ca}}{40.1\text{ g Ca}}\right) = 0.220\text{ mol Ca}$

(f) $316\text{ g Mg}\left(\frac{1\text{ mol Mg}}{24.3\text{ g Mg}}\right) = 13.0\text{ mol Mg}$

4.7 **(a)** $2.50\text{ mol Ne}\left(\frac{20.2\text{ g Ne}}{1\text{ mol Ne}}\right) = 50.5\text{ g Ne}$

(b) $0.452\text{ mol Na}\left(\frac{23.0\text{ g Na}}{1\text{ mol Na}}\right) = 10.4\text{ g Na}$

(c) $0.125\text{ mol K}\left(\frac{39.1\text{ g K}}{1\text{ mol K}}\right) = 4.89\text{ g K}$

(d) $0.742\text{ mol Hg}\left(\frac{200.6\text{ g Hg}}{1\text{ mol Hg}}\right) = 149\text{ g Hg}$

(e) $0.0531\text{ mol Au}\left(\frac{197.0\text{ g Au}}{1\text{ mol Au}}\right) = 10.5\text{ g Au}$

4.8 **(a)** $0.25\text{ mol He}\left(\frac{6.02 \times 10^{23}\text{ atoms He}}{1\text{ mol He}}\right)$

$= 1.5 \times 10^{23}\text{ atoms He}$

(b) $3.50 \text{ mol Fe}\left(\frac{6.02 \times 10^{23} \text{ atoms Fe}}{1 \text{ mol Fe}}\right) = 2.11 \times 10^{24} \text{ atoms Fe}$

(c) $0.0155 \text{ mol Au}\left(\frac{6.02 \times 10^{23} \text{ atoms Au}}{1 \text{ mol Au}}\right) = 9.33 \times 10^{21} \text{ atoms Au}$

(d) $6.35 \text{ g Cu}\left(\frac{1 \text{ mol Cu}}{63.5 \text{ g Cu}}\right)\left(\frac{6.02 \times 10^{23} \text{ atoms Cu}}{1 \text{ mol Cu}}\right) = 6.02 \times 10^{22} \text{ atoms Cu}$

(e) $253 \text{ g Na}\left(\frac{1 \text{ mol Na}}{23.0 \text{ g Na}}\right)\left(\frac{6.02 \times 10^{23} \text{ atoms Na}}{1 \text{ mol Na}}\right) = 6.62 \times 10^{24} \text{ atoms Na}$

(f) $10.0 \text{ g S}\left(\frac{1 \text{ mol S}}{32.1 \text{ g S}}\right)\left(\frac{6.02 \times 10^{23} \text{ atoms S}}{1 \text{ mol S}}\right) = 1.88 \times 10^{23} \text{ atoms S}$

4.9 (a) $1.56 \text{ g CO}\left(\frac{1 \text{ mol CO}}{28.0 \text{ g CO}}\right) = 0.0557 \text{ mol CO}$

(b) $15.0 \text{ g CO}_2\left(\frac{1 \text{ mol CO}_2}{44.0 \text{ g CO}_2}\right) = 0.341 \text{ mol CO}_2$

(c) $2.50 \text{ g NO}_2\left(\frac{1 \text{ mol NO}_2}{46.0 \text{ g NO}_2}\right) = 0.0543 \text{ mol NO}_2$

(d) $12.5 \text{ g KBr}\left(\frac{1 \text{ mol KBr}}{119.0 \text{ g KBr}}\right) = 0.105 \text{ mol KBr}$

(e) $20.0 \text{ g K}_2\text{CO}_3\left(\frac{1 \text{ mol K}_2\text{CO}_3}{138.2 \text{ g K}_2\text{CO}_3}\right) = 0.145 \text{ mol K}_2\text{CO}_3$

(f) $12.1 \text{ g SO}_3\left(\frac{1 \text{ mol SO}_3}{80.1 \text{ g SO}_3}\right) = 0.151 \text{ mol SO}_3$

4.10 (a) $2.50 \text{ mol CH}_4\left(\frac{16.0 \text{ g CH}_4}{1 \text{ mol CH}_4}\right) = 40.0 \text{ g CH}_4$

(b) $0.550 \text{ mol KCl}\left(\frac{74.6 \text{ g KCl}}{1 \text{ mol KCl}}\right) = 41.0 \text{ g KCl}$

(c) $0.315 \text{ mol NaOH}\left(\frac{40.0 \text{ g NaOH}}{1 \text{ mol NaOH}}\right) = 12.6 \text{ g NaOH}$

(d) $0.114 \text{ mol H}_2\text{SO}_4\left(\frac{98.1 \text{ g H}_2\text{SO}_4}{1 \text{ mol H}_2\text{SO}_4}\right) = 11.2 \text{ g H}_2\text{SO}_4$

4.11 (a) $5.0 \text{ g H}_2\left(\frac{1 \text{ mol H}_2}{2.0 \text{ g H}_2}\right) = 2.5 \text{ mol H}_2$

(b) $20.0 \text{ g Cl}_2\left(\frac{1 \text{ mol Cl}_2}{71.0 \text{ g Cl}_2}\right) = 0.282 \text{ mol Cl}_2$

(c) $11.3 \text{ g I}\left(\frac{1 \text{ mol I}}{126.9 \text{ g I}}\right) = 0.0890 \text{ mol I}$

(d) $11.3 \text{ g I}_2\left(\frac{1 \text{ mol I}_2}{253.8 \text{ g I}_2}\right) = 0.0445 \text{ mol I}_2$

(e) $1.40 \text{ g Ar}\left(\frac{1 \text{ mol Ar}}{39.9 \text{ g Ar}}\right) = 0.0351 \text{ mol Ar}$

(f) $17.0 \text{ g F}_2\left(\frac{1 \text{ mol F}_2}{38.0 \text{ g F}_2}\right) = 0.447 \text{ mol F}_2$

4.12 (a) $4.75 \text{ mol NH}_3\left(\frac{6.02 \times 10^{23} \text{ molecules NH}_3}{1 \text{ mol NH}_3}\right) = 2.86 \times 10^{24} \text{ molecules NH}_3$

(b) $0.450 \text{ mol CCl}_4\left(\frac{6.02 \times 10^{23} \text{ molecules CCl}_4}{1 \text{ mol CCl}_4}\right) = 2.71 \times 10^{23} \text{ molecules CCl}_4$

(c) $0.0135 \text{ mol F}_2\left(\frac{6.02 \times 10^{23} \text{ molecules F}_2}{1 \text{ mol F}_2}\right) = 8.13 \times 10^{21} \text{ molecules F}_2$

(d) $14.7 \text{ g C}_3\text{H}_6\text{O}\left(\frac{1 \text{ mol C}_3\text{H}_6\text{O}}{58.0 \text{ g C}_3\text{H}_6\text{O}}\right) \times \left(\frac{6.02 \times 10^{23} \text{ molecules C}_3\text{H}_6\text{O}}{1 \text{ mol C}_3\text{H}_6\text{O}}\right) = 1.53 \times 10^{23} \text{ molecules C}_3\text{H}_6\text{O}$

(e) $8.55 \text{ g N}_2\left(\frac{1 \text{ mol N}_2}{28.0 \text{ g N}_2}\right)\left(\frac{6.02 \times 10^{23} \text{ molecules N}_2}{1 \text{ mol N}_2}\right) = 1.84 \times 10^{23} \text{ molecules N}_2$

4.13 (a) $52.4 \text{ g K}\left(\frac{1 \text{ mol K}}{39.1 \text{ g K}}\right) = 1.34 \text{ mol K}; \quad \frac{1.34}{1.34} = 1$

$47.6 \text{ g Cl}\left(\frac{1 \text{ mol Cl}}{35.5 \text{ g Cl}}\right) = 1.34 \text{ mol Cl}; \quad \frac{1.34}{1.34} = 1$

empirical formula: KCl

(b) $16.2 \text{ g Na}\left(\frac{1 \text{ mol Na}}{23.0 \text{ g Na}}\right) = 0.704 \text{ mol Na};$

$\frac{0.704}{0.703} \approx 1$

$38.6 \text{ g Mn}\left(\frac{1 \text{ mol Mn}}{54.9 \text{ g Mn}}\right) = 0.703 \text{ mol Mn};$

$\frac{0.703}{0.703} = 1$

$45.2 \text{ g O}\left(\frac{1 \text{ mol O}}{16.0 \text{ g O}}\right) = 2.82 \text{ mol O}; \quad \frac{2.82}{0.703} \approx 4$

empirical formula: $NaMnO_4$

(c) $3.1 \text{ g H}\left(\frac{1 \text{ mol H}}{1.0 \text{ g H}}\right) = 3.1 \text{ mol H}; \quad \frac{3.1}{1.02} \approx 3$

$31.5 \text{ g P}\left(\frac{1 \text{ mol P}}{31.0 \text{ g P}}\right) = 1.02 \text{ mol P}; \quad \frac{1.02}{1.02} = 1$

$65.4 \text{ g O}\left(\frac{1 \text{ mol O}}{16.0 \text{ g O}}\right) = 4.09 \text{ mol O}; \quad \frac{4.09}{1.02} \approx 4$

empirical formula: H_3PO_4

(d) $56.6 \text{ g K}\left(\frac{1 \text{ mol K}}{39.1 \text{ g K}}\right) = 1.45 \text{ mol K}; \quad \frac{1.45}{0.72} \approx 2$

$8.7 \text{ g C}\left(\frac{1 \text{ mol C}}{12.0 \text{ g C}}\right) = 0.72 \text{ mol C}; \quad \frac{0.72}{0.72} = 1$

$34.7 \text{ g O}\left(\frac{1 \text{ mol O}}{16.0 \text{ g O}}\right) = 2.17 \text{ mol O}; \quad \frac{2.17}{0.72} \approx 3$

empirical formula: K_2CO_3

(e) $10.04 \text{ g C}\left(\frac{1 \text{ mol C}}{12.0 \text{ g C}}\right) = 0.837 \text{ mol C}; \quad \frac{0.837}{0.837} = 1$

$0.84 \text{ g H}\left(\frac{1 \text{ mol H}}{1.0 \text{ g H}}\right) = 0.84 \text{ mol H}; \quad \frac{0.84}{0.837} \approx 1$

$89.12 \text{ g Cl}\left(\frac{1 \text{ mol Cl}}{35.5 \text{ g Cl}}\right) = 2.51 \text{ mol Cl}; \quad \frac{2.51}{0.837} \approx 3$

empirical formula: $CHCl_3$

4.14 $1.67 \text{ g Ca}\left(\frac{1 \text{ mol Ca}}{40.1 \text{ g Ca}}\right) = 0.0416 \text{ mol Ca}; \quad \frac{0.0416}{0.0416} = 1$

$2.96 \text{ g Cl}\left(\frac{1 \text{ mol Cl}}{35.5 \text{ g Cl}}\right) = 0.0834 \text{ mol Cl}; \quad \frac{0.0834}{0.0416} \approx 2$

empirical formula: $CaCl_2$

4.15 $0.432 \text{ g C}\left(\frac{1 \text{ mol C}}{12.0 \text{ g C}}\right) = 0.0360 \text{ mol C}; \quad \frac{0.0360}{0.0180} = 2$

$0.090 \text{ g H}\left(\frac{1 \text{ mol H}}{1.0 \text{ g H}}\right) = 0.090 \text{ mol H}; \quad \frac{0.090}{0.0180} = 5$

$0.342 \text{ g F}\left(\frac{1 \text{ mol F}}{19.0 \text{ g F}}\right) = 0.0180 \text{ mol F}; \quad \frac{0.0180}{0.0180} = 1$

empirical formula: C_2H_5F

4.16 **(a)** $\frac{165 \text{ g}}{1.25 \text{ mol}} = 132 \text{ g/mol}$

(b) $\frac{14.5 \text{ g}}{0.320 \text{ mol}} = 45.3 \text{ g/mol}$

(c) $\frac{2.35 \text{ g}}{0.0155} = 152 \text{ g/mol}$

4.17 **(a)** Empirical formula mass for CH: $12.0 + 1.0 = 13.0$

$\frac{52.0}{13.0} = 4; \quad$ molecular formula: C_4H_4

(b) Empirical formula mass for CH: $12.0 + 1.0 = 13.0$

$\frac{78.0}{13.0} = 6; \quad$ molecular formula: C_6H_6

(c) Empirical formula mass for C_2H_3F: $2(12.0) + 3(1.0) + 19.0 = 46.0$

$\frac{92.0}{46.0} = 2; \quad$ molecular formula: $C_4H_6F_2$

(d) Empirical formula mass for NaO: $23.0 + 16.0 = 39.0$

$\frac{78.0}{39.0} = 2; \quad$ molecular formula: Na_2O_2

(e) Empirical formula mass for C_5H_5N: $5(12.0) + 5(1.0) + 14.0 = 79.0$

$\frac{79.0}{79.0} = 1; \quad$ molecular formula: C_5H_5N

(f) Empirical formula mass for C_2HNO_2: $2(12.0) + 1.0 + 14.0 + 2(16.0) = 71.0$

$\frac{213.0}{71.0} = 3; \quad$ molecular formula: $C_6H_3N_3O_6$

Chapter 5

5.1 **(a)** 4 u **(b)** +2

5.2 **(a)** 10 **(b)** 29 **(c)** 83 **(d)** 7 **(e)** 51 **(f)** 80

5.3 **(a)** Beryllium-9; $^{9}_{4}Be$ **(b)** Carbon-12; $^{12}_{6}C$
(c) Oxygen-18; $^{18}_{8}O$ **(d)** Chlorine-35; $^{35}_{17}Cl$
(e) Uranium-238; $^{238}_{92}U$ **(f)** Lawrencium-257; $^{257}_{103}Lr$
(g) Carbon-13; $^{13}_{6}C$ **(h)** Chlorine-37; $^{37}_{17}Cl$

5.4 Carbon-12 and carbon-13 are isotopes of one another, as are chlorine-35 and chlorine-37.

5.5

Name	Symbol	At. no.	p^+	e^-	n	Mass no.
Lithium	Li	3	3	3	4	7
Carbon	C	6	6	6	7	13
Nitrogen	N	7	7	7	8	15
Gold	Au	79	79	79	118	197
Lead	Pb	82	82	82	124	206
Uranium	U	92	92	92	143	235
Uranium	U	92	92	92	146	238
Neptunium	Np	93	93	93	144	237
Plutonium	Pu	94	94	94	148	242

5.6 **(a)** 3 : 1 **(b)** 75,770 **(c)** 40
(d) 1; nitrogen-14; yes
(e) Beryllium, fluorine, sodium, aluminum, and phosphorus each has only one naturally occurring isotope. For each of these elements, the atomic mass is approximately equal to the mass number of that naturally occurring isotope.
(f) Potassium-40 has 19 protons and 21 neutrons, whereas calcium-40 has 20 protons and 20 neutrons.
(g) Hydrogen-2 and helium-3 (1 each); beryllium-9 and boron-10 (5 each); carbon-13 and nitrogen-14 (7 each); oxygen-18, fluorine-19, and neon-20 (10 each); silicon-30 and phosphorus-31 (16 each); potassium-39 and calcium-40 (20 each)

5.7 Atomic mass = (0.6917)(62.9296 u) + (0.3083)(64.9278 u)
= 43.53 u + 20.02 u = 63.55 u

5.8 Atomic mass = (0.5184)(106.9051 u) + (0.4816)(108.9048 u)
= 55.42 u + 52.45 u = 107.87 u

5.9 **(a)** 47 protons, 46 electrons
(b) 26 protons, 23 electrons
(c) 30 protons, 28 electrons
(d) 50 protons, 46 electrons
(e) 34 protons, 36 electrons
(f) 15 protons, 18 electrons

5.10 **(a)** Hg^{2+} **(b)** Cu^{+} **(c)** Rb^{+} **(d)** Ni^{2+}
(e) Au^{3+} **(f)** Te^{2-}

Chapter 6

6.1 **(a)** T **(b)** F **(c)** T **(d)** T **(e)** F **(f)** T

6.2 **(a)** F **(b)** T **(c)** T **(d)** F **(e)** F

6.3 **(a)** 10 **(b)** 2 **(c)** 6 **(d)** 14 **(e)** 22 **(f)** 18 **(g)** 32

6.4 $3d \rightarrow 4p$; $5s \rightarrow 4d$; $6s \rightarrow 4f$; $4d \rightarrow 5p$; $4f \rightarrow 5d$

6.5 **(a)** Na: $1s^2 2s^2 2p^6 3s^1$ or $[Ne]3s^1$
(b) Mg: $1s^2 2s^2 2p^6 3s^2$ or $[Ne]3s^2$
(c) Al: $1s^2 2s^2 2p^6 3s^2 3p^1$ or $[Ne]3s^2 3p^1$
(d) Si: $1s^2 2s^2 2p^6 3s^2 3p^2$ or $[Ne]3s^2 3p^2$
(e) P: $1s^2 2s^2 2p^6 3s^2 3p^3$ or $[Ne]3s^2 3p^3$
(f) S: $1s^2 2s^2 2p^6 3s^2 3p^4$ or $[Ne]3s^2 3p^4$
(g) Cl: $1s^2 2s^2 2p^6 3s^2 3p^5$ or $[Ne]3s^2 3p^5$
(h) Ar: $1s^2 2s^2 2p^6 3s^2 3p^6$ or $[Ne]3s^2 3p^6$

6.6 **(a)** P: $1s^2 2s^2 2p^6 3s^2 3p^3$ or $[Ne]3s^2 3p^3$
(b) Ni: $1s^2 2s^2 2p^6 3s^2 3p^6 4s^2 3d^8$ or $[Ar]4s^2 3d^8$
(c) Sr: $1s^2 2s^2 2p^6 3s^2 3p^6 4s^2 3d^{10} 4p^6 5s^2$ or $[Kr]5s^2$
(d) Hg:
$1s^2 2s^2 2p^6 3s^2 3p^6 4s^2 3d^{10} 4p^6 5s^2 4d^{10} 5p^6 6s^2 4f^{14} 5d^{10}$
or $[Xe]6s^2 4f^{14} 5d^{10}$
(e) I: $1s^2 2s^2 2p^6 3s^2 3p^6 4s^2 3d^{10} 4p^6 5s^2 4d^{10} 5p^5$
or $[Kr]5s^2 4d^{10} 5p^5$
(f) Eu: $1s^2 2s^2 2p^6 3s^2 3p^6 4s^2 3d^{10} 4p^6 5s^2 4d^{10} 5p^6 6s^2 4f^7$
or $[Xe]6s^2 4f^7$

6.7 Filled = 2 electrons; half-filled = 1 electron; empty = 0 electrons.
(a) One filled orbital
(b) Two half-filled orbitals, one empty orbital
(c) Two filled orbitals, one half-filled orbital
(d) Five half-filled orbitals
(e) Three filled orbitals, two half-filled orbitals

6.8 **(a)** 2 **(b)** 2 **(c)** 1 **(d)** 2 **(e)** 3 **(f)** 3

6.9 **(a)**

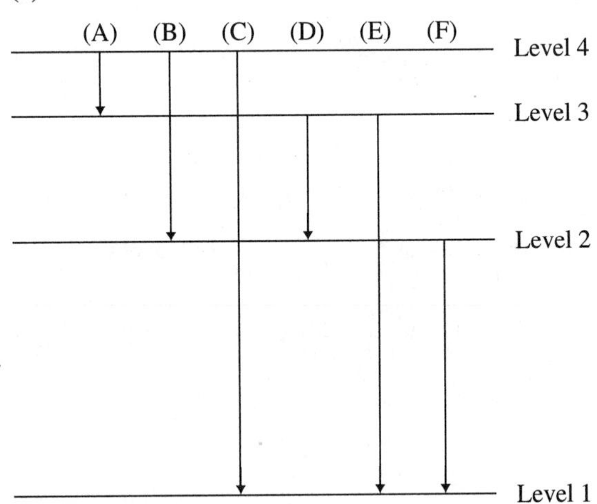

Six different frequencies of light are possible.

(b)

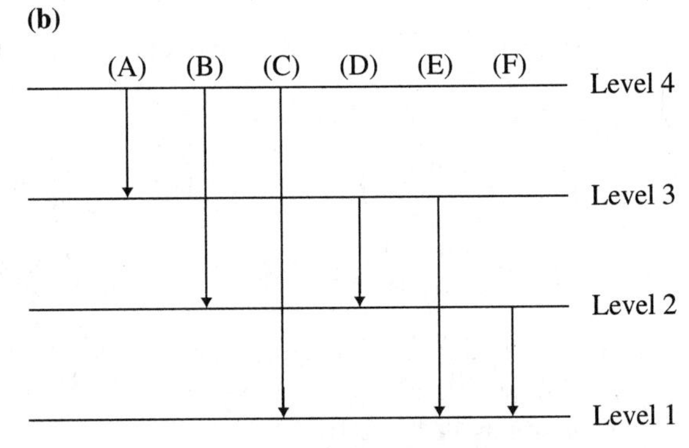

There are only three different frequencies of light. Transitions (D) and (F) are of the same energy as transition (A), and transition (E) is of the same energy as transition (B). Thus, transitions (A), (B), and (C) account for the three observed frequencies.

6.10 In the Bohr model, electrons orbit the nucleus in fixed radii. The higher the energy level, the greater the distance from the nucleus. For any energy level, the "average" distance equals the fixed radius. In the quantum-mechanical model, electrons exist in orbitals that describe the probability of finding an electron at given locations, relative to the nucleus. Unlike the Bohr model, the electrons in an orbital are not restricted to fixed distances from the nucleus; instead, they may be found over a wide range of distances. Nevertheless, for higher energy orbitals, there is a greater probability of finding an electron further from the nucleus. Thus, the average electronic distance from the nucleus increases with increasing energy in both the Bohr model and the quantum-mechanical model.

Chapter 7

7.1 **(a)** The atomic mass of an element represents the average mass of the isotopes that occur naturally. The average mass of the isotopes that make up tellurium is greater than that of the isotopes that make up iodine, even though each tellurium atom has one less proton than each atom of iodine. The greater average mass must mean that the average number of neutrons in a sample of tellurium must be greater than in a sample of iodine.

(b) Argon (atomic number 18) has a greater atomic mass than potassium (atomic number 19). Cobalt (atomic number 27) has a greater atomic mass than nickel (atomic number 28).

7.2 **(a)** Refer to the Glossary.

(b)
$$\underset{(19p^+,\ 19e^-)}{K} \rightarrow \underset{(19p^+,\ 18e^-)}{K^+} + \underset{(e^-)}{e^-}$$
$$\underset{(37p^+,\ 37e^-)}{Rb} \rightarrow \underset{(37p^+,\ 37e^-)}{Rb^+} + \underset{(e^-)}{e^-}$$

(c) Refer to the Glossary.

(d)
$$\underset{(35p^+,\ 35e^-)}{Br} + \underset{(e^-)}{e^-} \rightarrow \underset{(35p^+,\ 36\,e^-)}{Br^-}$$

7.3 **(a)** K^+ **(b)** F^- **(c)** Ca^{2+} **(d)** Se^{2-} **(e)** No ion **(f)** Cs^+ **(g)** O^{2-} **(h)** No ion **(i)** Te^{2-}

7.4 **(a)** Alkaline earth metal **(b)** Noble gas **(c)** Halogen **(d)** Alkali metal **(e)** Chalcogen **(f)** Noble gas **(g)** Halogen **(h)** Alkali metal **(i)** Alkaline earth metal **(j)** Alkali metal **(k)** Chalcogen

7.5 **(a)** 2, 3 **(b)** 1 **(c)** 2 **(d)** 5 **(e)** 1, 3 **(f)** 1 **(g)** 6 **(h)** 6

7.6 Refer to answer to Problem 6.6.

7.7 **(a)** Br_2 **(b)** Te and Po **(c)** Hg **(d)** F_2 and Cl_2 **(e)** B **(f)** He **(g)** I_2

7.8 **(a)** K **(b)** Cl **(c)** S **(d)** K **(e)** K

7.9 **(a)** He and Li^+ **(b)** F^- and Mg^{2+} **(c)** Ca^{2+} and P^{3-}

7.10 **(a)** K^+ **(b)** Br^- **(c)** S^{2-} **(d)** F^- **(e)** Ar **(f)** Cl^- **(g)** Na^+ **(h)** O^{2-}

7.11 Each of these ions has 18 electrons. However, the nuclear charge increases in the order $S^{2-} \rightarrow Cl^- \rightarrow K^+$. This draws the electrons closer to the nucleus, making K^+ the smallest of the three and S^{2-} the largest.

Chapter 8

8.1 Elements in group IA have one valence electron, those in group IIA have two valence electrons, and so forth. Those elements in group VIIIA (the noble gases) have eight valence electrons (with the exception of helium, which has two).

8.2 **(a)**
·F̤̈: [:F̤̈:]⁻
·Ȧl· + ·F̤̈: → Al^{3+} + [:F̤̈:]⁻
·F̤̈: [:F̤̈:]⁻

(b) Ṁg· + ·S̤̈· → Mg^{2+} + [:S̤̈:]²⁻

(c) Na· Na^+
+ ·Ö̤· → + [:Ö̤:]²⁻
Na· Na^+

(d)
·Ö̤· [:Ö̤:]²⁻
·Ȧl· Al^{3+}
+ ·Ö̤· → + [:Ö̤:]²⁻
·Ȧl· Al^{3+}
·Ö̤· + [:Ö̤:]²⁻

8.3 **(a)** Covalent **(b)** Ionic **(c)** Covalent **(d)** Covalent **(e)** Ionic **(f)** Covalent

8.4 **(a)** H:N̈:H with H below N
(b) H:C:H with H above and H below C
(c) H:S̈: with H below S
(d) H:P̈:H with H below P
(e) :F̤̈:N̈:F̤̈: with :F̤̈: below N
(f) :C̤̈l:C:C̤̈l: with :C̤̈l: above and :C̤̈l: below C
(g) H:Si:H with H above and H below Si
(h) :C̤̈l:C̤̈l:

(i)
```
:F:P:F:
  :F:
```
(j)
```
  :Cl:
H:C:Cl:
  :Cl:
```
(k)
```
  :F:
H:Si:H
  :F:
```

8.5 (a) :S::C::S: (b) H:C:::N:
(c) H:C:::C:H (d)
```
H:C::C:H
  H  H
```
(e)
```
  :O:
H:C:H
```

8.6 (a) H:O:O:H (b)
```
H:N:N:H
  H H
```
(c) :O::S:O: (d) O::C::O
(e) :C:::O: (f)
```
   :O:
H:O:S:O:H
   :O:
```
(g)
```
   :O:
H:O:N:O:
```
(h)
```
   :O:
H:O:Cl:O:
   :O:
```

8.7 From Problem 8.4:
(a)
```
H–N–H
  |
  H
```
(b)
```
  H
  |
H–C–H
  |
  H
```
(c)
```
H–S:
  |
  H
```
(d)
```
H–P–H
  |
  H
```
(e)
```
:F–N–F:
   |
  :F:
```
(f)
```
    :Cl:
     |
:Cl–C–Cl:
     |
    :Cl:
```
(g)
```
   H
   |
H–Si–H
   |
   H
```
(h) :Cl–Cl:
(i)
```
:F–P–F:
   |
  :F:
```
(j)
```
   :Cl:
    |
 H–C–Cl:
    |
   :Cl:
```
(k)
```
  :F:
   |
H–Si–H
   |
  :F:
```

From Problem 8.5:
(a) :S=C=S: (b) H–C≡N:
(c) H–C≡C–H (d)
```
H–C=C–H
  |  |
  H  H
```
(e)
```
  :O:
   ||
H–C–H
```

8.8 (a)
```
[   :O:    ]⁻
[ :O:N:O:  ]
```
(b) [:O:H]⁻
(c)
```
[   :O:    ]3–
[ :O:P:O:  ]
[   :O:    ]
```
(d)
```
[   :O:    ]2–
[ :O:C:O:  ]
```
(e)
```
[   H    ]⁻
[ H:B:H  ]
[   H    ]
```
(f) [:N::N::N:]⁻
(g)
```
[   :O:     ]⁻
[ :O:Cl:O:  ]
[   :O:     ]
```
(h)
```
[   :O:     ]⁻
[ :O:Cl:O:  ]
```

8.9 (a) C (b) O (c) O (d) Mg (e) P (f) Sb

8.10

Bond	*Electronegativity difference (More electronegative – Less electronegative)*
(a) $\overrightarrow{\text{H–Cl}}$	3.0 – 2.1 = 0.9
(b) $\overrightarrow{\text{Al–H}}$	2.1 – 1.5 = 0.6
(c) $\overrightarrow{\text{Mg–F}}$	4.0 – 1.2 = 2.8
(d) $\overleftarrow{\text{O–H}}$	3.5 – 2.1 = 1.4
(e) $\overrightarrow{\text{Be–S}}$	2.5 – 1.5 = 1.0
(f) $\overrightarrow{\text{P–S}}$	2.5 – 2.1 = 0.4
(g) $\overrightarrow{\text{Li–H}}$	2.1 – 1.0 = 1.1

(h) Hydrogen is the *less* electronegative element.
(i) Hydrogen is the *more* electronegative element.

8.11 In order of increasing polarity (electronegativity differences in parentheses):
F–F (0) < C–H (0.4) < C–N (0.5) < H–Cl (0.9) < Al–S (1.0) < Si–O (1.7) < Li–Cl (2.0)

8.12 (a) Covalent (b) Covalent (c) Ionic (d) Ionic (e) Covalent (f) Covalent (g) Ionic (h) Covalent

Chapter 9

9.1

	Cl^-	O^{2-}	N^{3-}
Na^+	$NaCl$	Na_2O	Na_3N
Ca^{2+}	$CaCl_2$	CaO	Ca_3N_2
Al^{3+}	$AlCl_3$	Al_2O_3	AlN

9.2 (a) $NaCl$ (b) KI (c) $MgBr_2$ (d) BaO (e) Li_2O (f) K_2S (g) Al_2S_3 (h) CaF_2 (i) SrO

9.3 (a) $AlPO_4$ (b) Li_2SO_4 (c) $Al(OH)_3$ (d) $Ba(NO_3)_2$ (e) $(NH_4)_2CO_3$ (f) $CaSO_4$ (g) $Al_2(CrO_4)_3$ (h) $Sr_3(PO_4)_2$ (i) $Mg_3(AsO_4)_2$ (j) $Na_2C_2O_4$

9.4 (a) $CuBr$ (b) $SnCl_2$ (c) $Pb(NO_3)_2$ (d) Co_2O_3 (e) $Fe(OH)_3$ (f) $Cr_2(SO_4)_3$ (g) Cu_2CO_3 (h) $CoCrO_4$ (i) HgF_2 (j) Hg_2F_2

9.5 An ionic compound is named by stating the name of the cation followed by the name of the anion.

9.6 (a) Barium sulfide (b) Calcium chloride (c) Silver nitride (d) Copper(II) fluoride (e) Mercury(II) oxide (f) Mercury(I) oxide (g) Cobalt(II) chloride (h) Cobalt(III) chloride (i) Iron(II) sulfide (j) Chromium(II) nitride (k) Zinc phosphide (l) Magnesium hydride

9.7 (a) Mercury(II) chloride, $HgCl_2$
(b) Cobalt(II) iodide, CoI_2
(c) Tin(II) bromide, $SnBr_2$
(d) Iron(II) phosphide, Fe_3P_2
(e) Mercury(I) iodide, Hg_2I_2
(f) Cobalt(III) nitride, CoN
(g) Lead(II) sulfide, PbS
(h) Tin(IV) hydride, SnH_4

9.8 (a) Potassium hydroxide (b) Copper(II) cyanide (c) Aluminum hydroxide (d) Sodium cyanide (e) Ammonium cyanide (f) Ammonium sulfide

9.9 (a) BrO^- (b) IO_4^- (c) IO_2^- (d) IO^-

9.10 (a) Copper(I) sulfate (b) Potassium nitrite (c) Lithium chlorate (d) Iron(II) sulfite (e) Magnesium nitrate (f) Sodium chlorite

9.11 (a) Sodium perchlorate
(b) Potassium hypobromite
(c) Calcium hypoiodite
(d) Sodium peroxide
(e) Potassium permanganate
(f) Silver hypochlorite

9.12 (a) Potassium hydrogen sulfate
(b) Lithium dihydrogen phosphate
(c) Magnesium hydrogen sulfate
(d Calcium hydrogen carbonate
(e) Magnesium monohydrogen phosphate

9.13 (a) Barium chloride dihydrate or barium chloride 2-hydrate
(b) Zinc sulfate heptahydrate or zinc sulfate 7-hydrate
(c) Copper(II) sulfate pentahydrate or copper(II) sulfate 5-hydrate
(d) Cobalt(II) chloride dihydrate or cobalt(II) chloride 2-hydrate

9.14 (a) CO_2 (b) PCl_3 (c) BCl_3 (d) N_2O_4 (e) $SeCl_2$ (f) SO_2 (g) As_2O_5 (h) NI_3 (i) NO_2 (j) XeF_4

9.15 (a) Selenium trioxide (b) Oxygen dichloride (c) Diphosphorus pentoxide (d) Nitrogen monoxide (e) Tellurium trioxide (f) Silicon tetrabromide (g) Nitrogen trifluoride (h) Dinitrogen monoxide (i) Xenon tetrafluoride (j) Sulfur hexafluoride

9.16 (a) Nitric acid (b) Phosphorous acid (c) Arsenic acid (d) Chromic acid (e) Nitrous acid (f) Acetic acid (g) Hydriodic acid (h) Hydrochloric acid (i) Hydrocyanic acid (j) Aluminum hydroxide (k) Lithium hydroxide (l) Magnesium hydroxide

Chapter 10

10.1 A chemical equation is balanced when every atom appearing on the reactant side also appears on the product side, and vice versa. This implies that all of the matter that enters into the reaction as reactants is found somewhere in the products. Hence, matter is neither created nor destroyed.

10.2 (a) $C_5H_{12} + 8\,O_2 \rightarrow 5\,CO_2 + 6\,H_2O$
(b) $H_2 + I_2 \rightarrow 2\,HI$
(c) $2\,Li + H_2 \rightarrow 2\,LiH$
(d) $S + O_2 \rightarrow SO_2$
(e) $2\,Al + Fe_2O_3 \rightarrow 2\,Fe + Al_2O_3$
(f) $2\,Fe_2S_3 + 9\,O_2 \rightarrow 2\,Fe_2O_3 + 6\,SO_2$
(g) $2\,KOH + H_2SO_4 \rightarrow K_2SO_4 + 2\,H_2O$
(h) $P_4 + 5\,O_2 \rightarrow P_4O_{10}$
(i) $Zn + 2\,HCl \rightarrow ZnCl_2 + H_2$
(j) $2\,Al + 3\,H_2SO_4 \rightarrow Al_2(SO_4)_3 + 3\,H_2$
(k) $C_6H_{12}O_6 + 6\,O_2 \rightarrow 6\,CO_2 + 6\,H_2O$
(l) $Mg_3N_2 + 3\,H_2O \rightarrow 3\,MgO + 2\,NH_3$
(m) $N_2H_4 + 3\,O_2 \rightarrow 2\,NO_2 + 2\,H_2O$

10.3 **(a)** Pentane + Oxygen → Carbon dioxide + Water

? C_5H_{12} + ? O_2 → ? CO_2 + ? H_2O

$C_5H_{12} + 8\,O_2 \rightarrow 5\,CO_2 + 6\,H_2O$

(b) Mercury(II) oxide → Mercury + Oxygen

? HgO → ? Hg + ? O_2

$2\,HgO \rightarrow 2\,Hg + O_2$

(c) Phosphoric acid + Potassium hydroxide → Potassium phosphate + Water

? H_3PO_4 + ? KOH → ? K_3PO_4 + ? H_2O

$H_3PO_4 + 3\,KOH \rightarrow K_3PO_4 + 3\,H_2O$

(d) Magnesium carbonate + Nitric acid → Magnesium nitrate + Water + Carbon dioxide

? $MgCO_3$ + ? HNO_3 → ? $Mg(NO_3)_2$ + ? H_2O + ? CO_2

$MgCO_3 + 2\,HNO_3 \rightarrow Mg(NO_3)_2 + H_2O + CO_2$

10.4 **(a)** Magnesium + Oxygen → Magnesium oxide

? Mg + ? O_2 → ? MgO

$2\,Mg + O_2 \rightarrow 2\,MgO$

(b) Barium oxide + Hydrochloric acid → Barium chloride + Water

? BaO + ? HCl → ? $BaCl_2$ + ? H_2O

$BaO + 2\,HCl \rightarrow BaCl_2 + H_2O$

(c) Sulfur trioxide + Water → Sulfuric acid

? SO_3 + ? H_2O → ? H_2SO_4

$SO_3 + H_2O \rightarrow H_2SO_4$

(d) Sodium + Water → Sodium hydroxide + Hydrogen

? Na + ? H_2O → ? NaOH + ? H_2

$2\,Na + 2\,H_2O \rightarrow 2\,NaOH + H_2$

(e) Sodium carbonate + Sulfuric acid → Sodium sulfate + Carbon dioxide + Water

? Na_2CO_3 + ? H_2SO_4 → ? Na_2SO_4 + ? CO_2 + ? H_2O

$Na_2CO_3 + H_2SO_4 \rightarrow Na_2SO_4 + CO_2 + H_2O$

(f) Diphosphorus pentoxide + Water → Phosphoric acid

? P_2O_5 + ? H_2O → ? H_3PO_4

$P_2O_5 + 3\,H_2O \rightarrow 2\,H_3PO_4$

(g) Aluminum + Oxygen → Aluminum oxide

? Al + ? O_2 → ? Al_2O_3

$4\,Al + 3\,O_2 \rightarrow 2\,Al_2O_3$

(h) Iron(III) oxide + Hydrogen → Iron + Water

? Fe_2O_3 + ? H_2 → ? Fe + ? H_2O

$Fe_2O_3 + 3\,H_2 \rightarrow 2\,Fe + 3\,H_2O$

10.5 **(a)** $C_5H_{12}(\ell) + 8\,O_2(g) \rightarrow 5\,CO_2(g) + 6\,H_2O(\ell)$

(b) $MgCO_3(s) + 2\,HNO_3(aq) \rightarrow Mg(NO_3)_2(aq) + CO_2(g) + H_2O(\ell)$

(c) $2\,Na(s) + 2\,H_2O(\ell) \rightarrow 2\,NaOH(aq) + H_2(g)$

(d) $2\,HgO(s) \rightarrow 2\,Hg(\ell) + O_2(g)$

(e) $2\,C(s) + O_2(g) \rightarrow 2\,CO(g)$

(f) $Pb(NO_3)_2(aq) + 2\,KI(aq) \rightarrow PbI_2(s) + 2\,KNO_3(aq)$

10.6 **(a)** 3 mol Cu : 8 mol HNO_3

(b) 3 mol Cu : 3 mol $Cu(NO_3)_2$ (or 1 mol Cu : 1 mol $Cu(NO_3)_2$)

(c) 8 mol HNO_3 : 2 mol NO (or 4 mol HNO_3 : 1 mol NO)

(d) 2 mol NO : 4 mol H_2O (or 1 mol NO : 2 mol H_2O)

(e) 3 mol Cu : 2 mol NO

(f) 3 mol $Cu(NO_3)_2$: 4 mol H_2O

10.7 **(a)** $C_5H_{12} + 8\,O_2 \rightarrow 5\,CO_2 + 6\,H_2O$

(b) $2.10\text{ mol } C_5H_{12}\left(\frac{8\text{ mol } O_2}{1\text{ mol } C_5H_{12}}\right) = 16.8\text{ mol } O_2$

(c) $2.10\text{ mol } C_5H_{12}\left(\frac{5\text{ mol } CO_2}{1\text{ mol } C_5H_{12}}\right) = 10.5\text{ mol } CO_2$

(d) $2.10\text{ mol } C_5H_{12}\left(\frac{6\text{ mol } H_2O}{1\text{ mol } C_5H_{12}}\right) = 12.6\text{ mol } H_2O$

10.8 **(a)** $2\,Al + 6\,HCl \rightarrow 2\,AlCl_3 + 3\,H_2$

(b) $0.450\text{ mol Al}\left(\frac{6\text{ mol HCl}}{2\text{ mol Al}}\right) = 1.35\text{ mol HCl}$

(c) $0.450\text{ mol Al}\left(\frac{2\text{ mol } AlCl_3}{2\text{ mol Al}}\right) = 0.450\text{ mol } AlCl_3$

(d) $0.450\text{ mol Al}\left(\frac{3\text{ mol } H_2}{2\text{ mol Al}}\right) = 0.675\text{ mol } H_2$

10.9 **(a)** $N_2H_4 + 3\,O_2 \rightarrow 2\,NO_2 + 2\,H_2O$

(b) $1.30\text{ mol } N_2H_4\left(\frac{3\text{ mol } O_2}{1\text{ mol } N_2H_4}\right) = 3.90\text{ mol } O_2$

(c) $1.30\text{ mol } N_2H_4\left(\frac{2\text{ mol } NO_2}{1\text{ mol } N_2H_4}\right) = 2.60\text{ mol } NO_2$

(d) $1.30\text{ mol } N_2H_4\left(\frac{2\text{ mol } H_2O}{1\text{ mol } N_2H_4}\right) = 2.60\text{ mol } H_2O$

10.10 **(a)** $2\,Fe_2S_3 + 9\,O_2 \rightarrow 2\,Fe_2O_3 + 6\,SO_2$

(b) $0.900\text{ mol } SO_2\left(\frac{2\text{ mol } Fe_2S_3}{6\text{ mol } SO_2}\right) = 0.300\text{ mol } Fe_2S_3$

(c) $0.900\text{ mol } SO_2\left(\frac{9\text{ mol } O_2}{6\text{ mol } SO_2}\right) = 1.35\text{ mol } O_2$

(d) $0.900\text{ mol } SO_2\left(\frac{2\text{ mol } Fe_2O_3}{6\text{ mol } SO_2}\right) = 0.300\text{ mol } Fe_2O_3$

10.11 **(a)** $14.0\text{ g CO}\left(\frac{1\text{ mol CO}}{28.0\text{ g CO}}\right) = 0.500\text{ mol CO}$

(b) $25.0\text{ g } CaCO_3\left(\frac{1\text{ mol } CaCO_3}{100.1\text{ g } CaCO_3}\right) = 0.250\text{ mol } CaCO_3$

(c) $30.0 \text{ g } Na_2SO_4\left(\frac{1 \text{ mol } Na_2SO_4}{142.1 \text{ g } Na_2SO_4}\right) = 0.211 \text{ mol } Na_2SO_4$

(d) $62.9 \text{ g } (NH_4)_3PO_4\left(\frac{1 \text{ mol } (NH_4)_3PO_4}{149.0 \text{ g } (NH_4)_3PO_4}\right) = 0.422 \text{ mol } (NH_4)_3PO_4$

10.12 **(a)** $2.00 \text{ mol } CO_2\left(\frac{44.0 \text{ g } CO_2}{1 \text{ mol } CO_2}\right) = 88.0 \text{ g } CO_2$

(b) $0.300 \text{ mol } SO_3\left(\frac{80.1 \text{ g } SO_3}{1 \text{ mol } SO_3}\right) = 24.0 \text{ g } SO_3$

(c) $0.835 \text{ mol } C_7H_{16}\left(\frac{100.0 \text{ g } C_7H_{16}}{1 \text{ mol } C_7H_{16}}\right) = 83.5 \text{ g } C_7H_{16}$

(d) $0.0175 \text{ mol } Ca(OH)_2\left(\frac{74.1 \text{ g } Ca(OH)_2}{1 \text{ mol } Ca(OH)_2}\right) = 1.30 \text{ g } Ca(OH)_2$

10.13 **(a)** $4\,Fe + 3\,O_2 \rightarrow 2\,Fe_2O_3$

(b) $5.58 \text{ g } Fe\left(\frac{1 \text{ mol } Fe}{55.8 \text{ g } Fe}\right)\left(\frac{2 \text{ mol } Fe_2O_3}{4 \text{ mol } Fe}\right) \times \left(\frac{159.6 \text{ g } Fe_2O_3}{1 \text{ mol } Fe_2O_3}\right) = 7.98 \text{ g } Fe_2O_3$

10.14 **(a)** $C_6H_{12}O_6 + 6\,O_2 \rightarrow 6\,CO_2 + 6\,H_2O$

(b) $36.0 \text{ g } C_6H_{12}O_6\left(\frac{1 \text{ mol } C_6H_{12}O_6}{180.0 \text{ g } C_6H_{12}O_6}\right) = 0.200 \text{ mol } C_6H_{12}O_6$

$0.200 \text{ mol } C_6H_{12}O_6\left(\frac{6 \text{ mol } O_2}{1 \text{ mol } C_6H_{12}O_6}\right) \times \left(\frac{32.0 \text{ g } O_2}{1 \text{ mol } O_2}\right) = 38.4 \text{ g } O_2$

(c) $0.200 \text{ mol } C_6H_{12}O_6\left(\frac{6 \text{ mol } CO_2}{1 \text{ mol } C_6H_{12}O_6}\right) \times \left(\frac{44.0 \text{ g } CO_2}{1 \text{ mol } CO_2}\right) = 52.8 \text{ g } CO_2$

(d) $0.200 \text{ mol } C_6H_{12}O_6\left(\frac{6 \text{ mol } H_2O}{1 \text{ mol } C_6H_{12}O_6}\right) \times \left(\frac{18.0 \text{ g } H_2O}{1 \text{ mol } H_2O}\right) = 21.6 \text{ g } H_2O$

(e) $36.0 \text{ g} + 38.4 \text{ g} = 74.4 \text{ g} = 52.8 \text{ g} + 21.6 \text{ g}$

10.15 **(a)** $N_2 + 3\,H_2 \rightarrow 2\,NH_3$

(b) $85.0 \text{ g } NH_3\left(\frac{1 \text{ mol } NH_3}{17.0 \text{ g } NH_3}\right) = 5.00 \text{ mol } NH_3$

$5.00 \text{ mol } NH_3\left(\frac{3 \text{ mol } H_2}{2 \text{ mol } NH_3}\right)\left(\frac{2.0 \text{ g } H_2}{1 \text{ mol } H_2}\right) = 15 \text{ g } H_2$

(c) $5.00 \text{ mol } NH_3\left(\frac{1 \text{ mol } N_2}{2 \text{ mol } NH_3}\right)\left(\frac{28.0 \text{ g } N_2}{1 \text{ mol } N_2}\right) = 70.0 \text{ g } N_2$

10.16 **(a)** $2\,C_8H_{18} + 25\,O_2 \rightarrow 16\,CO_2 + 18\,H_2O$

(b) $702 \text{ g } C_8H_{18}\left(\frac{1 \text{ mol } C_8H_{18}}{114.0 \text{ g } C_8H_{18}}\right) = 6.16 \text{ mol } C_8H_{18}$

$6.16 \text{ mol } C_8H_{18}\left(\frac{16 \text{ mol } CO_2}{2 \text{ mol } C_8H_{18}}\right)\left(\frac{44.0 \text{ g } CO_2}{1 \text{ mol } CO_2}\right) = 2170 \text{ g } CO_2\ (2.17 \times 10^3 \text{ g})$

(c) $6.16 \text{ mol } C_8H_{18}\left(\frac{25 \text{ mol } O_2}{2 \text{ mol } C_8H_{18}}\right)\left(\frac{32.0 \text{ g } O_2}{1 \text{ mol } O_2}\right) = 2460 \text{ g } O_2\ (2.46 \times 10^3 \text{ g})$

10.17 As in all stoichiometry problems, we must begin with a balanced equation.

$2\,Al + 3\,H_2SO_4 \rightarrow Al_2(SO_4)_3 + 3\,H_2$

(a) $455 \text{ g } Al_2(SO_4)_3\left(\frac{1 \text{ mol } Al_2(SO_4)_3}{342.3 \text{ g } Al_2(SO_4)_3}\right) = 1.33 \text{ mol } Al_2(SO_4)_3$

$1.33 \text{ mol } Al_2(SO_4)_3\left(\frac{2 \text{ mol } Al}{1 \text{ mol } Al_2(SO_4)_3}\right)\left(\frac{27.0 \text{ g } Al}{1 \text{ mol } Al}\right) = 71.8 \text{ g } Al$

(b) $1.33 \text{ mol } Al_2(SO_4)_3\left(\frac{3 \text{ mol } H_2SO_4}{1 \text{ mol } Al_2(SO_4)_3}\right) \times \left(\frac{98.1 \text{ g } H_2SO_4}{1 \text{ mol } H_2SO_4}\right) = 391 \text{ g } H_2SO_4$

(c) $1.33 \text{ mol } Al_2(SO_4)_3\left(\frac{3 \text{ mol } H_2}{1 \text{ mol } Al_2(SO_4)_3}\right)\left(\frac{2.0 \text{ g } H_2}{1 \text{ mol } H_2}\right) = 8.0 \text{ g } H_2$

10.18 Theoretical yield:

$3.28 \text{ g } C_7H_6O_3\left(\frac{1 \text{ mol } C_7H_6O_3}{138.0 \text{ g } C_7H_6O_3}\right)\left(\frac{1 \text{ mol } C_9H_8O_4}{1 \text{ mol } C_7H_6O_3}\right) \times \left(\frac{180.0 \text{ g } C_9H_8O_4}{1 \text{ mol } C_9H_8O_4}\right) = 4.28 \text{ g } C_9H_8O_4$

Percentage yield: $\frac{3.11 \text{ g}}{4.28 \text{ g}} \times 100\% = 72.7\%$

10.19 Theoretical yield:

$17.5 \text{ g } C_6H_6\left(\frac{1 \text{ mol } C_6H_6}{78.0 \text{ g } C_6H_6}\right)\left(\frac{1 \text{ mol } C_6H_5NO_2}{1 \text{ mol } C_6H_6}\right) \times \left(\frac{123.0 \text{ g } C_6H_5NO_2}{1 \text{ mol } C_6H_5NO_2}\right) = 27.6 \text{ g } C_6H_5NO_2$

Percentage yield: $\frac{22.6 \text{ g}}{27.6 \text{ g}} \times 100\% = 81.9\%$

10.20 Theoretical yield:

$$5.36 \text{ g } Pb(NO_3)_2\left(\frac{1 \text{ mol } Pb(NO_3)_2}{331.2 \text{ g } Pb(NO_3)_2}\right)$$

$$\times\left(\frac{1 \text{ mol } PbCl_2}{1 \text{ mol } Pb(NO_3)_2}\right)\left(\frac{278.2 \text{ g } PbCl_2}{1 \text{ mol } PbCl_2}\right) = 4.50 \text{ g } PbCl_2$$

Percentage yield: $\frac{3.21 \text{ g}}{4.50 \text{ g}} \times 100\% = 71.3\%$

10.21 $8.50 \text{ g Mg}\left(\frac{1 \text{ mol Mg}}{24.3 \text{ g Mg}}\right)\left(\frac{1 \text{ mol S}}{1 \text{ mol Mg}}\right)\left(\frac{32.1 \text{ g S}}{1 \text{ mol S}}\right)$

$$= 11.2 \text{ g S}$$

Only 9.00 g S are present, so there is not enough sulfur to react with all of the magnesium. Hence, sulfur is the limiting reagent.

$$9.00 \text{ g S}\left(\frac{1 \text{ mol S}}{32.1 \text{ g S}}\right)\left(\frac{1 \text{ mol MgS}}{1 \text{ mol S}}\right)\left(\frac{56.4 \text{ g MgS}}{1 \text{ mol MgS}}\right)$$

$$= 15.8 \text{ MgS}$$

10.22 $5.75 \text{ g } BaCl_2\left(\frac{1 \text{ mol } BaCl_2}{208.3 \text{ g } BaCl_2}\right)\left(\frac{1 \text{ mol } Na_2SO_4}{1 \text{ mol } BaCl_2}\right)$

$$\times\left(\frac{142.1 \text{ g } Na_2SO_4}{1 \text{ mol } Na_2SO_4}\right) = 3.92 \text{ g } Na_2SO_4$$

Because 4.25 g Na_2SO_4 are available, sodium sulfate is in excess and barium chloride is the limiting reagent.

$$5.75 \text{ g } BaCl_2\left(\frac{1 \text{ mol } BaCl_2}{208.3 \text{ g } BaCl_2}\right)\left(\frac{1 \text{ mol } BaSO_4}{1 \text{ mol } BaCl_2}\right)$$

$$\times\left(\frac{233.4 \text{ g } BaSO_4}{1 \text{ mol } BaSO_4}\right) = 6.44 \text{ g } BaSO_4$$

10.23 $4.55 \text{ g } Pb(NO_3)_2\left(\frac{1 \text{ mol } Pb(NO_3)_2}{331.2 \text{ g } Pb(NO_3)_2}\right)$

$$\times\left(\frac{2 \text{ mol KI}}{1 \text{ mol } Pb(NO_3)_2}\right)\left(\frac{166.0 \text{ g KI}}{1 \text{ mol KI}}\right) = 4.56 \text{ g KI}$$

Only 3.75 g KI are actually present, so there is not enough potassium iodide to react with all of the lead(II) nitrate. Thus, potassium iodide is the limiting reagent.

$$3.75 \text{ g KI}\left(\frac{1 \text{ mol KI}}{166.0 \text{ g KI}}\right)\left(\frac{1 \text{ mol } PbI_2}{2 \text{ mol KI}}\right)\left(\frac{461.0 \text{ g } PbI_2}{1 \text{ mol } PbI_2}\right)$$

$$= 5.21 \text{ g } PbI_2$$

10.24 **(a)** Endothermic **(b)** Exothermic
(c) Exothermic **(d)** Endothermic

10.25 $50.0 \text{ g HgO}\left(\frac{1 \text{ mol HgO}}{216.6 \text{ g HgO}}\right)\left(\frac{182 \text{ kJ}}{2 \text{ mol HgO}}\right) = 21.0 \text{ kJ}$

Chapter 11

11.1 The word *kinetic* refers to motion. Gas molecules are in constant motion, colliding with one another and with other objects in their paths.

11.2 $P_{gas} = 764 \text{ torr} + 7 \text{ torr} = 771 \text{ torr}$

11.3 **(a)** The pressure experienced underwater comes from the weight of the water above. Imagine that a column of water (much like the column of mercury in a barometer) exists between the surface of the pool and the level of the ear. The weight of water in this column exerts a force per unit area on the ear.
(b) The greater the depth, the taller the column of water above the ear, and hence, the greater the pressure.

11.4 **(a)** $30.0 \text{ mL}\left(\frac{750.0 \text{ torr}}{600.0 \text{ torr}}\right) = 37.5 \text{ mL}$

(b) $500.0 \text{ torr}\left(\frac{1.50 \text{ L}}{1.00 \text{ L}}\right) = 750 \text{ torr}$

(c) $45.0 \text{ mL}\left(\frac{1.00 \text{ atm}}{1.00 \text{ atm}}\right) = 45.0 \text{ mL}$

(d) $3.60 \text{ L}\left(\frac{1.20 \text{ atm}}{0.800 \text{ atm}}\right) = 5.40 \text{ L}$

(e) $700.0 \text{ torr}\left(\frac{0.0840 \text{ L}}{5.20 \text{ L}}\right) = 11.3 \text{ torr}$

(f) $76.0 \text{ mL}\left(\frac{532 \text{ torr}}{480 \text{ torr}}\right) = 84.2 \text{ mL}$

(g) $146 \text{ torr}\left(\frac{18.2 \text{ L}}{11.4 \text{ L}}\right) = 233 \text{ torr}$

11.5 **(a)** 300 K **(b)** 233 K **(c)** 546 K **(d)** −116°C
(e) 142°C **(f)** −273°C

11.6 **(a)** $205 \text{ torr}\left(\frac{636 \text{ K}}{212\text{K}}\right) = 615 \text{ torr}$

(b) $1.40 \text{ atm}\left(\frac{216 \text{ K}}{324 \text{ K}}\right) = 0.933 \text{ atm}$

(c) $575 \text{ torr}\left(\frac{400 \text{ K}}{300 \text{ K}}\right) = 767 \text{ torr}$

(d) $330 \text{ K}\left(\frac{3.00 \text{ atm}}{2.00 \text{ atm}}\right) = 495 \text{ K} = 222°\text{C}$

(e) $298 \text{ K}\left(\frac{1.00 \text{ atm}}{0.873 \text{ atm}}\right) = 341 \text{ K} = 68°\text{C}$

(f) $645 \text{ torr}\left(\frac{273 \text{ K}}{373 \text{ K}}\right) = 472 \text{ torr}$

(g) $195 \text{ K}\left(\frac{760 \text{ torr}}{555 \text{ torr}}\right) = 267 \text{ K} = -6°\text{C}$

11.7 **(a)** $50.0 \text{ mL}\left(\frac{400 \text{ K}}{300 \text{ K}}\right) = 66.7 \text{ mL}$

(b) $3.50 \text{ mL}\left(\frac{273 \text{ K}}{373 \text{ K}}\right) = 2.56 \text{ mL}$

(c) $200 \text{ K}\left(\frac{0.700 \text{ L}}{0.400 \text{ L}}\right) = 350 \text{ K}$

(d) $298\text{ K}\left(\frac{1020\text{ mL}}{68.0\text{ mL}}\right) = 4470\text{ K}\ (4.47 \times 10^3\text{ K})$

(e) $0.250\text{ L}\left(\frac{546\text{ K}}{273\text{ K}}\right) = 0.500\text{ L}$

11.8 **(a)** $75.0\text{ mL}\left(\frac{732\text{ torr}}{946\text{ torr}}\right)\left(\frac{273\text{ K}}{298\text{ K}}\right) = 53.2\text{ mL}$

(b) $310\text{ K}\left(\frac{60.0\text{ mL}}{55.0\text{ mL}}\right)\left(\frac{715\text{ torr}}{785\text{ torr}}\right) = 308\text{ K} = 35°\text{C}$

(c) $1.00\text{ atm}\left(\frac{4.20\text{ L}}{5.00\text{ L}}\right)\left(\frac{200\text{ K}}{415\text{ K}}\right) = 0.405\text{ atm}$

(d) $17.5\text{ L}\left(\frac{1.40\text{ atm}}{0.916\text{ atm}}\right)\left(\frac{273\text{ K}}{384\text{ K}}\right) = 19.0\text{ L}$

(e) $195\text{ K}\left(\frac{0.900\text{ L}}{0.0450\text{ L}}\right)\left(\frac{2.50\text{ atm}}{0.283\text{ atm}}\right)$

$= 34{,}500\text{ K}\ (3.45 \times 10^4\text{ K})$

(f) $463\text{ torr}\left(\frac{0.355\text{ L}}{0.500\text{ L}}\right)\left(\frac{173\text{ K}}{546\text{ K}}\right) = 104\text{ torr}$

11.9 $0.400\text{ L}\left(\frac{1.23\text{ atm}}{1.00\text{ atm}}\right)\left(\frac{273\text{ K}}{330\text{ K}}\right) = 0.407\text{ L}$

11.10 $72.0\text{ mL}\left(\frac{695\text{ torr}}{760\text{ torr}}\right)\left(\frac{273\text{ K}}{400\text{ K}}\right) = 44.9\text{ mL}$

11.11 $1.00\text{ atm}\left(\frac{6.00\text{ L}}{0.455\text{ L}}\right)\left(\frac{263\text{ K}}{350\text{ K}}\right) = 9.91\text{ atm}$

11.12 $295\text{ K}\left(\frac{50.0\text{ mL}}{40.0\text{ mL}}\right)\left(\frac{760\text{ torr}}{666\text{ torr}}\right) = 421\text{ K} = 148°\text{C}$

11.13 **(a)** $5.60\text{ L}\left(\frac{1\text{ mol}}{22.4\text{ L}}\right) = 0.250\text{ mol}$

(b) $V_{STP} = 3.20\text{ L}\left(\frac{273\text{ K}}{298\text{ K}}\right) = 2.93\text{ L}$

$2.93\text{ L}\left(\frac{1\text{ mol}}{22.4\text{ L}}\right) = 0.131\text{ mol}$

(c) $V_{STP} = 45.0\text{ mL}\left(\frac{745\text{ torr}}{760\text{ torr}}\right)\left(\frac{273\text{ K}}{228\text{ K}}\right) = 52.8\text{ mL}$

$= 0.0528\text{ L}$

$0.0528\text{ L}\left(\frac{1\text{ mol}}{22.4\text{ L}}\right) = 0.00236\text{ mol}$

(d) $V_{STP} = 5.00\text{ L}\left(\frac{855\text{ torr}}{760\text{ torr}}\right)\left(\frac{273\text{ K}}{516\text{ K}}\right) = 2.98\text{ L}$

$2.98\text{ L}\left(\frac{1\text{ mol}}{22.4\text{ L}}\right) = 0.133\text{ mol}$

(e) $V_{STP} = 18.0\text{ mL}\left(\frac{17.6\text{ atm}}{1.00\text{ atm}}\right)\left(\frac{273\text{ K}}{90\text{ K}}\right) = 960\text{ mL}$

$= 0.96\text{ L}$

$0.96\text{ L}\left(\frac{1\text{ mol}}{22.4\text{ L}}\right) = 0.043\text{ mol}$

11.14 $V_{STP} = 1.00\text{ mL}\left(\frac{1.00\text{ torr}}{760\text{ torr}}\right)\left(\frac{273\text{ K}}{1000\text{ K}}\right)$

$= 3.59 \times 10^{-4}\text{ mL}$

$3.59 \times 10^{-4}\text{ mL}\left(\frac{1\text{ L}}{1000\text{ mL}}\right)\left(\frac{1\text{ mol}}{22.4\text{ L}}\right)$

$\times \left(\frac{6.02 \times 10^{23}\text{ molecules}}{1\text{ mol}}\right) = 9.65 \times 10^{15}\text{ molecules}$

11.15 **(a)** $\frac{0.0821\text{ L}\cdot\text{atm}}{\text{mol}\cdot\text{K}}\left(\frac{1000\text{ mL}}{1\text{ L}}\right) = \frac{82.1\text{ mL}\cdot\text{atm}}{\text{mol}\cdot\text{K}}$

(b) $\frac{0.0821\text{ L}\cdot\text{atm}}{\text{mol}\cdot\text{K}}\left(\frac{760\text{ torr}}{1\text{ atm}}\right) = \frac{62.4\text{ L}\cdot\text{torr}}{\text{mol}\cdot\text{K}}$

(For Problems 11.16–11.19 use $PV = nRT$.)

11.16 $(745\text{ torr})(V) = 0.275\text{ mol}\left(\frac{62.4\text{ L}\cdot\text{torr}}{\text{mol}\cdot\text{K}}\right)(195\text{ K})$

$V = \frac{(0.275)(62.4)(195)\text{L}}{(745)} = 4.49\text{ L}$

11.17 $(P)(1.00\text{ L}) = 2.00\text{ mol}\left(\frac{0.0821\text{ L}\cdot\text{atm}}{\text{mol}\cdot\text{K}}\right)(295\text{ K})$

$P = \frac{(2.00)(0.0821)(295)\text{atm}}{(1.00)} = 48.4\text{ atm}$

11.18 $0.640\text{ g }O_2\left(\frac{1\text{ mol }O_2}{32.0\text{ g }O_2}\right) = 0.0200\text{ mol }O_2$

$(747\text{ torr})(575\text{ mL}) = 0.0200\text{ mol}$

$\times \left(\frac{6.24 \times 10^4\text{ mL}\cdot\text{torr}}{\text{mol}\cdot\text{K}}\right)(T)$

$\frac{(747)(575)\text{K}}{(0.0200)(6.24 \times 10^4)} = T = 344\text{ K} = 71°\text{C}$

11.19 $(1.15\text{ atm})(276\text{ mL}) = (n)\left(\frac{82.1\text{ mL}\cdot\text{atm}}{\text{mol}\cdot\text{K}}\right)(249\text{ K})$

$\frac{(1.15)(276)\text{mol}}{(82.1)(249)} = n = 0.0155\text{ mol}$

11.20 $6.14\text{ L }C_2H_6\left(\frac{4\text{ L }CO_2}{2\text{ L }C_2H_6}\right) = 12.3\text{ L }CO_2$

11.21 $6.14\text{ L }C_2H_6\left(\frac{7\text{ L }O_2}{2\text{ L }C_2H_6}\right) = 21.5\text{ L }O_2$

11.22 $4.50\text{ g }Ag_2O\left(\frac{1\text{ mol }Ag_2O}{231.8\text{ g }Ag_2O}\right)\left(\frac{1\text{ mol }O_2}{2\text{ mol }Ag_2O}\right)$

$\times \left(\frac{22.4\text{ L }O_2}{1\text{ mol }O_2}\right) = 0.217\text{ L }O_2$

11.23 $1.75\text{ L }C_4H_{10}\left(\frac{8\text{ L }CO_2}{2\text{ L }C_4H_{10}}\right)$

$= 7.00\text{ L }CO_2$ (at 298 K, 765 torr)

$$7.00\ \text{L}\left(\frac{528\ \text{K}}{298\ \text{K}}\right)\left(\frac{765\ \text{torr}}{395\ \text{torr}}\right) = 24.0\ \text{L}$$

11.24 $$V_{STP} = 1.35\ \text{L}\left(\frac{273\ \text{K}}{458\ \text{K}}\right)\left(\frac{635\ \text{torr}}{760\ \text{torr}}\right) = 0.672\ \text{L}$$

$$0.672\ \text{L}\left(\frac{1\ \text{mol O}_2}{22.4\ \text{L}}\right) = 0.0300\ \text{mol O}_2$$

$$0.0300\ \text{mol O}_2\left(\frac{1\ \text{mol S}}{1\ \text{mol O}_2}\right)\left(\frac{32.1\ \text{g S}}{1\ \text{mol S}}\right) = 0.963\ \text{g S}$$

11.25 $$5.00\ \text{g H}_2\text{O}_2\left(\frac{1\ \text{mol H}_2\text{O}_2}{34.0\ \text{g H}_2\text{O}_2}\right)\left(\frac{1\ \text{mol O}_2}{2\ \text{mol H}_2\text{O}_2}\right) = 0.0735\ \text{mol O}_2$$

$$0.0735\ \text{mol O}_2\left(\frac{22.4\ \text{L O}_2}{1\ \text{mol O}_2}\right) = 1.65\ \text{L O}_2\ \text{(at STP)}$$

$$V_f = 1.65\ \text{L}\left(\frac{296\ \text{K}}{273\ \text{K}}\right)\left(\frac{760\ \text{torr}}{712\ \text{torr}}\right) = 1.91\ \text{L}$$

11.26 $$V_{STP} = 6.00\ \text{L}\left(\frac{732\ \text{torr}}{760\ \text{torr}}\right)\left(\frac{273\ \text{K}}{373\ \text{K}}\right) = 4.23\ \text{L}$$

$$4.23\ \text{L}\left(\frac{1\ \text{mol}}{22.4\ \text{L}}\right) = 0.189\ \text{mol}$$

$$\text{Molar mass} = \frac{16.0\ \text{g}}{0.189\ \text{mol}} = 84.7\ \text{g/mol}$$

11.27 $$V_{STP} = 1.50\ \text{L}\left(\frac{777\ \text{torr}}{760\ \text{torr}}\right)\left(\frac{273\ \text{K}}{218\ \text{K}}\right) = 1.92\ \text{L}$$

$$1.92\ \text{L}\left(\frac{1\ \text{mol}}{22.4\ \text{L}}\right) = 0.0857\ \text{mol}$$

$$\text{Molar mass} = \frac{2.40\ \text{g}}{0.0857\ \text{mol}} = 28.0\ \text{g/mol}$$

11.28 $$V_{STP} = 0.500\ \text{L}\left(\frac{0.900\ \text{atm}}{1.00\ \text{atm}}\right)\left(\frac{273\ \text{K}}{283\ \text{K}}\right) = 0.434\ \text{L}$$

$$0.434\ \text{L}\left(\frac{1\ \text{mol}}{22.4\ \text{L}}\right) = 0.0194\ \text{mol}$$

$$\text{Molar mass} = \frac{2.67\ \text{g}}{0.0194\ \text{mol}} = 138\ \text{g/mol}$$

11.29 815 torr

11.30 Total number of moles = 1.00 mol

$$P_{He} = \frac{0.25}{1.00}(2.00\ \text{atm}) = 0.50\ \text{atm}$$

$$P_{Ar} = \frac{0.33}{1.00}(2.00\ \text{atm}) = 0.66\ \text{atm}$$

$$P_{Ne} = \frac{0.42}{1.00}(2.00\ \text{atm}) = 0.84\ \text{atm}$$

11.31 Total number of moles = 10.00 mol

$$P_{O_2} = \frac{1.00}{10.00}(1.20\ \text{atm}) = 0.120\ \text{atm}$$

$$P_{H_2} = \frac{2.00}{10.00}(1.20\ \text{atm}) = 0.240\ \text{atm}$$

$$P_{N_2} = \frac{7.00}{10.00}(1.20\ \text{atm}) = 0.840\ \text{atm}$$

11.32 $$2.80\ \text{g N}_2\left(\frac{1\ \text{mol N}_2}{28.0\ \text{g N}_2}\right) = 0.100\ \text{mol N}_2$$

$$6.40\ \text{g O}_2\left(\frac{1\ \text{mol O}_2}{32.0\ \text{g O}_2}\right) = 0.200\ \text{mol O}_2$$

Total moles = 0.100 mol + 0.200 mol = 0.300 mol

$$P_{N_2} = \frac{0.100}{0.300}(720\ \text{torr}) = 240\ \text{torr}$$

$$P_{O_2} = \frac{0.200}{0.300}(720\ \text{torr}) = 480\ \text{torr}$$

11.33 P_{H_2} = 770.0 torr − 21.1 torr = 748.9 torr

11.34 P_{N_2} = 758.8 torr − 19.8 torr = 739.0 torr

$$0.250\ \text{L}\left(\frac{739.0\ \text{torr}}{760.0\ \text{torr}}\right)\left(\frac{273\ \text{K}}{295\ \text{K}}\right) = 0.225\ \text{L}$$

$$0.225\ \text{L}\left(\frac{1\ \text{mol}}{22.4\ \text{L}}\right) = 0.0100\ \text{mol}$$

Chapter 12

12.1 **(a)** Trigonal planar **(b)** Angular **(c)** Trigonal pyramidal **(d)** Tetrahedral **(e)** Linear

12.2 **(a)** Nonpolar **(b)** Polar **(c)** Polar **(d)** Nonpolar **(e)** Polar **(f)** Nonpolar **(g)** Nonpolar **(h)** Polar

12.3 **(a)** 3 **(b)** 1 **(c)** 3 **(d)** 2 **(e)** 1 **(f)** 2 **(g)** 1

12.4 **(a)** 2 **(b)** 4 **(c)** 1 **(d)** 3

12.5 $CH_4 < CH_3Cl < CH_3OH$

12.6 **(a)** NaCl (ionic) is higher boiling than CH_3CH_2Cl (dipole–dipole), because ionic interactions are stronger than dipole–dipole interactions.
(b) CH_3OCH_3 (dipole–dipole) is higher boiling than $CH_3CH_2CH_3$ (London forces), because dipole–dipole interactions are stronger than London forces.
(c) $CH_3CH_2CH_2CH_3$ (London forces) is higher boiling than $CH_3CH_2CH_3$ (London forces), because it has a higher molecular mass.
(d) ICl (dipole–dipole) is higher boiling than Br_2 (London forces), because dipole–dipole interactions are stronger than London forces.

(e) $HOCH_2CH_2OH$ (hydrogen bonding) is higher boiling than $CH_3CH_2CH_2OH$ (hydrogen bonding), because it has two groups that hydrogen-bond, rather than one.

12.7, 12.8, and 12.9

20°C (Room Temperature)	−50°C	−100°C	100°C
(a) Liquid	Solid	Solid	Gas
(b) Solid	Solid	Solid	Solid
(c) Gas	Liquid	Liquid	Gas
(d) Liquid	Liquid	Liquid	Gas
(e) Liquid	Solid	Solid	Gas
(f) Liquid	Solid	Solid	Gas
(g) Liquid	Solid	Solid	Liquid
(h) Solid	Solid	Solid	Liquid
(i) Gas	Gas	Solid	Gas
(j) Gas	Liquid	Solid	Gas

12.10 $55.0\text{ g}\left(\frac{4.184\text{ J}}{\text{g}\cdot°\text{C}}\right)53.0°\text{C} = 12{,}200\text{ J} = 12.2\text{ kJ}$

12.11 $155\text{ g}\left(\frac{0.449\text{ J}}{\text{g}\cdot°\text{C}}\right)78.0°\text{C} = 5430\text{ J} = 5.43\text{ kJ}$

12.12 $75.0\text{ g}\left(\frac{0.897\text{ J}}{\text{g}\cdot°\text{C}}\right)83.0°\text{C} = 5580\text{ J} = 5.58\text{ kJ}$

12.13 $17.4\text{ g}\left(\frac{0.334\text{ kJ}}{1\text{ g}}\right) = 5.81\text{ kJ}$

12.14 $7.82\text{ mol}\left(\frac{6.01\text{ kJ}}{1\text{ mol}}\right) = 47.0\text{ kJ}$

12.15 $525\text{ g}\left(\frac{2.26\text{ kJ}}{1\text{ g}}\right) = 1190\text{ kJ} (= 1.19 \times 10^3\text{ kJ})$

12.16 $6.73\text{ g}\left(\frac{2.26\text{ kJ}}{1\text{ g}}\right) = 15.2\text{ kJ}$

12.17 **(a)** The substance with the highest vapor pressure is the one with the lowest boiling point: *A* (53°C).
(b) The substance with the lowest vapor pressure is the one with the highest boiling point: *B* (165°C).

12.18 **(a)** The one with the lowest vapor pressure has the highest boiling point: *E* (13 torr).
(b) At room temperature, *F* is a gas, because its vapor pressure (960 torr) exceeds atmospheric pressure.
(c) We do not have enough information to determine whether any are solids.

12.19 Water will be on top, because it is less dense.

12.20 Since water is a molecular substance, a snowflake must be a molecular crystal.

12.21 When water freezes, a rigid network of hydrogen bonds forms, creating considerable space between the molecules. Because density and volume are inversely proportional, this expansion of volume produces a decrease in density. Hence, the solid is less dense than the liquid.

Chapter 13

13.1 3 mol; 1 mol $MgCl_2$ separates into 1 mol of Mg^{2+} and 2 mol Cl^-.

13.2 $32.6\text{ g} - 11.2\text{ g} = 21.4\text{ g}$ (dissolves in 40.0 mL H_2O)

$$100\text{ mL H}_2\text{O}\left(\frac{21.4\text{ g}}{40.0\text{ mL H}_2\text{O}}\right) = 53.5\text{ g}$$

The solubility is 53.5 g/100 mL H_2O.

13.3 AgCl *(i)*, $Ca(OH)_2$ *(ss)*, $HgCl_2$ *(s)*, NaCl *(vs)*, KI *(vs)*, $C_{12}H_{22}O_{11}$ *(vs)*

13.4 **(a)** $444\text{ g NaOH}\left(\frac{1\text{ mol NaOH}}{40.0\text{ g NaOH}}\right) = 11.1\text{ mol NaOH}$

$$\frac{11.1\text{ mol NaOH}}{5.00\text{ L soln}} = 2.22\text{ M NaOH}$$

(b) $4.04\text{ g KNO}_3\left(\frac{1\text{ mol KNO}_3}{101.1\text{ g KNO}_3}\right) = 0.0400\text{ mol KNO}_3$

$$\frac{0.0400\text{ mol KNO}_3}{0.200\text{ L soln}} = 0.200\text{ M KNO}_3$$

(c) $39.0\text{ g NaBr}\left(\frac{1\text{ mol NaBr}}{102.9\text{ g NaBr}}\right) = 0.379\text{ mol NaBr}$

$$\frac{0.379\text{ mol NaBr}}{0.500\text{ L soln}} = 0.758\text{ M NaBr}$$

(d) $2.22\text{ g CaCl}_2\left(\frac{1\text{ mol CaCl}_2}{111.1\text{ g CaCl}_2}\right) = 0.0200\text{ mol CaCl}_2$

$$\frac{0.0200\text{ mol CaCl}_2}{0.0800\text{ L soln}} = 0.250\text{ M CaCl}_2$$

(e) $0.385\text{ g MgBr}_2\left(\frac{1\text{ mol MgBr}_2}{184.1\text{ g MgBr}_2}\right)$

$$= 0.00209\text{ mol MgBr}_2$$

$$\frac{0.00209\text{ mol MgBr}_2}{0.00500\text{ L soln}} = 0.418\text{ M MgBr}_2$$

(f) $7.50\text{ g KI}\left(\frac{1\text{ mol KI}}{166.0\text{ g KI}}\right) = 0.0452\text{ mol KI}$

$$\frac{0.0452\text{ mol KI}}{0.0400\text{ L soln}} = 1.13\text{ M KI}$$

(g) $4.36\text{ g Na}_2\text{SO}_4\left(\frac{1\text{ mol Na}_2\text{SO}_4}{142.1\text{ g Na}_2\text{SO}_4}\right)$

$$= 0.0307\text{ mol Na}_2\text{SO}_4$$

$$\frac{0.0307 \text{ mol } Na_2SO_4}{0.350 \text{ L soln}} = 0.0877 \text{ M } Na_2SO_4$$

(h) $12.3 \text{ g } Ca(NO_3)_2\left(\frac{1 \text{ mol } Ca(NO_3)_2}{164.1 \text{ g } Ca(NO_3)_2}\right) = 0.0750 \text{ mol } Ca(NO_3)_2$

$$\frac{0.0750 \text{ mol } Ca(NO_3)_2}{0.640 \text{ L soln}} = 0.117 \text{ M } Ca(NO_3)_2$$

(i) $2.65 \text{ g } CuSO_4\left(\frac{1 \text{ mol } CuSO_4}{159.6 \text{ g } CuSO_4}\right) = 0.0166 \text{ mol } CuSO_4$

$$\frac{0.0166 \text{ mol } CuSO_4}{0.0450 \text{ L soln}} = 0.369 \text{ M } CuSO_4$$

(j) $6.25 \text{ g } NH_4Cl\left(\frac{1 \text{ mol } NH_4Cl}{53.5 \text{ g } NH_4Cl}\right) = 0.117 \text{ mol } NH_4Cl$

$$\frac{0.117 \text{ mol } NH_4Cl}{0.0350 \text{ L soln}} = 3.34 \text{ M } NH_4Cl$$

13.5 **(a)** $3.00 \text{ L soln}\left(\frac{4.00 \text{ mol KOH}}{\text{L soln}}\right) = 12.0 \text{ mol KOH}$

(b) $1.50 \text{ L soln}\left(\frac{2.40 \text{ mol } LiNO_3}{\text{L soln}}\right) = 3.60 \text{ mol } LiNO_3$

(c) $0.500 \text{ L soln}\left(\frac{0.600 \text{ mol } CaCl_2}{\text{L soln}}\right) = 0.300 \text{ mol } CaCl_2$

(d) $0.0450 \text{ L soln}\left(\frac{1.60 \text{ mol } FeCl_3}{\text{L soln}}\right) = 0.0720 \text{ mol } FeCl_3$

(e) $0.150 \text{ L soln}\left(\frac{0.460 \text{ mol } CuSO_4}{\text{L soln}}\right) = 0.0690 \text{ mol } CuSO_4$

(f) $0.750 \text{ L soln}\left(\frac{0.0440 \text{ mol } MgBr_2}{\text{L soln}}\right) = 0.0330 \text{ mol } MgBr_2$

(g) $0.00350 \text{ L soln}\left(\frac{3.00 \text{ mol } Na_2CrO_4}{\text{L soln}}\right) = 0.0105 \text{ mol } Na_2CrO_4$

(h) $0.0250 \text{ L soln}\left(\frac{0.645 \text{ mol } Ba(NO_3)_2}{\text{L soln}}\right) = 0.0161 \text{ mol } Ba(NO_3)_2$

(i) $0.0020 \text{ L soln}\left(\frac{12 \text{ mol HCl}}{\text{L soln}}\right) = 0.024 \text{ mol HCl}$

(j) $0.025 \text{ L soln}\left(\frac{18 \text{ mol } H_2SO_4}{\text{L soln}}\right) = 0.45 \text{ mol } H_2SO_4$

13.6 See the answers to Problem 13.5 for the calculation of the moles of solute. To prepare each solution, dissolve the calculated mass of solute shown here *in sufficient water to achieve the desired total volume.*

(a) $12.0 \text{ mol KOH}\left(\frac{56.1 \text{ KOH}}{1 \text{ mol KOH}}\right) = 673 \text{ g KOH (in 3.00 L)}$

(b) $3.60 \text{ mol } LiNO_3\left(\frac{68.9 \text{ g } LiNO_3}{1 \text{ mol } LiNO_3}\right) = 248 \text{ g } LiNO_3 \text{ (in 1.50 L)}$

(c) $0.300 \text{ mol } CaCl_2\left(\frac{111.1 \text{ g } CaCl_2}{1 \text{ mol } CaCl_2}\right) = 33.3 \text{ g } CaCl_2 \text{ (in 0.500 L)}$

(d) $0.0720 \text{ mol } FeCl_3\left(\frac{162.3 \text{ g } FeCl_3}{1 \text{ mol } FeCl_3}\right) = 11.7 \text{ g } FeCl_3 \text{ (in 45.0 mL)}$

(e) $0.0690 \text{ mol } CuSO_4\left(\frac{159.6 \text{ g } CuSO_4}{1 \text{ mol } CuSO_4}\right) = 11.0 \text{ g } CuSO_4 \text{ (in 0.150 L)}$

(f) $0.0330 \text{ mol } MgBr_2\left(\frac{184.1 \text{ g } MgBr_2}{1 \text{ mol } MgBr_2}\right) = 6.08 \text{ g } MgBr_2 \text{ (in 0.750 L)}$

(g) $0.0105 \text{ mol } Na_2CrO_4\left(\frac{162.0 \text{ g } Na_2CrO_4}{1 \text{ mol } Na_2CrO_4}\right) = 1.70 \text{ g } Na_2CrO_4 \text{ (in 3.50 mL)}$

(h) $0.0161 \text{ mol } Ba(NO_3)_2\left(\frac{261.3 \text{ g } Ba(NO_3)_2}{1 \text{ mol } Ba(NO_3)_2}\right) = 4.21 \text{ g } Ba(NO_3)_2 \text{ (in 25.0 mL)}$

13.7 **(a)** $0.50 \text{ mol NaOH}\left(\frac{1 \text{ L soln}}{6.0 \text{ mol NaOH}}\right) = 0.083 \text{ L soln (or 83 mL soln)}$

(b) $2.5 \text{ mol } NH_3\left(\frac{1 \text{ L soln}}{15 \text{ mol } NH_3}\right) = 0.17 \text{ L soln (or } 1.7 \times 10^2 \text{ mL soln)}$

(c) $0.125 \text{ mol KCl}\left(\frac{1 \text{ L soln}}{2.45 \text{ mol KCl}}\right) = 0.0510 \text{ L soln (or 51.0 mL soln)}$

(d) $0.0575 \text{ mol } H_2C_2O_4\left(\frac{1 \text{ L soln}}{0.785 \text{ mol } H_2C_2O_4}\right) = 0.0732 \text{ L soln (or 73.2 mL soln)}$

(e) $0.00350 \text{ mol } CaCl_2\left(\frac{1 \text{ L soln}}{0.120 \text{ mol } CaCl_2}\right) = 0.0292 \text{ L soln (or 29.2 mL soln)}$

13.8 **(a)** $V_{con} = 500 \text{ mL soln}\left(\frac{3.0 \text{ M}}{12 \text{ M}}\right) = 125 \text{ mL soln}$

Dilute 125 mL of 12 M HCl to 500 mL.

(b) $V_{con} = 250\text{ mL soln}\left(\frac{1.0\text{ M}}{18\text{ M}}\right) = 14\text{ mL soln}$

Dilute 14 mL of 18 M H_2SO_4 to 250 mL.

(c) $V_{con} = 75\text{ mL soln}\left(\frac{0.60\text{ M}}{3.0\text{ M}}\right) = 15\text{ mL soln}$

Dilute 15 mL of 3.0 M NaOH to 75 mL.

(d) $V_{con} = 150\text{ mL soln}\left(\frac{0.10\text{ M}}{3.0\text{ M}}\right) = 5.0\text{ mL soln}$

Dilute 5.0 mL of 3.0 M KI to 150 mL.

(e) $V_{con} = 5.0\text{ L soln}\left(\frac{0.25\text{ M}}{15\text{ M}}\right) = 0.083\text{ L soln}$

$= 83\text{ mL soln}$

Dilute 83 mL of 15 M NH_3 to 5.0 L.

13.9 **(a)** $M_{dil} = \frac{(75\text{ mL soln})(18\text{ M})}{(350\text{ mL soln})} = 3.9\text{ M}$

(b) $M_{dil} = \frac{(25\text{ mL soln})(2.4\text{ M})}{(750\text{ mL soln})} = 0.080\text{ M}$

(c) $M_{dil} = \frac{(10.0\text{ mL soln})(1.25\text{ M})}{(60.0\text{ mL soln})} = 0.208\text{ M}$

(d) $M_{dil} = \frac{(0.0350\text{ L soln})(2.00\text{ M})}{(0.500\text{ L soln})} = 0.140\text{ M}$

(e) $M_{dil} = \frac{(5.00\text{ mL soln})(6.00\text{ M})}{(1250\text{ mL soln})} = 0.0240\text{ M}$

13.10 **(a)** $\frac{5.0\text{ mL}}{30.0\text{ mL}} \times 100\% = 17\%$

(b) $\frac{15\text{ mL}}{60\text{ mL}} \times 100\% = 25\%$

(c) $\frac{3.0\text{ g}}{60.0\text{ g}} \times 100\% = 5.0\%$

(d) $\frac{14\text{ g}}{250\text{ g}} \times 100\% = 5.6\%$

13.11 **(a)** 7.0% of 500 mL: (0.070)(500 mL) = 35 mL
Add 35 mL of acetic acid to enough water to make 500 mL of solution.

(b) 7.0% of 500 g: (0.070)(500 g) = 35 g
Add 35 g of acetic acid to 465 g of water.

(c) 12% of 250 mL: (0.12)(250 mL)
$= 30\text{ mL } (3.0 \times 10^1\text{ mL})$
Add 30 mL of ethyl alcohol to enough water to make 250 mL of solution.

(d) 6.0% of 750 g: (0.060)(750 g) = 45 g
Add 45 g of H_3BO_3 to 705 g of water.

(e) 2.0% of 25 g: (0.020)(25 g) = 0.50 g
Add 0.50 g of $NaHCO_3$ to 24.5 g of water.

13.12 $\frac{2.1 \times 10^{-3}\text{ g}}{15 \times 10^3\text{ g}} \times 10^6\text{ ppm} = 0.14\text{ ppm}$

No. This level exceeds the recommended limit of 0.05 ppm.

13.13 $0.0645\text{ L soln}\left(\frac{0.125\text{ mol } H_2SO_4}{\text{L soln}}\right)\left(\frac{1\text{ mol } BaSO_4}{1\text{ mol } H_2SO_4}\right)$

$\times \left(\frac{233.4\text{ g } BaSO_4}{1\text{ mol } BaSO_4}\right) = 1.88\text{ g } BaSO_4$

13.14 $0.0550\text{ L soln}\left(\frac{0.250\text{ mol } AgNO_3}{\text{L soln}}\right)\left(\frac{1\text{ mol } Ag_2S}{2\text{ mol } AgNO_3}\right)$

$\times \left(\frac{247.9\text{ g } Ag_2S}{1\text{ mol } Ag_2S}\right) = 1.70\text{ g } Ag_2S$

13.15 $3.55\text{ g } CaCO_3\left(\frac{1\text{ mol } CaCO_3}{100.1\text{ g } CaCO_3}\right)\left(\frac{2\text{ mol } HNO_3}{1\text{ mol } CaCO_3}\right)$

$= 0.0709\text{ mol } HNO_3$

$0.0709\text{ mol } HNO_3\left(\frac{1\text{ L soln}}{0.275\text{ mol } HNO_3}\right) = 0.258\text{ L soln}$

$= 258\text{ mL soln}$

13.16 $0.0400\text{ L soln}\left(\frac{0.350\text{ mol } Na_3PO_4}{\text{L soln}}\right)\left(\frac{3\text{ mol } Pb(NO_3)_2}{2\text{ mol } Na_3PO_4}\right)$

$= 0.0210\text{ mol } Pb(NO_3)_2$

$0.0210\text{ mol } Pb(NO_3)_2\left(\frac{1\text{ L soln}}{0.125\text{ mol } Pb(NO_3)_2}\right)$

$= 0.168\text{ L soln } (= 168\text{ mL soln})$

13.17 **(a)** $0.0160\text{ L soln}\left(\frac{0.150\text{ mol HCl}}{\text{L soln}}\right)\left(\frac{1\text{ mol } H_2}{2\text{ mol HCl}}\right)$

$\times \left(\frac{22.4\text{ L } H_2}{1\text{ mol } H_2}\right) = 0.0269\text{ L } H_2 \text{ } (= 26.9\text{ mL } H_2)$

(b) $26.9\text{ mL } H_2\left(\frac{308\text{ K}}{273\text{ K}}\right)\left(\frac{1.00\text{ atm}}{1.15\text{ atm}}\right) = 26.4\text{ mL } H_2$

13.18 $0.0350\text{ L soln}\left(\frac{0.440\text{ mol HCl}}{\text{L soln}}\right)\left(\frac{1\text{ mol NaOH}}{1\text{ mol HCl}}\right)$

$= 0.0154\text{ mol NaOH}$

$\frac{0.0154\text{ mol NaOH}}{0.0250\text{ L soln}} = 0.616\text{ M NaOH}$

13.19 $0.0250\text{ L soln}\left(\frac{0.210\text{ mol KOH}}{\text{L soln}}\right)\left(\frac{1\text{ mol } H_3PO_4}{3\text{ mol KOH}}\right)$

$= 0.00175\text{ mol } H_3PO_4$

$\frac{0.00175\text{ mol } H_3PO_4}{0.0140\text{ L soln}} = 0.125\text{ M } H_3PO_4$

13.20 **(a)** $\Delta T = \left(\frac{1.86°\text{C}}{m}\right)1.20\, m = 2.23°\text{C}$

$T_f = 0.00°\text{C} - 2.23°\text{C} = -2.23°\text{C}$

(b) $\Delta T = \left(\frac{3.9°C}{m}\right)2.34\ m = 9.1°C$

$T_f = 16.6°C - 9.1°C = 7.5°C$

(c) $\Delta T = \left(\frac{40.0°C}{m}\right)0.670\ m = 26.8°C$

$T_f = 17.6°C - 26.8°C = -9.2°C$

(d) $\Delta T = \left(\frac{4.9°C}{m}\right)0.23\ m = 1.1°C$

$T_f = 5.5°C - 1.1°C = 4.4°C$

(e) $\Delta T = \left(\frac{6.8°C}{m}\right)0.78\ m = 5.3°C$

$T_f = 80.2°C - 5.3°C = 74.9°C$

13.21 (a) The water will flow to the right.
(b) Because 0.9% NaCl is isotonic with 5% glucose and 5% NaCl is more concentrated than 0.9% NaCl, water will flow to the left.

13.22 It does not settle out upon standing, and it is milky in appearance.

Chapter 14

14.1 (a) Combustion (b) Double-replacement
(c) Decomposition (d) Synthesis
(e) Single-replacement

14.2 (a) $Pb(NO_3)_2(aq) + BaCl_2(aq) \rightarrow PbCl_2(s) + Ba(NO_3)_2(aq)$
(b) $H_2SO_4(aq) + 2\ KOH(aq) \rightarrow K_2SO_4(aq) + 2\ H_2O(\ell)$

14.3 (a) Nonelectrolyte (b) Electrolyte
(c) Weak electrolyte (d) Electrolyte
(e) Weak electrolyte (f) Nonelectrolyte

14.4 $HBr(g) + H_2O(\ell) \rightarrow H_3O^+(aq) + Br^-(aq)$

14.5 1:1 ratio

14.6 (a) $HCl + KOH \rightarrow KCl + H_2O$
(b) $HNO_3 + NaOH \rightarrow NaNO_3 + H_2O$
(c) $2\ HBr + Ca(OH)_2 \rightarrow CaBr_2 + 2\ H_2O$
(d) $H_3PO_4 + 3\ KOH \rightarrow K_3PO_4 + 3\ H_2O$
(e) $H_2SO_4 + Ca(OH)_2 \rightarrow CaSO_4 + 2\ H_2O$

14.7 (a) $HI + NaOH \rightarrow NaI + H_2O$
(b) $HBr + LiOH \rightarrow LiBr + H_2O$
(c) $2\ HCl + Ca(OH)_2 \rightarrow CaCl_2 + 2\ H_2O$
(d) $H_3PO_4 + 3\ NaOH \rightarrow Na_3PO_4 + 3\ H_2O$
(e) $2\ HI + Ba(OH)_2 \rightarrow BaI_2 + 2\ H_2O$

14.8 (a) $HgO + 2\ HCl \rightarrow HgCl_2 + H_2O$
(b) $Na_2O + 2\ HI \rightarrow 2\ NaI + H_2O$
(c) $2\ HBr + CaCO_3 \rightarrow CaBr_2 + H_2O + CO_2$
(d) $2\ HNO_3 + MgCO_3 \rightarrow Mg(NO_3)_2 + H_2O + CO_2$
(e) $HNO_3 + NaHCO_3 \rightarrow NaNO_3 + H_2O + CO_2$
(f) $HI + LiHCO_3 \rightarrow LiI + H_2O + CO_2$

14.9 (a) $Zn^{2+}(aq)$ and $Cl^-(aq)$ in a 1:2 ratio
(b) $Li^+(aq)$ and $SO_4^{2-}(aq)$ in a 2:1 ratio
(c) $Fe^{3+}(aq)$ and $NO_3^-(aq)$ in a 1:3 ratio
(d) $NH_4^+(aq)$ and $CO_3^{2-}(aq)$ in a 2:1 ratio
(e) $Ba^{2+}(aq)$ and $I^-(aq)$ in a 1:2 ratio

14.10 (a) $HClO_4(aq) + KOH(aq) \rightarrow KClO_4(aq) + H_2O(\ell)$
$H^+(aq) + ClO_4^-(aq) + K^+(aq) + OH^-(aq) \rightarrow K^+(aq) + ClO_4^-(aq) + H_2O(\ell)$
$H^+(aq) + OH^-(aq) \rightarrow H_2O(\ell)$
(b) $2\ HBr(aq) + Ba(OH)_2(aq) \rightarrow BaBr_2(aq) + 2\ H_2O(\ell)$
$2\ H^+(aq) + 2\ Br^-(aq) + Ba^{2+}(aq) + 2\ OH^-(aq) \rightarrow Ba^{2+}(aq) + 2\ Br^-(aq) + 2\ H_2O(\ell)$
$H^+(aq) + OH^-(aq) \rightarrow H_2O(\ell)$
(c) $2\ HI(aq) + Ca(OH)_2(aq) \rightarrow CaI_2(aq) + 2\ H_2O(\ell)$
$2\ H^+(aq) + 2\ I^-(aq) + Ca^{2+}(aq) + 2\ OH^-(aq) \rightarrow Ca^{2+}(aq) + 2\ I^-(aq) + 2\ H_2O(\ell)$
$H^+(aq) + OH^-(aq) \rightarrow H_2O(\ell)$

14.11 (a) $AgNO_3(aq) + NH_4Cl(aq) \rightarrow AgCl(s) + NH_4NO_3(aq)$
$Ag^+(aq) + NO_3^-(aq) + NH_4^+(aq) + Cl^-(aq) \rightarrow AgCl(s) + NH_4^+(aq) + NO_3^-(aq)$
$Ag^+(aq) + Cl^-(aq) \rightarrow AgCl(s)$
(b) $2\ AgNO_3(aq) + Na_2CO_3(aq) \rightarrow Ag_2CO_3(s) + 2\ NaNO_3(aq)$
$2\ Ag^+(aq) + 2\ NO_3^-(aq) + 2\ Na^+(aq) + CO_3^{2-}(aq) \rightarrow Ag_2CO_3(s) + 2\ Na^+(aq) + 2\ NO_3^-(aq)$
$2\ Ag^+(aq) + CO_3^{2-}(aq) \rightarrow Ag_2CO_3(s)$
(c) $Sr(NO_3)_2(aq) + K_2SO_4(aq) \rightarrow SrSO_4(s) + 2\ KNO_3(aq)$
$Sr^{2+}(aq) + 2\ NO_3^-(aq) + 2\ K^+(aq) + SO_4^{2-}(aq) \rightarrow SrSO_4(s) + 2\ K^+(aq) + 2\ NO_3^-(aq)$
$Sr^{2+}(aq) + SO_4^{2-}(aq) \rightarrow SrSO_4(s)$

14.12 (a) $K_2CO_3(aq) + 2\ HClO_4(aq) \rightarrow 2\ KClO_4(aq) + H_2O(\ell) + CO_2(g)$
$2\ K^+(aq) + CO_3^{2-}(aq) + 2\ H^+(aq) + 2\ ClO_4^-(aq) \rightarrow 2\ K^+(aq) + 2\ ClO_4^-(aq) + H_2O(\ell) + CO_2(g)$
$2\ H^+(aq) + CO_3^{2-}(aq) \rightarrow H_2O(\ell) + CO_2(g)$
(b) $Mg(s) + 2\ HNO_3(aq) \rightarrow Mg(NO_3)_2(aq) + H_2(g)$
$Mg(s) + 2\ H^+(aq) + 2\ NO_3^-(aq) \rightarrow Mg^{2+}(aq) + 2\ NO_3^-(aq) + H_2(g)$
$Mg(s) + 2\ H^+(aq) \rightarrow Mg^{2+}(aq) + H_2(g)$

(c) $2\,Al(s) + 6\,HCl(aq) \rightarrow 2\,AlCl_3(aq) + 3\,H_2(g)$
$2\,Al(s) + 6\,H^+(aq) + 6\,Cl^-(aq) \rightarrow 2\,Al^{3+}(aq) + 6\,Cl^-(aq) + 3\,H_2(g)$
$2\,Al(s) + 6\,H^+(aq) \rightarrow 2\,Al^{3+}(aq) + 3\,H_2(g)$

(d) $BaCO_3(s) + 2\,HNO_3(aq) \rightarrow Ba(NO_3)_2(aq) + H_2O(\ell) + CO_2(g)$
$BaCO_3(s) + 2\,H^+(aq) + 2\,NO_3^-(aq) \rightarrow Ba^{2+}(aq) + 2\,NO_3^-(aq) + H_2O(\ell) + CO_2(g)$
$BaCO_3(s) + 2\,H^+(aq) \rightarrow Ba^{2+}(aq) + H_2O(\ell) + CO_2(g)$

14.13 (a) $Cu^{2+}(aq) + 2\,NO_3^-(aq) + Mg(s) \rightarrow Mg^{2+}(aq) + 2\,NO_3^-(aq) + Cu(s)$
$Cu^{2+}(aq) + Mg(s) \rightarrow Mg^{2+}(aq) + Cu(s)$

(b) $2\,Ag^+(aq) + 2\,NO_3^-(aq) + Zn(s) \rightarrow Zn^{2+}(aq) + 2\,NO_3^-(aq) + 2\,Ag(s)$
$2\,Ag^+(aq) + Zn(s) \rightarrow Zn^{2+}(aq) + 2\,Ag(s)$

(c) $3\,Pb^{2+}(aq) + 6\,NO_3^-(aq) + 2\,Al(s) \rightarrow 2\,Al^{3+}(aq) + 6\,NO_3^-(aq) + 3\,Pb(s)$
$3\,Pb^{2+}(aq) + 2\,Al(s) \rightarrow 2\,Al^{3+}(aq) + 3\,Pb(s)$

14.14 (a) $NaNO_2(aq) + HCl(aq) \rightarrow NaCl(aq) + HNO_2(aq)$
$Na^+(aq) + NO_2^-(aq) + H^+(aq) + Cl^-(aq) \rightarrow Na^+(aq) + Cl^-(aq) + HNO_2(aq)$
$H^+(aq) + NO_2^-(aq) \rightarrow HNO_2(aq)$

(b) $KF(aq) + HNO_3(aq) \rightarrow KNO_3(aq) + HF(aq)$
$K^+(aq) + F^-(aq) + H^+(aq) + NO_3^-(aq) \rightarrow K^+(aq) + NO_3^-(aq) + HF(aq)$
$H^+(aq) + F^-(aq) \rightarrow HF(aq)$

(c) $KCN(aq) + HCl(aq) \rightarrow KCl(aq) + HCN(aq)$
$K^+(aq) + CN^-(aq) + H^+(aq) + Cl^-(aq) \rightarrow K^+(aq) + Cl^-(aq) + HCN(aq)$
$H^+(aq) + CN^-(aq) \rightarrow HCN(aq)$

14.15 (a) $2\,KI(aq) + Pb(NO_3)_2(aq) \rightarrow PbI_2(s) + 2\,KNO_3(aq)$

(b) $2\,HClO_4(aq) + Ca(OH)_2(aq) \rightarrow Ca(ClO_4)_2(aq) + 2\,H_2O(\ell)$

(c) No reaction occurs; the ions mix.

(d) $HBr(aq) + KC_2H_3O_2(aq) \rightarrow HC_2H_3O_2(aq) + KBr(aq)$

(e) $2\,HI(aq) + Li_2CO_3(aq) \rightarrow 2\,LiI(aq) + H_2O(\ell) + CO_2(g)$

(f) $NH_4Cl(aq) + NaOH(aq) \rightarrow NaCl(aq) + H_2O(\ell) + NH_3(aq)$

Chapter 15

15.1 (a) $[OH^-] = 0.30$ M **(b)** $[H^+] = 0.50$ M
(c) $[OH^-] = 0.65$ M **(d)** $[H^+] = 0.73$ M

15.2 (a) $[OH^-] = \dfrac{1.0 \times 10^{-14}}{1.0 \times 10^{-10}} = 1.0 \times 10^{-4}$ M

(b) $[H^+] = \dfrac{1.0 \times 10^{-14}}{1.0 \times 10^{-6}} = 1.0 \times 10^{-8}$ M

(c) $[OH^-] = \dfrac{1.0 \times 10^{-14}}{2.0 \times 10^{-5}} = 5.0 \times 10^{-10}$ M

(d) $[H^+] = \dfrac{1.0 \times 10^{-14}}{6.5 \times 10^{-4}} = 1.5 \times 10^{-11}$ M

15.3 (a) pOH = 7.00; neutral
(b) pOH = 6.50; weakly basic
(c) pOH = 8.20; weakly acidic
(d) pOH = 2.30; strongly basic
(e) pOH = 11.40; strongly acidic

15.4 (a) $pH = -\log[1.0 \times 10^{-4}] = -(-4.00) = 4.00$; pOH = 10.00
(b) $pOH = -\log[1.0 \times 10^{-4}] = -(-4.00) = 4.00$; pH = 10.00
(c) $pH = -\log[1.0 \times 10^{-2}] = -(-2.00) = 2.00$; pOH = 12.00
(d) $pOH = -\log[1.0 \times 10^{-5}] = -(-5.00) = 5.00$; pH = 9.00
(e) $pH = -\log[1.0 \times 10^{0}] = -(0.00) = 0.00$; pOH = 14.00
(f) $pH = -\log[6.8 \times 10^{-2}] = -(-1.17) = 1.17$; pOH = 12.83
(g) $pH = -\log[3.1 \times 10^{-4}] = -(-3.51) = 3.51$; pOH = 10.49
(h) $pH = -\log[5.8 \times 10^{-5}] = -(-4.24) = 4.24$; pOH = 9.76
(i) $pOH = -\log[2.2 \times 10^{-2}] = -(-1.66) = 1.66$; pH = 12.34
(j) $pOH = -\log[7.4 \times 10^{-3}] = -(-2.13) = 2.13$; pH = 11.87

15.5 (a) $0.63\text{ g HNO}_3\left(\dfrac{1\text{ mol HNO}_3}{63.0\text{ g HNO}_3}\right) = 0.010\text{ mol HNO}_3$

$\text{Molarity} = \dfrac{0.010\text{ mol HNO}_3}{1000\text{ L soln}} = 1.0 \times 10^{-5}\text{ M HNO}_3$

$[H^+] = 1.0 \times 10^{-5}$ M
$pH = -\log[1.0 \times 10^{-5}] = -(-5.00) = 5.00$

(b) $4.0\text{ g NaOH}\left(\dfrac{1\text{ mol NaOH}}{40.0\text{ g NaOH}}\right) = 0.10\text{ mol NaOH}$

$\text{Molarity} = \dfrac{0.10\text{ mol NaOH}}{100\text{ L soln}} = 1.0 \times 10^{-3}\text{ M NaOH}$

$[OH^-] = 1.0 \times 10^{-3}$ M
$pOH = -\log[1.0 \times 10^{-3}] = -(-3.00) = 3.00$
pH = 11.00

(c) $0.080\text{ g NaOH}\left(\dfrac{1\text{ mol NaOH}}{40.0\text{ g NaOH}}\right) = 0.0020\text{ mol NaOH}$

$$\text{Molarity} = \frac{0.0020 \text{ mol NaOH}}{0.200 \text{ L soln}}$$

$= 1.0 \times 10^{-2}$ M NaOH

$[OH^-] = 1.0 \times 10^{-2}$ M

pOH $= -\log[1.0 \times 10^{-2}] = -(-2.00) = 2.00$

pH $= 12.00$

(d) $73 \text{ g HCl}\left(\frac{1 \text{ mol HCl}}{36.5 \text{ g HCl}}\right) = 2.0 \text{ mol HCl}$

$$\text{Molarity} = \frac{2.0 \text{ mol HCl}}{20.0 \text{ L soln}} = 1.0 \times 10^{-1} \text{ M HCl}$$

$[H^+] = 1.0 \times 10^{-1}$ M

pH $= -\log[1.0 \times 10^{-1}] = -(-1.00) = 1.00$

(e) $0.051 \text{ g HCl}\left(\frac{1 \text{ mol HCl}}{36.5 \text{ g HCl}}\right) = 1.4 \times 10^{-3} \text{ mol HCl}$

$$\text{Molarity} = \frac{1.4 \times 10^{-3} \text{ mol HCl}}{15 \text{ L soln}}$$

$= 9.3 \times 10^{-5}$ M HCl

$[H^+] = 9.3 \times 10^{-5}$ M

pH $= -\log[9.3 \times 10^{-5}] = -(-4.03) = 4.03$

(f) $0.10 \text{ g NaOH}\left(\frac{1 \text{ mol NaOH}}{40.0 \text{ g NaOH}}\right)$

$= 2.5 \times 10^{-3}$ mol NaOH

$$\text{Molarity} = \frac{2.5 \times 10^{-3} \text{ mol NaOH}}{5.0 \text{ L soln}}$$

$= 5.0 \times 10^{-4}$ M NaOH

$[OH^-] = 5.0 \times 10^{-4}$ M

pOH $= -\log[5.0 \times 10^{-4}] = -(-3.30) = 3.30$

pH $= 10.70$

(g) $0.084 \text{ g KOH}\left(\frac{1 \text{ mol KOH}}{56.1 \text{ g KOH}}\right)$

$= 1.5 \times 10^{-3}$ mol KOH

$$\text{Molarity} = \frac{1.5 \times 10^{-3} \text{ mol KOH}}{0.250 \text{ L soln}}$$

$= 6.0 \times 10^{-3}$ M KOH

$[OH^-] = 6.0 \times 10^{-3}$ M

pOH $= -\log[6.0 \times 10^{-3}] = -(-2.22) = 2.22$

pH $= 11.78$

(h) $0.59 \text{ g HBr}\left(\frac{1 \text{ mol HBr}}{80.9 \text{ g HBr}}\right) = 7.3 \times 10^{-3} \text{ mol HBr}$

$$\text{Molarity} = \frac{7.3 \times 10^{-3} \text{ mol HBr}}{4.0 \text{ L soln}}$$

$= 1.8 \times 10^{-3}$ M HBr

$[H^+] = 1.8 \times 10^{-3}$ M

pH $= -\log[1.8 \times 10^{-3}] = -(-2.74) = 2.74$

15.6 **(a)** $[H^+] = 1.0 \times 10^{-4}$ M
(b) $[H^+] = 1.0 \times 10^{-11}$ M
(c) pH $= 12.00$; $[H^+] = 1.0 \times 10^{-12}$ M
(d) $[H^+] = 3.6 \times 10^{-9}$ M
(e) $[H^+] = 1.3 \times 10^{-3}$ M
(f) pH $= 3.67$; $[H^+] = 2.1 \times 10^{-4}$

15.7 **(a)** $ClO^- + H_2O \rightleftharpoons HClO + OH^-$
(b) $NO_2^- + H_2O \rightleftharpoons HNO_2 + OH^-$
(c) $C_7H_5O_2^- + H_2O \rightleftharpoons HC_7H_5O_2 + OH^-$

15.8 **(a)** Br^- **(b)** PH_4^+ **(c)** $HC_2O_4^-$ **(d)** HS^- **(e)** S^{2-} **(f)** $HC_2O_4^-$ **(g)** $H_2C_2O_4$

15.9 **(a)** H_2SO_4 **(b)** SO_4^{2-}

15.10

	Acid	Base
(a)	HCN	OH^-
(b)	H_2SO_4	NH_3
(c)	HNO_3	H_2O
(d)	H_2O	F^-
(e)	HSO_4^-	HPO_4^{2-}

15.11

	Acid	Base
(a)	H^+	PH_3
(b)	Cu^+	NH_3
(c)	BF_3	NH_3
(d)	$FeCl_3$	Cl^-

15.12 **(a)** No **(b)** Yes

15.13 **(a)** Will work as a buffer
(b) Will not work as a buffer
(c) Will work as a buffer
(d) Will not work as a buffer

15.14 **(a)** 3 **(b)** 1 **(c)** 3 **(d)** 1 **(e)** 2 **(f)** 2

15.15 **(a)** $0.250 \text{ mol HCl}\left(\frac{1 \text{ equiv HCl}}{1 \text{ mol HCl}}\right) = 0.250 \text{ equiv HCl}$

(b) $0.550 \text{ mol Ba(OH)}_2\left(\frac{2 \text{ equiv Ba(OH)}_2}{1 \text{ mol Ba(OH)}_2}\right)$

$= 1.10 \text{ equiv Ba(OH)}_2$

(c) $2.50 \text{ mol } H_3PO_4\left(\frac{3 \text{ equiv } H_3PO_4}{1 \text{ mol } H_3PO_4}\right)$

$= 7.50 \text{ equiv } H_3PO_4$

(d) $0.0140 \text{ mol } H_2SO_4\left(\frac{2 \text{ equiv } H_2SO_4}{1 \text{ mol } H_2SO_4}\right)$

$= 0.0280 \text{ equiv } H_2SO_4$

(e) $0.630 \text{ mol Al(OH)}_3\left(\frac{3 \text{ equiv Al(OH)}_3}{1 \text{ mol Al(OH)}_3}\right)$

$= 1.89 \text{ equiv Al(OH)}_3$

15.16 **(a)** $0.150 \text{ equiv } H_3PO_4\left(\frac{1 \text{ mol } H_3PO_4}{3 \text{ equiv } H_3PO_4}\right)$

$= 0.0500 \text{ mol } H_3PO_4$

(b) $0.650 \text{ equiv HBr}\left(\frac{1 \text{ mol HBr}}{1 \text{ equiv HBr}}\right) = 0.650 \text{ mol HBr}$

(c) $1.55 \text{ equiv Ca(OH)}_2\left(\frac{1 \text{ mol Ca(OH)}_2}{2 \text{ equiv Ca(OH)}_2}\right) = 0.775 \text{ mol Ca(OH)}_2$

(d) $3.25 \text{ equiv KOH}\left(\frac{1 \text{ mol KOH}}{1 \text{ equiv KOH}}\right) = 3.25 \text{ mol KOH}$

(e) $0.0750 \text{ equiv H}_2\text{C}_2\text{O}_4\left(\frac{1 \text{ mol H}_2\text{C}_2\text{O}_4}{2 \text{ equiv H}_2\text{C}_2\text{O}_4}\right) = 0.0375 \text{ mol H}_2\text{C}_2\text{O}_4$

15.17 **(a)** $3.65 \text{ g HCl}\left(\frac{1 \text{ mol HCl}}{36.5 \text{ g HCl}}\right)\left(\frac{1 \text{ equiv HCl}}{1 \text{ mol HCl}}\right) = 0.100 \text{ equiv HCl}$

(b) $0.0200 \text{ g NaOH}\left(\frac{1 \text{ mol NaOH}}{40.0 \text{ g NaOH}}\right)\left(\frac{1 \text{ equiv NaOH}}{1 \text{ mol NaOH}}\right) = 5.00 \times 10^{-4} \text{ equiv NaOH}$

(c) $7.41 \text{ g Ca(OH)}_2\left(\frac{1 \text{ mol Ca(OH)}_2}{74.1 \text{ g Ca(OH)}_2}\right) \times \left(\frac{2 \text{ equiv Ca(OH)}_2}{1 \text{ mol Ca(OH)}_2}\right) = 0.200 \text{ equiv Ca(OH)}_2$

(d) $0.162 \text{ g HBr}\left(\frac{1 \text{ mol HBr}}{80.9 \text{ g HBr}}\right)\left(\frac{1 \text{ equiv HBr}}{1 \text{ mol HBr}}\right) = 0.00200 \text{ equiv HBr}$

(e) $21.0 \text{ g H}_2\text{C}_2\text{O}_4\left(\frac{1 \text{ mol H}_2\text{C}_2\text{O}_4}{90.0 \text{ g H}_2\text{C}_2\text{O}_4}\right)\left(\frac{2 \text{ equiv H}_2\text{C}_2\text{O}_4}{1 \text{ mol H}_2\text{C}_2\text{O}_4}\right) = 0.467 \text{ equiv H}_2\text{C}_2\text{O}_4$

(f) $2.94 \text{ g H}_2\text{SO}_4\left(\frac{1 \text{ mol H}_2\text{SO}_4}{98.1 \text{ g H}_2\text{SO}_4}\right)\left(\frac{2 \text{ equiv H}_2\text{SO}_4}{1 \text{ mol H}_2\text{SO}_4}\right) = 0.0599 \text{ equiv H}_2\text{SO}_4$

(g) $0.392 \text{ g H}_3\text{PO}_4\left(\frac{1 \text{ mol H}_3\text{PO}_4}{98.0 \text{ g H}_3\text{PO}_4}\right)\left(\frac{3 \text{ equiv H}_3\text{PO}_4}{1 \text{ mol H}_3\text{PO}_4}\right) = 0.0120 \text{ equiv H}_3\text{PO}_4$

15.18 **(a)** $1.00 \text{ g NaOH}\left(\frac{1 \text{ mol NaOH}}{40.0 \text{ g NaOH}}\right)\left(\frac{1 \text{ equiv NaOH}}{1 \text{ mol NaOH}}\right) = 0.0250 \text{ equiv NaOH}$

$\text{Normality} = \frac{0.0250 \text{ equiv NaOH}}{0.0500 \text{ L soln}} = 0.500 \text{ N NaOH}$

(b) $0.500 \text{ g H}_2\text{SO}_4\left(\frac{1 \text{ mol H}_2\text{SO}_4}{98.1 \text{ g H}_2\text{SO}_4}\right)\left(\frac{2 \text{ equiv H}_2\text{SO}_4}{1 \text{ mol H}_2\text{SO}_4}\right) = 0.0102 \text{ equiv H}_2\text{SO}_4$

$\text{Normality} = \frac{0.0102 \text{ equiv H}_2\text{SO}_4}{0.250 \text{ L soln}} = 0.0408 \text{ N H}_2\text{SO}_4$

(c) $2.00 \text{ g H}_3\text{PO}_4\left(\frac{1 \text{ mol H}_3\text{PO}_4}{98.0 \text{ g H}_3\text{PO}_4}\right)\left(\frac{3 \text{ equiv H}_3\text{PO}_4}{1 \text{ mol H}_3\text{PO}_4}\right) = 0.0612 \text{ equiv H}_3\text{PO}_4$

$\text{Normality} = \frac{0.0612 \text{ equiv H}_3\text{PO}_4}{0.150 \text{ L soln}} = 0.408 \text{ N H}_3\text{PO}_4$

(d) $0.350 \text{ g Mg(OH)}_2\left(\frac{1 \text{ mol Mg(OH)}_2}{58.3 \text{ g Mg(OH)}_2}\right) \times \left(\frac{2 \text{ equiv Mg(OH)}_2}{1 \text{ mol Mg(OH)}_2}\right) = 0.0120 \text{ equiv Mg(OH)}_2$

$\text{Normality} = \frac{0.0120 \text{ equiv Mg(OH)}_2}{2.50 \text{ L soln}} = 0.00480 \text{ N Mg(OH)}_2$

(e) $9.87 \text{ g HBr}\left(\frac{1 \text{ mol HBr}}{80.9 \text{ g HBr}}\right)\left(\frac{1 \text{ equiv HBr}}{1 \text{ mol HBr}}\right) = 0.122 \text{ equiv HBr}$

$\text{Normality} = \frac{0.122 \text{ equiv HBr}}{1.80 \text{ L soln}} = 0.0678 \text{ N HBr}$

15.19 **(a)** $2.00 \text{ L soln}\left(\frac{0.300 \text{ equiv H}_2\text{SO}_4}{\text{L soln}}\right) = 0.600 \text{ equiv H}_2\text{SO}_4$

(b) $0.500 \text{ L soln}\left(\frac{1.20 \text{ equiv HCl}}{\text{L soln}}\right) = 0.600 \text{ equiv HCl}$

(c) $0.350 \text{ L soln}\left(\frac{0.750 \text{ equiv NaOH}}{\text{L soln}}\right) = 0.262 \text{ equiv NaOH}$

(d) $0.175 \text{ L soln}\left(\frac{0.180 \text{ equiv H}_3\text{PO}_4}{\text{L soln}}\right) = 0.0315 \text{ equiv H}_3\text{PO}_4$

(e) $0.0500 \text{ L soln}\left(\frac{0.125 \text{ equiv H}_2\text{C}_2\text{O}_4}{\text{L soln}}\right) = 0.00625 \text{ equiv H}_2\text{C}_2\text{O}_4$

15.20 $N_a = \frac{V_b \cdot N_b}{V_a} = \frac{(45.0 \text{ mL soln})(0.200 \text{ N})}{55.0 \text{ mL soln}} = 0.164 \text{ N}$

15.21 $N_b = \frac{V_a \cdot N_a}{V_b} = \frac{(25.0 \text{ mL soln})(0.450 \text{ N})}{35.0 \text{ mL soln}} = 0.321 \text{ N}$

15.22 $N_b = \frac{V_a \cdot N_a}{V_b} = \frac{(23.14 \text{ mL soln})(0.1200 \text{ N})}{21.20 \text{ mL soln}} = 0.1310 \text{ N}$

15.23 $N_a = \frac{V_b \cdot N_b}{V_a} = \frac{(40.00 \text{ mL soln})(0.1000 \text{ N})}{52.00 \text{ mL soln}} = 0.07692 \text{ N}$

15.24 $N_b = \frac{V_a \cdot N_a}{V_b} = \frac{(50.9 \text{ mL soln})(0.0400 \text{ N})}{42.4 \text{ mL soln}} = 0.0480 \text{ N}$

Chapter 16

16.1 (a) Oxidation (b) Reduction
(c) Oxidation (d) Reduction

16.2 (a) A loss of electrons (b) A gain of electrons
(c) Either the oxidation or the reduction portion of an oxidation–reduction
(d) A substance that causes oxidation
(e) A substance that causes reduction

16.3 Key: oa = oxidizing agent; ra = reducing agent
(a) oa = Cu^{2+}; ra = Mg (b) oa = Br_2; ra = I^-
(c) oa = Cl_2; ra = Na

16.4 (a) H = 0 (b) Al = +3
(c) H = +1; P = +5; O = −2
(d) N = +2; O = −2 (e) N = +5; O = −2
(f) H = +1; N = +3; O = −2
(g) Cr = +6; O = −2 (h) Mn = +4; O = −2
(i) Mn = +7; O = −2 (j) C = +3; O = −2

16.5 (a) F = 0 (b) C = +4; O = −2
(c) Br = +3; O = −2 (d) K = +1; I = −1
(e) Cu = +2; Br = −1
(f) Li = +1; N = +5; O = −2
(g) Na = +1; S = +6; O = −2
(h) Al = +3; S = +6; O = −2

16.6 (a) $\overset{0}{Zn} + \overset{+1+6-2}{H_2SO_4} \rightarrow \overset{+2+6-2}{ZnSO_4} + \overset{0}{H_2}$
Yes. Zinc is oxidized; hydrogen is reduced.

(b) $\overset{+1-1}{KI} + \overset{+2\,+5-2}{Pb(NO_3)_2} \rightarrow \overset{+2-1}{PbI_2} + \overset{+1+5-2}{KNO_3}$
No changes.

(c) $\overset{0}{Mn} + \overset{+1-1}{HCl} \rightarrow \overset{+2\,-1}{MnCl_2} + \overset{0}{H_2}$
Yes. Manganese is oxidized; hydrogen is reduced.

(d) $\overset{+2-2}{HgO} + \overset{+1+5-2}{HNO_3} \rightarrow \overset{+2\,+5-2}{Hg(NO_3)_2} + \overset{+1-2}{H_2O}$
No changes.

16.7 $\overset{0}{Cu} + \overset{+1+5-2}{HNO_3} \rightarrow \overset{+2\,+5-2}{Cu(NO_3)_2} + \overset{+2-2}{NO} + \overset{+1-2}{H_2O}$
Copper is oxidized from 0 to +2. Nitrogen is reduced from +5 to +2. However, only a portion of the nitrogen is reduced.

16.8 (a) $\overset{0}{Sn} + 4\,\overset{+5}{HNO_3} \rightarrow \overset{+4}{SnO_2} + 4\,\overset{+4}{NO_2} + 2\,H_2O$
Each Sn atom loses 4 e^-; each N atoms gains 1 e^-.

(b) $2\,\overset{-1}{HBr} + \overset{+6}{H_2SO_4} \rightarrow \overset{+4}{SO_2} + \overset{0}{Br_2} + 2\,H_2O$
Each Br atom loses 1 e^-; each S atom gains 2 e^-.

(c) $2\,\overset{+5}{HNO_3} + 6\,\overset{-1}{HCl} \rightarrow 2\,\overset{+2}{NO} + 3\,\overset{0}{Cl_2} + 4\,H_2O$
Each Cl atom loses 1 e^-; each N atom gains 3 e^-.

(d) $Na_2\overset{+6}{Cr_2}O_7 + 6\,\overset{+2}{Fe}Cl_2 + 14\,HCl \rightarrow 2\,\overset{+3}{Cr}Cl_3 + 2\,NaCl + 6\,\overset{+3}{Fe}Cl_3 + 7\,H_2O$
Each Fe atom loses 1 e^-; each Cr atom gains 3 e^-.

(e) $\overset{0}{I_2} + 5\,\overset{0}{Cl_2} + 6\,H_2O \rightarrow 2\,\overset{+5}{HIO_3} + 10\,\overset{-1}{HCl}$
Each I atom loses 5 e^-; each Cl atom gains 1 e^-.

(f) $\overset{0}{Zn} + 4\,\overset{+5}{HNO_3} \rightarrow \overset{+2\,+5}{Zn(NO_3)_2} + 2\,\overset{+4}{NO_2} + 2\,H_2O$
Each Zn atom loses 2 e^-; two N atoms gain 1 e^- each; two N atoms do not change.

16.9 (a) $Sn^{2+} \rightarrow Sn^{4+} + 2\,e^-$ (oxidation)
(b) $MnO_4^- + 8\,H^+ + 5\,e^- \rightarrow Mn^{2+} + 4\,H_2O$ (reduction)
(c) $C_2O_4^{2-} \rightarrow 2\,CO_2 + 2\,e^-$ (oxidation)
(d) $2\,I^- \rightarrow I_2 + 2\,e^-$ (oxidation)
(e) $Cr_2O_7^{2-} + 14\,H^+ + 6\,e^- \rightarrow 2\,Cr^{3+} + 7\,H_2O$ (reduction)

16.10 (a)
$3(Cu \rightarrow Cu^{2+} + 2\,e^-)$ oxidation
$2(3\,e^- + 4\,H^+ + NO_3^- \rightarrow NO + 2\,H_2O)$ reduction
$3\,Cu + 2\,NO_3^- + 8\,H^+ \rightarrow 3\,Cu^{2+} + 2\,NO + 4\,H_2O$

(b) $3(4\,H_2O + S^{2-} \rightarrow SO_4^{2-} + 8\,H^+ + 8\,e^-)$ oxidation
$8(3\,e^- + 4\,H^+ + NO_3^- \rightarrow NO + 2\,H_2O)$ reduction
$8\,NO_3^- + 3\,S^{2-} + 8\,H^+ \rightarrow 8\,NO + 3\,SO_4^{2-} + 4\,H_2O$

(c) $5(2\,H_2O + NO \rightarrow NO_3^- + 4\,H^+ + 3\,e^-)$ oxidation
$3(5\,e^- + 8\,H^+ + MnO_4^- \rightarrow Mn^{2+} + 4\,H_2O$ reduction
$3\,MnO_4^- + 5\,NO + 4\,H^+ \rightarrow 3\,Mn^{2+} + 5\,NO_3^- + 2\,H_2O$

(d) $6(Ag \rightarrow Ag^+ + e^-)$ oxidation
$1(6\,e^- + 14\,H^+ + Cr_2O_7^{2-} \rightarrow 2\,Cr^{3+} + 7\,H_2O)$ reduction
$6\,Ag + Cr_2O_7^{2-} + 14\,H^+ \rightarrow 6\,Ag^+ + 2\,Cr^{3+} + 7\,H_2O$

(e) $1(2\,I^- \rightarrow I_2 + 2\,e^-)$ oxidation
$1(2\,e^- + 2\,H^+ + H_2O_2 \rightarrow 2\,H_2O)$ reduction
$H_2O_2 + 2\,I^- + 2\,H^+ \rightarrow I_2 + 2\,H_2O$

(f)

$$\begin{array}{rlr} 3(2\,Cl^- & \rightarrow Cl_2 + 2\,e^-) & \text{oxidation} \\ 2(3\,e^- + 4\,H^+ + NO_3^- & \rightarrow NO + 2\,H_2O) & \text{reduction} \\ \hline 2\,NO_3^- + 6\,Cl^- + 8\,H^+ & \rightarrow 2\,NO + 3\,Cl_2 + 4\,H_2O & \end{array}$$

16.11 Key: soa = stronger oxidizing agent;
woa = weaker oxidizing agent;
sra = stronger reducing agent;
wra = weaker reducing agent

(a) $Ba + Cd^{2+} \rightarrow Ba^{2+} + Cd$
(sra) (soa) (woa) (wra)
Yes. Reaction occurs.

(b) $3\,Cu + 2\,Al^{3+} \rightarrow 3\,Cu^{2+} + 2\,Al$
(wra) (woa) (soa) (sra) No reaction.

(c) $Mg + 2\,H^+ \rightarrow Mg^{2+} + H_2$
(sra) (soa) (woa) (wra)
Yes. Reaction occurs.

(d) $2\,Au + 6\,H^+ \rightarrow 2\,Au^{3+} + 3\,H_2$
(wra) (woa) (soa) (sra) No reaction.

(e) $Pb + 2\,H^+ \rightarrow Pb^{2+} + H_2$
(sra) (soa) (woa) (wra)
Yes. Reaction occurs.

16.12 (a) An electrochemical cell is an arrangement of two half-reactions such that the electrons transferred in the oxidation–reduction reaction must pass through an external circuit from one electrode to another.

(b) A voltaic cell is an electrochemical cell that operates in the spontaneous direction. An electrolytic cell is connected to an external source of energy and runs in the nonspontaneous direction.

(c) Oxidation occurs at the anode.

(d) Reduction occurs at the cathode.

(e) The porous partition prevents the solutions in the two half-cells from mixing, but it enables ions to migrate in order to maintain the electrical neutrality in each half-cell.

(f) Anions migrate toward the anode.

16.13 (a) $Cu \rightarrow Cu^{2+} + 2\,e^-$

(b) The spoon is the anode.

(c) The spoon dissolves.

16.14 $2\,PbSO_4(s) + 2\,H_2O(\ell) \rightarrow Pb(s) + PbO_2(s) + 2\,H_2SO_4(aq)$

During recharging, the battery operates as an electrolytic cell.

Chapter 17

17.1 When a flammable liquid achieves a high enough temperature, there is a chance that some high-energy molecules will collide with sufficient activation energy to undergo combustion. Once the reaction is initiated in this fashion, the heat that is liberated provides more energy to continue combustion in other molecules.

17.2 As the temperature is increased, so are the velocities of the molecules. This leads to an increased frequency of collision and an increase in rate.

17.3 (a) When $[A]$ is doubled, the rate doubles. When $[B]$ is doubled, the rate doubles.

(b) When $[A]$ is doubled, the rate doubles. When $[B]$ is doubled, the rate quadruples.

(c) When $[A]$ is doubled, the rate doubles. When $[B]$ is doubled, the rate does not change.

17.4 Because a catalyst is not consumed in a reaction, it is not included with the reactants. Nevertheless, to show its presence, we place it over the arrow in the equation.

17.5 The iodine color will increase. No hydrogen is introduced into the flask, so the only reaction that takes place at first is $2\,HI \rightarrow H_2 + I_2$. Because this reaction produces I_2, the intensity of the iodine color increases until equilibrium is established.

17.6 (a) $K_{eq} = \dfrac{[SO_3]^2}{[SO_2]^2[O_2]}$ **(b)** $K_{eq} = \dfrac{[NCl_3]^2}{[N_2][Cl_2]^3}$

(c) $K_{eq} = \dfrac{[NOBr]^2}{[NO]^2[Br_2]}$ **(d)** $K_{eq} = \dfrac{[CO_2][H_2]}{[CO][H_2O]}$

(e) $K_{eq} = \dfrac{[CH_4][H_2O]}{[CO][H_2]^3}$ **(f)** $K_{eq} = \dfrac{[PCl_3][Cl_2]}{[PCl_5]}$

(g) $K_{eq} = \dfrac{[O_3]^2}{[O_2]^3}$ **(h)** $K_{eq} = \dfrac{[COCl_2]}{[CO][Cl_2]}$

(i) $K_{eq} = \dfrac{[HF]^2}{[H_2][F_2]}$

17.7 (a) $K_{eq} = \dfrac{[NO]^2}{[N_2][O_2]} = \dfrac{(0.10)^2}{(1.00)(0.10)} = 0.10$

(b) $K_{eq} = \dfrac{[I_2][Cl_2]}{[ICl]^2} = \dfrac{(0.25)(0.20)}{(0.67)^2} = 0.11$

(c) $K_{eq} = \dfrac{[PCl_3][Cl_2]}{[PCl_5]} = \dfrac{(0.20)(0.10)}{(0.40)} = 0.050$

17.8 The equilibrium mixture will be almost entirely reactants. A very small equilibrium constant means that the relative amount of product formed is very small.

17.9 (a) Shift to right **(b)** Shift to left
(c) Shift to left **(d)** Shift to right
(e) Shift to left

17.10 (a) Shift to right **(b)** No effect
(c) Shift to left **(d)** Shift to left

17.11 (a) Shift to right **(b)** No effect
(c) Shift to left

17.12 **(a)** Shift to left **(b)** Shift to right
(c) Shift to right **(d)** Shift to left

17.13 $K_a = \frac{[H^+][F^-]}{[HF]} = \frac{(0.026)(0.026)}{(0.97)} = 7.0 \times 10^{-4}$

17.14 $K_a = \frac{[H^+][ClO^-]}{[HClO]} = \frac{(5.6 \times 10^{-4})(5.6 \times 10^{-4})}{(0.10)}$
$= 3.1 \times 10^{-6}$

17.15

	$HZ(aq)$ ⇌	$H^+(aq)$ +	$Z^-(aq)$
Initial concentration	0.10	0	0
Dissociates	−0.05	+0.05	+0.05
Equilibrium concentration	0.05	0.05	0.05

$K_a = \frac{[H^+][Z^-]}{[HZ]} = \frac{(0.05)(0.05)}{(0.05)} = 5 \times 10^{-2}$

$pH = -\log[0.05] = -\log[5 \times 10^{-2}]$
$= -(-1.3) = 1.3$

17.16 **(a)** $AgBr(s) \rightleftharpoons Ag^+(aq) + Br^-(aq)$
$K_{sp} = [Ag^+][Br^-]$
$= (7.1 \times 10^{-7})(7.1 \times 10^{-7}) = 5.0 \times 10^{-13}$
(b) $BaCO_3(s) \rightleftharpoons Ba^{2+}(aq) + CO_3^{2-}(aq)$
$K_{sp} = [Ba^{2+}][CO_3^{2-}]$
$= (4.0 \times 10^{-5})(4.0 \times 10^{-5}) = 1.6 \times 10^{-9}$
(c) $CaF_2(s) \rightleftharpoons Ca^{2+}(aq) + 2\,F^-(aq)$
$K_{sp} = [Ca^{2+}][F^-]^2$
$= (3.5 \times 10^{-4})(7.0 \times 10^{-4})^2 = 1.7 \times 10^{-10}$

Appendix A

A.1 **(a)** 71.8 **(b)** 6.437 **(c)** 0.84 **(d)** 418
(e) 0.27 **(f)** 13.8 **(g)** 450 **(h)** 1.0010
(i) 7.6 **(j)** 14.5 **(k)** 7.66 **(l)** 10.000

A.2 **(a)** 0.375 **(b)** 0.312 **(c)** 0.647 **(d)** 0.600

A.3 **(a)** 3^5 **(b)** 6^7 **(c)** 7^3 **(d)** 10^1

A.4 **(a)** 4^8 **(b)** 7^4 **(c)** 3^{-3}

A.5 **(a)** 375 **(b)** 42,100 **(c)** 0.00739
(d) 0.832 **(e)** 5,930,000 **(f)** 0.00000593

A.6 **(a)** 4.76×10^4 **(b)** 5.41×10^2
(c) 9.43×10^{-2} **(d)** 3.15×10^{-4}
(e) 2.19×10^6 **(f)** 2.19×10^{-6}

A.7 **(a)** $(4 \times 10^6)(2 \times 10^{-4}) = 8 \times 10^2$

(b) $\frac{2.2 \times 10^3}{5.5 \times 10^{-3}} = 0.40 \times 10^6 = 4.0 \times 10^5$

(c) $\frac{(4.36 \times 10^2)(5.39 \times 10^{-3})}{(6.20 \times 10^{-2})} = 3.79 \times 10^1$

(d) $\frac{(3.18 \times 10^{-4})(2.17 \times 10^1)}{(4.83 \times 10^3)} = 1.43 \times 10^{-6}$

A.8 **(a)** 5.48×10^{-4} **(b)** 7.95×10^{-5} **(c)** 2.45×10^{-2}

A.9 **(a)** $\frac{16\text{ g}}{225\text{ g}} \times 100\% = 7.1\%$

(b) $(250\text{ g})(0.18) = 45\text{ g}$

A.10 **(a)** $\frac{1.2}{44.0} \times 100\% = 2.7\%$

(b) $\frac{14}{602} \times 100\% = 2.3\%$

ANSWERS TO PHOTO QUIZ QUESTIONS

1. Both compounds that produce red contain strontium. A plausible hypothesis is: Strontium will produce a red color in fireworks.

2. Chemical change; the substances in the test tube before and after heating have clearly different properties. A red substance (before) produced a silver substance and a gas (after).

3. There are six liquid layers. A seventh gaseous layer (air) is also present.

4. The number of moles and atoms in each of the beakers is as follows:

$$\begin{aligned} 1.00 \text{ lb } C_{12}H_{22}O_{11} &= 454 \text{ g } C_{12}H_{22}O_{11} \\ &= 1.33 \text{ moles } C_{12}H_{22}O_{11} \\ &= 3.60 \times 10^{25} \text{ atoms} \\ 1.00 \text{ lb NaCl} &= 454 \text{ g NaCl} \\ &= 7.76 \text{ moles NaCl} \\ &= 9.34 \times 10^{24} \text{ atoms} \\ 1.00 \text{ lb } NaHCO_3 &= 454 \text{ g } NaHCO_3 \\ &= 5.40 \text{ moles } NaHCO_3 \\ &= 1.95 \times 10^{25} \text{ atoms} \end{aligned}$$

   Thus, 1 lb of table salt has the greatest number of moles, and 1 lb of table sugar has the greatest number of atoms.

5. There are approximately 4×10^7 atoms along a 1-cm length. Thus, an area of 1 cm^2 would contain approximately 1.6×10^{15} atoms.

6. The electrons in atoms exist only in certain discrete energy levels. When electrons fall from higher to lower energy levels, light is emitted with frequencies (colors) that correspond to the differences between the levels. Thus, only certain colors are given off, corresponding to the various energy differences that are possible. Each element has its own unique set of colors, as shown in the various spectra. The orange glow of a mercury lamp corresponds to the combination of colors given off as its electrons fall from higher to lower levels.

7. Electrons from the Van de Graf generator flow to all parts of the woman, including her hair. Since like charges repel, the strands of her hair are repelled from her body and from one another, causing them to stand on end.

8. As a general rule, the number of bonds an atom forms corresponds to its number of unpaired valence electrons: C, O, H, and Cl typically form 4, 2, 1, and 1 bond, respectively. Thus, the atoms in ball-and-stick model kits have holes drilled corresponding to these numbers. Since silicon is in the same chemical family as carbon, it also forms four bonds. Thus, a carbon ball could be used to substitute for silicon in a model of SiH_4.

9. Ammonium nitrate is NH_4NO_3, ammonium phosphate is $(NH_4)_3PO_4$, and potassium sulfate is K_2SO_4. Ammonium nitrate and ammonium phosphate would provide nitrogen. Ammonium phosphate would provide phosphorus. Potassium sulfate would provide sulfur. The name *iron sulfate* does not specify the oxidation state of iron, thereby leaving the exact chemical formula of the compound ambiguous.

10. A 12.5-g sample of $(NH_4)_2Cr_2O_7$ (0.0496 mole) will produce 7.54 g of Cr_2O_3 (0.0496 mole).

11. Gases expand upon heating, thereby decreasing in density. Since less dense substances rise to the top of more dense substances, the hotter, less dense air inside the balloon rises above the cooler, more dense surrounding air, carrying the balloonist aloft. Air tends to be coolest at night and in the early morning, having its greatest density during these hours. These conditions provide the greatest density difference between the hot air in the balloon and the surroundings. Thus, a hot air balloon rises most easily in the morning.

12. Whereas most substances contract upon cooling, water is unusual in its property of expanding as it freezes. This expansion results in a solid that is less dense than the liquid form. Thus, ice floats in liquid water. Paraffin, which exhibits the more typical behavior, contracts as it freezes, giving a solid that is more dense than its liquid form. Thus, solid paraffin sinks in liquid paraffin.

13. The solubilities of most solids are greater in solvents at high temperature than at low temperature. Sodium acetate in water exhibits this behavior. Thus, all of the solid dissolves in hot water. However, as the solution is allowed to cool, the solubility decreases. When the solubility of the substance

drops below the concentration actually present in solution, the excess solute is forced from the solution, forming an insoluble solid. In this case, slow cooling has permitted the solid to form beautiful needle-like crystals.

14. Acid rain is a severe problem in some parts of the world, including the United States. Carbonates react with acids to form a salt, water, and carbon dioxide. In the case of insoluble carbonates such as marble ($CaCO_3$), the salt produced may be soluble, thereby dissolving in the acid. Thus, as acid rain washes over a marble statue, the statue dissolves slowly, as observed in the "before" and "after" shots.

15. The more acidic a solution, the lower the pH. Thus, vinegar, with a pH of 2.5, is the most acidic.

16. The following oxidation–reduction is occurring: $Cu + 2\,Ag^+ \rightarrow Cu^{2+} + 2\,Ag$. The metal formed from the reduction of silver ion is plating out on the copper coil. The blue color is from the copper(II) ion that forms as the copper metal is oxidized.

17. These photos illustrate Le Chatelier's principle. The reaction given in the question describes the equilibrium between the blue $CoCl_4^{2-}$ ion and the pink $Co(H_2O)_6^{2+}$ ion. Addition of water drives the equilibrium to the right, converting blue $CoCl_4^{2-}$ ion to pink $Co(H_2O)_6^{2+}$. Addition of chloride ion drives the equilibrium to the left, converting pink $Co(H_2O)_6^{2+}$ to blue $CoCl_4^{2-}$ ion.

18. Exposure to beta particles can cause damage to living tissue. Thus, beta emitters are not generally desirable for diagnostic purposes, where the goal is to assess the patient's condition. However, the destruction of living cells is often the goal of treatment, as is the case in reducing the size of an enlarged thyroid gland. Thus, a physician would select the milder iodine-123 for diagnosis, using iodine-131 only if the goal of treatment is to kill some of the cells present.

19. Organic compounds are carbon-containing covalent compounds, generally characterized by relatively low melting points. In contrast to this, inorganic substances are often ionic compounds with relatively high melting points. The substance on the left melts and then chars, leaving a carbon residue. Thus, it must be the organic compound. The substance on the right never melts under the same conditions. Thus, it must be the inorganic compound.

20. Celery is a polysaccharide, composed of repeating glucose units. Table sugar is a disaccharide, composed of a glucose and a fructose unit. Thus, both of these foods are carbohydrates. By contrast, nonfat milk is a protein, and butter is a lipid.

Photo Credits:
Photos 1, 2, 3, 7, 8, 9, 11, 12, 13, 15, 17, 19: Richard Megna/Fundamental Photographs. Photos 4, 10: Joseph P. Sinnot/Fundamental Photographs. Photo 5: Courtesy of Materials Analysis Group at Philips Semiconductor. Photo 6: Paul Silverman/Fundamental Photographs & Wabash Instrument Corp. Photo 14: NYC Parks Photo Archive & Joseph P. Sinnot/Fundamental Photographs. Photo 16: Peticolas/Megna/Fundamental Photographs. Photo 18: Courtesy of Dr. James Kowalski. Photo 20: Michael Dalton/Fundamental Photographs.

INDEX

Page numbers in **boldface type** indicate pages where terms are defined.

Useful Conversion Factors

Conversion factors

Length:	1 m = 1000 mm = 100 cm
	2.54 cm = 1 in.
	1.00 km = 0.621 mile
Mass:	1 kg = 1000 g
	1 g = 1000 mg
	454 g = 1.00 lb
	1.00 kg = 2.20 lb
Volume:	1 L = 1000 mL
	946 mL = 1.00 qt
	3.785 L = 1.000 gal
Energy:	4.184 J = 1 cal
Temperature:	$T_F = 1.8T_C + 32$
	$T_K = T_C + 273$
Gases:	1 mol = 22.4 L (at STP)
Avogadro's number:	6.02×10^{23}
Universal gas constant:	$R = \frac{0.0821\ \text{L} \cdot \text{atm}}{\text{mol} \cdot \text{K}}$

Important SI prefixes

Greater than 1:

kilo (k) means 10^3 or 1000

mega (M) means 10^6 or 1,000,000

Less than 1:

deci (d) means 10^{-1} = 0.1

centi (c) means 10^{-2} = 0.01

milli (m) means 10^{-3} = 0.001

micro (μ) means 10^{-6} = 0.000001

nano (n) means 10^{-9} = 0.000000001

pico (p) means 10^{-12} = 0.000000000001